Elementary Plane Surveying

ELEMENTARY PLANE SURVEYING

Fourth Edition

RAYMOND E. DAVIS, M.S., C.E., D.ENG.
Professor of Civil Engineering, Emeritus
University of California, Berkeley

JOE W. KELLY, B.S.
Professor of Civil Engineering, Emeritus
University of California, Berkeley

McGRAW-HILL BOOK COMPANY
New York San Francisco Toronto London Sydney St. Louis

Elementary Plane Surveying

Library of Congress Catalog Card Number 66–26577

15771

1234567890 MP 7432106987

PREFACE

For the fourth edition of *Elementary Plane Surveying* the entire text has been critically reviewed in detail, many sections have been rewritten, and a number of additions have been made. The new edition maintains its logical arrangement and thoroughness and its characteristic emphasis throughout on precision and accuracy. As before, it presents surveying not only for its own sake but also as a medium for studying the principles of measurement and the adjustment of measurements in general. So far as practicable it parallels the more comprehensive and detailed text *Surveying: Theory and Practice,* which may be used for reference or for special assignments.

Modern instruments and methods are described including levels, transits, plane-table alidades, electronic distance-measuring instruments, photogrammetric plotting instruments, and industrial surveying. Recently adopted classifications and standards of precision for surveys are given, and the model map symbols are now shown in color. In the chapter on route surveying relatively greater emphasis is given to highways, yet the applicability to other route structures is retained. The brief chapter on photogrammetric surveying includes a sequence of operations and a classification of plotting instruments.

Material has been added on precise leveling, adjustment of a level net by successive approximations, circular curves, a spiral curve, strength of figure in triangulation, reduction of distance to equivalent at sea level, and map projection.

The astronomy examples and field notes have been updated, as well as the references to more detailed publications. The numerical problems have been reexamined and many revised. Numerous cross references relate various aspects of topics, and an ample index is provided.

Logarithms are described briefly and six-place logarithmic tables are given because in many situations computers are not available. Principles and precision of computations are retained. Field problems are included for direct use by classes and because they contain many details on procedure not covered in the text.

Even though new types of levels and transits are described, conventional types and their adjustment are retained because they will continue to be used in ordinary surveying for many years. The newer types are used in a manner similar to the older. Errors and precision of level and transit work are evaluated.

Account has been taken throughout of the many suggestions offered by users of the book over many years, and the authors are grateful to them. Professors Sturla Einarsson, F. S. Foote, Bruce Jameyson, and Milos Polivka, colleagues of the authors at the University of California, and Mr. Sidney W. Smith of the U.S. Geological Survey rendered valuable suggestions and counsel.

Many of the illustrations and tables have been taken or adapted from publications or directly furnished material of Federal agencies, especially the U.S. Coast and Geodetic Survey, U.S. Geological Survey, and U.S. Bureau of Land Management. Use has also been made of illustrations and tabular information from the California Division of Highways, Federal Board of Surveys and Maps, U.S. Air Force, U.S. Army Corps of Engineers, and U.S. Naval Observatory.

Illustrative material was furnished by the following manufacturers of surveying equipment: Fairchild Space and Defense Systems, Keuffel & Esser Co., and The A. Lietz Company.

Thanks are due John Wiley & Sons, Inc., for permission to use Table III.

Raymond E. Davis
Joe W. Kelly

CONTENTS

LIST OF GENERAL TABLES

LIST OF FIELD AND OFFICE PROBLEMS

1
CONCEPTS; FIELD WORK; COMPUTATIONS

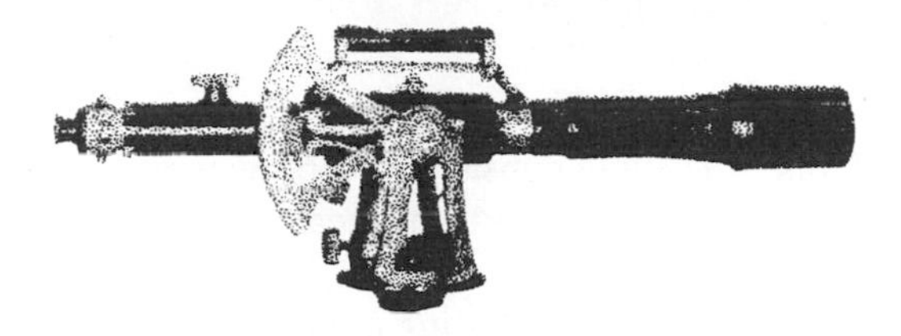

CONCEPTS

1-1 Surveying. Surveying is the art of determining horizontal distances, differences in elevation, directions, angles, locations, areas, and volumes on or near the surface of the earth. It comprises the *field work* of measuring and the *office work* of computing and drawing. Surveys are made for the purpose (1) of establishing the boundaries of land, (2) of providing information necessary to the construction of engineering works, or (3) of portraying land and water forms for purposes of navigation, mining, construction, and general use.

1-2 Plane and Geodetic Surveying. In surveying that extends over a relatively small part of the earth's surface, generally the spheroidal shape of the earth is neglected and the surface of the earth is considered as a plane. This is called *plane surveying.* With regard to horizontal distances and directions, a level line is considered to be mathematically straight, the direction of the plumb line is considered to be the same at all points within the limits of the survey, and all angles are considered to be plane angles. The usual procedure for determining vertical distances, called *leveling,* takes into account the curvature of the earth; but—since the operation is simple and in most cases automatic—leveling is usually considered to be a division of plane surveying. This book deals chiefly with the methods of plane surveying.

In *geodetic surveying* the shape of the earth is taken into account, either approximately by considering the earth as a true sphere or more precisely by considering the earth as an oblate spheroid of revolution. The polar axis of this spheroid is less than the equatorial axis by $\frac{1}{297}$, or about $\frac{1}{3}$ per cent. The values adopted in 1924 by the International Geodetic and Geophysical Union are as follows: polar axis, 41,711,940 ft; equatorial axis, 41,852,860 ft. Geodetic surveys are of high precision and usually extend over large areas; they are usually conducted only by government agencies. They furnish points of reference to which the multitude of surveys of lower precision may be tied.

For each state, a system of plane coordinates has been devised, to which all points in the state can be referred without significant error in distance or direction arising from the difference between the reference surface and the actual mean surface of the earth.

1-3 Operations of Surveying. *Land surveying* consists in subdividing land into parcels, reestablishing old or obliterated land lines, calculating areas, constructing land maps, and writing descriptions for deeds. *Topographic surveying* consists in securing field data for plotting a map which shows the configuration of the earth's surface and the location of natural and artificial objects. *Route surveying* consists in determining the ground configuration and location of objects along a proposed line such as that of a highway, railway, canal, or pipeline; establishing the line on the ground; and calculating volumes of earthwork.

Hydrographic surveying has reference to surveying bodies of water for purposes of navigation, water supply, or subaqueous construction; it usually comprises a topographic survey along the shore line, taking and plotting soundings, and observing the change in level of bodies of water or the discharge of streams. *Mine surveying* makes use of the principles of land, topographic, and route surveying in establishing mineral claims, locating the mine structures and underground workings, determining geological formations, and calculating volumes removed. *Cadastral surveying,* a practically obsolete term, has particular reference to extensive urban or rural surveys made for the purpose of locating property lines and improvements in detail, primarily for use in connection with the extent, value, ownership, and transfer of land; the term is sometimes applied to the primary control of public-land surveys.

City surveying is a term frequently applied to the surveying operations within a municipality with regard to mapping its area, laying out new streets and lots, and constructing streets, sewers and other public utilities, and buildings; it differs from ordinary surveying only in the precision of measurement and in certain details of procedure. The term *city survey* has come to mean a coordinated topographical survey of the area in and near a city for use in planning city improvements and for a variety of other purposes.

Photogrammetric surveying is the application to surveying of the science of measurement by means of photographs taken with specially designed cameras either from airplanes or from ground stations. The principles of perspective are employed in the projection of details from the photographs onto maps drawn to scale. In conjunction with limited ground surveys to locate control points visible from the air, *aerial photogrammetry* is employed on many topographic surveys because of its outstanding advantages of speed, wealth of detail, economy, applicability to difficult or inaccessible terrain, and usefulness for a variety of purposes. Aerial photogrammetry is replacing or supplementing the more laborious ground methods on many surveys.

The American Society of Civil Engineers has adopted the following general classification of kinds of surveying and mapping, with specific subdivisions of each. The first four items are classified as part of civil engineering:

1 Land or property surveying (cadastral).
2 Engineering surveys for design and construction.
3 Geodetic surveying, geodetic engineering, or geodesy.
4 Cartographic surveying, cartographic engineering, or map and chart surveying.
5 Aerial survey services.
6 Cartography (not requiring original surveys).

With regard to complete surveys, this book deals principally with land surveying, topographic surveying, and route surveying.

1-4 Units of Measurement. The units of angular measurement are the *degree*, *minute*, and *second*.

In English-speaking countries the common units of linear measurement are the *yard*, *foot*, and *inch*. The *Gunter's chain*, employed in the subdivision of the United States public lands, is 66 ft long and is divided into 100 links each 7.92 in. long. 1 mile = 80 chains = 320 rods = 5,280 ft.

Many other countries employ the *meter* as the unit of length. 1 m = 39.370 in. = 3.2808 ft. The meter is the unit of length employed by the U.S. Coast and Geodetic Survey.

The *vara* is a Spanish unit of linear measurement used in Mexico and several other countries falling under early Spanish influence. Commonly 1 vara = 32.993 in. (Mexico), 33 in. (California), or 33⅓ in. (Texas).

In the United States the units of area commonly used are the *square foot* and the *acre*. Formerly the *square rod* and the *square Gunter's chain* were also used. 1 acre = 10 sq Gunter's chains = 160 sq rods = 43,560 sq ft.

The units of volumetric measurement are the *cubic foot* and *cubic yard*.

1-5 Principles and Practice. The principles of plane surveying involve a working knowledge of geometry, trigonometry, physics, astronomy, and theory of probability. The practice of surveying requires skill in the art of observing and in field and office procedures. Traits of character required for a successful surveyor or engineer include reliability, thoroughness, initiative, resourcefulness, logical reasoning, sound judgment, and an impartial scientific attitude toward each measurement and result.

FIELD WORK

1-6 Student Field Practice. Field work consists in (1) adjusting instruments and caring for field equipment, (2) measuring horizontal and vertical distances and angles, and (3) recording the field measurements usually in the field notebook. The field problems herein give procedures and useful techniques; in practice they might become parts of extended surveys.

Members of the student field parties should from day to day alternately assume the various duties involved in the field work.

Before going into the field, the student should understand exactly what he is to do and why he is to do it. He should note the object of the problem, critically examine the various steps involved, and list the necessary equipment. All equipment should be examined as it is issued, and any defect or injury should be reported immediately.

In the two succeeding articles are given some general suggestions regarding the care, handling, and adjustment of surveying instruments. Other details of field practice are given in connection with the individual field problems.

1-7 Care and Handling of Instruments. The following suggestions apply to such instruments as the transit, level, surveyor's compass, and plane table.

1 Handle the instrument with care, especially when removing it from the box.

2 See that it is securely fastened to the tripod head.

3 Avoid carrying the instrument on the shoulder while passing through doorways or beneath low-hanging branches; carry it under the arm, with the head of the instrument in front.

4 Before climbing over a fence or similar obstacle, place the instrument on the other side, with the tripod legs well spread.

5 Whenever the instrument is being carried or handled, the clamp screws should be clamped lightly so as to allow the parts to move if the instrument is struck.

6 Protect the instrument from impact and vibration.

7 If the instrument is to be shipped, pack paper or cloth around it in the case; and pack the case, well padded, in a larger box.

8 Never leave the instrument unattended while it is set up in the street, on the sidewalk, near construction work, in fields where there are livestock, or in any other place where there is a possibility of accident.

9 Just before setting up the instrument, adjust the wing nuts controlling the friction between tripod legs and head, so that each leg when placed horizontally will barely fall from its own weight.

10 Do not set the tripod legs too close together, and see that they are firmly planted. Push *along* the leg, not vertically downward. So far as possible, select solid ground for instrument stations. On soft or yielding ground, do not step near the feet of the tripod.

11 While an observation is being made, do not touch the instrument except as necessary to make a setting; and do not move about.

12 In tightening the various clamp screws, adjusting screws, and leveling screws, bring them only to a firm bearing. *The general tendency is to tighten these screws far more than necessary.* Such a practice may strip the threads, twist off the screw, bend the connecting parts, or place undue stresses in the instrument so that the setting may not be stable. Special care should be taken not to strain the small screws which hold the cross-hair ring.

13 For the plumb-bob string, learn to make a sliding bowknot that can be easily undone. Hard knots in the string indicate an inexperienced or slovenly instrumentman.

14 Before observations are begun, focus the eyepiece on the cross-hairs, and (by moving the eye slightly from side to side) see that no parallax is present.

15 *When the magnetic needle is not in use, see that it is raised off the pivot.* While the needle is resting on the pivot, impact is apt to blunt the point of the pivot or to chip the jewel and thus to cause the needle to be sluggish.

16 Always use the sunshade. Attach or remove it by a clockwise motion, in order not to loosen the objective.

17 If the instrument is to be returned to its box, put on the dust caps for the objective and the eyepiece, and wipe the instrument clean and dry.

18 Have the rain hood available when the instrument is in use. If caught without it, place the dust cap on the objective and point the telescope up.

19 Remove grit from exposed movable parts such as the threads of tangent screws by wiping them with an oiled rag. If the threads or slides work hard, clean them in gasoline or kerosene, and oil lightly. Use no abrasives.

20 Use only the best quality of clock or watch oil. Oil sparingly, and wipe off the excess oil. In cold weather, it may be necessary to use graphite for lubrication.

21 Never rub the objective or the eyepiece of a telescope with the fingers or with a rough cloth. Use a camel's-hair brush to remove dust, or use clean chamois or lint-free soft cloth if the dust is caked or damp. Occasionally the lenses may be cleaned with a mixture of equal parts of alcohol and water. Keep oil off the lenses.

22 Never touch the graduated circles and verniers with the fingers. Do not wipe them more than necessary, and particularly do not rub the edges.

23 Do not touch the level vials nor breathe on them, as unequal heating of the level tube will cause the bubble to move out of its correct position.

24 If the level vial becomes loose, it can be reset with plaster of paris, or wedged lightly with strips of paper or with toothpicks.

25 Minor repairs such as the placement of a level vial or broken cross hairs may be made in the field, but whenever possible, repairs should be made by an experienced instrument mechanic in the shop or factory.

26 It is advisable to carry a few spare parts such as a level vial, cross-hair ring, foot screw, and tangent screw.

27 In cold weather, the instrument should not be exposed to sudden changes in temperature (as by bringing it indoors); mittens should be worn over the instrumentman's gloves; and the observer should be careful not to breathe on the eyepiece. Films of ice which may form on the lenses may be removed with a pointed piece of wood.

Chaining Equipment. Keep the tape straight when in use; any tape will break when kinked and subjected to a strong pull. Steel tapes rust readily and for this reason should be wiped dry after being used. The method of doing up a tape which is not on a reel is given in Art. 3-3. Use special care when working near electric power lines. Fatal accidents have resulted from throwing a metallic tape over a power line.

Do not use the flagpole as a bar to loosen stakes or stones; such use bends the steel point and soon renders the point unfit for lining purposes.

To avoid losing pins, tie a piece of colored cloth (preferably bright red) through the ring of each.

Leveling Rod. Do not allow the metal shoe on the foot of the rod to strike against hard objects, as this, if continued, will round off the foot of the rod and thus introduce a possible error in leveling. Keep the foot of the rod free from dirt. When not in use, long rods should be either placed upright or supported for their entire length; otherwise they are likely to warp. When not in use, jointed rods should have all clamps loosened to allow for possible expansion of the wood.

Safety. In addition to the precautions previously mentioned, on highway surveys provision should be made against the hazards of passing traffic.

1-8 Adjustment of Instruments. By "adjustment" of a surveying instrument is meant the bringing of the various fixed parts into proper relation with one another, as distinguished from the ordinary operations of leveling the instrument, alining the telescope, etc. It is important that the surveyor:

1 Understand the principles upon which the adjustments are based.

2 Learn the method by which nonadjustment is discovered.

3 Know how to make the adjustments.

4 Appreciate the effect of one adjustment upon another.

5 Know the effect of each adjustment upon the use of the instrument.

6 Learn the order in which adjustments may most expeditiously be performed.

The frequency with which adjustments are required depends upon the particular adjustment, the instrument and its care, and the precision with which measurements are to be taken. Often in a good instrument, well cared for, the adjustments will be maintained with sufficient precision for ordinary surveys over a period of months or even years. On the other hand, blows that may pass unnoticed are likely to disarrange the adjustments at any time. On ordinary surveys, it is good practice to test the critical adjustments once each day, especially on long surveys where frequent checks on the accuracy of the field data are impossible. Failure to observe this simple practice sometimes results in the necessity of retracing lines which may represent the work of several days. Testing the adjustments with reasonable frequency lends confidence to the work and is a practice to be strongly commended. The instrumentman should, if possible, make the necessary tests at a time that will not interfere with the general progress of the survey party. Some adjustments may be made with little or no loss of time during the regular progress of the work.

The adjustments are made by tightening or loosening certain screws. Usually these screws have capstan heads which may be turned by a pin called an *adjusting pin.* Following are some general suggestions:

1 The adjusting pin should be carried in the pocket and not left in the instrument box. Disregard of this rule frequently leads to loss of valuable time.

2 The adjusting pin should fit the hole in the capstan head. If the pin is too small, the head of the screw is soon ruined.

3 Preferably make the adjustments with the instrument in the shade.

4 Before adjusting the instrument, see that no parts (including the objective) are loose. When an adjustment is completed, always check it before using the instrument.

5 When several interrelated adjustments are necessary, time will be saved by first making an approximate or rough series of adjustments and then by repeating the series to make finer adjustments. In this way, the several disarranged parts are gradually brought to their correct position. This practice does not refer to those adjustments which are in no way influenced by others.

1-9 Signals. Some of the common hand signals are as follows:

Right or *Left*. The corresponding arm is extended in the direction of the desired movement. A long, slow, sweeping motion of the hand indicates a long movement; a short, quick motion indicates a short movement.

Up or *Down*. The arm is extended upward or downward, with wrist straight. When the desired movement is nearly completed, the arm is moved toward the horizontal.

All Right. Both arms are extended horizontally, and the forearms waved vertically.

Plumb the Flagpole or *Plumb the Rod*. The arm is held vertically and moved in the direction in which the flagpole or rod is to be plumbed.

Give a Foresight. The instrumentman holds one arm vertically above his head.

Establish a Turning Point or *Set a Hub*. The instrumentman holds one arm above his head and waves it in a circle.

Turning Point or *Bench Mark*. In profile leveling, the rodman holds the rod horizontally above his head and then brings it down on the point.

Give Line. The flagman holds the flagpole horizontally in both hands above his head and then brings it down and turns it to a vertical position.

If he desires to set a hub, he waves the flagpole (with one end of it on the ground) from side to side.

Wave the Rod. The levelman holds one arm above his head and moves it from side to side.

Pick Up the Instrument. Both arms are extended outward and downward, then inward and upward, as they would be in grasping the legs of the tripod and shouldering the instrument.

1-10 Field Notes. Field notes reflect the competency of the surveyor to an even greater extent than does his handling of the instrument. They should be regarded as a permanent record and not merely as memoranda for immediate use. They may be needed later in court proceedings or if the survey is to be rerun or extended. Their value will depend largely upon the clearness and completeness with which they are recorded.

All field notes should be recorded *in the field book* at the time the work is being done. Nothing should be left to memory or copied from temporary notes. The field notes consist of numerical data, explanatory notes, and sketches. Also, the record of every survey or student problem should include the date, the weather conditions, the names and duties of the members of the party, and a title indicating the location of the survey and its nature or purpose.

Before any survey is made, the necessary data to be collected should be considered carefully; and in the field all such data should be obtained, but no more. Sketches should be used freely. The recorder should put himself in the place of one who is not on the ground at the time the survey is made.

Convenient forms of notes in common use are shown in this book, but in many cases it will be necessary for the surveyor to devise his own form of record.

In locating details for mapping, the field notes may be supplemented by photographs taken with an ordinary camera.

1-11 Notebook. The field notebook should be of good-quality rag paper, with stiff cover, and of pocket size. Special field notebooks for particular kinds of notes are sold by engineering supply companies. For general surveying or for students in field work where the problems to be done are general in character, an excellent form of notebook has the right-hand page divided into small rectangles with a red line running up the middle and has the left-hand page divided into several columns; both pages have the same horizontal ruling. In general, tabulated numerical values are written on the left-hand page; sketches and explanatory notes on the right. This type is called a *field book* (see Fig. 10-3). Another common form, used in leveling, has both pages ruled in columns and has wider horizontal spacing than the field book; this is called a *level book* (see Fig. 5-2).

The field notebook may be bound in any of three ways: conventional, ring or spiral, and loose-leaf. Loose-leaf notebooks are increasing in use because of the following advantages: (1) Only one book need be carried; (2) sheets can be withdrawn for use in the field office while the survey is being continued; (3) carbon copies can be made in the field, for use in the field office or headquarters office, and such copies are a protection against loss of data; (4) files can be made consecutive and are not bulky; and (5) the cost of binders is less than that for bound books. The disadvantages are that (1) sheets may be lost or misplaced; (2) sheets may be substituted for other sheets—an undesirable practice; and (3) there may be difficulty in establishing the identity of the data in court, as compared with a bound book. When loose-leaf books are used, *each sheet* should be fully identified by date, serial number, and location.

1-12 Recording Data. A 4H pencil, well pointed, should be used. Because of its simplicity, Reinhardt's style of slope lettering (Fig. 14-3*a*) is commonly considered to be the best form of lettering for taking notes rapidly and neatly. Office entries of reduced or corrected values should be made in red ink, to avoid confusion with the original data.

The figures used should be plain; one figure should never be written over another. In general, *numerical data should not be erased;* if a number is in error, a line should be drawn through it and the corrected value written above. Portions of sketches and explanatory notes may be erased if there is a good reason.

In tabulating numbers, the recorder should place all figures of the tens column, etc., in the same vertical line. Where decimals are used, the decimal point should never be omitted. The number should always show with what degree of precision the measurement was taken; thus a rod reading taken to the nearest 0.01 ft should be recorded not as 7.4 ft but as 7.40 ft. Notes should not be made to appear either more precise or less precise than they really are.

Sketches are rarely made to exact scale, but in most cases they are made approximately to scale. They are made freehand and of liberal size. The recorder should decide in advance just what the sketch is to show. A sketch crowded with unnecessary data is often confusing, even though all necessary features are included. Many features may be most readily shown by conventional symbols (Art. 14-8); special symbols may be adopted for the particular organization or job.

Explanatory notes are employed to make clear what the numerical data and sketches fail to do. Usually they are placed on the right-hand page in the same line with the numerical data that they explain. If sketches are used, the explanatory notes are placed where they will not interfere with other data and as near as possible to that which they explain.

If a page of notes is abandoned, either because it is illegible or because it contains erroneous or useless data, it should be retained, and the word "void" written in large letters diagonally across the page. The page number of the continuation of the notes should be indicated.

1-13 Student Field Notes. A neat title should be made either on the flyleaf or on the cover, showing the owner of the notebook, the number and name of the course, and the year in which the notes are taken. Two or three pages should be left in the front of the book for a table of contents, and the table of contents should always be kept up to date. The remaining pages of the book should be numbered, with one number assigned to each two facing pages, or "spread."

In the field, the recorder should always record at once the uncorrected readings of the instrument and apply any necessary computed corrections afterward. All written field calculations should be made in the field notebook, usually on the right-hand page.

COMPUTATIONS

1-14 General. Herein *computations* are usually considered to be direct mathematical operations (such as multiplication) on given numerical data and yielding a definite result. *Calculations* are considered to be relatively broad general operations which may involve not only computation but also the exercise of judgment in such matters as the organization of related computations and the formulation of any necessary assumptions. The terms necessarily overlap somewhat in meaning.

Computations are made *algebraically* through the use of the simple arithmetical processes, logarithms, and the trigonometric functions; *graphically* by accurately scaled drawings and nomographic charts; or *mechanically* by devices such as the slide rule and computing machines.

The ability to make mental computations quickly is desirable. The student should review the elementary relations of trigonometry; those which are most likely to be needed in surveying are included in Table XII.

Before making calculations of any great importance or extent, the computer should carefully plan a clear and orderly arrangement, using tabular forms so far as practicable. All computations should be preserved in a notebook for that purpose, preferably 8½ by 11 in., having cross-ruled pages which facilitate the making of tabulations and sketches. The pages should be numbered, and a table of contents made. Each problem should have a clear title for identification. Usually enough of the field notes should be transcribed to make computations possible without further reference to the field notebook.

1-15 Checking. In practice, *no confidence is placed in results that have not been checked,* and important results are preferably checked by more than one method. Rarely is a lengthy computation made without some mistake. *Each student should depend upon himself;* comparing each step with that of other students as the work proceeds is not true checking and would not be countenanced in practice. Approximate checks to discover large arithmetical mistakes can be obtained in many cases by the use of such mechanical devices as the slide rule, the planimeter, and the protractor. Often graphical methods may be used as an approximate check; they generally take less time than arithmetical or logarithmic solutions, and possible incorrect assumptions in the precise solution may be detected.

In many cases, large mistakes such as faulty placing of the decimal point can be detected by inspection of the value to see whether it appears reasonable.

1-16 Significant Figures. The term *significant figures* is used to refer to those digits in a number which have meaning; that is, those digits the values of which are known. Measured quantities are not exact, and the number of digits that have meaning is limited by the precision with which the measurement has been made. If we measure roughly a given distance with a steel tape, we may find it to be 732 ft; but if we measure it more carefully, we may find it to be 732.4 ft; or by still greater refinements we may determine it to be 732.38 ft. But we have not yet reached an exact number, nor can we, for there will always be an error of indeterminable amount. The number of significant figures in the three foregoing results is 3, 4, and 5, respectively.

It is not always easy to determine just the degree of uncertainty with which a measurement has been made, but in some cases it can be estimated or calculated with some precision and is expressed by a number called the *probable error* (Art. 2-7). We say, in such cases, that each digit is a significant figure until we reach that one for which the probable error equals or exceeds 5 units. Thus the number 623.58 ± 0.02 has five significant figures, but the number 623.58 ± 0.08 has only four significant figures and should properly be written as 623.6.

In a decimal the number of significant figures is not necessarily the number of decimal places, as the following examples illustrate:

0.0000065 contains two significant figures.
0.00000650 contains three significant figures.
10.00000650 contains ten significant figures.
0.08000650 contains seven significant figures.

In a number ending with one or more ciphers and having no decimal point, the number of significant figures is not definite but can be made so by using powers of 10 as a factor. Thus if 65,000 is written as 65.0×10^3, it is clear that there are three significant figures. If 0.0000065 is written as 6.5×10^{-6}, it is clear that there are two significant figures.

If the maximum allowable error in a value is 1 per cent, the value must be at least 100 and therefore must have at least three significant figures; and so on for other degrees of precision.

The demarcation between successive numbers of significant figures is not a sharp one. Thus, considering whole numbers, 999 has three significant figures and 1,001 has four; but an error of 1 in the last digit of 999 is practically the same in percentage as that for 1,001. For purposes of computation, numbers in the upper range of those having a given number of significant figures may appropriately be used with numbers in the lower range of those having one additional significant figure.

Certain quantities are mathematically exact, and the corresponding numbers can be considered to have as many significant figures as necessary. For example, if a quantity containing several significant figures is to be taken twice, or multiplied by 2, the number 2 is mathematically exact, or *absolute*. Also mathematically exact are the number of degrees about a point and the sum of the angles of a closed figure. Numbers obtained by counting are absolute.

In computations it is advisable to carry out the intermediate results to one figure more than that desired in the final result.

The relation of significant figures to precision of computations is discussed in Art. 1-22.

1-17 Use of Logarithms. For multiplying, dividing, squaring, cubing, or taking the roots of numbers, arithmetic cannot be used advantageously beyond three or perhaps four significant figures. For four or more significant figures, there is a decided advantage in the use of *logarithms,* both to save time and to lessen the likelihood of mistakes.

The logarithm of a number is the power to which some base must be raised to produce the number. In computations made by the surveyor the *common* system of logarithms is employed, for which the base is 10.* Hence,

$$\log 10 = \log 10^1 = 1 \qquad \log 100 = \log 10^2 = 2$$
$$\log 1{,}000{,}000 = \log 10^6 = 6 \qquad \log 1 = \log 10^0 = 0$$
$$\log 0.1 = \log\frac{1}{10} = \log\frac{10^0}{10^1} = 0 - 1 = -1 = 9 - 10$$

For any number (except 1) that is not a power of 10, the logarithm is a fractional quantity. For example,

$$\log 1.5 = \log 10^{0.17609} = 0.17609$$
$$\log 15 = \log 10^{1.17609} = 1.17609$$
$$\log 0.015 = \log\frac{15}{1000} = \log\frac{10^{1.17609}}{10^3} = 1.7609 - 3$$
$$= 8.17609 - 10 = \bar{2}.17609$$

The whole number of the logarithm is called the *characteristic;* the decimal is called the *mantissa.* For a number greater than 1, the logarithm is a positive quantity, and the value of the characteristic is one less than the number of places in the integer of the number.

For a number less than 1, the logarithm is a negative quantity, and to determine the characteristic the common practice is to deduct from 10 a number equal to one more than the number of ciphers to the right of the decimal point. When a logarithm is written in this manner, it is *understood* that 10 is to be deducted from it. Instead of writing the logarithm in this manner, the characteristic is often shown as a quantity one more than the number of ciphers to the right of the decimal point, and a negative sign is placed over it. The logarithm of 0.0435 may therefore appear either as 8.63849 −10 or as $\bar{2}.63849$.

For the same sequence of figures, the mantissa remains unchanged regardless of the position of the decimal point. Thus the logarithm of 4,350 is 3.63849, and the logarithm of 0.00435 is 7.63849 −10 or 3.63849.

The same considerations govern the number of places to be used in logarithmic computations as govern those of arithmetic. The number of significant figures in the final result should be consistent with its purpose or with the precision of the given data, as discussed in Art. 1-16. The usual practice is to use a number of places of logarithms one more than the number of places desired in the final result. In the ordinary work of the surveyor five places are usually sufficient, but sometimes six places are required.

Tables VII and VIII give logarithms of numbers and of the functions of angles to six places, but in using these tables the last figure should be dropped if only five places are required. In tables of logarithms of num-

* For *natural* logarithms, the base is 2.71828+. The natural logarithm of any number is the common logarithm multiplied by 2.30258+.

bers (see Table VII) only the mantissa is shown, and the characteristic must be supplied by the computer. Tables of the logarithmic functions of angles (see Table VIII) show both the characteristic and the mantissa.

Finding Logarithm. The process of finding the logarithm of a number from the tables is best illustrated by an example.

example 1. Find the logarithm of 6,458.6 correct to the sixth place.

In Table VII opposite the number 645 the mantissa in the column headed 8 is 810098, and in the column headed 9 is 810165. The difference between the two is 67. The last figure in the given number is 6, and hence the desired mantissa is $810098 + 0.6 \times 67$. To facilitate the multiplication, the table of proportional parts at the bottom of the page is given. Opposite 67 the quantity in the 6 column is 40.2. Hence $0.6 \times 67 = 40.2$. The mantissa is, therefore, $810098 + 40 = 810138$. The number has four digits to the left of the decimal point, and hence the characteristic is 3 and the logarithm is 3.810138. Note that all logarithms between two adjacent horizontal broken lines have the same figures in their first two places, these figures being shown in the column headed 0.

Finding Antilogarithm. The process of finding an antilogarithm, or number the logarithm of which is known, is the reverse of that just described.

example 2. Find the number whose logarithm is 2.688544. The number is to have five significant figures.

By Table VII it is seen that the mantissa of the logarithm of 4881 is 688509 and that of the logarithm of 4882 is 688598. The difference between these two mantissas is $688598 - 688509 = 89$; the difference between the given mantissa and that for 4881 is $688544 - 688509 = 35$. The required number is therefore $4881 + {}^{35}\!/_{89}$. By the table of proportional parts opposite 89 find the number nearest 35 (it is 35.6). At the head of the column the corresponding number is seen to be 4. The characteristic is 2, and the number is therefore 488.14.

If six places were required, it would be necessary to interpolate between values given in the table of proportional parts. Thus, for the preceding example, the difference between 35 and 26.7 is 8.3, and the difference between 35.6 and 26.7 is 8.9. But $8.3/8.9 = 0.9$ (approximately). Therefore the digit in the sixth place is 9, and the number is 488.139. This interpolation may also be made by using the table of proportional parts directly, merely moving the decimal point one place.

Multiplication. The product of two numbers is determined by adding their logarithms.

example 3

$15 \times 12 = 10^{1.18} \times 10^{1.08} = 10^{2.26} =$ number whose log is $2.26 = 180$

Division. One number is divided by another by subtracting the logarithm of the divisor from that of the dividend.

example 4. $^{180}\!/_{12} = 10^{2.26}/10^{1.08} =$ number whose log is $1.18 = 15$.

Powers. A number is raised to a power by multiplying its logarithm by that power.

example 5. $12^4 = 10^{1.08 \times 4} =$ number whose log is $4.32 = 21{,}000$ (to two significant figures).

Following is an example of a logarithmic process of raising a number less than 1 to a fractional power:

example 6

$0.6324^{1.7180} = 10^{(9.8010-10)1.7180} =$ number whose log is $(9.8010 - 10)1.7180$

$$\log 9.8010 + \log 1.7180 = 0.99127 + 0.23502 - 1.22629$$

$$9.8010 \times 1.7180 = 16.838$$

$$0.6324^{1.7180} = \text{number whose log is } 16.838 - (10 \times 1.7180)$$

$$= \text{number whose log is } \bar{9}.658$$

$$= 0.455$$

In the preceding example the logarithm of a logarithm has been determined; for short this is called the *log log.* Similarly for brevity, the antilogarithm is often called the *antilog.*

Trigonometric Functions. When the logarithm of a trigonometric function is to be found from Table VIII, it should be noted that there are four angular values on each page, one at each corner of the table. Each angular value is used with the line of headings and the column of minutes which are nearest to it, as illustrated by the following example:

example 7. From Table VIII,

log sin 22°12′ is 9.577309 − 10

log sin 67°12′ is 9.964666 − 10

log sin 112°12′ is 9.966550 − 10

log sin 157°12′ is 9.588289 − 10

Functions of an angle α greater than 45° can be determined conveniently by use of the corresponding complement $(90° - \alpha)$ or supplement $(180° - \alpha)$ of the angle, with due regard to the sign of the corresponding function.

1-18 Graphical and Mechanical Methods. Frequently, combined graphical and mechanical methods may be employed in conjunction with algebraic processes. For example, earthwork cross-sections may be plotted to scale (graphical), the area of the cross-sections may be measured with the planimeter (mechanical), and the volume of earthwork may be determined by arithmetic.

The ordinary 10-in. *slide rule* (Art. 1-19) greatly facilitates computations involving no more than three significant figures and is in every way the equivalent of a three-place table of logarithmic functions.

For results of more than three significant figures the *computing machine* is coming largely to replace other methods of computing; the use of logarithms is largely limited to computations in the field. The details of operation vary with the type of machine, but they are simple and can be mastered in a few minutes of instruction.

Electronic computers are used in surveying for repetitive computations, such as those for traverses (Chaps. 15 and 16), cross-sections and volumes of earthwork (Chaps. 6 and 7), and applications of photogrammetry to highway location and construction (Chap. 21). It is not necessary for the surveyor to own or to operate a computer, as standarized programs and commercial computing services are available.

The *polar planimeter* (Art. 7-6) is of great value in finding the area of figures plotted to scale and is the most efficient means of determining the area of figures with irregular or curved boundaries. In general, results may be determined to three significant figures.

1-19 Use of the Slide Rule. Special books of instruction for each of the several varieties of slide rules are issued by the manufacturers. Herein some of the more frequent computations which may be performed on all rules of the Mannheim type will be described briefly.

In Fig. 1-1 is shown one simple style. The two scales on the body of the rule are lettered *A* and *D*, and those on the slide are lettered *B* and *C*. The rectangular glass runner may be moved to any position along the rule, its setting being indicated by a fine line etched on the glass at right angles to the axis of the rule. The *C* and *D* scales are exactly alike and are graduated with numbers from 1 to 10. The *A* and *B* scales are similar to the *C* and *D* scales, but the corresponding graduations are only one-

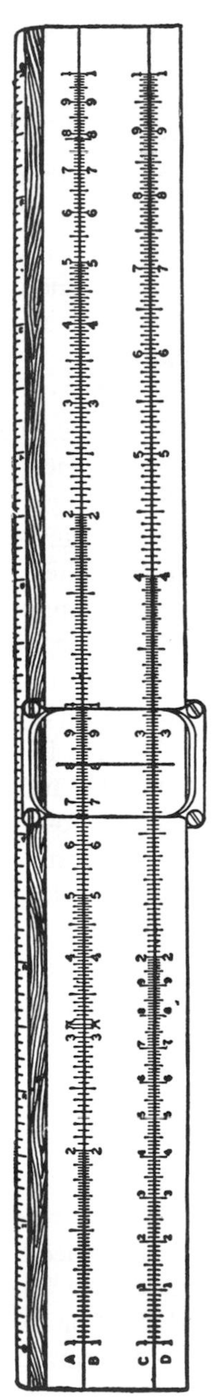

Fig. 1-1. Slide rule (Mannheim type).

half as great. All four of the scales are logarithmic, but the figures shown are for the numbers, not the logarithms.

If for any computation the student is uncertain as to which scales should be employed, he may make a trial computation with simple numbers or with known values of the logarithms or trigonometric functions.

Multiplication. In logarithmic computations two numbers are multiplied together by adding their logarithms. This operation can be performed mechanically by using the slide rule as illustrated by the following examples:

example 1. Multiply 4 by 2.

Set the runner at 2 on scale *D;* move the slide until its left index is at the runner. (Graphically the mantissa of 2 has now been laid off.) Move the runner to 4 on scale *C.* (Graphically the mantissa of 4 has been added to that of 2.) On scale *D* read 8.

example 2. Multiply 8.2 by 7.3.

Set runner to 8.2 on scale *D;* right index to runner; runner to 7.3 on scale *C;* read answer 59.9 on scale *D.* The position of the decimal point is determined by mental computation, using round numbers.

Settings may be made with either index, that one being chosen which will bring the final result within the length of the *D* scale.

Division. Division of one number by another is accomplished by finding the difference between their logarithms. On the slide rule the subtraction is performed mechanically, as follows:

example 3. Divide 8 by 4.

Set the runner at 8 on scale *D.* Set 4 on scale *C* to the runner. Set the runner to the left index on scale *C.* Read the answer 2 on scale *D.*

Squares. Scales *A* and *B* may be used for solving multiplications and divisions in exactly the same manner as are the *C* and *D* scales, but are more commonly employed in conjunction with the *C* and *D* scales for finding the squares and square roots of numbers. The logarithmic scale for *A* and *B* is one half of that for *C* and *D.* If the runner is set to a given number on the *D* scale, the square of the number is given by the runner reading on the *A* scale.

example 4. Square 6.23.

Set runner to 6.23 on scale *D;* read answer 38.8 on scale *A.* Or set runner to 6.23 on scale *C;* read answer 38.8 on scale *B.*

Square Roots. Square root is obtained by setting the runner to the number on the *A* (or *B*) scale and reading the root on the *D* (or *C*) scale. If the *integer* of the number contains an odd number of places (as 4.83; 125; 17,536), the runner is set on the left scale; if it contains an even number of places (as 16; 42.8; 1,174), the runner is set on the right scale. If the number is a decimal without ciphers between the decimal point and the first finite figure (as 0.428; 0.87) or with an even number of ciphers between the decimal point and the first finite figure (as 0.0087; 0.000064), the runner is set to the number on the right scale; if the number is a decimal with an odd number of ciphers between the decimal point and the first finite figure (as 0.0426; 0.0000065), the runner is set to the number on the left scale. If in doubt, the correct scale to use may be found by rough trial, using a round number which is a perfect square and which is near to the number under consideration.

example 5. Find the square root of 16.4.
Set runner to 16.4 on right *A;* read answer 4.05 on *D*.

example 6. Find the square root of 1.64.
Set runner to 1.64 on left *A;* read answer 1.28 on *D*.

Trigonometric Functions. On the back of the slide is a scale *S* for sines and a scale *T* for tangents. If the slide is removed, turned over, and replaced so that the indexes of the *S* and *T* scales coincide with those of the *A* and *D* scales, respectively, the values of the natural sines of angles between 0°34′ and 90° and of natural tangents between 5°43′ and 45° are read directly from the *A* and *D* scales, respectively. Thus by readings on the *A* scale the sine of 0°34′ is seen to be 0.0100, the sine of 5°44′ is seen to be 0.100, the sine of 30° is seen to be 0.500, etc.; and by readings on the *D* scale the tangent of 5°43′ is seen to be 0.100, the tangent of 30° is seen to be 0.577, etc. The tangent of any angle less than 5°43′ may be considered to equal the sine, within the precision of the slide rule. Sines of angles less than 0°34′ may be assumed to be proportional to the angle. On this assumption, since the sine of 0°34′ is 0.0100, the sine of 0°05′ would be $5/34 \times 0.0100 = 0.00147$. The correct value is 0.00145.

The values of other trigonometric functions may be obtained by the simple relations of trigonometry.

Logarithms. Almost every slide rule has a scale *L* divided into 100 equal major parts. On the rule shown in Fig. 1-1, this scale is on the back of the slide and its graduations are numbered from the right end. From this scale the logarithm of a number may be found as follows: Set the runner on the number on the D scale. Move the slide until the left index is at the number. Read the mantissa from the scale of equal parts.

1-20 Drafting. It is assumed that the student is familiar with the use of the ordinary drafting instruments and with the elements of mechanical drawing. The drawings of surveying consist of maps, profiles, cross-sections, and, to some extent, graphical computations. The usefulness of these drawings is dependent largely upon the accuracy with which points and lines are projected upon paper; few dimensions are shown. Much of the drafting calls for a degree of skill and precision unnecessary on dimensional plans, and the student should exercise great care in plotting.

The details of drafting that pertain to the drawings of surveying are described in Chap. 14.

PRECISION OF MEASUREMENTS AND COMPUTATIONS

1-21 Precision of Measurements. The degree of precision of a given measurement depends upon the methods and instruments employed and upon other conditions surrounding the survey. Although it is desirable that all measurements be made with high precision, the fact must be considered that increasing the precision usually increases the time and labor in more than direct proportion. It therefore becomes the duty of the surveyor to maintain a degree of precision as high as justified by the purpose of the survey, but not higher. Before undertaking a series of measurements he should consider the sources of error, the instruments and methods employed to reduce these errors within allowable limits, the methods of checking, and the most efficient organization of the work.

It is common practice before beginning a survey to fix the permissible error of linear measurement, this being expressed as a ratio, for example, 1/1,000; sometimes it is desired to locate a given point within a specified distance of its true location. If measurements are to be consistent, the precision of angles should correspond to the precision of related distances—in other words, the error in location of a point on account of error in angle should not be greatly different from its error in location on account of error in distance. With the ratio of precision of linear measurements and the average length of sighting to determine angles known, the permissible limit of error in angles can be determined by simple trigonometry. As an illustration, the tangent (or sine) of 01′ is 0.000291; hence the linear error at a distance of 1,000 ft for an angular error of 01′ is 0.291 ft, and the corresponding ratio of precision in linear measurement is 1/3,440. For other distances from the vertex of the angle the linear error is in direct proportion.

In some surveys in which distances are determined only roughly, it is possible to measure the angles with greater than the required precision without increased effort or loss of time. For example, in rough taping the ratio of precision might be 1/1,000, corresponding to an angular error of 03′; but with the ordinary transit the angles could be determined to the nearest 01′ as readily as to the nearest 03′.

Often field measurements are made the basis of computations involving the trigonometric functions. As these functions are not proportional to the size of the angle, theoretically the precision with which an angle is determined should be made to depend upon the size of the angle and upon the function to be used in the computations. This procedure is not practicable in every case, but attention to the foregoing principle will often enable the surveyor to avoid the use of angles which would result in large errors in computation. For example, in triangulation the computations involve the use of sines; if a precision not lower than 1/4,000 is desired and angles are measured to the nearest minute (possible error 30″), no angle used in the computations should be less than 30°. Charts are available giving the ratios of precision of linear measurements corresponding to various angular errors for sines, cosines, tangents, and cotangents (Ref. 8 in the list following Chap. 23).

1-22 Precision of Computations. A proper regard for consistency between measured values and the computed results based upon them requires an understanding of the effects of the errors of measurement when combined in the operations of arithmetical computations.

Two important principles are as follows: in addition (or subtraction), the precision of the values is governed by the number of *places of figures,* whereas in multiplication (or division), the precision is governed by the number of *significant figures* (Art. 1-16). Application of these principles will be discussed in the following paragraphs.

Addition. Suppose that it is desired to add two (or more) quantities of earthwork which have been measured with precision appropriate to the values 37.2 and 468 cu yd. The sum, 505.2, cannot properly be expressed to tenths because one (or more) of the quantities has not been measured to tenths. Therefore, the sum should be written as 505 cu yd.

To illustrate the fact that significant figures do not control the precision of addition, suppose that the taped distances 104.32 ft and 0.64 ft are to be added together, yielding the sum 104.96 ft. The result is properly expressed to hundredths of feet and contains five significant figures, even though one of the quantities used in the computation has only two significant figures.

Further, the amount of the total error in a sum may be shown by supposing that a considerable number of earthwork quantities are to be added, as shown, and that the probable error in each quantity is know to be ±0.3 cu yd. The sum is 501.7 ± cu yd, but this number is affected by the probable error of each quantity of which it is composed. These separate probable errors, assumed here to be ±0.3 cu yd, are accidental in nature (Art. 2-4). Hence they will probably combine in the sum as the square root of the number of times which they occur. In this example, then, since there are 10 numbers, the total probable error will be ±0.3 cu yd $\times \sqrt{10}$, or about ±1.0 cu yd. But when the last digit is in doubt by more than 5 units, it has ceased to be a significant figure, and the result should more properly be written as 502 cu yd. Whether it is so written or as 501.7 cu yd, it has essentially three significant figures and no more.

37.2
45.6
53.1
63.2
43.7
45.2
63.8
72.1
36.4
41.4
501.7

Multiplication. The relative error in a product is equal to the sum of the relative errors in the factors. It follows that on a relative basis the probable error of a product cannot be less than that of the least precise factor.

An application of this principle is shown by the triangle (Fig. 1-2) whose sides $a = 680.8$ ft, $b = 75.30$ ft, and angle $C = 132°02'$ are obtained by field measurement. If it is desired to compute the area of the triangle (area = ½ $ab \sin C$) with a precision of four significant figures, then a, b, and C must be measured with such precision as to yield four significant figures. Side a must be measured to tenths, whereas side b must be measured to hundredths.

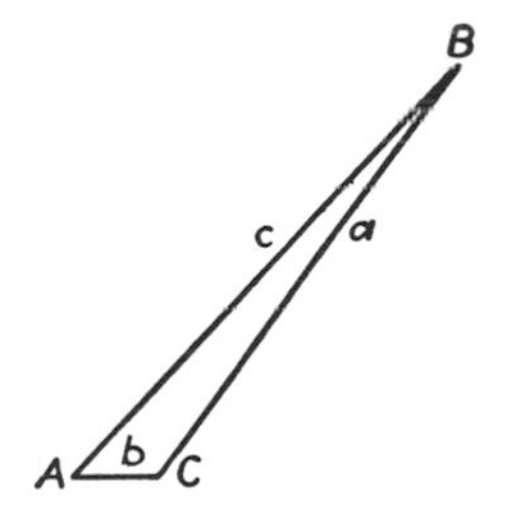

Fig. 1-2.

Four significant figures are used when the ratio of precision lies between 1/1,000 and 1/10,000. In the example just given, assuming that the permissible error is about 1/4,000, the angle C should be measured with such precision that the use of sin C will not introduce an error greater than about 1/4,000. For the angle shown, this allowable error is slightly less than 01′.

Units. The precision of a result is independent of the unit in which it is expressed. Thus, in the case of the triangle mentioned above, the area is 19,040 sq ft, which is properly expressed in acres as 0.4371 acre. There are four significant figures in each value.

1-23 Numerical Problems

1 How many feet are there in 3½ rods? In 6 chains and 8 links?

2 How many acres are there in a rectangular field 18.25 rods wide and 920.0 ft long?

3 How many significant figures are there in each of the following numbers?

a. 0.208
b. 76248.0
c. 1.4800
d. 30.0×10^6
e. 0.006
f. 1.04 ± 0.02
g. 28.46 ± 0.08

4 Find the product of each of the following series of numbers, using logarithms:

a. $2.64 \times 79.18 \times 0.0767 \times 1.0028$
b. $(3.415)^2 \times \sqrt[3]{72.849}$
c. $(6.618)^{1/3} \times (68.627)^{1/2}$

5 Find the quotient for each of the following series of numbers, using logarithms:

a. $76.21 \div 49.20$ (5-place logs)
b. $0.4092 \div 0.006314$
c. $1.4823 \div 16.4917$ (6-place logs)
d. $8.472 \div \tan 42°21'$

6 Compute the log of 27, $\sqrt{15}$, and $\sqrt[3]{12}$ without reference to tables, using only the logarithms of the following numbers: log 2 = 0.30103, log 3 = 0.47712, log 5 = 0.69897.

2
ERRORS

ERRORS

2-1 General. Every observed or measured quantity contains errors of unknown magnitude due to a variety of causes, and hence a measurement is never exact. One of the important functions of the surveyor is to secure measurements which are correct within certain limits of error prescribed by the nature and purpose of the survey. This requires that he know the sources of errors, understand the effect of the various errors upon the observed quantities, and be familiar with the procedure necessary to maintain a required precision.

In dealing with measurements, it is important to distinguish between *accuracy* and *precision*. As defined by the American Society of Civil Engineers, accuracy is "absolute nearness to the truth" whereas precision is "relative or apparent nearness to the truth." (Formerly ASCE defined precision as "degree of fineness of reading in a measurement, or, the number of places to which a computation is carried.") As defined by the U.S. Coast and Geodetic Survey, accuracy is "degree of conformity with a standard" whereas precision is "degree of refinement in the performance of an operation or in the statement of a result." From these mutually consistent definitions it follows that a measurement may be accurate without being precise, and *vice versa*. For example, a distance may be measured very carefully with a tape, to thousandths of a foot, and still be in error of several hundredths of a foot because of erroneous length of tape; the measurement is precise but not accurate.

2-2 Sources of Error. In the measurements of surveying, *instrumental errors* arise from imperfections or faulty adjustment of the devices with which measurements are taken; *personal errors* occur through the observer's inability to manipulate or to read the instruments exactly; and *natural errors* or *external errors* occur from variations in the phenomena of nature such as temperature, humidity, wind, gravity, refraction, and magnetic declination.

2-3 Kinds of Error. A *mistake,* or *blunder,* is an unintentional fault of conduct arising from poor judgment or from confusion in the mind of the observer. Mistakes have no place in a discussion of the theory of errors. They are detected and eliminated by checking all work.

The *resultant error* is the difference between a measurement and the true value. It is made up of individual errors from a variety of sources. If the measurement is too large, the error is said to be positive; if too small, the error is said to be negative.

A *discrepancy* is the difference between two measurements of the same quantity.

A *systematic error* is one that, so long as conditions remain unchanged, always has the same magnitude and the same algebraic sign (which may be either positive or negative). If conditions do not change during a series of measurements, the error is termed a *constant* systematic error; for example, a line may be measured with a tape which is too short. If conditions change, resulting in corresponding changes in the magnitude of the error, it is termed a *variable* systematic error; for example, a line may be chained during a period in which the temperature varies. A systematic error always follows some definite mathematical or physical law, and a correction can be determined and applied. The error may be instrumental, personal, or natural.

An *accidental error* is an error due to a combination of causes beyond the ability of the observer to control and for which it is impossible to make correction; for each observation the magnitude and algebraic sign of the accidental error are matters of chance and hence cannot be computed as can the magnitude and algebraic sign of a systematic error. However, accidental errors taken collectively obey the law of probability (see Art. 2-5). Since each accidental error is as likely to be positive as negative, a certain compensative effect exists; and accidental errors are sometimes called "compensating" errors. Accidental errors are also termed "irregular errors" or "erratic errors." As an example of the occurrence of an accidental error, in chaining it is impossible to set the chaining pin exactly at the proper graduation on the tape. Accidental errors remain after mistakes have been eliminated by checking and systematic errors have been eliminated by correction.

2-4 Systematic and Accidental Errors Compared. The total systematic error in any given number of measurements is the algebraic sum of the individual errors of the individual measurements. According to the mathematical theory of probability, accidental errors tend to increase in proportion to the *square root* of the number of opportunities for error. Thus if the accidental error in measuring one tape length were ± 0.02 ft, the chances would be even that the total accidental error due to measuring 100 tape lengths would not exceed $\pm 0.02 \times \sqrt{100} = \pm 0.20$ ft. A systematic error of the same magnitude would produce a total error of $0.02 \times 100 = 2.00$ ft. It is thus seen that for any connected series of observations of independent but related quantities the accidental errors are of relatively small importance as compared with systematic errors *of the same magnitude*. However, for many observations the order of procedure is such that the systematic errors are either eliminated or at least reduced to a negligible quantity. In general, the more refined the methods used, the smaller the systematic errors as compared with the accidental errors.

If the discrepancy between a check measurement and the original measurement is small, it is an indication that no mistakes have been made and that the accidental errors are small, but it is not an indication that the systematic errors are small.

2-5 The Theory of Probability. For surveys of higher precision special effort is made to eliminate systematic errors, and the precision of a measured quantity is governed by the accidental error which it contains. To form a judgment of the probable value or the probable precision of a quantity, *from which systematic errors have been eliminated,* it is necessary to rely upon the theory of probability, which deals with accidental errors of a series of either like or related measurements. It is assumed that small errors are more numerous than large errors, that no very large errors occur, that errors are as likely to be positive as negative, and that the true value of a quantity is the mean of an infinite number of like observations. In practice it is impossible either to eliminate systematic errors entirely or to take an infinite number of observations; hence the value of a quantity is never known exactly. However, in the discussions to follow, it is assumed that systematic errors are so far eliminated as to be a negligible factor.

A thorough understanding of the law of probability may be obtained only by the study of a text on least squares, but a few of its simpler applications will be stated in the following articles.

OBSERVATIONS OF EQUAL RELIABILITY

2-6 Probable Value. The *most probable value* of a quantity is a mathematical term used to designate that adjusted value which, according to the principles of least squares, has more chances of being correct than has any other. Determination of the most probable value from a series of measurements is the principal use which the surveyor makes of the theory of probability.

Same Quantity. For a series of measurements of the same quantity made under identical conditions, the most probable value is the mean of the measurements.

example 1. After all systematic errors have been eliminated, the several measured lengths of a line are 1,012.36, 1,012.35, 1,012.38, 1,012.32, 1,012.33, and 1,012.30 ft. The most probable value is the mean of the measurements, or 1,012.34 ft.

Related Quantities. For related measurements taken under identical conditions, the sum of which should equal a mathematically exact quantity, the most probable values are the observed values corrected by an equal part of the total error. (This situation can arise only in the case of angles about a point or angles in a closed figure.) The correction is in proportion to the *number* of related measurements and not to the *magnitude* of the individual measurements.

example 2. The angles about a point have the following observed values: 130°15′20″, 142°37′30″, and 87°07′40″. The sum of these measurements is 360°00′30″; therefore the total error is 30″. Since there are three angles, the error is assumed to be 10″ for each measurement. The most probable values are

$$\begin{array}{rcrcr} 130°15'20'' & - & 10'' & = & 130°15'10'' \\ 142°37'30'' & - & 10'' & = & 142°37'20'' \\ 87°07'40'' & - & 10'' & = & 87°07'30'' \\ \hline 360°00'30'' & - & 30'' & = & 360°00'00'' \end{array}$$

For related measurements taken under identical conditions, the sum of which should equal a single measurement taken under the same conditions, the most probable values are obtained by dividing the discrepancy equally among all the measurements, including the sum. If the correction is added to each of the related measurements, it is subtracted from the measurement representing their sum, and *vice versa*.

example 3. Measurements of three angles about a point O are AOB = 12°31′50″, BOC = 37°29′20″, and COD = 47°36′30″. The measurement of the single angle AOD is 97°37′00″. The discrepancy between the sum of the three measured angles and the measurement of the angle representing their sum is 40″. Since the size of the errors is independent of the size of the angle, the discrepancy is divided into equal parts: $^{40}\!/_{4} = 10''$. This correction is to be subtracted from measurements of each of the angles *AOB, BOC,* and *COD* and is to be added to the measurement of the angle *AOD*. The most probable values are

$$
\begin{aligned}
AOB &= 12°31'50'' - 10'' = 12°31'40'' \\
BOC &= 37°29'20'' - 10'' = 37°29'10'' \\
COD &= 47°36'30'' - 10'' = 47°36'20'' \\
\text{Sum} \quad & \ 97°37'40'' - 30'' = 97°37'10'' \\
AOD &= 97°37'00'' + 10'' = 97°37'10''
\end{aligned}
\left.\vphantom{\begin{aligned}a\\b\end{aligned}}\right\}\textit{check}
$$

In adjusting several measurements the sum of which should equal a single measurement, there is a distinction between observations like those cited in Example 3 and those for which several operations are involved in the measurement of a single quantity, for example, the length of a long line. If the case illustrated by Example 3 involved linear instead of angular measurements, application of the method would be limited to distances not greater than one tape length.

2-7 Probable Error. *Probable error* is a plus or minus quantity within which limits the actual accidental error is as likely as not to fall. Thus if 6.23 represents the mean of several measurements and 0.11 represents the probable error of the mean value, the chances are even that the true value lies between the limits $6.23 - 0.11 = 6.12$ and $6.23 + 0.11 = 6.34$. The quantity would be written 6.23 ± 0.11. The *probable ratio of precision* of the measurement is $0.11 \div 6.23 = 1/57$ (approximately). Throughout this chapter the discussion of probable errors may, in the case of linear measurements, be applied to probable ratios of precision by converting one method of expressing the precision into terms of the other method as desired.

Another measure of the precision of a series of observations is *standard deviation*, or *standard error*. It is computed, as indicated by Eq. (1) and Example 1, by the principles of least squares. Probable error is simply a special case of standard deviation which corresponds to a 50 per cent probability that the accidental error will lie within (or without) the stated limits. The two measures are related by a numerical constant (probable error = 0.6745 standard deviation). For the purposes of this book the probable error, or 50 per cent error, is adequate and simple, and the discussions of this chapter are based on probable error rather than on standard deviation.

In the adjustment of observations, the probable error of the most probable value of each quantity can be estimated from a series of measurements of that quantity, and the probable errors can then be used in computing weights and/or the corrections to be applied to related quantities. Consideration of probable error is also useful in choosing methods of surveying to produce desired degrees of precision.

Same Quantity. It has been stated that the mean of a series of observations of a single quantity is the most probable value. For the purpose of determining the probable error, this mean value is mathematically regarded as being the most likely value (based on this series of observations), and the difference between each of the individual measurements and the mean value is determined. These differences are termed *residuals* or *deviations*. The theory of least squares demonstrates that the probable error is a function of the square root of the sum of the squares of the residuals.

For the ordinary measurements of surveying, the relations stated below may be applied to series of perhaps 10 or more measurements.

The probable error of a *single observation* is calculated by the equation

$$E = 0.6745 \sqrt{\frac{\Sigma v^2}{n - 1}} = 0.6745\sigma \tag{1}$$

in which Σv^2 is the sum of the squares of the residuals, n is the number of observations, and σ is the standard deviation.

The probable error of a single observation may be calculated approximately by the equation

$$E = 0.845\bar{v} \text{ (approx.)} \tag{2}$$

in which $\bar{v}$ is the mean value of the residuals without regard to signs. The term $\bar{v}$ is also called the *average deviation.*

Since the determination of probable error is at best an approximation, in many cases it is permissible (and usually conservative) to take the probable error of a single observation as being roughly equal to the average deviation, or

$$E = \bar{v} \text{ (rough)} \tag{3}$$

The probable error of the *mean* of a number of observations of the same quantity is inversely proportional to the square root of the number of observations, or

$$E_m = \frac{E}{\sqrt{n}} \tag{4}$$

example 1. Following is a series of 10 rod readings which were taken with a wye level under identical conditions.

Rod reading, ft	*v, ft*	v^2
2.467	0.002	0.000004
2.460	0.005	0.000025
2.469	0.004	0.000016
2.465	0.000	0.000000
2.471	0.006	0.000036
2.461	0.004	0.000016
2.463	0.002	0.000004
2.466	0.001	0.000001
2.460	0.005	0.000025
2.468	0.003	0.000009
	$\Sigma v = 0.032$	$\Sigma v^2 = 0.000136$
Mean = 2.465	$\bar{v} = 0.0032$	$\sigma = \pm\sqrt{\dfrac{0.000136}{9}} = \pm 0.00389$

From the values in the tabulation, the probable error of a single observation is

By Eq. (1), $E = \pm 0.6745\sqrt{\dfrac{0.000136}{9}} = \pm 0.00262$ ft

By Eq. (2), $E = \pm 0.845 \times 0.0032 = \pm 0.00270$ ft

By Eq. (3), $E = \pm 0.00320$ ft

By Eq. (4), using the value of E determined by Eq. (1), the probable error of the mean is

$$E_m = \pm\frac{0.00262}{\sqrt{10}} = \pm 0.00083 \text{ ft}$$

As soon as the residuals are computed, they should be examined in comparison with the average residual. Values corresponding to any unduly large residuals, say four or five times the average residual (excluding the doubtful value), should be rejected and the computation continued with the remaining values.

Related Quantities. The probable error of the sum of observations each having the same probable error is equal to the probable error of a single observation multiplied by the square root of the number of observations (opportunities for error), or

$$E_s = E\sqrt{n} \tag{5}$$

Equation (5) corresponds to a special case of Eq. (10), Art. 2-10. See also "Addition" in Art. 1-22.

example 2. If the probable error in measuring one tape length were ± 0.01 ft, the probable error in the measured length of a line 1 mile long (assuming full 100-ft tape lengths, without breaking tape) would be $\pm 0.01 \times \sqrt{53} = \pm 0.07$ ft.

OBSERVATIONS OF DIFFERENT RELIABILITY

2-8 Weight. In the foregoing discussion it has been assumed that all observations are equally reliable. Frequently in surveying, however, it is required to combine the results of measurements which are not made under similar conditions and which therefore have different degrees of reliability. In such cases it is necessary to consider the degree of reliability, or *weight* (as nearly as it can be determined), that applies to each of the separate measurements. For example, suppose that an angle has been measured perhaps at different times and perhaps by different observers but presumably with equal care; and suppose that the results are as follows:

47°37′40″ (one measurement)
47°37′22″ (four measurements)
47°37′30″ (nine measurements)

It is assumed that weights are proportional to the number of observations; thus in the example the second value has four times the reliability of the first, and the third value has nine times the reliability of the first. For convenience, usually a weight of unity is assigned to the least precise value. Weights are relative or comparative, not absolute; thus the numbers 2, 8, and 18 would represent the weights as well as the numbers 1, 4, and 9.

Often weights will be assigned to observations, not according to the number of observations, but arbitrarily according to the judgment of the observer. For example, he might judge the value of an elevation secured from a line of levels run on a calm, temperate day as being two or three times as reliable as that secured from another line of levels run over the same route but on a windy, cold day.

If the probable error is known instead of the number of observations, the weight can be computed as follows: For observations (of a given quantity) made with equal care, it has been stated that (1) weights vary directly with the number of observations and (2) probable errors (of the mean value) vary inversely with the square root of the number of observations. It follows that *weights are inversely proportional to the square of the corresponding probable errors,* or

$$\frac{W_1}{W_2} = \frac{E_2^2}{E_1^2} \quad \text{or} \quad W_1E_1^2 = W_2E_2^2 = W_3E_3^2 \cdots \tag{6}$$

2-9 Adjustment of Weighted Observations. With the weights known, the most probable values can be determined as follows:

Same Quantity. The most probable value of a quantity for which measurements of different reliability have been made is the *weighted mean.* The weighted mean is computed by multiplying each value by its weight, adding the products, and dividing by the sum of the weights.

example 1. It is desired to determine the most probable value of the angle discussed in the preceding article. For each value the number of observations is given, and hence the weight is known. In the following computation the labor is reduced by employing only the seconds, which represent the *differences* between the observed values and the common value 47°37′.

47°37′40″ ×	1 =	47°37′40″
22″ ×	4 =	88″
30″ ×	9 =	270″
Sum =	14	398″

Weighted mean = 47°37′28″, most probable value

example 2. The length of a line has been determined by three different methods, with results (after correction for systematic errors) as follows:

Measurement	*Observed length, ft*
a	1,021.05 ± 0.02
b	1,021.37 ± 0.04
c	1,021.62 ± 0.06

Since the probable errors are given, the weights can be computed from Eq. (6).

$$W_a(0.02)^2 = W_b(0.04)^2 = W_c(0.06)^2$$

or

$$W_a = 4W_b = 9W_c$$

Let $W_a = 1$; then $W_b = \frac{1}{4}$ and $W_c = \frac{1}{9}$.

1,021.05 ×	1	=	1,021.05
1,021.37 ×	¼	=	255.34
1,021.62 ×	⅑	=	113.51
Sum =	49/36		1,389.90

Weighted mean = 1,021.15 ft, most probable value

The labor of computation could have been reduced by weighting the differences between the observed values and some common value such as 1,021.00.

Probable Error of Weighted Mean. By the principles of least squares, it is known that the probable error of the weighted mean is

$$E_{mw} = 0.6745 \sqrt{\frac{\Sigma(Wv^2)}{(\Sigma W)(n-1)}} \tag{7}$$

Equation (7) is considered sufficiently precise for most purposes of surveying, although in other fields a method of "propagation of error" would be used.

example 3. To determine the probable error of the weighted mean computed in the preceding example, the computation is as indicated by the successive columns of the accompanying tabulation.

Observed length, ft	v	v^2	W	Wv^2
1,021.05	0.10	0.0100	1	0.0100
1,021.37	0.22	0.0484	¼	0.0121
1.021.62	0.47	0.2209	⅑	0.0245
1,021.15, weighted mean			$\Sigma W = 49/36$	$\Sigma(Wv^2) = 0.0466$

Then by Eq. (7),

$$E_{wm} = 0.6745 \sqrt{\frac{0.0466}{(49/36)(3-1)}} = 0.09 \text{ ft}$$

Related Quantities. When the sum of measured values having different weights must equal a known value, either measured or exact, the most probable values are the observed values each corrected by an appropriate portion of the discrepancy or of the total error. *The corrections to be applied are inversely proportional to the weights.* As before, the weights may be determined from the number of observations, from the probable errors, or arbitrarily.

$$\frac{C_1}{C_2} = \frac{W_2}{W_1} \qquad \text{or} \qquad C_1W_1 = C_2W_2 = C_3W_3 \cdots \tag{8}$$

example 4. Two angles *AOB* and *BOC* and the single angle *AOC* are measured about a point *O* under identical conditions, with results as given in the following tabulation. It is desired to determine the most probable values.

Angle	*Observed value*	*No. of measurements*
AOB	23°46′00″	1
BOC	59°14′27″	4
AOC	83°01′07″	6

The discrepancy between the sum of angles *AOB* and *BOC* and the angle *AOC* is 40″. The weights are 1, 4, and 6, respectively; hence the comparative corrections are 1, ¼, and ⅙, respectively. The sum of the comparative corrections is equal to $^{24}/_{24} + {}^{6}/_{24} + {}^{4}/_{24} = {}^{34}/_{24}$; in such cases it is said that there are 34 *parts* of the total correction. The total correction in seconds is divided among the angles in proportion to the individual comparative corrections (parts); thus the individual corrections are

$$C_{AOB} = {}^{24}/_{34} \times 40'' = 28''$$
$$C_{BOC} = {}^{6}/_{34} \times 40'' = 07''$$
$$C_{AOC} = {}^{4}/_{34} \times 40'' = 05''$$

For angles *AOB* and *BOC* whose sum was smaller than *AOC,* the correction is to be added; for *AOC* the correction is to be subtracted. The most probable values are

$$\begin{aligned} AOB &= 23°46'00'' + 28'' = 23°46'28'' \\ BOC &= 59°14'27'' + 07'' = \underline{59°14'34''} \\ &\qquad\qquad \text{Sum} = 83°01'02'' \\ AOC &= 83°01'07'' - 05'' = 83°01'02'' \end{aligned} \Bigg\} \textit{check}$$

Since corrections are inversely proportional to weights, and since weights are inversely proportional to the square of the corresponding probable errors, it follows that *corrections are directly proportional to the square of the corresponding probable errors,* or

$$\frac{C_1}{C_2} = \frac{E_1^2}{E_2^2} \qquad \text{or} \qquad \frac{C_1}{E_1^2} = \frac{C_2}{E_2^2} = \frac{C_3}{E_3^2} \cdots \tag{9}$$

Direct use of this relation obviates the determination of weights when probable errors are given, and thus simplifies the computations.

example 5. Three angles about a point are measured by a series of observations. The mean values with their probable errors are given in the following tabulation.

$$\begin{aligned} AOB &= 130°15'20'' \pm 02'' \\ BOC &= 142°37'30'' \pm 04'' \\ COA &= \underline{\ 87°07'40''} \pm 06'' \\ \text{Sum} &= 360°00'30'' \end{aligned}$$

The total error, and therefore the total correction to be made, is 30″; that is, $C_{AOB} + C_{BOC} + C_{COA} = 30''$. The successive columns of the accompanying tabulation show the steps in computing the corrections.

	Probable error E			*Correction C*		
Angle	*Absolute*	*Comparative*	E^2	*Comparative*	*Absolute*	
AOB	02″	1	1	1	$^1/_{14} \times 30'' = 02''$	
BOC	04″	2	4	4	$^4/_{14} \times 30'' = 09''$	
COA	06″	3	9	9	$^9/_{14} \times 30'' = 19''$	
Sum				14	$^{14}/_{14}$	30″

The most probable values are therefore

$$
\begin{aligned}
AOB &= 130°15'20'' - 02'' = 130°15'18'' \\
BOC &= 142°37'30'' - 09'' = 142°37'21'' \\
COA &= \ \ 87°07'40'' - 19'' = \ \ 87°07'21'' \\
\text{Sum} &= 360°00'30'' - 30'' = 360°00'00''; \textit{ check}
\end{aligned}
$$

2-10 Errors in Computed Quantities. The probable error of the *sum* of independent measurements $Q_1, Q_2, \ldots, Q_n$ for which the probable errors are $E_1, E_2, \ldots, E_n$, respectively, is

$$E_s = \sqrt{E_1^2 + E_2^2 + \cdots + E_n^2} \tag{10}$$

If $E_1, \ldots, E_n$ are all equal, this equation reduces to $E_s = E\sqrt{n}$.

The probable error of the *difference* between two independent measurements Q_1 and Q_2 for which the probable errors are E_1 and E_2, respectively, is

$$E_d = \sqrt{E_1^2 + E_2^2} \tag{11}$$

The probable error of the *product* of a constant or known quantity K and a measured quantity Q for which the probable error is E is

$$E_p = KE \tag{12}$$

If E_1 and E_2 represent, respectively, the probable errors of lengths L_1 and L_2, the probable error of the area representing the *product* of these two lengths is

$$E_a = \sqrt{L_1^2 E_2^2 + L_2^2 E_1^2} \tag{13}$$

2-11 Summary of Principal Relations. In each of the four cases which arise in practice, the most probable value is as shown in the following tabulation:

	Most probable value	
Measurements	*Same quantity*	*Related quantities*
Of equal reliability	Mean	Each observed value corrected equally
Of different reliability	Weighted mean	Each observed value corrected by an amount inversely proportional to its weight

The corresponding probable errors are as follows:

Measurements	Probable error of most probable value	
	Same quantity	*Related quantities*
Of equal reliability	$E_m = 0.6745\sqrt{\frac{\Sigma v^2}{n(n-1)}} = \frac{E}{\sqrt{n}}$	$E_s = E\sqrt{n}$
Of different reliability	$E_{wm} = 0.6745\sqrt{\frac{\Sigma(Wv^2)}{(\Sigma W)(n-1)}}$	$E_s = \sqrt{E_1^2 + E_2^2 + \cdots + E_n^2}$

Weights to be used in the adjustment of weighted observations are determined (1) as being proportional to the number of like observations of a given quantity ($W \propto n$); (2) as being inversely proportional to the square of corresponding probable errors ($W \propto 1/E^2$); or (3) by arbitrary assignment.

For the adjustment of weighted observations of related quantities, the corrections are taken as being inversely proportional to the corresponding weights ($C \propto 1/W$).

2-12 Numerical Problems

1 The following values were observed in a series of rod readings under identical conditions. What is the most probable value? Its probable error? What is the probable error of a single measurement (*a*) as nearly as can be determined and (*b*) as determined by the various approximate relations?

Rod readings, ft

3.187	3.181	3.186	3.181
3.182	3.184	3.183	3.188
3.179	3.176	3.178	3.179

2 Adjust the following angles measured at station *O*:

Angle	*Observed value*
AOB	46°14′45″
BOC	74°32′29″
COD	85°54′38″
AOD	206°41′28″

3 The interior angles of a triangle are observed to be: $A = 28°53'58''$, $B = 61°05'50''$, and $C = 90°00'00''$. What is the most probable value of each of these angles?

4 The difference in elevation between two points is determined to be 117.843 ft, by leveling over a route in which 18 setups are required. It is estimated that the probable error of the difference in elevation determined at each setup is 0.003 ft. What is the probable error of the total difference in elevation?

5 The difference in elevation between two points is observed by three independent measurements, with results as follows:

214.38 ± 0.09 ft
214.19 ± 0.06 ft
213.86 ± 0.15 ft

What is the most probable value of the difference in elevation? Its probable error?

6 Adjust the angles of problem 2, if weights of 6, 1, 3, and 5, respectively, are assigned to the four angles.

7 Adjust the angles of problem 3 if weights of 1, 1½, and 3, respectively, are assigned to angles *A*, *B*, and *C*.

8 A base line is measured in three sections with probable errors of ±0.014, ±0.022, and ±0.016 ft, respectively. What is the probable error of the total length?

9 The sides of a rectangular field are 1193.6 ± 0.6 and 582.7 ± 0.4 ft, respectively. What is the probable error of the computed area?

3

DISTANCE; TAPING

METHODS

3-1 General. In surveying, the distance between two points is understood to mean the *horizontal* distance. Though frequently slope distances are measured, they are reduced to their equivalent on the horizontal projection for use. Horizontal distances may be converted (by proportion) into equivalent distances at another elevation such as that of the state plane-coordinate system.

Various methods of determining distance are useful, depending upon the precision required, the cost, and other conditions (see Table 3-1). On some surveys, a combination of methods may be used to advantage.

The common method of determining distance is by direct measurement with the tape (Arts. 3-2 to 3-11); the operation is called *taping* or *chaining*. (Herein the two terms are used interchangeably.) The precision of taping varies widely with the purpose of the survey, ordinarily ranging from 1/1,000 to 1/30,000; in precise base-line measurement, the probable error may be as low as 1/1,000,000.

The *stadia* method, described in detail in Chap. 11, offers a rapid indirect means of determining distance and is particularly useful in topographic surveying. It is used extensively. Under average conditions, it will yield a precision between $\frac{1}{300}$ and 1/1,000. The telescope of the transit, level, or plane-table alidade is equipped with two horizontal hairs, one above and the other an equal distance below the horizontal cross hair; the distance from the instrument to a given point is indicated by the intercept between these stadia hairs as shown on a graduated rod held vertically at the point.

Distance can be measured indirectly by sighting with the transit through a *small angle* at a distant scale transverse to the line of sight and in the plane of the angle.

The method of *pacing* furnishes a rapid means of approximately checking more precise measurements of distance. It is used on reconnaissance surveys and, in small-scale mapping, for locating details and traversing with the plane table. Under average conditions, a person of

Table 3-1 Methods of measuring distance

Method	*Usual precision*	*Use*	*Instrument for measuring angles with corresponding precision*
Pacing	1/100 to 1/200	Reconnaissance; small-scale mapping; checking tape measurements	Hand compass; peep-sight alidade
Stadia	1/300 to 1/1,000	Location of details for mapping; rough traverses; checking more precise measurements	Transit; telescopic alidade of plane table; surveyor's compass
Ordinary taping	1/1,000 to 1/5,000	Traverses for land surveys and for control of route and topographic surveys; ordinary construction work	Transit (angles doubled)
Precision taping	1/10,000 to 1/30,000	Traverses for city surveys; base lines for triangulation of intermediate precision; precise construction work	Transit (angles by repetition); theodolite
Base-line taping	1/100,000 to 1/1,000,000	Triangulation of high precision for large areas, city surveys, or long bridges and tunnels	Theodolite; direction instrument
Electronic measurements	± 0.04 ft ± 1/300,000	Traverses for control of precise surveys; base lines or triangle sides for triangulation	Theodolite

experience will have little difficulty in pacing with a precision of $\frac{1}{100}$. Either the natural pace (about $2\frac{3}{4}$ ft), a $2\frac{1}{2}$-ft pace, or a 3-ft pace may be employed, according to the preference of the surveyor, who standardizes his pace before beginning a survey. 2 paces = 1 stride. Paces or strides are usually counted by means of a tally register operated by hand or by means of a pedometer which is attached to the leg. In hilly country, rough corrections for slope may be applied.

Distance may be measured by observing the number of revolutions of a wheel of a vehicle. The *mileage recorder* attached to the ordinary automobile speedometer registers distance to 0.1 mile and may be read by estimation to 0.01 mile. Special speedometers are available reading to 0.01 or 0.002 mile. The *odometer* is a simple device which can be attached to any vehicle and which registers directly the number of revolutions of the wheel. With the circumference of the wheel known, the relation between revolutions and distance is fixed. The distance indicated by either the mileage recorder or the odometer is somewhat greater than the true horizontal distance, but in hilly country a rough correction for slope may be applied.

Distances are sometimes roughly estimated by *time interval of travel* for person at walk, saddle animal at walk, or saddle animal at gallop. By *mathematical* or *graphical* methods, unknown distances may be determined through their relation to known distances. These methods are used in triangulation and plane-table work.

Electronic Measurements. Several portable instruments are available for accurate measurement of distances by electronic measurement through the use of light waves or radio waves. The maximum inherent instrumental error is approximately 0.04 ft $\pm$ 1/300,000. The proportional accuracy is greater for longer distances. The measurements are subject to error due to atmospheric conditions, and in some cases it is necessary to apply corrections. Reflections may result in errors in radio measurements.

The *Geodimeter* (Fig. 3-1) utilizes light waves, and its use is based upon the known velocity of light. Its basic principle is the indirect determination of the time required for a light beam to travel between two stations. The instrument, set up on one station, emits a modulated light beam to a passive reflector set up on the other end of the line to be measured. The reflector, acting as a mirror, returns the light pulse to the instrument, where a phase comparison is made between the projected and reflected pulses. A clear line of sight is required, and observations cannot be made if conditions do not permit intervisibility between the two stations. The maximum range for measurements is 2 to 3 miles in daylight and 15 to 20 miles at night, depending upon atmospheric conditions.

The *Tellurometer* and the *Electrotape* both consist of two interchangeable instruments, one being set up on each end of the line to be measured. The sending instrument transmits a series of microwaves which are run through the circuitry of the receiving unit and are retransmitted back to the original sending unit, which measures the time required. Distances are computed on the basis of the velocity of the radio waves. A clear line of sight between the two instruments is necessary. However, intervisibility is not required, and therefore observations can be made in fog and other unfavorable weather conditions. Distances up to 40 or 50 miles can be measured under favorable conditions.

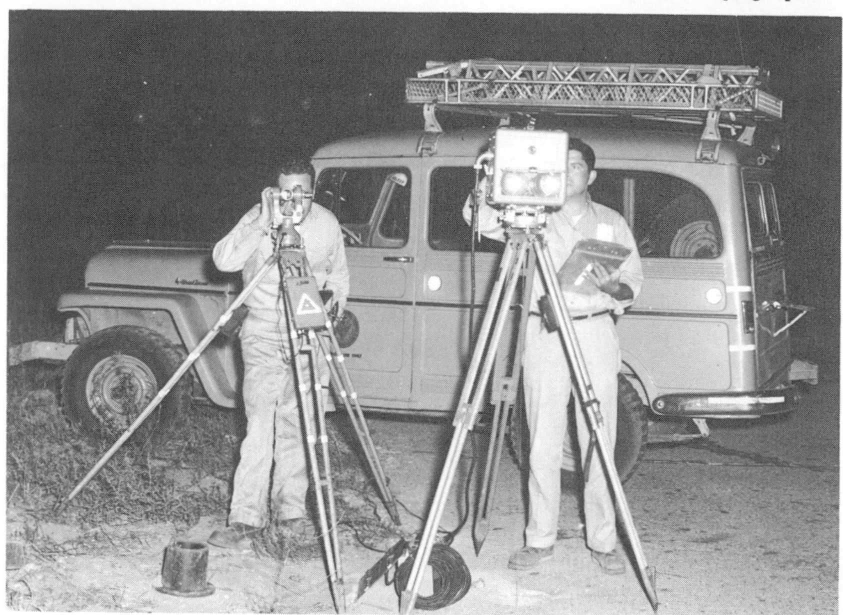

Fig. 3-1. Night measurement of distance with Geodimeter (right) and of vertical angle with theodolite. (*California Division of Highways.*)

Distances measured by use of electronic instruments are *slope* distances, and they must be converted to horizontal distances for use. The usual practice is to measure vertical angles with a transit or a theodolite (Fig. 3-1).

TAPING

3-2 Equipment. The tapes commonly used by the surveyor are the heavy steel tape, called the engineer's tape, surveyor's tape, or chain tape, and the metallic tape. The link chain is obsolete.

The *steel tape* usually employed is 100 ft long and is graduated in feet, with the first and last foot graduated in tenths or hundredths of feet. Other lengths and various types of graduation are available. The graduations may be etched or may be stamped on babbitt metal or on brass sleeves. Some tapes have an extra graduated foot at one or both ends; some have ends graduated so that corrections for slope can be applied directly; and some have ends graduated so that temperature corrections can be applied directly. The ordinary widths of 100-ft tape are $\frac{1}{4}$ and $\frac{5}{16}$ in.; longer tapes are usually narrower. A steel tape commonly used in land surveying is 66 ft long and is graduated in 100 "links" each 7.92 in. long, corresponding to the old Gunter's chain. One common type of metric tape is graduated in half meters, with the end meters graduated in decimeters.

The *metallic tape* is a ribbon of waterproofed fabric into which are woven small brass or bronze wires to prevent its stretching. It is usually 50 or 100 ft long and is graduated in feet, tenths, and half-tenths; it is usually $\frac{5}{8}$ in. wide. It is used principally in earthwork cross-sectioning, in location of details, and in similar work where a light flexible tape is desirable and where small errors in length are not of consequence. Recently a nonmetallic tape, which is a nonconductor of electricity, has been developed for use near power lines.

For very precise measurements the *invar tape* has come into extensive use. Invar is a composition of steel and nickel which is affected but little by temperature changes; the thermal expansion of commercial invar tapes is about one-tenth that of steel tapes. It is a soft metal, and the tape must be handled very carefully to avoid bends and kinks. This property and its high cost make the invar tape impracticable for ordinary use.

Steel *chaining pins,* also called taping arrows or surveyor's arrows, are commonly employed to mark the ends of the tape during the process of chaining between two points more than a tape length apart. They are usually 10 to 14 in. long. A set consists of 11 pins (see Fig. 3-2). For more precise chaining or for future reference, nails may be driven into the earth. On paved surfaces, tape lengths and other points may be marked with keel, pencil, tape, or spray paint.

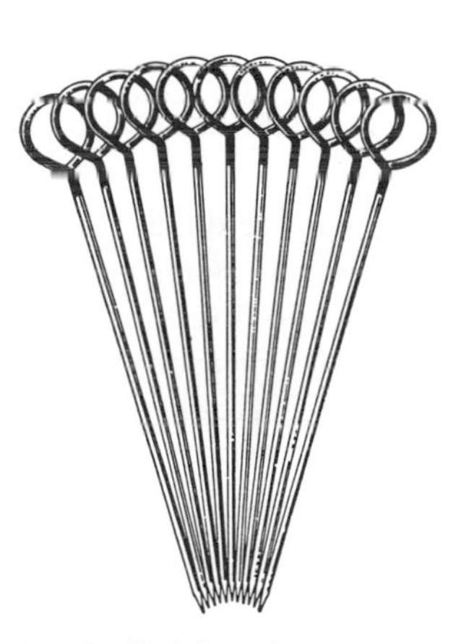

Fig. 3-2. Chaining pins.

Metal, wooden, or fiberglass *range poles,* also called flags, flagpoles, or lining rods, are used as temporary signals to indicate the location of points or the direction of lines. They are of octagonal or circular cross-section and are pointed at the lower end. Wooden and fiberglass range poles are shod with a steel point. The common length is 8 ft. Usually the pole is painted with alternate bands of red and white 1 ft long (see Fig. 3-3).

The type of *plumb bob* used in surveying is relatively slender so that the point can be seen from almost directly above; usually it consists of a brass body weighing 10 to 16 oz, a replaceable tip of wear-resistant alloy steel, and a device to which the plumb-bob string can be attached centrally.

3-3 Handling the Steel Tape. The steel tape will break if kinked and subjected to a strong pull. Some tapes are wound on a reel, but usually the tape is done up in 5-ft lengths into a figure 8 and then "thrown" into the form of a circle with diameter about 10 in., as follows:

Stand beside the zero end of the tape, take the end of the tape in the left hand, and—allowing the tape to slide loosely through the right hand—extend the arms. As the 5-ft mark is reached, grasp it with the right hand. Bring the hands together, and lay the 5-ft mark of the tape in the fingers of the left hand without permitting the tape to turn over. Then grasp this loop with the left hand, and again extend the arms for another 5-ft length, and so on. When the last mark is reached, tie the loop tightly where the ends of the tape come together, by means of the rawhide thongs. Grasp the loop with the right hand, and at the opposite point with the left hand. Twist the loop in such a manner that it will be thrown into circular form, with diameter half that of the loop.

To undo the tape, reverse the operation of throwing; untie the thongs; remove the first loop in such a way as not to twist the tape; and walk in the direction of measurement, removing one loop at a time and watching for kinks.

3-4 Taping on Smooth Level Ground. The following represents the usual method of taping between two fixed points over smooth level ground when the measurements are of ordinary precision, say 1/5,000. Errors and mistakes in taping are discussed in Arts. 3-8 to 3-11.

The rear chainman *with one pin* stations himself at the point of beginning. The head chainman, with the zero (graduated) end of the tape and 10 pins, advances toward the distant point. When the head chainman has gone nearly 100 ft, the rear chainman calls "chain" or "tape," a signal for the head chainman to halt. The rear chainman holds the 100-ft mark at the point of beginning and, by hand signals or by voice, lines in a chaining pin (held by the head chainman) with the range pole or other signal marking the distant point. During the lining-in process, the rear chainman is in a kneeling position on the line, facing the distant point ahead; the head chainman is in a kneeling position to one side, facing the line so that he can hold the tape steady and so that

Fig. 3-3. Range pole.

the rear chainman will have a clear view. The head chainman with one hand sets the pin vertically on line and a short distance to the rear of the zero mark. With his other hand he then pulls the tape taut and, making sure that it is straight, brings it in contact with the pin. The rear chainman, when he observes that the 100-ft mark is at the point of beginning, calls "stick" or "all right." The head chainman pulls the pin and sticks it at the zero mark of the tape, with the pin sloping away from the line (Fig. 3-4, right). As a check, he again pulls the tape taut and notes that the zero point coincides with the pin at its intersection with the ground. He then calls "stuck" or "all right"; the rear chainman releases the tape; the head chainman moves forward as before; and so the process is repeated.

Fig. 3-4. Taping on smooth level ground.

As the rear chainman leaves each intermediate point, he pulls the pin. Thus there is always one pin in the ground, and the number of pins held by the rear chainman at any time indicates the number of hundreds of feet, or *stations,* from the point of beginning to the pin in the ground. At the end of each 10 full stations, the head chainman signals for pins, the rear chainman comes forward and gives the 10 pins to the head chainman, both check the tally, and the head chainman records it. The procedure is then repeated. The count of pins is important, as the number of tape lengths is easily forgotten.

When the end of the course is reached, the head chainman halts, and the rear chainman comes forward to the last pin set. The head chainman holds the zero mark at the terminal point. The rear chainman pulls the tape taut and observes the number of *whole feet* between the last pin and the end of the line. For a tape having the end foot graduated, he then holds the next *larger* foot mark at the pin, and the head chainman pulls the tape taut and reads the decimal by means of the finer graduations of the end foot. The decimal is counted from the 1-ft mark; thus the distance between the last pin and the end of the line is 1 ft *less* than that indicated at the pin, plus the decimal read by the head chainman. The chainmen should agree on some system of checking to prevent mistakes.

For tapes having an extra graduated foot beyond the zero point, it is not necessary to subtract 1 ft as just described. The rear chainman holds the next *smaller* foot mark at the pin, and the head chainman reads the decimal.

Where the transit is set up on line, the transitman usually directs the head chainman in placing the pins on line.

On some surveys it is required that stakes be set on line at short intervals, usually 100 ft. Sometimes stakes are driven by the rear chainman after he has pulled the pin. On surveys of low precision no pins are used, and the head chainman sets the stakes in the manner previously described for pins, measuring the distance between centers of stakes at their junction with the ground. On more precise surveys, measurements are carried forward by setting a tack in the head of each stake.

3-5 Horizontal Taping over Sloping or Overgrown Ground. The measurements are carried forward much as just described for smooth level ground. The tape is held horizontal, and a plumb line is used by either, or at times by both, chainman for projecting from tape to pin, or *vice versa.* For rough work, plumbing can be accomplished with the range pole.

Where the slope is less than 5 or 6 ft in 100, the head chainman advances a full tape length at a time. If the course is downhill, he holds the plumb line at the zero point of the tape, with the tape horizontal and the plumb bob a few inches from the ground, then pulls the tape taut and is directed to line by the rear chainman. When the plumb bob comes to rest, he lowers it carefully to the ground (see Fig. 3-5) and then sets a pin in its place. As a check, the measurement is repeated. If the course is uphill, the rear chainman holds a plumb line suspended from the 100-ft mark and signals the head chainman to give or take until the rear chainman's plumb bob comes to rest over the pin; the head chainman then sets a pin.

Where the course is steeper and is downhill, the head chainman advances a full tape length and then returns to an intermediate point from which he can hold the tape horizontal. He suspends the plumb line at a foot mark, is lined in by the rear chainman, and sets a pin at the indicated point. The rear chainman comes forward, *gives the head chainman a pin,* and at the pin in the ground holds the tape at the foot mark from which the plumb line was previously suspended. The head chainman proceeds

Fig. 3-5. Plumbing at downhill end of horizontal tape

to another point from which he can hold the tape horizontal, and so the process is repeated until the head chainman reaches the zero mark on the tape. At each *intermediate* point of a tape length the rear chainman gives the head chainman a pin, but not at the point marking the full tape length. In this manner the tape is always advanced a full length at a time, and the rear chainman's count of pins is not confused. The process, called "breaking chain" or "breaking tape," is illustrated in Fig. 3-6, and one form of field notes is given in Fig. 3-13.

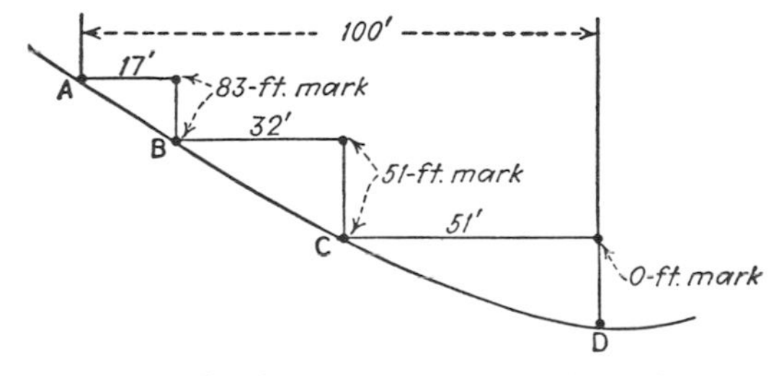

Fig. 3-6. Horizontal measurements on steep slope.

Some surveyors prefer to measure distances less than the tape length individually and add these measurements. However, this practice requires recording and may lead to mistakes in addition.

Usually the tape is estimated to be horizontal by eye. This practice commonly results in the downhill end's being too low, sometimes causing a significant error in horizontal measurement. The safe procedure in rough country is to use a hand level.

In horizontal measurements over uneven or sloping ground, the tape sags between supports and becomes effectively shorter. The effect of sag can be eliminated by standardizing the tape, by applying a computed correction, or by using the normal tension (Art. 3-8). In breaking chain, or when the tape is supported for part of its length, the difference in effect of sag as between a full tape length and the unsupported length can be taken into account roughly by varying the pull.

In measurements over crops, vegetation, or other low obstructions, if the tape is partially supported between the chainmen, account should be taken of the effect of this support upon sag.

3-6 Slope Taping.

Where the ground is fairly smooth, measurements on the slope may sometimes be made more accurately and quickly than horizontal measurements; and slope measurements are generally preferred. Some means of determining either the slope or the difference in elevation between successive 100-ft points or breaks in slope is required. For surveys of ordinary precision, either the clinometer for measuring slope or the hand level for measuring difference in elevation may be used to advantage: If only the distance between the ends of the line is required, the procedure of chaining is the same as on level ground, but a record is kept either of the slope or of the difference in elevation for each tape length (or less at breaks in slope). The horizontal distances are then computed from the distances measured on the slope.

Where stakes are to be placed every 100 ft, corrections to the slope distances may be applied as the chaining progresses, either by mental calculation or by use of the slide rule. Corrections are much more readily calculated than are the horizontal distances themselves.

3-7 Corrections for Slope.

Where the slopes are measured with sufficient precision so to warrant, the corresponding horizontal distances may be calculated by exact trigonometric relations. For measurements of ordinary precision where the slope is not greater than about 20 in 100, it is simpler and sufficiently precise to subtract an approximate correction from the slope distance.

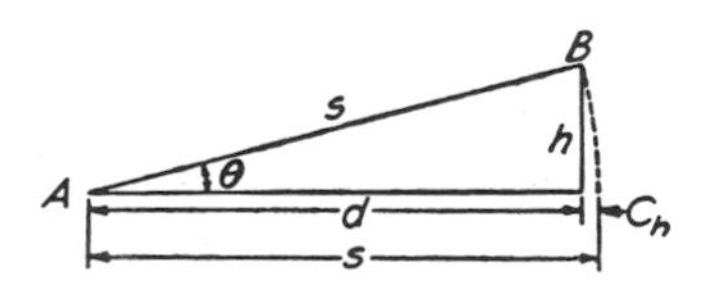

Fig. 3-7. Slope correction.

In Fig. 3-7 let s represent the slope distance between two points A and B, h the difference in elevation, and d the horizontal distance, all in feet. The correction is $C_h = s - d$. Then

$$h^2 = s^2 - d^2 = (s - d)(s + d)$$

Where the slope is not large, $s + d = 2s$ (approx.); making this substitution,

$$h^2 = 2s(s - d) \qquad \text{(approx.)}$$

and the correction is

$$C_h = s - d = \frac{h^2}{2s} \qquad \text{(approx.)} \tag{1}$$

For the usual case, where $s = 100$ ft, this formula can be solved mentally, and it is worth remembering.

For steep slopes, higher precision in making the slope correction can be obtained by using Eq. (1*a*) which includes the first two terms of a power series.

$$C_h = \frac{h^2}{2s} + \frac{h^4}{8s^3} \qquad \text{(approx.)} \tag{1a}$$

Where the *angle* of slope is determined, as when using the clinometer, Eq. (1) may still be used readily if it is remembered that for small angles the difference in elevation per 100 ft is about 1.75 ft times the slope angle in degrees ($\tan 1° = 0.0175$). Hence, if θ is the slope angle (Fig. 3-7) in degrees and S is the slope distance in 100-*ft stations,*

$$C_h = \frac{(1.75S\theta°)^2}{200S} = 0.015S(\theta°)^2 \qquad \text{(approx.)} \tag{2}$$

The distance S is expressed in 100-ft stations rather than in feet in order to reduce the number of decimal places in the formula.

Having given the slope distance, to find its horizontal projection the correction is *added.*

For the case in which it is desired to set stakes at a given horizontal distance (as 100 ft) apart or to compute a slope distance when the horizontal projection is known, the corresponding slope distance is given *approximately* by *adding* the same correction C_h to the required horizontal distance.

3-8 Errors in Taping; Corrections. Errors in taping may be caused by (1) incorrect length of tape, (2) imperfect alinement of tape, (3) imperfections of observing, (4) variations in temperature, (5) variations in tension, or (6) sag in tape. The precision of transit-tape surveys is discussed in Art. 10-11.

Table 3-2 summarizes the various errors and the procedures which can be adopted to eliminate or at least to reduce their effect, as discussed in the following paragraphs. It is remarkable that the apparently simple operation of linear measurement by taping is affected by so many factors.

In applying corrections to the observed length of a line measured with a tape that is too long, the correction is *added*. In laying out a required distance with a tape that is too long, the correction is *subtracted* from the required distance to determine the distance to be laid out. For a tape that is too short, of course, the corrections are opposite in direction to those just stated.

1 Tape Not Standard Length. The 100-ft tape as received from the manufacturer is usually close to the standard length when subjected to the standard pull of 10 lb at 68°F, the tape being horizontal and supported throughout its length. However, from time to time it is well to refer the tape to a standard of length. Many cities have such standards. For a small fee the National Bureau of Standards, Washington, D.C., will compare a tape with the official standard for any specified pull, temperature, and conditions of support of the tape.

If the tape is not of standard length under the specified conditions, a systematic error is produced in measurements. To eliminate the error, a correction (per tape length) determined by verifying the tape is applied to the observed lengths. Small differences from the correct length can be compensated by varying the pull, using a spring scale.

2 Imperfect Alinement of Tape. The head chainman is likely to set the pin to one or the other side of the line, producing a positive variable systematic error. The error can be reduced to a negligible quantity by care in lining. Usually it is the least important of the errors of taping, and extreme care in lining is not justified.

In taping through grass and brush or when the wind is blowing, it is impossible to have all parts of the tape in perfect alinement with its ends, and a positive variable systematic error is produced. If the head chainman is careful to stretch the tape taut and to observe that it is straight, the error is not of consequence.

If the tape is not horizontal for a supposedly horizontal measurement, a positive variable systematic error is produced similar to that due to imperfect alinement. Often slopes are deceptive, even to experienced men; the tendency is to hold the downhill end of the tape too low. In ordinary taping this is one of the largest of contributing errors. It can be reduced to a negligible amount by leveling the tape by means of either a hand level or a clinometer.

Table 3-2 Errors and corrections in taping

NOTE: In *measuring* a distance with a tape that is *too long*, *add* the correction.

Error	*Source*		*Amount*	*Error of 0.01 ft per 100 ft tape length caused by*	*Makes tape too*	*Importance in ordinary taping*	*Procedure to eliminate or reduce*
Systematic	Erroneous length		....	......	Long or short	Usually small, but should be checked	Standardize tape and apply computed correction
	Temperature		$C_x = 0.0000065L(T - T_0)$	15°F	Long or short	Of consequence only in hot or cold weather	Measure temperature and apply computed correction. In precise work, tape at favorable times and/or use invar tape
	Pull or tension (change in)		$C_p = \frac{(P - P_0)L}{AE}$	15 lb (for 1½-lb tape)	Long or short	Negligible	Apply computed correction. In precise work, use spring scale. May use normal tension, $P_n = \frac{0.204W\sqrt{AE}}{\sqrt{P_n - P_0}}$
	Sag		$C_s = \frac{w^2L^3}{24P^2} = \frac{W^2L}{24P^2}$	0.6 ft	Short	Large, especially with heavy tape	Apply computed correction. May be avoided by taping on slope
	Slope		$C_h = h^2/2s$ (approx.) $= 0.015S(\theta^\circ)^2$ (") $= s$ vers θ (exact)	1.4 ft	Short	....	At breaks in slope, determine difference in elevation or slope angle; apply computed correction
	Imperfect alinement	Horizontal	Same as slope	1.4 ft	Short	Not serious	Use reasonable care in sighting. Keep tape taut and reasonably straight
		Grass	Twice slope	0.7 ft	Short	Not serious	
		Wind	Same as sag	0.6 ft			
		Vertical	Same as slope	1.4 ft	Short	Often large	Level tape
Accidental	Manipulation	Plumbing	In rough country, breaking tape, 0.05 to 0.10 ft per tape length		±	Large, but accidental	May avoid by slope taping. Use reasonable care to reduce
		Marking ends of tape; reading graduations	0.01 ft per tape length		±	Not serious	....
		Error in pull	Same as pull	15 lb (1½-lb tape)	±	Not serious	....
		Error in determination of slope	Amount varies with slope		±	Not serious	....

3 Imperfections of Observing. Errors in plumbing, reading the tape, and setting the pins are accidental errors; hence the probable error tends to vary as the square root of the number of tape lengths. Only the error due to plumbing is of real importance; it can be avoided by taping on the slope. Although the errors in reading the tape and setting the pins cannot be eliminated, their effect upon the resultant error is usually not large. All the accidental errors can be kept reasonably low by care in taping.

4 Variations in Temperature. The tape expands as the temperature rises and contracts as the temperature falls. Steel has a coefficient of thermal expansion about 0.00000645 per 1°F; hence a 100-ft steel tape will change in length about 0.01 ft for a change in temperature of 15°F. Even for measurements of ordinary precision, the error due to thermal expansion becomes of consequence in extremely cold or hot weather.

If the tape is standard at a temperature of T_0 degrees and measurements are taken at a temperature of T degrees, the correction C_x for change in length is given by the formula

$$C_x = 0.00000645L(T - T_0) \tag{3}$$

in which L is the measured length.

Errors due to variations in temperature are greatly reduced by using an invar tape. If a steel tape is used, one or more plastic thermometers should be taped to it and corrections applied.

5 Variations in Tension. The tape, being elastic, stretches when tension is applied. If the pull is greater or less than that for which the tape is standardized, the tape is correspondingly too long or too short. For the medium (1½-lb) 100-ft steel tape, a change in tension of 1½ lb changes the length of the tape about 0.001 ft. The error is systematic. It is of consequence only for light tapes and is negligible except in precise work.

The correction for variation in tension in a steel tape is given by the formula

$$C_p = \frac{(P - P_0)L}{AE} \tag{4}$$

in which C_p = correction per distance L, in feet
P = applied tension, in pounds
P_0 = tension for which the tape is standardized, in pounds
L = length, in feet
A = cross-sectional area, in square inches
E = elastic modulus of the steel, in pounds per square inch

The elastic modulus of steel can be taken as 30 million lb per sq in. with sufficient precision for many purposes. The cross-sectional area of the tape can be computed from the weight and dimensions, since steel weighs approximately 490 lb per cu ft. For the 100-ft tape, it is usually sufficiently precise to take A, in square inches, as being equal to 0.003 times the weight of the tape in pounds.

6 *Sag.* If the tape is supported at the ends or at intervals rather than throughout its full length, it sags between the points of support. If a heavy 100-ft tape weighing 3 lb, standardized while supported throughout its length, is supported at the ends under a tension of 10 lb, the positive systematic error due to sag alone is about 0.37 ft per tape length. This example illustrates the serious disadvantage of using a very heavy tape for horizontal measurements over sloping ground.

The correction for sag is given with sufficient precision for most purposes by the formula

$$C_s = \frac{w^2L^3}{24P^2} = \frac{W^2L}{24P^2} \tag{5}$$

in which C_s = correction between points of support, in feet
w = weight of tape, in pounds per foot
W = total weight of tape *between supports,* in pounds
L = distance *between supports,* in feet
P = applied tension, in pounds

The correction is seen to vary directly as the cube of the unsupported length and inversely as the square of the pull. It may be applied without error of consequence to a tape held on a slope up to approximately 10°.

7 *Normal Tension.* By equating the right-hand members of Eqs. (4) and (5), the elongation due to increase in tension is made equal to the shortening due to sag. The pull that will produce this condition, called *normal tension* P_n, is given by the formula

$$P_n = \frac{0.204W\sqrt{AE}}{\sqrt{P_n - P_0}} \tag{6}$$

This equation is solved by trial, after all terms except P_n have been converted into numerical values and the numerator reduced to a single value.

3-9 Combined Corrections. Whenever corrections for several effects such as tension, temperature, and sag are to be applied, for convenience they may be combined as a single net correction per tape length. Since the corrections are relatively small, the value of each is not appreciably affected by the others and each may be computed on the basis of the nominal tape length. For example, even if the verified length of a tape were 100.21 ft, the correction for temperature (within the required precision) would be found to be the same whether computed for the exact length or for a nominal length of 100 ft.

Further, the method of *adding* (or subtracting) the small corrections encountered in taping is far more convenient than would be that of *multiplying* (or dividing) by correction factors, since fewer figures are required.

3-10 Resultant Error. In ordinary taping, the systematic errors are likely to be much larger than the accidental errors. Hence the resultant error varies as the number of tape lengths or as the length measured.

When every possible method is employed to eliminate or to correct for the systematic errors, the resultant error is almost entirely accidental; hence the resultant error may be expected to vary as the square root of the number of tape lengths or as the square root of the distance.

3-11 Mistakes in Taping. Some of the mistakes commonly made by inexperienced chainmen are:

1 Adding or dropping a full tape length, through failure to handle or count the pins properly.

2 Adding a foot, usually in measuring the fractional part of a tape length at the end of the line.

3 Other points incorrectly taken as 0 or 100-ft marks on tape.

4 Reading numbers incorrectly, as "6" for "9." As a check, good practice is to observe the numbers on each side of the one indicating the measurement.

5 Calling numbers incorrectly or not clearly. Decimal points and zeros should be called, and the recorder should repeat a number as it is recorded.

Often large mistakes will be prevented or discovered if the chainmen form the habit of pacing distances or of estimating them by eye. If a transit is being used to give line, distances can be checked by reading the approximate stadia interval on a flagpole.

SURVEYS WITH TAPE

3-12 Surveys. The survey of a field with the tape is accomplished by dividing the field into triangles and obtaining sufficient measurements of the sides, altitudes, and/or angles of the triangles to permit the computation of remaining sides and angles required for plotting and for the calculation of areas.

Where the measurement of angles is involved, surveying with the tape alone is too slow to be used to any great extent except on surveys covering small areas. However, often it is convenient to measure an angle or erect a perpendicular with the tape.

3-13 Measurement of Angles. Angles are measured by the chord method, as follows: With the vertex A of the angle α as a center (Fig. 3-8), the tape is swung, and pins are set at points a and b where the arc intersects the sides AB and AC of the angle. The chord distance ab is measured. Then

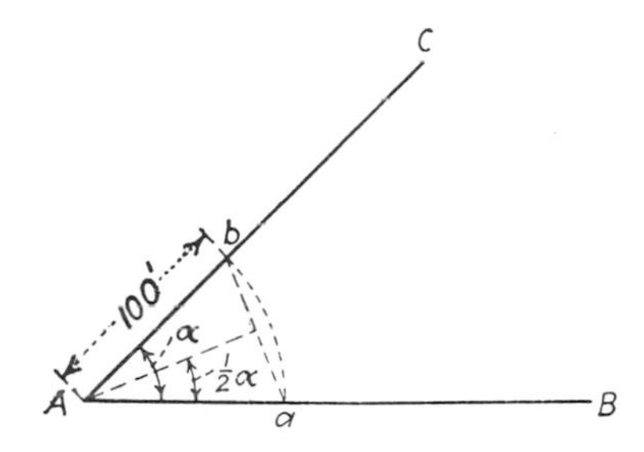

Fig. 3-8. Angle with tape.

$$\sin \frac{1}{2} \alpha = \frac{ab}{200} \tag{7}$$

An angle is laid off in a similar manner, making use of the chord relation of Eq. (7).

Angles measured by the tape method are usually considered as being less accurate than those measured with the transit, but for very small angles the reverse is likely to be true.

3-14 Erecting Perpendicular to Line. For the purpose of locating the altitude of a triangle or of laying out a right angle as for a building corner, it is necessary to erect on the ground a perpendicular to an established line. This is usually done either by the 3:4:5 method or by the chord-bisection method. The 3:4:5 method requires less time, but the chord-bisection method is more precise. A prismatic sighting device for erecting perpendiculars is available.

3:4:5 *Method.* To erect a perpendicular to the line AB (Fig. 3-9a) that will include point C, a point a on line AB is assumed to be on the perpendicular, and a pin is set at a. With sides 3, 4, and 5 ft or a multiple

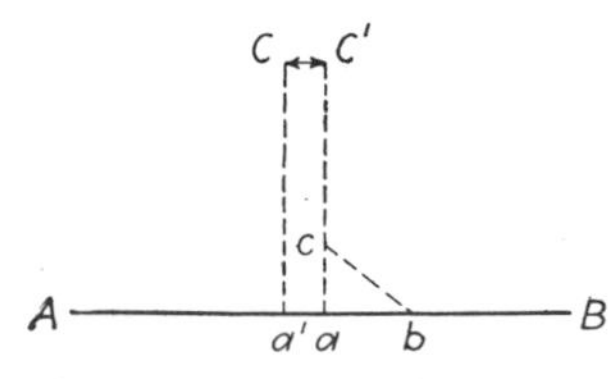

Fig. 3-9a. Perpendicular by 3:4:5 method.

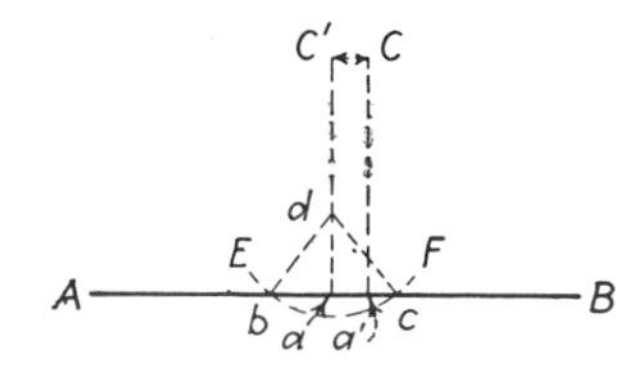

Fig. 3-9b. Perpendicular by chord-bisection method.

of these numbers, such as 24, 32, and 40 ft, a right triangle *abc* is constructed as follows: A pin is set on line *AB* at *b*, 32 ft from *a*. The zero end of the tape is fixed with a pin at *a*, and the 100-ft end at *b*. The head chainman moves to *c* and holds the 24-ft and the 60-ft marks of the tape in one hand, with the tape between these marks laid out so as to avoid kinking. He then sets a pin at *c*. The rear chainman moves from *a* to *b* as necessary to check the position of the tape at these points as *c* is established. He then sights along *ac* to *C′* beside *C*. If *C* does not lie on this prolongation, the perpendicular distance from *C* to the line *aC′* is measured, and the foot *a* of the perpendicular *aC′* is moved along the line *AB* by an equal amount, to the point *a′*. If the trial perpendicular *aC′* fails to include the point *C* by several feet, the process is repeated for *a′*, the new point; otherwise the location of *a′* may be assumed as correct.

If ground conditions are favorable, the point *c* may be established by striking arcs on the ground with $ac = 24$ ft and $bc = 40$ ft as radii, using a chaining pin. The point *c* lies at the intersection of the arcs. This procedure avoids either fastening or bending the tape.

A good way of finding the approximate location of the perpendicular is to stand on the line *AB* with arms extended horizontally along the line. With eyes closed, bring the arms to the front, palms together; then sight along the line of the hands.

Chord-bisection Method. To erect a perpendicular to the line *AB* (Fig. 3-9*b*) that will include point *C*, the location of the perpendicular is estimated, and a pin is set at *d* on this estimated perpendicular, somewhat less than one tape length from the line *AB*. With *d* as center and length of tape as radius, the head chainman describes the arc *EF* of a circle, setting pins at the intersections *b* and *c* of the arc with the line *AB*. The rear chainman stationed at *A* or *B* determines the location of the intersections *b* and *c* on line. The point *a* is established midway between *b* and *c*. The line *ad* is prolonged to *C′* beside *C*, and the point *a* is moved if necessary, as described for the 3:4:5 method.

3-15 Irregular Boundary. Where a boundary is irregular or curved, as along a shoreline or a winding road, the usual procedure of locating the boundary is by means of perpendicular offsets from a straight line run as near the boundary as practicable. For straight portions of the boundary, offsets need be taken only at the ends. Where a curved boundary has many changes in direction, offsets should be taken at short intervals, which will usually be irregular. However, for convenience in calculating areas (Arts. 16-4 to 16-6), so far as possible the offsets are taken at regular intervals.

If the distance from the line to the boundary is not more than about

50 ft and the boundary is fairly regular, usually it is sufficiently precise to erect the perpendiculars by estimation with the eye. For more precise work, the perpendicular may be erected with the tape, the transit, or a prismatic sighting device.

3-16 Obstructed Distances. Often it becomes necessary to determine the distance between two points where direct taping is impossible. If the points are intervisible, the distance may be determined by swing offsets, parallel lines, or similar triangles. If the points are not intervisible, the methods employing parallel lines are impossible.

1 Swing Offsets. To find the distance AB (Fig. 3-10) by the swing-offset method, the head chainman attaches the end of the tape to one end of the line as at B and describes an arc with center B and radius 100 ft. The rear chainman stationed at A lines in the end of the tape with some distant object as O and directs the setting of pins at points a and b where the end of the tape crosses line AO. A point C midway between a and b lies on the perpendicular CB. The line ab is bisected; a pin is set at C; and the distances BC and CA are measured to obtain the necessary data for computing the length of AB.

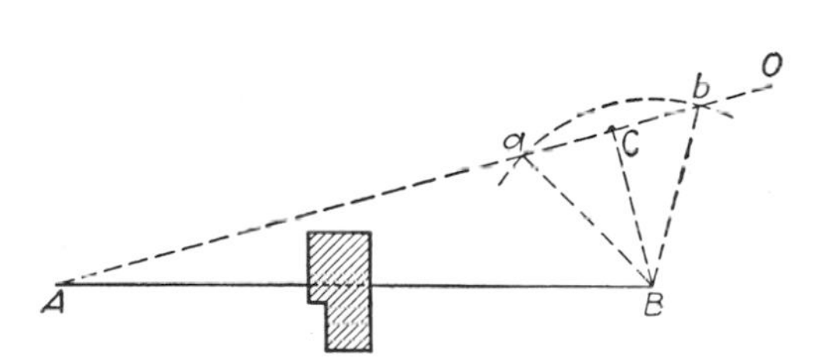

Fig. 3-10. Swing offset with tape.

2 Parallel Lines. If the necessary offset distance from the line AB is *short*, perpendiculars $AA' = BB'$ are erected by either method of Art. 3-14 to clear the obstacle. The line $A'B'$ is then chained, and its length is taken as that of AB.

If a *long* offset is necessary, the method just described will be inaccurate because of the uncertainty of right angles measured with the tape. In such a case, a point C (Fig. 3-11) is established to clear the obstacle, such that the estimated value of α is less than 45°. The chord length of α for a radius of 100 ft is determined. AC and CD are measured, CD being roughly perpendicular to AB. At C the angle α is laid off so that CE will be parallel to AB. CE is measured, E being any convenient point such that β will be less than 45°. EB and EF are meas-

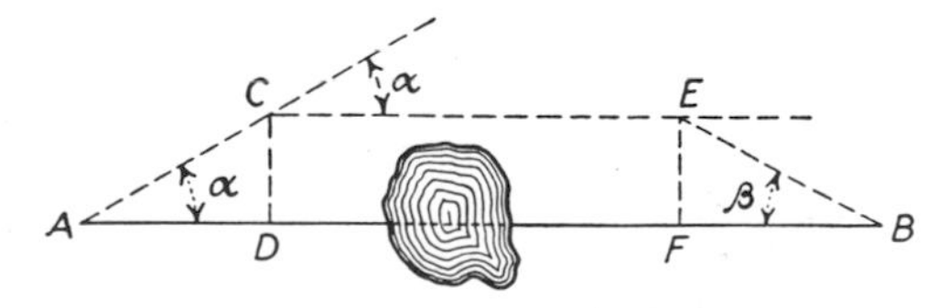

Fig. 3-11. Obstructed distance (long offset).

ured, EF being roughly perpendicular to AB. The angle β is measured by determining its chord length for a radius of 100 ft. The right-angle triangles ADC and BFE are solved for AD and FB. Then $AB = AD + CE + FB$.

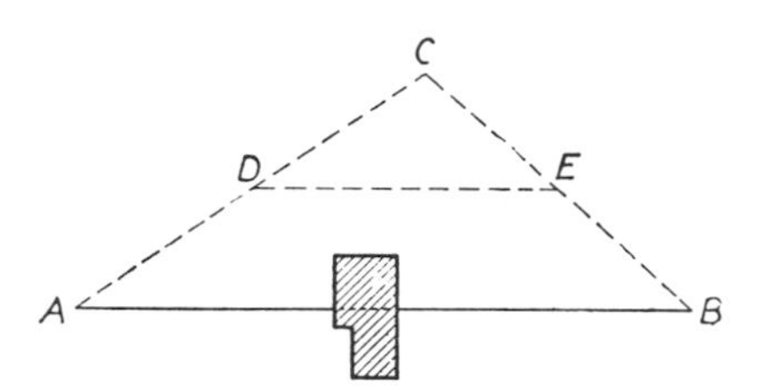

Fig. 3-12. Obstructed distance (similar triangles).

3 Similar Triangles. Let C (Fig. 3-12) be a point from which A and B are visible. AC and BC are measured. CD and CE are laid off so that CD will bear the same relation to CA that CE bears to CB. The triangles ACB and DCE are similar. DE is measured, and AB is computed.

3-17 NUMERICAL PROBLEMS

1 The length of a line as measured with a 100-ft steel tape is 1,012.3 ft. Afterward the tape is compared with the standard and is found to be 0.03 ft too long. Compute the length of the line.

2 A building 80.00 by 160.00 ft is to be laid out with a 50-ft tape which is 0.016 ft too long. What ground measurements should be made?

3 The actual distance between two marks at the City Hall is known to be 100.080 ft. When a field tape is held on this line, the observed distance is 100.03 ft. What is the actual length of the tape?

4 The slope measurement of a line is 800.0 ft. The differences in elevation between successive 100-ft points, as measured with a hand level, are 1.0, 1.5, 2.5, 3.8, 4.6, 5.0, 7.5, and 6.2 ft. Determine the horizontal distance.

5 The slope measurement of a line is 957.2 ft. Slope angles measured with a clinometer are as shown below. Determine the horizontal distance.

Chainage, ft	0		300		800		1,000		1,246.5
Slope angle, degrees		½		1¼		2½		4	

6 Two points at a slope distance of about 100 ft apart have a difference in elevation of 12.0 ft. What slope distance should be laid off to establish a horizontal distance of 100.00 ft? Compute by exact and approximate methods.

7 Compute the effect of sag *per tape length* for a 100-ft tape weighing 1½ lb, under a pull of 20 lb, (*a*) supported at the ends only; (*b*) supported at ends and quarter-points.

8 A 100-ft tape weighing 2 lb is of standard length under a tension of 12 lb and supported for full length. A line on smooth level ground is measured with the tape under a tension of 35 lb and found to be 4,863.5 ft long. $E =$ 29,000,000 lb per sq in.; 3.53 cu in. of steel weighs 1 lb. Make the correction for increase in tension.

9 Compute the normal tension for the tape of problem 8, the tape being supported at its ends.

10 A line along a paved street having a grade of 2.5 per cent is measured on the slope to be 1,320.64 ft long. The applied tension is 12 lb and the mean temperature is 87.4°F. The 100-ft steel tape used for the measurements is standardized at 70°F, is supported for its full length under a tension of 12 lb, and is found to be 0.004 ft too short. Determine the horizontal length of the line.

11 A line through rough country is taped by horizontal measurements and found to be 2,450 ft long. On the average, it was necessary to use the plumb line every 50 ft. If the probable error of plumbing from the end of the tape to the ground is ±0.03 ft in the direction of the line, compute the probable error due to inaccurate plumbing.

12 A hedge along the line AB makes direct measurement impossible. A point C is established at an offset distance of 25 ft from the line AB and roughly equidistant from A and B. The distances AC and CB are then taped; $AC = 1{,}287.2$ ft and $CB = 1{,}353.0$ ft. By an approximate method, using the "slope correction," compute the length of the line AB.

3-18 FIELD PROBLEMS

Problem 1. Taping over Level Ground

Object. To measure with the 100-ft steel tape over an approximately level course about 1,000 ft long and to check the distance by taping in the opposite direction.

Procedure. Set a flagpole at each end of the line. Follow the procedure indicated in Art. 3-4, reading the distance to tenths of feet and estimating to hundredths. Record the two lengths, and compute the ratio of discrepancy to length. Measurements should check within 1/5,000. As a rough check, standardize the pace, and measure the distance by pacing.

Hints and Precautions. (1) The rear chainman should not hold the end of the tape as he moves forward from station to station. (2) Be careful not to disturb the "stuck" pin by allowing the tape to press against it. (3) Check every measurement.

Problem 2. Taping over Uneven Ground

Object. To standardize the 100-ft steel tape; to find the horizontal length of an assigned course about 800 ft long over uneven ground by two methods; and to correct for the error in length of tape as determined by the standardization tests.

Procedure. (1) Standardize the tape before and after the field work, by comparing it with the official standard of length. Maintain the required pull by means of a spring scale. Determine the error with a finely divided scale. (2) Tape the course by horizontal measurement after the manner described in Art. 3-5. For measurements in which the line slopes more than 5 or 6 ft in 100, "break tape." (3) Correct for the error in length of tape. Record the data in a form similar to that of Fig. 3-13. (4) Tape the course by measurement on the slope, recording the difference in elevation (to the nearest foot) of each 100-ft tape length as determined by a hand level. If sharp breaks in slope occur between 100-ft stations, treat each distance between breaks in the same manner as a full station. Correct the measurements for error in length

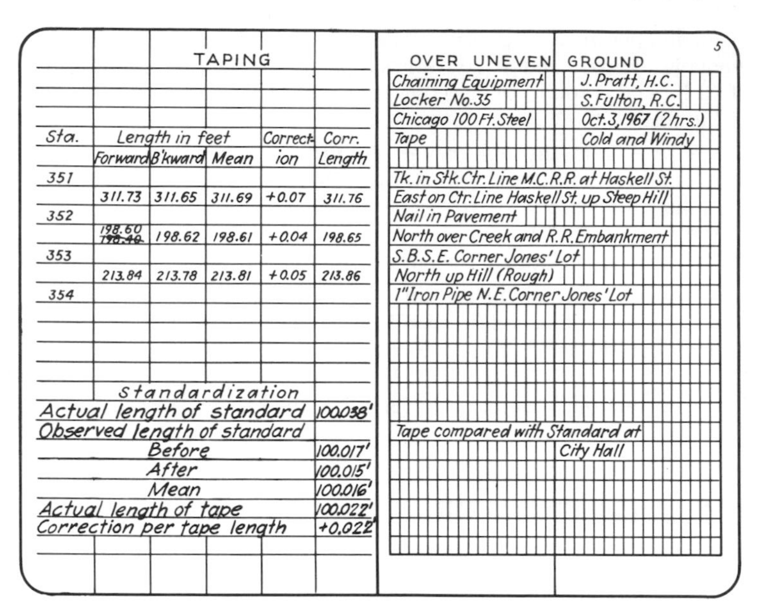

Fig. 3-13. Notes for taping over uneven ground with tape horizontal.

of tape. Correct the measurements for slope, and determine the horizontal distance. (5) Compare the results obtained by the two methods of measurement.

Hints and Precautions. (1) In taping with the tape horizontal, avoid the general tendency to hold the downhill end of the tape too low by (*a*) comparing the tape with some level line, (*b*) having the two ends in line with the horizon, or (*c*) estimating the angle between tape and plumb line. (2) It is usually more accurate to tape downhill, as the rear end of the tape can be held firmly on the ground.

Problem 3. Survey of Field with Tape

Object. To collect sufficient data for calculating the area of a field having one or more irregular boundaries, including data for calculation of partial areas by more than one method. The data may be used in office problem 39 (Chap. 16).

Procedure. (1) Divide into triangles as much of the field as its shape will permit, avoiding very acute angles wherever possible. Lay out the triangles adjacent to irregular boundaries so that no long offsets will be necessary. (2) Measure the sides, the altitude, and one angle of each triangle. (3) From the triangle sides nearest the boundary, take offsets at such intervals as will ensure sufficient precision for plotting and computation. (4) Include a simple sketch with the recorded notes.

Hints and Precautions. (1) For measurement of the triangles, the following form of notes is suggested:

Station	Distance	Perpendicular	Chord length	sin ½ angle	Angle

For measurement of the offsets, the following form is suggested:

Station	Distance	Offset left	Offset right

(2) To take the offsets correctly and quickly, set several pins at the desired intervals along the side of the triangle in advance. (3) Do not measure the offsets with unnecessary precision; for example, if the triangle sides are measured to the nearest 0.01 ft, measure the offsets no closer than the nearest 0.1 ft.

Problem 4. Determining Obstructed Distance with Tape

Object. To determine an obstructed distance between two points.

Procedure. (1) On an assigned course about 800 ft long, assume that there is some obstacle which makes direct measurement and intervision impossible, and find the distance by the swing-offset method. (2) Find the distance by similar triangles. (3) Assume that the points are intervisible but that direct taping is impossible, and determine the distance by the method of parallel lines, using offsets of 20 ft and of 100 ft. (4) As a check, measure the distance directly. (5) Compare the results.

4

LEVELING: METHODS AND INSTRUMENTS

METHODS

4-1 Definitions. The *elevation* of a point near the surface of the earth is its vertical distance above or below an arbitrarily assumed *level surface,* or curved surface every element of which is normal to the plumb line. The level surface (real or imaginary) used for reference is called the *datum.* Usually the datum is taken at mean sea level, but it may be taken at any convenient elevation below the lowest point of the survey. A *level line* is a line in a level surface.

The *difference in elevation* between two points is the vertical distance between the two level surfaces in which the points lie. *Leveling* is the operation of measuring vertical distances, either directly or indirectly, in order to determine differences in elevation.

A *horizontal line* is a line, in surveying taken as straight, tangent to a level surface.

A *vertical angle* is an angle between two intersecting lines in a vertical plane. In surveying it is commonly understood that one of these lines is horizontal.

4-2 Curvature and Refraction. In leveling, it is necessary to consider the effects of (1) curvature of the earth and (2) atmospheric refraction, which affects the line of sight. Usually these two effects are considered together.

In Fig. 4-1 is shown a horizontal line tangent at A to a level line near the surface of the earth. The vertical distance between the horizontal line and the level line is a measure of the earth's curvature. It varies approximately as the square of the distance from the point of tangency.

Owing to the phenomenon of atmospheric refraction, rays of light transmitted along the surface of the earth are bent downward slightly. Thus, as viewed from point A (Fig. 4-1), an object actually at B would appear to be at C; the actual line of sight is along a curve AB which is concave downward. The amount of refraction is variable, but it is relatively small compared with the earth's curvature; therefore, for ordinary work it is taken as constant.

Let K be the distance, in miles, from the point of tangency (station of the observer); M the same distance, in thousands of feet; and h' the

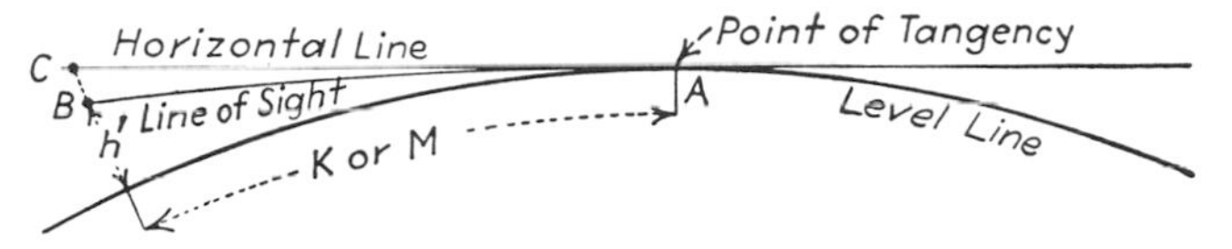

Fig. 4-1. Earth's curvature and atmospheric refraction.

combined effect of the earth's curvature and atmospheric refraction, in feet. Then

$$h' = 0.57K^2 = 0.021M^2 \qquad \text{(approx.)} \tag{1}$$

The effect of the earth's curvature alone is about $0.66K^2$ or $0.024M^2$. The effect of refraction alone is about $0.09K^2$ or $0.003M^2$ in the opposite direction.

4-3 Methods. Difference in elevation may be measured by the following methods:

1 *Direct* or *spirit leveling,* by measuring vertical distances directly. Direct leveling is the most precise method of determining elevations and is the one most commonly used (Art. 4-4).

2 *Indirect* or *trigonometric leveling,* by measuring vertical angles and horizontal distances (Art. 4-5).

3 *Barometric leveling,* by measuring the difference in atmospheric pressure at various stations, by means of a barometer (Art. 4-6).

Differential leveling (Chap. 5) is the operation of determining differences in elevation of points some distance apart. Usually differential leveling is accomplished by direct leveling. *Precise leveling* is a precise form of differential leveling.

Profile leveling (Chap. 6) is the operation—usually by direct leveling—of determining elevations of points at short measured intervals along a definitely located line such as the center line for a highway or a sewer.

4-4 Direct Leveling. In Fig. 4-2, A represents a point of known elevation and B a point the elevation of which is desired. In the method of direct or spirit leveling, the level is set up at some intermediate point as L, and the vertical distances AC and BD are observed by holding a leveling rod first at A and then at B, the line of sight of the instrument being horizontal. (Owing to refraction the line of sight is slightly curved as explained previously.)

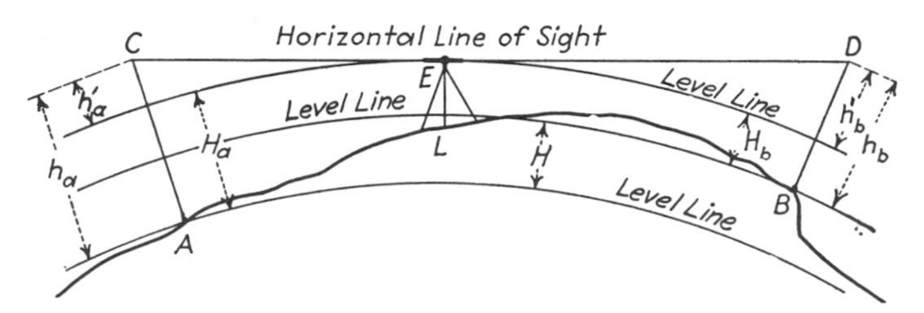

Fig. 4-2. Direct leveling. (Owing to refraction, the line of sight is slightly curved.)

If the difference in elevation between the points A and E is designated as H_a and between E and B as H_b; if the vertical distances read at A and B, respectively, are designated as h_a and h_b; and if the effects of curvature and refraction for the horizontal distances LA and LB, respectively, are designated h'_a and h'_b, then

$$H_a = h_a - h'_a \qquad \text{and} \qquad H_b = h_b - h'_b \tag{2}$$

The difference in elevation H between A and B is then

$$H = H_a - H_b = h_a - h_b - h'_a + h'_b \tag{3}$$

If the backsight distance LA is equal to the foresight distance LB, then $h'_a = h'_b$ and

$$H = h_a - h_b \tag{4}$$

Thus if backsight and foresight distances are balanced, the difference in elevation between two points is equal to the difference between the rod readings taken to the two points, and no correction for curvature and refraction is necessary. In direct leveling, usually the work is so conducted that the effect of curvature and refraction is reduced to a negligible amount.

The procedure of direct leveling is described in Arts. 4-21 and 4-22 and in Chaps. 5 and 6.

4-5 Indirect Leveling. In Fig. 4-3, A represents a point of known elevation and B a point the elevation of which is desired. In the method of indirect or trigo-

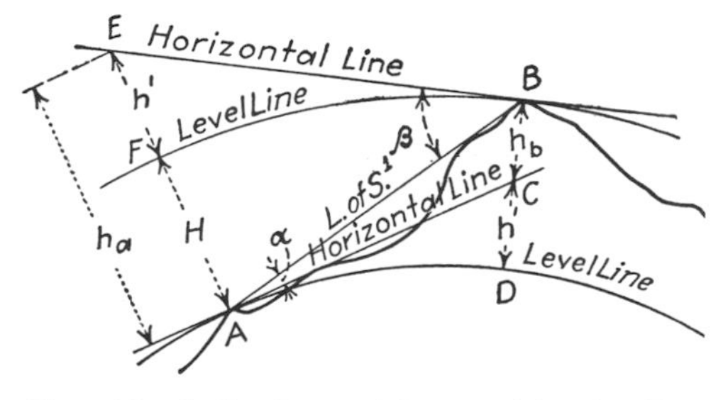

Fig. 4-3. Indirect, or trigonometric, leveling. (Owing to refraction, the line of sight is slightly curved.)

nometric leveling, the verical angle α a A is measured, and the distance AD is determined by some method of measurement. Within the limits of ordinary practice $AD = AC$ and $\angle BCA = 90°$. Therefore,

$$h_b = AC \tan \alpha \tag{5}$$

The correction h' for curvature and refraction is, by Eq. (1),

$$h' = 0.021M^2 = 0.021 \left(\frac{AC}{1{,}000}\right)^2$$

in which AC is in feet. The difference in elevation H is therefore

$$H = h_b + h' = AC \tan \alpha + 0.021 \left(\frac{AC}{1{,}000}\right)^2 \tag{6}$$

If the vertical angle β is now taken from B to A, by a similar course of reasoning

$$H = h_a - h' = AC \tan \beta - 0.021 \left(\frac{AC}{1{,}000}\right)^2 \tag{7}$$

When the vertical angle is upward or positive, the curvature and refraction correction is added; and when downward or negative, the correction is subtracted. Combining Eqs. (6) and (7),

$$H = \frac{AC}{2} (\tan \alpha + \tan \beta) \tag{8}$$

Thus, if vertical angles are measured from (and to) each of the two points, the effect of curvature and refraction is eliminated. This procedure is employed in precise indirect leveling.

In ordinary surveying, indirect leveling furnishes a rapid means of determining the elevations of points in rolling or rough country. On reconnaissance surveys, angles may be measured with the clinometer and distances may be obtained by pacing. On more precise surveys, angles are measured with the transit and distances by the stadia. Indirect leveling is used extensively in plane-table work.

The errors of indirect leveling are chiefly accidental. Under average conditions when the transit is used, the error may be expected to be not greater than 0.4 ft times the square root of the distance in miles.

4-6 Barometric Leveling. Since the pressure of the earth's atmosphere varies inversely with the elevation, the barometer may be employed for making observations of difference in elevation. Barometric leveling is employed principally on exploratory or reconnaissance surveys where differences in elevation are large, as in hilly or mountainous country.

Since atmospheric pressure may vary over a considerable range in the course of a day or even an hour, elevations determined by one ordinary barometer carried from one elevation to another may be several feet in error. However, by means of extremely sensitive barometers and special techniques, elevations can be determined within a foot or even less.

Usually barometric observations are taken at a fixed station during the same period that observations are made on a second barometer which is carried from point to point in the field. This procedure makes it possible to correct the readings of the portable barometer for atmospheric disturbances.

The mercurial barometer is accurate, but it is cumbersome and is suitable only for observations at a fixed station. For field use, an aneroid barometer is commonly used because it is light and is easily transported. The usual type has a dial about 3 in. in diameter, graduated both in inches of mercury and in feet of altitude (elevation); it is compensated for temperature. At a point of known altitude, the pointer can be set at the corresponding reading on the scale in order to place the instrument in adjustment. The aneroid barometer can be calibrated against a mercurial barometer by comparing values at a given station over a range of temperature.

In use, the barometer should be given time to reach the temperature of the air before an observation is made.

A single aneroid barometer is sometimes used by topographers on small-scale surveys where the contour interval is large. Stops are made at frequent intervals during the day, and the rate of change in atmospheric conditions is observed; suitable corrections are thus determined and are applied to the observed values. Where distances permit, it is preferable to return to the starting point and to correct the intermediate readings in proportion to the change in atmospheric pressure during the interval between observations.

One type of sensitive barometer used in topographic surveying employs two of the instruments at fixed bases and one or more instruments carried from point to point over the area being surveyed. One fixed instrument is located at a point of known elevation near the highest elevation of the area, and one near the lowest elevation; these instrument stations are called the *upper base* and *lower base,* respectively. A third instrument is carried to the point whose elevation is desired, and a reading is taken. Readings on the fixed instruments are taken either simultaneously (as determined by signaling) or at fixed intervals of time; in the latter case the readings at the desired instant are determined by proportion. The elevation of the portable instrument is then determined by interpolation. The horizontal location of each point at which a reading is taken is determined by conventional methods.

Another procedure is to employ one barometer at a fixed base and one barometer which is carried to points whose elevation is desired, simultaneous readings being taken. The carried barometer is finally brought either back to the starting point or to another point of known elevation; the computation of the elevation of each point then takes account of the corresponding change in atmospheric pressure at the fixed base.

INSTRUMENTS FOR DIRECT LEVELING

4-7 Kinds of Levels. Any instrument commonly used for direct leveling has as its essential features a *line of sight* and a *level tube* mounted with its axis parallel to the line of sight and used as a guide in making the line of sight horizontal.

The instrument used principally in the United States is the *engineer's level,* either the *dumpy level* for which the telescope tube is permanently fastened to the supporting bar or the *wye level* for which the telescope is removable and rests in Y-shaped supports. The *architect's level* or *construction level,* a modified form of the engineer's level but less sensitive, is used in establishing grades for buildings. Modern American and European levels include *tilting* and *self-leveling* or *automatic* levels. The *hand level* is a simple and useful device for rough leveling. The *transit* and the telescopic alidade of the *plane table* are not designed primarily for direct leveling, but are frequently used for this purpose. The measurements of difference in elevation are made by sighting on a graduated rod called a *leveling rod.*

4-8 Engineer's Level. The principal parts of the engineer's level are shown in Figs. 4-6 and 4-7, as described in Arts. 4-11 and 4-12. The details of the instrument are constructed quite differently by the different manufacturers. Essentially the level consists of a *telescope* mounted on a *level bar* which is rigidly fastened to a vertical *center spindle.* Attached to the telescope or the level bar and parallel to the telescope is a *level tube.* The spindle fits into the cone-shaped bearing of a *leveling head,* so that the level is free to revolve about the spindle as a vertical axis. At the lower end of the spindle is a ball-and-socket joint which makes a flexible connection between the movable leveling head and a stationary *foot plate.* Any desired relation between leveling head and foot plate is maintained by means of vertical *leveling screws,* usually four. The foot plate is screwed to a metal *tripod head* which is supported by wooden tripod legs. In the tube of the telescope are *cross hairs* supported on a metal ring; the intersection of these hairs defines a point on the line of sight.

4-9 Level Tube. The level tube is a glass vial with the inside ground barrel-shaped, so that a longitudinal line on its inner surface is the arc of a circle. The tube is nearly filled with sulfuric ether or with alcohol, the remaining space being occupied by a bubble of air. A longitudinal line tangent to the curved inside surface of the tube at its upper mid-point is called the *axis of the level tube;* when the bubble is centered, the axis of the level tube is horizontal. The metal housing of the tube is attached to the instrument by means of screws that permit adjustment, as shown in Figs. 4-6 and 4-7.

The longer the radius of the circle to which the level tube is ground, the more sensitive the tube. However, the more sensitive the tube, the longer the time required to center the bubble. The sensitivity of the tube on any given telescope should be such that the first noticeable movement of the bubble will be accompanied by a barely apparent movement of the line of sight on an object at an average sighting distance from the instrument.

The level tube is graduated so that the bubble can be centered; the length of a division is 2 mm on some instruments and 0.1 in. (2.5 mm) on others. The sensitivity of a level tube is usually expressed in seconds of the central angle whose arc is one division of the tube, being inversely proportional to the number of seconds; it is directly proportional to the radius of the arc. For the better grade of engineer's levels the radius of curvature is about 68 ft; in this case the bubble will be displaced one 2-mm division by a rotation amounting to 20 seconds of arc.

4-10 Telescope. Figure 4-4*a* shows the principal parts of an *external-focusing* telescope. Rays of light emanating from an object within the field of view of the telescope are caught by the objective lens *A* and are brought to a focus and form an image in the plane of the cross hairs *B.*

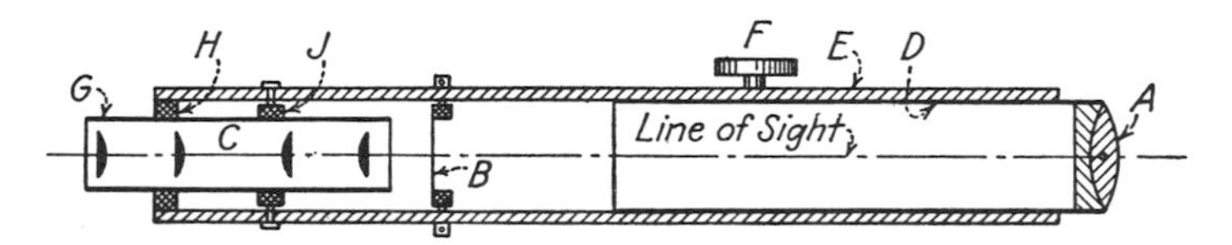

Fig. 4-4*a*. Longitudinal section of external-focusing telescope.

The lenses of the eyepiece *C* form a microscope which is focused on the image at the cross hairs. The objective lens is screwed in the outer end of the objective slide *D* which fits in the telescope tube *E*. The objective lens is focused by the screw *F*, at the inner end of which is a pinion that engages the teeth of a rack fixed to the objective slide. The eyepiece slide *G* is held in position transversely by rings *H* and *J*, through which it may be moved in a longitudinal direction for focusing. By means of screws the ring *J* may be moved transversely so that the intersection of the cross hairs will appear in the center of the field of view.

The line of sight is defined by the intersection of the cross hairs and the optical center of the objective lens (defined later in this article).

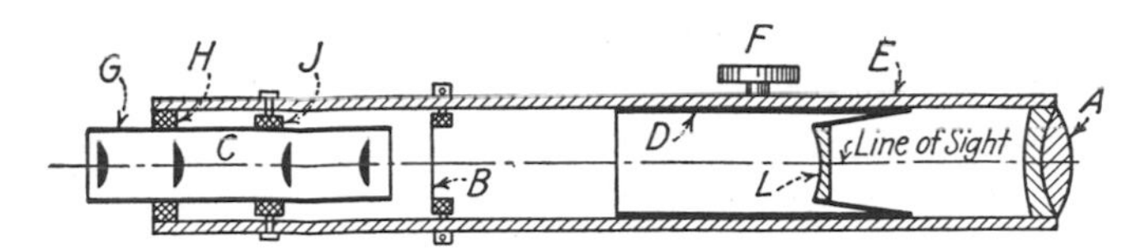

Fig. 4-4*b*. Longitudinal section of internal-focusing telescope.

Another type, called the *internal-focusing* telescope, has increased greatly in use. It is shown in section in Fig. 4-4*b*. Its arrangement and operation are similar to those just described, except that the objective lens *A* is fixed in the end of the telescope tube and that the slide carries a focusing lens *L*. It has the advantages that the focusing slide is practically free from grit and that the interior is practically free from dust and moisture; the telescope tends to balance well; and in stadia work an instrumental constant is eliminated and the computations thus simplified. Its disadvantages are that the required extra lens reduces the illumination and that the interior of the telescope is not so accessible for field cleaning or repairs.

The discussions hereinafter refer to the external-focusing type of telescope, except as specifically stated. With modifications in detail, they apply also to the internal-focusing type.

1 Focusing. When the telescope is to be used, the eyepiece is first moved in or out until the cross hairs appear distinct. This adjustment of the eyepiece should be tested frequently, as the observer's eye becomes tired.

When an object is sighted, the objective slide is moved in or out until the image appears clear, when the image should be in the plane of the cross hairs. If a slight movement of the eye from side to side produces an apparent movement of the cross hairs over the image, the plane of the image and the plane of the cross hairs do not coincide, and *parallax* is said to exist. Since parallax is a source of error in observations, it should be eliminated by refocusing the objective, the eyepiece, or both until further trial shows no apparent movement. The objective lens must be focused for each distance sighted.

The telescope cannot be focused on objects closer than about 6 ft from the center of the instrument.

The strain on the observer's eyes will be reduced if he learns to keep both eyes open while sighting.

2 *Objective.* The principal function of the telescope objective is to form an image for sighting purposes. For accuracy of measurements the objective should produce an image that is well lighted, accurate in form, distinct in outline, and free from discolorations.

The *optical center* of the objective is that point in the lens through which any ray of light will pass without permanent change in direction. The *optical axis* passes through the optical center and the centers of curvature of the lens.

The *principal focus* is a point on the optical axis back of the objective where rays entering the telescope parallel with the optical axis are brought to a focus. The *focal length* of the objective is the distance from its optical center to the principal focus. When the telescope is focused on a distant point, the focal length is very nearly the distance from the optical center of the objective to the plane of the cross hairs.

3 *Objective Slide.* Particular care should be taken to protect the objective slide from dust, water, and other foreign matter. If the slide is lubricated at all, only a drop of the finest watch oil should be used, and all excess oil should be removed with a soft cloth.

4 *Cross Hairs.* Threads from the cocoon of the brown spider are usually used as cross hairs. In some instruments the cross hairs are made of very fine platinum wire.

One vertical and one horizontal hair are fastened to a metal ring called the *cross-hair ring,* or *reticule,* as shown in Fig. 4-5. The ring is smaller than the telescope tube. It is held in position by four capstan-headed screws by means of which it may be moved either horizontally or vertically or may be rotated through a small angle about the axis of the telescope. To rotate the cross-hair ring about the axis of the telescope without disturbing its centering, loosen any two adjacent screws, tap one lightly, then tighten the same two screws. To move the cross hair vertically or horizontally, loosen one screw and then tighten the opposite screw.

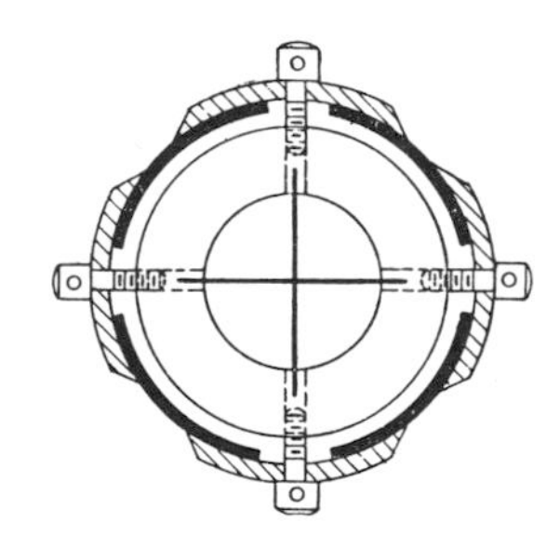

Fig. 4-5. Cross hairs.

In modern instruments usually the reticule consists of a glass plate on which are etched fine vertical and horizontal lines which serve as cross hairs.

5 *Stadia Hairs.* Many telescopes are also equipped with two horizontal *stadia hairs,* one above and the other an equal distance below the horizontal cross hair, for use in measuring distances by stadia. Usually the stadia hairs are mounted in the same plane as the cross hairs.

6 *Eyepiece.* The image formed by the objective is inverted. The eyepiece commonly used, called the *erecting* or *terrestrial* eyepiece, reinverts

the image so that the object appears to the eye in its normal position. Usually it consists of four plano-convex lenses placed in the eyepiece slide (*G*, Figs. 4-4*a* and 4-4*b*).

The *inverting* or *astronomical* eyepiece simply magnifies the image without reinverting it. It consists of two plano-convex lenses.

The advantage of the erecting eyepiece lies in the fact that objects appear in their natural position. The principal advantage of the inverting eyepiece is that objects are more brilliantly illuminated, as there are fewer lenses to absorb light; also the telescope is relatively short. Most American engineers and surveyors prefer the erecting eyepiece, but the use of the inverting eyepiece is increasing.

7 *Magnifying Power.* The magnifying power of the telescope is the ratio of the apparent length of the image to that of the object. For the better grade of engineer's levels it is about 30 diameters. For transit telescopes it is commonly 18 to 24 diameters. The greater the magnifying power, the less the illumination and the smaller the field of view.

4-11 Dumpy Level. Figure 4-6 shows the details of a conventional dumpy level with erecting eyepiece. The telescope *A* is rigidly attached to the level bar *B*, and the instrument is so constructed that the optical axis of the telescope is perpendicular to the axis of the center spindle. The level tube *C* is permanently placed so that its axis lies in the same

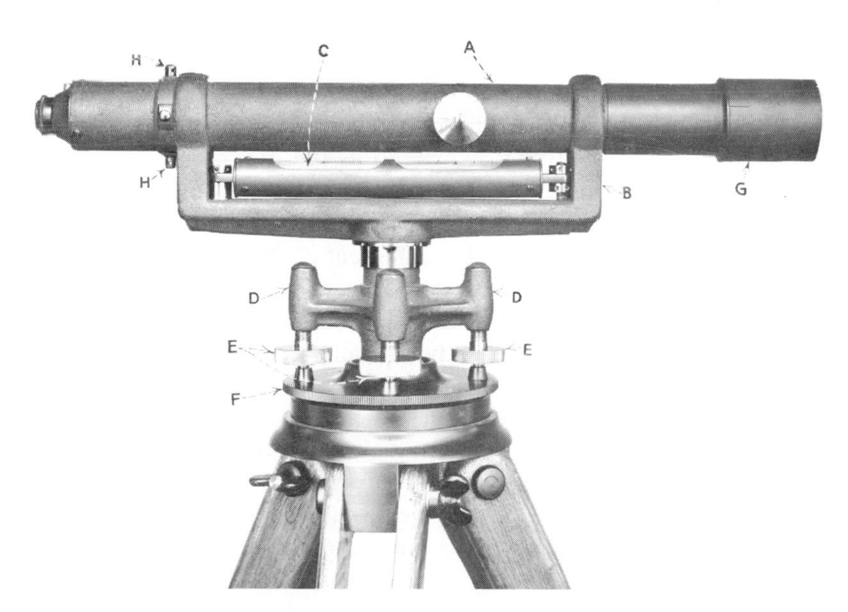

Fig. 4-6. Engineer's dumpy level.

vertical plane as the optical axis, but it is adjustable in altitude by means of a capstan-headed screw at one end. The spindle revolves in the socket of the leveling head *D,* which is controlled in position by the four leveling screws *E.* At the lower end of the spindle is a ball-and-socket joint which makes a flexible connection between the instrument proper and the foot plate *F.* When the leveling screws are turned, the level is moved about this joint as a center. The sunshade *G* protects the objective from the direct rays of the sun. The adjusting screws *H* for the cross-hair ring are near the eyepiece end of the telescope.

The name "dumpy level" originated from the fact that formerly this level was usually equipped with an inverting eyepiece and therefore was shorter than a wye level of the same magnifying power. Its advantages over the wye level are that it is simpler in construction, has fewer parts subject to wear, requires fewer adjustments, and stays in adjustment better. It is the type most commonly employed.

The *precise levels* of the U.S. Geological Survey and of the U.S. Coast and Geodetic Survey are refined forms of the dumpy level.

4-12 Wye Level. Although the wye level is passing out of use, so many are continuing in service in the United States that the characteristics and adjustments of this type of level are described herein.

Figure 4-7 shows the details of a wye level with erecting eyepiece. The telescope rests in Y-shaped bearings called the *wyes.* By means of capstan nuts (30 and 31) on one of the wye legs, the wye can be raised or lowered. When the wye clips or stirrups (32) are raised, the telescope may be revolved in the wyes, or it may be lifted from the wyes and turned end for end. The enlarged cylindrical portions (39) of the telescope barrel which rest in the wyes are called the *rings* or *collars.* The line joining the centers of the rings is defined as the *axis of the collars.* The *axis of the wyes* is a general term used to denote a reference line—sometimes actually the axis of the collars and sometimes actually the axis of the supports—which represents the alinement of the telescope tube in the wye supports.

The level tube is attached to the telescope and is adjustable in both a horizontal and a vertical plane. Other details are much the same as for the dumpy level previously described.

The distinguishing characteristics of the wye level are (1) that the telescope may be revolved about its own axis in the wyes and (2) that the telescope may be lifted from the wyes and turned end for end. These features are of no particular advantage in the work of leveling, but they facilitate the making of adjustments, provided the bearings are not worn. (The wye level can be adjusted by one man whereas the dumpy level

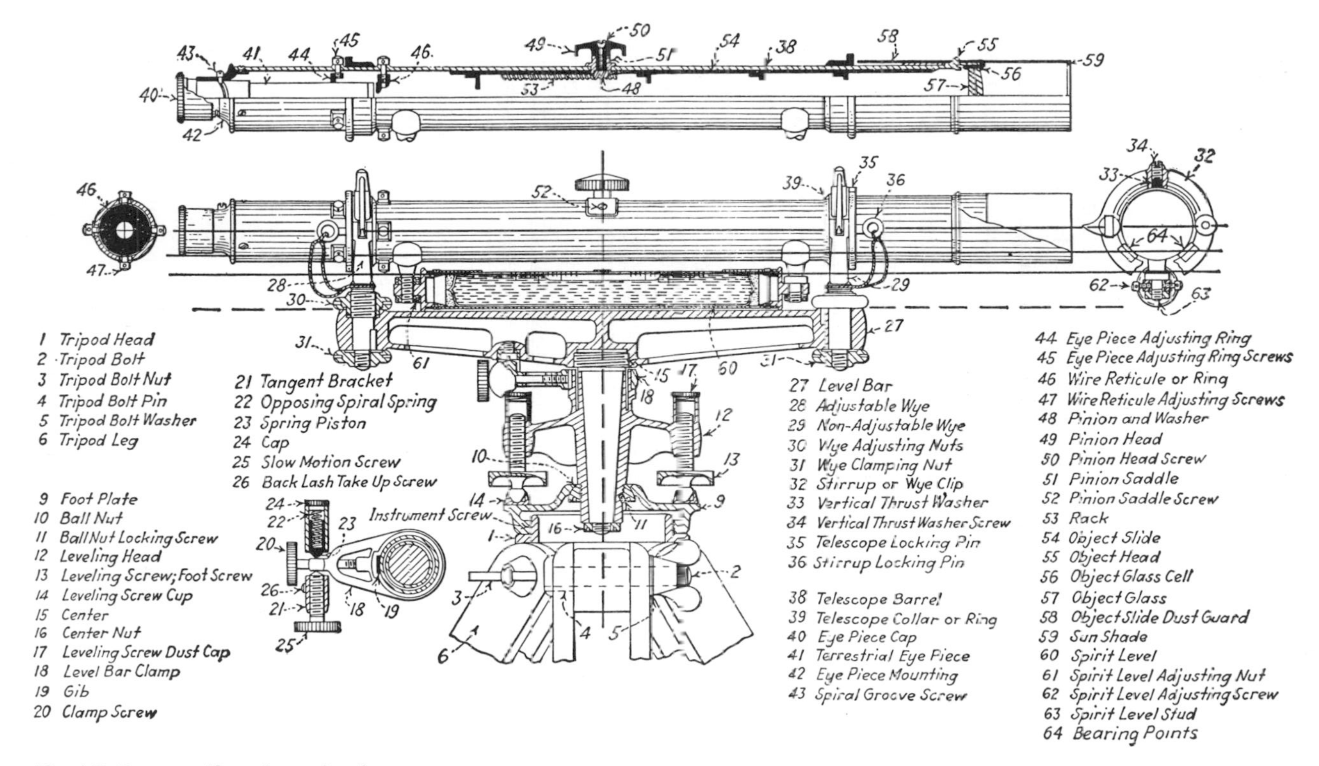

Fig. 4-7. Cross-section of wye level.

requires two men.) Each collar is in contact with the wye at two points (64, Fig. 4-7). At these points the collars become worn and flattened in use so that they are no longer cylindrical, nor are they likely to be of the same size or shape; furthermore, the bearing points of the wyes may become worn unevenly. Under these conditions it is impossible to adjust the instrument correctly by the usual methods, and the adjustments are the same as for the dumpy level.

4-13 Modern Types. Modern levels, although essentially of the "dumpy" type, incorporate several features which make them quite different from the conventional dumpy level described in Art. 4-11. They are small, light, and very accurate and permit of rapid setting up and observing. They are usually light in color to minimize temperature effects of sunlight. They need little or no adjustment in the field. They are equipped with a circular spirit level for approximate preliminary leveling. They may have either three or four leveling screws; one type has a clamping ball-and-socket arrangement instead of leveling screws. The telescope is internal-focusing and has the crosslines, including stadia lines, etched on a glass diaphragm. The telescope level bubble is easily and accurately centered by bringing the images of the ends of the bubble into coincidence as viewed from the eyepiece end of the telescope through a system of prisms. Electric illumination can be provided for night observations. Some instruments are equipped with an optical micrometer, which raises or lowers the line of sight slightly without changing its direction and thus permits readings to be taken at even graduations on the rod; the micrometer reading is applied to the rod reading. Some levels are equipped with a horizontal circle, with graduations on glass, for observing directions approximately.

Tilting Level. The distinctive feature of a *tilting level* (Fig. 4-8) is that the telescope is mounted on a transverse fulcrum at the vertical axis and on a micrometer screw at the eyepiece end of the telescope. After the instrument has been leveled approximately in the usual manner, perhaps by use of a circular spirit level, the telescope is pointed in the direction desired and is then "tilted" or rotated slightly in the vertical plane of its axis by turning the micrometer screw, until the sensitive telescope-level bubble is centered. The line of sight is then horizontal, even though the instrument as a whole is not exactly level. On the micrometer screw is mounted a graduated drum which may be used as a gradienter.

Self-leveling Level. A *self-leveling,* or *automatic,* level keeps its line of sight horizontal by means of some form of pendulum and a system of prisms and mirrors. The pendulum is damped magnetically, and its oper-

Fig. 4-8. K & E tilting level.

ation is quick and accurate. The level has no tilting arrangement and no spirit-level tube; at each setup the instrument is leveled approximately by use of a circular spirit level, and the pendulum then maintains the line of sight horizontal. The telescope may be provided with an optical micrometer for measuring the difference in elevation between the line of sight and the nearest graduation on the leveling rod. One form of self-leveling level is shown in Fig. 4-9.

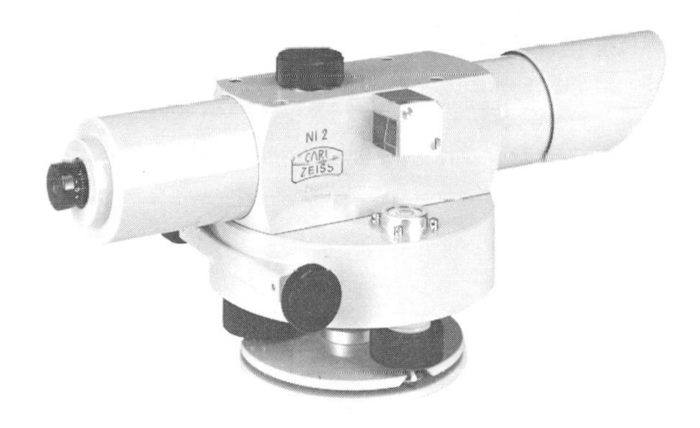

Fig. 4-9. Zeiss self-leveling level. (*Keuffel & Esser Co.*)

4-14 Locke Hand Level. The Locke hand level is widely used for rough leveling. It consists of a metal sighting tube about 6 in. long on which is mounted a level vial (Fig. 4-10). In the tube beneath the vial is a prism which reflects the image of the bubble to the eye end of the level. Just beneath the level vial is an adjustable cross wire. The eyepiece consists of a peephole mounted in the end of a slide which fits inside the tube and is held in a given position by friction. Mounted on the right half of the eyepiece slide is a semicircular convex lens which magnifies the image of the bubble and cross wire as reflected by the prism.

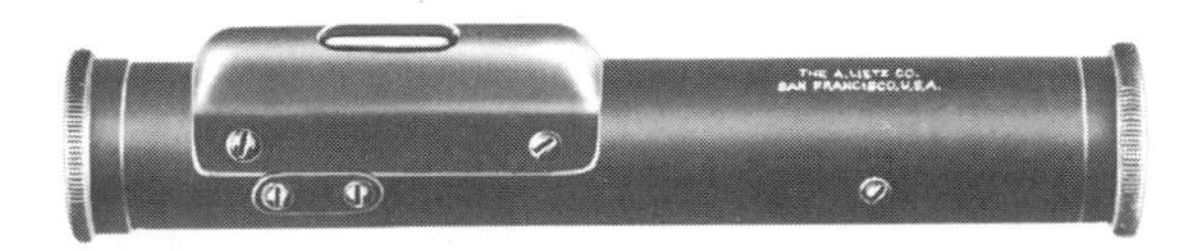

Fig. 4-10. Locke hand level. (*The A. Lietz Co.*)

In using the level, the object is viewed directly through the left half of the sighting tube, without magnification, while the position of the bubble is observed in the right half of the field of view. The level is held with the level vial uppermost and is tipped up or down until the cross wire bisects the bubble, when the line of sight is horizontal.

4-15 Abney Hand Level and Clinometer. This instrument is suitable both for direct leveling and for measuring the angles of slopes. The instrument shown in Fig. 4-11 is graduated both in degrees and in percentage of slope, or grade. When it is used as a level, the index of the vernier is set at zero, and it is then used in the same way as the Locke hand level. When it is used as a clinometer, the object is sighted, and the level tube is caused to rotate about the axis of the vertical arc until the cross wire

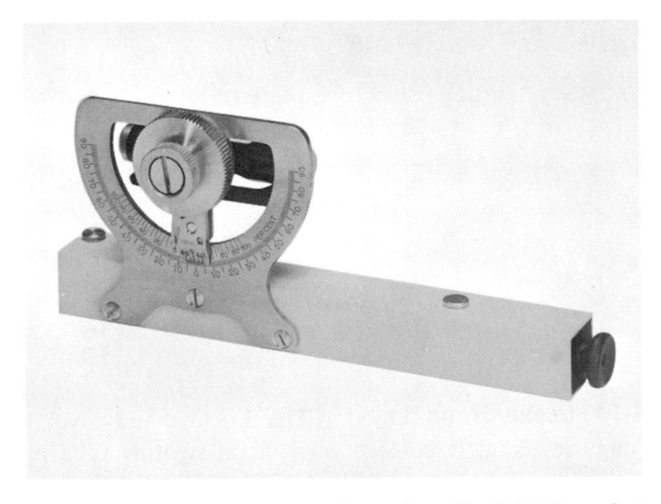

Fig. 4-11. Abney hand level and clinometer. (*Keuffel & Esser Co.*)

bisects the bubble as viewed through the eyepiece. Either the slope angle or the slope percentage is then read on the vertical arc.

4-16 Leveling Rods. These are graduated wooden rods of rectangular cross section by means of which difference in elevation is measured. The lower end is shod with metal.

The rod is held vertical, and hence the reading of the rod as indicated by the horizontal cross hair of the level is a measure of the vertical distance between the point on which the rod is held and the line of sight.

Rods are obtainable in a variety of types, patterns, and graduations, and are either in single pieces or in sections. Common lengths are 12 and 13 ft. In the United States, ordinarily rods are graduated in hundredths of a foot. On some government surveys the rods are graduated in decimals of the meter or the yard.

The two general classes of leveling rods are (1) *self-reading* rods, which may be read directly by the leveler and (2) *target* rods, for which a target sliding on the rod is set by the rodman as directed by the leveler. Under ordinary conditions, observations with the self-reading rod can be made with nearly the same precision and much more rapidly. The self-reading rod is the one commonly employed, even for precise leveling.

4-17 Self-reading Rods. Of the many types of self-reading rods, the *Philadelphia rod* (Fig. 4-12) is the most widely used. It is equipped with a target and is therefore also a target rod. It is usually in two sections or strips, the strips being held in contact by two brass sleeves. By means of a screw attached to the upper sleeve the two strips may be clamped together in any relative position desired. For readings of 7 ft or less (on a 13-ft rod) the back strip is clamped in its normal position. For greater readings, the rod is extended its full length; the graduations on the front face of the back strip are then a continuation of those on the front strip. When thus extended, the rod is called a "long" rod.

The background is white with graduations 0.01 ft wide painted in black as shown in Fig. 4-12. The readings are made on the edges of the graduations. The numbers indicating feet are in red, and those indicating

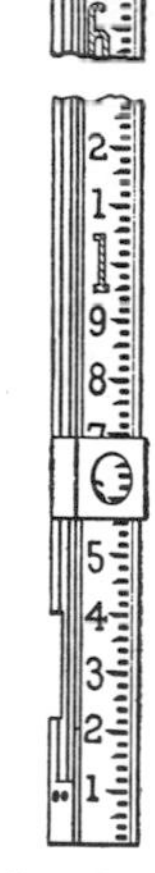

Note:- Cross-hatched portions indicate red

Fig. 4-12. Philadelphia rod.

tenths of feet are in black. This style of graduation is satisfactory for self-reading when the length of sight is less than 400 or 500 ft. Observations closer than the smallest division on the rod are made by estimation.

The *Chicago rod* is graduated similarly to the Philadelphia rod, but is made in three sections with slip joints. The *Florida rod* is a one-piece rod 10 ft long, graduated with alternate bands of red and white 0.1 ft wide.

The *topographer's rod* is graduated in half-feet, with zero either at the bottom of the rod or near the middle, at the height of the observer's eye; in the latter case the graduations extend both upward and downward from the zero mark. Usually the topographer's rod is homemade.

Graduated flexible *ribbons* of enameled and waterproofed fabric are obtainable. Such a ribbon attached to a plain wooden strip makes a serviceable and accurate leveling rod.

Some rods are provided with a graduated strip of invar steel in order to minimize the effect of changes in temperature and of changes in length of the wooden rod due to changes in humidity. The invar strip is fastened at the ends only and is kept taut by means of a spring. Invar scales may be mounted on conventional rods.

4-18 Target Rods. The usual target (Fig. 4-13, left) is a circular or elliptical disk about 5 in. in diameter, with horizontal and vertical lines formed by the junction of alternate quadrants of white and red (shown cross-hatched in the figure). A rectangular opening in the front of the target exposes a portion of the rod to view so that readings can be taken. The attached *vernier* (Art. 4-19) fits closely to the rod, its zero point or index being at the horizontal line of the target. The leveler signals the rodman to slide the target up or down until it is bisected by the line of sight. The target is then clamped, and the rodman, leveler, or both observe the indicated reading. Where very long sights are taken, where the rod is partly obscured from view, or where a number of points are to be

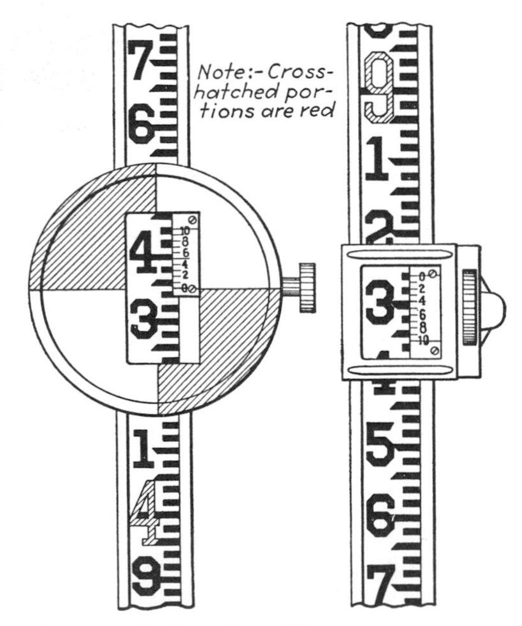

Fig. 4-13. Direct-vernier settings.

established all at the same elevation, the use of the target greatly facilitates the work. However, under ordinary conditions its use retards progress without adding much, if anything, to the precision. Its principal advantage is that mistakes are less likely to occur.

The Philadelphia rod (Fig. 4-12) may be used as a target rod. For readings on its lower half, the reading is made to thousandths of feet by means of the vernier attached to the target. For readings greater than can be taken with the "short" rod, the target is clamped at the same graduation on the face of the rod as the reading of the vernier on the back of the upper sleeve of the rod. The rod is then extended until the target is bisected by the line of sight. The vertical distance from the foot of the rod to the target is then indicated by the reading of the vernier on the back of the rod (Fig. 4-13, right).

The *New York rod,* used to some extent in building construction, is similar to the Philadelphia rod but has graduations consisting of short, fine lines at intervals of 0.01 ft and longer, numbered lines at intervals of 0.1 ft. It is not a self-reading rod.

The *Detroit rod* has pointed black graduations alternately 0.02 and 0.01 ft wide and alternately of two lengths. The points of the long wide graduations indicate tenths of feet, and the points of the short wide graduations indicate half-tenths of feet. It is made in three sections for convenience in transporting.

The *architect's rod* sometimes used in building construction is similar to the New York rod but is graduated in 1/8 in. and is equipped with verniers reading to 1/64 in.

4-19 Verniers. A vernier is a short auxiliary scale placed parallel and in contact with the main scale, by means of which fractional parts of the least division of the main scale can be measured precisely; the length of one space on the vernier scale differs from that on the main scale by the amount of one fractional part. The precision of the vernier depends on the fact that the eye can determine more closely when two lines coincide than it can estimate the distance between two parallel lines. The scale may be either straight (as on a leveling rod) or curved (as on the horizontal and vertical circles of a transit). The zero of the vernier scale is the index for the main scale.

The *least count,* or fineness of reading, of a vernier is equal to the difference in length between one scale space and one vernier space. Furthermore, the number n of spaces on the vernier is equal to the number of equal parts into which one space s on the main scale can be subdivided by reading the vernier. It follows that

$$\text{Least count} = \frac{s}{n} \qquad (9)$$

Verniers are of two types: (1) the *direct* vernier which has spaces slightly shorter than those of the main scale and (2) the *retrograde* vernier which has spaces slightly longer than those of the main scale. On the direct vernier, n spaces on the vernier are equal in length to $n - 1$ spaces on the main scale; on the retrograde vernier, $(n + 1)$ spaces. The use of the two types is identical, and they are equally sensitive and equally easy to read. Since they extend in opposite directions, however, one or the other may be preferred because it permits a more advantageous location of the vernier on the instrument. Both types are in common use.

Direct Vernier. Figure 4-14*a* represents a scale graduated in hundredths of feet and a direct vernier having each space 0.001 ft shorter than a 0.01-ft space on the main scale; thus each vernier space is equal to 0.009 ft, and 10 spaces on the vernier are equal to 9 spaces on the scale. The index, or zero of the vernier, is set at 0.400 ft on the scale. If the vernier were moved upward 0.001 ft, its graduation numbered 1 would coincide with a graduation (0.41 ft) on the scale and the index would be at 0.401 ft, and so on. It is thus seen that the position of the index is determined to thousandths of feet without estimation, simply by noting which graduation on the vernier coincides with one on the scale. Note that the coinciding graduation on the main scale does *not* indicate the main-scale reading.

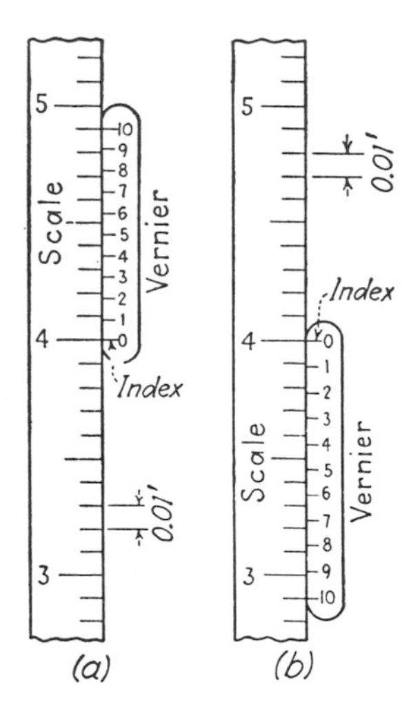

Fig. 4-14. (*a*) **Direct vernier.** (*b*) **Retrograde vernier.**

Retrograde Vernier. On the retrograde vernier shown in Fig. 4-14*b* each space on the vernier is 0.001 ft *longer* than a 0.01-ft space on the main scale, and 10 spaces on the vernier are equal to 11 spaces on the scale. As before, the index is set at 0.400 ft on the scale. If the vernier were moved upward 0.001 ft, its graduation numbered 1 would coincide with a graduation (0.39 ft) on the scale, and so on. It is seen that, from the index, the retrograde vernier extends backward along the main scale and that the vernier graduations are also numbered in reverse order; however, the retrograde vernier is read in the same manner as the direct vernier.

Reading the Vernier. Figure 4-13 illustrates settings of direct verniers on the target (at left) and on the back (at right) of a Philadelphia leveling rod. The rod reading indicated by the target (4.347 ft in figure) is determined by first observing the position of the vernier index on the scale to hundredths of feet (4.34 in figure), next by observing the number of spaces *on the vernier* from the index to the coinciding graduations (7 spaces in figure), and finally by adding the vernier reading (0.007 ft in figure) to the scale reading (4.34 ft). On the back of the rod (Fig. 4-13, right) both the main scale and the direct vernier read *down* the rod. The scale reading is 9.26 ft and the vernier reading is 0.004 ft; hence the rod reading is 9.264 ft.

4-20 Rod Levels. The rod level is an attachment for indicating the verticality of the leveling rod. One type (Fig. 4-15) consists of a circular or metal angle or bracket which is either metal angle or bracket which is either attached by screws to the side of the rod or held against the rod as desired. Another type consists of a hinged casting on each wing of which is mounted a level tube which is held parallel to a face of the rod.

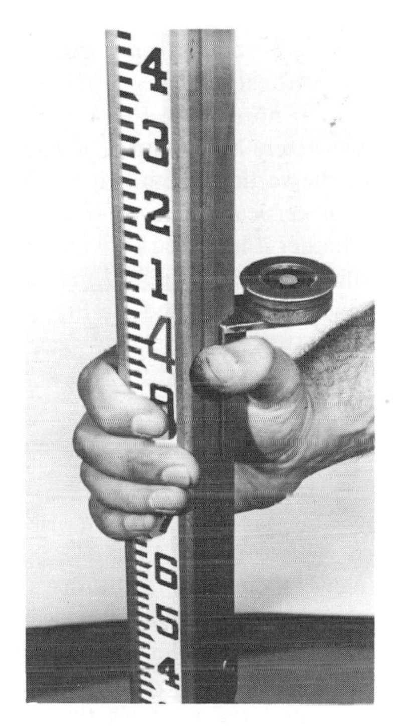

Fig. 4-15. Rod level.

USE OF LEVEL AND ROD

4-21 Setting Up Engineer's Level. The engineer's level is placed in a desired location with the tripod legs well spread and firmly pressed into the ground, with the tripod head nearly level, and with the telescope at a convenient height for sighting. If the setup is on a slope, it is preferable to orient the tripod so that one of its legs extends up the slope. If the leveling head has four screws, the telescope is brought over one pair of opposite leveling screws and the bubble is centered approximately; then the process is repeated with the telescope over the other pair. By repetition of this procedure the leveling screws are manipulated until the bubble remains centered, or nearly so, for any direction in which the telescope is pointed. If the instrument is in adjustment, the line of sight is then horizontal. If the leveling head has three screws, the process is similar, the telescope being brought successively over each of the screws. Suggestions for the care and handling of the level are given in Art. 1-7.

4-22 Reading Rods. For observations to hundredths or thousandths of feet, the rod is held vertical on some well-defined point of a stable object. The leveler sights at the rod, focuses the objective, and carefully centers the bubble. If the self-reading rod is used, the leveler records the reading of the cross hair on the rod. As a check he again observes the bubble and the rod. If the target rod is used, the procedure is identical except that the target is set by the rodman as directed by the leveler.

For leveling of lower precision, as when rod readings for ground points are determined to the nearest 0.1 ft, the observations usually are not checked, and proportionately less care is exercised in keeping the rod vertical and the bubble centered.

If no rod level is used, in calm air the rodman can plumb the rod accurately by balancing it upon the point on which it is held. By means of the vertical cross hair the leveler can determine when the rod is held in a vertical plane passing through the instrument, but he cannot tell whether it is tipped forward or backward in this plane. If it is in either of these positions, the rod reading will be greater than the true vertical distance, as illustrated by Fig. 4-16. To eliminate this error, the rodman

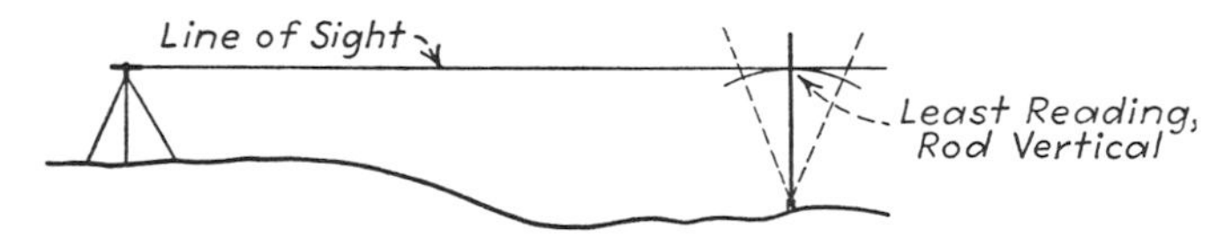

Fig. 4-16. Waving the rod.

waves the rod, or tilts it forward and backward as indicated by the figure, and the leveler takes the least reading, which occurs when the rod is vertical. The larger the rod reading, the larger the error due to the rod's being held at a given inclination; hence it is more important to wave the rod for large readings than for small readings. Further, whenever the rod is tipped backward about any support other than the front edge of its base, the graduated face rises and an error is introduced; for small readings this error is likely to be greater than that caused by not waving the rod.

ADJUSTMENT OF DUMPY AND WYE LEVELS

4-23 Adjustment of Dumpy Level. The axis of the objective slide and the optical axis are permanently fixed perpendicular to the vertical axis by the manufacturer. In addition, for a dumpy level in perfect adjustment the following relations should exist (see Fig. 4-17):

1 The axis of the level tube should be perpendicular to the vertical axis.

2 The horizontal cross hair should lie in a plane perpendicular to the

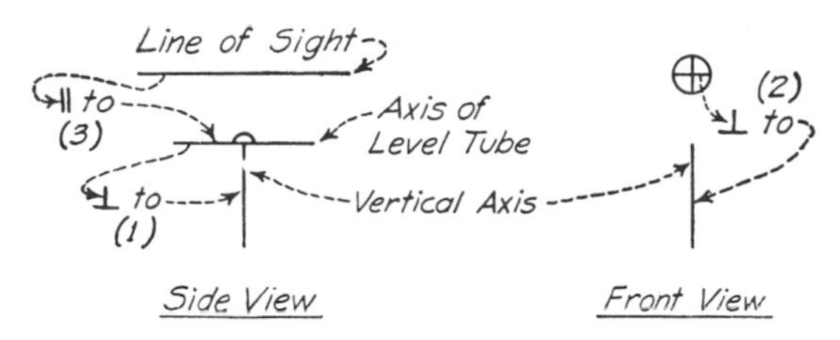

Fig. 4-17. Desired relations among principal lines of dumpy level.

vertical axis, so that it will lie in a horizontal plane when the instrument is level.

3 The line of sight should be parallel to the axis of the level tube.

The parts capable of and requiring adjustment are the cross hairs and the level tube. The basis for adjustments is the vertical axis. The adjustments are as given in the following paragraphs. (For one form of student field notes, see Fig. 4-24.)

1 To Make the Axis of the Level Tube Perpendicular to the Vertical Axis. Approximately center the bubble over each pair of opposite leveling screws; then carefully center the bubble over one pair. Rotate the level end for end about its vertical axis. If the bubble is displaced, bring it back halfway to the center by means of the capstan nuts at one end of the level tube. Relevel the instrument with the leveling screws, and repeat the process until the adjustment is perfected.

This adjustment involves the principle of *reversion*. If the tube is not in adjustment, the displacement of the bubble indicates double the actual error, as shown by Fig. 4-18. If $(90° - \alpha)$ represents the angle between

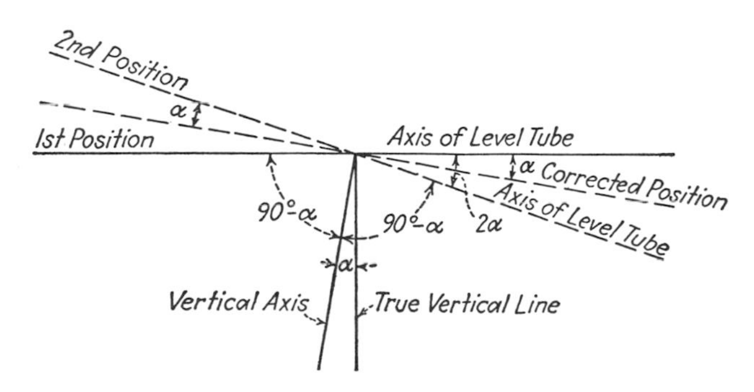

Fig. 4-18. Adjustment of axis of level tube of dumpy level.

the vertical axis and the axis of the level tube, then when the bubble is centered the vertical axis makes an angle of α with the true vertical. When the level is reversed, the bubble is displaced through the arc whose angle is 2α. Hence the correction is the arc whose angle is α.

2 To Make the Horizontal Cross Hair Lie in a Plane Perpendicular to the Vertical Axis (and Thus Horizontal When the Instrument Is Level). Sight the horizontal cross hair on some clearly defined point (as *A*, Fig. 4-19) and rotate the instrument slowly about its vertical axis. If the point appears to depart from the cross hair and takes some position as *A'* on the opposite side of the field of view, loosen two adjacent capstan screws and rotate the cross-hair ring until by further trial the point appears to travel along the cross hair. Tighten the same two screws. The instrument need not be level when the test is made.

3 To Make the Line of Sight Parallel to the Axis of the Level Tube (Two-peg Test). Method A. Set two pegs 200 to 300 ft apart on approximately level ground. Set up and level the instrument in a location such that the eyepiece is ½ in. or less in front of the rod held on one of the pegs as at *A* (Fig. 4-20). With the rod held at *A*, take a rod reading *a* by sighting through the objective end of the telescope (with the eyepiece next to the rod). The center of the field of view may be determined within one or two thousandths of a foot by holding the point of a pencil on the rod.

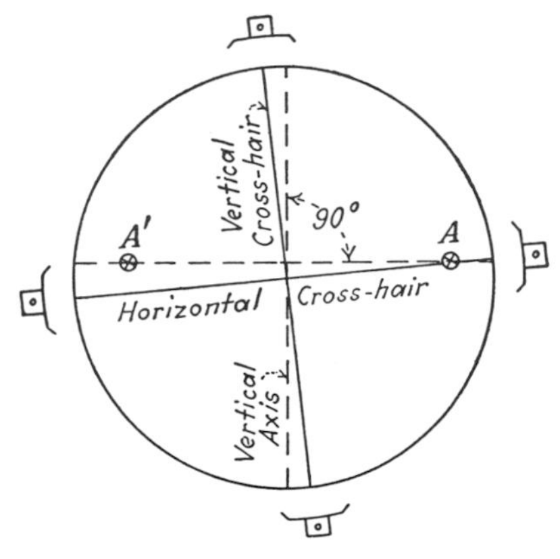

Fig. 4-19. Adjustment of horizontal cross hair.

Move the rod to the other peg *B*, and take a rod reading *b* with level at *A*, in the usual manner.

Move the instrument to *B*, set up as before, and take rod readings *c* and *d*.

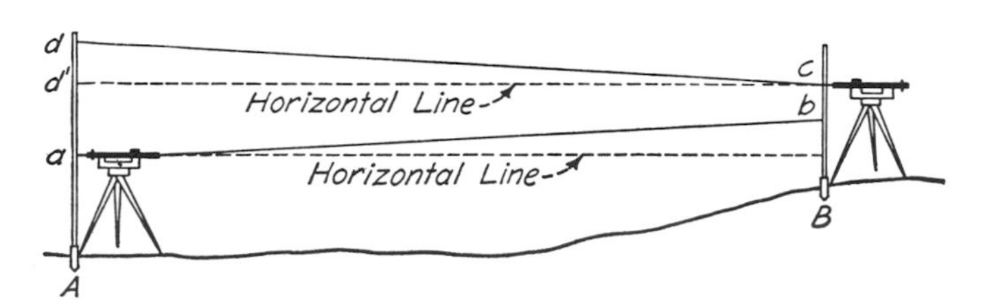

Fig. 4-20. Two-peg test, Method A.

If the two differences in elevation thus determined are equal, that is, if $(a - b) = (d - c)$, the line of sight is in adjustment. If not, the correct rod reading at *A* for instrument with position unchanged at *B* is

$$d' = c + \text{correct diff. el.} = c + \frac{(a-b)+(d-c)}{2} \tag{10}$$

Equation (10) must be solved with due regard to signs. Strictly speaking, the effect of curvature and refraction (Art. 4-2) should be included; but for ordinary leveling the quantities are negligible. To show the readings and distances in their true relation, a sketch should always be drawn.

The adjustment is made by moving the cross-hair ring vertically until the line of sight cuts the rod at d'. The preceding steps are then repeated as a check.

example 1. With level at A, observed readings are $a = 4.086$ and $b = 2.705$; with level at B, $c = 3.871$ and $d = 5.542$. Then, by Eq. (10), the correct difference in elevation is $\frac{(4.086 - 2.705) + (5.542 - 3.871)}{2} = 1.526$ ft, with B indicated as being higher than A. The correct rod reading for a horizontal line of sight with instrument still at B would be $3.871 + 1.526 = 5.397$ ft.

Instead of viewing the near rod through the objective end of the telescope, the level may be set up a short distance (say, 6 or 8 ft) beyond each near rod and the near rod reading observed in the customary manner. The adjustment should be considered as a first approximation and should be repeated for precise results.

The advantages of Method A are (1) the computations are relatively simple and (2) the objective slide is in the same position for the two sights; thus a possible error in sighting is eliminated.

Method B. Set two pegs 200 to 300 ft apart on approximately level ground, and designate as A the peg near which the second setup will be made (Fig. 4-21); call the other peg B. Set up and level the instrument at any point M equally distant from A and B, that is, in a vertical plane bisecting the line AB. Take rod readings a on A and b on B; then $(a - b)$

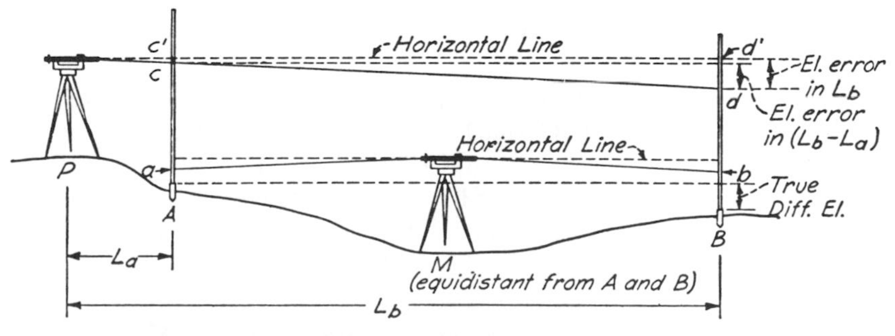

Fig. 4-21. Two-peg test, Method B.

will be the correct difference in elevation, since any error would be the same for the two equal sight distances L_m. Due account must be taken of signs throughout the test.

Move the instrument to a point P near A, preferably but not necessarily on line with the pegs; set up as before, and measure the distances L_a to A and L_b to B. Take rod readings c on A and d on B. Then $(c - d)$, taken in the same order as before, is the indicated difference in elevation; if $(c - d) = (a - b)$, the line of sight is parallel to the axis of the level tube and the instrument is in adjustment. If not, $(c - d)$ is called the "false" difference in elevation, and the inclination (error) of the line of sight in the net distance $(L_b - L_a)$ is equal to $(a - b) - (c - d)$. By proportion, the error in the reading on the far rod is $\dfrac{L_b}{L_b - L_a}[(a - b) - (c - d)]$. Subtract algebraically the amount of this error from the reading d on the far rod to obtain the correct reading d' at B for a horizontal line of sight with the position of the instrument unchanged at P. Set the target at d' and bring the line of sight on the target by moving the cross-hair ring vertically.

example 2. With level at M, the rod reading a is 0.970 and b is 2.986; the correct difference in elevation $(a - b)$ is then $0.970 - 2.986 = -2.016$ ft, with B thus indicated as being lower than A. With level at P, the rod reading c is 5.126 and d is 7.018; the false difference in elevation $(c - d)$ is then $5.126 - 7.018 = -1.892$, with B again indicated as being lower than A. The distance L_a is observed to be 30 ft, and L_b to be 230 ft. Inclination of the line of sight in $(230 - 30 = 200)$ ft is $(-2.016) - (-1.892) = -0.124$ ft. The error in elevation of the line of sight at the far rod is $(^{230}/_{200}) \times (-0.124) = -0.143$ ft. The correct rod reading d' for a horizontal line of sight is $7.018 - (-0.143) = 7.161$ ft.

As a partial check on the computations, the correct rod reading c' at A may be computed by proportion; the difference in elevation computed from the two corrected rod readings c' and d' should be equal to the correct difference in elevation observed originally at M.

example 3. In the preceding example, the error in elevation of the line of sight at the near rod is $(^{30}/_{200}) \times (-0.124) = -0.019$ ft. The correct rod reading c' is $5.126 - (-0.019) = 5.145$ ft. The "false" difference in elevation is $5.145 - 7.161 = -2.016$ ft, which is equal to the correct difference in elevation; hence the computations are checked to this extent.

As in Method A, a sketch should always be drawn. Also, theoretically a correction for earth's curvature and atmospheric refraction should be added numerically to the final rod reading d', although in practice it is usually considered negligible.

Some surveyors prefer to set up at P within 6 to 8 ft of A and to consider $[(a - b) - (c - d)]$ as being the *total* error in elevation, to be subtracted directly from d. This serves as a first approximation; the procedure for the setup at P is then repeated. The amount of computation is thus reduced, but the amount of field work is increased.

The advantages of Method B are that in no case is it necessary to take a rod reading by projecting from the eyepiece and that it is not necessary to set up on a line through AB.

4-24 Adjustment of Wye Level.

For a wye level in perfect adjustment, the following relations should exist (see Fig. 4-22):

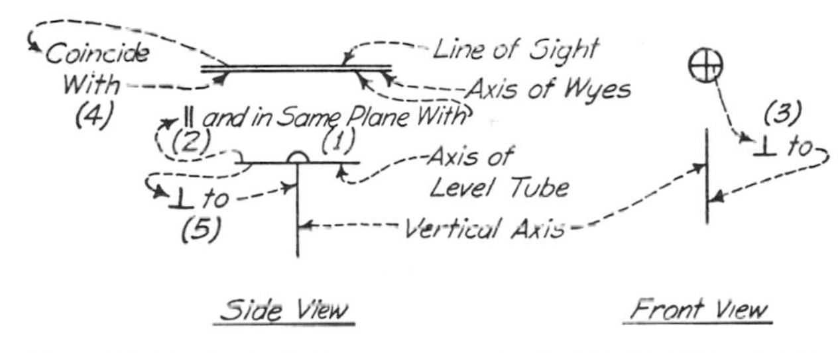

Fig. 4-22. Desired relations among principal lines of wye level.

1 The axis of the level tube should lie in the same plane with the axis of the wyes.

2 The axis of the level tube should be parallel to the axis of the wyes.

3 The horizontal cross hair should lie in a plane perpendicular to the vertical axis, so that it will be horizontal when the instrument is level.

4 The line of sight should coincide with the axis of the wyes, so that it will be parallel to the axis of the level tube.

5 For convenience in leveling, the axis of the wyes (and hence the axis of the level tube) should be perpendicular to the vertical axis.

The optical axis and the axis of the objective slide are fixed in the proper relation by the manufacturer, and they are presumed to remain in coincidence with the axis of the wyes (and hence with the line of sight) without further attention. The remaining relations are established by the following adjustments.

1 To Make the Axis of the Level Tube Lie in the Same Plane with the Axis of the Wyes. Raise the wye clips; level the instrument; and rotate the telescope a few degrees in the wyes. If the bubble is displaced, bring it back to the center by means of the lateral adjusting screws at one end of the level tube.

2 To Make the Axis of the Level Tube Parallel to the Axis of the Wyes. Raise the wye clips; level the instrument carefully; lift the telescope from the wyes, and turn it end for end. If the bubble is displaced, bring it back halfway to the center by means of the vertical adjusting nuts at one end of the level tube. Relevel the instrument by means of the leveling screws, and repeat the process until the adjustment is perfected.

3 To Make the Horizontal Cross Hair Lie in a Plane Perpendicular to the Vertical Axis (and Thus Horizontal When the Instrument Is Level). This adjustment is the same as adjustment 2 for the dumpy level. For some instruments the adjustment may be made by rotating the telescope in the wyes instead of rotating the cross-hair ring in the telescope.

4 To Make the Line of Sight Coincide with the Axis of the Wyes (and Thus Parallel to the Axis of the Level Tube). Raise the wye clips, sight the intersection of the cross hairs on some well-defined point, and clamp the vertical axis. Revolve the telescope 180° (about its own axis) in the wyes. If the line of sight is displaced from the point, adjust the cross-hair ring until the line of sight is midway between its two former positions (Fig. 4-23). Loosen only one screw at a time, and tighten the opposite screw. Repeat the process until the proper relation is obtained.

5 To Make the Axis of the Level Tube (and Thus the Axis of the Wyes) Perpendicular to the Vertical Axis. (This adjustment does not add to the precision of observations, but makes it possible to level the instrument so that the bubble will remain centered for any direction in which the telescope may be pointed.) Level the instrument, and rotate the telescope end for end about the vertical axis. If the bubble is displaced, bring it back halfway to the center by means of the capstan nuts controlling the vertical position of one of the wyes.

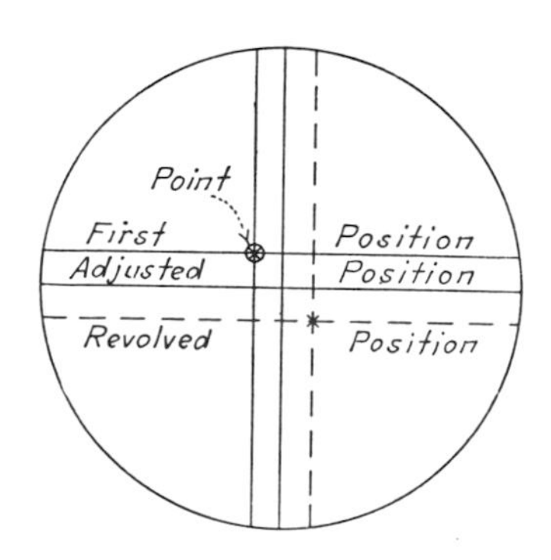

Fig. 4-23. Adjustment of line of sight of wye level.

Worn Collars. When either the collars or the bearing points become worn, the adjustments just described are inadequate, and the two-peg test must be made as for the dumpy level. The common procedure is first to perform the adjustments for the cross hairs in the usual manner, making the line of sight coincide as nearly as may be with the axis of the wyes. The true difference in elevation between two points having been determined by the two-peg test, the line of sight is then set to the proper rod reading for a horizontal line by means of the leveling screws, and the bubble is centered by means of the capstan nuts at one end of the level tube.

4-25 Adjustment of Hand Level. The simplest procedure is to hold the hand level alongside an engineer's level that has been leveled and sighted at some well-defined point. The line of sight of the hand level should strike the same point when the bubble is centered. The Locke hand level is adjusted by means of a screw which moves the cross wire. The Abney hand level is adjusted by means of a screw which raises or lowers one end of the level tube.

The hand level, even if not in adjustment, may be used to establish a horizontal line by employing the principle of the two-peg test (Art. 4-23, adjustment 3).

4-26 NUMERICAL PROBLEMS

1 What is the combined effect of the earth's curvature and mean atmospheric refraction in a distance of 300 ft? 3,000 ft? 6 miles? 60 miles?

2 An observer standing on a beach can just see the top of a lighthouse 15 miles away. The eye height of the observer above sea level is 5.7 ft. What is the height of the lighthouse above sea level?

3 Two points, *A* and *B*, are each distant 2,000 ft from a third point *C*, from which the measured vertical angle to *A* is $+3°21'$ and that to *B* is $+0°32'$. What is the difference in elevation between *A* and *B*?

4 Two points, *A* and *B*, are 1,000 ft apart. The elevation of *A* is 615.03 ft. A level is set up on the line between *A* and *B* and at a distance of 250 ft from *A*. The rod reading on *A* is 9.15 and that on *B* is 2.07. What is the elevation of *B* (*a*) considering curvature and refraction and (*b*) neglecting curvature and refraction?

5 Through the telescope of a level a magnified 0.2 ft on the rod apparently covers the unmagnified image between 2.1 and 8.3. What is the magnifying power of the telescope?

6 Design a direct vernier and a retrograde vernier, both reading to $\frac{1}{32}$ in., for an architect's rod graduated to $\frac{1}{8}$ in. Draw a neat sketch of each vernier and a portion of the rod for a reading of $2\frac{19}{32}$ in., with graduations shown and labeled.

7 If the rod were inclined forward 0.4 ft in a length of 13 ft, what error would be introduced in a rod reading of 5.0 ft? Compute by an approximate "slope" method and by an exact method, and compare results.

8 In the two-peg test of a dumpy level by Method A, the following observations are taken:

	Instrument at A	*Instrument at B*
Rod reading on *A*	4.937	3.077
Rod reading on *B*	6.736	4.752

What is the correct difference in elevation between the two points? With the instrument in the same position at *B*, to what rod reading on *A* should the line of sight be adjusted? What is the error in the line of sight for the distance *A* to *B*?

9 In the two-peg test of a dumpy level by Method B, the following observations are taken:

	Instrument at M	*Instrument at P*
Rod reading on *A*	3.612	1.862
Rod reading on *B*	3.248	0.946

M is equidistant from *A* and *B; P* is 40 ft from *A* and 240 ft from *B*. What is the correct difference in elevation between the two points? With the level in the same position at *P*, to what rod reading on *B* should the line of sight be adjusted? What is the corresponding rod reading on *A* for a horizontal line of sight? Check these two rod readings against the correct difference in elevation, previously determined.

4-27 FIELD PROBLEM

Problem 5. Adjustment of Engineer's Level

Object. To make the field adjustment of the engineer's level.

Procedure. (1) Proceed as outlined in Art. 4-23 for the dumpy level or Art. 4-24 for the wye level. (2) Record all field observations in the field book. For each adjustment separately, state such items as the desired relation, the test, the condition of the instrument with regard to the adjustment, the method of adjusting, and the final test. Include sketches if necessary to make the method clear. One form of student field notes is shown in Fig. 4-24. In practice, knowledge of the amount by which an instrument is out of adjustment may be important as a basis for deciding whether prior work needs to be repeated.

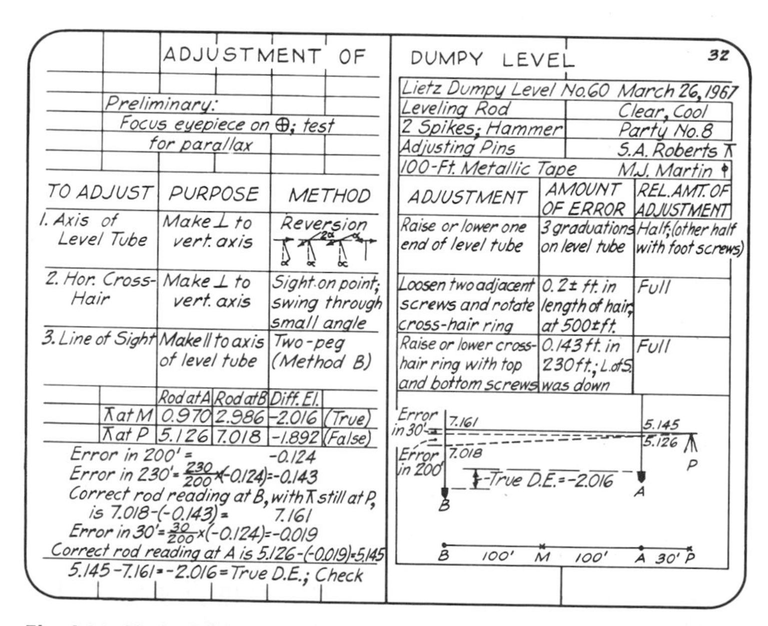

Fig. 4-24. Student field notes for adjustment of dumpy level.

5

DIFFERENTIAL LEVELING

5-1 Definitions. *Differential leveling* is the operation of determining the elevations of points some distance apart. Usually this is accomplished by direct leveling. Differential leveling requires a series of setups of the instrument along the general route and, for each setup, a rod reading back to a point of known elevation and forward to a point of unknown elevation.

A *bench mark* (B.M.) is a definite point on an object of more or less permanent character, the elevation and location of which are known. Bench marks serve as points of reference for levels in a given locality. Their elevations are established by differential leveling. Permanent bench marks have been established throughout the United States by the U.S. Geological Survey and the U.S. Coast and Geodetic Survey; these consist of bronze plates set in stone or concrete and marked with the elevation above mean sea level. Other agencies have established permanent monuments. For any survey or construction enterprise, local bench marks are established, based on some selected datum and employing natural or artificial objects such as stones, pegs, spikes in trees or pavements, and marks painted or chiseled on street curbs.

In some areas, the elevation of bench marks may be altered by earth movements such as those caused by earthquakes, slides, lowering of water tables, pumping from oil fields, mining, or construction.

A *turning point* (T.P.) is an intervening point between two bench marks, upon which point foresight and backsight rod readings are taken. The nature of the turning point is usually indicated in the notes, but no record is made of its location unless it is to be reused. A bench mark may be used as a turning point.

A *backsight* (B.S.) is a rod reading taken on a point of known elevation. It is sometimes called a *plus sight.*

A *foresight* (F.S.) is a rod reading taken on a point the elevation of which is to be determined. It is sometimes called a *minus sight.*

The *height of instrument* (H.I.) is the elevation of the line of sight of the telescope when the instrument is leveled.

In surveying with the transit, the terms backsight, foresight, and height of instrument have meanings different from those here defined.

5-2 Procedure. In Fig. 5-1, B.M.$_1$ represents a point of known elevation (bench mark) and B.M.$_2$ represents a bench mark to be established some distance away. It is desired to determine the elevation of B.M.$_2$. The rod is held at B.M.$_1$, and the level is set up in some convenient location, as L_1, along the general route but not necessarily on the direct line joining B.M.$_1$ and B.M.$_2$. A backsight is taken on B.M.$_1$. The rodman then goes forward and, as directed by the leveler, chooses a turning point T.P.$_1$ at some convenient spot within the range of the telescope along the

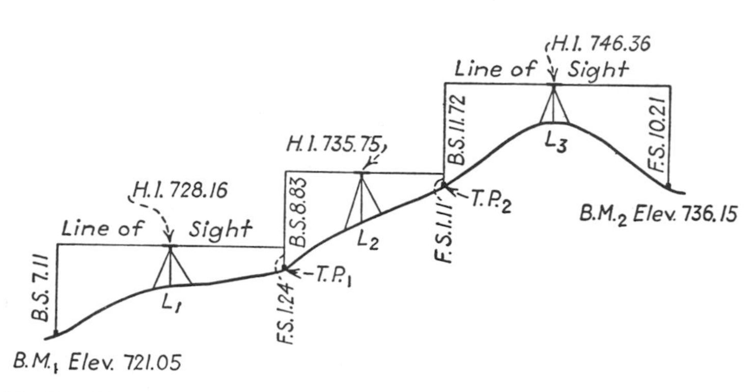

Fig. 5-1. Differential leveling.

general route B.M.$_1$ to B.M.$_2$. It is desirable, but not necessary, that each foresight distance, as L_1-T.P.$_1$, be made approximately equal to its corresponding backsight distance, as B.M.$_1$-L_1. The rod is held on the turning point, and a foresight is taken. The leveler then sets up the instrument at some favorable point, as L_2, and takes a backsight to the rod held on the turning point; the rodman goes forward to establish a second turning point T.P.$_2$; and so the process is repeated until finally a foresight is taken on the terminal point B.M.$_2$.

It is seen in Fig. 5-1 that a *backsight added to the elevation* of a point on which the backsight is taken gives the height of instrument, and that a *foresight subtracted from the height of instrument* determines the elevation of the point on which the foresight is taken. Also, the difference between the backsight taken on a given point and the foresight taken on the following point is equal to the difference in elevation between the two points. It follows that the difference between the sum of all backsights and the sum of all foresights gives the difference in elevation between the bench marks.

Sometimes in leveling for a tunnel or a building it is necessary to take rod readings on points which are at a higher elevation than that of the H.I. In such cases the rod is held inverted, and in the field notes each such backsight is indicated with a minus sign and each foresight with a plus sign.

When several bench marks are to be established along a given route, each intermediate bench mark is made a turning point in the line of levels. Elevations of bench marks are checked usually by continuing the line of levels back to the initial bench mark. A line of levels that ends at the point of beginning is called a *level circuit*. If the circuit checks within the prescribed limits of error, it is considered that the elevations of all turning points are correct within prescribed limits, but such closure does not serve as a check on side shots taken from the circuit.

5-3 Balancing Backsight and Foresight Distances. In Art. 4-4 it has been shown that, if a foresight distance were equal to the corresponding backsight distance, any error in readings due to earth's curvature and to atmospheric refraction (under uniform conditions) would be eliminated.

In ordinary leveling no special attempt is made to balance *each* foresight distance against the preceding backsight distance. Whether or not such distances are approximately balanced between bench marks will depend upon the desired precision. The effect of earth's curvature and atmospheric refraction is slight unless there is an abnormal difference between backsight and foresight distances. The effect of instrumental errors is likely to be of considerably greater consequence with regard to the balancing of these distances. The chances are that there is not absolute parallelism between the line of sight and the axis of the level tube. The error in a rod reading due to this imperfection of adjustment is proportional to the distance from the instrument to the rod and is of the same sign for a backsight as for a foresight. Since backsights are added and foresights are subtracted, this instrumental error is eliminated if, between bench marks, the *sum* of the foresight distances is made equal to the *sum* of the backsight distances.

In leveling of ordinary precision, *with an instrument in good adjustment,* the backsight and foresight distances are not measured, and no special attempt is made to equalize them even by estimation. Normally they tend to balance in the long run. However, a succession of very long backsight distances and very short corresponding foresight distances, or *vice versa,* as would be likely for leveling between two points having a large difference in elevation, would produce a systematic error of appreciable magnitude.

For leveling of moderately high precision it is necessary to equalize backsight and foresight distances between bench marks. In less refined leveling, distances are usually determined by pacing; in precise leveling, they are usually measured with the stadia or the gradienter.

In leveling uphill or downhill a balance between foresight and backsight distances can be obtained with a minimum number of setups by following a zigzag course.

The effect of earth's curvature and atmospheric refraction cannot be entirely eliminated by making the *sum* of the foresight distances equal to that of the backsight distances; rather it would be necessary that *each* foresight distance be made equal to the corresponding backsight distance.

5-4 Differential-level Notes. For ordinary differential leveling when no special effort is made to equalize backsight and foresight distances, usually the record of field work is kept in the form indicated by Fig. 5-2, in which the levels from B.M.$_1$ to B.M.$_2$ are the same as those shown by

LEVELS FOR BENCH MARKS ALONG RIDGE ROAD — Dec. 31, 1965, Fair — J.G. Sutter, ⊼; W.R. Knowles, Rod

Sta.	B.S.	H.I.	F.S.	Elev.		Remarks
B.M.$_1$	7.11	728.16		721.05		Top of Hydrant Cor. Oak St.
T.P.$_1$	8.83	735.75	1.24	726.92		Curb
T.P.$_2$	11.72	746.36	1.11	734.64		
B.M.$_2$	4.32	740.47	10.21	736.15		Spike in Pole North of Williams House Marked B.M 736.15
T.P.$_3$	3.06	733.57	9.96	730.51		
T.P.$_4$	2.74	727.40	8.91	724.66		Stone
T.P.$_5$	0.81	716.59	11.62	715.78		
B.M.$_3$			12.42	704.17		Concrete Monument No. of Road; County Line
ΣB.S.=	38.59	ΣF.S.=	55.47	721.05		
			38.59			
		Diff=	16.88 =	16.88	; ck.	

Fig. 5-2. Differential-level notes.

Fig. 5-1. The heights of instrument and the elevations are computed as the work progresses. Usually at the foot of each page of level notes the *computations* are checked by comparing the difference between the sum of the backsights and the sum of the foresights with the difference between the initial and the final elevation, as illustrated at the bottom of Fig. 5-2. However, this procedure does not check against mistakes in observing or recording.

When backsight and foresight distances are to be balanced, usually they are recorded in the last column of the left page of the field notebook, and the cumulative excess or deficiency of foresight over backsight distances is noted for each turning point and bench mark.

5-5 Mistakes in Leveling. Some of the mistakes commonly made in leveling are:

1 Confusion of numbers in reading the rod. To avoid this mistake, the observer should notice the numbers on both sides of the observed reading.

2 Recording backsights in foresight column, and *vice versa.*

3 Faulty additions and subtractions; adding foresights and subtracting backsights. As a check, the difference between the sum of the backsights and the sum of the foresights should be computed for each page or between bench marks.

4 Rod not held on same point for both foresight and backsight.

5 Not having the Philadelphia rod fully extended when reading the long rod.

6 Wrong reading of vernier when the target rod is used.

7 When the long target rod is used, not having the vernier on the target set to read exactly the same as the vernier on the back of the rod when the rod is short.

5-6 Errors in Ordinary Leveling. In ordinary leveling, errors of consequence may be due to the following causes (see also Table 5-1):

1 Imperfect Adjustment of Instrument. Insofar as results are concerned, the only essential relation is that the line of sight should be parallel to the axis of the level tube. The effect of imperfect adjustment is minimized by balancing backsight and foresight distances.

2 Parallax. This condition produces an accidental error. It can be practically eliminated by careful focusing.

3 Earth's Curvature and Atmospheric Refraction. In ordinary leveling, when backsight and foresight distances roughly balance each other, any error due to earth's curvature and atmospheric refraction is accidental in nature and is usually negligible in amount. If heat waves are apparent, the line of sight should be well above the ground (say, at least 2 ft), and the sight distances should be short.

4 Rod Not Standard Length. If the error is distributed over the length of the rod, a systematic error is produced which varies directly as the difference in elevation and bears no relation to the length of the line over which levels are run. The error can be eliminated by comparing the rod with a standard length and applying the necessary corrections. The case is analogous to measurements of distance with a tape that is too long or too short.

If the rod is worn uniformly at the bottom, an erroneous height of instrument is shown at each setup, but the error in backsight is balanced by that in the following foresight, and no error results in the elevation of the foresight point.

Table 5-1 Errors in leveling

Source	*Type*	*Cause*	*Remarks*	*Procedure to eliminate or reduce*
Instrumental	Systematic	Line of sight not parallel to axis of level tube	Error of each sight proportional to distance [a]	Adjust instrument, or balance sum of backsight and foresight distances
		Rod not standard length (throughout length) [b]	May be due to manufacture, moisture, or temperature. Error usually small	Standardize rod and apply corrections, same as for tape
Personal	Accidental	Parallax	...	Focus carefully
		Bubble not centered at instant of sighting	Error varies as length of sight	Check bubble before making each sight
		Rod not held plumb	Readings are too large. Error of each sight proportional to square of inclination [a]	Wave the rod, or use rod level
		Faulty reading of rod or setting of target	...	Check each reading before recording. For self-reading rod, use fairly short sights
		Faulty turning points	...	Choose definite and stable points
Natural	Accidental	Temperature	May disturb adjustment of level	Shield level from sun
		Earth's curvature	Error of each sight proportional to square of distance [a]	Balance *each* backsight and foresight distance, or apply computed correction
		Atmospheric refraction	Error of each sight proportional to square of distance [a]	Same as for earth's curvature; also take short sights, well above ground, and take backsight and foresight readings in quick succession
	Systematic	Settlement of tripod or turning points	Observed elevations are too high	Choose stable locations; take backsight and foresight readings in quick succession

[a] The error of *each sight* is systematic, but the resultant error is the difference between the systematic error for foresights and that for backsights; hence the resultant error tends to be accidental.

[b] Uniform wear of the bottom of the rod causes no error.

5 Rod Not Held Plumb. This condition produces rod readings which are too large. The error varies directly as the first power of the rod reading and directly as the square of the inclination. It can be eliminated by waving the rod or by using a rod level.

6 Faulty Turning Points. An accidental error results when turning points are not well defined. If either the level tripod or the turning point settles between readings, as in soft ground, the error will be systematic and the observed elevations will be too high.

7 Bubble Not Exactly Centered at Instant of Sighting. This produces an accidental error which tends to vary as the distance from instrument to rod. The longer the sight, the greater the care that should be observed in leveling the instrument.

8 Inability of Observer to Read the Rod Exactly or to Set the Target Exactly on the Line of Sight. This inability causes an accidental error, which can be confined within reasonable limits through proper choice of length of sight.

Summary. If the proper leveling procedure is observed, the important errors are accidental. Hence the resultant error may be expected to vary as the square root of the number of setups or as the square root of the distance. It is customary to express limiting errors of leveling in terms of the square root of the distance, usually in miles.

5-7 Precision of Differential Leveling. The conditions affecting the precision of differential leveling are many and variable, but experience indicates that under average conditions and with a level in good adjustment the *maximum* error may be kept within the limits shown below. Usually the *average* error will be materially less.

1 Rough leveling, as practiced on preliminary surveys. Sights up to 1,000 ft in length. Rod readings to tenths of feet. No particular attention paid to balancing backsight and foresight distances. Maximum error, in feet, $\pm 0.4 \sqrt{\text{distance in miles}}$.

2 Ordinary leveling, as necessary in connection with most engineering works. Sights up to 500 ft in length. Rod readings to hundredths of feet. Backsight and foresight distances roughly balanced only when running for long distances uphill or downhill. Turning points on solid objects. Maximum error, in feet, $\pm 0.1 \sqrt{\text{distance in miles}}$.

3 Excellent leveling for the principal bench marks on extensive surveys. Sights up to 300 ft in length. Rod readings to thousandths of feet. Backsight and foresight distances measured by pacing and approximately balanced between bench marks. Rod waved for large readings. Bubble centered carefully before each sight. Turning points on well-defined points of solid objects. Tripod set on firm ground. Maximum error, in feet, $\pm 0.05 \sqrt{\text{distance in miles}}$.

4 Precise leveling for establishing bench marks with great accuracy at widely distributed points. High-grade level equipped with stadia hairs and with sensitive level tube. Adjustments carefully tested daily. Rod standardized frequently. Sights up to 300 ft in length. Rod readings of three horizontal hairs to thousandths of feet. Level protected from sun. Turning points on metal pin or plate. Two rodmen. Backsights and following foresights taken in quick succession. Bubble very carefully centered and under observation at instant of taking sight. Rod plumbed with rod level. Backsight and foresight distances balanced between bench marks by stadia readings. Level set up securely on firm ground. Levels not run when the air is boiling badly nor during high winds. Maximum error in feet, $\pm 0.02 \sqrt{\text{distance in miles}}$.

5-8 Precise Leveling. The subject of precise leveling as practiced on government surveys is not to be considered in detail here, as it is thoroughly covered in several U.S. Coast and Geodetic Survey manuals. However, it is appropriate to call attention here to certain refinements by means of which a relatively high degree of precision may be obtained with the ordinary wye or dumpy level and the self-reading rod.

For work of this nature the rod should be treated in some manner to prevent expansion or contraction through change in moisture content, and at intervals it should be compared with a standard length. Rods with a graduated strip of invar steel are preferable. The rod should have an attached rod level for plumbing. It is particularly important that turning points be on solid objects with rounded tops so that the base of the rod can be held in the same location for both backsight and foresight.

The level should be equipped with stadia hairs in addition to the regular cross hairs or with the micrometer tilting device which can be used as a gradienter. To prevent unequal thermal expansion, the level should be protected from the sun's rays. It should be set up very firmly so that no settlement will occur. It is desirable that the shortest possible time elapse between a backsight and the succeeding foresight. The backsight and foresight distances are determined preferably by stadia or gradienter but sometimes by pacing and are balanced very closely between bench marks. For stadia readings all three horizontal hairs should be read by estimation to thousandths of feet, and the readings should be recorded; the mean of the readings for the three hairs is taken as the correct rod reading for each sight. The interval between the reading of the upper hair and that of the lower hair is a measure of the distance from instrument to rod.

Two Rods. Excellent results have been obtained by employing two rods and two rodmen, each occupying alternate turning points (of the same set). In order further to eliminate possible systematic errors, the order of readings (backsight and foresight) may be interchanged at alternate setups of the level.

In another method, which requires only one rodman and which permits checks to be made as the leveling progresses, readings are taken not only on the front face of the rod but also on the back face, on which is a scale with a different zero or different units.

Double-rodding with Two Sets of Turning Points. This method was formerly used on some government surveys where two rods and two rodmen were usually employed. Two sets of turning points are established so that at each setup of the level two independent backsights and two independent foresights are taken. The turning points on one line are usually a foot or more higher than corresponding points on the other line so as to eliminate the possibility of making the same mistake in reading the foot marks on both rods. One rodman gives readings for points along the "high" line and the other for points along the "low" line.

5-9 Geodetic Leveling. Geodetic leveling is direct or spirit leveling of a high order of accuracy and precision. It is based on the irregular surfaces associated with the geoid rather than the regular surface of a spheroid or an ellipsoid representing the surface of the earth. The geoid is the figure of the earth considered as a sea-level surface extended continuously through the continents. At every point, the geodetic surface is perpendicular to the plumb line; it is determined by observing deflections of the vertical.

Geodetic leveling is of first-order precision (see Art. 20-3) and is usually conducted in connection with triangulation over large areas to furnish vertical control for topographic and other surveys.

The high order of precision required calls for instruments and methods which are ordinarily employed only by governmental agencies such as the U.S. Coast and Geodetic Survey and the U.S. Geological Survey. Following are given briefly some features of the field work of first-order leveling, principally as followed by the Coast Survey.

The unit of vertical distance used by the Coast Survey is the meter. (The Geological Survey uses the yard.) An extremely sensitive level of special design is used. The leveling rod is of the self-reading type. The front of the wooden rod is graduated in meters and decimeters; along a groove in the face of the rod extends a strip of invar metal which is graduated in centimeters and which permits readings to be taken (by estimation) to millimeters. The invar strip is fastened rigidly only at the bottom of the rod and is kept taut by means of a spring at the top; thus it is free to expand or contract independently of the wooden rod. For each sighting on the front of the rod three readings are taken—one of the horizontal cross hair and one of each stadia hair. The back of the rod is graduated in feet and tenths; after each sighting on the front face, the horizontal cross hair is read to 0.01 ft on the back face as a check against mistakes. As the eyepiece of the telescope is inverting, the graduations of the rod are inverted. The rod is equipped with a rod level and a thermometer. Where other stable objects are not available as turning points, a steel pin driven into the ground is used.

Two rods are employed in order that corresponding backsight and foresight readings can be taken as closely together as possible. At one setup of the level, the backsight is observed before the foresight; at the next setup, this order is reversed. Each foresight distance is kept within 10 m of the corresponding backsight distance, and the cumulative difference between backsight and foresight distances is kept within 20 m. The maximum length of sight is 150 m. Between bench marks, the lines of levels are run both forward and backward; if these do not agree within specified limits (± 0.017 ft $\sqrt{\text{distance in miles}}$), the line is rerun until two runnings in opposite directions do agree. The instrument is shaded from the direct rays of the sun even while it is being carried from setup to setup; and it is shielded from strong winds.

5-10 Reciprocal Leveling. On occasion it becomes necessary to determine accurately the relative elevations of two intervisible points a considerable distance apart, between which points levels cannot be run in the ordinary manner. For example, it may be desired to transfer levels from one side to the other of a deep canyon, or from bank to bank of a wide stream. With certain modifications the method employed in getting the true difference in elevation between two points prior to the adjustment of the line of sight of the dumpy level (the two-peg test) may be utilized in situations of this kind.

If A and B are two such points, then the level is set up near A and one or more rod readings are taken on both A and B. Then the level is set up in a similar location near B, and rod readings to near and distant points are taken as before. The mean of the two differences in elevation thus determined is taken to be the true difference between the two points. Usually the distance between points is large (often a half mile or more) so that it is necessary to use a target on the distant rod. If precise results are desired, a series of foresights is taken on the distant rod and sometimes also a series of backsights on the near rod, the bubble being recentered and the target reset after each observation. The difference in elevation is then computed by using the mean of the backsights and the mean of the foresights. The three-wire method described in Art. 5-8 can be used here.

When one point cannot be quickly reached from the other, the effect of variation in refraction may be eliminated by taking simultaneous observations with two instruments, one being set up near each point. The instruments are then interchanged, and simultaneous readings on near and far points are taken as before.

5-11 Adjustment of Elevations. *Intermediate Bench Marks.* When a line of levels makes a complete circuit, the final elevation of the initial bench mark as computed from the level notes will not agree with the initial elevation of this point. The difference is called the *error of closure.* The elevations of intermediate bench marks will also be in error, and it is necessary to correct them. Since the probable error in leveling varies as the square root of the distance and since corrections to related quantities are proportional to the square of the corresponding probable errors (Art. 2-9), it follows that the corrections will be directly proportional to the distances from the point of beginning. Thus if E_c is the error of closure of a level circuit of length L and if $C_a, C_b, \ldots, C_n$ are the respective corrections to be applied to observed elevations of bench marks $A, B, \ldots, N$ whose respective distances from the point of beginning are $a, b, \ldots, N$, *then*

$$C_a = -\frac{a}{L}E_c \qquad C_b = -\frac{b}{L}E_c \ \ldots \qquad \text{and} \qquad C_n = -\frac{n}{L}E_c \tag{1}$$

The foregoing principles apply also to the adjustment of bench marks on a line of levels run between two points whose difference in elevation has previously been determined by more accurate methods and is assumed to be correct.

Levels over Different Routes. When the elevation of a bench mark is established by lines of levels run over several routes, there will be as many observed elevations as there are lines terminating at the point. Under the assumption that the probable error of each of the observed values is proportional to the square root of the length of the line of levels (or of the number of setups) by means of which that determination is secured, the weight to be applied to a given observed elevation will vary inversely as the length of the corresponding line (or of the number of setups). The most probable value of the elevation will then be the weighted mean of the observed values.

Level Net. Where elevations of bench marks in an interconnecting network of level circuits are to be adjusted, the method of least squares may be employed. A simpler and equally precise method is that of successive approximations. It consists in adjusting each separate figure in the net in turn, with the adjusted values for each circuit used in the adjustment of adjacent circuits; the process is repeated for as many cycles as necessary to balance the values for the whole net. Within each circuit the error of closure is normally distributed to the various sides in proportion to their length (or number of setups), as previously explained. The order in which the various circuits and lines are taken is immaterial, although it is advisable to begin with the circuit having the largest error of closure. The computations may be based on *corrections* rather than on differences in elevation, and given sides or circuits may be weighted as desired.

example. Figure 5-3 represents a level net made up of the circuits *BCDEB*, *AEDA*, and *EABE*. Along each side of the circuit is shown the length in miles and the observed difference in elevation in feet between terminal bench marks; the sign of the difference in elevation corresponds with the direction indicated by the arrows. Within each circuit are shown its length and the error of closure computed by summing up the differences in elevation in a clockwise direction.

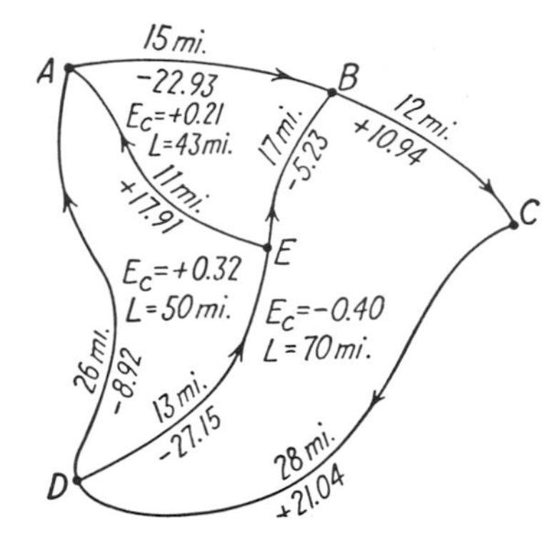

Fig. 5-3. Adjustment of level net.

Table 5-2 shows the computations required to balance the net. For each circuit are listed the sides, the distances (expressed in miles and in percentages of the total), and the differences in elevation. For circuit *BCDEB* the error of closure is −0.40 ft. This is distributed among the lines in proportion to their lengths; thus for the line *BC* the correction is $12/70 \times 0.40$ or $0.17 \times 0.40 = 0.07$ ft, with sign opposite to that of the error of closure. The corrections are applied to the differences in elevation to obtain the values of "corrected difference in elevation" shown in the seventh column. The line *DE* in circuit *BCDEB* is the same as the line *ED* in circuit *AEDA*. Hence, in listing the differences in elevation for circuit *AEDA*, the difference in elevation for *ED* is taken not as the observed value (27.15), but as the adjusted value (27.08) from circuit *BCDEB*, with opposite sign. The error of closure for circuit *AEDA* is then +0.25 ft. The error is distributed as before. Similarly, in circuit *EABE* the differences in elevation listed for *EA* and *BE* are the adjusted values from the previous circuits. In Cycle II the process for Cycle I is repeated, always listing the latest values from previously adjusted circuits before computing the new error of closure. And so the cycles are continued until the corrections become zero or negligible.

Table 5-2 Adjustment of level net by successive approximations

Circuit	Side	Distance: Miles	Distance: Per cent	Cycle I: DE [a]	Cycle I: Corr. [b]	Cycle I: Corr. DE [c]	Cycle II: DE [a]	Cycle II: Corr. [b]	Cycle II: Corr. DE [c]	Cycle III: DE [a]	Cycle III: Corr. [b]	Cycle III: Corr. DE [c]	Cycle IV: DE [a]	Cycle IV: Corr. [b]	Cycle IV: Corr. DE [c]
BCDEB	*BC*	12	17	+10.94	+0.07	+11.01	+11.01	−0.02	+10.99	+10.99	−0.01	+10.98	+10.98	0	+10.98
	CD	28	40	+21.04	+0.16	+21.20	+21.20	−0.05	+21.15	+21.15	−0.01	+21.14	+21.14	−0.01	+21.13
	DE	13	19	−27.15	+0.07	−27.08	−27.02	−0.03	−27.05	−27.03	−0.01	−27.04	−27.03	0	−27.03
	EB	17	24	− 5.23	+0.10	− 5.13	− 5.06	−0.03	− 5.09	− 5.07	−0.01	− 5.08	− 5.08	0	− 5.08
	Total	70	100	− 0.40	+0.40	0	+ 0.13	−0.13	0	+ 0.04	−0.04	0	+ 0.01	−0.01	0
AEDA	*AE*	11	22	−17.91	−0.06	−17.97	−17.93	−0.01	−17.94	−17.93	0	−17.93	−17.93		
	ED	13	26	+27.08	−0.06	+27.02	+27.05	−0.02	+27.03	+27.04	−0.01	+27.03	+27.03		
	DA	26	52	− 8.92	−0.13	− 9.05	− 9.05	−0.04	− 9.09	− 9.09	−0.01	− 9.10	− 9.10		
	Total	50	100	+ 0.25	−0.25	0	+ 0.07	−0.07	0	+ 0.02	−0.02	0	0		
EABE	*EA*	11	26	+17.97	−0.04	+17.93	+17.94	−0.01	+17.93	+17.93	0	+17.93	+17.93		
	AB	15	35	−22.93	−0.06	−22.99	−22.99	−0.01	−23.00	−23.00	−0.01	−23.01	−23.01		
	BE	17	39	+ 5.13	−0.07	+ 5.06	+ 5.09	−0.02	+ 5.07	+ 5.08	0	+ 5.08	+ 5.08		
	Total	43	100	+ 0.17	−0.17	0	+ 0.04	−0.04	0	+ 0.01	−0.01	0	0		

[a] Difference in elevation.
[b] Correction.
[c] Corrected difference in elevation.

5-12 Numerical Problems

1 A line of differential levels was run between two bench marks 20 miles apart, and the measured difference in elevation was found to be 2,163.4 ft. Later the rod whose nominal length was 13 ft was found to be 0.003 ft too short, the error being distributed over its full length. Correct the measured difference in elevation for erroneous length of rod.

2 Suppose that the levels of problem 1 had been run by using a rod which was 0.003 ft too short owing to wear on the lower end. What would have been the error?

3 Suppose that the line of levels of problem 1 were continued to form a circuit closing on the initial bench mark. What error of closure due to erroneous length of rod would be expected?

4 Differential levels were run from B.M.$_1$ (el. 470.07 ft) to B.M.$_2$, a distance of 100 miles. On the average, the backsight distances were 400 ft in length and the foresight distances were 200 ft in length. The elevation of B.M.$_2$, as computed from the level notes, was 3,652.74 ft. Compute the error due to earth's curvature and atmospheric refraction, and correct the elevation of B.M.$_2$.

5 Complete the differential-level notes shown below. Perform the customary check.

Station	*B.S.*	*H.I.*	*F.S.*	*El.*
B.M.$_1$	6.11		...	416.23
T.P.$_1$	9.25		7.36	
T.P.$_2$	11.48		3.12	
T.P.$_3$	8.30		2.98	
B.M.$_2$	12.29		4.37	
T.P.$_4$	7.73		5.16	
T.P.$_5$	8.24		3.38	
T.P.$_6$	10.66		0.47	
B.M.$_3$	....		4.33	

6 Complete the differential-level notes shown below. Determine the error of closure of the level circuit and adjust the elevations of B.M.$_2$ and B.M.$_3$, assuming that the error is a constant per setup.

Station	*B.S.*	*H.I.*	*F.S.*	*El.*
B.M.$_1$	4.127		.....	100.000
T.P.$_1$	3.831		9.346	
T.P.$_2$	4.104		10.725	
T.P.$_3$	2.654		12.008	
B.M.$_2$	4.368		7.208	
T.P.$_4$	6.089		6.534	
T.P.$_5$	8.863		4.736	
B.M.$_3$	12.356		2.100	
T.P.$_6$	10.781		3.662	
T.P.$_7$	12.365		4.111	
B.M.$_1$	.....		9.059	

7 Lines of differential levels are run from B.M.$_1$ to B.M.$_2$ over three different routes. Following are the lengths of the routes and the observed elevations of B.M.$_2$. Determine the most probable value of the elevation of B.M.$_2$.

Route	*Length, miles*	*El. of B.M.$_2$*
a	10	742.81
b	16	742.58
c	40	743.27

8 The following data are for a level net whose perimeter (reading clockwise) is *ABCDEFA*. Within the net, a line of levels extends from *B* to *F* and from *C* to *E*. The elevation of *A* is 100.00 ft. Adjust the elevations by the method of successive approximations.

Circuit	*From*	*To*	*Distance, miles*	*Diff. el., ft*
	A	*B*	40	+17.47
ABFA	*B*	*F*	35	−10.87
	F	*A*	52	− 6.26
	B	*C*	33	+11.88
BCEFB	*C*	*E*	16	− 8.48
	E	*F*	26	−14.01
	F	*B*	35	+10.87
	C	*D*	27	−16.36
CDEC	*D*	*E*	34	+ 7.59
	E	*C*	16	+ 8.48

5-13 Field Problems

Problem 6. Differential Leveling with Engineer's Level and Self-reading Rod

Object. Given the elevation of an initial bench mark, to determine the elevations of points in an assigned level circuit.

Procedure. Follow the procedure outlined in Art. 5-2. Keep notes in the form of the sample notes shown in Fig. 5-2, but estimate rod readings to thousandths of feet. Check each rod reading by taking a second observation. Close the circuit, and compute the error of closure.

Hints and Precautions. (1) Sight in the direction of the rod before centering the bubble exactly. (2) Center the bubble just before each reading, and check its position immediately afterward. (3) Test for parallax (see Art. 4-10). (4) Choose the turning points with an eye to simplicity of field operations, but roughly balance the backsight and foresight distances between bench marks; keep no record of these distances. (5) Use signals (Art. 1-9). (6) Keep the foot of the rod free from dirt. (7) Be sure that the rod is held vertical while a sight is being taken. When the "long rod" is used, wave the rod, and before each reading check the position of the back vernier. (8) Describe each bench mark in the field notes. (9) Check the computations.

Problem 7. Differential Leveling with Engineer's Level and Target Rod

Object. To determine the elevations of points in an assigned level circuit.

Procedure. The procedure differs from that of the preceding problem only in that the rodman sets the target as directed by the leveler. Both men read the rod from the attached vernier to the nearest 0.001 ft. The field notes are kept by the leveler as explained in the preceding problem. In more precise leveling, the rodman in a separate book records the backsight and foresight distances (in paces) and the rod readings and, by observing the cumulative excess or deficiency of foresight distances over backsight distances as he goes along, roughly balances these distances between bench marks. The circuit is then closed, and the error of closure computed. The error of closure and the average time required per setup are compared with those for the preceding problem.

Hints and Precautions. (1) The record of backsight and foresight distances is kept in the right-hand column of the left-hand page of the field notebook. The column is headed "Dist." and is subdivided into "B.S." and "F.S." A cumulative excess of 11 paces, for example, of foresight distances over backsight distances is noted as "+11" directly above each foresight distance. (2) When the long rod is used, the target must be clamped to read exactly the reading of the vernier on the back of the rod when the rod is short.

6

PROFILE LEVELING; CROSS-SECTIONS; GRADES

PROFILE LEVELING

6-1 General. During the location and construction of highways, railroads, canals, and sewers, stakes or other marks are placed at regular intervals along an established line, usually the center line. Ordinarily the interval between stakes is 100 ft, 50 ft, or 25 ft. The 100-ft points, reckoned from the beginning of the line, are called *full stations,* and all other points are called *plus stations.* Thus a stake set at 1,600 ft from the point of beginning is numbered "16 + 00," and one set at 1,625 ft from the point of beginning is numbered "16 + 25." Elevations by means of which the profile may be constructed are obtained by taking level-rod readings on the ground at each stake and at intermediate points where marked changes in slope occur.

Figure 6-1 illustrates in plan and elevation the steps in leveling for profile. In this case stakes are set every 100 ft. The instrument is set up in some convenient location not necessarily on the line (as at L_1), the rod is held on a bench mark (B.M. 28, el. 564.31); a backsight (1.56)

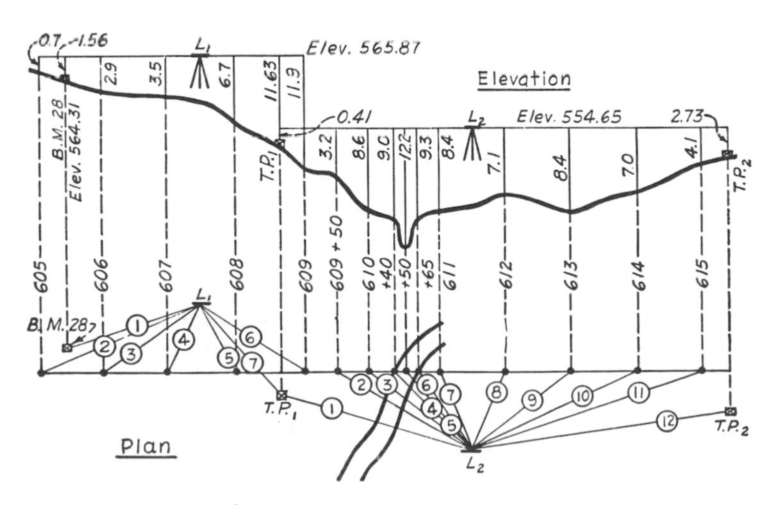

Fig. 6-1. Profile leveling.

is taken, and the height of instrument (565.87) is obtained as in differential leveling. Readings are then taken with the rod held on the ground at successive stations along the line. These rod readings are foresights, frequently designated as *intermediate foresights* to distinguish them from foresights taken on turning points or bench marks. The intermediate foresights (0.7, 2.9, . . . , 11.9) subtracted from the H.I. (565.87) give ground elevations of stations. When the rod has been advanced to a point beyond which further readings to ground points cannot be observed, a turning point (T.P.$_1$) is selected, and a foresight (11.63) is taken to establish its elevation. The level is set up in an advanced position (L_2), and a backsight (0.41) is taken on the turning point (T.P.$_1$) just established. Rod readings on ground points are then continued as before. The rodman observes where changes of slope occur (as 609 + 50, . . . , 610 + 65), and readings are taken to these intermediate stations. The "plus," or distance from the preceding full station to the intermediate point, is measured by pacing or with a tape or the rod according to the precision required. Usually the backsights and foresights are read to hundredths of feet, and intermediate foresights to tenths of feet only.

6-2 Profile-level Notes. The notes for profile leveling may be recorded as shown in Fig. 6-2. The values shown in the notes are the same as those illustrated in Fig. 6-1. The notes for turning points are kept in the same manner as for differential leveling. Elevations of ground points are obtained by subtracting the corresponding intermediate foresights from the preceding height of instrument and are recorded only to the number of decimal places in the intermediate foresights.

CROSS-SECTIONS

6-3 Leveling for Earthwork. Four general situations arise in connection with field measurements to determine volumes of earthwork.

1 Excavation to Predetermined Surface. A given area is to be cut or filled to a predetermined surface, for example, in excavating the basement for a building or in grading a piece of land. Cross-sections may be taken at short intervals in the manner described in Art. 6-4. When the grade of the finished surface has been established, the cut or fill at each station will be known and the volume of earthwork can be calculated.

2 Excavation for Trench. A trench is to be excavated, as when a sewer or pipeline is to be laid. Profile levels are run along the proposed line. When the grade of the bottom of the trench has been fixed, the cut at

PROFILE LEVELS FOR						I. N. RY. LOCATION	
							J.C. Brown, ⊼
	Cox Brook to Big Forks					Buff Dumpy Level	F. Graham, Rod
							Sept. 16, 1967
Sta.	B.S.	H.I.	I.F.S.	F.S.	Elev.		Fair
B.M.28	1.56	565.87			564.31	On spruce root 50 ft. lt. Sta. 605.	
605			0.7		565.2		
606			2.9		563.0		
607			3.5		562.4		
608			6.7		559.2		
609			11.9		554.0		
T.P.	0.41	554.65		11.63	554.24	On stone.	
609+50			3.2		551.5		
610			8.6		546.1		
+40			9.0		545.7	Bank Cox Brook.	
+50			12.2		542.5	Ctr. " "	water 1.5 ft. deep
+65			9.3		545.4	Bank " "	
611			8.4		546.3		
612			7.1		547.6		
613			8.4		546.3		
614			7.0		547.7		
615			4.1		550.6		
T.P.	8.02	559.94		2.73	551.92	On plug.	
616			9.7		550.2		
+40			6.3		553.6	Ctr. highway to St. Leonards.	
	9.99		564.31	14.36			
			559.94	9.99			
			4.37 =	4.37	ck.		

Fig. 6-2. Profile-level notes.

each station can be computed. With the necessary width of trench at top and bottom and also its depth at each station known, the volume of excavation can be calculated.

3 *Borrow-pit Cross-sections.* An irregular mass of unknown volume is to be excavated at a given site, as, for example, to furnish material for a highway fill. Sufficient data for the calculation of volume can be obtained by taking cross-sections of the site before and after the material is removed. Usually a base line is staked out near one side, and crosslines are established at regular intervals. Levels are run over the crosslines. When the material has been removed, levels are rerun over the crosslines. The difference between the original cross-section and the final cross-section shows the area cut at each crossline, from which the volume can be determined.

4 *Road or Canal Cross-sections.* Earth must be cut or filled to a given grade line along some route, as a highway, railroad, or canal, and must have a prescribed shape of cross-section (see Arts. 6-6 to 6-9).

6-4 Cross-section Levels. Frequently the shape of the surface of a piece of land is obtained by staking out the area into a system of 100-ft, 50-ft, or 25-ft squares and then determining the elevations of the corners and of other points where changes in slope occur. Directions of the lines may be obtained with either the tape or the transit, distances may be laid off either with the tape or by stadia, and elevations may be determined with either the engineer's level or the hand level, all depending upon the required precision. The elevations are determined as in profile leveling. Figure 6-3 illustrates a suitable form of notes. The data may be employed in the construction of a contour map (see Chap. 19).

SURVEY OF PROPOSED RICE FARM, C.K. PERRY

(For Contour Map)
Scale 1"=50' Contour Int. = 1 ft.

Sta.	B.S.	H.I.	F.S.	Elev.	
B.M.1	5.31	105.31		100.00	(Assumed)
A-8			7.3	98.0	
B-8			6.0	99.3	
C-8			4.1	101.2	
D-8			2.2	103.1	
E-8			1.4	103.9	
E-7			2.9	102.4	
D-7			4.2	101.1	
C-7			5.0	100.3	
(B+50)-7			5.6	99.7	
B-7			4.8	100.5	
A-7			3.2	102.1	
A-(7+60)			6.2	99.1	
A-6			2.5	102.8	
(A+50)-6			0.4	104.9	
B-6			1.9	103.4	
C-6			3.6	101.7	
D-6			4.2	101.1	
E-6			3.5	101.8	
E-(5+40)			6.7	98.6	
E-5			6.1	99.2	

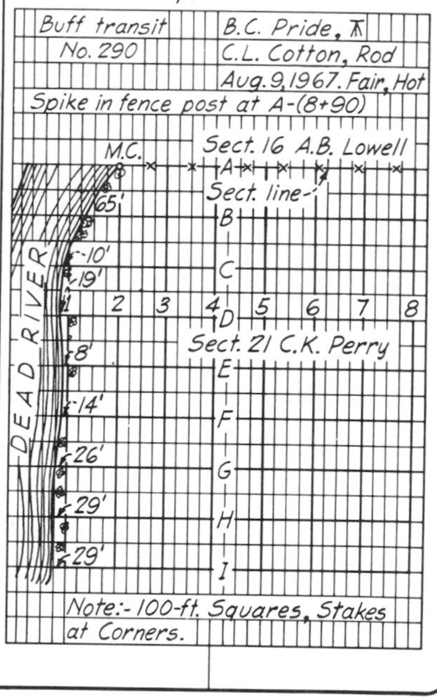

Buff transit No. 290
B.C. Pride, ⊼
C.L. Cotton, Rod
Aug. 9, 1967. Fair, Hot
Spike in fence post at A-(8+90)
M.C.
Sect. 16 A.B. Lowell
A
Sect. line
B
C
D
E
F
G
H
I
1 2 3 4 5 6 7 8
65'
10'
19'
8'
14'
26'
29'
29'
DEAD RIVER
Sect. 21 C.K. Perry
Note:- 100-ft. Squares, Stakes at Corners.

Fig. 6-3. Cross-section (checkerboard) notes.

6-5 Preliminary Route Cross-sections. Preliminary surveys for railroads, highways, and canals are often made by running a taped traverse line along the proposed route, stations being established by stakes set every 100 ft. The elevations of the stations are then determined by profile leveling, as already described. To furnish data for location studies and for estimating volumes of earthwork, it is customary to determine the shape of the ground on both sides of the traverse line by running levels over crosslines at right angles to the traverse, usually at each station. Usually the elevations are determined with the hand level in rough country and with the engineer's level in flat country. For each crossline the height of instrument is established by a backsight on the ground at the center stake. The rod is then held on the crossline at breaks in the surface slope, and distances from the traverse line to these points are measured with the metallic tape. The direction of short crosslines is laid off by eye; that of long crosslines by means of a compass, transit, or right-angle mirror.

Notes may be kept in the form shown in Fig. 6-4. The center line of the right-hand page represents the traverse line, and to the right and left of this line are recorded the observed distances and rod readings and the computed elevations.

PRELIMINARY CROSS-SECTIONS C. & R. EXTENSION						O.H. Ellis, ⊼ C.O. Lord, Rod J.A. Crum, Tape		Jan. 20, 1967 Cold, Snow
Sta.	B.S.	H.I.	F.S.	Elev.		Left	℄	Right
405	12.4	633.0		620.6	(Dist.) (Elev.) (Rod)	300 210 123 80 632.1 630.8 627.0 626.7 0.9 2.2 6.0 6.3	620.6	
405	0.6	621.2		620.0				50 160 250 350 617.0 612.3 610.0 609.7 4.2 8.9 11.2 11.5
406	12.1	628.9		616.8		90 50 628.2 624.3 0.7 4.6	616.8	
406	11.5	639.7	0.7	628.2		280 200 120 638.3 635.6 632.2 1.4 4.1 7.5		
406	1.9	618.7		616.8				60 100 200 300 615.5 611.2 610.9 609.3 3.2 4.5 7.8 9.4
407	4.7	615.9		611.2		300 180 100 615.6 614.7 612.6 0.3 1.2 3.3	611.2	75 155 270 606.2 604.5 602.9 9.7 11.4 13.0
408	10.6	615.9		605.3		280 200 100 611.6 609.2 607.7 4.3 6.7 8.2	605.3	100 200 300 604.1 603.2 603.0 11.8 12.7 12.9

Fig. 6-4. Preliminary route cross-section notes.

When the route is located definitely, final cross-sections are taken as described in the following articles.

6-6 Final Road Cross-sections. Figures 6-7 and 6-8 illustrate typical highway or railroad cross-sections in fill and in cut. The rough subgrade is usually a plane surface, transversely level but on highway curves perhaps superelevated. On a given road the subgrade is of uniform width in cut and of uniform but usually a smaller width in fill; still a third width may be used where the section is partly in cut and partly in fill. The finished cross-section may be sloped variously to provide shoulders, drainage, and rounded corners, as illustrated by Fig. 6-5.

The side slopes are plane surfaces of constant slope for a given material of excavation. The rate of the side slope (as 2:1) is stated in terms of the number of units measured horizontally (as 2 ft) to one unit measured vertically (as 1 ft).

The preliminary survey is made, and the profile is plotted. With the profile and the data of the preliminary cross-sections as a guide, the grade line of the subgrade is established on the sheet with the plotted profile. The center cut or fill to be made at each station is then equal to the difference between the observed elevation of the ground line and the established elevation of the grade line.

Prior to actual construction, final cross-sections are taken, and *slope stakes* marking the intersection of the side slopes with the natural ground surface are set opposite each center stake (Art. 6-9).

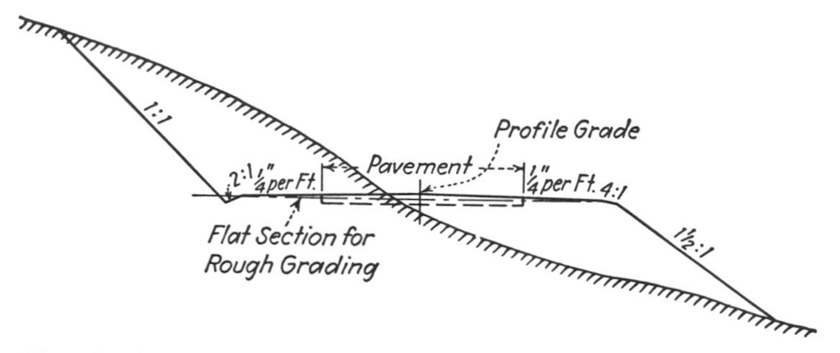

Fig. 6-5. Sidehill cross-section for highway.

Usually level readings for final cross-sections are taken with the engineer's level, and distances to the right or left of the center stake are measured with the metallic tape, all to tenths of feet. Alternatively, slope distances and vertical angles may be measured. At each station a foresight is taken, the ground elevation is checked against that obtained by profile leveling, and the computed cut or fill is marked on the back of the center stake. A crossline is established, and the slope stakes are set as later described.

If the ground is level in a direction transverse to the center line, the cross-section is called a *level section*. When rod readings are taken at each slope stake in addition to the reading taken at the center, as will normally be done where the ground is sloping, the cross-section is called a *three-level section*. When rod readings are taken at the center stake, the slope stakes, and points on each side of the center at a distance of half the width of the roadbed, the cross-section is called a *five-level section*. A cross-section for which observations are taken to points between center and slope stakes at irregular intervals is called an *irregular section*. Where the cross-section passes from cut to fill, it is called a *sidehill section* (Fig. 6-5).

6-7 Final Cross-section Notes.

Figure 6-6 illustrates a suitable form of final cross-section notes. The left-hand page is essentially the same as for profile leveling except that a column is added for grade elevations. Some engineers prefer to have the notes read *up* the page. The notes

CROSS-SECTIONS FOR I.N.RY. FINAL LOCATION

Cox Brook to Big Forks — Dec. 4, 1967 — F.F. Smith, ⊼

Roadbed. 20 ft. in Cut, 16 ft. in Fill — Cloudy — J. Richie, Rod

Slope 1½:1 — O. Byram, Tape

Sta.	B.S.	H.I.	F.S.	Elev.	Grade	Left	Ctr.	Right	Remarks
B.M.28	2.67	566.98		(564.31)					50 ft. Lt. Sta. 605
605			1.9	565.1	556.00	C8.6/22.9	C9.1	C11.2/26.8	Gravel in this hill.
606			4.0	563.0	555.60	C5.0/17.5	C7.4	C8.4/22.6	
607			4.5	562.5	555.20	C4.6/16.9	C7.3	C8.0/22.0	
608			8.0	559.0	554.80	C1.8/12.7	C4.2	C5.1/17.7	
+25			9.2	557.8	554.70	0.0/10.0	C3.1	C2.6/13.9	
T.P.	1.94	557.19	(11.73)	(555.25)					On plug.
+90			2.8	554.4	554.44	F1.8/10.7	0.0	C2.4/13.6	
609			3.4	553.8	554.40	F3.2/12.8	F0.6	0.0/4.0 C1.0/11.5	
+20			5.6	551.6	554.32	F3.6/13.4	F2.7	0.0/8.0	
610			11.1	546.1	554.00	F8.2/20.3	F7.9	F6.3/17.5	
+40			11.2	546.0	553.84	F8.0/20.0	F7.8	F7.8/19.7	Top of bank Cox Brook.
+45			14.6	542.6	553.82	F11.2/24.0	F11.2	F11.2/24.8	In brook.
+60			14.5	542.7	553.76	F11.1/24.7	F11.1	F11.1/24.7	" "
+65			11.6	545.6	553.74	F8.2/20.3	F8.1	F8.0/20.0	Top of bank.
611			10.9	546.3	553.60	F7.8/19.7	F7.3	F7.2/18.8	
612			9.7	547.5	553.20	F7.6/19.4	F5.7	F3.1/12.7	
613			10.8	546.4	552.80	F7.8/19.7	F6.4	F2.0/11.0	
T.P.	11.96	559.69	(9.46)	(547.73)					On stump.
	16.57	564.31	21.19						
		559.69	16.57						
		4.62–4.62 ck.							

Fig. 6-6. Cross-section notes for final location of roadway.

shown are for a portion of the line for which profile-level notes are shown in Fig. 6-2. The values in columns marked "Left" or "Right" are for points at which the slope stakes are driven; for each such point the upper number is the cut or fill, and the lower number is the distance out from center. The slope is such that three-level sections are adequate. Cross-sections are taken where the left edge, center, and right edge of the roadbed pass from cut to fill. The cross-sections at 608 + 90 and 609 are sidehill sections.

6-8 Cuts and Fills.

6-8 Cuts and Fills. Figure 6-7 shows at *A* the engineer's level in position above grade, for taking rod readings at a section in fill. The height

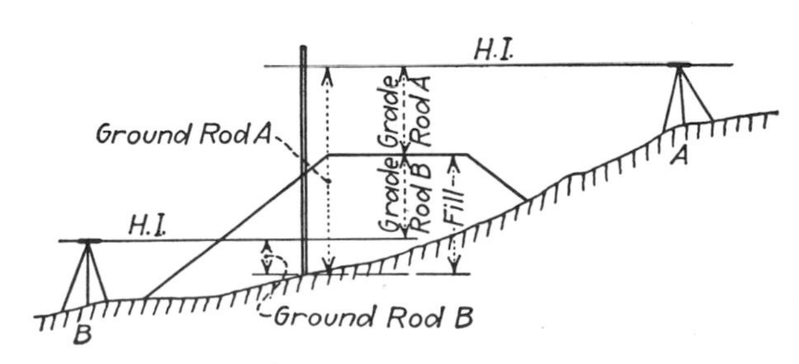

Fig. 6-7. Road cross-section in fill.

of instrument has been determined; the elevation of grade at the particular station is known. The leveler computes the difference between the H.I. and the grade elevation, a difference known as the *grade rod;* that is, H.I. − el. of grade = grade rod. The rod is held at any point for which the fill is desired, and a reading called the *ground rod* is taken. The difference between the grade rod and the ground rod is equal to the fill. Similarly, for a section in cut this difference is equal to the cut.

If the H.I. is *below* grade, as at *B* in Fig. 6-7, the fill is the *sum* of the grade rod and the ground rod.

6-9 Setting Slope Stakes.

If w is the width of roadbed, d the measured distance from center to slope stake, s the side-slope ratio (ratio of horizontal distance to drop or rise), and h the cut (or fill) at the slope stake, then, by Fig. 6-8, when the slope stake is in the correct position (at C),

$$d = \frac{w}{2} + hs \qquad (1)$$

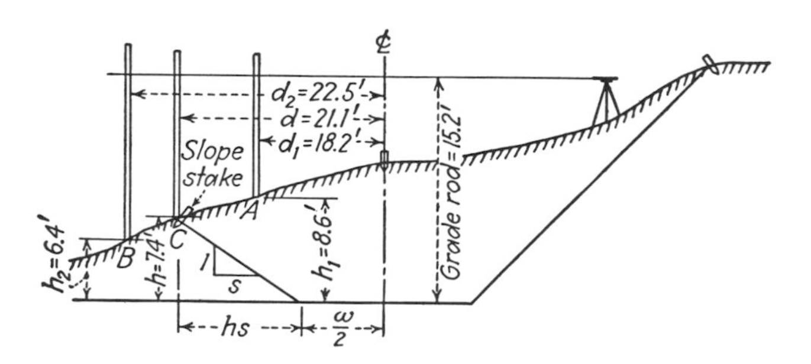

Fig. 6-8. Setting slope stakes.

example. The following numerical example for a cut illustrates the steps involved in establishing the correct location for a slope stake in the field; the same procedure is followed in fill.

Let w = 20 ft; side slope = 1½ to 1; grade rod = 15.2 ft. As a first trial the rod is held at A (Fig. 6-8); ground rod = 6.6 ft;

$$h_1 = \text{grade rod} - \text{ground rod} = 15.2 - 6.6 = 8.6 \text{ ft}$$

The computed distance for this value of h_1 is $w/2 + h_1 s = 10.0 + 8.6 \times 3/2 = 22.9$ ft. Measurement from the center stake shows d_1 to be 18.2 ft; hence the rodman should go farther out. A second trial is made perhaps at *B*, and by similar computation it is found that the rod is too far out.

Eventually, by trial, the rod will be held at *C;* ground rod = 7.8 ft; $h = 15.2 - 7.8 = 7.4$ ft. The computed distance for this value of h is $w/2 + hs = 10.0 + 7.4 \times 3/2 = 21.1$ ft. The measured value of d is also 21.1 ft; hence this is the correct location for the slope stake. In the notes the coordinates of the slope stake are given by the expression $c7.4/21.1$ in the form of a fraction, but the trial observations are not recorded.

Slope stakes are set side to the line, sloping outward in fill and inward in cut. On the back of the stake is marked the station number. On the front (side nearest the center line) are marked the cut or fill at the stake, and sometimes the distance from center to slope stake. The numbers read down the stake.

In cut, some organizations set the slope stakes at a fixed distance, say 2 ft, back from the edge of the slope. The cut marked on the stake applies to the elevation of the ground at the stakes thus offset.

If cuts and fills are only a few feet deep, sometimes the slope stakes are omitted, and the stakes used for alinement are also employed as reference elevations for grade.

GRADES

6-10 Setting Grades. In surveying, the term *grade* or *gradient* is used to denote the slope, or rate of regular ascent or descent, of a line. It is usually expressed in per cent; for example, a 4 per cent grade is one which rises or falls 4 ft in a horizontal distance of 100 ft. The term *grade* is also used to denote an established line on the profile of an existing or a proposed roadway. In such expressions as "at grade" or "to grade" it denotes the elevation of a point either on a grade line or at some established elevation as in construction work.

The operation of setting grades is similar to profile leveling. The grade rod to be employed in setting a given grade stake to grade is computed by subtracting the established grade elevation (taken from the profile) from the H.I. The rodman starts the stake and holds the rod on its top. The leveler reads the rod and calls out the approximate distance which the stake must be driven to reach grade. The rodman drives the stake nearly the desired amount, and a second rod reading is taken; and so the process is continued until the rod reading is made equal to the grade rod. Sometimes the rod is held alongside of the stake, and the position of grade is indicated by a crayon mark or a nail driven into the stake at the foot of the rod. Usually grade elevations are determined to hundredths of feet. The notes are kept as in profile leveling except that the right-hand column of the left-hand page is for grade elevations.

The procedure of setting grades for typical works of construction is described in Chaps. 20 to 22.

6-11 Vertical Curves. On highways and railroads, adjacent segments of differing gradient are connected by a curve in a vertical plane, called a *vertical curve*. Usually the vertical curve is the arc of a parabola.

The length of a vertical curve cannot be less than the algebraic difference in gradient between the two segments connected, divided by the maximum allowable change in grade per station (usually established by specifications). Usually it is some convenient whole number of feet in highway work or an even number of stations in railway work.

The station and plus of the vertex, or point of intersection of the two segments, and the elevations of stations along the uniform grade lines are determined from the grade profile. The length of vertical curve is then computed or is taken at some convenient value which meets the specification requirements; and the stations and elevations of the beginning and end of curve are calculated. Then the offsets from the uniform gradients to the curve are computed, and the grade elevations at stations along the curve are thus determined. In the field the vertical curve is laid out by setting grade stakes at these stations, just as along a uniform grade.

One method of calculation for a vertical curve is as follows: The elevation of the mid-point of the "long chord" (Fig. 6-9) connecting the points of beginning and end of the vertical curve is computed. As the curve is a parabola, the elevation of the mid-point of the vertical curve is the mean of the elevation of the vertex and the elevation of the mid-point of the long chord. The tangent offsets to various points along the curve are then computed, employing the known property of a parabola that the tangent offset varies as the square of the distance from the tangent point.

example. On a railroad a +0.8 per cent grade meets a −0.4 per cent grade at station 90 + 00 and at elevation 100.00 (Fig. 6-9). The maximum allowable change in grade per station is 0.2. It is desired to establish a vertical curve connecting the two grades.

The algebraic difference in gradient is $+0.8 - (-0.4) = 1.2$ per cent. The minimum length of curve is then $1.2/0.2 = 6$ stations, or 600 ft.

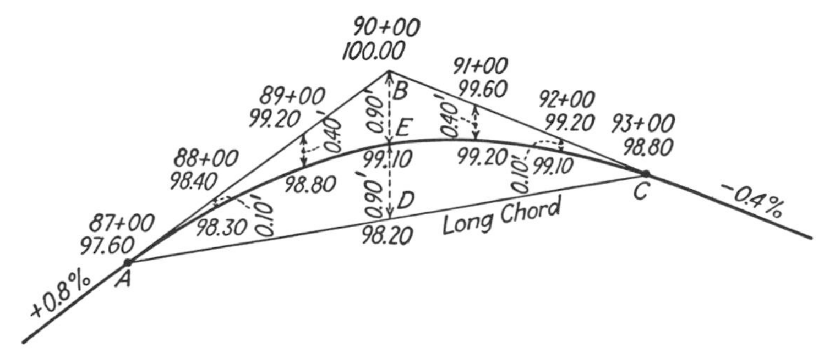

Fig. 6-9. Vertical curve.

The length on either side of the vertex ($AB = BC$) is $^{600}\!/_{2} = 300$ ft. The station at A is therefore $90 - 3 = 87$, and the station at C is $90 + 3 = 93$. The elevation of A is $100.00 - 3 \times 0.80 = 97.60$, and the elevation of C is $100.00 - 3 \times 0.40 = 98.80$ ft.

The elevation of the mid-point D of the long chord AC is the mean of the elevations A and C:

$$\tfrac{1}{2}\,(97.60 + 98.80) = 98.20 \text{ ft}$$

The mid-point E of the vertical curve is midway between D and the vertex B:

$$\tfrac{1}{2}\,(98.20 + 100.00) = 99.10 \text{ ft}$$

The offset from vertex to curve is

$$100.00 - 99.10 = 0.90 \text{ ft}$$

The tangent offsets at stations 89 and 91 are

$$\frac{2^2}{3^2} \times 0.90 = 0.40 \text{ ft}$$

and the offsets at stations 88 and 92 are

$$\frac{1^2}{3^2} \times 0.90 = 0.10 \text{ ft}$$

The elevations of points on the curve are shown in Fig. 6-9.

6-12 Numerical Problems

1 Complete the profile-level notes shown below.

Sta.	*B.S.*	*H.I.*	*I.F.S.*	*F.S.*	*El.*
B.M. 10	6.32		....	.....	836.76
179	....	...	10.1	.....	
180	....	...	7.8	.....	
+35	....	...	12.6	.....	
181	....	...	4.7	.....	
182	....	...	3.4	.....	
T.P. 38	7.32		....	2.11	
183	....	...	8.5	.....	
+40	....	...	4.6	.....	
184	....	...	7.2	.....	
185	....	...	10.6	.....	
T.P. 39	5.93		....	11.49	
186	....	...	4.2	.....	

2 The width of roadbed of a proposed railroad is 24 ft in cut, and the side slopes are 1½ to 1. At a given station the elevation of grade is 515.75. For obtaining the cross-section at the station the H.I. of the level is 528.32.

The ground rod at the center stake is 6.5. Compute the grade rod and the center cut. The rod reading at the right slope stake is 1.2 and at the left slope stake is 10.9. Compute the cut and the distance out to each slope stake.

3 Make a page of cross-section notes (similar to Fig. 6-6) for a highway running from cut into fill. The grade of the highway is 4.0 per cent, width of roadbed 24 ft in cut and 18 ft in fill, and side slopes 1½ to 1. Show observations at center and slope stakes and at grade points.

4 Make a page of notes for establishing grades of top of rail from station 750 to station 762. Elevation of grade at station 750 is 381.60; grade from station 750 to station 758 is −0.6 per cent; grade from station 758 to station 762 is −0.4 per cent; elevation of bench mark near station 750 is 378.47. Do not consider the vertical curve.

5 On a highway a −6.0 per cent grade meets a +4.0 per cent grade at station 67 + 50 and at elevation 516.32. The maximum allowable change in grade per station is 2.5. Compute the elevations of stations at 50-ft intervals along a vertical curve connecting the two grades.

6-13 Field Problems

Problem 8. Profile Leveling for a Roadway

Object. To determine the elevations necessary for plotting the profile of a line. It is assumed that the center line has already been laid out with numbered stakes every 100 ft [see field and office problems 10, 11 (Chap. 7), and 26 (Chap. 10)].

Procedure. Select a bench mark, with datum below every point of the proposed route. Adapt the procedure indicated in Arts. 6-1 and 6-2 to the field conditions encountered. Keep the notes in the form of the sample notes (Fig. 6-2).

Hints and Precautions. (1) Read the rod carefully to the nearest 0.01 ft on bench marks and turning points, and quickly to the nearest 0.1 ft on ground points. (2) Take rod readings on the ground at all full stations and at such other points on line (plus stations) as are necessary to obtain a sufficiently accurate profile. In general, these plus stations will be at points where the slope of the ground changes noticeably and at highways, railways, and streams. (3) If the line is long, establish bench marks every 1,500 to 2,000 ft, and mark each bench mark. (4) Make the computations as the work progresses, and check each page of notes.

Problem 9. Profile Leveling for a Pipeline

Object. To prepare the line of a proposed sewer or water main for construction. It is assumed that the line has already been run and that center stakes marked with station and plus have been set every 25 or 50 ft. Ground pegs are to be set to give line and grade for ditchers and pipelayers.

Procedure. (1) Opposite each stake on the line and far enough from the line to ensure its not being disturbed by the excavation, drive a short peg (or spike) flush with the ground, and beside this peg drive a stake marked (on the side away from the line) with the station number of the center stake and the offset of peg from center stake. (2) Start from a bench mark as in the preceding field problem, and take profile readings on the ground pegs to the nearest 0.01 ft. Keep notes in the form of the sample notes shown in Fig. 6-2, except that additional columns are required for offsets of pegs from center line, grade elevations, and cuts. Complete the level work as in the preceding field problem. (3) Roughly plot the profile; then fix the grade of the bottom of the trench, and determine the amount of cut at each station. (4) Mark the cut, expressed in feet and inches to the nearest ⅛ in., on the front of each side stake (facing the line).

Hints and Precautions. (1) Take rod readings with greater care on the turning points than on the ground pegs. (2) Mark all stakes to read down. (3) In paved streets or hard roads, spikes driven flush with the surface, chisel marks, or paint marks are used instead of stakes; their location is recorded.

Problem 10. Setting Slope Stakes; Cross-sections

Object. To prepare a proposed highway or railroad for grading and to obtain data for calculating earthwork (see office problem 13, Chap. 7).

Procedure. (1) From the level notes of field problem 8 plot a profile and fix a grade such that the amount of cut will approximately balance the amount of fill. (2) Keep field notes in the form of Fig. 6-6. (3) Drive short pegs flush with the ground against the center stakes, on the side farthest from the beginning of the line. Run profile levels over the line of pegs, checking the elevations obtained with those of field problem 8; and mark on the back of each center stake the cut or fill at that point, as *C* 3.9 or *F* 4.7. (4) Assume a roadway 20 ft wide with side slopes of 1½ to 1. Opposite each center stake, at right angles to and on both sides of the line, set and mark slope stakes as described in Art. 6-9. (5) Drive ground pegs at "grade points" where the center line and each edge of roadbed pass from cut to fill; mark the location of these grade pegs by stakes marked "grade."

7

PLOTTING PROFILES AND CROSS-SECTIONS; EARTHWORK

PROFILES AND CROSS–SECTIONS

7-1 Plotting Profiles. The profile is plotted from profile-level notes or from elevations taken from a topographic map. Usually it is plotted on regular *profile paper,* which is ruled with vertical lines at intervals of $\frac{1}{4}$ or $\frac{1}{2}$ in. and with horizontal lines at intervals of $\frac{1}{20}$ or $\frac{1}{10}$ in. Figure 7-1 illustrates to reduced scale a portion of the profile for a proposed railroad; in the original drawing the vertical scale was 1 in. = 20 ft, and the horizontal scale was 1 in. = 400 ft. Other scales may be used, depending upon the purpose of the profile. The vertical scale is exaggerated because the vertical distances on the ground are relatively small as compared with the horizontal distances.

The station numbers increase from left to right. The ground line is drawn freehand, usually as the elevations are plotted. The profile should not be unduly rounded at summits and depressions.

Notes on the profile show the station and plus of important objects, as streams and roads, crossed by the line (Fig. 7-1). Usually an alinement diagram is drawn near the bottom of the sheet, with points on the diagram directly below corresponding points on the profile. The alinement diagram is not a true plan view, except where the line is straight. In some cases general drawings show the profile on the same sheet with the map or plan of the line (Fig. 21-1). The profile is finished in ink.

7-2 Fixing Grades. The ground profile furnishes the basis for the study of economic grade elevation. For road location, the elevation of grade is fixed at certain controlling points as at terminals and at stream and roadway crossings. In addition, maximum permissible rates of grade are established by considerations of traffic. Conforming to these limitations, the grade of the proposed road is fitted to the ground until so far as possible the volume of earthwork in cuts will balance that in adjacent fills. For sewers and drains certain minimum permissible rates of grade are established by considerations of flow, and these with the profile of the ground determine the grade between controlling points.

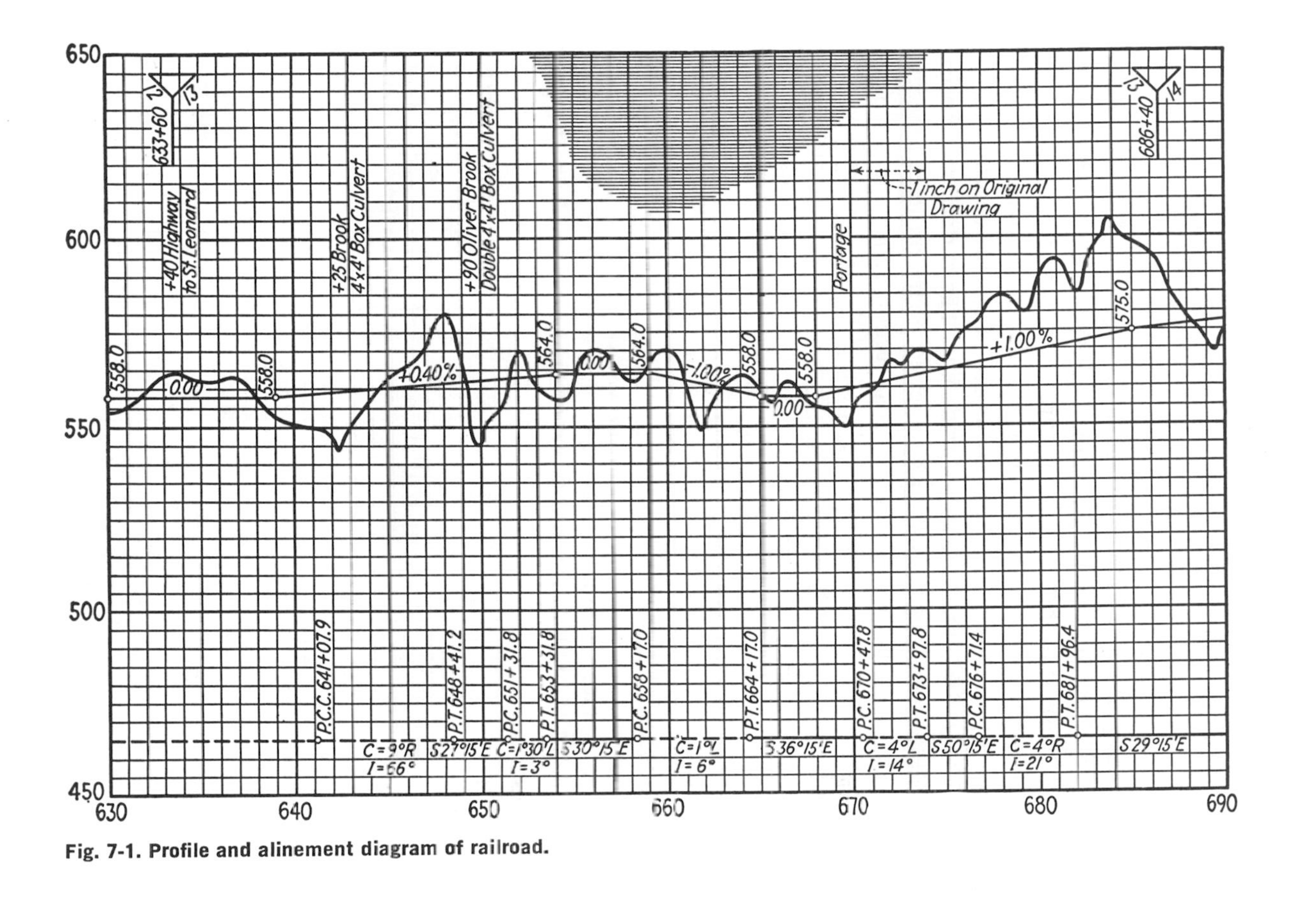

Fig. 7-1. Profile and alinement diagram of railroad.

Grade lines are established on the profile until by trial a satisfactory solution is obtained. The elevations of points of change are noted on the profile, and the rate of grade is noted just above each segment of the grade line (Fig. 7-1).

7-3 Plotting Cross-sections. Irregular cross-sections for earthwork are commonly drawn to scale on cross-section paper which is ruled usually with 10 divisions to the inch in both directions. Governing points of the cross-section are plotted either from the cross-section notes or from the data of a topographic map. These points are essentially coordinates, with the origin on the center line at the grade line. The surface may be indicated either by an irregular line or by a series of straight lines connecting the points; the fixed portions of the cross-section either may be plotted directly or may be drawn by means of a templet. For large cross-sections a scale of 1 in. = 10 ft is common. If the cross-sections are shallow, sometimes the vertical scale is exaggerated. Usually cross-sections are plotted in pencil only. If volumes of earthwork are to be determined by means of a computer, cross-sections need not be plotted.

The cross-section for the first station of the route is usually placed in the upper left corner of the sheet, and successive cross-sections are placed one below the other (Fig. 7-2). Below each cross-section is shown its station number. Within each cross-section is shown its computed area in square feet, and between successive cross-sections is shown the com-

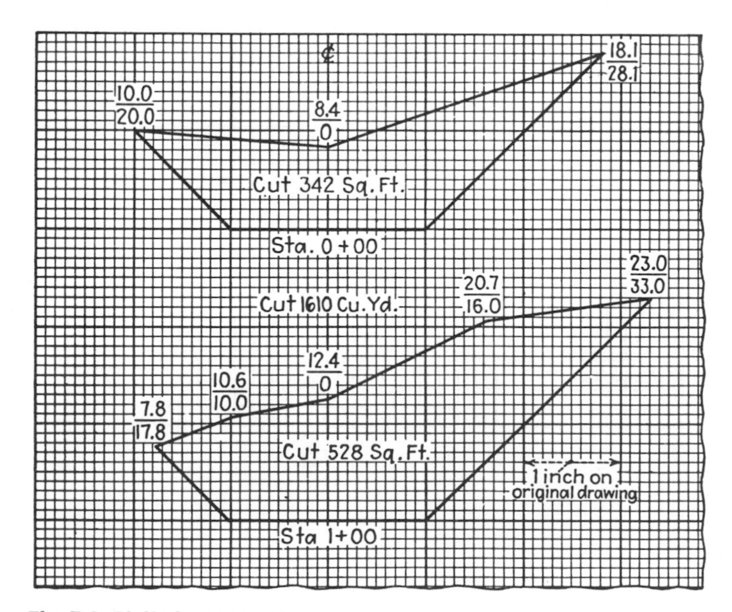

Fig. 7-2. Plotted cross-sections.

puted volume of earthwork in cubic yards. The coordinates may or may not be noted on the sheet. Some engineers place the first cross-section in the *lower* left corner of the sheet, show the elevation of the ground line at each station, and place the notations at locations other than those shown in the figure.

AREAS OF CROSS-SECTIONS

7-4 Regular Cross-sections. Areas of regular cross-sections are readily determined by numerical computations without plotting.

For a trench the cross-sectional area at any point is determined by multiplying the average of the top and bottom widths by the depth.

The same method may be applied to level cross-sections for highways and railroads; if d is the distance to either slope stake from the center, w is the width of the roadbed, and c is the center cut or fill, then the area A of the level section is

$$A = c\left(d + \frac{w}{2}\right) \tag{1}$$

A three-level section may be divided into four triangles, as shown in Fig. 7-3. Then from the figure the area A is

$$A = \frac{w}{4}(h_l + h_r) + \frac{c}{2}(d_l + d_r) \tag{2}$$

7-5 Irregular Road Cross-sections. Areas of irregular sections may be computed by plotting the sections and dividing them into trapezoids and triangles, but the computations are tedious except for sidehill sections. The method of coordinates (Chap. 16) may be adapted in simplified form for the particular case. If the cross-sections are bounded by curved lines or are very irregular, usually they are plotted and the area is determined by traversing the perimeter of each plotted cross-section with a polar planimeter.

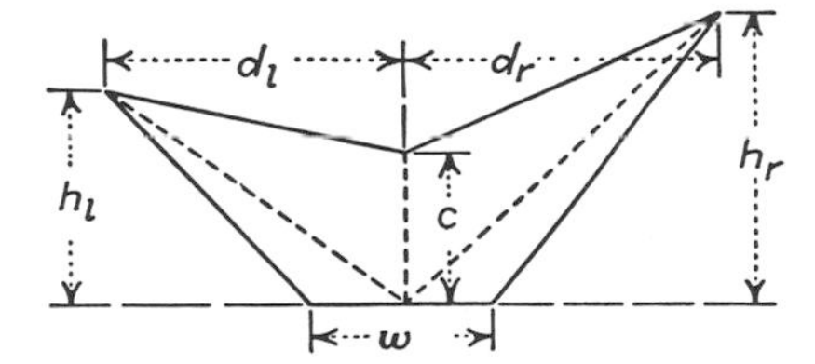

Fig. 7-3. Three-level section.

7-6 Polar Planimeter. Figure 7-4 shows a polar planimeter with adjustable tracing arm and adjustable pole arm. The planimeter is supported at the anchor point or pole P, the roller R, and the tracing point T.

The arm carrying the anchor point is hinged to the frame of the planimeter. On the adjustable tracing arm *A* are graduations which give known relations between the readings of the planimeter and the area. The circumference of the drum *D* of the roller *R* is graduated into 100 parts. At *E* is a vernier for the roller. By means of a worm, the roller when revolving turns the graduated disk *F* in the ratio 10:1. The whole number of revolutions of the roller is read on the disk *F* by means of an index; hundredths of a revolution of the roller are indicated by the drum reading at the index of the vernier *E;* and thousandths are estimated by reading the vernier.

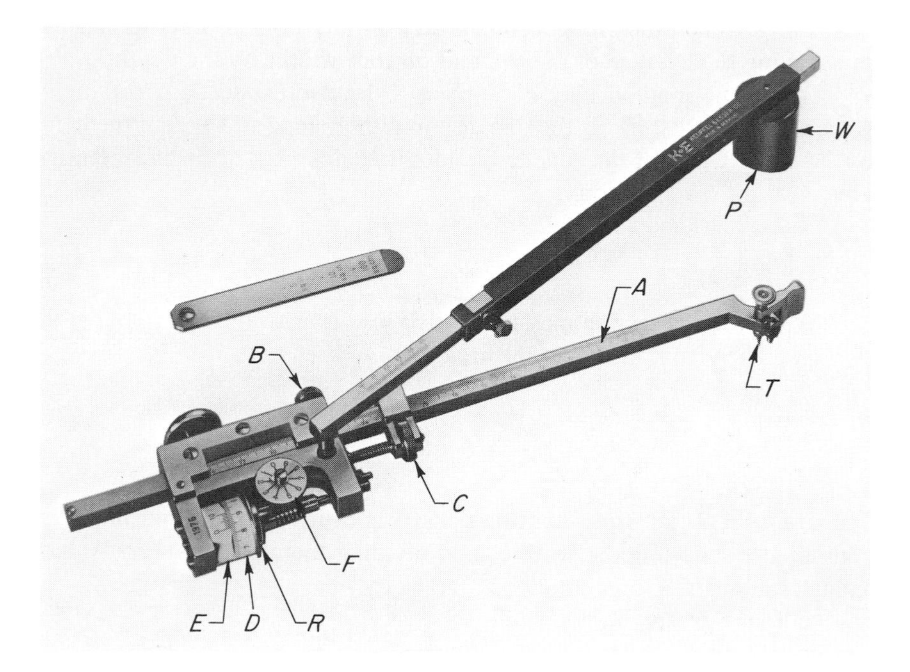

Fig. 7-4. Polar planimeter. (*Keuffel & Esser Co.*)

An alternative feature of the polar planimeter is a tracing lens having an index mark at the center, which serves as the tracing point. The lens magnifies the line to be followed.

When a plotted area is to be determined, the needle tip of the anchor point is pressed into the paper in a convenient location and is held down by the weight *W*. The tracing point is set at a definite point on the perimeter of the figure, and an initial reading is taken. The perimeter is then completely traversed until the tracing point is brought to its original location, and a final reading is taken. The difference between the initial reading and the final reading is the net number n of revolutions of the roller:

$$n = \text{final reading} - \text{initial reading} \tag{3}$$

in which n is positive if the net rotation of the roller is forward and negative if the net rotation is backward. Account must be taken of the net number of times the zero graduation of the disk F may have passed the index. The area of the figure is computed as described in the following paragraphs.

For small areas, the anchor point is placed *outside* the figure. If the figure is larger than can be traced in one operation with the anchor point outside, it may be divided into smaller figures; however, if there are many large areas, faster progress can be made by placing the anchor point *inside* the figure.

In order to simplify the discussion, herein the direction of motion of the tracing point around the figure is taken as *always clockwise.* For counterclockwise traversing, the rotation of the roller would be the opposite of that for clockwise traversing.

Area with Anchor Point Outside Figure. The planimeter is so constructed that, when the anchor point is outside the figure and the perimeter is traversed clockwise, the final reading will be greater than the initial reading, and n will be positive. The area A of the figure is directly proportional to the number of revolutions, or

$$A = Cn \tag{4}$$

in which C, called the *planimeter constant,* indicates the area per revolution of roller. It can be shown that the value of C is equal to the product of the length of the tracing arm and the circumference of the roller. If the length of the tracing arm is fixed, as on many planimeters, C is usually 10.00 sq in., and the value is stated either on the top of the tracing arm or in the planimeter case. On some instruments, C is stated in metric units.

example 1. The roller of a fixed-arm planimeter having a planimeter constant of 10.00 sq in. is set at zero, and the perimeter of a figure is traversed clockwise with anchor point outside. The final reading is 2.367. Then the area of the figure is $10.00 \times (2.367 - 0.000) = 23.67$ sq in.

Determination of Constant. The value of the planimeter constant can by determined by traversing the perimeter of a figure of known area, with anchor point outside. Preferably several trials should be made, and the average computed for use.

example 2. The length of the tracing arm of a planimeter is so set that the roller registers 0.893 revolution when the perimeter of a 2 by 5-in. rectangle is traversed with anchor point outside. From Eq. (4),

$$C = \frac{A}{n} = \frac{2.00 \times 5.00}{0.893} = 11.20 \text{ sq in.}$$

It is not necessary to determine the instrumental constant. All that is necessary is to determine the difference in planimeter readings for a known area. Then by proportion, any required area is to the corresponding difference in readings as the figure of known area is to its difference in readings. However, the computation of the constant is so simple that it is usually made.

Area with Anchor Point Inside Figure. When the tracing arm is held in such a position relative to the anchor arm that the plane of the roller passes through the anchor point, the tracing point can be made to describe completely the circumference of a circle without there being any revolution of the roller. This is called the *zero circle*. It can be shown that when the perimeter of a figure is traversed with the anchor point inside the figure, the indicated area ($A' = Cn'$) is equal to the difference between the area A of the figure and the area Z of the zero circle. The planimeter is so constructed that, for clockwise traversing with the anchor point inside the figure, the net rotation of the roller will always be $\left\{\begin{matrix}\text{forward}\\ \text{backward}\end{matrix}\right\}$ if the area of the figure is $\left\{\begin{matrix}\text{greater}\\ \text{less}\end{matrix}\right\}$ than that of the zero circle; hence the final reading will be $\left\{\begin{matrix}\text{greater}\\ \text{less}\end{matrix}\right\}$ than the initial reading, and n' will be $\left\{\begin{matrix}\text{positive}\\ \text{negative}\end{matrix}\right\}$. It follows that the area of the figure is

$$A = Cn' + Z \tag{5}$$

with due regard to the sign of n'.

example 3. The perimeter of a cross-section is traversed clockwise with the anchor point inside the figure, with the length of the tracing arm so set that the planimeter constant is 10.00 and the area of the zero circle is 132.16 sq in. The initial reading is 1.234, and the final reading is 8.703, the net rotation of the roller being forward. Then $n' = 8.703 - 1.234 = +7.469$; and, from Eq. (5),

$$A = [10.00 \times (+7.469)] + 132.16 = 206.85 \text{ sq in.}$$

example 4. Conditions as in the preceding example, except that it was observed on the disk that the net rotation of the disk was *backward* and that the zero graduation of the disk had passed the index once. Then

$$n' = 8.703 - (10 + 1.234) = -2.531$$

and, from Eq. (5),

$$A = [10.00 \times (-2.531)] + 132.16 = 106.85 \text{ sq in.}$$

Area of Zero Circle. The area of the zero circle can be determined by traversing the perimeter of a figure, once with the anchor point outside the figure and once with the anchor point inside. The first determination gives the area of the figure ($A = Cn$), and the second gives an indicated area Cn' representing the difference between the area of the figure and the area of the zero circle; it follows from Eq. (5) that

$$Z = Cn - Cn' = C(n - n') \tag{6}$$

in which n' is $\left\{\begin{matrix}\text{positive}\\\text{negative}\end{matrix}\right\}$ if the area of the figure is $\left\{\begin{matrix}\text{greater}\\\text{less}\end{matrix}\right\}$ than that of the zero circle, as indicated by the direction of rotation of the roller.

If the length of the tracing arm is fixed, the area of the zero circle is usually stated either on top of the tracing arm or in the planimeter case.

example 5. A given planimeter has a constant of 10.00 sq in. A figure is traversed clockwise, first with anchor point outside and then with anchor point inside; the observed differences in planimeter readings are 2.124 and −9.537, respectively. Then, by Eq. (6), the area of the zero circle is

$$Z = 10.00 \times [2.124 - (-9.537)] = 116.61 \text{ sq in.}$$

Figure Plotted at Other than Full Scale. If a figure is plotted to scale other than full size, the required area is computed by multiplying the actual area by the product of the horizontal and vertical scaled relationships. For example, if a profile is plotted to the scale of 1 in. = 400 ft (horizontal) and 1 in. = 20 ft (vertical), each square inch on the paper represents 400 × 20 = 8,000 sq ft.

7-7 Precision of Planimeter Measurements. If the relation between revolutions and area is established accurately, the errors of planimeter measurement are accidental. Areas should be plotted to a scale consistent with the desired accuracy. Ordinarily, planimeter measurements of small areas may be expected to be correct within 1 per cent, and measurements of figures of considerable size may be correct within perhaps 0.1 or 0.2 per cent. In general, an area determined directly from a difference between initial and final planimeter readings may be determined to three (or the lower range of four) significant figures; summations of such areas (to a given number of decimal places) may have a greater number of significant figures.

In conformity with the precision of planimeter work, observations on the roller should be made to 0.001 revolution, and values of area (C, A, and Z) should be determined to 0.01 sq in. The constants C and Z should be the mean of several observations.

VOLUMES OF EARTHWORK

7-8 General. Volumes of earthwork are calculated by a variety of methods. If cross-sections have been taken along a route, their areas are determined and the volumes of the successive prismoids are determined either by the method of average end areas (Art. 7-10) or by the prismoidal formula (Art. 7-11). The same procedure may be followed for borrow pits and similar excavations; or if elevations are observed at the same points before and after excavating, the volume may be computed by dividing it into vertical truncated prisms. Volumes of earthwork may be determined by the use of contours (see Chap. 19).

Because of the repetitive nature of computations of earthwork, including the areas of cross-sections, machine computation is highly desirable.

Total volumes are almost invariably expressed in cubic yards.

7-9 Borrow Pit. The common method of determining the volume of a borrow pit is to cross-section the area before and after excavating (Arts. 6-3 and 6-4). The pit is then plotted in plan, and its area is divided into rectangles, triangles, and trapezoids, thus dividing the volumes into truncated prisms. Actually the upper and lower surfaces of the prisms are warped, but for earthwork computations they are assumed to be plane.

The volume V of a triangular truncated prism of horizontal sectional area A is

$$V = \frac{A}{3}(h_1 + h_2 + h_3) \tag{7}$$

in which h_1, h_2, and h_3 are the corner heights.

Any rectangular prism may be divided into two triangular prisms by either of two diagonal planes; and where differences in corner heights are large, the proper division into triangles should be indicated in the field notes. Where differences in corner heights are not large, the volume of earthwork in rectangular prisms is computed with sufficient precision by multiplying the average of the corner heights by the horizontal sectional area.

7-10 Average End Areas. The common method of determining volumes of excavation along a line is that of *average end areas*. It is assumed that the volume between successive cross-sections is the average of their areas multiplied by the distance between them, or

$$V = \frac{l}{2}(A_1 + A_2) \qquad (8)$$

in which V is the volume (cubic feet) of the prismoid of length l (feet) between cross-sections having areas (square feet) A_1 and A_2.

The formula is exact only when $A_1 = A_2$ but is sufficiently precise for ordinary earthwork. The maximum error (50 per cent) occurs as one of the end areas approaches zero, but in this case the volume is usually computed as a pyramid; that is,

$$\text{Volume} = \tfrac{1}{3}\ \text{area of base} \times \text{length}$$

7-11 Prismoidal Formula. It can be shown that the volume of a prismoid is

$$V = \frac{l}{6}(A_1 + 4A_m + A_2) \qquad (9)$$

in which V, l, A_1, and A_2 are the same as in Art. 7-10 and A_m is the area halfway between the end sections. A_m is determined by averaging the corresponding linear dimensions of the end sections and not by averaging the end areas.

The use of the prismoidal formula in earthwork computations is justified only if cross-sections are taken at short intervals, if small surface deviations are observed, and if the areas of successive cross-sections differ widely. Usually it yields smaller values than those computed from average end areas. For excavation under contract, the basis of computation should be understood in advance; otherwise the contractor will usually claim (and obtain) the benefit of the common method of average end areas.

7-12 Road Profiles. Preliminary estimates of earthwork for highways, railroads, and canals are based upon the preliminary profile. As the side slopes are inclined, the area representing cut or fill on the profile cannot be taken directly as a measure of volume, as would be the case for a trench.

For very rough estimates, the profile area of any given cut or fill may be measured with the planimeter and divided by the length to obtain the average depth of cut (or fill) at the center line. The area of a level section having this average cut or fill is then computed, and the area is multiplied by the length of cut (or fill) to obtain the volume of earthwork.

For less approximate calculations the cut or fill at each full station is scaled from the profile, and the corresponding volume per station is computed for a level section whose depth is the scaled cut (or fill). The level section at each station is assumed to extend halfway to both adjacent stations. The total volume for a given cut (or fill) is obtained by summing up the volumes per station. The work may be simplified by the use of graphs or tables. Tables of volumes (cubic yards) per 100 ft for various widths of roadbed, side slopes, and depths of cut (or fill) are given in texts on highway and railroad surveying.

7-13 Precision of Determining Volumes. Under the usual conditions of leveling for earthwork, the principal errors are accidental and are due to holding and reading the tape, to reading the rod, and particularly to variations in the ground surface. The *percentage* of error in horizontal dimensions is much less than in vertical dimensions; hence errors in computed volumes result for the most part from errors in cuts and fills. The percentage of error in the final result is greater for small cuts and fills than for large ones, ranging from perhaps 1 per cent for an average roadway cut of 2 ft to perhaps 0.2 per cent for an average cut of 12 ft. Hence it will be consistent to carry one decimal place in intermediate computations of areas and volumes; but it is absurd to record values of volumes beyond the last whole unit.

7-14 Numerical Problems

1 Plot a profile for data given in numerical problem 1 of Chap. 6, between stations 179 and 186. Use a horizontal scale of 1 in. = 100 ft and a vertical scale of 1 in. = 4 ft.

2 Following are the notes for cross-sections at stations 109 and 110. The width of the roadbed is 24 ft, and the side slopes are 2 to 1. Compute the areas of the two sections.

Station	*Cross-section*		
109	$\frac{c2.4}{16.8}$	$\frac{c1.2}{0.0}$	$\frac{c0.4}{12.8}$
110	$\frac{c12.2}{36.4}$	$\frac{c9.2}{0.0}$	$\frac{c4.8}{21.6}$

3 Following are the notes for an irregular road cross-section. The width of the roadbed is 24 ft and the side slopes are 1½ to 1. Determine the cross-sectional area by computing the areas of triangles and rectangles and by planimeter. Compare the results.

c4.2	c6.8	c11.2	c14.4	c16.8	c18.4
18.3	12.0	0	10.0	25.0	39.6

4 Compute the volume in cubic yards between stations 109 and 110 of problem 2. Use both the average-end-area method and the prismoidal formula. Note the discrepancy in percentage between volumes as determined by the two methods.

5 What error in volume between station 109 and station 110 of problem 4 would be introduced if the recorded cuts at centers and slope stakes were 0.1 ft too great? What is the error in terms of percentage of the volume by average end areas?

6 In plan, a borrow pit is 75 by 135 ft. Before and after excavation, levels are run and offsets are measured from stations along one of the 135-ft sides. The computed cuts are shown in the following table:

	Cut, ft						
Offsets	*Sta.* 0	*Sta.* 0 + 30	*Sta.* 0 + 50	*Sta.* 0 + 75	*Sta.* 1 + 00	*Sta.* 1 + 15	*Sta.* 1 + 35
0	0.0	1.5	0.0	4.5	6.2	4.7	0.0
25	1.2	2.9	10.6	9.7	7.9	8.4	2.5
50	2.5	3.7	8.7	8.7	9.4	8.4	3.6
75	0.0	0.0	1.9	7.6	6.8	6.3	0.0

Compute the volume by the method of Art. 7-9.

7 The circumference of a circle 4 in. in diameter is traversed clockwise, with the anchor point of the planimeter outside the figure. The initial reading is 5.637 and the final reading is 6.939. What is the planimeter constant?

8 For a given planimeter set so that the instrumental constant is 10.22, the area of the zero circle is 116.24 sq in. A figure is traversed clockwise with anchor point inside. The initial reading is 1.085 and the final reading is 9.632, the net rotation being backward and the zero graduation of the disk having passed the index once. What is the area of the figure?

9 The tracing arm of a planimeter is set so that the roller reads 0.583 revolution for 10.00 sq in. The perimeter of an area is traversed clockwise first with anchor point outside and then with anchor point inside. The corresponding differences in readings are 2.095 and −7.786. What is the planimeter constant *C*? What is the area of the zero circle?

10 With the tracing arm of the planimeter set as in the preceding problem, a figure for which the vertical scale is 1 in. = 8 ft and the horizontal scale is 1 in. = 20 ft is traversed clockwise with anchor point outside. The difference in planimeter readings is 1.932. What is the actual area in square inches? What area in square feet does it represent?

11 A given fill for a railroad is 1,350 ft long. The profile is plotted to a horizontal scale of 1 in. = 400 ft and to a vertical scale of 1 in. = 20 ft. The perimeter of the area between ground line and grade line is traversed clockwise with anchor point outside, employing a planimeter set so that 1 revolution of the roller is equal to 10 sq in. on the paper. The difference in readings is 0.269. What is the average depth of the fill in feet? Estimate roughly the volume of the fill in cubic yards, assuming a level section of the average depth, roadbed 18 ft wide, and side slopes 1½ to 1.

7-15 Office Problems

Problem 11. Plotting Profile

Object. To plot a profile from level notes (field problem 8, Chap. 6) and to fix the grade line for a road or a pipeline.

Procedure. (1) Choose a horizontal and vertical scale in keeping with the purpose of the profile. (2) Examine the field notes to determine the range between points of maximum and minimum elevation. Number each of the heaviest horizontal lines of the profile paper with its elevation. Number each of the heaviest vertical lines with its station number. (3) From the profile-level notes plot the profile. Through the plotted points draw a freehand curve (see Fig. 7-1). Show the names of streams and roads crossed. Check the profile, and ink it in black. (4) Fix the grade line, and ink it in red. Show elevations of points of change in grade and rates of grade. (5) Near the bottom of the sheet indicate the horizontal alinement, using a scheme similar to that shown in the figure. (6) Make an appropriate title.

Hints and Precautions. (1) Avoid the common mistake of reading elevations of turning points and bench marks as ground elevations, by enclosing the elevations of turning points and bench marks in the field notebook with a circle. (2) In checking the profile, read stations and elevations back from the profile. (3) Ink the profile freehand, preferably with a ruling pen; do not round off the summits and depressions unduly.

Problem 12. Area with Planimeter

Object. With the polar planimeter, to determine the area of a figure plotted to scale (see office problem 13).

Procedure. (1) Set the tracing arm so that one revolution of the roller will bear some simple relation to the given scale and unit of measurement. (2) Test the accuracy of the setting by traversing a figure of known area, say a 2 by 5-in. rectangle, three times. Record the readings to 0.001 revolution. If necessary, adjust the tracing arm until the desired relation is obtained. (3) With the anchor point outside, traverse clockwise the perimeter of the figure whose area is to be determined. Check the operation. (4) Convert the difference between readings into terms of area. (5) Determine the area of the zero circle as described in Art. 7-6. (6) Measure the given figure with anchor point *inside,* and compute the area.

Hints and Precautions. (1) See that the paper is flat and free from wrinkles. (2) Locate the anchor point so that the roller will stay on the paper as the tracing point is moved about the figure. Preferably have the tracing arm and the anchor arm nearly at right angles when the tracing point is near the center of the area, and never allow the area between the arms to become small. (3) See that the contact edge of the roller is free from dirt.

Problem 13. Plotting Cross-sections; Quantities of Earthwork

Object. To plot cross-sections of the roadway from field notes and to compute quantities of earthwork. It is assumed that the cross-section notes give cut and fill (see field problem 10, Chap. 6).

Procedure. (1) Beginning near the upper left corner of a sheet of cross-section paper, choose convenient heavy horizontal and vertical lines as grade and center lines. With these as coordinates plot the cross-section notes of the first station (see Fig. 7-2). Mark the plotted points with dimensions identical with those of corresponding points in the notes. Draw straight lines showing roadbed and side slopes of cut or fill and the original ground, thus enclosing the section. Below the cross-section, mark its station number. (2) At a convenient distance below and on the same center line, plot the next section in similar manner. When the bottom of the sheet is reached, plot the next section a little farther to the right and at the top of the sheet; and in this way continue until all plotting is done. (3) Compute the area of each section, and show its value within the section (as 123 sq ft). Irregular sections may be planimetered. (4) Compute volumes by the average-end-area method and by the prismoidal formula. Show the volume of each prismoid (as 97 cu yd) between its end sections. (5) By each method find the total yardage in each cut and fill, and mark these totals conspicuously. (6) Make an appropriate title.

8

ANGLES AND DIRECTIONS; THE COMPASS

8-1 General. This chapter treats only of angles and directions in the horizontal plane. The horizontal angle between two points is understood to mean the angle between the projections in the horizontal plane of two lines passing through the two points and converging at a third point.

The location of a point is fixed if measurements are made of (1) its direction and distance from a known point, (2) its direction from two known points, (3) its distance from two known points, or (4) its direction from one known point and its distance from another.

The direction of any line (as fixed by two points) is defined by the horizontal angle that it makes either with an adjacent line of the survey or with some real or imaginary reference line of fixed direction called a *meridian*. If the meridian is arbitrarily chosen it is called an *assumed meridian*. If it is a true north-and-south line passing through the geographical poles of the earth, it is called a *true meridian*. If it lies parallel with the magnetic lines of force of the earth it is called a *magnetic meridian*.

Methods of determining the meridian are described in Art. 8-5 and Chap. 17.

8-2 Magnetic Meridian. The direction of the magnetic meridian is that taken by a freely suspended magnetic needle. The magnetic poles are at some distance from the true geographic poles; hence in general the magnetic meridian is not parallel to the true meridian. The location of the magnetic poles is constantly changing; hence the direction of the magnetic meridian is not constant. However, the magnetic meridian is employed as a line of reference on rough surveys where a magnetic compass is used and often is employed in connection with more precise surveys in which direct angular measurements are checked approximately by means of the compass. It was formerly used extensively for land surveys.

8-3 Magnetic Declination. The angle between the true meridian and the magnetic meridian is called the *magnetic declination,* or *variation.*

If the north end of the compass needle points to the east of the true meridian, the declination is said to be east (Fig. 8-7); if it points to the west of the true meridian, the declination is said to be west.

If a true north-and-south line is established, the mean declination of the needle for a given locality can be determined by compass observations extending over a period of time. The declination may be estimated with sufficient precision for most purposes from an *isogonic chart* published by the U.S. Coast and Geodetic Survey; the chart shows lines of equal magnetic declination for the date of issue, and indicates the rate of change in magnetic declination from year to year.

The isogonic chart of the continental United States shown in Fig. 8-1 applies to January 1, 1960. The solid lines are lines of equal magnetic declination, or *isogonic lines.* East of the heavy solid line of zero declination, or *agonic line,* the north end of the compass needle points west of north; west of that line it points east of north. The north end of the compass needle is moving eastward over the area of eastward annual change and westward elsewhere over the chart at an annual rate indicated by the lines of equal annual change.

The magnetic declination changes more or less systematically in cycles over periods of (1) approximately 300 years, (2) 1 year, and (3) 1 day, as follows:

1 Secular Variation. This amounts to several degrees in a half-cycle of approximately 150 years. In Fig. 8-1 are shown by dash lines the annual rates of change in the secular variation for the year 1960. On account of its magnitude, the secular variation is of considerable importance to the surveyor, particularly in retracing lines the directions of which are referred to the magnetic meridian as it existed years previously. When *variation* is mentioned without further qualification, it is taken to mean the secular variation.

2 Annual Variation. This is a small annual swing distinct from the secular variation. For most places in the United States it amounts to less than 01′.

3 Daily Variation. This variation, also called *solar-diurnal* variation, occurs during each day. For points in the United States the north end of the needle reaches its extreme easterly swing at about 8 or 9 A.M. and its extreme westerly swing at about 1 or 2 P.M. The needle usually reaches its mean position between 10 and 11 A.M. and between 7 and 11 P.M. The average range for points in the United States is less than 08′, a quantity so small as to need no consideration for most of the work for which the compass needle is employed. However, in the United States in summer, a line run 1,000 ft by compass at 8 A.M. would end as much as 3 ft to the right of the point where it would end if run at 1 P.M.

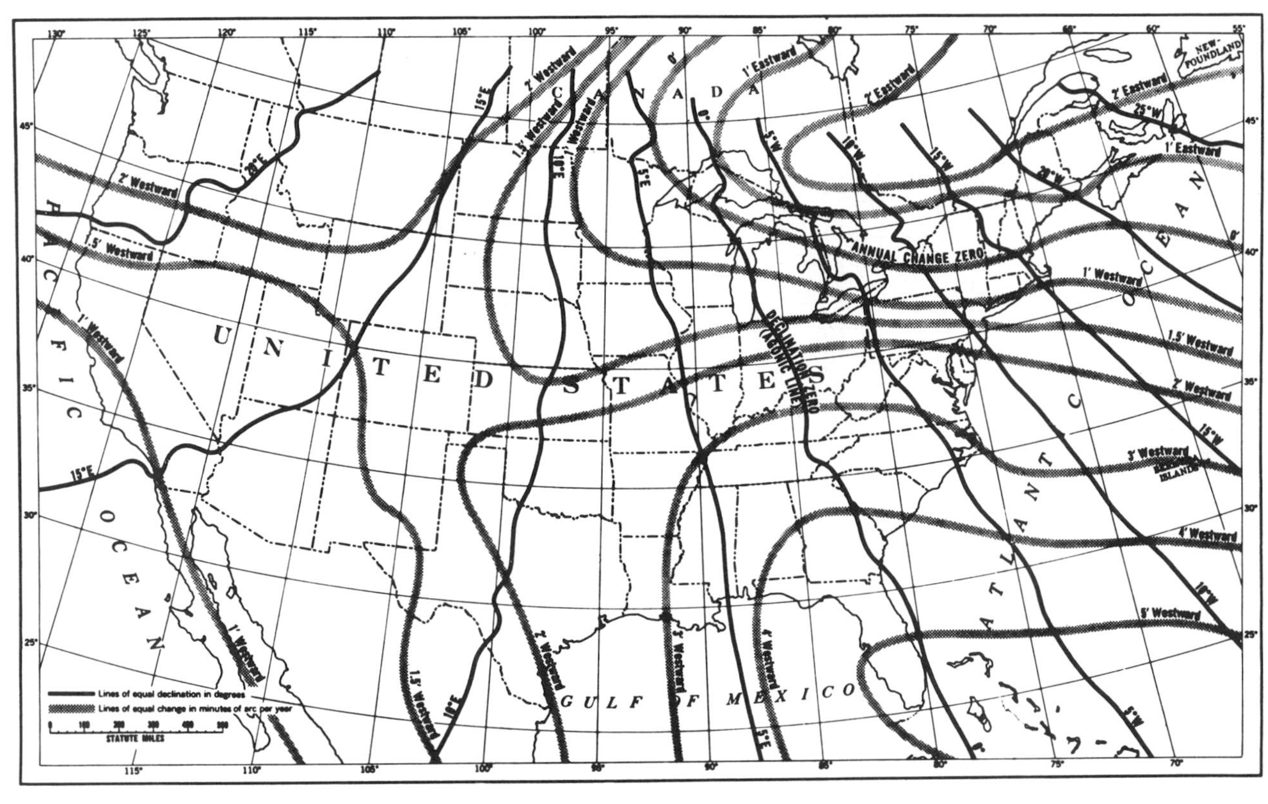

Fig. 8-1. Isogonic chart of the United States for 1960. (*U.S. Coast and Geodetic Survey.*)

4 *Irregular Variations*. These are due to magnetic disturbances and are most likely to occur during magnetic storms. They may amount to a degree or more, particularly at high latitudes.

8-4 Local Attraction. Objects of iron or steel, some kinds of iron ore, and currents of direct electricity alter the direction of the lines of magnetic force in their vicinity and hence are likely to cause the compass needle to deviate from the magnetic meridian. The deviation arising from such local sources is called *local attraction* or *local disturbance*. In certain localities, particularly in cities, its effect is so pronounced as to render the magnetic needle of no value for determining directions. It is not likely to be the same at one point as at another, and is even affected by such objects as the steel tape, chaining pins, axe, and small objects of iron or steel that are on the person. Usually its magnitude can be determined, and directions observed with the compass can be corrected accordingly (Art. 8-16).

8-5 Establishing Meridian. The *true* meridian is established by astronomical observations as described in Chap. 17. For surveys of ordinary precision the transit is employed.

On compass surveys, in order to determine the magnetic declination, sometimes the true meridian is established by ranging two plumb lines with Polaris (the North Star), usually when the star is at elongation (farthest east or farthest west). If the time is accurately known, the observations are sometimes made when the star is at culmination (directly above or below the pole and hence on the meridian). The time of elongation or culmination may be taken from published tables. If the line between the plumb lines is established at elongation, the true meridian is established by laying off an angle equal to the horizontal angle between the star and the meridian, as given in published tables. If care is exercised in setting the ground points and if the plumb lines are at least 15 ft apart, the error need not exceed 05′.

A *magnetic* meridian can be established by setting up the compass over any convenient point and then sighting to set a point on a stake or other object that marks another point on the meridian. Several sights should be taken during the setup. The mean of the points thus established is assumed to be on the magnetic meridian, provided the observations are taken at a time of day when the declination is approximately at its mean value.

United States by the U.S. Coast and Geodetic Survey, reference lines of

At many of the triangulation stations established throughout the known true direction have been established for use by local surveyors.

8-6 Angles and Directions. Angles and directions may be defined by means of *bearings, azimuths, deflection angles, angles to the right,* or *interior angles,* as described in the following articles. These quantities are said to be *observed* when obtained directly in the field and *calculated* when obtained indirectly by computation. Conversion from one means of expressing angles and directions to another means is a simple matter if a sketch is drawn to show the existing relations.

8-7 Bearings. The direction of any line with respect to a given meridian may be defined by the *bearing*. Bearings are called *true* (*astronomic*) *bearings, magnetic bearings,* or *assumed bearings* according as the meridian is true, magnetic, or assumed. The bearing of a line is indicated by the quadrant in which the line falls and the acute angle which the line makes with the meridian in that quadrant. Thus in Fig. 8-2*a* the bearing of the

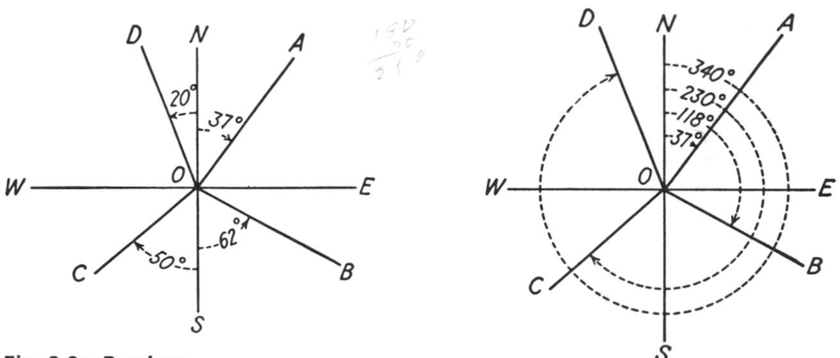

Fig. 8-2*a*. Bearings.

Fig. 8-2*b*. Azimuths from north.

line *OA* is read north 37° east and is written N37°E. The bearings of *OB, OC,* and *OD* are, respectively, S62°E, S50°W, and N20°W. In all cases, values of bearing angles lie between 0° and 90°.

8-8 Azimuths. The *azimuth* of a line is its direction as given by the angle between the meridian and the line, measured in a clockwise direction usually from the south branch of the meridian. In astronomical observations azimuths are generally reckoned from the true south; in surveying, some surveyors reckon azimuths from the south and some from the north branch of whatever meridian is chosen as a reference, but on any given survey the direction of zero azimuth is either always south or always north. Herein, unless otherwise stated, azimuths are reckoned from south. Azimuths are called *true azimuths, magnetic azimuths,* or *assumed azimuths* according as the meridian is true, magnetic, or assumed. Azimuths may have values between 0° and 360°.

In some special cases, the term "azimuth" is used in the sense of a bearing and therefore may be taken either clockwise or counterclockwise, as in "azimuth of Polaris" (astronomy, Chap. 17) or "azimuth of the secant" (land surveying, Chap. 18).

In Fig. 8-2*b* the positions of the lines are the same as in Fig. 8-2*a*, with azimuths from north indicated. Az. $OA = 37°$; Az. $OB = 118°$; Az. $OC = 230°$; Az. $OD = 340°$.

8-9 Deflection Angles. The angle between a line and the prolongation of the preceding line is called a *deflection angle*. Deflection angles are recorded as *right* or *left* according as the line to which measurement is taken lies to the right (clockwise) or left (counterclockwise) of the prolongation of the preceding line. Thus in Fig. 8-3 the deflection angle at *B* is

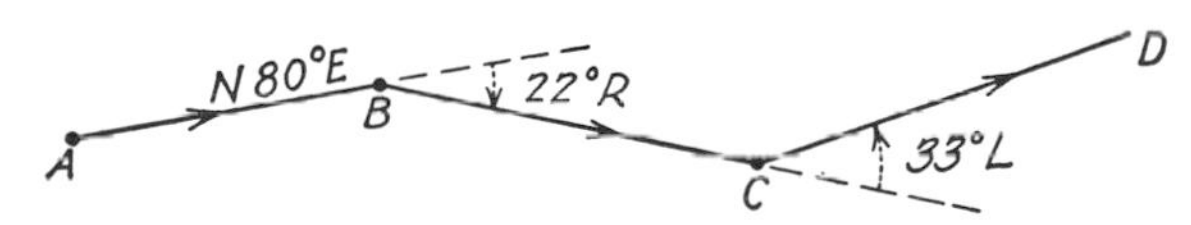

Fig. 8-3. Deflection angles.

22°R, and at *C* is 33°L. Deflection angles may have values between 0° and 180°, but usually they are not employed for angles greater than 90°. In any closed polygon the algebraic sum of the deflection angles (considering right deflections as of sign opposite to left deflections) is 360°.

8-10 Angles to Right. Angles may be determined by clockwise measurements from the preceding to the following line, as illustrated by Fig. 8-4. Such angles are called *angles to right* or *azimuths from back line*.

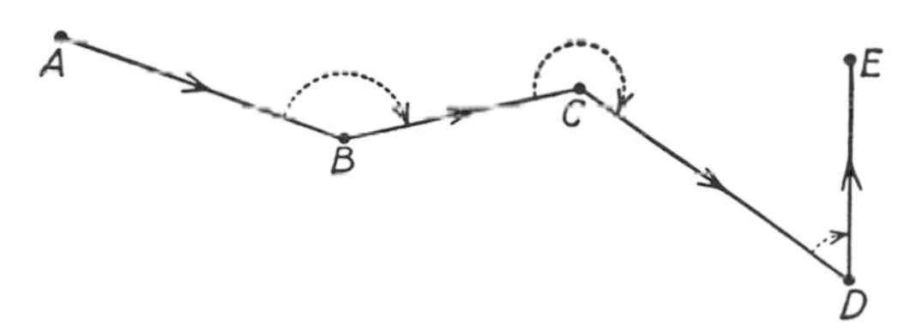

Fig. 8-4. Angles to right.

8-11 Interior Angles. In a closed polygon the angles inside the figure between adjacent lines are called *interior angles*. If n is the number of sides in a closed polygon, the sum of the interior angles is $(n - 2)180°$.

8-12 Traversing. The succession of straight lines connecting a succession of established points along the route of a survey is called a *traverse* or *traverse line*. The points defining the traverse line are called *traverse stations* or *traverse points*. Distances along the line between successive points are determined either by direct measurement or by stadia. At each point where the traverse changes direction an angular measurement is taken. If the traverse forms a closed figure, as, for example, the boundary of a parcel of land, it is called a *closed traverse;* if it does not form a closed figure, as, for example, the line for a highway, it is called an *open traverse* or *continuous traverse*. Traverses are also designated according to the purpose of the survey (as preliminary), the field instrument or method employed (as transit-tape), or the kind of angular measurements observed (as azimuth).

For details of the method of traversing, see Chap. 10. Traversing is the method of surveying in most common use where favorable routes are available.

8-13 Triangulation. Where the lines of a survey form triangular figures whose angles are measured and whose distances are determined by trigonometric computations, the operation of making the necessary field observations is called *triangulation*. The simplest case is that of a single triangle, one of whose sides is of known length; if any two angles of the triangle are measured, sufficient data are obtained for computing the length of the other two sides. Furthermore, if the third angle is measured, the angular measurements may be checked.

A triangulation system is made up of a series of triangles so connected that, having measured the angles of the triangles and the length of one line, the length of other lines may be computed. The line of known length, upon which all computed distances are based, is called a *base line*.

For details of the method of triangulation, see Chap. 12. The advantage of triangulation over traversing lies in the small number of linear measurements that are necessary; the disadvantage lies in the greater amount of computing required. Triangulation is superior to traversing where the terrain offers many obstacles (such as hills, vegetation, or marsh) to traverse work.

COMPASS

8-14 General. Angles and directions may be determined by means of the *tape* (Art. 3-13), the *transit* (Art. 9-9), the *plane table* (Chap. 13), the *sextant,* and the magnetic *compass*. The compass is useful alone in making rough surveys and retracing early land surveys and is useful on the transit as a means of approximately checking horizontal angles measured by more precise methods.

The essential features of the magnetic compass used by the surveyor are (1) a compass box with circle graduated from 0° to 90° in both directions from the N and S points and usually having the E and W points interchanged as illustrated in Fig. 8-5, (2) a line of sight in the direction of the SN points of the compass box, and (3) a magnetic needle. When the line of sight is pointed in a given direction, the compass needle (when pivoted and brought to rest) gives the magnetic bearing. Thus in the figure the bearing of *AB* is N60°E. If the N point of the compass box is nearest the object sighted, the bearing is read by observing the north end of the needle.

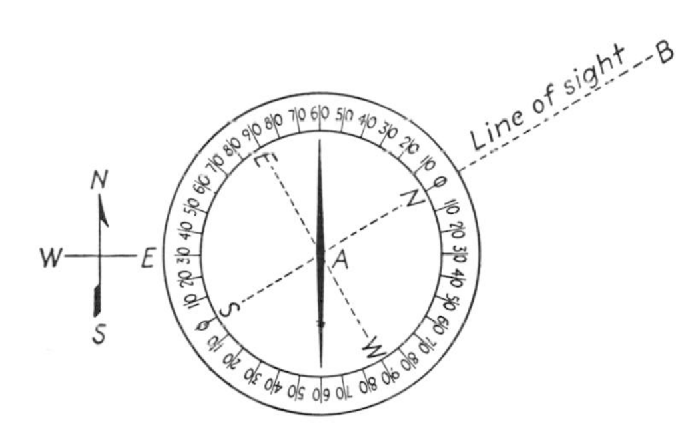

Fig. 8-5. Features of magnetic compass used in surveying.

The varieties of compasses are (1) various *pocket compasses* which are generally held in the hand; (2) the *surveyor's compass* (Art. 8-15) which is mounted usually on a light tripod or sometimes on a Jacob's staff (a pointed stick about 5 ft long); and (3) the *transit compass* (Art. 9-1), a compass box similar to that of the surveyor's compass, mounted on the upper or vernier plate of the engineer's transit and often used to check horizontal angles.

8-15 Surveyor's Compass. The *surveyor's compass* consists of a compass box to opposite sides of which are fastened vertical sight vanes (Fig. 8-6). The box is rigidly connected to a vertical spindle which is free to revolve in a conical socket. Below the spindle is a leveling head consisting of a ball-and-socket joint by means of which the compass may be leveled; the upper portion of the socket is a thumb nut which is tightened until the ball is held securely by friction. The leveling head may be screwed onto a wooden tripod or a Jacob's staff. Within the compass box is a circular, or "bull's eye," level. The compass is provided with a screw for lifting and clamping the needle and with a screw for clamping the vertical spindle. To counteract the effect of magnetic dip so that the needle will be horizontal while in use, an adjustable counterweight of fine brass wire is attached to the south end of the needle. The compass circle is graduated usually in half degrees, and bearings may be read by estimation to 05′ or 10′.

In order that *true* bearings may be read directly, some compasses, as the one shown in the illustration, are so designed that the compass circle may be rotated with respect to the box in which it is mounted. If the magnetic declination is set off by means of the circle, the observed bearings will be true (Fig. 8-8).

When the direction of a line is to be determined, the compass is set up on line and is leveled. The needle is released, and the compass is rotated about its vertical axis until a range pole or other object on line is viewed through the slits in the two sight vanes. When the needle comes to rest, the bearing is read. Ordinarily the sight vane at the end of the compass box marked "S" is held next to the eye; in this case the bearing is given by the north end of the needle.

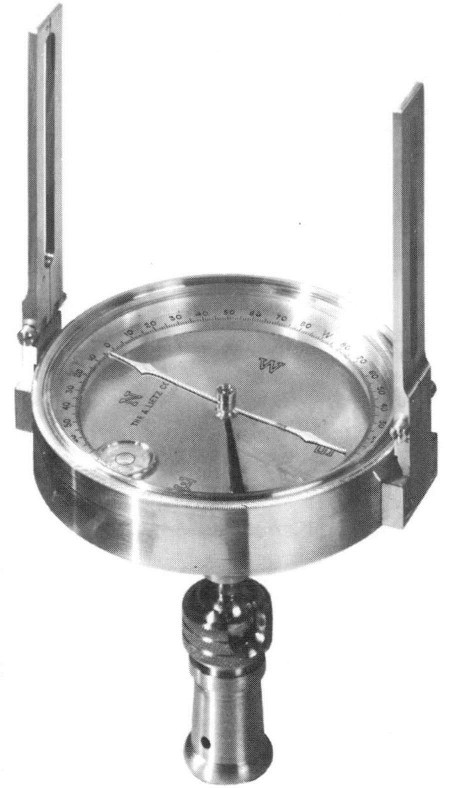

Fig. 8-6. Surveyor's compass. (*The A. Lietz Co.*)

The following suggestions apply to compass observations: At each observation the compass box should be tapped lightly as the needle comes to rest, so that the needle may swing freely. Since the precision with which angles may be read depends on the delicacy of the needle, special care should be taken to avoid any jar between the jewel bearing of the needle and the pivot point. *Never move the instrument without making certain that the needle is lifted and clamped.*

Sources of magnetic disturbance such as chaining pins and axe should be kept away from the compass while a reading is being taken. Care should be taken not to produce static charges of electricity by rubbing the glass; a moistened finger pressed against the glass will remove such charges. Ordinarily the amount of metal about the person of the instrumentman is not large enough to deflect the needle appreciably, but a change of position between two readings should be avoided.

Surveying with the compass is usually by traversing (Art. 8-12). Only alternate stations need be occupied, but a check is secured and local attraction is detected if both a backsight and a foresight are taken from each station. Unlike a transit traverse, in which an error in any angle affects the observed or computed directions of all following lines, an error in the observed bearing of one line in a compass traverse has no effect upon the observed *directions* of any of the other lines. This is an important advantage. Another advantage of the compass is that obstacles

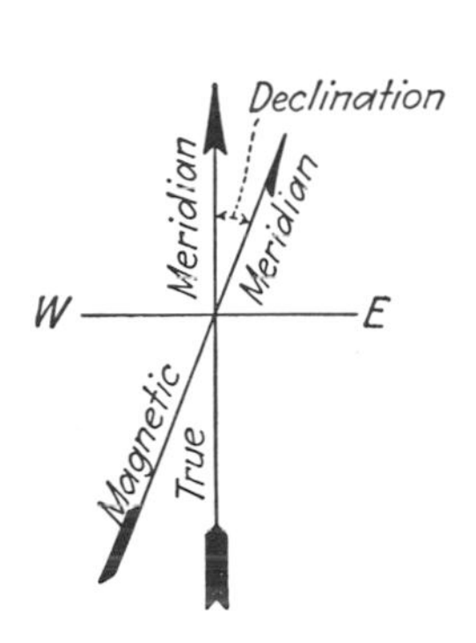

Fig. 8-7. Declination east.

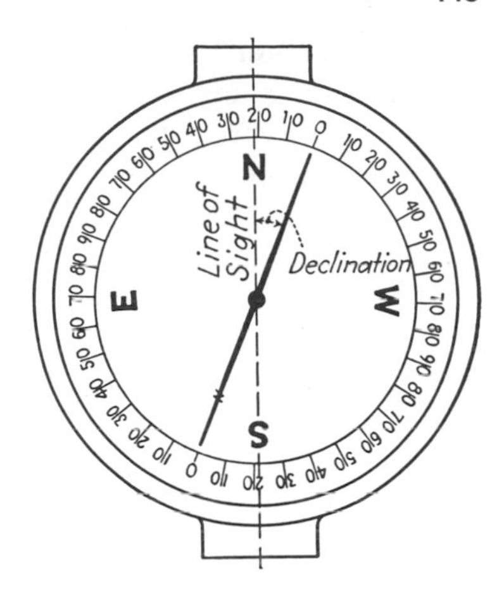

Fig. 8-8. Declination set off on compass circle.

such as trees can be passed readily by offsetting the instrument a short measured distance from the line.

Field notes for a compass traverse are kept in a form similar to that of Fig. 8-9. The declination was set off on the compass so that bearings were referred to the true meridian.

SURVEY OF WOOD LOT OF R.D. FLY, BEMIS, ME.

With Surveyor's Compass and 66-ft. Tape

Sta. At	Sta. To	Dist., Chains	Obs. Bear.	Int. Angle Comp.	Int. Angle Corr.	Corr. Bear.
A	E		N28°00'W			
	B	24.93	S30°40'W	238°40'	238°45'	S30°40'W
B	A		N30°40'E			
	C	37.56	S83°50'E	65°30'	65°35'	S83°45'E
C	B		N84°30'W			
	D	48.42	N 2°00'W	82°30'	82°35'	N 1°10'W
D	C		S 2°15'E			
	E	35.26	S89°30'W	91°45'	91°50'	N89°20'W
E	D		East			
	A	25.77	S28°50'E	61°10'	61°15'	S28°05'E
			Sum	539°35'	540°00'	

11

N.E. Dunn ⊼
Gurley Vernier Compass No. 89. Declin. 20°15' Set off with Vernier
F. Arsneault, Chain
Nov. 13, 1967
Snow

Spruce tree 18"○ blazed B.M.Co./R.D.F.
Cedar stump 12"○ " " / "
Rough stone, d.h.
Ledge, d.h. and cross
Cedar stake 4"○ 4' high

E B.M.Co. D
N
R.D. Fly
A
B
B.M.Co.
C
P.&F.R.R. right of way

Note: Bearings are referred to the true meridian

Fig. 8-9. Notes for compass survey.

8-16 Correction for Local Attraction. If local attraction from a fixed source exists at any station in a traverse, both the back and the forward bearings taken from that station will be affected by the same amount. Disregarding for the time being the accidental errors due to observing, it is probable that the terminal points of any line, as *AB*, are free from local attraction if the back bearing from *B* is the reverse of the forward bearing from *A*. Keeping in mind that the computed *angle* between the forward and back lines from any station can be determined correctly from the observed bearings taken from that station regardless of whether or not the needle is affected locally, the direction free from local attraction may be chosen as a basis, the traverse angles may be computed from observed bearings, and—starting from the unaffected line—the correct bearings of successive lines may be computed.

example 1. The following observations were made on an open compass traverse:

Line	*Forward bearing*	*Back bearing*
AB	N45°E	S45°W
BC	S60°E	N62°W
CD	S31°W	N30°E

Since the back and forward bearings of *AB* are in agreement, it is assumed that stations *A* and *B* are free from local attraction. Hence the correct forward bearing of *BC* is S60°E. The angle at *C*, computed from the observed bearings, is $180 - 62 - 31 = 87°$; and this value of the angle is correct (excluding errors of observation) regardless of local attraction. The correct forward bearing of *CD* is therefore $180 - 60 - 87 =$ S33°W.

Some surveyors find it more expedient to consider the magnitude and direction of the error due to local attraction and then to make corrections to observed bearings without computing the traverse angles.

example 2. For the observed bearings of Example 1 it is seen that the *correct* back bearing of *BC* is N60°W and that the *observed* back bearing is N62°W. The local attraction at *C* is therefore 2° clockwise (as may be seen from a sketch), and the correction to any observed bearing taken with the compass at *C* is 2° counterclockwise. The observed forward bearing of *CD* is S31°W, and the corrected forward bearing of *CD* is therefore $31 + 2 =$ S33°W. In similar manner the local attraction at *D* is found to be 3° clockwise.

Owing to errors of observation, there are likely to be discrepancies between the observed forward and back bearings of lines, even though no local attraction exists. If the discrepancies are small and apparently not of a systematic character, it is reasonable to assume that the errors are due to causes other than local attraction.

8-17 Adjustment of Closed Compass Traverse. When the compass traverse forms a closed figure, the angle method of Example 1 may be extended to include the effect of observational errors, as follows: The interior angle at each station is computed from the observed bearings; the computed value will be free from local attraction as previously described. The sum of the interior angles should equal $(n - 2)180°$, in which n is the number of sides in the traverse. Since the error of observing a bearing is accidental, the error of closure of the traverse (as indicated by the sum of the computed interior angles) is assumed to be distributed equally, and the interior angles are corrected accordingly. The bearings are then adjusted by starting from some line whose observed bearing is assumed to be correct and by computing the bearings of successive lines by means of the corrected interior angles.

example. The observed bearings and computed interior angles for the compass traverse of Fig. 8-9 are shown in Fig. 8-10*a*, in which the short vertical lines represent the compass needle. The sum of the interior angles is 25′ less than the correct value of 540°00′; hence 05′ is to be added to each of the five interior angles to correct for observational errors. The corrected interior angles are shown in Fig. 8-10*b*, in which the short vertical lines represent the true meridian. Since line *AB* had the same observed back bearing as observed forward bearing, both ends of that line are assumed to be free from local attraction, and the bearing of *AB* is taken as being correct. Using the corrected interior angle at *B*, the corrected forward bearing of *BC* is then 180°00′ − 30°40′ − 65°35′ = S83°45′E. The corrected back bearing of *BC* must be the same as the forward bear-

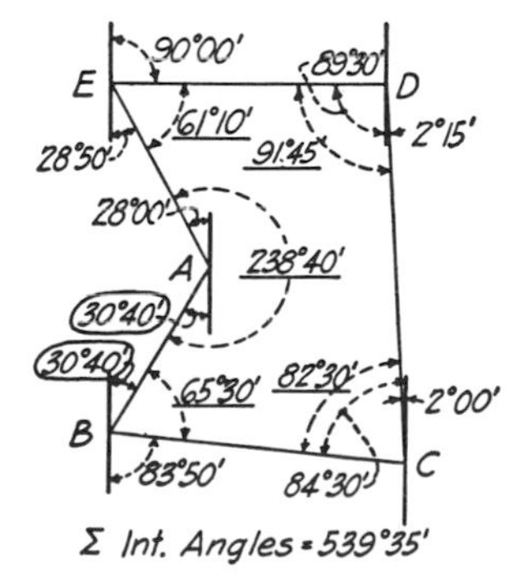

Fig. 8-10*a*. Observed bearings and computed interior angles.

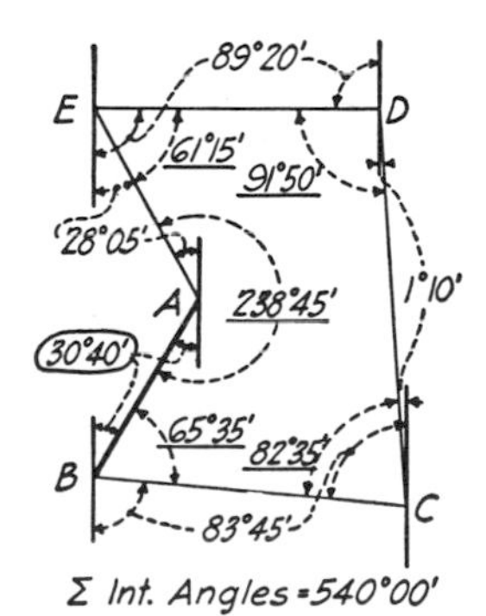

Fig. 8-10*b*. Corrected bearings computed from *AB* and corrected interior angles.

ing, just computed. At C, the corrected forward bearing of CD is 85°45′ − 82°35′= N1°10′W. In this manner the computations are continued around the traverse. As a check, the forward bearing of the initial line AB is computed from the corrected back bearing AE of the preceding line and the corrected interior angle at A.

If the error in the sum of the interior angles is greater than 05′ or 10′ times the number of angles, it is probable that a mistake in reading the compass has occurred, and the field measurements should be repeated. If the error is within permissible limits but cannot be divided equally among the angles in amounts of 05′ or 10′, the greater corrections (in multiples of 05′) should be applied arbitrarily to those angles for which the conditions of observing were estimated to be the least favorable. The precision of compass measurements does not justify computations with a precision closer than multiples of 05′.

If two or more of the traverse lines appear to be free from local attraction, as indicated by the agreement between forward and back bearings, one of these lines is arbitrarily chosen as the "best line," and the computation of corrected bearings is referred to this line. If none of the lines is free from local attraction, that line is chosen which has the least discrepancy between forward and back bearings; and its forward bearing is assumed to be correct.

8-18 Sources of Error; Adjustment of Compass

1 Needle Bent. A bent needle introduces a constant error which can be eliminated by reading both ends of the needle and averaging the two angular values. The needle can be straightened with pliers.

2 Pivot Bent. A bent pivot introduces a variable systematic error the magnitude of which depends upon the direction of sighting. The error can be eliminated by reading both ends of the needle and averaging the two values. The instrument can be corrected by bending the pivot until the end readings of the needle are 180° apart for any direction of pointing.

3 Plane of Sight Not Vertical, or Graduated Circle Not Horizontal. This misalinement introduces a systematic error, but it is usually so small as to be of no consequence. However, the sight vanes may become bent so that, even though the instrument is leveled, an appreciable error is introduced particularly if the line of sight is steeply inclined when taking a bearing. The vanes may be tested by leveling the compass and sighting at a plumb line. The adjustment of the level tubes may be tested by reversal, as described for the transit.

4 Sluggish Needle. The lag of a sluggish or "weak" needle produces an accidental error, often large, which can be reduced by remagnetizing the needle or sharpening the pivot, as needed. As the needle comes nearly to rest, tapping the glass lightly will tend to prevent the needle from sticking to the pivot.

5 Reading Needle. Inability of the observer to determine exactly the point on the graduated circle at which the needle comes to rest is usually the source of the largest accidental error in compass work. The needle should be level, and the eye of the observer should be above the coinciding graduation and in line with the needle.

6 Magnetic Variations. Undetected deviations of the magnetic needle from whatever cause are the source of the largest and most important systematic errors in compass work.

8-19 Numerical Problems

1 The magnetic bearing of a line is S47°30′W and the magnetic declination is 12°10′W. What is the true bearing of the line?

2 In an old survey made when the declination was 2°10′W, the magnetic bearing of a given line was N35°15′E. The declination in the same locality is now 3°15′E. What are the true bearing and the present magnetic bearing that would be used in retracing the line?

3 Following are the observed magnetic bearings of a compass traverse: *AB*, N37°15′E; *BC*, N81°30′E; *CD*, S66°10′E; *DE*, S79°00′E; *EF*, N55°15′E. Compute the deflection angles.

4 Following are deflection angles of traverse *A* to *F*: *B*, 37°21′L; *C*, 12°39′L; *D*, 63°31′R; *E*, 14°07′L. The true bearing of *AB* is S37°56′E. Compute the bearings of the remaining lines.

5 For the traverse of problem 3, the declination is 7°15′E. Compute the true azimuths reckoned from the north point.

6 For the traverse of problem 4, compute the true azimuths reckoned from the south point.

7 The interior angles of a five-sided closed traverse are as follows: *A*, 117°36′; *B*, 96°32′; *C*, 142°54′; *D*, 132°18′. The angle at *E* is not measured. Compute the angle at *E*, assuming the given values to be correct.

8 (*a*) What are the deflection angles of the traverse of problem 7? (*b*) What are the computed bearings if the bearing of *AB* is due north?

9 Following are the deflection angles of a closed traverse: *A*, 85°20′L; *B*, 10°11′R; *C*, 83°32′L; *D*, 63°27′L; *E*, 34°18′L; *F*, 72°56′L; *G*, 30°45′L. Compute the error of closure. Adjust the angular values on the assumption that the error is the same for each angle.

10 In triangulating across a river a base line *AB* of the triangle *ABC* has a measured length of 536.27 ft, and the angles at *A* and *B* are respectively 87°32′ and 68°48′. Compute the distance *AC*.

11 The following are bearings taken on a closed compass traverse. Compute the interior angles and correct them for observational errors. Assuming the observed bearing of the line *AB* to be correct, adjust the bearings of the remaining sides.

Line	*Forward bearing*	*Back bearing*
AB	S37°30′E	N37°30′W
BC	S43°15′W	N44°15′E
CD	N73°00′W	S72°15′E
DE	N12°45′E	S13°15′W
EA	N60°00′E	S59°00′W

8-20 Field Problem

Problem 14. Survey of Field with Surveyor's Compass and Tape

Object. To find the true bearing and length of each side of an assigned field, using the surveyor's compass and 66-ft chain tape. The data may be used in office problem 37 or 38 (Chap. 15).

Procedure. (1) On the compass, set off the magnetic declination in order that true bearings may be read directly; if this cannot be done, observe the magnetic bearings and convert them to true bearings later. (2) Set up at one corner *A* of the field, and observe the forward bearing of the line *AB* to the nearest 05′. (3) Take a back bearing from *A*. (4) Tape the line *AB*, and record the distance in chains to the nearest link. (5) Set up the compass at *B*, and proceed in the same manner as at *A*. (6) Continue around the field, taking both back and forward bearings from each point and chaining the lines. (7) Compute the interior angles of the field, and correct the observed bearings for local attraction and/or errors of observation. (8) If magnetic bearings have been observed, apply the magnetic declination to convert them to true bearings.

Hints and Precautions. (1) Keep notes in a form similar to that of Fig. 8-9; add a column for true bearing if desired. (2) Be sure to set off the declination in the correct direction. (3) Have the end of the compass box marked "S" next to the eye. (4) As the needle comes nearly to rest, tap the compass lightly with a pencil. When the needle becomes stationary, read the north end, estimating the bearing to the nearest 05′.

9
ENGINEER'S TRANSIT

INSTRUMENTS

9-1 Conventional Transit. The transit may be employed for measuring and laying off horizontal angles, directions, vertical angles, differences in elevation, and distances, and for prolonging lines.

Some of the modern types of transit differ considerably in design and construction from those long in use, but their essential features do not differ greatly, and their use not at all. Inasmuch as thousands of the conventional type are continuing in use, and as the conventional type requires field adjustment whereas some modern types require little or none, herein the conventional transit is described in detail in this article and other types in less detail (Arts. 9-5 and 9-6). For a given type of transit, the details of design and construction differ somewhat as among the various instrument makers; however, their essential features are similar.

Figures 9-1 and 9-2 show a conventional engineer's transit of the type in common use. It is seen to consist of an *upper,* or *vernier, plate* to which are attached standards supporting the telescope and a *lower plate* to which is fixed a horizontal graduated circle. The upper and lower plates are fastened, respectively, to vertical inner and outer spindles, the two axes of rotation being coincident with and at the geometric center of the graduated circle. The outer spindle is seated in the tapered socket of the leveling head. Near the bottom of the leveling head is a ball-and-socket joint which secures the instrument to the foot plate yet permits rotation of the instrument about the joint as a center.

The outer spindle carrying the lower plate may be clamped in any position by means of the lower *clamp screw.* Similarly, the inner spindle carrying the upper plate may be clamped to the outer spindle by means of the upper clamp screw. After either clamp has been tightened, small movements of the spindle may be made by turning the corresponding *tangent screw.* The axis about which the spindles revolve is called the *vertical axis* of the instrument.

Level tubes, called *plate levels,* are mounted at right angles to each

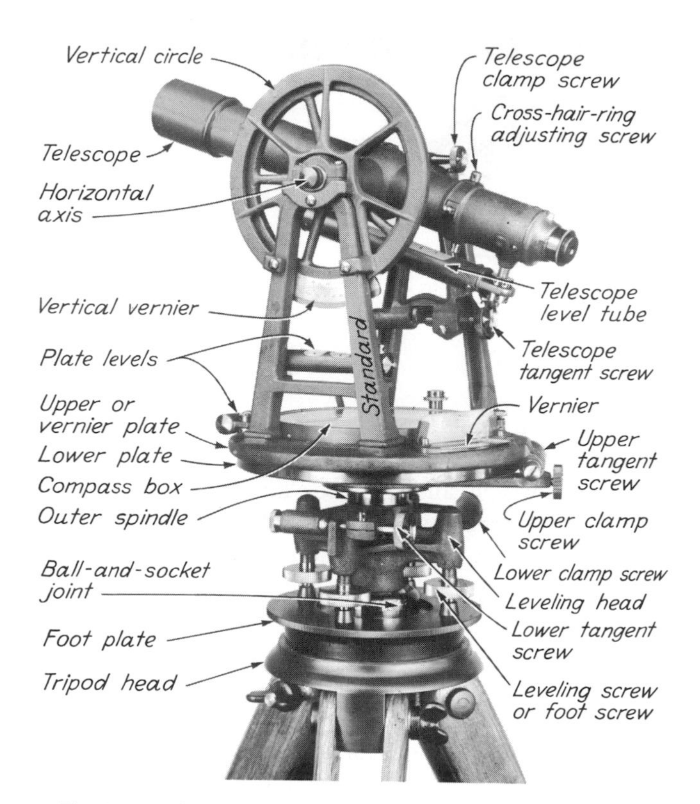

Fig. 9-1. Engineer's transit (conventional type).

other on the upper plate. Four *leveling screws,* or foot screws, are threaded into the leveling head and bear against the foot plate; when the screws are turned, the instrument is tilted about the ball-and-socket joint. When all four screws are loosened, pressure between the sliding plate and the foot plate is relieved, and the transit may then be shifted laterally with respect to the foot plate. From the end of the spindle is suspended a chain with hook for the plumb line. The instrument is mounted on a tripod by screwing the foot plate onto the tripod head.

The *telescope* is fixed to a transverse *horizontal axis* which rests in bearings on the standards. The telescope may be rotated about this horizontal axis and may be fixed in any position in a vertical plane by means of the telescope clamp screw; small movements about the horizontal axis may then be secured by turning the telescope tangent screw. Fixed to the horizontal axis is the *vertical circle,* and attached to one of the standards is the vertical vernier. Beneath the telescope is the *telescope level tube.*

Attached to the upper plate is the *compass box,* the details of which are the same as for the surveyor's compass described in Art. 8-15.

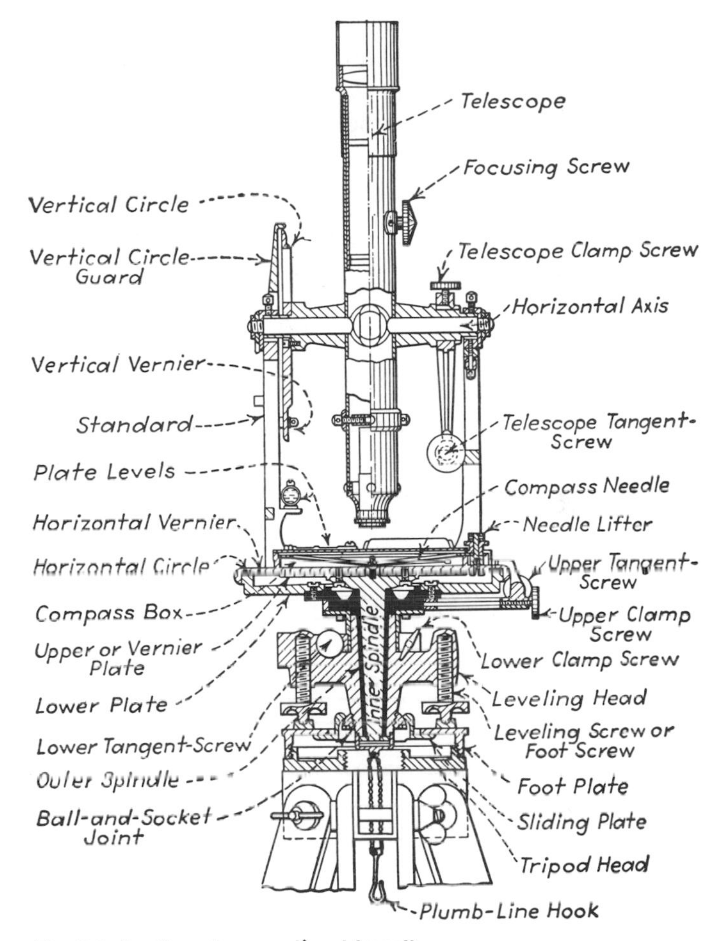

Fig. 9-2. Section of conventional transit.

9-2 Level Tubes; Telescope. On the ordinary transit reading to 01′, the sensitivity of the plate levels is about 75″ per 2-mm graduation of the level tube, and the sensitivity of the telescope level is about 30″ per 2-mm graduation.

The telescope of the transit is similar to that of the engineer's level (Art. 4-10). When the transit is used as an instrument for either direct or trigonometric leveling, any point on the horizontal cross hair may be used in sighting; when the transit is used for establishing lines, measuring angles, or taking bearings, any point on the vertical cross hair may be used. Most instruments are equipped with stadia hairs. The magnifying power is usually 18 to 24 diameters. Usually the transit is equipped with an erecting eyepiece. Some telescopes are of the internal-focusing type (Art. 4-10).

9-3 Graduated Circles. The vertical circle has two opposite zero points and is graduated usually in half degrees, the numbers increasing to 90° in both directions from the zero points as illustrated in Fig. 9-3*a*. When the telescope is level, the index of the vernier is at 0°.

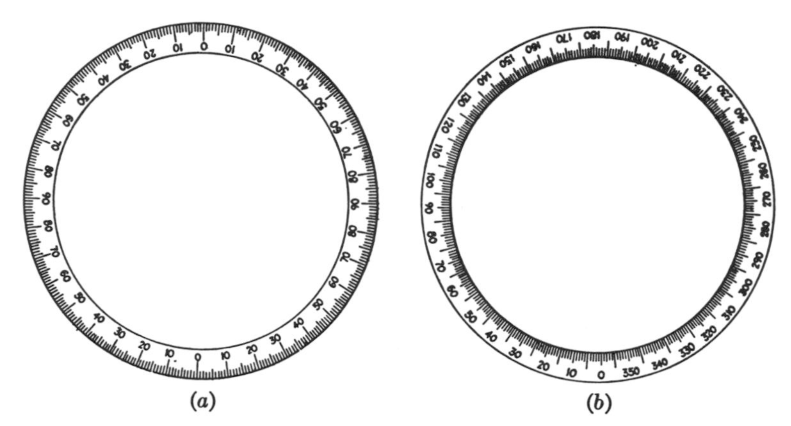

Fig. 9-3. Numbering of circles. (*a*) **Vertical circle numbered in quadrants.** (*b*) **Horizontal circle numbered 0 to 360.**

The horizontal circle is graduated usually in half degrees but sometimes in third degrees or quarter degrees. It may be numbered from 0° to 360° clockwise (Fig. 9-3*b*), 0° to 360° clockwise and 0° to 90° in quadrants, or 0° to 360° in both directions. Usually the numbers slope in the direction of reading.

On the conventional transit, both vertical and horizontal angles can be read to 01′ by means of the verniers.

9-4 Verniers. The verniers employed for reading the horizontal and vertical circles are identical in principle with those for the target rod, described in Art. 4-19. Practically all transit verniers are of the direct type.

Figure 9-4 shows the usual type of double direct vernier. The vernier on the left of the index is for reading clockwise angles, while that on the right is for reading counterclockwise angles. The circle is graduated in half degrees, or 30′. Each space on the vernier is 01′ less than a space on the circle, and 30 spaces on the vernier are equal to 29 spaces on the circle; the least count is 30′/30 = 01′. Considering counterclockwise

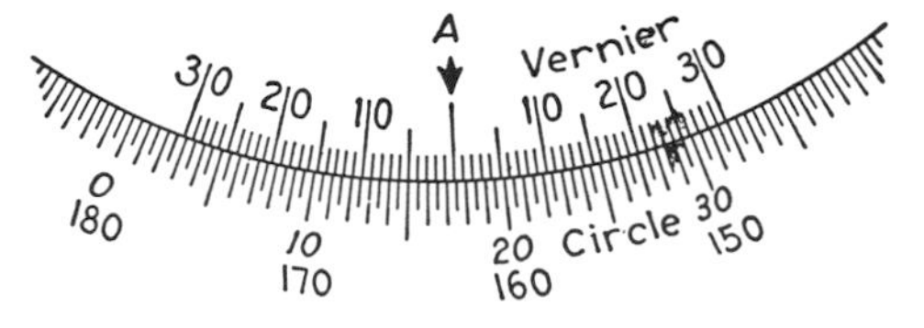

Fig. 9-4. Double direct-vernier reading to minutes.

angles, the reading of the right vernier is seen to be 17°25′. Considering clockwise angles, the reading of the left vernier is 162°35′.

Transits have two verniers, called the *A* and *B* verniers, for reading the horizontal circle, their indexes being 180° apart. Failure of the two verniers to register readings exactly 180° apart on the circle may be due to either or both of two following causes:

1 Eccentricity of Verniers. The verniers may have become displaced, so that a line joining their indexes does not pass through the center of rotation of the upper plate. The error will be the same for all parts of the horizontal circle.

2 Eccentricity of Centers. The spindles may have become worn or otherwise damaged so that the center of rotation of the upper plate does not coincide with the geometrical center of the graduated horizontal circle. The error varies according to the position of the verniers with respect to the horizontal circle.

Taking the mean of the two vernier readings (*A* and *B* verniers) eliminates errors due to either or both types of eccentricity. Further, if the verniers only are eccentric, no error is introduced in an angle as long as the same vernier is used for making the final reading as for making the initial setting.

9-5 Other Transits and Features. A transit without vertical circle and telescope level tube is called a *plain transit*. One without compass and having U-shaped, one-piece standards instead of the A-shaped separate standards shown in Fig. 9-1 is often called a *city transit*. A *mountain transit* is relatively small and light. A *mining transit* has one or more of the following features: a small auxiliary telescope attached either to one end of the horizontal axis or to the top of the main telescope, the vertical circle graduated on the edge instead of the side, a vertical arc of 180° instead of a full circle, and a reversion telescope level which makes it possible to level the telescope with the level tube above as well as below the telescope.

Some transits have two opposite vertical verniers which may be rotated as a unit about the horizontal axis and which are equipped with a level tube for control. When the level bubble is centered, a line through the zeros of the two verniers is horizontal. This device enables vertical angles to be read correctly whether or not the plate-level bubbles are centered. Some transits are provided with a sensitive *striding level* which rests on the horizontal axis and ensures that the line of sight will revolve in a vertical plane; this feature is particularly important in astronomical observations.

9-6 Theodolites. A *theodolite* is essentially a transit of high precision. A *repeating theodolite* is so designed that successive measures of a horizontal angle may be accumulated on the graduated circle and the total arc passed through then divided by the number of repetitions, as described in Art. 9-10. Usually the precision of a repeating theodolite corresponds to that used in third-order traversing or triangulation (Arts. 20-3 and 20-4).

A *direction theodolite* is so designed that the graduated horizontal circle remains fixed during a series of observations, the direction to each point being read on the circle only once, by means of micrometer microscopes, in terms of the angle from the zero of the circle. Several sets of observations are made, with the circle shifted between sets so that each direction is measured on different parts of the circle. The direction technique, the precision in reading the circle, the high precision of manufacture of the circle, and—for American instruments—the use of a relatively large circle result in greater precision of angular measurement with the direction instrument than with the repeating instrument. The direction instrument is mounted on a relatively stable base.

American Repeating Theodolite. A repeating theodolite of American manufacture is similar to the conventional engineer's transit but is larger and is of a higher grade of workmanship. Its horizontal circle is larger and more finely graduated, and its levels are more sensitive. Attached magnifying glasses are used for reading the verniers, which usually read to 10″. Ordinarily the instrument has no compass, and an optical plummet is used instead of a plumb line.

American Direction Theodolite. The Parkhurst direction theodolite was developed and has long been used by the U.S. Coast and Geodetic Survey. The 9-in. horizontal plate has a single tangent motion, comparable to the upper motion of the ordinary transit, except that in the direction theodolite the graduated circle can be rotated while the telescope is clamped in a fixed direction. When angles are to be measured about a point, an initial circle reading is made when the instrument has been pointed on the first distant signal. This initial reading is a measure of the azimuth, or direction, of the first object sighted with respect to some (undefined) reference meridian the direction of which depends entirely on the chance position of the plate when it is fixed before sighting. The directions of all distant stations are then read successively without disturbing the horizontal circle, and from these readings the values of the angles (from one station to the next) are computed. Instead of a vernier, a micrometer microscope mounted on each side of the instrument is used to read the subdivisions of the graduated circle. The micrometer may usually be read directly to 01″ and by estimation to 0.1″.

European Theodolites. Theodolites of European manufacture combine a high degree of precision with facility of operation and lightness of weight. Their use is increasing in the United States not only for precise work but also for the ordinary operations of surveying.

In general construction and appearance, European direction theodolites are similar to repeating theodolites of the same manufacturer. The difference lies principally in the mounting of the horizontal circle; in the direction type the horizontal circle is fixed, whereas in the repeating type it can be rotated. Furthermore, usually the direction type is designed for measurements of higher precision than those made with the repeating type. Two forms of European theodolites are shown in Figs. 9-5*a* and 9-5*b*.

Although there are differences as among the various manufacturers,

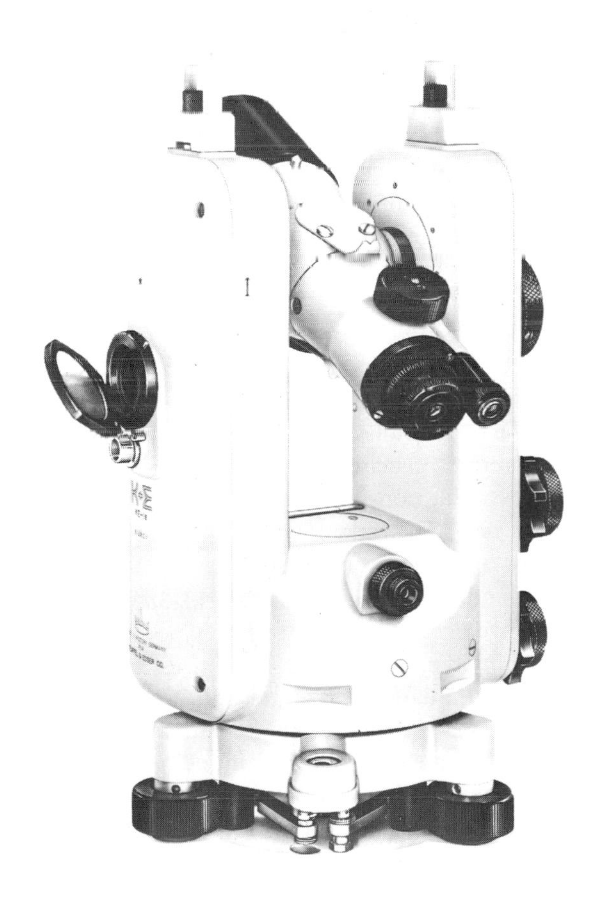

Fig. 9-5*a*. Askania repeating theodolite. (*Keuffel & Esser Co.*)

certain general features of European theodolites are fairly common. The weight (without tripod) is about 10 lb as compared with about 15 lb for the conventional American transit; the difference is due both to compact size and to the use of light materials in manufacture. The finish is light in color in order to minimize temperature effects of sunlight. There is no vernier plate, and the center is cylindrical and rotates on ball bearings. The plate level is circular and of moderate sensitivity, and the telescope level is tubular and of high sensitivity. There are three leveling screws. An optical plummet is used instead of a plumb line. All motions are enclosed, so that the instrument is dustproof and moistureproof. Interior lighting is accomplished by means of mirrors and prisms, and provision is made for night lighting. The only field adjustments are those of the levels, which adjustments are similar to those for the conventional American transit.

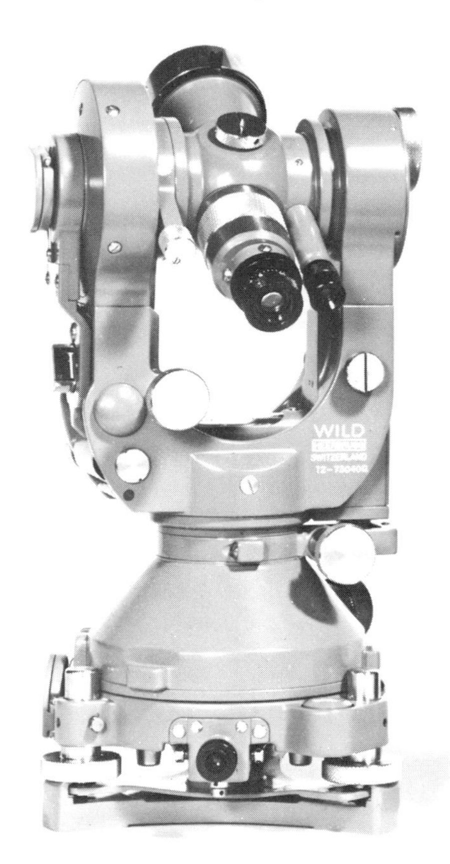

Fig. 9-5*b*. Wild direction theodolite.

The telescope of the European theodolite is short and can be rotated completely about the horizontal axis. It is of the internal-focusing type, so that the stadia constant (Art. 11-3) is zero. The reticule is of glass with etched crosslines, and the stadia interval factor (Art. 11-3*a*) is fixed at 100.0. For European use the eyepiece is of the inverting type; for American use it is usually of the erecting type.

The graduated horizontal and vertical circles are of glass and are relatively small; they are viewed simultaneously from the eye end of the telescope by means of a system of microscopes, prisms, and mirrors. Likewise the levels are viewed from the eye end of the instrument; it is not necessary for the observer to move around during an observation. The horizontal circle is read by means of an optical micrometer, in accordance with systems devised by the various manufacturers. Each reading is obtained by one observation which is the average of two readings at opposite sides of the circle and which is therefore free from errors due to eccentricity. Direction theodolites are usually read directly to 01″ and repeating theodolites to 20″ or 01′ and are read by estimation to one-tenth of the corresponding direct reading.

USE OF THE TRANSIT

9-7 General. The succeeding articles describe the elementary processes employed in running lines and in measuring horizontal and vertical angles with the transit. Transit surveys are considered in detail in the following chapter.

The process of taking magnetic bearings with the transit is the same as with the surveyor's compass. The transit may be employed for running direct levels in the same manner as with the engineer's level, the telescope level bubble being centered each time a rod reading is taken.

The operation of reversing the telescope by rotating it about the horizontal axis is called "plunging the telescope." When the telescope level tube is below, the telescope is said to be in the *normal,* or *direct,* position; when the level tube is above, the telescope is said to be in the *inverted,* or *reversed,* position.

Signals generally applicable to transit work are given in Art. 1-9. Suggestions for the care and handling of the transit are given in Art. 1-7.

9-8 Setting Up Transit. Ordinarily the transit is set over a definite point, such as a tack in a stake. First the transit is placed approximately over the point. Each tripod leg is then moved as required to bring the plumb bob within about ¼ in. of being over the tack, with the foot plate nearly level and with the shoe of each tripod leg pressed firmly into the ground. The instrument is leveled approximately by means of the leveling screws. Then two adjacent leveling screws are loosened, and the instrument is shifted laterally until the plumb bob is exactly over the tack. The length of the plumb line is changed as necessary to make the bob just clear the tack. The leveling screws are tightened to a firm, but not tight, bearing. The instrument is leveled by means of the leveling screws and the plate levels, each level tube being first brought parallel to a pair of opposite leveling screws. Both bubbles are brought approximately to center, and then each bubble is centered carefully. The telescope is tested for parallax (see Focusing, Art. 4-10) before observations are begun.

The operation of setting up and leveling the transit expeditiously requires on the part of the instrumentman a skill that is acquired only with practice.

Just before the transit is taken up, the instrument is centered on the foot plate, the leveling screws are roughly equalized, the upper motion is clamped, the lower motion is either unclamped or is clamped lightly, and the telescope is pointed vertically up and is clamped lightly.

9-9 Measuring Horizontal Angle. If a horizontal angle, as *AOB,* is to be measured, the transit is set up over *O*. The upper motion is clamped, with one of the horizontal verniers near zero, and by means of the upper tangent screw one vernier is set at 0°. The telescope is sighted approximately to *A*, the lower motion is clamped, and by turning the lower tangent screw the line of sight is set exactly on a range pole or other object marking the point. The upper clamp is loosened, and the telescope is turned until the line of sight cuts *B*. The upper clamp is tightened, and the line of sight is set exactly on *B* by turning the upper tangent screw. The reading of the vernier which was initially set at 0° gives the value of the angle.

A horizontal angle is laid off in similar manner, except that after the sight is taken on *A* the angle is set off on the circle, and the point *B* is established on the line of sight.

Following is a list of suggestions:

1 Make reasonably close settings by hand so that the tangent screws will not need to be turned through more than one or two revolutions.

2 Make the last movement of the tangent screw clockwise, thus compressing the opposing spring (see parts 20 to 26, Fig. 4-7).

3 When reading the vernier, have the eye directly over the coinciding graduation, to avoid parallax. It is also helpful to observe that the graduations on both sides of those coinciding fail to concur by the same amount.

4 As a check on the reading of one vernier, the other vernier may be read also.

5 The plate bubbles should be centered before measuring an angle, but between initial and final settings of the line of sight the leveling screws should not be disturbed. When an angle is being measured by repetition (Art. 9-10), the plate may be releveled after each turning of the angle before again sighting on the initial point.

6 When a number of angles are to be observed from one point without moving the horizontal circle, the instrumentman should sight at some clearly defined object that will serve as a reference mark and should observe the angle. If occasionally the angle to the reference mark is read again, any accidental movement of the horizontal circle will be detected.

7 Whenever an angle is doubled, if the instrument is in adjustment the two readings should not differ by more than the least count of the vernier.

A greater discrepancy, if confirmed by repeating the measurement, will indicate that the instrument is out of adjustment.

Common Mistakes. In measuring horizontal angles, mistakes often made are:

1 Turning the wrong tangent screw.

2 Failing to tighten the clamp.

3 Reading numbers on the horizontal scale from the wrong row.

4 Reading angles in the wrong direction.

5 Dropping 30′, with a circle graduated to 30′; for example, calling an angle 21°14′ when it is actually 21°44′.

6 Reading the vernier in the wrong direction.

7 Reading the wrong vernier.

9-10 Measuring Angle by Repetition. By means of the transit, a horizontal angle may be mechanically multiplied and the product can be read with the same precision as the single value. The precision increases directly with the number of repetitions up to six or eight; beyond this number the precision is not appreciably increased by further repetition on account of lost motion in the instrument and on account of accidental errors such as those due to setting the line of sight.

To repeat an angle, as *AOB*, the transit is set up at *O* and the single value of the angle is observed as previously described. The vernier setting is left unaltered, the instrument is turned on its lower motion, and a second sight is taken to the first point, as *A*. The upper clamp is loosened, and the telescope is again sighted to *B*. The angle has now been doubled. In this way the process is continued until the angle has been multiplied the desired number of times. The vernier is read, and the value of the angle is determined by dividing the difference between initial and final readings by the number of times the angle was turned.

Usually the angle is multiplied four to ten times, half the observations being made with the telescope normal and half with it inverted. Both verniers are read, and the mean values are used in the computations.

Sample notes for measuring the angles about a point by repetition are shown in Fig. 9-6. For each angle, five "repetitions" are taken with telescope normal and five with telescope inverted, always measuring clockwise. The vernier is set at zero at the beginning but not thereafter; the error of closure (called the "horizon closure") is thus obtained directly as a check on the computations, and errors in setting the vernier are avoided. Rough computations on the right-hand page serve as a check on the number of repetitions and detect appreciable mistakes in turning the

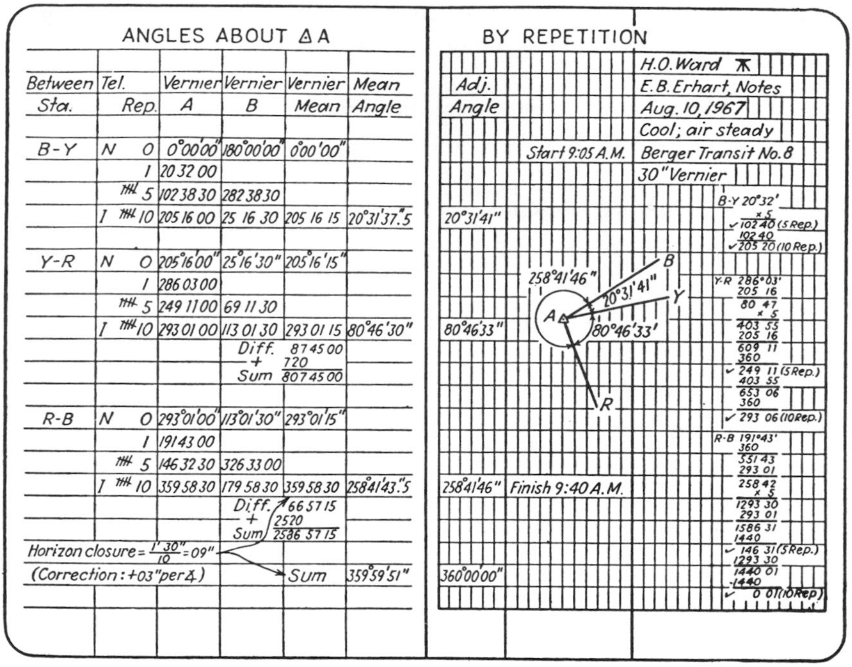

ANGLES ABOUT △A

Between Sta.	Tel. Rep.	Vernier A	Vernier B	Vernier Mean	Mean Angle
B-Y	N 0	0°00'00"	180°00'00"	0°00'00"	
	1	20 32 00			
	𝍸 5	102 38 30	282 38 30		
	I 𝍸 10	205 16 00	25 16 30	205 16 15	20°31'37".5
Y-R	N 0	205°16'00"	25°16'30"	205°16'15"	
	1	286 03 00			
	𝍸 5	249 11 00	69 11 30		
	I 𝍸 10	293 01 00	113 01 30	293 01 15	80°46'30"
			Diff.	87 45 00	
			+	720	
			Sum	807 45 00	
R-B	N 0	293°01'00"	113°01'30"	293°01'15"	
	1	191 43 00			
	𝍸 5	146 32 30	326 33 00		
	I 𝍸 10	359 58 30	179 58 30	359 58 30	258°41'43".5
			Diff.	66 57 15	
			+	2520	
			Sum	2586 57 15	
Horizon closure = 1' 30"/10 = 09"					
(Correction: +03" per ∡)				Sum	359°59'51"

BY REPETITION

Adj. Angle	
20°31'41"	
80°46'33"	
258°41'46"	Finish 9:40 A.M.
360°00'00"	

Start 9:05 A.M.

H.O. Ward ⊼
E.B. Erhart, Notes
Aug. 10, 1967
Cool; air steady
Berger Transit No. 8
30" Vernier

258°41'46"
20°31'41"
80°46'33'
A B Y R

B-Y 20°32'
× 5
✓ 102 40 (5 Rep.)
102 40
✓ 205 20 (10 Rep.)

Y-R 286°03'
205 16
80 47
× 5
403 55
205 16
609 11
360
✓ 249 11 (5 Rep.)
403 55
653 06
360
✓ 293 06 (10 Rep.)

R-B 191°43'
360
551 43
293 01
258 42
× 5
1293 30
293 01
1586 31
1440
✓ 146 31 (5 Rep.)
1293 30
1440 01
-1440
✓ 0 01 (10 Rep.)

Fig. 9-6. Notes for measuring angles by repetition.

wrong tangent screw. The recorded value for five repetitions is used only as a check; and the *B* vernier is used only as a check, except with regard to the number of seconds. The final adjusted values of the angles (to the nearest second) are recorded on the sketch for ready reference in further computations.

9-11 Laying Off Angle by Repetition.

If it is desired to establish an angle *AOB* from a fixed line *OA* with a precision greater than that possible by a single observation, the transit is set up at *O*, and the angle is laid off by a single observation to establish a trial point *B'*. The angle *AOB'* is then measured by repetition, and the line *OB'* is measured. The angle *AOB'* must be corrected by an angular amount *B'OB* to establish the correct angle *AOB*. The correction, which is too small to be laid off accurately by angular measurement, is applied by offsetting the distance $B'B = OB' \tan$ (or sin) $B'OB$, thus establishing the point *B* beside *B'*. As a check, the angle *AOB* is measured by repetition.

9-12 Measuring Vertical Angle.

The vertical angle to a point is its angle of elevation (+) or depression (−) from the horizontal. The transit is set up and leveled as when measuring horizontal angles.

For a transit having a fixed vertical vernier, the plate bubbles should be centered carefully. The telescope is sighted approximately at the point, and the horizontal axis is clamped. The horizontal cross hair is set exactly on the point by turning the telescope tangent screw, and the angle is read by means of the vertical vernier.

For a transit having a movable vertical vernier with control level, the telescope is sighted on the point as described above, the vernier control bubble is centered, and the angle is read.

In ordinary trigonometric leveling, vertical angles are taken by sighting usually at a leveling rod, the line of sight being directed at a rod reading equal to the height of the horizontal axis of the transit above the station over which the transit is set up. Sights may be taken to points defined by signals erected at the distant stations.

For leveling with the transit, for astronomical observations, or for measurement of horizontal angles requiring steeply inclined sights, usually it is desired to level the transit with greater precision than that which is possible through the use of the plate levels. In such cases the vertical axis is made truly vertical by means of the telescope level, the plate levels being disregarded.

9-13 Double-sighting For a transit having a full vertical circle, sights to determine vertical angles can be taken with the telescope either normal or inverted. The method of *double-sighting* consists in reading once with the telescope normal and once with it inverted, and taking the mean of the two values thus obtained. It eliminates the effect of certain instrumental errors and reduces the personal error of observation.

The method of double-sighting is used, for example, in astronomical observations and in similar measurements of vertical angles to distant objects. In traversing, a similar result is obtained by measuring the vertical angle of each traverse line from each end, with the telescope the *same side up* for the two observations, and taking the mean of the two values.

9-14 Index Error. *Index error* is the error in an observed angle due to (1) lack of parallelism between the line of sight and the axis of the telescope level, (2) displacement (lack of adjustment) of the vertical vernier, and/or (3) for a transit having a fixed vertical vernier, inclination of the vertical axis. If the instrument were in perfect adjustment and were leveled perfectly for each observation, there would be no index error; however, in practice these conditions seldom exist.

The effect of index errors due to lack of adjustment of the instrument can be eliminated either by double-sighting for each observation or by applying to each observation a correction determined (by double-sighting) for the instrument in its given condition of adjustment. For the common type of transit having a fixed vertical vernier, the effect of imperfect leveling cannot be eliminated by double-sighting, but—provided the line of sight is in adjustment—for each direction of pointing a correction can be determined (as described later) and applied. Often it is more convenient to apply the correction than to ensure that the instrument is perfectly adjusted and leveled.

The index correction is equal in amount but opposite in sign to the index error. Thus, if the observed vertical angle is $+12°14'$ and if the index error is determined to be $+02'$, the correct value of the angle is

$$+12°14' - 02' = +12°12'$$

Methods of determining the index error (and, therefore, the correction) are given in the following paragraphs.

1 Lack of Parallelism between Line of Sight and Axis of Telescope Level. If the axis of the telescope level is not parallel to the line of sight and if the vertical vernier reads zero when the bubble is centered (Fig. 9-7*a*), an error in vertical angle results. This error can be rendered

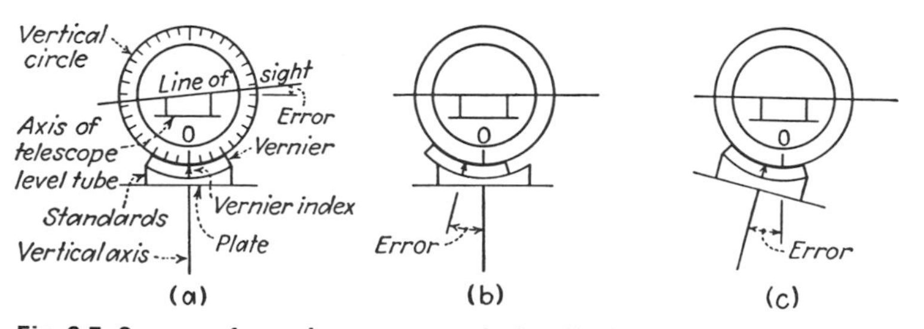

Fig. 9-7. Sources of error in measurement of vertical angles.

negligible for ordinary work by careful adjustment of the instrument (Art. 9-23, adjustment 5). The combined error due to this cause and to displacement of the vertical vernier (see following paragraph) can be eliminated by double-sighting. The index error due to the two causes can be determined by comparing a single reading on any given point with the mean of the two readings obtained by double-sighting to the same point. Thus, if the observed vertical angle to a point is $+2°58'30''$ with telescope normal and is $+2°55'30''$ with telescope reversed, the index error for readings with telescope normal is

$$(2°58'30'' - 2°55'30'')/2 = +1'30''$$

2 Displacement of Vertical Vernier. Displacement of the vertical vernier (Fig. 9-7*b*) introduces a constant index error. The error can be rendered negligible by careful adjustment (Art. 9-23, adjustments 6 and 6*a*). The combined index error due to this cause and to lack of parallelism between the line of sight and the axis of the telescope level can be eliminated by double-sighting; or the combined error can be determined as described in the preceding paragraph. For a transit having a fixed vertical vernier, the error due to displacement of the vertical vernier alone can be determined—provided the line of sight is in adjustment—by leveling the transit carefully, leveling the telescope, and reading the vertical vernier. For a transit having a movable vertical vernier with control level, the error due to displacement of the vertical vernier alone can be determined by leveling both the telescope level and the vernier level, and reading the vertical vernier.

3 *Inclination of Vertical Axis.* For a transit having a fixed vertical vernier, any inclination of the vertical axis (Fig. 9-7*c*) due to erroneous leveling of the instrument introduces an index error which varies with the direction in which the telescope is pointed and which is equal in amount to the angle through which the fixed vertical vernier is displaced about the horizontal axis while the instrument is directed toward the point. This index error can be rendered negligible by careful leveling of the transit before each observation, making sure that the plate-level bubbles remain in position for any direction of pointing. It is not eliminated by double-sighting, since the condition causing the error is not changed by reversal (and plunging) of the instrument (see Fig. 9-7*c*). If the line of sight and the vertical vernier are in adjustment, the index error due to inclination of the vertical axis alone can be determined for each direction of pointing by leveling the telescope and reading the vertical vernier.

For a transit having a movable vertical vernier with control level, any moderate inclination of the vertical axis does not introduce an appreciable error in vertical angles, provided the instrument is in adjustment and provided the vernier control-level bubble is centered each time an observation is made. On topographic surveys or simliar work where many horizontal and vertical angles are to be observed, the use of the movable vertical vernier with control level results in a considerable saving of time.

9-15 Prolonging Straight Line.

A straight line as *AB* (Fig. 9-8) may be prolonged by any of the following three methods. The second is most common and usually most convenient. Lines may also be prolonged by means of a prismatic sighting device.

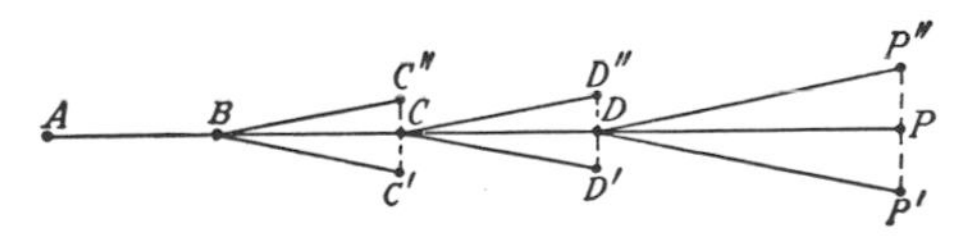

Fig. 9-8. Double-sighting to prolong line.

Method 1. The transit is set up at A, a sight is taken to B, and a point C is established on line beyond B. The transit is moved to B, a point D is set in similar manner beyond C, and so on until the desired distance is traversed.

Method 2. The transit is set up at B, and a backsight is taken to A. With both upper and lower motions clamped, the telescope is plunged, and a point C is set on line. The transit is moved to C, a backsight is taken to B, a point D is set in similar manner, and so on.

If the line of sight is not perpendicular to the horizontal axis of the transit, C will not lie on the true prolongation of AB but to one side, as at C'; the angular deflection of BC' from the correct prolongation of AB will be double the error of adjustment of the line of sight. If the line is extended by this method and all backsights are taken with the telescope in one position (either normal or inverted), the points established will lie along a curve instead of a straight line. To correct approximately for such a condition, backsights may be taken with the telescope alternately normal and inverted.

Method 3. This method, known as "double-sighting," or "double-centering," is employed if the instrument is in poor adjustment or if it is desired to establish the line with high precision. The transit is set up at B, and a backsight is taken to A with the telescope in, say, its *normal* position. The telescope is plunged, and a point C' is set on line. The transit is then revolved about its vertical axis, and a second backsight is taken to A with the telescope *inverted*. The telescope is plunged, and a point C'' is established on line beside C'. Midway between C' and C'' a point C is set defining a point on the correct prolongation of AB. The line may be prolonged farther in similar manner.

9-16 Prolonging Line Past Obstacle. Figure 9-9 illustrates one method of prolonging a line AB past an obstacle where the offset space is limited. The transit is set up at A, a right angle is turned, and a point C is estab-

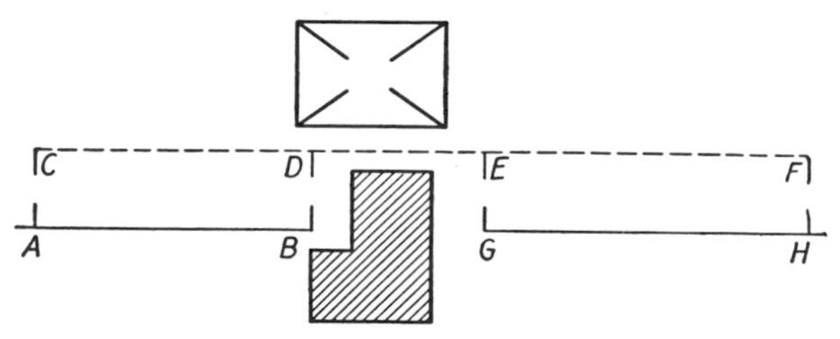

Fig. 9-9. Prolonging line past obstacle by perpendicular offsets.

lished at a convenient distance from A. Similarly the point D is established, the distance BD being made equal to AC. The line CD, which is parallel to AB, is prolonged; and points E and F are established in convenient locations beyond the obstacle. From E and F right-angle offsets are made, and G and H are set as were C and D; GH then defines the prolongation of AB. If precise results are desired, it is necessary to erect the perpendiculars and measure the offset distances carefully, and that the distances between the two offsets on each side of the obstacle be as long as practicable.

Another method of prolonging a line past an obstacle is to follow a zigzag course, preferably symmetrical, until it is possible to resume traveling on the direct prolongation of the line. This method requires less measurement but more computation than the method of perpendicular offsets. If the deflection angles are small, say not greater than a degree or so, it will be sufficiently precise to take the distance along the main line as equal to that along the auxiliary lines.

9-17 Running Straight Line between Two Points. Various methods are employed to establish intervening points on the straight line joining two fixed terminals A and B.

Case 1. Terminals Intervisible. The transit is set up at A, a sight is taken to B, and intervening points are established on line. Sometimes the intervening points are set by double-sighting.

Case 2. Terminals Not Intervisible, but Visible from an Intervening Point on Line. The location of the line at the intervening point C is determined by trial, as follows: The transit is set up on the estimated location of the line near C, a backsight is taken to A, the telescope is plunged, and the location of the line of sight at B is noted. The amount that the transit must be shifted laterally is estimated; and the process is repeated until, when the telescope is plunged, the line of sight falls on the point at B. This process is known as "balancing in." The location of the transit should then be tested by double-sighting; the test will also disclose whether or not the line of sight and the horizontal axis are in adjustment.

Case 3. Terminals Not Visible from Any Intermediate Point. By one of the methods previously explained, a straight line, called a *random line*, is run from A in the estimated direction of B. In Fig. 9-10, AX is such a

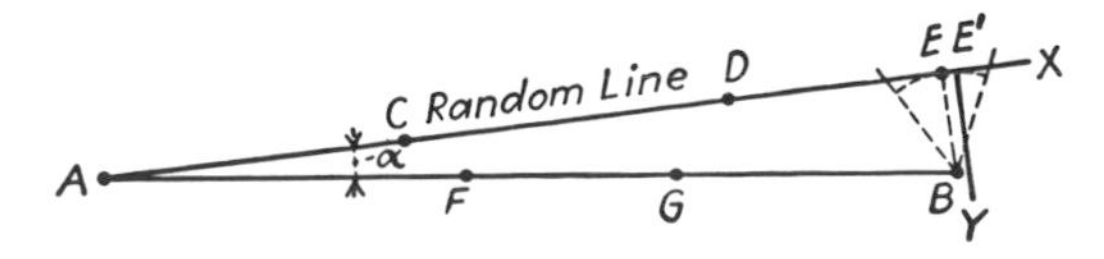

Fig. 9-10. Random line.

random line, and C and D are stations established during the process of extending it. The transit is set up at D and sighted toward X. The tape is swung about B as a center, and the offset distance BE from the point B to the line DX is determined by sighting at the tape through the transit and taking the least reading (that is, by a swing offset).

The length AE is then measured. If the offset distance BE is relatively short, the location of E on the line may be estimated with sufficient precision. If the offset distance is long, a perpendicular $E'Y$ is erected at the estimated location of E, and E is located by trial (Art. 3-14) so that a perpendicular at E will pass through B.

The angle α which the random line makes with AB may be computed by the equation $\tan \alpha = BE/AE$. The transit is again set up at A; a sight is taken to C; the computed value of α is laid off; and the line is run toward B, intermediate stations as F and G being established at desired points. With the transit at G, a backsight is taken to F, the telescope is plunged, and the linear offset error at B is noted. If the error is sufficiently large to be of importance, the points at F and G are corrected by linear measurements so as to place them on the true line, the correction being made proportional to the distance from the terminal A to the point. Thus the offset correction at F is to AF as the error at B is to AB.

It is not necessary to compute the angle α, as the direction of AB can be established by an offset from the random line at any convenient distance from A. As α is usually very small, either of the approximate slope formulas (Art. 3-7) will usually yield sufficiently precise values of the length of AB.

9-18 Determining Inaccessible Distance. This determination involves triangulation (see Art. 8-13). The usual case is illustrated in Fig. 9-11, in which AB cannot be measured directly and in which B can be seen from A and vicinity. The transit is set up at A, a sight is taken to B, an angle of 90° is laid off, and C is set at any convenient point. The line AC is measured. The transit is set up at C, and the angle ACB is observed. Then $AB = AC \tan ACB$.

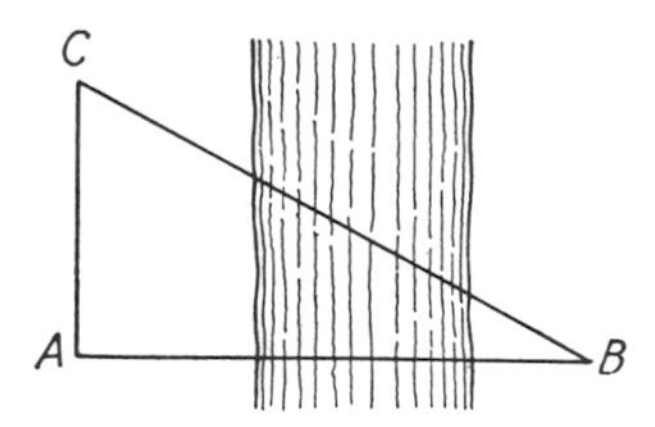

Fig. 9-11. Inaccessible distance.

9-19 Intersection of Lines. The point of intersection of two lines as ab and cd (Fig. 9-12) is established as follows: One of the lines ab is prolonged (Art. 9-15), and points e and f are established a short distance on opposite sides of the estimated location of the prolongation of *cd*. A string is stretched between *e* and *f*. The line *cd* is prolonged until it intersects the string at *g*.

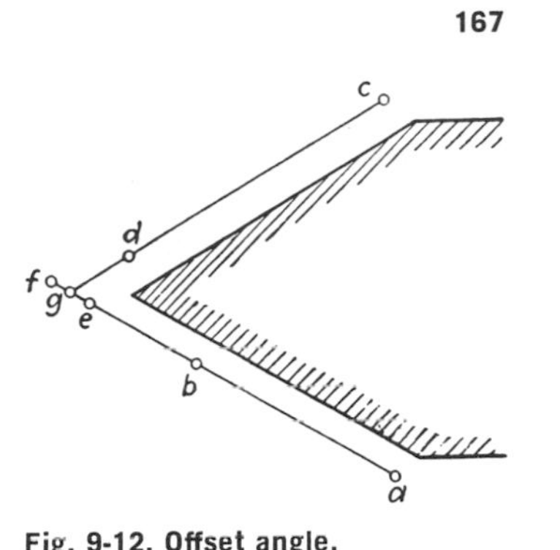

Fig. 9-12. Offset angle.

9-20 Setting Monument. Often it is desired to set a subsurface monument to mark permanently a point on a transit survey. In such a case the location of the surface point established by the survey is well referenced (Art. 10-12), preferably by the intersection of two lines. When the monument is set and the subsurface mark is to be established, a string is stretched along each of the two reference lines. The location of the mark on the monument is projected below the surface by plumbing from the intersection of the two strings. If desired, a batter board or other frame may be set over the surface mark for the purpose of plumbing from a surface mark to a subsurface mark, or *vice versa*.

9-21 Measuring Angle When Transit Cannot Be Set at Vertex. Figure 9-12 illustrates a typical case where it is desired to determine the angle between walls of a building or between fence lines.

The point *a* is established at any convenient distance from the wall. The perpendicular distance from the wall is determined by holding the tape on point *a* and swinging the end of the tape through an arc, varying the radius until the arc becomes tangent to the wall (that is, by a swing offset). Similarly a second point *b* is established at the same distance from the wall as *a;* then *ab* is parallel to the wall. In similar manner points *c* and *d* are established. The point of intersection *g* of lines *ab* and *cd* is determined as described in Art. 9-10. The transit is set up at *g,* and the angle *agc,* which is equal to the angle between the walls, is measured in the usual manner.

ADJUSTMENT OF CONVENTIONAL TRANSIT

9-22 Desired Relations. Much of Arts. 4-23 and 4-24 concerning adjustment of the level applies equally well to the transit. Some general suggestions regarding adjustments are given in Art. 1-8.

For a transit in perfect adjustment the relations shown in Fig. 9-13 should exist, as well as the following (see adjustments 6 and 6*a,* later in this article): (6) If the transit has a fixed vernier for the vertical circle, the vernier should read zero when the plate bubbles and telescope bubble are centered; and (6*a*) If the vertical vernier is movable and has a control level, the axis of the control level should be parallel to that of the telescope level when the vernier reads zero. The transit adjustments commonly made are adjustments 1 to 6*a*, given herein.

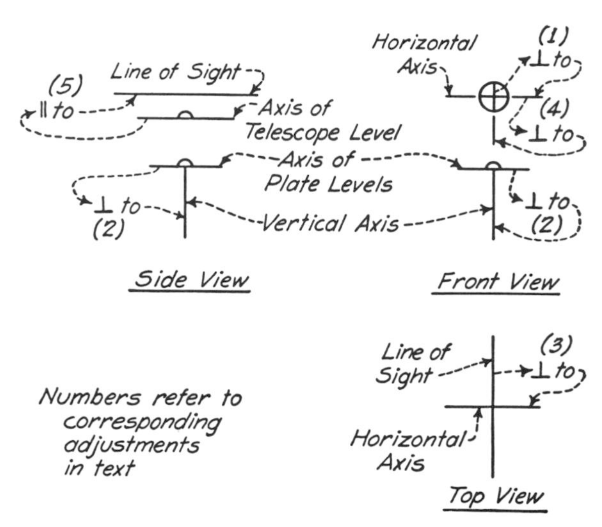

Fig. 9-13. Desired relations among principal lines of transit.

Still other relations are implied by the four additional adjustments which are listed at the end of this article. Herein, however, it is assumed that the line of sight coincides with the optical axis; that the objective slide is permanently fixed in the telescope tube with respect to transverse motion; that the optical axis and the axis of the objective slide coincide and are perpendicular to the horizontal axis; that the intersection of the cross hairs is in the center of the field of view of the eyepiece; and that the transit is not equipped with a striding level.

9-23 Adjustments

1 To Make the Vertical Cross Hair Lie in a Plane Perpendicular to the Horizontal Axis. Sight the vertical cross hair on a well-defined point not less than 200 ft away, and swing the telescope through a small vertical angle. If the point appears to depart from the cross hair, loosen two adjacent capstan screws and rotate the cross-hair ring in the telescope tube until the point traverses the entire length of the hair. Tighten the same two screws. This adjustment is similar to adjustment 2 of the dumpy level (Art. 4-23), with the terms *vertical* and *horizontal* interchanged.

2 *To Make the Axis of Each Plate Level Lie in a Plane Perpendicular to the Vertical Axis.* Rotate the instrument about the vertical axis until each level tube is parallel to a pair of opposite leveling screws. Center the bubbles by means of the leveling screws. Rotate the instrument end for end about the vertical axis. If the bubbles become displaced, bring them *halfway* back by means of the adjusting screws. This is the method of reversion (Art. 4-23, adjustment 1).

3 *To Make the Line of Sight Perpendicular to the Horizontal Axis.* Level the instrument. Sight on a point *A* (see Fig. 9-14) about 500 ft

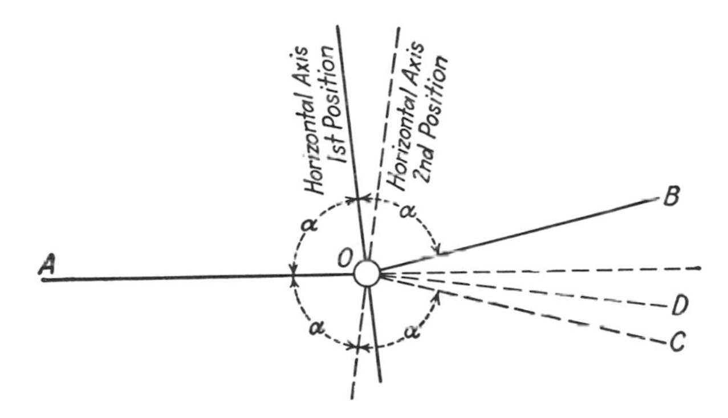

Fig. 9-14. Adjustment of line of sight of transit.

away, with telescope normal. With both horizontal motions of the instrument clamped, plunge the telescope and set another point *B* on the line of sight and about the same distance away on the opposite side of the transit. Unclamp the upper motion, rotate the instrument end for end about the vertical axis, and again sight at *A* (with telescope inverted). Clamp the upper motion. Plunge the telescope as before; if the line of sight does not fall on *B*, set a point *C* on the line of sight beside *B*. Mark a point *D*, *one quarter* of the distance from *C* to *B*, and adjust the cross-hair ring (by means of the two opposite horizontal screws) until the line of sight passes through *D*. The points sighted should be at about the same elevation as the transit.

4 *To Make the Horizontal Axis Perpendicular to the Vertical Axis.* Set up the transit near a building or other object on which is some well-defined point *A* at a considerable vertical angle. Level the instrument very carefully, thus making the vertical axis truly vertical. Sight at the high point *A* (see Fig. 9-15) and, with the horizontal motions clamped, depress the telescope and set a point *B* on or near the ground. If the horizontal axis is perpendicular to the vertical axis, *A* and *B* will be in the same vertical plane. Plunge the telescope, rotate the instrument end for end about the vertical axis, and again sight on *A*. Depress the telescope as before; if the line of sight does not fall on *B*, set a point *C* on the

line of sight beside *B*. A point *D*, halfway between *B* and *C*, will lie in the same vertical plane with *A*. Sight on *D;* elevate the telescope until the line of sight is beside *A;* loosen the screws of the bearing cap; and raise or lower the adjustable end of the horizontal axis until the line of sight is in the same vertical plane with *A*.

5 *To Make the Axis of the Telescope Level Parallel to the Line of Sight.* Proceed the same as for the two-peg adjustment of the dumpy level (Art. 4-23, adjustment 3), except as follows: With the line of sight set on the rod reading established for a horizontal line, the correction is made by raising or lowering one end of the telescope level tube until the bubble is centered.

6 (For Transit Having a Fixed Vertical Vernier) *To Make the Vertical Circle Read Zero When the Telescope Bubble Is Centered.* With the plate bubbles centered, center the telescope bubble. If the vertical vernier does not read zero, loosen it and move it until it reads zero. Care should be taken that the vernier will not bind on the vertical circle.

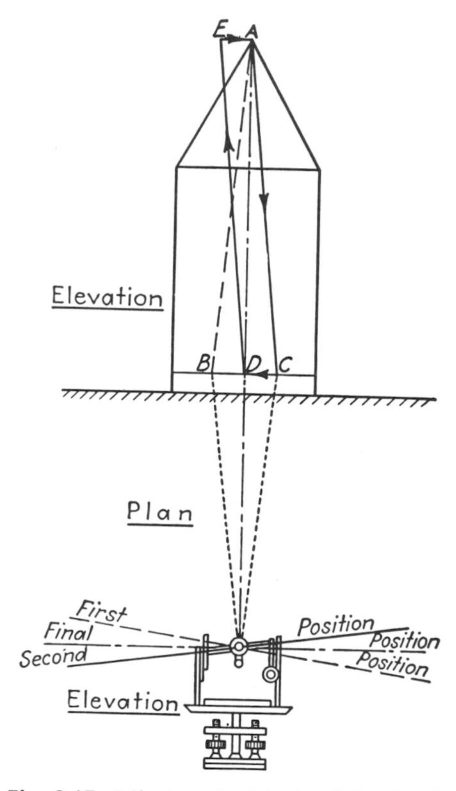

Fig. 9-15. Adjustment of horizontal axis of transit.

6a (For Transit Having a Movable Vertical Vernier with Control Level) *To Make the Axis of the Auxiliary Level Parallel to the Axis of the Telescope Level When the Vertical Vernier Reads Zero.* Center the telescope bubble, and by means of the vernier tangent screw move the vertical vernier until it reads zero. By means of the capstan screws adjust the level tube attached to the vertical vernier until the bubble is centered.

Other Adjustments. The following additional adjustments are described in more detailed texts on surveying, such as Ref. 10:

7 To make the line of sight, insofar as defined by the horizontal cross hair, coincide with the optical axis.

8 (For transit having an adjustable objective slide) To make the axis of the objective slide perpendicular to the horizontal axis.

9 (For transit having an adjustable eyepiece slide) To center the eyepiece slide.

10 (For transit equipped with striding level) To make the axis of the striding level parallel to the horizontal axis.

ERRORS IN TRANSIT WORK

9-24 Errors. Except in field astronomy, a measured angle is always closely related to a measured distance. On surveys of ordinary precision it usually requires much more care to keep *linear* errors within prescribed limits than to maintain a corresponding degree of *angular* precision. Often undue attention is paid to securing precision in angular measurements, and at the same time large and important errors in the measurement of distances are overlooked.

Errors in transit work may be instrumental, personal, or natural. Although a detailed discussion of the errors in transit work is beyond the scope of this book, the following observations may be made:

1 *Instrumental Errors*. The adjustments, even though carefully made, are never exact. Likewise the graduations are not perfect, and the centers are not absolutely true.

Errors due to instrumental imperfections and/or nonadjustment are all systematic, and they can be either eliminated or reduced to a negligible amount by proper methods of procedure.

a. Horizontal angles. When the bubbles of plate levels in nonadjustment are centered, the vertical axis is inclined, and hence measured angles are not truly horizontal angles. Also the horizontal axis is inclined to a varying degree depending upon the direction in which the telescope is sighted; hence a variable systematic error in horizontal angle results. For a given direction of sighting, the larger the vertical angle, the greater the error in direction. The error cannot be eliminated by double-sighting. Where the vertical angles are large, even for surveys of ordinary precision, it is important that the vertical axis be made truly vertical, either by proper adjustment and use of the plate levels or by means of the telescope level.

If the line of sight is not perpendicular to the horizontal axis, an error in horizontal angle results. If the telescope is plunged between backsight and foresight, the resultant error is doubled. The angular error E in the observed direction of any line along with a sight is taken depends both on the angle e by which the line of sight departs from the perpendicular to the horizontal axis and on the observed vertical angle a to the point sighted. It can be shown (Ref. 10) that

$$E = e \sec a \text{ (approx.)} \tag{1}$$

The error, which is of consequence for steeply inclined sights or when the telescope is plunged, can be eliminated by double-sighting.

If the horizontal axis is not perpendicular to the vertical axis, an error in horizontal angle results (except when the points sighted are at the same vertical angle). The angular error θ in the observed direction of any line along which a sight is taken depends both on the angle e' by which the horizontal axis departs from the perpendicular to the vertical axis and on the observed vertical angle a to the point sighted. It can be shown (Ref. 10) that

$$\theta = e' \tan a \text{ (approx.)} \tag{2}$$

The error, which becomes large for steeply inclined sights, can be eliminated by double-sighting.

Errors due to eccentricity of verniers and/or eccentricity of centers can be eliminated by taking the means of readings indicated by opposite verniers.

b. Vertical angles. Instrumental errors in vertical angle result whenever (1) the line of sight is not parallel to the axis of the level tube, (2) the vertical vernier is not in adjustment, and/or (3) the plate levels are not in adjustment. Methods of reducing or eliminating these errors are described in Arts. 9-12 to 9-14.

2 *Personal Errors.* Personal errors arise from the limitations of the human eye in setting up and leveling the transit and in making observations. The transit may not be set up exactly over the station; the plate bubbles may not be centered exactly; the verniers may not be set or read precisely; parallax may exist in focusing; and the line of sight may not be directed exactly on the point. These personal errors form a large part of the resultant error in transit work. All the personal errors are accidental and hence cannot be eliminated. They can be kept within reasonable limits by care in observing. However, time should not be wasted by needless care in such matters as setting up exactly over a station when sights are long or centering the plate bubbles exactly when sights are nearly horizontal and horizontal angles only are being observed.

3 *Natural Errors.* Sources of natural errors are (*a*) settlement of the tripod; (*b*) unequal atmospheric refraction; (*c*) unequal expansion of parts of the telescope due to temperature changes; and (*d*) wind, producing vibration of the transit or making it difficult to plumb correctly.

In general, the errors resulting from natural causes are not large enough to affect appreciably the measurements of ordinary precision. However, when the transit is set up on boggy or thawing ground, large errors are likely to arise from settlement, usually accompanied by horizontal and angular displacement. Errors due to adverse atmospheric conditions can usually be rendered negligible by choosing appropriate times for observing.

9-25 Precision of Angular Measurements. Many factors influence the angular precision of transit work, and no rigid rules can be formulated to ensure a required precision. The values given below represent, in a general way, the *maximum* error likely to occur in measuring a horizontal angle under average conditions of practice, instruments being in fair condition and in fair adjustment except as otherwise stated. The *average* angular error will be materially less. Also, as the errors are largely accidental, the resultant error in the sum of a series of measured angles may be expected to vary as the square root of the number of angles involved.

Case 1. Short sights, point indicated by range pole obscured near ground. Range pole plumbed by eye. Single observation of angle. Maximum error 02′ to 04′.

Case 2. Long sights, but otherwise as stated for case 1. Maximum error 01′ to 02′.

Case 3. Unobscured but steeply inclined sights; no special attention given to making horizontal axis truly horizontal; single measurement of angle. Maximum error 01′ to 02′.

Case 4. Unobscured sights on well-defined points; sights not steeply inclined. Single observation of angle, vernier reading to minutes. Maximum error 30″ to 01′.

Case 5. As for case 4, but transit in excellent condition and in good adjustment. Angles estimated to ½ min. Maximum error 20″ to 30″.

Case 6. As for case 4 but angle doubled, the telescope being plunged between sights. Maximum error 15″ to 30″.

Case 7. Unobscured sights on well-defined points. Sights not steeply inclined. Verniers reading to 30″. Single observation of angle represented by mean of readings of both verniers. Transit in excellent condition and in good adjustment. Maximum error 15″ to 30″.

Case 8. As for case 7, but verniers reading to 10″. Also instrument set up with great care. Maximum error 10″ to 15″.

Case 9. Unobscured sights on well-defined points. Instrument set up with great care. Sights not steeply inclined. Transit in excellent condition and adjustment. Verniers reading to 30″. Angles repeated six times with telescope normal and six times with it inverted. Maximum error 02″ to 04″.

Case 10. As above, but transit reading to 10″. Observations taken at favorable times. Maximum error 01″ to 02″.

9-26 Numerical Problems

1 Thirty spaces on a transit vernier are equal to 29 spaces on the graduated circle, and 1 space on the circle is 15′. What is the least count of the vernier?

2 A transit for which the circle is graduated 0° to 360° clockwise is used to measure an angle by 10 clockwise "repetitions," 5 with telescope normal and 5 with telescope inverted. Compute the most probable value of the angle from the following data:

Telescope	*Reading*	*Vernier A*	*Vernier B*
Normal	Initial	48°46′	228°46′
Normal	After first turning	161°09′	......
Inverted	After tenth turning	92°41′	272°42′

3 In laying out the lines for a building, a 90° angle was laid off as precisely as possible with a 01′ transit. The angle was then measured by repetition and found to be 89°59′40″. What offset should be made at a distance of 250 ft from the transit to establish the true line?

4 The following observations were made to determine an index correction: Vertical angle to point $A = +7°16'$ with telescope direct and $+7°14'$ with telescope inverted. Compute the index correction for observations with telescope direct.

5 A vertical angle measured by a single observation is $-12°02'$, and the index error is determined to be $+06'$. What is the correct value of the angle?

6 A line *AB* is prolonged to *F* by setting up the transit at succeeding points *B, C, D,* and *E;* backsighting to *A, B, C,* and *D,* respectively; and plunging the telescope. If the line of sight made an angle of 10″ with the normal to the horizontal axis and the procedure were such that each backsight was taken with the telescope normal, what would be the angular error in the segment *EF?* What would be the offset error (approximate) in the position of *F* if the segments *AB, BC,* etc., were each 400 ft long?

7 Two points *A* and *B,* 5,280 ft apart, are to be connected by a straight line. A random line run from A in the general direction of B is found by computation to deviate 03′18″ from the true line. On the random line at a distance 1,250.6 ft from A an intermediate point C is established. What must be the offset from C to locate a corresponding point D on the true line?

8 In Fig. 9-10, a straight line AX is run at random from A in the general direction of B, point B not being visible from A. A swing offset is measured from B to line AX and found to be 63.40 ft. The transit is set up at E', and $E'Y$ (perpendicular to DX) is erected. The swing offset from B to $E'Y$ is 1.1 ft. Also, the distance AE' is 2,633.9 ft. Compute the angle α which must be laid off from the random line in order to establish points on the straight line AB and determine the length AB. What must be the precision of α in order that the line established shall fall within 0.1 ft of the point B?

9 Given the data of problem 8, it is proposed to establish points on the line AB by perpendicular offsets from C and D. What must these offsets be if $AC = 937.6$ ft and $AD = 1{,}932.0$ ft?

10 In Fig. 9-11, suppose that the distance AC is 317.2 ft and the angle ACB is 67°13′. What is the distance AB?

11 What error would be introduced in the measurement of a horizontal angle, with sights taken to points at the same elevation as the transit, if, through nonadjustment, the horizontal axis was inclined (*a*) 03′? (*b*) 3°? (*c*) If the horizontal axis was inclined 03′, what error would be introduced if both sights were inclined at angles of +30°? (*d*) If one sight was inclined at +30° and the other at −30°?

12 In measuring a horizontal angle the error of setting up the transit is 0.03 ft, the direction of displacement being such as to produce a maximum angular error. What error is introduced in a 60° angle if the length of sights is (*a*) 50 ft? (*b*) 1,000 ft?

13 If the ratios of linear precision to be maintained on the various parts of a survey are 1/1,000, 1/5,000, 1/20,000, and 1/40,000, about how closely should the corresponding horizontal angles be observed in order that a consistent relation may exist between precision of angles and precision of distances?

9-27 Field Problems

Problem 15. Measurement of Horizontal Angles with Transit

Object. To measure several angles about a point with the transit, and to check the values of the angles by the use of magnetic bearings.

Procedure. (1) Set up and level the instrument at any point O. (2) Set four chaining pins at about 150 ft from the transit, forming four angles about the station O. The pins may be plumbed by the vertical cross hair of the transit. (3) According to the procedure of Art. 9-9, measure each of the angles, using the A vernier only and resetting the vernier on each backsight. (4) The sum of the measured angles should not differ from 360° by more than ±02′. If this difference is exceeded, remeasure the angles until the sum falls within the stated limits. (5) Release the compass needle, sight on each point, and according to the method of Art. 8-15 read and record the magnetic bearing to each pin. (6) Compute the angles by bearings, and compare with the transit angles. The discrepancy between any transit angle and the same angle by bearings should not exceed 30′.

Problem 16. Measurement of Vertical Angles with Transit

Object. To determine the height of a building above the water table, by measurement of vertical angles with the transit.

Procedure. (1) Set up the transit at A, at a distance from the building approximately twice the height of the building. Level the telescope, and note the location of the horizontal hair on the building; mark the point sighted, and measure the distance h_a above or below the water table of the building. (2) By double-sighting, determine the index error of the vertical circle. Sight on the high point T of the building, and record the vertical angle. (3) Depress the telescope, and set a point B in the same vertical plane with A and T, about halfway between A and the building. (4) Set up the transit at B, and measure h_b and the vertical angle to point T, as at A. (5) Measure the horizontal distance AB. (6) Draw a sketch, and compute the difference in elevation (a) between T and the horizontal line of sight from either transit station and (b) between T and the water table.

Hints and Precautions. The index error may be read either before or after observing a vertical angle; while this reading is being taken, the line of sight should be in the same vertical plane as the point sighted.

Problem 17. Prolongation of Line by Double-sighting with Transit

Object. To prolong a straight line with precision, setting stakes at intervals of about 300 ft (see Art. 9-15).

Procedure. (1) Set two points about 300 ft apart in such location as to afford an open view for 1,000 ft or more in advance. (2) Set up the instrument on the forward point. Backsight with the telescope inverted. (3) Plunge the telescope, and set a stake on the line 100 paces in advance. Mark a point on the stake exactly on line. (4) Take a second backsight on the rear stake in the same manner but with the telescope normal. Plunge the telescope again, and mark a point on the advance stake. (5) If this point does not coincide with the first point set, a point midway between them is on the line. (6) Set up the transit over this point, and advance by the same process, backsighting on the nearest point in the rear. Continue in this way for the desired distance. (7) Check the work by setting the instrument over the first point, sighting carefully on the next point, and then noting the linear error of the points set by double-sighting, without moving either horizontal motion of the instrument.

Problem 18. Prolongation of Line Past Obstacle

Object. To prolong a line AB past an obstacle where the conditions are such as to limit the lengths of the offsets.

Procedure. (1) As outlined in the first paragraph of Art. 9-16. (2) The lengths of offsets should be measured very carefully. If the instrument is in good adjustment, the points E and F may be set by a single reversal; or if a clear sight can be obtained, the transit may be set up at C and the points E and F established without reversal. (3) If the obstacle is imaginary, check the accuracy of the work by setting the transit at A and locating G and H by the direct prolongation of AB.

Problem 19. Running Straight Line between Two Points Not Intervisible

Object. To establish points along a straight line joining two given points not intervisible.

Procedure. (1) As outlined in Art. 9-17, cases 2 and 3. (2) Under case 2, take two points on opposite sides of a hill. To check the located position of *C*, set up the transit at that station, and by the method of double-sighting prolong the line *AC* to *B*. Note the error at *B*. (3) Under case 3, determine offsets from the intermediate points on the line *AX* (Fig. 9-10) to the line *AB*, and establish the corresponding points on *AB* by tape measurements. Then lay off the angle α from *AX*, and establish a second set of points on the line *AB* by the method of double-sighting. Note the discrepancies.

Problem 20. Determination of Inaccessible Distance

Object. To obtain the distance between two points on opposite sides of a river.

Procedure. As outlined in Art. 9-18. The transit points should be tacked stakes. If the river is imaginary, after the distance has been computed check it by direct measurement.

Problem 21. Intersection of Lines with Transit

Object. To bisect the three angles of a triangle and to mark the point of concurrence of the bisectors.

Procedure. (1) Drive three stakes *A, B, C* at the vertices of a roughly equilateral triangle having sides about 300 ft in length. Tack each stake. (2) Set up the transit, and measure the angle at *A*. Lay off one half of the measured amount, thus establishing the bisector of angle *A*. On the bisecting line of sight and on an estimated bisector of angle *B* drive a stake *o*, and drive a tack halfway. Set two more tacked stakes *m* and *n* on the bisecting line of sight about 10 ft from and on opposite sides of *o*. (3) Set up the transit at *B*, and locate the position of the bisector as at *A*. Drive a stake on this line and under a cord stretched from *m* to *o* or *n* to *o*, as the conditions require. Tack the exact point of intersection *p*. (4) Set up the transit at *C*, measure the angle, and bisect as at *A* and *B*. (5) Measure the discrepancy between this bisector and the point of intersection of the first two bisectors at *p*, to hundredths of feet. Also measure the angular discrepancy; it should not exceed 02′.

Problem 22. Measurement of Angle When Transit Cannot Be Set at Vertex

Object. To measure the angle between two walls of a building.

Procedure. As outlined in Art. 9-21. The points *a, b,* etc., should be tacks in stakes driven at appropriate locations.

Problem 23. Measurement of Angles by Repetition

Object. To obtain a more precise determination of the horizontal angles between various stations about a point than would be possible by a single measurement (see Art. 9-10).

Procedure. (1) Set up the transit very carefully over the point. (2) Set the *A* vernier at zero, read the *B* vernier, and record the readings. (3) Keep notes in a form similar to that of Fig. 9-6. (4) With the telescope *normal,* measure one of the angles clockwise, and record both vernier readings to the least reading of the vernier. (5) Leaving the upper motion clamped, again set on the first point and again measure the angle clockwise (thus doubling the angle). (6) Continue until five "repetitions" (observations) have been secured. Record both vernier readings and the total angle turned. (7) In like manner, without resetting the vernier, measure the angle (five repetitions) with the telescope *inverted,* always measuring clockwise. (8) Go through the same process for all other angles about the point. (9) Compute the value of each of the angles for the 10 repetitions, and compare with the single measurement. (10) For a transit reading to single minutes, the error of horizon closure should not exceed $10'' \sqrt{\text{number of angles}}$. (11) Adjust the angles so that their sum will equal 360° by distributing the error equally among the mean values.

Hints and Precautions. (1) Level the transit very carefully before each repetition, but do not disturb the leveling screws while a measurement is being made. (2) Be careful not to loosen the wrong clamp screw. (3) When the lower motion is being turned, the hands should be in contact with the *lower* plate, not the upper motion. (4) Do not walk around the transit to read the second vernier; rotate it to you, always turning the instrument clockwise in order to avoid possible errors due to slackness and twist.

Problem 24. Laying Off Angle by Repetition

Object. To lay off a given horizontal angle more precisely than is possible with a single setting of the vernier (see Art. 9-11).

Procedure. (1) Drive and tack two stakes about 500 ft apart. (2) Carefully set up the transit over one end of the line. Sight at the point at the other end, and lay off the given angle. (3) Set a stake on the line of sight about 500 ft from the instrument (distance by pacing), and carefully set a tack. (4) By repetition measure the angle laid off, as in the previous problem, making five "repetitions" with telescope normal and five with it inverted. (5) Find the difference between the angle laid off and the required angle, and by trigonometry compute the linear distance that the tack must be moved perpendicular to the line of sight. (6) Set the tack accordingly.

Problem 25. Adjustment of Transit

Object. To make the field adjustments of the conventional engineer's transit.

Procedure. As outlined in Art. 9-23.

10
TRANSIT-TAPE SURVEYING

10-1 General. The practices described in this chapter apply in general to most surveying work of ordinary precision, excluding leveling.

The transit party is usually composed of a *transitman,* a *head chainman,* and a *rear chainman.* The transitman directs the party, operates and cares for the transit, and keeps the notes. The chainmen give line, set stakes, and assist in clearing the line. In wooded country, axemen may be needed to clear the line. Where sights are long, a rear flagman may be employed to give backsights to the transitman. Where many observations are taken, the notes may be kept by a recorder, who may also act as the chief of party.

The equipment of the transit party usually consists of a transit, 100-ft steel tape, two range poles, stakes, tacks, axe or hammer, two or three plumb bobs, field notebook, chaining pins, and marking crayon. Also included are devices for marking stations such as ordinary or special nails, a cold chisel, spray paint, and colored plastic flagging or tape. A light tripod with plumb bob is useful for sighting on transit stations where vision is obscured near the ground. Portable sending and receiving radios, called "walkie-talkies," may be used to expedite communication.

Any temporary or permanent point of reference over which the transit is set up is called a *transit station.* Usually the transit station is a peg, called a *hub,* driven flush with the ground and having a tack driven in its top. The location of a hub is usually indicated by a flat *guard stake* extending above the ground and driven sloping so that its top is over the hub. The number of the station is marked with keel or lumber crayon on the underside of the guard stake and reads down the stake. A transit station may also be a driven nail or a cross cut in the pavement or curb.

Lines connecting transit stations are called *transit lines.* Traverses and triangulation systems are formed by transit lines. In the survey notes and on maps a traverse transit station is indicated by the symbol ⨀, and a triangulation transit station by the symbol ◬.

For many open traverses where lengths are measured with the tape, stakes are set every 100 ft. These 100-ft points are called *full* stations, and intermediate points are called *plus* stations. A full station at, say

1,200 ft from the initial point would be numbered 12 + 00 or simply 12; and a plus station at, say, 1,927.2 ft from the initial point would be numbered 19 + 27.2. The stations intermediate between transit stations are marked usually by flat stakes driven vertically, with the number on the side toward the initial point and with the number reading down the stake. Where stakes cannot be driven, nails or marks may be employed.

10-2 Transit Surveys. The field work of surveying with the transit may be divided as follows:

1 Establishing transit stations and lines by angular and linear measurements. The transit lines may be said to form the skeleton of the survey and are called the *control* or *horizontal control.*

2 Locating objects and points with respect to the transit lines, thus furnishing the *details* with which the transit lines are clothed (see Art. 10-13).

The details may be observed either while the transit lines are being run or after they have been established and checked. Instruments and methods employed in collecting details may differ from those employed in establishing the control.

Transit surveys may be made by the method of *radiation* (Art. 10-3), of *intersection* (Art. 10-4), of *traversing* (Arts. 10-5 to 10-11), or of *triangulation* (Chap. 12). On surveys of ordinary precision, transit traverses in one form or another probably make up more than nine-tenths of the systems of control. The methods of radiation and intersection are employed on surveys of small areas and in the collection of details from transit traverses. Triangulation is not often used on surveys of ordinary precision except for hydrographic surveying or for topographic surveying in rough open country; but it could be employed to advantage more generally than at present.

10-3 Radiation. The transit is set up at any convenient station from which can be seen all points that it is desired to locate. The distance from the transit station to each of the points is measured, and the horizontal angle is observed. The angles between successive points may be measured; or the true, magnetic, or assumed bearing or azimuth of each of the lines joining the points with the transit station may be observed.

Where points only are to be located, the method is excellent. Where lengths and directions of lines other than the radial lines are to be determined, trigonometric computations are necessary. The method is not commonly used for property-line surveys.

10-4 Intersection. In Fig. 10-1 let the points *A, B, C,* etc. represent objects which it is desired to locate, and let *OP* be a convenient line from both ends of which the unknown points are visible. The length of the base line *OP* is measured with the tape. The transit is set up at *O,* and angles to the unknown points are observed; these may be expressed as azimuths, as bearings, or as angles between successive points. A similar series of observations is made with the transit at *P.* In this manner each of the unknown points becomes the vertex of a triangle of which the base line *OP* is the side of measured length and in which the angles adjacent thereto are observed values; the locations of the unknown points are thus defined.

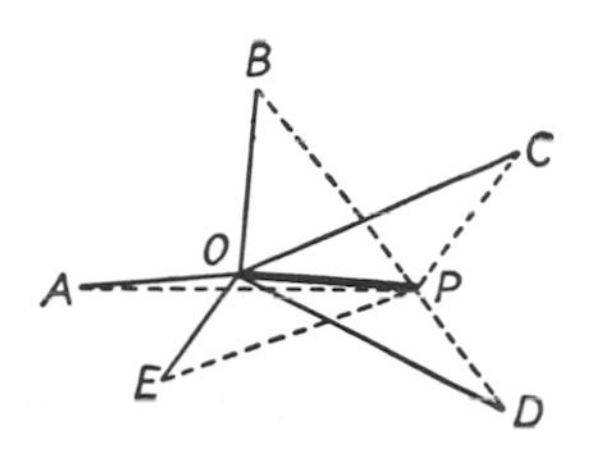

Fig. 10-1. Intersection.

The method of intersection just described is a form of triangulation. It is useful in locating details for traverses, each unknown point being observed from two successive stations of the traverse. Angles taken to some landmark from several successive stations provide a means of checking the traverse. The method is not well-suited to property-line surveys because of the large amount of computing necessary and because of the uncertainties of the computed values when triangles are weak.

Sights may be taken simultaneously by means of two transits set up at opposite ends of a measured base line. In hydrographic surveying this method of intersection is useful in locating soundings. For the construction of bridges and dams, the method is used to establish points for piers and other structural parts which are difficult of access.

A variation of the method of intersection is the method of *resection,* by means of which the transit may be set up at a station of unknown location and its location determined by sighting on points of known location. The principle of resection employing the transit is as described for the plane table in Art. 13-11.

10-5 Traversing. Following is a general description of the work of running a *closed* traverse, the transit stations being established in advantageous locations as the survey progresses, and distances being measured between successive transit stations. In Fig. 10-2 let *A* and *B* be selected locations for transit stations marking the first line of a traverse. Hubs are set defining the points. The transit is set up at *B,* the horizontal vernier is set to a given angular value, a backsight is taken on a range pole at *A,* and the lower motion is clamped. The line

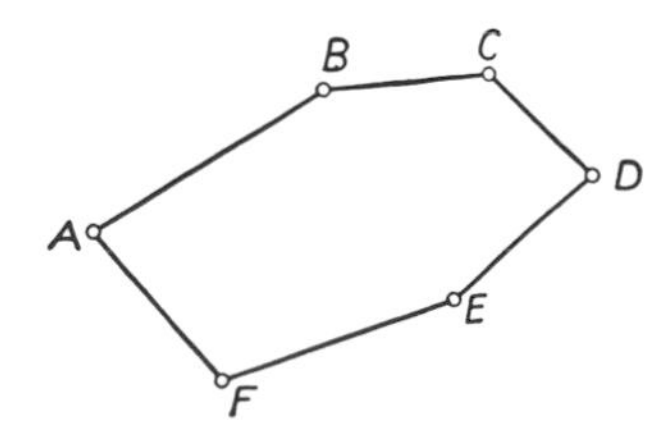

Fig. 10-2. Closed traverse.

AB is then taped, the head chainman being lined in by the transitman. The distance *AB* is recorded. The location of *C* is selected, and the transit point is established. The transit is turned on the upper motion until a foresight is secured on *C*. The upper motion is clamped, and the angular value is read and recorded. The distance *BC* is taped and recorded in a manner similar to that for *AB*. The transit is moved forward to *C;* a backsight is taken to *B;* the point *D* is chosen; a foresight is taken to *D;* and the angle is read. The line *CD* is taped. The process is repeated for point *E,* etc., until the traverse is finally brought to closure on the initial point *A*. As a check, usually the initial station is occupied by the transit and the angle between the first and last lines of the traverse is measured; in this way the angular error of closure is determined.

An *open* or continuous traverse may be run in exactly the same manner, except that of course there is no closure. The cumulative angular error cannot be determined except by astronomical observations (taking account of convergency of meridians) or by beginning and closing on previously established lines the directions and locations of which are known.

The angles of the traverse may be measured by observing *deflection angles* (Art. 8-9), *azimuths* (Art. 8-8), *angles to right* (Art. 8-10), or *interior angles* (Art. 8-11) as desired. As a check against mistakes or large errors, usually the magnetic bearings are observed on both backsights and foresights and are compared with bearings computed from the measured angles at each station. For rough traverses the angles may be measured by means of the transit compass alone, the bearings of the lines being observed.

One or more transit stations may be necessary between points at which angles are turned. In such cases the line is prolonged by one of the methods of Art. 9-15.

It is customary to show a sketch of the traverse on the right-hand page of the field notebook, as shown for a closed azimuth traverse in Fig. 10-4. For most short closed traverses the notes read down the page. For long open or closed traverses usually the center line of the right-hand page is taken to represent the traverse, and the notes and sketch read *up* the page (Fig. 10-3); thus objects on the sketch appear in the same relative position as they appear to the recorder proceeding along the line.

10-6 Deflection-angle Traverse. Successive transit stations are occupied, and at each station a backsight is taken with the *A* vernier set at zero and the telescope inverted. The telescope is then plunged, the foresight is taken by turning the instrument about the vertical axis on its upper motion, and the deflection angle is observed. The angle is recorded as right *R* or left *L*, according to whether the upper motion is turned clockwise or counterclockwise.

Figure 10-3 shows the notes for a portion of an open deflection-angle traverse where stakes are set every 100 ft. The notes are typical of the form used on highway, railroad, and other route surveys. Magnetic bearings have been observed as a check. The center line of the right-hand page represents the traverse, and the notes read up the page.

PRELIMINARY SURVEY OF I.N. RY., C. TO ST. LEONARDS

(Defl. Ang. Traverse)

Sta.	Defl.Ang.	Mag.B.	Cal.B.	Remarks
12				
+82				Water Line
+17				" "
11				
10				
9				
○8+36.3	23°14'L	N 35°W	N34°47'W	
8				
+99.8				
+68				℄ Due E to St.L.
+33.1				
7				
6+00				Ctr. of Brook
5				
4				
3				
2				
+23.6				
1				
○0+00		N11°W	N11°33'W	on C.L. C.P.Ry. 523'N. M'l. Post No. 26

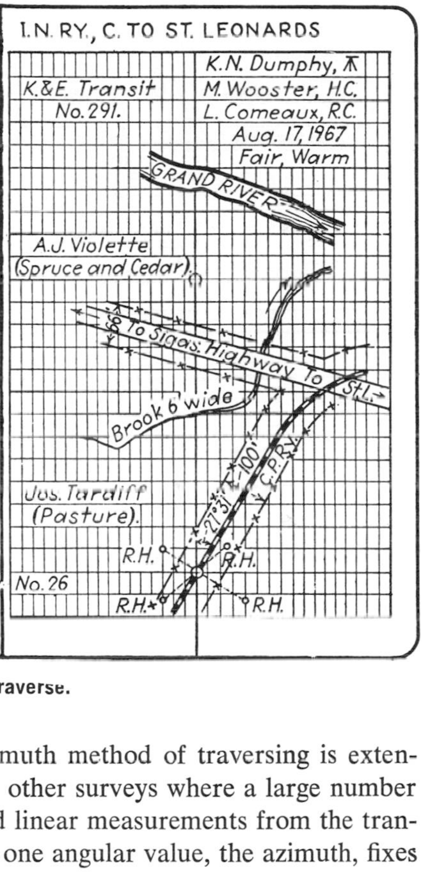

K.&E. Transit No. 291.

K.N. Dumphy, ⊼
M. Wooster, H.C.
L. Comeaux, R.C.
Aug. 17, 1967
Fair, Warm

Fig. 10-3. Notes for open deflection-angle traverse.

10-7 Azimuth Traverse.

The azimuth method of traversing is extensively employed on topographic and other surveys where a large number of details are located by angular and linear measurements from the transit stations. The simple statement of one angular value, the azimuth, fixes the direction of the line to which it refers. The reference meridian may be either true or assumed.

Successive stations are occupied, beginning with a line of known or assumed azimuth. At each station the transit is "oriented" by setting the *A* vernier to read the back azimuth (forward azimuth $\pm 180°$) of the preceding line and then backsighting to the preceding transit station. The instrument is then turned on the upper motion, and a foresight on the following transit station is secured. The reading indicated by the *A* vernier is the azimuth of the forward line. Traverse angles are checked by magnetic bearings.

Figure 10-4 illustrates the notes for a short closed traverse for which the azimuths were observed as just described.

AZIMUTH TRAVERSE AT HIGH-WATER LINE

Proposed Mill Pond, El. 741.36
Silver Creek, Penn.
(For Land Damage Est.)

Sta.	Obj.	Dist.	Azimuth	Mag. B.	Cal. Bear.
1	5		270°28′	N 80½° W	N 89°32′ W
	2	689.32	350°30′	N	N 9°30′ W
2	1		170°30′	S	S 9°30′ E
	3	509.66	303°05′	N 48° W	N 56°55′ W
3	2		123°05′	S 48¼° E	S 56°55′ E
	4	678.68	236°13′	S 65½° W	S 56°13′ W
4	3		56°13′	N 65¼° E	N 56°13′ E
	5	572.50	177°58′	S 7° W	S 2°02′ E
5	4		357°58′	N 7¼° E	N 2°02′ W
	1	1082.71	90°29′	S 80° E	S 89°31′ E
			Error =	01′	

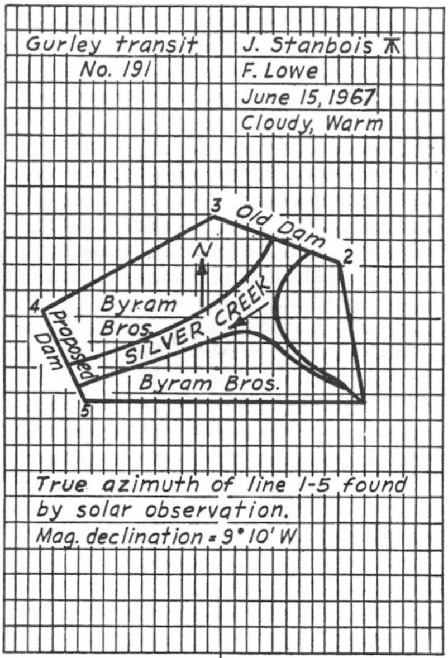
Gurley transit
No. 191
J. Stanbois ⊼
F. Lowe
June 15, 1967
Cloudy, Warm
3
Old Dam
2
N
4
Byram
Bros.
SILVER CREEK
Proposed
Dam
Byram Bros.
5
True azimuth of line 1-5 found
by solar observation.
Mag. declination = 9° 10′ W

Fig. 10-4. Notes for short closed azimuth traverse.

Another method, faster but more liable to mistake, consists in leaving the vernier setting unchanged between a foresight and the following backsight and plunging the telescope between each backsight and the corresponding foresight.

10-8 Traverse by Angles to Right.

This method is similar to the azimuth method, except that at each station the backsight to the preceding transit station is taken with the *A* vernier set at zero. The instrument is turned on the upper motion, a foresight is taken to the following station, and the clockwise angle is read on the *A* vernier. Angles to the right are often called *azimuths from back line*. Notes are kept in much the same form as for deflection angles. The method is used chiefly on open traverses, particularly where many details are located from the traverse stations. For such work the method lessens the chances of confusion.

10-9 Interior-angle Traverse. At each station the vernier is set at zero, and a backsight to the preceding transit station is taken. The instrument is then turned on its upper motion, a foresight is taken on the following station, and the interior angle is read. Either the notes may be kept in the form of a sketch or the numerical values may be tabulated.

10-10 Checking Traverses. Magnetic bearings offer an excellent means of checking observed angles against mistakes or large errors. On important traverses often the angular values are checked by doubling the angles and the linear measurements are checked by taping forward and back over each line.

For a *closed* traverse it is possible by known geometrical relations to determine the angular error of closure. The linear error of closure due to errors in both angles and distances can be determined by comparing the coordinates of the initial point as originally assumed and as computed by successive angles and distances around the traverse. (However, a closed traverse may close perfectly even though the tape is not of the correct length.)

For an *open* traverse no means are available for checking the measurements as a whole. Angular errors can be determined closely by taking astronomical observations at intervals, taking the convergency of meridians (Art. 18-24) into account if the distances are considerable. Occasionally conditions are favorable to establishing *cutoff lines* as the traverse progresses (Fig. 10-5). These lines together with the intermediate traverse lines essentially form closed traverses covering parts of the survey. Another check is obtained by observing the angle to some distant landmark from each of several stations.

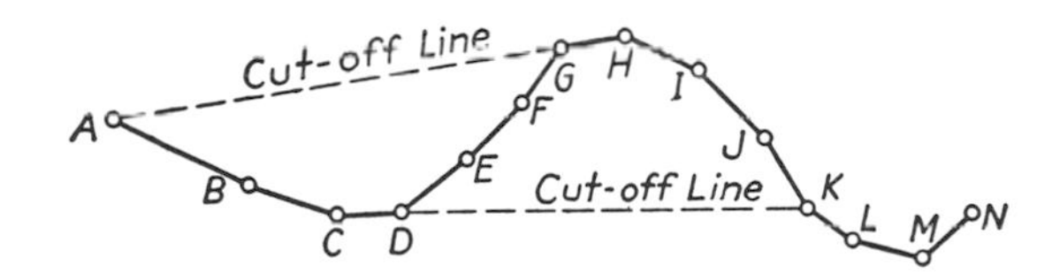

Fig. 10-5. Cutoff lines.

In some cases the open traverse begins and ends at points the locations of which have been accurately determined by previous field operations (for example, the triangulation stations of the U.S. Coast and Geodetic Survey or the monuments set in connection with state plane-coordinate systems). In such cases the error of closure of traverse on the known point may be determined by comparing its established or accepted coordinates with those computed from the traverse observations. The error of closure obtained by this method contains the combined effect of angular and linear errors.

10-11 Precision; Specifications for Traversing. The precision of transit-tape traverses is affected by both linear and angular errors of measurement (see Arts. 3-8, 9-24, and 9-25), and it can be expressed only in very general terms. On surveys of ordinary precision, the important linear errors are likely to be systematic and the important angular errors are largely accidental. The precision is generally influenced much more by the systematic linear errors than by the accidental angular errors; and it is usually found to vary approximately as the length of the traverse lines.

The following specifications give approximately the *maximum* linear and angular errors to be expected when the methods stated are followed. If the surveys are executed by well-trained men, with instruments in good adjustment, and under average field conditions, in general the error of closure should not exceed *half* the specified amount. The specifications apply to traverses of considerable length. It is assumed that a standardized tape is used.

Class 1. Precision sufficient for many preliminary surveys, for horizontal control of surveys plotted to intermediate scale, and for land surveys where the value of the land is low.

Transit angles read to the nearest minute. Sights taken on a range pole plumbed by eye. Distances measured with a 100-ft steel tape. Pins or stakes set within 0.1 ft of end of tape. Slopes under 3 per cent disregarded. On slopes over 3 per cent, distances either measured on the slope and corrections roughly applied, or measured with the tape held level and with an esimated standard pull.

Angular error of closure not to exceed $1'30'' \sqrt{n}$, where n is the number of observations. Total linear error of closure not to exceed 1/1,000.

Class 2. Precision sufficient for most land surveys and for location of highways, railroads, etc. By far the greater number of transit traverses fall into this class.

Transit angles read carefully to the nearest minute. Sights taken on a range pole carefully plumbed. Pins or stakes set within 0.05 ft of end of tape. Temperature corrections applied to the linear measurements if the temperature of air differs more than 15°F from standard. Slopes under 2 per cent disregarded. On slopes over 2 per cent, distances either measured on the slope and corrections roughly applied, or measured with the tape held level and with a carefully estimated standard pull.

Angular error of closure not to exceed $1' \sqrt{n}$. Total linear error of closure not to exceed 1/3,000.

Class 3. Precision sufficient for much of the work of city surveying, for surveys of important boundaries, and for the control of extensive topographic surveys.

Transit angles read twice with the telescope plunged between observations. Sights taken on a plumb line or on a range pole carefully plumbed. Pins set within 0.05 ft of end of tape. Temperature of air determined within 10°F, and corrections applied to the linear measurements. Slopes determined within 2 per cent and corrections applied. If tape is held level, the pull kept within 5 lb of standard and corrections for sag applied.

Angular error of closure not to exceed $30'' \sqrt{n}$. Total linear error of closure not to exceed 1/5,000.

Class 4. Precision sufficient for precise surveying in cities and for other especially important surveys.

Transit angles read twice with the telescope plunged between readings, each reading being taken as the mean of both *A* and *B* vernier readings. Verniers reading to 30″. Instrument in excellent adjustment. Sights taken with special care. Pins set within 0.02 ft of end of tape. Temperature of tape determined within 5°F and corrections applied. Slopes determined within 1 per cent and corrections applied. If tape is held level, the pull kept within 3 lb of standard and corrections for sag applied.

Angular error of closure not to exceed $15'' \sqrt{n}$. Total linear error of closure not to exceed 1/10,000.

10-12 Referencing Transit Stations.

A transit station is said to be *referenced* when it is so tied by linear and/or angular measurements to nearby objects that it can be replaced readily. (Sometimes such a station is said to be "witnessed," but this term is incorrect.) The reference points may be wooden *reference hubs* or may be objects of a more or less permanent character. Stations likely to be disturbed or destroyed should always be referenced, and the measurements recorded by means of a sketch.

The station may be referenced by linear measurements only (Fig. 10-6) to two reference hubs which either may be on line with the station (as R.H. *A* and R.H. *B*) or may not be on line (as R.H. *A* and R.H. *D*). If either of the two reference hubs is disturbed or destroyed, the station cannot be relocated.

If four reference hubs are placed as shown in Fig. 10-6, with the transit station at the intersection of the line *AB* and the line *CD*, any two reference hubs may be destroyed, and still the station can be relocated.

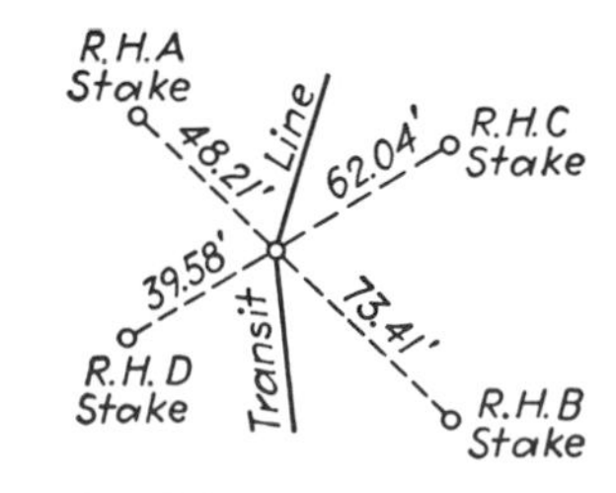

Fig. 10-6.

Figure 10-7 is typical of the manner in which corners are referenced in land surveying. Both angular and linear ties are employed.

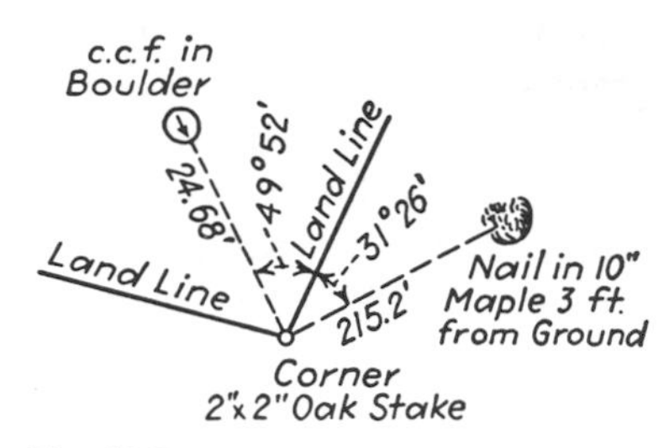

Fig. 10-7.

10-13 Details from Transit Lines.

On nearly all transit surveys certain *details,* or natural and artificial features of the terrain, are located with respect to the transit lines. The amount of detail required may be small, as on property-line surveys, or large, as on topographic surveys.

The precision with which details are located depends upon the purpose of the survey. In retracing property lines, the actual lines may be obstructed by hedges or buildings, so that the actual corners must be located by measurements from other transit lines; such measurements should be taken with a precision as great as that for the transit line. If details are located solely for map-making purposes, usually the required precision of measurements to details is less than that for the transit lines.

All well-defined objects should be correctly shown within the scale of the map, bearing in mind that points cannot be plotted within less than perhaps 0.01 in. Thus if the map scale were 1 in. = 1,000 ft there would be no particular advantage in taking measurements closer than the nearest 10 ft from a transit line to details; but if the scale were 1 in. = 10 ft measurements should be taken to 0.1 ft.

Angular measurements to details are usually made with the transit, with angles read to minutes. Often the angles used in mapping details are estimated to the nearest 05′ without the aid of the vernier in order to save time. Where details are located with respect to stations intermediate between transit stations, often some hand instrument is employed to measure the angles or directions.

Linear measurements to details are made with the 100-ft steel tape, with the 50-ft metallic tape, by stadia (Chap. 11), or sometimes by pacing, according to the precision required and the convenience of measuring.

On most surveys the details are located as the traverse progresses. The observations with the transit, called *side shots,* are made after the foresight to the following station has been taken.

10-14 Locating Details.

Following are descriptions of the common methods of locating details with the transit and tape. The method or combination of methods is chosen which requires the least time in the particular case. In general, the *dimensions* of an object (as a building) should be determined by direct measurement. Sketches are used freely. Where details are numerous, each observed point is given a number on the sketch, and the angles and distances are tabulated.

1 *By Angle and Distance from Transit Station.* As illustrated by Fig. 10-8, any given point on the object is located by an angle and a distance. The method is widely used where details are close to transit stations.

2 *By Angles from Two Transit Stations.* As shown by Fig. 10-9, the point is located with respect to the transit line by angles taken from two

Fig. 10-8.

Fig. 10-9.

stations; that is, by the method of intersection. No linear measurements are required. The method is useful for distant or inaccessible points.

3 *By Distances from Two Stations.* If the details are close to the transit line yet distant from the nearest transit station, they are conveniently located by linear ties from two traverse stations. Thus the corners of the middle building of Fig. 10-10 are located by ties to stations 12 + 00 and 13 + 00, neither of which is a transit station.

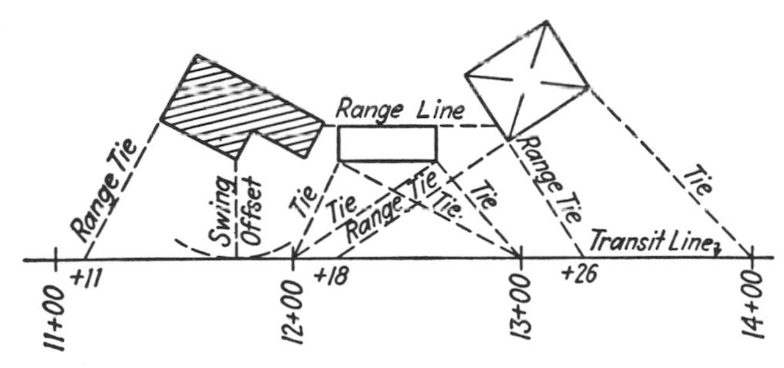

Fig. 10-10. Details by ranges, ties, and offsets.

4 *By Angle from One Station and Distance from Another.* Occasionally angles are measured from a transit station and distances are measured from intermediate stations. Care must be taken to secure good intersections.

5 *By Ranges, Range Ties, and Swing Offsets.* If the features to be located are buildings, the work of location may be facilitated by *ranging,*

or sighting along one or more sides of the building and finding the points of intersection of lines thus defined with other lines such as the transit line, a fence, or the side of another building. The station and plus of such a point of intersection with a transit line is called a *range,* and a range together with the distance along the range line to a corner of the building is called a *range tie.*

A *swing offset,* or perpendicular distance from a transit line to a point, is determined by swinging the tape about the point as a center and taping the least distance from the point to the transit line.

Figure 10-10 shows how a group of buildings near a traverse line may be located by tape measurements only. The building at the right is completely located by two range ties, one intersecting the traverse line at 12 + 18 and the other at 13 + 26. The location of the building is checked (assuming that the lengths of sides are measured) by the check tie to station 14 + 00. The location of the middle building of the group is established by ties to station 12 + 00 and station 13 + 00, three ties being sufficient to locate the building and the fourth tie being taken to check the location. The location of the building on the left is fixed by the range tie intersecting the traverse line at 11 + 11 and by the swing offset shown. The locations are checked by the range line tying the three buildings together.

6 By Perpendicular Offsets from Transit Line. This method is adapted to the location of irregular or curved boundaries, streams, and roads that closely parallel the transit lines. As indicated by Fig. 10-11, a point is

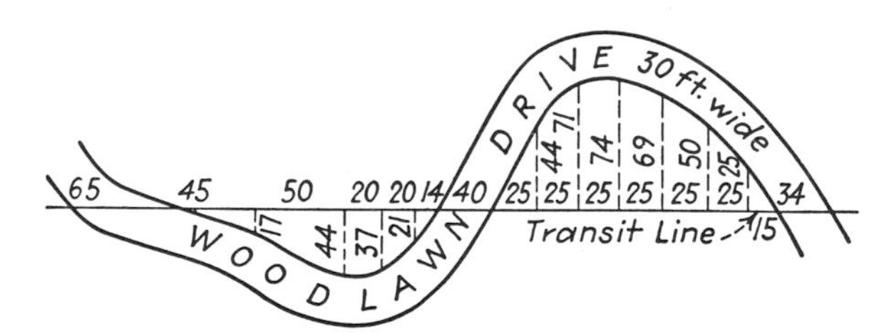

Fig. 10-11. Details by perpendicular offsets.

located by measuring the distance along the transit line to the foot of a perpendicular offset through the point and then measuring the length of the offset. Usually the direction of the perpendiculars is estimated by eye except where offsets are long; in this case the tape or one of the sighting instruments is employed.

CURVES

10-15 Circular Curves. In highway and railway location the horizontal curves employed at points of change in direction are arcs of circles. The straight lines connecting these *circular curves* are tangent to them and are therefore called *tangents*. For the completed line, the transition from tangent to circular curve and from circular curve to tangent may be accomplished gradually by means of a segment in the form of a *spiral* (Art. 10-22). On railway work spirals are used almost invariably. On highway work spirals are usually used only on the sharper curves of primary roads.

Vertical curves (Art. 6-11) are usually arcs of parabolas. Horizontal parabolic curves are occasionally employed in route surveying and in landscaping; they are similar to vertical curves and will not be discussed herein.

The subject of route curves is extensive and is covered by separate textbooks. Herein are discussed only some of the simpler relationships.

The stationing of a route progresses around a curve in the same manner as along a tangent. The point where a circular curve begins is commonly called the *point of curve*, written P.C.; that where the curve ends is called the *point of tangent*, written P.T.; and that where two tangents produced intersect is called the *point of intersection* or the *vertex*, written P.I. or V. A point on the curve is written P.O.C. Other notations are also used; for example, the point of curve may be written T.C. signifying that the route changes from tangent to circular curve, whereas the point of tangent is written C.T. Or the beginning of curve may be written B.C. and the end of curve E.C. The point of change from tangent to spiral is written T.S., and the point of change from spiral to circular curve S.C.

In the field the distances from station to station (usually 100 ft) on a curve are necessarily measured in straight lines, so that essentially the curve consists of a succession of 100-ft chords. Where the curve is of long radius, as in railroad practice, the distances along the arc of the curve are considered to be the same as along the chords. In highway practice and along curved property boundaries usually the distances are considered to be along the *arcs;* either a correction is applied for the difference between arc length and chord length or the chords are made so short as to reduce the error to a negligible amount.

The sharpness of curvature may be expressed in any of three ways:

1 Radius. By stating the length of the radius. This method is often employed in highway work, with the radius for a given curve taken as a multiple of 100 ft.

2 Degree of Curve, Arc Basis. By stating the "degree of curve," or the angle subtended at the center by an *arc* 100 ft long. This method is usually followed in highway practice. Thus, if the degree of curve is D and the radius is R,

$$R = \left(\frac{360^\circ}{D^\circ}\right)\left(\frac{100}{2\pi}\right) \tag{1}$$

From this relation the radius may be found if the degree of curve is known, and *vice versa.*

3 Degree of Curve, Chord Basis. By stating the degree of curve as the angle subtended by a *chord* of 100 ft. This method is followed in railroad practice. Thus, in Fig. 10-12,

$$R = \frac{50}{\sin \frac{1}{2}D} \tag{2}$$

On the arc basis the radius of curvature varies inversely as the degree of curve; for example, the radius of a 1° curve is 5,729.58 ft and the radius of a 10° curve is 572.96 ft. On the chord basis the radius of a 1° curve is 5,729.65 ft and the radius of a 10° curve is 573.68 ft. The difference per 100 ft in length between the chord and the arc is for a 1° curve less than 0.01 ft, for a 5° curve 0.03 ft, and a 10° curve 0.13 ft.

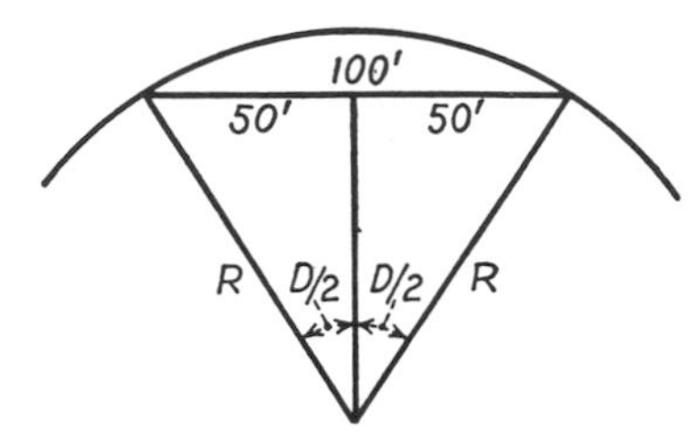

Fig. 10-12. Degree of curve.

Except as specifically stated, hereinafter the discussions refer to the circular curve and to the chord basis for expressing curvature.

10-16 Curve Formulas. In the following discussion the geometrical fact is employed that either an inscribed angle or an angle formed by a tangent and a chord is measured by one-half its intercepted arc and is therefore equal to one-half the corresponding central angle.
From the triangle AVF, in which $\angle VAF = \frac{1}{2}I$ and $AF = \frac{1}{2}C$,
field the intersection angle I between the two tangents is measured. The radius of the curve is selected to fit the topography and the proposed operating conditions on the line. The line OV bisects the angles at V and at O, bisects the chord AB and the arc ADB, and is perpendicular to the chord AB at F. From the figure, $\angle AOB = I$ and $\angle AOV = \angle VOB = \frac{1}{2}I$.

The chord $AB = C$ from beginning to end of curve is called the *long chord.* The distance $AV = BV = T$ from vertex to P.C. or P.T. is called the *tangent distance.* The distance $DF = M$ from mid-point of arc to mid-point of chord is called the *middle ordinate.* The distance $DV = E$ from mid-point of arc to vertex is called the *external distance.*

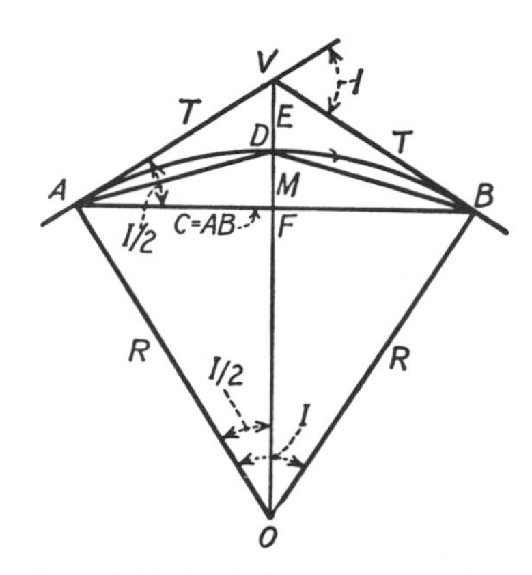

Fig. 10-13. Basis for curve formulas.

Given the radius of the curve $OA = OB = R$ and the intersection angle I, then in the triangle OAV

$$T = R \tan \tfrac{1}{2}I = \text{tangent distance} \tag{3}$$
$$E = R \sec \tfrac{1}{2}I - R = R \operatorname{exsec} \tfrac{1}{2}I = \text{external distance} \tag{4}$$

From the triangle AOF, in which $AF = \tfrac{1}{2}C$,

$$C = 2R \sin \tfrac{1}{2}I = \text{long chord} \tag{5}$$
$$M = R - R \cos \tfrac{1}{2}I = R \operatorname{vers} \tfrac{1}{2}I = \text{middle ordinate} \tag{6}$$

From the triangle AVF, in which $\angle VAF = \tfrac{1}{2}I$ and $AF = \tfrac{1}{2}C$,

$$C = 2T \cos \tfrac{1}{2}I \tag{7}$$

From the triangle ADF, in which $\angle DAF = \tfrac{1}{4}I$,

$$M = \tfrac{1}{2}C \tan \tfrac{1}{4}I \tag{8}$$

10-17 Length of Curve. As the arc length corresponding to a given radius varies in direct proportion to the central angle subtended by the arc, the length of arc for any central angle I is

$$\text{Arc} = \left(\frac{I^\circ}{360^\circ}\right)2\pi R \tag{9}$$

in which the angle I° is expressed in degrees. If the curvature is expressed on the *arc* basis, from Eqs. (1) and (9) the length of curve L_a is

$$L_a = 100\frac{I}{D} \tag{10}$$

If the curvature is expressed on the *chord* basis, the length of curve is considered to be the sum of the lengths of the chords, normally each 100 ft long. In this case also, the length of curve (on the chords) is

$$L_c = 100\frac{I}{D} \tag{11}$$

which is somewhat less than the actual arc length. Thus if the central angle I of the curve AD (Fig. 10-14) is equal to three times the degree of curve D, as shown, then there are three 100-ft chords between A and D, and the length of "curve" on this basis is 300 ft.

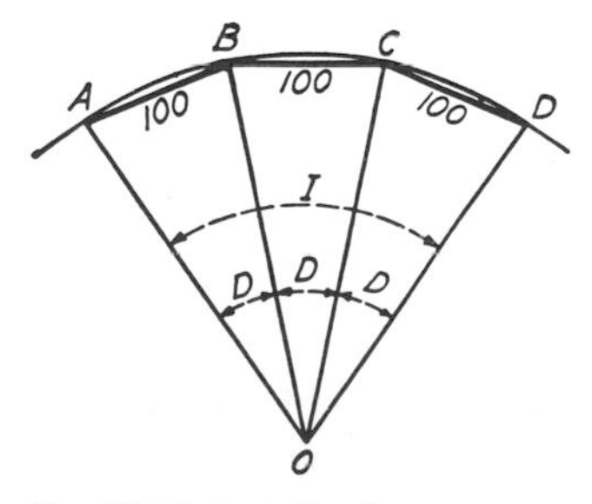

Fig. 10-14. Length of curve.

10-18 Laying Out Curve by Deflection Angles. Curves are staked out usually by the use of deflection angles turned at the P.C. from the tangent to stations along the curve together with the use of chords measured from station to station along the curve. The method is illustrated in Fig. 10-15, in which *ABC* represents the curve, *AX* the tangent to the curve at *A*, and angles *XAB* and *XAC* the deflection angles from the tangent to the chords *AB* and *AC*.

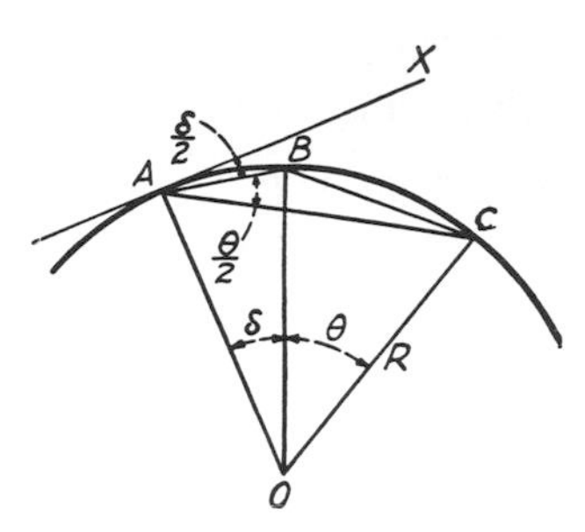

Fig. 10-15. Curve by deflection angles.

Assume the transit to be set up at *A*. Given *R*, δ, θ. Required to locate *B* and C. Considering the point *B*,

$$\angle XAB = \tfrac{1}{2}\delta \qquad (12)$$
$$AB = 2R \sin \tfrac{1}{2}\delta \qquad (13)$$

In the field, *B* is located as follows: The deflection of $XAB = \frac{1}{2}\delta$ is set off from the tangent, the distance *AB* is measured from *A*, and the forward end of the tape at *B* is lined in with the transit.

Considering the point *C*,

$$\angle BAC = \tfrac{1}{2}\theta \qquad (14)$$
$$BC = 2R \sin \tfrac{1}{2}\theta \qquad (15)$$
$$\angle XAC = \tfrac{1}{2}(\delta + \theta) \qquad (16)$$

In the field, *C* is located as follows: With the transit still at *A*, the deflection angle *XAC* is set off from the tangent, the distance *BC* is measured from *B*, and the forward end of the tape at *C* is lined in with the transit sighted along the line *AC*. Succeeding stations on the curve are located in similar manner.

Should the chord lengths be given instead of the central angles, then the angles are computed.

If *B* is at a full station and $BC = 100$ ft, then $\theta = D =$ degree of curve, and $\angle BAC = \frac{1}{2}D$.

A curve is located in the field normally as follows: The P.C. and P.T. are marked on the ground. The deflection angle from the P.C. is computed for each full station on the curve and for any intermediate stations that are to be located. The transit is set up at the P.C., a backsight is taken along the tangent with telescope inverted, the telescope is plunged, and each point on the curve is located by deflection angle and by distance measured from the preceding *full station*. Figure 10-16 illustrates the usual form of field notes.

10-19 Transit Setups on Curve. Often it is impracticable or impossible to run all of a given curve with the transit at the P.C., and one or

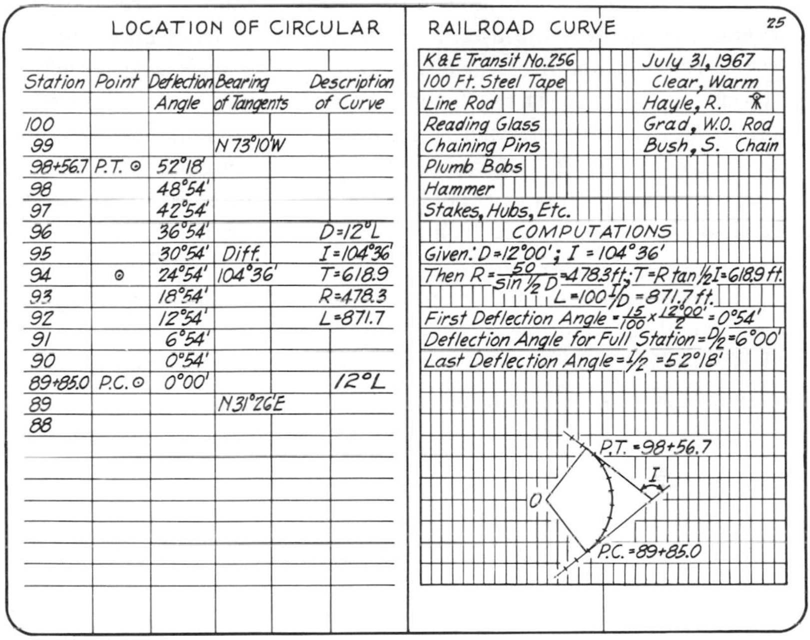

LOCATION OF CIRCULAR RAILROAD CURVE 25

Station	Point	Deflection Angle	Bearing of Tangents	Description of Curve
100				
99			N 73°10′W	
98+56.7	P.T. ⊙	52°18′		
98		48°54′		
97		42°54′		
96		36°54′		D=12°L
95		30°54′	Diff.	I=104°36′
94	⊙	24°54′	104°36′	T=618.9
93		18°54′		R=478.3
92		12°54′		L=871.7
91		6°54′		
90		0°54′		
89+85.0	P.C. ⊙	0°00′		12°L
89			N31°26′E	
88				

K&E Transit No. 256	July 31, 1967
100 Ft. Steel Tape	Clear, Warm
Line Rod	Hayle, R. ⊼
Reading Glass	Grad, W.O. Rod
Chaining Pins	Bush, S. Chain
Plumb Bobs	
Hammer	
Stakes, Hubs, Etc.	

COMPUTATIONS

Given: $D = 12°00'$; $I = 104°36'$

Then $R = \frac{50}{\sin \frac{1}{2} D} = 478.3$ ft; $T = R \tan \frac{1}{2} I = 618.9$ ft.

$L = 100 \, I/D = 871.7$ ft.

First Deflection Angle $= \frac{15}{100} \times \frac{12°00'}{2} = 0°54'$

Deflection Angle for Full Station $= D/2 = 6°00'$

Last Deflection Angle $= I/2 = 52°18'$

P.T. = 98+56.7

I

O

P.C. = 89+85.0

Fig. 10-16. Field notes for circular curve.

more setups are required along the curve. In any case, all of the deflection angles are computed as if for use at the P.C. (see Fig. 10-16).

The transit is set up at any point along the curve, and a backsight (with telescope inverted) is taken on the last preceding transit station with the vernier reading the computed deflection angle for the point sighted. If the backsight is to the P.C., the vernier is set at zero. The telescope is plunged, and points ahead of the transit are then set by using the deflection angles previously computed for these points.

10-20 Laying out Curve by Tape Alone. Often it is convenient or necessary to lay out a circular curve by means of the tape alone. Where the angle of intersection of the two tangents is small, the various points on the curve may be established by perpendicular offsets from the tangents, in other cases by perpendicular offsets from chords. The necessary lengths of the offsets are computed from the traverse data and the properties of a circular curve.

10-21 Stringlining of Curves. Railroad track, particularly on curves, is eventually thrown out of alinement by the action of trains. *Stringlining* is a simple method of determining and applying the amounts by which the track must be moved laterally at various points to restore proper curvature. It involves the use of middle ordinates from chord to curve; it is described in detail in various texts on route surveying. Briefly, the method is as follows: At regular intervals along the outer rail, a cord of length equal to two intervals is stretched, and the middle ordinate is measured with a scale and is recorded. For the circular portion of the curve, all middle ordinates should be equal; for the portion along which a gradual transition is made from curve to tangent, the middle ordinates should be progressively smaller by uniform increments. Irregularities in the tabulated values of middle ordinate are noted, and for each point of measurement the amount necessary to move the track is computed. Stakes are set in the ballast to serve as reference points, and the track is moved to conform with computed values.

10-22 Spiral Curves. *Railway.* On railway lines where trains are to be operated at high speed, it is common practice to insert between circular curve and tangent a curve of varying radius, called a *spiral,* in order that the degree of curvature and centrifugal force may be developed gradually. At the end of the spiral adjacent to the tangent its radius is very long; along the curve it decreases gradually until at the point where the spiral joins the circular curve the radii of the two are equal. Spiral curves are also called *easement curves* or *transition curves.*

In order to provide room for the spiral, the circular curve is offset from the main tangent, as to the position *AFGB* of Fig. 10-17. If the two spirals *EF* and *GH* are of equal length, the offsets *AC* and *BN* are equal, and the distance $VC = VN = (R + o) \tan \frac{1}{2}I$, in which o is the length of the offset.

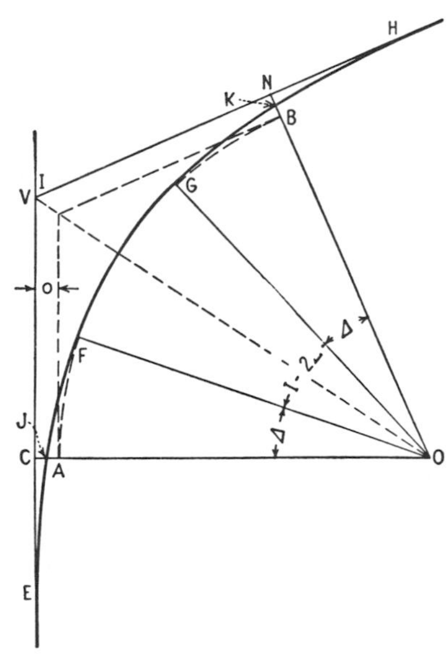

Fig. 10-17. Railway spiral.

Many mathematical solutions of the spiral are available, and the reader is referred to these for exact values (Ref. 4). The following approximate and empirical solution is not greatly in error.

1 The central angle I and the degree of circular curve D are known.

2 The length of spiral L' is selected; for curves likely to limit train speed L' should be not less than 240 ft; for minor curves, L' may be 100 ft or even less.

3 The length of the offset $o = AC = BN$ is computed. This may be assumed to be 6.50 ft for $D = 10°$ and $L' = 300$ ft, varying directly as the degree of curve and as the square of the length of spiral. Thus for $D = 5°$ and $L' = 200$ ft,

$$o = \frac{5}{10} \times \frac{(2)^2}{(3)^2} \times 6.50 = 1.44 \text{ ft}$$

4 $VC = VN = (R + o) \tan \frac{1}{2}I$.

5 $EC = NH$ = one half of the spiral length minus a correction. For curves of dimensions common in railroad practice this correction has approximately the following values: for spiral angle $\Delta = 5°$, 0.06 ft; for $\Delta = 10°$, 0.25 ft; for $\Delta = 15°$, 0.50 ft. For exact formula, see Ref. 4.

6 Spirals bisect the offsets AC and BN so that $CJ = \frac{1}{2}AC$ and $NK = \frac{1}{2}BN$.

7 Between E and J, perpendicular offsets from the tangent to the spiral vary in proportion to the cubes of the distances from E; between J and F, radial offsets from the circular curve to the spiral vary as the cubes of the distances from F; similarly for the other spiral GH.

8 Angle AOF = angle $BOG = \Delta = DL'/200$.

9 In the field, the points N, B, C, and A are located, and the direction of each offset tangent is established by means of another and equal offset from the main tangent. The simple curve $AFGB$ is located.

The necessary offsets are made to points on the spirals. For construction surveys it is usually sufficient to offset the circular curve, leaving the staking of the spirals to be done after the line is graded.

10 The alinement with spirals is along the line $EJFGKH$.

Highway. The procedure of computing and laying out a highway spiral is similar to that for railway spirals to which reference has just been made. Important simplifications have been made in the procedure through the publication of tables which give (1) values of the various functions involved, over a wide range; and (2) recommendations for superelevations, minimum transition lengths, safe maximum curvatures for various speeds, and widening of the pavement at the curve.

10-23 Numerical Problems

1 Given: $I = 34°30'$, D (chord basis) $= 3°00'$, and P.C. = station 74 + 30.0. Required: R, L, T, and E; also deflection angles arranged in notebook form for staking out this curve, using 100-ft stations.

2 Given: $I = 92°30'$, $T = 425.00$ ft, and P.C. = station 25 + 10.00. Required: R, D (arc basis), C, E, M, and L; also deflection angles arranged in notebook form for staking out this curve, using 50-ft stations.

3 Given: $I = 60°40'$, $E = 125.5$ ft. Required: R, D (arc basis), C, T, M, and L.

4 Given the data of problem 1. Make the necessary computations for the insertion of a spiral of length 250 ft at each end of the curve.

10-24 Field Problems

Problem 26. Open Deflection-angle Traverse with Transit and Tape

Object. To locate a section of an assigned route, setting stakes at full stations. This problem may be employed in connection with field problems 8 (Chap. 6) and 50 (Chap. 21). For plotting, see office problems 37 and 38 (Chap. 15).

Procedure. (1) Stake out a route perhaps 1,500 ft long with three or four changes in direction. (2) Set up the transit over the first stake marking a change in direction. With the *A* vernier set at zero and with the telescope inverted, sight on the stake at the beginning of the line. Read and record the magnetic bearing of this backsight. (3) Tape the line from the beginning (station 0 + 00) to the transit station, and record the length to the nearest 0.01 ft. The transitman lines in the head chainman. As the rear chainman pulls each pin, he replaces it with a stake on which he has written the station number. (4) Plunge the telescope, unclamp the upper motion, and sight on the next forward transit station. Record the reading of the *A* vernier and *R* or *L* to indicate whether the deflection is right or left from the prolongation of the preceding line. Also record the magnetic bearing of the forward line (5) Before the transit is moved, compute the deflection angle from the magnetic bearings taken at the station, and compare with the deflection angle indicated on the horizontal circle, as a rough check. (6) Tape the forward line. (7) Set up the transit at succeeding stations, and observe the deflection angle at each. Give line for the chainmen on each forward line. (8) Keep notes in the form of the sample notes (Fig. 10-3). Include sketches of streams, roadways, fences, etc., crossed by the line. (9) Assume as correct the magnetic bearing of some course the back and forward bearings of which have the same angular value, and compute the forward bearing of each of the other courses.

Problem 27. Closed Azimuth Traverse with Transit and Tape

Object. To collect data sufficient for plotting the boundaries and determining the area of a field, employing the azimuth method of traversing. The data may be used later, as in field problem 28 and in office problem 37 or 38 (Chap. 15).

Procedure. (1) Stake out an irregular field having perhaps five sides and containing an acre or more. (2) Measure the sides with the steel tape, to the nearest 0.01 ft. (3) Set up the transit at one corner. Set the *A* vernier at 0°, and turn the instrument on the lower motion to sight along the magnetic meridian either north or south according as azimuths are to be reckoned from north or south. Clamp the lower motion in this position. (4) Unclamp the upper motion, and turn the instrument to sight the next corner forward. The angle turned off in a clockwise direction and read from the *A* vernier is the azimuth of the forward line. Record the azimuth. (5) Record the magnetic bearing. Compare this with the bearing computed from the azimuth of the line. (6) Compute the back azimuth of the line by adding 180° to the forward azimuth. (7) Set up the instrument on the next corner forward, and with the *A* vernier set on the back azimuth of the line, backsight on the corner previously occupied. The instrument is now oriented. (8) Turn the instrument on the upper motion, and sight on the next corner forward; the azimuth of the forward line is then indicated by the *A* vernier. (9) Proceed in this manner until each corner has been occupied. Also set up again at the first corner, and take readings as for the other corners. (10) Keep notes in the form of the sample notes (Fig. 10-4). (11) Note the angular error of closure, which should not exceed $30'' \sqrt{\text{number of sides}}$. Distribute the error equally among the angles.

Problem 28. Details with Transit and Tape

Object. To obtain sufficient data for plotting a detailed map of a portion of the campus. The data may be used in office problem 37 or 38 (Chap. 15).

Procedure. (1) Run a closed azimuth traverse, as described in the preceding problem, through the area to be mapped. Locate the lines and corners of the traverse so that linear and angular measurements necessary to fix the position of details with respect to the traverse may be taken with the least labor. Reference the corner hubs. (2) Obtain the location of buildings, streets, walks, hydrants, fences, etc., by the methods of Arts. 10-13 and 10-14. (3) Sketch indefinite details such as trees, streams, and shrubbery without measurements other than by pacing. (4) When the traverse is plotted (office problem 37 or 38), plot the details according to the manner in which they were secured.

Hints and Precautions. (1) Details may be taken as the traverse is run. (2) Determine the angular error of closure of the traverse. If this exceeds $30'' \sqrt{\text{number of sides}}$ rerun the traverse until the mistake is discovered. (3) The location of important details should be checked, preferably by a different method. (4) Sketches should not be overcrowded with measurements. Many measurements can be tabulated on the left-hand page and referred to on the sketch by a single letter or number.

Problem 29. Laying Out a Circular Curve

Object. To lay out a circular curve, as for the curb line of a driveway, by the use of deflection angles.

Procedure. (1) From the assigned central angle and degree of curve, compute the tangent distance and the length of curve. (2) Assume that the P.C. is station 9 + 83.2, and compute deflections for each full station and +50. Prepare notes in the form shown in Fig. 10-16. (3) In the field, locate two tangents making the assumed angle. Locate the P.C. and the P.T. (4) Set up the transit at the P.C., orient the instrument, and stake out the full and +50 stations by the method described in Art. 10-18. Report the error observed at the P.T. (5) Set up the transit at the P.T., and check the angle. (6) If transit setups on the curve are required, follow the method of Art. 10-19.

11
STADIA SURVEYING

11-1 Tacheometry. Distances can be measured indirectly by sighting through a small angle at a distant scale transverse to the line of sight and in the plane of the angle. The angle may be fixed and the length it subtends on the scale measured, or a length on the scale may be fixed and the angle measured. The general process is called *tacheometry* or *tachymetry*.

In European practice usually a horizontal rod of fixed length, called a *subtense bar,* is used. Sighting marks usually 2 m apart are supported on a bar which is mounted on a tripod. The tripod is centered over the station to which it is desired to measure, the bar is leveled, and the bar is oriented perpendicular to the line by sighting through an attached telescope toward the angle-measuring station. The horizontal angle subtended by the 2-m length is then read precisely by means of theodolite or a transit, and the distance is computed by trigonometry. The horizontal angle is independent of the inclination of the line of sight; however, for most purposes the measured inclined distance must be reduced to the equivalent horizontal distance.

Although the subtense bar is used to some extent in the United States, common practice is to employ the *stadia* method of measuring distances indirectly. In this method the sighting telescope of the transit or plane-table alidade has a fixed small angle defined in the vertical plane, and the length which this angle subtends is read on a graduated rod held vertical (whether or not the line of sight is inclined) on the distant station.

Stadia. The stadia method is far more rapid than taping, and under certain conditions is as precise. It is a useful means of checking more precise measurements. It is employed extensively in transit surveying, plane-table surveying, and leveling.

The equipment for stadia measurements consists of (1) a telescope with two horizontal *stadia hairs,* one above and the other an equal distance below the horizontal cross hair, and (2) a *stadia rod* graduated usually in feet and decimals. Any self-reading leveling rod may be used as a stadia rod, but for long sights special graduations are employed; some of these are shown in Fig. 11-1.

The process of taking a stadia measurement consists in observing through the telescope the apparent locations of the two stadia hairs on the rod, which is held vertical. The interval between the rod readings, called the *stadia interval* or *stadia reading,* is a direct function of the distance from instrument to rod. For most instruments the ratio of distance to stadia interval is 100.

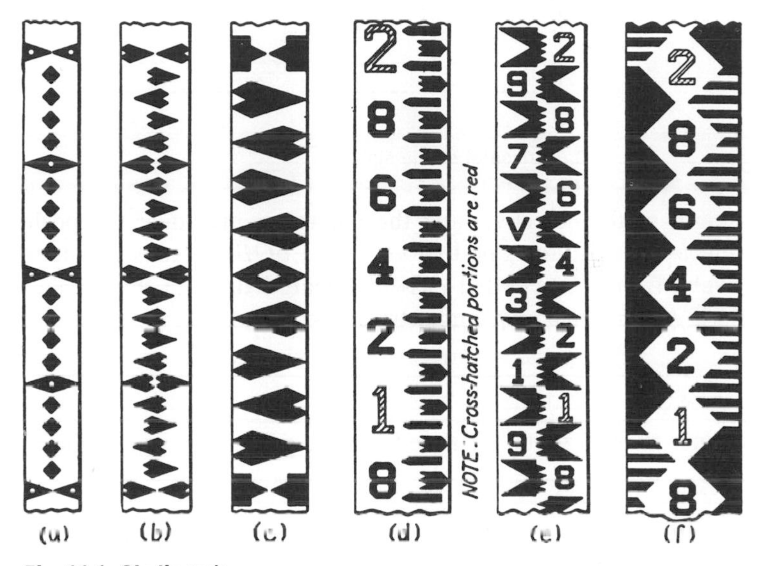

Fig. 11-1. Stadia rods.

For convenience and to lessen the chance of mistake, on transit or plane-table surveys the stadia interval is usually determined by setting the lower stadia hair on a foot mark and then reading the location of the upper stadia hair.

Whenever the stadia interval is in excess of the length of the rod, the separate half-intervals are observed and their sum is taken.

11-2 Principle of the Stadia. In Fig. 11-2 the line of sight is horizontal and the stadia rod is vertical. The stadia hairs are indicated by the points a and b; the distance between the stadia hairs is i. The stadia interval is s.

In optics it is shown that a ray of light passing through the optical center of a lens remains undeviated in direction and, further, that rays which are parallel on one side of the lens are all brought to a focus at a fixed point on the optical axis. This point is called the *principal focus,* and its distance from the optical center is called the *focal length* of the lens.

Imagine that aa' and bb' in the figure are parallel rays emanating from the stadia hairs a and b. Then F is the principal focus, f is the focal

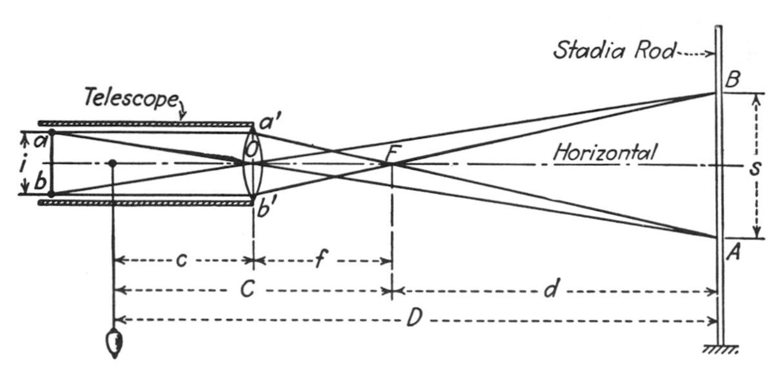

Fig. 11-2. Horizontal stadia sight.

length of the objective, and the emerging rays take the positions $a'FA$ and $b'FB$. Also imagine that aOA and bOB are rays emanating, respectively, from a and b that pass undeviated through the optical center O.

As $ab = a'b'$, by similar triangles

$$\frac{f}{i} = \frac{d}{s}$$

Hence the horizontal distance from the principal focus to the rod is

$$d = (f/i)s = Ks$$

in which $K = f/i$ is a coefficient called the *stadia interval factor* which for a particular instrument is a constant so long as conditions remain unchanged. Thus for a horizontal sight the distance from principal focus to rod is obtained by multiplying the stadia interval factor by the stadia interval. The horizontal distance from center of instrument to rod is then

$$D = Ks + (f + c) = Ks + C \qquad (1)$$

in which C is the distance from center of the instrument to principal focus. This formula is employed in computing horizontal distances from stadia intervals when sights are horizontal.

11-3 Stadia Constants. The focal distance f is a constant for a given instrument. It can be determined with all necessary accuracy by focusing the objective on a distant point and then measuring the distance from the cross-hair ring to the objective. The distance c, though a variable depending upon the position of the objective, may for all practical purposes be considered a constant. Its mean value can be determined by measuring the distance from the vertical axis to the objective when the objective is focused for an average length of sight.

Usually the value of $C = f + c$ is determined by the manufacturer and is stated on the inside of the instrument box. For external-focusing telescopes, under ordinary conditions C may be considered as 1 ft without error of consequence. Internal-focusing telescopes are so constructed that C is zero or nearly so; this is an important advantage of internal-focusing telescopes for stadia work.

11-3a Stadia Interval Factor. The nominal value of the stadia interval factor $K = f/i$ is usually 100. The interval factor can be determined by observation. The usual procedure is to set up the instrument in a location where a horizontal sight can be obtained and with a tape to lay off, from a point distant $C = (f + c)$ in front of the center of the instrument, distances of 100 ft, 200 ft, etc., up to perhaps 1,000 ft, stakes being set at the points thus established. The stadia rod is then held on each of the stakes, and the stadia interval is read. The stadia interval factor is computed for each sight, and the mean is taken as the most probable value. To overcome prejudicial tendencies in sighting, the stakes may be set at random distances which are measured later.

For use on long sights, where the full stadia interval would exceed the length of the rod, the stadia interval factor may be determined separately for the upper stadia hair and horizontal cross hair, and for the lower stadia hair and horizontal cross hair.

11-4 Inclined Sights. In stadia surveying most sights are inclined, and usually it is desired to find both the horizontal and the vertical distances from instrument to rod. For convenience in field operations the rod is always held vertical.

Figure 11-3 illustrates an inclined line of sight, AB being the stadia interval on the vertical rod and $A'B'$ being the corresponding projection normal to the line of sight. The length of the inclined line of sight from center of instrument is

$$D_i = \frac{f}{i} A'B' + C \tag{2}$$

For all practical purposes the angles at A' and B' may be assumed to be 90°. Let $AB = s$; then $A'B' = s \cos \alpha$. Making this substitution in Eq. (2) and letting $K = f/i$, the inclined distance is

$$D_i = Ks \cos \alpha + C \tag{3}$$

The horizontal component of this inclined distance is

$$H = Ks \cos^2 \alpha + C \cos \alpha \tag{4}$$

which is the general equation for determining the horizontal distance from center of instrument to rod, when the line of sight is inclined.

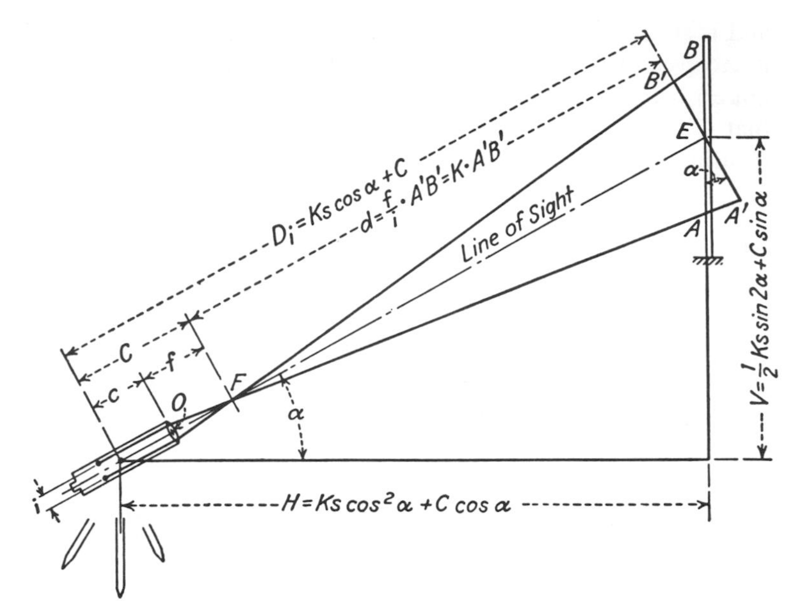

Fig. 11-3. Inclined stadia sight.

The vertical component of the inclined distance is

$$V = Ks\cos\alpha\sin\alpha + C\sin\alpha$$

The equivalent of $\cos\alpha\sin\alpha$ is conveniently expressed in terms of double the angle α, or

$$V = \tfrac{1}{2}Ks\sin 2\alpha + C\sin\alpha \qquad (5)$$

which is the general equation for determining the difference in elevation between the center of the instrument and the point where the line of sight cuts the rod. To determine the difference in ground elevations, the height of instrument and the rod reading of the line of sight must be considered.

Equations (4) and (5) are known as the *stadia formulas for inclined sights.*

11-5 Permissible Approximations. More approximate forms of the stadia formulas are sufficiently precise for most stadia work. Usually distances are computed only to feet and elevations to tenths of feet. For side shots where vertical angles are less than 3°, Eq. (4) may properly be reduced to the form

$$H = Ks + C \qquad (6)$$

which is the same as for horizontal sights (Art. 11-2).

Owing to unequal refraction and to accidental inclination of the rod, observed stadia intervals are in general slightly too large. To offset the systematic errors from these sources, frequently on surveys of ordinary precision the constant C is neglected. Hence in any ordinary case Eq. (4) may with sufficient precision be expressed in the form

$$H = Ks \cos^2 \alpha \text{ (approx.)} \tag{7}$$

Also Eq. (5) may often be expressed with sufficient precision for ordinary work in the form

$$V = \tfrac{1}{2}Ks \sin 2\alpha \text{ (approx.)} \tag{8}$$

except where the vertical angle is large.

Equations (7) and (8) are simple in form and are most generally employed.

When K is 100, the common practice is to multiply mentally the stadia interval by 100 at the time of observation and to record this value Ks in the field notebook. Thus, if the stadia interval were 7.37 ft, the *stadia distance* recorded would be 737 ft.

For external-focusing telescopes, the degree of approximation in using Eqs. (7) and (8) may be greatly reduced either by adding 0.01 ft to the observed stadia interval s or—when K is 100—by adding 1 ft to the observed stadia distance Ks. It is convenient to add the correction mentally and to record the corrected value in the field notebook. The notes should state that corrected values are recorded.

11-6 Stadia Reductions. Ordinarily in practice the horizontal distances and the differences in elevation are not computed by actually solving the stadia formulas but are obtained by the use of a table, diagram, stadia slide rule, or stadia arc on the vertical circle of the transit or alidade, all of these devices being based upon the formulas.

The precision of stadia surveying is such that ordinarily horizontal distances are determined to the nearest foot and vertical distances to the nearest 0.1 ft. Readings and computations are usually taken to three significant figures and in the lower range of four significant figures; slide-rule computations are sufficiently precise.

Table III gives, for each 02′ of vertical angle up to 30°, the horizontal distances (principal focus to rod) and differences in elevation for $Ks = 100$ ft, computed from the equations $H = Ks \cos^2 \alpha$ and $V = \frac{1}{2}Ks \sin 2\alpha$ [see Eqs. (4), (5), (7), and (8)]. For any other value of Ks, the tabular quantities are to be multiplied by the value of Ks in hundreds of feet. The table also gives the horizontal distances and differences in elevation for three values of $C = (f + c)$, indicated as C in the table. Tables arranged differently will be found in various other publications.

Diagrams showing graphically the quantities $Ks \cos^2 \alpha$ and $\frac{1}{2}Ks \sin 2\alpha$ for all ordinary distances are published in a variety of forms. It is a simple matter to prepare such a diagram.

The *stadia slide rule* is constructed like the ordinary slide rule, except that on the slide are given values of $\cos^2 \alpha$ and $\frac{1}{2} \sin 2\alpha$ to logarithmic scale. When the index of the slide is set at the observed value of the stadia distance Ks on the main scale, the horizontal distance from principal focus to rod ($Ks \cos^2 \alpha$) is found by setting the runner at the observed vertical angle on the "$\cos^2 \alpha$" scale, and the corresponding difference in elevation is found by setting the runner to this same angle on the "$\frac{1}{2} \sin 2\alpha$" scale.

11-7 Beaman Stadia Arc. The Beaman stadia arc, in modified form known also as the *stadia circle,* is a specially graduated arc on the vertical circle of the transit or the plane-table alidade. It is used to determine distances and differences in elevation by stadia without reading vertical angles and without the use of tables, diagrams, or stadia slide rule. The stadia arc has no vernier, but settings are read by an index mark.

Horizontal Distance. In the type of stadia arc shown in Fig. 11-4, the graduations for determining distances are at the left, inside the vertical circle. When the telescope is level (vertical vernier reading zero as shown) the reading of the arc is 100, indicating that the horizontal distance is 100 per cent of the observed stadia distance. When an inclined sight is taken, the observed stadia distance is multiplied by the reading of the "Hor." stadia arc, expressed as a percentage, to obtain the horizontal distance from principal focus to rod. For example, if the stadia distance is 411 ft and the reading of the stadia arc is 99, the horizontal distance is $411 \times 0.99 = 407$ ft. The ordinary slide rule is sufficiently precise for the multiplication.

Another type of stadia arc is graduated to give the *correction,* in per cent, to be subtracted from the observed stadia distance. Thus, for the foregoing example the reading of the stadia arc would be 1, and the horizontal distance would be $411 - (0.01 \times 411) = 407$ ft. Since the value of the correction is small, the multiplication can be performed mentally; but the computation involves both a multiplication and a subtraction.

Difference in Elevation. In Fig. 11-4, the graduations for determining differences in elevation are at the right, inside the vertical circle. When the telescope is level (vertical vernier reading zero as shown), the reading of the arc is zero. When an inclined sight is taken, first the stadia distance is observed in the usual manner, that is, with the lower stadia hair on a

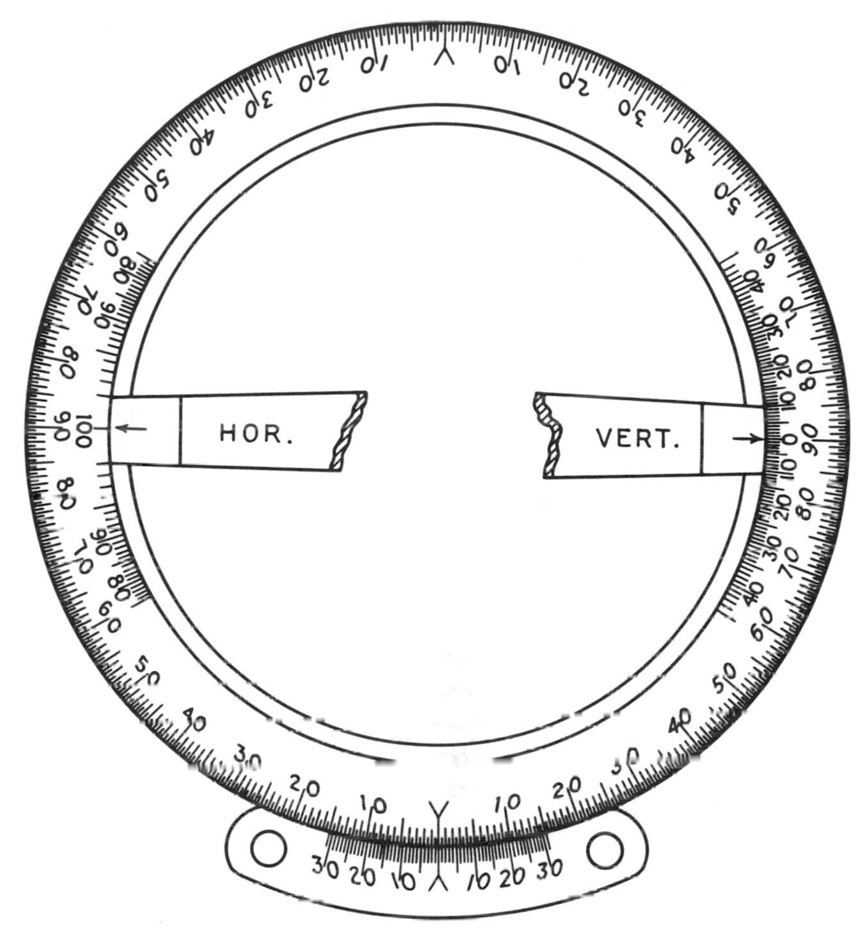

Fig. 11-4. Stadia arc (multiplier-type).

foot mark of the rod. The telescope is then either elevated or depressed slightly until the nearest graduation on the "Vert." scale of the stadia arc coincides with the index of the arc (in order to avoid interpolation), and a rod reading is taken at the point where the line of sight strikes the rod. The observed stadia distance is multiplied by the reading of the "Vert." stadia arc, expressed as a percentage, to obtain the vertical distance from center of the instrument to point last sighted on the rod. This difference in elevation, combined with the height of instrument and the rod reading, gives the difference in elevation between the instrument station and the point on which the rod is held.

example. The observed stadia interval with the rod held on a given point is 4.11 ft; with the index of the stadia arc at −9 (the minus sign indicating that the line of sight is depressed) the line of sight falls at 3.3

ft on the rod; the height of instrument (vertical distance from instrument station to center of telescope) is 4.5 ft. The difference in elevation between the instrument station and the point on which the rod is held is to be found.

The difference in elevation between center of the telescope and point sighted on the rod is

$$-9 \times 4.11 = -36.99, \text{ say, } -37.0 \text{ ft}$$

The difference in elevation between instrument station and point on which the rod is held is

$$4.5 - 37.0 - 3.3 = -35.8 \text{ ft}$$

One type of stadia arc is so graduated that when the telescope is level the reading of the "Vert." stadia arc is 50 instead of 0; when the telescope is elevated, the reading is greater than 50, and when the telescope is depressed, the reading is less than 50. In all cases, 50 is subtracted from the reading of the stadia arc, and the remainder (positive or negative as determined by the subtraction) is multiplied by the observed stadia distance to obtain the vertical distance from center of the telescope to point sighted on the rod. This arrangement of the scale on the stadia arc avoids mistakes of reading a positive value for a negative value, or *vice versa;* but it introduces an additional step (subtraction) in the computations.

General. Observations with the Beaman stadia arc do not include the effect of the instrumental constant $C = f + c$. If, for an external-focusing telescope, more precise results are desired than are yielded by the approximate formulas [Eqs. (7) and (8)], the observed values should be corrected, particularly if the vertical angles are large. The simplest method is to add 0.01 ft to the observed stadia interval.

11-8 Uses of the Stadia in Leveling.

In *differential leveling* the backsight and foresight distances are balanced conveniently if the level is equipped with stadia hairs.

In *profile leveling* and *cross-sectioning,* the stadia is a convenient means of finding distances from level to points on which rod readings are taken.

In rough *indirect leveling* with the transit, the stadia method is more rapid than any other. In running a line of levels by this method, the transit is set up in a convenient location. A backsight is taken on the rod held at the initial bench mark, first by observing the stadia interval and then by measuring the vertical angle to some arbitrarily chosen mark on the rod. A turning point is then established in advance of the transit, and similar observations are taken. The transit is moved to a new location in advance of the turning point, and the process is repeated. So long as the index mark to which vertical angles are taken is unchanged, the difference in elevation computed from the stadia formula is also the difference in elevation between the two points on which the rod is held. If the chosen index reading cannot be sighted, the vertical angle is measured to some other graduation, and this rod reading is given in the notes alongside the corresponding value of vertical angle. The notes are kept in a form similar to the following:

	Backsight		Foresight		
Station	Observations	Diff. el.	Observations	Diff. el.	El.

In the columns headed "Observations," both the vertical angle and the stadia distance (or the stadia interval) are recorded, one above the other, in the line opposite each station. Horizontal sights are taken wherever possible.

11-9 Transit-stadia Surveying. Where only the horizontal location of objects and lines is desired—as for certain reconnaissance surveys, preliminary surveys, rough surveys for the location of boundaries, and detailed surveys for maps—the transit-stadia method is sufficiently precise and considerably more rapid and economical than corresponding surveys made with transit and tape. The field party consists of a transitman, one to three rodmen, and usually a recorder. In general, the surveying procedure parallels that when the tape is used. Stadia intervals and horizontal angles (or directions) are observed as each point is sighted. Vertical angles, however, are observed only if large enough to make the horizontal distance appreciably different from the stadia distance (say, when greater than 3°), and then are estimated without reading the vernier.

On topographic and similar surveys, both the elevation of each point and its horizontal location are desired. As each point is sighted, both the horizontal and the vertical angle are measured, and the stadia interval is observed. This method may be employed merely for the location of details, the horizontal and vertical control being established by other means; or it may be employed for establishing control as well as for details.

If details only are to be located, the transit is set up at a traverse or triangulation station the elevation and location of which are known. The height of the instrument (H.I.) above the station is measured with a rod or a tape, or by swinging the plumb bob alongside a scale which has been laid off on a leg of the tripod. (As used here, the term "height of instrument" has a meaning different from that for direct leveling.) The transit is oriented by backsighting along a line the azimuth of which is known, this azimuth having been set off on the horizontal circle. The upper motion is unclamped, and sights to desired points are taken.

Where the required precision is not high, the control may be established by transit-stadia traverse, and the details may be located at the same time. For the traverse it is customary to observe the stadia interval and the vertical angle both forward and back from each setup of the transit, employing the mean value in computations. In measuring vertical angles it is customary, wherever practicable, to sight at a rod reading equal to the height of instrument above the station over which the transit is set up. Horizontal sights are taken wherever possible.

Figure 11-5 shows a page of notes for a stadia traverse for which side shots are taken as the work of running the traverse progresses. The elevation of station *P*49 has previously been determined as 785.1. Directions of the traverse lines are determined by azimuths and roughly checked by observed magnetic bearings, and stadia distances are recorded rather than the rod intervals. The backsight from station *P*50 to *P*49 checks reasonably close with the foresight from station *P*49 to *P*50. The sights to points 502 and *P*51 are horizontal; the rod reading is shown in the notes, and the difference in elevation is determined by direct leveling. In topographic surveying a sketch is included in the notes, the points to which sights are taken being numbered in the sketch.

PRELIMINARY (STADIA) SURVEY OF I. N. RY., BRIGHTON TO CAMBY

Obj.	Az.	Mag. B.	Stad. Dist.	Vert. Ang.	Hor. Dist.	Diff. Elev.	Elev.	Nov. 27, 1967 J.C. Clark, ⊼
	Inst. at Sta. P49; H.I. = 4.7						785.1	Cold T.N. Tillman, Notes
P48	169°34'	S10°30'E	637	-2°27'	636	-27.2	757.9	W.W. & H.H., Rods
P50	38°21'	N38°15'E	681	+1°14'	681	+14.6	799.7	On slope.
491	151°10'		366	-7°21'	360	-45.6	739.5	West bank Green River.
492	126°35'		418	-5°59'	413	-43.3	741.8	East " " "
493	78°05'		385	-5°36'	381	-37.4	747.7	West " " "
494	81°20'		387	-5°40'	383	-38.0	747.1	East " " "
495	298°55'		214	+6°34'	211	+24.3	809.4	Top Slope.
	Inst. at Sta. P50; H.I. = 4.9						799.7	
P49	218°21'	S38°30'W	683	-1°13'	683	-14.5	785.2	
501	294°40'		415	+4°38'	412	+33.4	833.1	Top slope.
502	16°00'		308	0° on 2.1	308	+ 2.8	802.5	On slope.
503	137°35'		374	-6°36'	369	-42.8	756.9	West bank Green River.
504	136°10'		486	-5°52'	481	-49.4	750.3	East " " "
505	5°45'		322	+7°36'	316	+41.8	841.5	Top slope.
P51	59°38'	N59°30'E	529	0° on 10.1	529	- 5.2	794.5	On slope.
506	94°25'		487	-3°36'	485	-30.5	769.2	West bank Green River.

Fig. 11-5. Stadia notes for preliminary route survey.

11-10 Errors in Stadia Surveying. Many of the errors of stadia surveying are those common to all similar operations of measuring horizontal angles and differences in elevation, previously discussed. Sources of error in horizontal and vertical distances computed from observed stadia intervals are as follows:

1 *Stadia Interval Factor Not That Assumed.* This condition produces a systematic error in distances, the error being proportional to that in the stadia interval factor.

2 *Rod Not Standard Length.* In stadia work of ordinary precision, errors from this source are usually of no consequence.

3 *Incorrect Stadia Interval.* An accidental error occurs owing to the inability of the instrumentman to observe the stadia interval exactly. This is the principal error affecting the precision of computed distances. It can be kept to a minimum by proper focusing to eliminate parallax, by care in observing, and by taking observations at favorable times.

4 *Rod Not Plumb.* This condition produces a small error in the vertical angle. It also produces an appreciable error in the observed stadia interval and hence in computed distances, this error being greater for larger vertical angles. It can be eliminated by using a rod level.

5 *Unequal Refraction.* The sight on the lower stadia hair, being nearer the earth's surface, is affected by refraction more than the sight on the upper hair; hence a positive systematic error is produced. In ordinary stadia surveying the error is of no consequence; but whenever atmospheric conditions are unfavorable, the sights should not be taken near the bottom of the rod.

Effect of Error in Vertical Angles. Errors in vertical angles have a relatively unimportant effect upon the precision of computed *horizontal distances* but a relatively important effect upon the precision of corresponding *differences in elevation.* For example, in sighting on a point 300 ft away and within the usual range of vertical angles, an error of 01′ in vertical angle produces no appreciable error in horizontal distance but produces an error in elevation of nearly 0.1 ft.

To maintain a given precision in computed values of difference in elevation, stadia intervals must be observed with much greater refinement where vertical angles are large than where they are small.

11-11 Precision of Stadia Surveying. An important advantage of transit-stadia surveying over transit-tape surveying is that, in determining distances and differences in elevation by stadia, the principal errors are largely accidental whereas, in chaining, the principal errors are largely systematic.

Many factors influence the precision of stadia surveying, and no definite statement of the precision for a given procedure can be made. Following are estimates believed to be fairly representative of several classes of stadia work, these estimates being based upon experience.

1 For side shots where a single observation is taken with sights steeply inclined and with no particular care taken to ensure the rod's being plumb, horizontal distances may have a precision lower than $1/100$, and individual differences in elevation may be in error 2 ft or more per 1,000 ft of horizontal distance.

2 Under the same conditions as in (1) but with small vertical angles and reasonable care used in approximately plumbing the rod and with lengths of sight between 200 and 1,500 ft, the precision of horizontal distances should be not lower than $1/200$; differences in elevation per 1,000 ft of horizontal distance need not be in error more than 0.3 ft if vertical angles are observed to 01′, or more than 1 ft if vertical angles are estimated to 05′.

3 For a rapid stadia traverse of considerable length run through rough country with numerous long sights, angles being measured to minutes but without special precaution to eliminate systematic errors, the error of closure may be as low as 25 ft per mile in plan and 3 ft per mile in elevation.

4 For conditions as in (3) but for country fairly level so that all vertical angles are small, the error of closure ought not to exceed 15 ft per mile in plan and 0.5 ft $\sqrt{\text{distance in miles}}$ in elevation.

5 For rough country with vertical angles up to 15°, angles to minutes, rod standardized, rod plumbed with level, sights limited to 1,500 ft and taken forward and backward from each transit station, and interval factor carefully determined, the error of closure may be less than 15 ft $\sqrt{\text{distance in miles}}$ in plan and 1 ft $\sqrt{\text{distance in miles}}$ in elevation.

6 For conditions as in (5) but for level country so that all vertical angles are small, the error of closure may be as small as 6 ft $\sqrt{\text{distance in miles}}$ in plan and 0.3 ft $\sqrt{\text{distance in miles}}$ in elevation.

7 For conditions as in (5) but stadia intervals determined by use of a target rod with two targets and observations made during cloudy days, the error of closure will probably not exceed 4 ft $\sqrt{\text{distance in miles}}$ in plan and 0.5 ft $\sqrt{\text{distance in miles}}$ in elevation.

11-12 Numerical Problems

1 With line of sight horizontal, a stadia reading is taken on a rod held at a taped distance of $600.0 + C$ ft from the transit station. The rod reading of the lower stadia hair is 0.82 ft and of the upper stadia hair is 6.77 ft. What stadia interval factor is indicated by this observation?

2 A stadia interval of 6.31 ft is observed with a transit for which the stadia interval factor is 98.5 and C is 1.00 ft. The vertical angle is $+7°42'$. Determine the horizontal distance and difference in elevation by means of (a) the exact stadia formulas for inclined sights, (b) the approximate formulas, and (c) Table III.

3 In determining the difference in elevation and the distance between two points A and B, a transit equipped with a stadia circle is set up at A and the following data are obtained: $V = +10$, $H = 98.0$, stadia interval $= 3.50$ ft, H.I. $= 4.5$ ft, line of sight at 4.5 ft on rod. The instrumental constants are $K = 100.0$ and $C = 0$ (internal-focusing telescope). The stadia circle has index marks of $H = 100$ and $V = 0$ for a horizontal line of sight. Compute the distance and difference in elevation between the two points A and B.

4 In determining the elevation of point B and the distance between two points A and B, a transit equipped with a stadia arc is set up at A and the following data are obtained: $V = 38$, $H = 3.0$, stadia interval $= 4.30$ ft, H.I. $= 4.2$ ft, line of sight at 8.6 ft on rod. The instrument constants are $K = 100.0$ and $C = 1.00$ ft. The stadia arc has index marks of $H = 0$ and $V = 50$ for a horizontal line of sight. The elevation of point A is 125.6 ft. Compute the distance AB and the elevation of point B.

5 Following are the notes for a line of stadia levels. The elevation of B.M.$_1$ is 637.05 ft. The stadia interval factor is 100.0, and $C = 1.25$ ft. Rod readings are taken at height of instrument. By use of Table III determine the elevations of remaining points.

	Backsight		*Foresight*	
Station	*Stadia interval, ft*	*Vertical angle*	*Stadia interval, ft*	*Vertical angle*
B.M.$_1$	4.26	$-3°38'$	....	......
T.P.$_1$	2.85	$-1°41'$	3.18	$+2°26'$
T.P.$_2$	3.30	$+0°56'$	2.71	$-4°04'$
T.P.$_3$	2.66	$+2°09'$	4.45	$-0°38'$
B.M.$_2$	....	......	3.09	$+7°27'$

6 Following are stadia intervals and vertical angles for a transit-stadia traverse. The elevation of station A is 418.6 ft. The stadia interval factor is 100.0, and $C = 1.00$ ft. Rod readings are taken at height of instrument. Compute the horizontal lengths of the courses and the elevations of the transit stations, using Table III.

Station	*Object*	*Stadia interval, ft*	*Vertical angle*
B	A	8.50	$+0°48'$
	C	4.37	$+8°13'$
C	B	4.34	$-8°14'$
	D	12.45	$-2°22'$
D	C	12.41	$+2°21'$
	E	7.18	$-1°30'$

7 Following are stadia intervals and vertical angles taken to locate points from a transit station the elevation of which is 415.7 ft. The height of instrument above the transit station is 4.6 ft, and rod readings are taken at 4.6 ft, except as noted. The stadia interval factor is 100.0, and $C = 1.00$ ft. Compute the horizontal distances and the elevations.

Object	*Stadia interval, ft*	*Vertical angle*
43	7.04	−0°58′
45	7.56	−0°44′ on 9.2
47	3.72	−5°36′

8 A transit equipped with a stadia arc is used in locating points from a transit station the elevation of which is 765.7 ft. The stadia arc has index marks of $H = 100$ and $V = 50$ for a horizontal line of sight. The instrument constants are $K = 100.0$ and $C = 0$ (internal-focusing telescope). The height of instrument above the transit station is 4.5 ft. Compute the horizontal distances and the elevations.

Object	*Stadia interval, ft*	*Rod reading, ft*	*Stadia arc readings*	
			V	*H*
114	3.26	3.6	18	88.3
115	7.84	5.8	35	97.7
116	2.18	4.7	39	98.8
117	1.66	4.3	76	92.6
118	8.14	6.4	69	96.2

11-13 Field Problems

Problem 30. Determination of Stadia Interval Factor

Object. To determine the stadia interval factor $K = f/i$ of transit or level.

Procedure. (1) As described in Art. 11-3*a*, employing a line about 800 ft long. Determine f and c by measurement (Art. 11-3). (2) For each observation, read the rod for lower, middle, and upper hairs. (3) Compute K for each distance for lower half-interval, upper half-interval, and full interval; take the mean of all computed values as the factor for the instrument. Discard any readings that differ widely from the others.

Hints and Precautions. (1) On fair days the line of sight defined by the lower hair should be at least 2 ft above the ground. (2) It is convenient to set the lower hair on the nearest foot mark, and this may be done without appreciable error.

Problem 31. Preliminary Traverse of Route with Transit and Stadia

Object. To obtain data for plotting a topographic map of a proposed highway route between two governing points. The data may be used in office problem 45 (Chap. 19).

Procedure. (1) Run a stadia azimuth traverse between the two points, establishing stadia stations at advantageous points near where it appears that the line will eventually be placed. (2) Determine the distance between adjacent stations by observing the stadia interval on both backsights and foresights. (3) Observe the vertical angle between instrument stations on both backsights and foresights. Record the H.I. at each setup. (4) Make the available checks before moving the transit. (5) While running the traverse, take side shots 200 to 600 ft on each side of the traverse line, as necessary to define the configuration of the land and the location of objects that might affect the proposed line. (6) Note the type of soil, any indications of rock near the surface, and the type of cover. (7) Keep notes in the form of the sample notes (Fig. 11-5).

Hints and Precautions. (1) In determining the differences in elevation and the horizontal distances between traverse points, use the mean of the two vertical angles and the mean of the two stadia readings taken along the line joining these points. (2) Before taking side shots about a station occupied, set the next stadia station in advance. (3) In running the traverse, the magnetic bearing of each line should be recorded and immediately compared with the bearing computed from the azimuth of the line. (4) Inclined distances with the vertical angles less than 3° may be considered as horizontal without appreciable error. (5) Observe vertical angles to the nearest minute. Observe azimuths of traverse lines to the nearest minute, and azimuths of sights to details to the nearest 05′ without the use of the vernier. Read the rod intercept to the nearest 0.01 ft. (6) Many shots can be taken with the telescope leveled as in direct leveling. (7) The observer should form the habit of judging distances by eye, in order to avoid large mistakes. The middle cross hair should not be mistaken for one of the stadia hairs.

Problem 32. Traverse and Location of Details with Transit and Stadia

Object. To collect sufficient data for making a topographic map of an assigned tract. The data may be used in office problem 45 (Chap. 19).

Procedure. (1) Make a rapid reconnaissance of the tract, selecting the most advantageous points for instrument stations from which areas comprising the entire area can be observed. (2) Run a closed azimuth traverse through the selected points, observing the stadia intervals and vertical angles. The allowable angular error of closure should not exceed $01' \sqrt{\text{number of sides}}$. The error of closure in elevation should not exceed $0.3 \text{ ft} \sqrt{\text{distance in thousands of feet}}$. (3) Occupy each of the traverse stations, and with the instrument correctly oriented observe the azimuth, stadia distance, and vertical angle to all changes in ground slope and to other natural and artificial features which are within range of the instrument. (4) Include in the notes a sketch drawn approximately to scale. (5) By means of Table III, a stadia slide rule, or a stadia reduction diagram, determine the horizontal distance and the elevation of each side shot.

Hints and Precautions. (1) See Hints and Precautions of the preceding field problem. (2) If the elevation of a point is not required for mapping, often the point can be located advantageously by the method of intersection, the azimuth being observed from two or more traverse stations. (3) It is sometimes advantageous, particularly if there are a large number of details, to plot the map in the field as the work progresses.

12

TRIANGULATION OF ORDINARY PRECISION

12-1 General. Triangulation is employed extensively as a means of control for topographic and similar surveys. A *triangulation system* consists of a series of triangles in each of which one or more sides are also sides of adjacent triangles. The lines of a triangulation system tie together the *triangulation stations* at the vertices of the triangles.

In a triangulation system only two angles in each triangle need be measured, but it is customary to measure all angles in order that the angular errors may be determined and adjusted. Only one line in a triangulation system need be measured, as the length of the remaining lines can be computed from this length and the measured angles. However, as a check it is desirable that two or more lines in each system be measured. These measured lines are called *base lines.*

The recent development of electronic distance-measuring devices (Chap. 3) has made feasible the method of *trilateration,* or the measurement of all sides of the triangles. Use of the method has been limited, principally because of the complexity of the computations for adjustment of a system of triangles, but it is increasing. Except for the measurement of lines instead of angles, the field procedures of trilateration are similar to those discussed herein for triangulation.

There is a quality of triangulation corresponding to every degree of precision used in traversing. Thus, triangulation may be used for a simple topographic survey covering but a few acres or it may be used to extend control of the highest order across the continent. The relative merits of the triangulation method and the traverse method are based upon the character of the terrain and not upon the degree of precision to be attained. If favorable routes are available, the method of traversing is superior to the method of triangulation; but if the terrain offers many obstacles to traverse work (such as hills, vegetation, or marsh), triangulation is superior.

The most notable example of triangulation is the precise transcontinental system established by the U.S. Coast and Geodetic Survey. The system is being developed to form a network to establish a control for the entire domain of the United States. A permanent reference point for the datum, called the "North American Datum of 1927," has been established at Meade's Ranch in Osborne County, Kansas, and to this point the precise surveys of the United States, Canada, and Mexico are referred.

Triangulation systems are classified with respect to precision according to (1) the average angular error of closure in the triangles of the system and (2) the discrepancy between the measured length of a base line and its length as computed through the system from an adjacent base line.

The Federal Board of Surveys and Maps has classified triangulation for the extensive surveys of the United States Government as shown in Table 12-1.

First-order and second-order triangulation call for methods of high precision not often necessary except on very extensive surveys. Third-order triangulation establishes points of horizontal control at short intervals in advantageous locations for detail mapping. This order is often employed in intermediate-scale and large-scale surveys of limited extent. It calls for methods of intermediate precision, although the requirements may sometimes be met by methods of ordinary precision. Fourth-order triangulation, not included in the Federal classification, calls for methods of ordinary precision. Herein the discussion is limited to triangulation of ordinary and, to some extent, intermediate precision.

12-2 Triangulation Figures. In a narrow triangulation system a chain of figures is employed, consisting of *single triangles, polygons, quadrilaterals,* or combinations of these figures. A wide system consists of similar combinations, irregularly overlapping and intermingling so that several routes exist for computing the length of triangle sides and the length of each base line from an adjacent base line.

1 Chain of Triangles. In the chain of single triangles (Fig. 12-1) there is but one route by which distances can be computed through the chain.

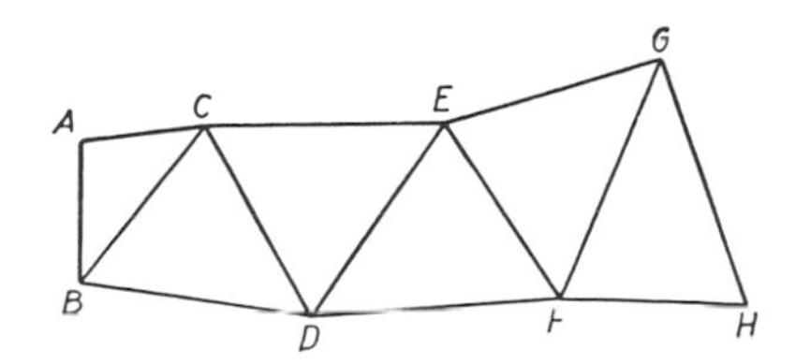

Fig. 12-1. Chain of single triangles.

As the sum of the measured angles in each triangle normally will not equal exactly 180°, the angles are adjusted to satisfy this requirement before the distances are computed.

2 Chain of Polygons. In triangulation, a polygon, or "central-point figure," is composed of a group of triangles, the figure being bounded by three or more sides and having within it a triangulation station at a vertex

Table 12-1 Classification and standards of precision for triangulation

	First order			*Second order*		
	Class I (special)	*Class II (optimum)*	*Class III (standard)*	*Class I*	*Class II*	*Third order*
Principal uses	Urban surveys, scientific studies	Basic network of U.S.	All other (state, private)	Area networks and supplemental cross arcs in national net	Coastal areas, inland waterways, and engineering surveys	Topographic mapping
Base measurement:						
Probable error not to exceed 1 part in	1,000,000	1,000,000	1,000,000	1,000,000	500,000	250,000
Triangle closure:						
Average error not to exceed	1″	1″	1″	1.5″	3″	5″
Length closure:						
Discrepancy between measured and computed length of base line not to exceed 1 part in	100,000	50,000	25,000	20,000	10,000	5,000

common to all the triangles. Figure 12-2a illustrates a chain composed of a six-sided polygon $FEGJKI$ and two five-sided polygons.

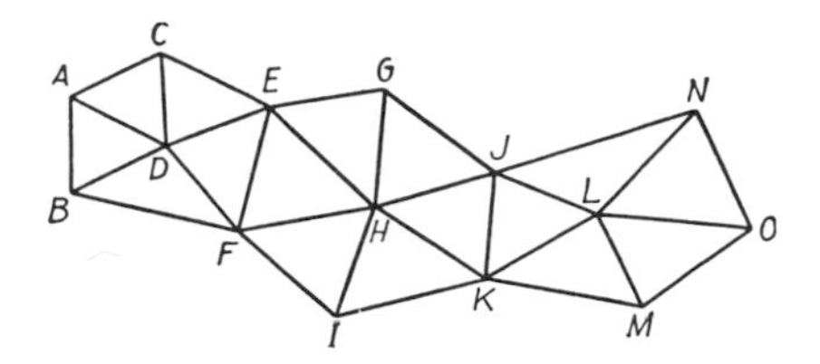

Fig. 12-2a. Chain of polygons.

The sum of the measured angles in each triangle of the polygon should equal 180°; also, the sum of the angles about the central point should equal 360°. Further, the length of any side may be computed by two routes, and these two computed lengths should agree. The observed angles are so adjusted by computation that these three conditions exist.

3 *Chain of Quadrilaterals.* A quadrilateral differs from a four-sided polygon in that there is no triangulation station within the figure. In Fig. 12-2b, consider one of the quadrilaterals, as $ACDB$. The measured angles give values for four triangles ACD, CBA, DBA, and BCD, in each of which the sum of the angles should equal 180°. In addition, the length of any line should be the same when computed by one route as when computed by another.

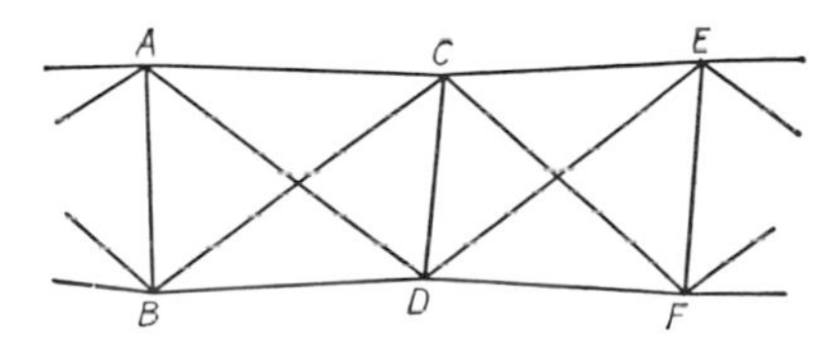

Fig. 12-2b. Chain of quadrilaterals.

Choice of Figure. The chain of single triangles is the simplest but does not afford so many checks as the other forms; hence for a given precision the base lines would need to be placed closer together. The chain of triangles is satisfactory for work of low or—in some cases—ordinary precision. For more precise work, usually quadrilaterals or polygons are used; quadrilaterals are best adapted to long narrow systems, and polygons to wide systems.

12-3 Strength of Figure. Values computed from the sine of angles near 0° or 180° are subject to large ratios of error. Since in triangulation computations nearly always the sine is used, it follows that angles near 0° and 180° are undesirable. It has been found in practice that satisfactory results can be secured for most purposes if the angles *used in the computations* fall between 30° and 150°. However, many angles measured in the field are not used in computing the length of the sides in the system. Such angles may be near 0° or 180° without impairing the excellence of the system as a whole.

This and other principles may be made clear by reference to Fig. 12-3. In the figure, let AB represent a side the length of which is known.

This side and all others in the system the lengths of which are desired are shown by heavy lines. The sine law, used in computing the lengths, states that in any triangle the sides are proportional to the sines of the angles opposite; accordingly, the angles affecting the computed lengths of sides in each triangle are those opposite the known and computed sides.

Consider the quadrilateral *CEFD*. The length of the side *CD* results from calculations carried through the quadrilateral *ACDB*. Then the length of *CF*, in triangle *CDF*, is computed by use of the known side *CD* and the angles 78° and 88° (13° + 75°); and *EF*, in triangle *CEF*, may be computed by use of the known side *CF* and the angles 93° and 72° (60° + 12°). In these two computations involving the small angles (12° and 13°) it is seen that neither one is used separately and so neither, by itself, has any effect upon the length of the side *EF*. Similarly, the side *ED*, in triangle *CED*, is computed by the use of the side *CD*, and again, neither of the small angles (14° and 15°) is used separately in computing the length of *EF*. Thus it is seen that the side *EF* can be computed by two independent series of computations neither of which is affected detrimentally by the small angles involved. As a matter of fact, the quadrilateral *CEFD* is a stronger figure than is *ACDB* in which no angle less than 36° occurs.

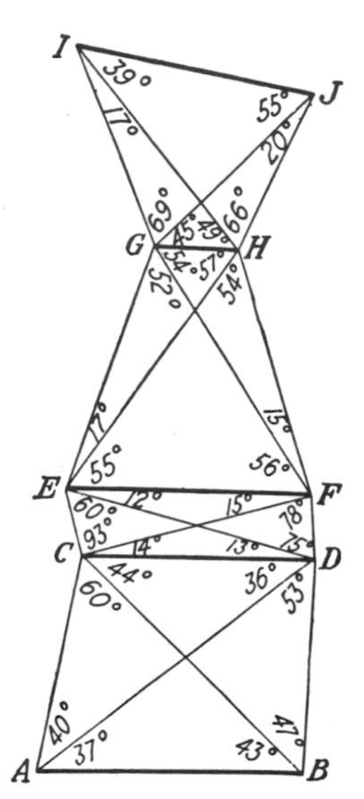

Fig. 12-3. Strength of figure.

By a similar analysis of the quadrilateral *EGHF*, it will be found, however, that it is impossible to compute the length of the side *GH* without making use in one series of computations of the angle 15° separately and in the other of the angle 17° separately. Therefore, by any possible means, the computed length of the side *GH* must be affected by the large ratios of error resulting from the use of small angles separately. The large degree of uncertainty thus introduced into the computed length of the side *GH* will be effective in all dependent values computed therefrom, as, for example, the length of the side *IJ* in the system shown.

As an aid in checking which of several alternative figures (or chains of figures) is to be used in triangulation, the relative strengths of the figures can be determined by computations based upon the size of the angles, the number of directions to be observed, and the number of geometric conditions that must be satisfied. Considerations of economy may render one figure more desirable than another even though it may be the weaker of the two. For example, in Fig. 12-3 the quadrilateral *ACDB* may be more desirable than *CEFD* because the progress of the work is advanced more rapidly by the former than by the latter, the ratio of progress being about that of the lengths BD/DF.

Computation of R. The relative strength of figure can be evaluated quantitatively in terms of a factor R based on the theory of probability; *the lower the value of R, the stronger the figure.* Strength of figure is a factor to be considered in establishing a triangulation system for which the computations can be maintained within a desired degree of precision. For example, for third-order triangulation it is desirable that R for a single figure not exceed 25 and that R between two base lines not exceed 125. In some cases it may not be necessary to occupy all the stations of the system or to observe all the lines in both directions. Furthermore, by means of computed strengths of figure, alternative routes of computation (chains of elemental triangles) can be compared and the best route chosen. The methods are described in detail in Ref. 19. The following brief treatment gives the essential relations for computing R.

Let C = number of conditions to be satisfied in figure

n = total number of lines in figure, including known line

n' = number of lines observed in both directions, including known line if observed

s = total number of stations

s' = number of occupied stations

D = number of directions observed (forward and/or back), excluding those along known line

δ_A, δ_B = respective logarithmic differences of the sines, expressed in units of the sixth decimal place, corresponding to a change of 1 sec in the "distance angles" A and B of a triangle. The distance angles are those opposite the known side and the side required

$\Sigma(\delta_A^2 + \delta_A\delta_B + \delta_B^2)$ = summation of values for the particular chain of triangles through which the computation is carried from the known line to the line required. Values of $(\delta_A^2 + \delta_A\delta_B + \delta_B^2)$ for a triangle are given in Table XI

Then

$$C = (n' - s' + 1) + (n - 2s + 3) \tag{1}$$

$$R = \frac{D - C}{D} \Sigma (\delta_A^2 + \delta_A\delta_B + \delta_B^2) \tag{2}$$

example. It is desired to compute the strength of the quadrilateral

ACDB in Fig. 12-3 for computation of the side *CD* from the known side *AB* when all lines are observed in both directions. From Eq. (1)

$$C = (6 - 4 + 1) + (6 - 8 + 3) = 4$$

$$\frac{D - C}{D} = \frac{10 - 4}{10} = 0.60$$

The computation may be carried through any of four chains of triangles, as indicated in the accompanying tabulation.

Common side	*Chain of triangles*	*Distance angles, deg.*	$(\delta_A^2 + \delta_A\delta_B + \delta_B^2)$ *Each*	Σ	*R*
AC	*ACB*	60; 43	9.8	32.0	19
	ACD	40; 36	22.2		
AD	*ADB*	90; 53	2.4	7.6	5
	ACD	104; 40	5.2		
BC	*BAC*	77; 60	2.0	5.7	3
	BCD	89; 47	3.7		
BD	*BAD*	53; 37	15.2	28.0	17
	BCD	47; 44	12.8		

It is seen that the strongest chain consists of triangles *BAC* and *BCD* and that the relative strength of the quadrilateral is 3.

By similar computations, for the remaining quadrilaterals, the least values of *R* are found to be *CEFD*, 0; *EGHF*, 29; and *GIJH*, 20. Therefore, the strongest quadrilateral is *CEFD* and the weakest is *EGHF*, as previously discussed. The strength of the figure as a whole (for *IJ* computed from *AB*) is represented by a value of *R* of 52, which is the sum of the lowest values for the four consecutive quadrilaterals in the chain.

12-4 Base Nets. In practice, for economic reasons usually the base lines are much shorter than the average length of the sides of the triangles in the main triangulation system. In order to obtain the required precision in the computed length of the sides of the main triangles, it is necessary to expand the base line through a group of smaller triangles called the *base net*. Figure 12-4 shows examples of base nets affording quick and accurate expansion of the base line to the longer sides of the system. In triangulation of ordinary precision, base lines are placed perhaps 20 to 60 triangles apart, the distance depending upon the strength of the figures.

METHODS

12-5 General. The work of triangulation consists of the following steps:

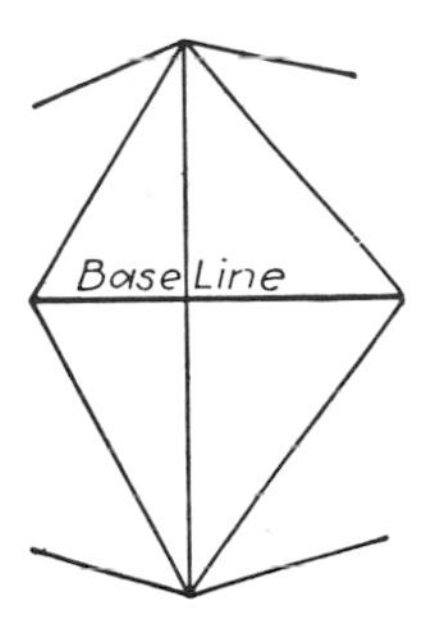

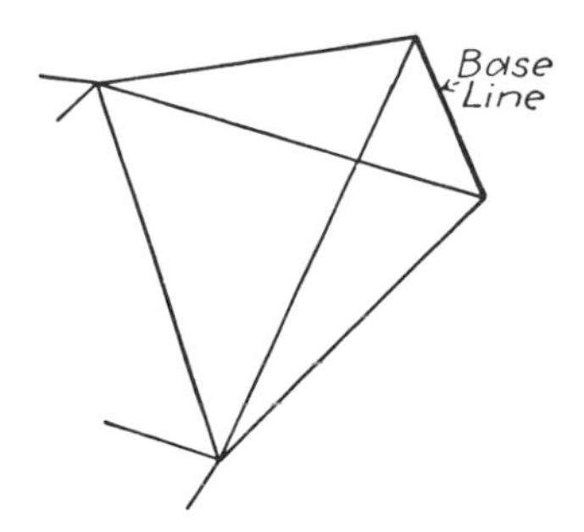

Fig. 12-4. Base nets.

1 Reconnaissance, to select the location of stations.

2 Erection of signals and, in some cases, tripods or towers for elevating the signals and/or the instrument.

3 Measurement of angles between the sides of triangles (for trilateration, measurement of lengths of the triangle sides).

4 In most cases, astronomical observations at one or more triangulation stations, in order to determine the true meridian to which azimuths are referred (Chap. 17); also in extensive systems to determine the geographical coordinates (latitude and longitude) of all points in the system.

5 Measurement of the base lines.

6 Computations, including the adjustment of the observations, the computation of the length of each triangle side, and the computation of the coordinates of the stations.

Herein the description of methods will be concerned principally with triangulation of ordinary precision (corresponding roughly to third-order or fourth-order triangulation). Triangulation of high or low precision differs from that of ordinary and intermediate precision as follows:

1 *Triangulation of High Precision.* The reconnaissance may amount to a preliminary survey. Extensive use is made of tall towers and signals and of signaling devices for reflecting sunlight or for night work. Angles are measured with either the repeating theodolite or the direction instrument. The angles of a system are adjusted by the method of least squares, and account is taken of spherical excess. The computations for latitude and longitude of the various stations take into account the curvature of the earth.

2 *Triangulation of Low Precision.* There is practically no reconnaissance, and often the stations are selected as the work progresses. The stations are marked with a stake, pole, or portable tripod. The base line is measured by the ordinary methods of taping or sometimes even by stadia. The angles of the triangles are not necessarily adjusted to meet the known geometric and trigonometric conditions. No correction is made when the instrument is not set up exactly over the station. No astronomical observations are made. Often the method of graphical triangulation with the plane table is employed (Art. 13-10).

12-6 Reconnaissance. Because of its influence on the accuracy and economy of the work, the reconnaissance is of the greatest importance. Reconnaissance consists in the selection of stations, and it determines the size and shape of the resulting triangles, the number of stations to be occupied, and the number of angles to be measured. In this connection are considered the intervisibility and accessibility of stations, the usefulness of stations in later work, the strength of figures, the cost of necessary signals, and the convenience of base-line measurements.

After a preliminary study of available maps and information, the chief of the party examines the terrain, choosing the most favorable sites for stations. If the information is not available from maps, angles and distances to other stations are estimated or measured roughly en route, so that the suitability of the system as a whole can be examined before the detailed work is begun. Angles are determined either directly by use of the prismatic compass or similar hand instrument or graphically by use of the plane table. Distances are determined either directly by pacing or odometer or graphically by use of the plane table. Where forest growth is present, the observer must make use of standing trees or guyed ladders or poles to establish visibility with adjacent stations.

12-7 Signals and Instrument Supports. At an instrument station it is desirable to have a signal of a type that will permit placing the instrument directly over the station when angles are to be measured. In a small triangulation system with triangle sides only a few hundred feet in length, and with but few angles to be measured, portable signals such as poles or light tripods with a plumb line may be used. In larger systems, often the signal is a permanent tripod such as that shown in Fig. 12-5.

Where the instrument must be

Fig. 12-5. Tripod signal. (*U.S. Coast and Geodetic Survey.*)

elevated to secure visibility, a combined observing tower and signal may be built. A central tripod supports the instrument; around this is a three-sided or four-sided structure supporting the platform upon which the observer stands. Thus the instrument tower is free from the vibrations caused by movements of the observing party.

In addition to the *major stations* from which observations are taken, often *minor stations* for local control are established by observations from the major stations. Minor stations may be signals erected at desirable locations, or they may be such objects as lone trees, spires, flagstaffs, and chimneys.

The best time for observing is in the late afternoon or at night. For night observations, an electric lamp is used as a signal.

12-8 Angle Measurements. For triangulation of intermediate precision, usually the angles in the system are measured by means of a repeating theodolite (Art. 9-6). For triangulation of ordinary and low precision the ordinary transit may be used. In triangulation of ordinary precision, angles are measured by the method of repetition described in Art. 9-10, the number of repetitions and procedure employed depending upon the required precision. The instrument should be protected from sun and wind; the air should be clear; and great care should be taken in setting up the instrument and in observing. In triangulation of low precision, angles are usually doubled, with a reversal of the telescope between measurements.

Azimuth Determinations. In computing the coordinates of triangulation stations, a meridian of reference, either true or assumed, is used, and azimuths of all lines in the system are computed from corrected angles. For an extensive system, the true meridian is employed, and account must be taken of the convergency of meridians. The direction of the true meridian or of the true azimuth of any line can be determined by astronomical observations, as described in Chap. 17.

BASE-LINE MEASUREMENT

12-9 General. For base-line measurements of ordinary precision either the steel tape or the invar tape may be employed, but for measurements of higher precision the invar tape is always used. Often a "long tape" (length 50 m to 500 ft) is used. It is desirable that the tape be compared with a standard under the conditions of tension and support that will be employed in the field.

Where the base line is along a paved highway or a railroad, usually measurements are made with the tape supported over its entire length and at a time when the temperature of the supporting surface (highway or rail) is not appreciably different from that of the surrounding air.

Where the base line is over uneven ground, end supports for the tape are provided, usually by substantial posts, perhaps 2 by 4 in., driven firmly into the ground. These are placed on a transit line at intervals of one tape length, as nearly as can be determined by careful preliminary measurements. A strip of copper or zinc is tacked to the top of the post to receive the markings. Portable tripods are also used to some extent as tape supports. Profile levels are run over the top of the end supports to determine the gradient from support to support. The tape is usually supported at one, two, or three points between the end supports. These intermediate points are placed accurately on the grade line by driving nails at grade in stakes placed on line at the proper intervals.

The equipment for base-line measurement of ordinary or intermediate precision includes at least one verified tape; two stretcher devices for applying tension, one of which is equipped with a spring scale (Fig. 12-6) or a weight and pulley; two or three thermometers; a finely divided pocket scale; dividers; and a needle or a marking awl.

The party consists of four to six men whose duties are indicated by the following description of the procedure. The proper tension is applied to the tape by means of the stretchers, with the spring scale (or weight and pulley) at the forward end of the tape (Fig. 12-6). When the rear end of the tape is observed to coincide with the previously established mark, and when the proper tension is applied, the position of the forward end of the tape is marked by a fine line engraved by means of a needle or marking awl on the metal strip on top of the post (Fig. 12-6). Ther-

Fig. 12-6. Base-line measurement: making forward contact.

mometers fastened to the tape, one near each end, are read at the time that the tape length is marked on the forward post.

The tape is then carried forward without allowing it to drag on the ground, and the process is repeated. After a few measurements, the end of the tape will probably fall either beyond or short of the limits of the metal strip of the next forward post; accordingly it will be necessary to use either *set backs* or *set forwards* (also called *setups*), which are measurements of small distances made by means of a finely divided pocket scale and a pair of dividers. The conditions of measurement are recorded in detail, and notes are kept in the following form:

From stake No.	To stake No.	Temperature, °F			Set forward	Set back
		Forward	Middle	Rear		

12-10 Errors in Base-line Measurement; Corrections. The various errors and corrections in ordinary measurements with the tape are discussed in Art. 3-8. In the measurement of base lines the effect of temperature is the most serious source of error; hence in more precise work it is customary to use an invar tape and to measure the base line on cloudy days or at night, when the air and the ground are at nearly the same temperature. Corrections are made for length of tape, temperature, and slope. They are also made for sag and tension when conditions of use make such corrections necessary. Errors due to imperfect marking of the tape lengths on the metal strips are reduced to a minimum by careful manipulation.

Following is an example showing the corrections applied to the measured length of a base line:

example. The length of a base line is recorded as 3,243.063 ft, and the average observed temperature is 59.7°F. In the field the tape is supported and the tension is maintained the same as when the tape was compared with the standard. The standardization data are as follows: Length at 68°F = 100.0214 ft (tape supported at 0, 50, and 100-ft marks; tension = 10 lb); coefficient of thermal expansion = 0.00000645 per degree Fahrenheit. Corrections (including corrections for slope) are as follows:

	Feet
Recorded length	3,243.063
Length correction	+0.694
Total set forwards	+0.364
Total set backs	−0.158
Temperature correction	−0.174
Total slope correction	−0.364
Length of base	3,243.425

12-11 Reduction to Sea Level. It is sometimes necessary to reduce the length of the base line to the equivalent length at mean sea level. The correction C_l to be subtracted from the actual length is given by the equation

$$C_l = \frac{LA}{R} \tag{3}$$

in which L is the length of the base line, A is the mean altitude of the base line above sea level, and R is the radius of the earth (mean $R =$ 20,889,000 ft, log $R = 7.31992$). For convenience, the correction may be determined from published tables. For a distance of 1,000 ft and a difference in elevation of 1,000 ft, it is 0.0487 ft, and as shown by Eq. (3) it is directly proportional both to distance and to difference in elevation.

Conversely, to compute the length at the elevation of a survey corresponding to a given length at sea level, Eq. (3) may be solved with the sea-level length as L, and the correction added.

COMPUTATIONS

12-12 General. In triangulation of low precision, the measured angles and base line may be used, without correction or adjustment, for computation of the lengths of the remaining sides. In triangulation of ordinary and higher precision, however, the observed angles are corrected before the lengths of the sides are computed. If sights have been taken from, or to, any point which is not exactly at a triangulation station, the angles at that point are corrected for such eccentricity by a procedure called *reduction to center*. The angles about each station are adjusted to total 360°. In precise work involving large triangles, the angles of each triangle are corrected for *spherical excess*, which amounts to about 1 sec for each 75 sq miles of area. The system of triangles or quadrilaterals is adjusted to make the angles meet the known geometric and trigonometric conditions. The lengths of the triangle sides are then computed from the corrected angles and the base line, and the coordinates (plane or geographic) of the stations are computed.

12-13 Adjustment of Chain of Triangles. A single chain of triangles is adjusted in two steps: (1) the *station adjustment* or *local adjustment*, to make the sum of the angles about each point equal 360° and (2) the *figure adjustment*, to make the sum of the three angles in each triangle equal 180° (plus the spherical excess).

In precise triangulation the station adjustment and the figure adjustment are made in one operation by the method of least squares, but the following approximate solution yields results sufficiently precise for most cases of triangulation of ordinary precision.

To make the sum of the angles about each point equal 360°, the observed angles are added together and the sum is subtracted from 360°. The resulting difference is divided by the number of angles, and the quantity so found is added algebraically to each angle. To make the sum of the angles in each triangle equal 180°, a similar plan is followed, using the values obtained by the station adjustment; that is, the three angles of each triangle are added together, and their sum is subtracted from 180°. One third of the difference is added algebraically to each of the three angles. If certain angles are measured with a higher precision than others, the observations may be weighted (Arts. 2-8 and 2-9).

example. In the accompanying tabulation are given the observed and adjusted angles of a triangle *ABC* for which the adjusted values are shown in Fig. 12-7. In succeeding columns of the tabulation are shown the adjusted values from the station adjustment and finally from the figure adjustment. If the triangulation were one of a chain, the number of angles around each station would of course be greater than two as shown in this example, but the procedure of adjustment would be the same.

			Station adjustment	*Figure adjustment*	
Station	*Angle (clockwise)*	*Observed value*	*Adjusted value*	*Triangle ABC*	*Adjusted value*
A	*BAC*	84°32′41″	84°32′38″	84°32′38″	84°32′40″
	CAB	275°27′25″	275°27′22″		
	Sum	360°00′06″	360°00′00″		
B	*CBA*	48°13′56″	48°13′57″	48°13′57″	48°13′59″
	ABC	311°46′02″	311°46′03″		
	Sum	359°59′58″	360°00′00″		
C	*ACB*	47°13′20″	47°13′19″	47°13′19″	47°13′21″
	BCA	312°46′43″	312°46′41″		
	Sum	360°00′03″	360°00′00″		
			Sum	179°59′54″	180°00′00″

12-14 Adjustment of Quadrilateral. As in the case of the chain of triangles, the angles around each station of a quadrilateral are adjusted to total 360° before the figure adjustment is made. In the figure adjustment, two conditions are considered: (1) the *geometric condition* that the sum of the interior angles of a plane rectilinear figure is equal to $(n - 2)\ 180°$, in which n is the number of sides of the figure and (2) the *trigonometric condition* that in any triangle the sines of the angles are proportional to the lengths of the sides opposite.

First the station adjustment is made; then the geometric condition is satisfied by adjustment of the angles of the four overlapping triangles forming the quadrilateral. Then the trigonometric condition is satisfied by means of computations involving the sines of the angles, the angles being adjusted so that the computed length of an unknown side opposite a known side will be the same regardless of which of the four possible routes is used.

12-15 Adjustment of Chain between Two Base Lines. If two base lines are measured, the length of each side in the connecting chain of triangles or quadrilaterals must be the same when computed from one base line as when computed from the other. An exact solution is possible only by the method of least squares, but the following approximate methods may be used in triangulation of ordinary precision in the case of a single chain of figures.

The figures (triangles or quadrilaterals) are adjusted individually as previously described. The lengths of the sides are then computed from each base line to a common line about midway between them. This common side may then be corrected to reconcile the two computed values of its length, with equal or different weights being assigned to the two computed values as desired, based upon the known conditions. The effect of this correction may then be carried back through each half of the chain, as follows:

If the precision of the *angular* measurements is relatively high as compared with that of the linear measurements, the lines of each half of the chain are corrected in proportion to their lengths as compared with the length of the common line, leaving the angles unchanged.

If, however, the precision of the *linear* measurements is relatively high, the correction is tapered off from the full amount at the common line to zero at each base line, the correction to each line being not only proportional to the length of the line but also roughly proportional to the relative distance of the given line from the base line. This procedure changes the values of the angles, and the new values of the angles are used in further computations.

Between these two extremes, the procedure depends upon the relative precision of the angular and the linear measurements, and weights may be assigned accordingly.

12-16 Computation of Triangles and Coordinates. In computing the lengths of the sides and the coordinates of the stations in a triangulation system, it is desirable to follow an orderly procedure to expedite the work and to avoid mistakes. Convenient arrangements for these computations for plane triangulation are given in Figs. 12-7 and 12-8.

Station or line	Angle or distance	Logarithm	Figure
c	1,432.58 ft	3.156119	Given:
C	47°13'21"	0.134306 (colog)	
A	84°32'40"	9.998028	
B	48°13'59"	9.872657	
a	1,942.94 ft	3.288453	
b	1,455.74 ft	3.163082	

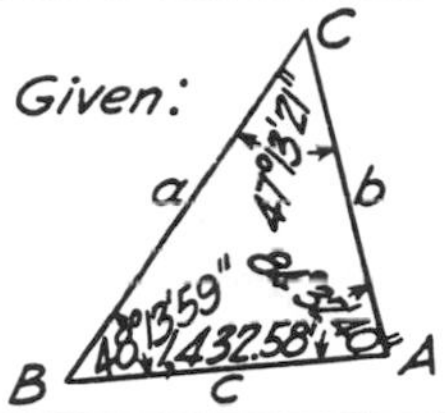

Fig. 12-7. Computation of triangle.

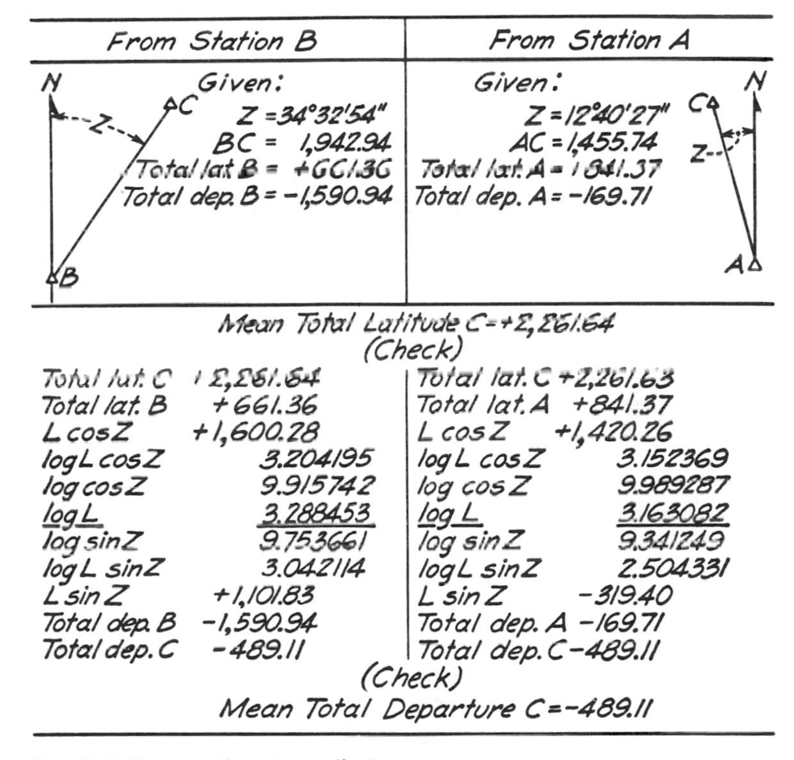

Fig. 12-8. Computation of coordinates.

Triangles. A sketch of the figure is drawn (Fig. 12-7), and the vertices are lettered as A, B, and C in a clockwise direction, beginning with the side the length of which is known. The sides opposite the vertices are indicated by the corresponding lower-case letters, as a, b, and c. The sine relation states that

$$b = c\,\frac{\sin B}{\sin C} \qquad \text{or} \qquad \log b = \log c - \log \sin C + \log \sin B \tag{4}$$

Accordingly, if the logarithms are recorded in the column of logarithms in the order log *c*, colog sin *C*, log sin *A*, and log sin *B*, then log *a* is found by covering log sin *B* with a narrow strip of paper and adding the other three values. Also log *b* is found by covering log sin *A* and adding the other three values. Finally, the distances *a* and *b* are found as the numbers corresponding to their respective logarithms. An example of the computations is shown in the tabulation of Fig. 12-7, for which the known data are given on the sketch of the triangle.

Coordinates. Figure 12-8 shows a form and example for computing the coordinates of a station *C* from each of the stations *B* and *A*. The known plane coordinates (total latitude and total departure) of *B* and *A* and the known bearings and lengths of *BC* and *AC* are shown at the top of the figure. The computation is carried out as indicated, with due regard to signs. Beginning with log *L* in the tabulation, computations for total latitude are made reading upward, and computations for total departure are made reading downward.

STATE PLANE COORDINATES

12-17 State Systems. One activity of the Federal agencies engaged in triangulation is the establishment throughout the country of monuments whose geographic coordinates, or latitude and longitude, are known. It is desirable to refer local surveys, which employ plane coordinates, to such monuments in order as far as possible to avoid discrepancies at the edges of adjacent surveys and in order to coordinate the work of surveying as a whole.

For many years, information has been available whereby the geographic coordinates of available triangulation stations may be converted into plane coordinates of a local system, and *vice versa*. The projection is on a plane tangent to the spheroid representing the earth, and its use is limited to areas not farther than about 20 miles from the origin of the local system. The tangent-plane projection has been employed on surveys of several large cities but not to a great extent elsewhere.

A far greater opportunity for use of the national triangulation system has come about through the adoption of state systems of plane coordinates, whereby one set of plane rectangular coordinates is made to serve the whole area of a small state or a portion (usually half) of the area of a large state. Many additional triangulation stations have been established and monumented by the U.S. Coast and Geodetic Survey. A map projection has been chosen for the state or portion thereof such that the errors of projection will not exceed 1/10,000 in a width of 158 miles and, therefore, will be negligible for most local surveys. For states of greater extent east and west a Lambert conformal conic projection (Art. 15-23) is used, while for states of greater extent north and south a transverse Mercator projection (Art. 15-25) is employed. Both projections are conformal; that is, at any point on the projection the angles are the correct angles. Along the two standard parallels of the Lambert projection and along the two standard meridional lines of the Mercator projection the scale of the projection is exact. Variations from exact scale with distance from the standard lines are given in published tables, and full allowance may be made for these scale errors. Also, distances at the elevation of the survey can be converted to equivalent sea-level distances (Art. 12-11) on which the coordinate systems are based, or *vice versa*. Reference axes for each zone are such that the x and y coordinates of all points within the area will be positive; the location of any point can be designated by simply stating these two coordinates, as illustrated in Art. 18-10.

At a distance of ¼ to 2 miles from each triangulation station is established an *azimuth mark*, or monument, which can be sighted from the station. The plane coordinates of the station and the plane azimuth to the mark are published for the information of engineers and surveyors. (Care must be taken not to confuse true or geodetic azimuths, which take account of the convergency of meridians, with plane or grid azimuths, which are referred to a single meridian for the zone.) In addition to the monuments for horizontal control, a system of bench marks for vertical control has been established throughout the country.

To make use of the state system for a local survey, the surveyor sets up the transit at a nearby station of the triangulation system, orients it on the line of known plane azimuth, and runs a survey (by traversing or triangulation) to the area under consideration. The coordinates of any point in the local survey can then be conveniently computed in terms of the state system by the ordinary methods of plane surveying. Preferably the survey should be checked by traversing either back to the original station or to another triangulation station. If elevations are determined, they are referred to the established system of bench marks.

An important advantage of the state-wide systems of coordinates is that the location of obliterated monuments can be reestablished with certainty and checked from various control points. Already most of the states have legalized the use of the state coordinate system for establishing and describing the monuments which mark the boundaries of land. For general mapping purposes, the coordinate system ensures reasonable agreement between maps of adjacent or overlapping areas. For extensive surveys such as those for routes, waterways, or municipal areas, the coordinate system facilitates checking, unifies the surveys of various portions of the project, and permits economies to be made in the conduct of the work. The use of state coordinates is simple and should be more widely adopted by surveyors.

12-18 Numerical Problems

1 For the measurement of a base line the following data are given: The Bureau of Standards certificate states that the tape has a length of 99.942 ft at 68°F when supported at the 0 and 100-ft points and under a tension of 10 lb; the coefficient of thermal expansion of the tape is 0.00000645 per degree Fahrenheit; the tape weighs 1½ lb. The field records give the measured length as 1,418.314 ft; the average temperature was 63.6°F; the stakes were set on a 2 per cent grade; the sum of the set forwards was 0.234 ft; the sum of the set backs was 0.114 ft. The interval and tension were the same as those used for the standard comparison. Compute the length of the line.

2 For the conditions given in problem 1, compute the normal tension.

3 For the conditions given in problem 1, assume that the interval between supports in the field is 50 instead of 100 ft. Compute the corresponding change in the distance between end marks of the tape.

4 For a given triangle *ABC*, the observed angles are as follows:

Station	*Interior angle*	*Other angles about station*
A	78°30′28″	281°29′36″
B	54°17′30″	78°45′03″, 95°06′11″, 131°51′12″
C	47°12′16″	110°27′15″, 202°20′32″

Determine the most probable value of the interior angles at *A, B,* and *C.*

12-19 Field Problem

Problem 33. Measurement of Base Line

Object. To measure a short base line with the steel or invar tape. It is assumed that the base-line site has already been chosen and that permanent marks have been established at the end points.

Procedure. (1) Install strips of zinc or copper, perhaps ½ by 5 in., along the base line at intervals of the length of tape with which measurements are to be taken. If the base line is not to be measured along a smooth surface, as a paved highway or a railroad track, install substantial posts and intermediate stakes at grade as described in Art. 12-9. Line in the strips with the transit. (2) If the base line is to be measured on posts, build a substantial table over each end of the base line, and tack a metal strip directly above the end point. Carefully project each end point of the base line to the strip as follows: Set up the transit at about 25 ft from the table, and sight to the end point on the ground. Elevate the telescope, and mark two points on the line of sight a few inches apart on the metal strip above. Scratch a straight line between these two points. Repeat this procedure with the transit set up so that the line of sight is approximately at a right angle with the first line of sight. (3) Run levels over the line, determining the elevation of all marking strips and intermediate supports. (4) Measure the line, following the procedure of Art. 12-9. At the end of the line, usually there will be a fractional part of a tape length; mark on the strip at the tape division that falls nearest the end point, and measure the remaining distance with a finely divided scale. (5) In the same manner measure the base line at least four times. (6) Make the necessary reductions to determine the correct length.

Hints and Precautions. (1) Measurements should not be taken with the steel tape in sunlight, or with a suspended steel or invar tape when a cross wind is blowing. (2) If measurements are to be taken with the invar tape in sunlight, at no place should the grade line come closer to the ground than 1 ft. (3) When the required tension has been applied, the tape should be set in vibration long enough to allow the amount of tension to become uniform throughout its length. The tapeman should take particular care to see that the device for applying tension is free from friction; and the device should be held at such a height that the tape will barely rest on the adjacent support. (4) Unless the spring scale is already adjusted for weighing in a horizontal position, it should be so calibrated.

13
PLANE-TABLE SURVEYING

13-1 Plane Table. A plane table consists essentially of (1) a *drawing board* mounted on a tripod and (2) an *alidade* having the vertical plane of the line of sight fixed parallel to a straightedge which rests upon but is not attached to the board (Fig. 13-1*a*). A sheet of drawing paper, called a *plane-table sheet,* is fastened to the board.

The location of any object is determined as follows (Fig. 13-2): With the straightedge through the plotted point *o* representing the station *O* occupied by the instrument, the line of sight is directed to the object *A,* and a line *oa* is drawn along the straightedge on the plane-table sheet; this line represents the direction from station to object. The measured distance *OA* between station and object is then plotted to scale from *o,* thus locating *A* on the map at *a.*

The term "plane table" is somewhat ambiguous, being used sometimes to designate only the board with its supporting tripod and sometimes (more generally) both the table proper and its accompanying alidade.

In the construction of maps the plane table has wide application, especially for securing the details; but as no record of numerical values is secured, the instrument is useful for mapping only. It is a valuable and commonly used means of completing the compilation of maps from aerial photographs (Chap. 23), especially where the ground is hidden by overgrowth.

13-2 Tables. The *Coast Survey table* (Fig. 13-1*a*) is the most stable of the three types in common use. The drawing board, 24 by 31 in., is attached to a metal casting which is so arranged that the board can be leveled accurately by means of three leveling screws. By means of a clamp and tangent screw the board can be fixed in any position in azimuth. The plane-table sheet is held in position by metal spring clamps, thus permitting the use of a sheet larger than the board.

The drawing board of the *Johnson table* is either 18 by 24 in. or 24 by 31 in. It is attached to the tripod head by means of a ball-and-socket joint having two clamp screws. When one clamp is loosened, the ball is free to rotate in the socket and the board can be leveled; this clamp is then tightened. When the other clamp is loosened, the board can be rotated about the vertical axis and thus can be oriented. The plane-

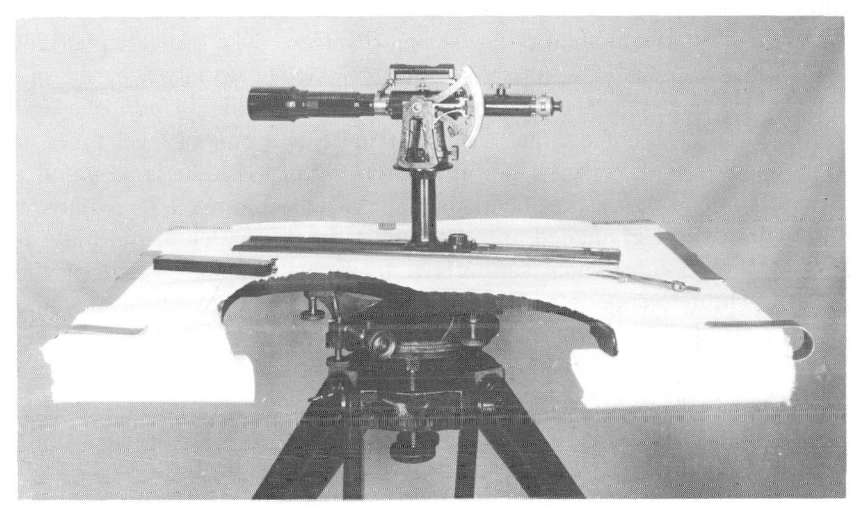

Fig. 13-1*a*. U.S. Coast and Geodetic Survey plane table and alidade.

table sheet is held in position by countersunk screws in the top of the board.

The *traverse table* consists of a small drawing board, usually 15 by 15 in., mounted on a light tripod in such manner that the board can be rotated about the vertical axis and can be clamped in any position. The table is leveled by adjusting the tripod legs, usually by estimation with the eye. A compass is fixed into a recess in the board. Ordinarily a peep-sight alidade is used with the traverse table. The traverse table is suitable for military reconnaissance sketches, traverses for small-scale maps, and the mapping of relatively inaccessible areas to fill in a topographic map being drawn on a larger sheet.

13-3 Alidades.

A plane-table alidade is a combined sight and straight-edge ruler.

Peep-sight Alidade. One type of alidade consists of a peep sight mounted on a ruler. The peep sight is formed by two vertical sight vanes, either fixed or folding, similar to those employed on the surveyor's compass. The ruler usually consists of a brass plate, 6 to 10 in. long, one edge of which is beveled and graduated to a suitable scale. The peep-sight alidade is convenient for sighting details while sketches are being made and is often employed by topographers as an auxiliary to the telescopic alidade.

Conventional Telescopic Alidade. The telescopic alidade is designed to afford greater precision in the control of the table and especially to make possible the stadia method of measuring distances. The base, or plate, of the alidade consists of a brass ruler or straightedge usually 3 by 18 in., beveled on one edge and chrome-plated on the bottom. Upon one end of the plate is mounted either a circular level or a pair of level tubes at right angles to each other. Upon the other end of the plate is mounted a trough compass consisting of a magnetic needle mounted in a narrow box with a short graduated arc at the end. In the center of the plate is mounted a column which supports a telescope similar to that of the transit. The telescope is equipped with stadia hairs, a vertical arc, and either a striding level or an attached level. In addition, many instruments are provided with a Beaman stadia arc, a vernier-control level, and a gradienter as previously described for the transit.

The telescope tube is fitted into a cylindrical sleeve which is rigidly attached to the horizontal axis. The telescope can be turned about its axis in the sleeve much as the telescope of the wye level can be turned in its wyes. On the telescope are turned two shoulders perhaps 5 in. apart, upon which rests a striding level.

The vernier of the vertical arc may be fixed or movable, or there may be an auxiliary level tube attached to the vernier arm as with the transit. Because the plane table is relatively unstable as compared with the transit, a control level mounted on the movable vernier arm greatly facilitates the measurement of vertical angles, rendering unnecessary an initial reading and index correction.

Self-indexing Alidade. Recently the U.S. Geological Survey has been active in the development of a self-indexing alidade, in which a damped pendulum automatically brings the index of the vertical arc to the correct scale reading even if the plane-table board is not quite level. The form shown in Fig. 13-1*b* incorporates other modern features, as follows: All three scales (degree, H, and V) of the vertical arc are visible simultaneously through a microscope; the degree scale can be read to ½′ by estimation and requires no vernier. The telescope is internal-focusing; therefore the stadia constant $(f + c)$ is zero. The reticule is fixed in the telescope, and the telescope is fixed to its axis; thus neither the line of sight nor the axis of the telescope needs adjustment. No striding level or collars are required. The design is simple, and most parts are enclosed. A telltale circle on the bull's-eye level indicates when the blade is tilted beyond the range of the pendulum. A detachable elbow eyepiece is provided. The only field adjustment is to zero the index of the vertical arc when the line of sight is horizontal, as follows: With the line of sight on the rod reading established for a horizontal line by means of the two-peg test (Art. 4-23, adjustment 3), the index is brought to the zero mark (or equivalent scale graduation) by means of a capstan-head screw which controls its setting.

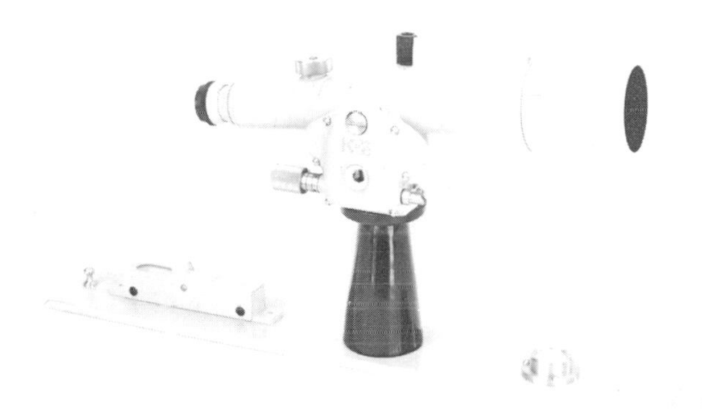

Fig. 13-1*b*. Self-indexing alidade. (*Keuffel & Esser Co.*)

13-4 Plane-table Sheet. As the plane-table sheet is exposed to outdoor conditions, specially prepared papers are required to avoid undue expansion or shrinkage. Only the best drawing papers should be used. The paper can be seasoned, that is, rendered more resistant to changes in humidity, by exposing it alternately to very moist and very dry atmospheres for a number of cycles.

The drawing paper can be mounted on muslin or on each side of a sheet of muslin with the grain of the paper of one sheet laid transverse to the grain of the other. These forms of plane-table sheet are excellent but are not sufficiently flexible to be rolled under the edge of the plane-table board if a sheet larger than the board is desired. For accurate work, such as graphical triangulation, the drawing paper may be mounted on a thin aluminum sheet. A sheet of celluloid with roughened surface is sometimes used for work in light rains; the details thus plotted are later transferred to the regular sheet. Cellulose-acetate or drafting film sheets are used in compilation of data from and in connection with aerial photographs.

If the plotting is to be continued for several days, the map sheet is protected by a cover sheet of tough paper which is torn away piece by piece to expose the map sheet as the work progresses.

Sharply pointed, hard (6H to 9H) pencils are used for drawing lines and plotting details, and a fine needle is used for plotting control stations. Special care should be taken not to smear the drawing; the alidade should be lifted instead of slid into position.

13-5 Setting Up and Orienting the Table.

The plane table is set up approximately waist-high, so that the topographer can bend over the board without resting against it. The tripod legs are spread well apart and planted firmly in the ground. The board is leveled, but no special attempt is made to see that it is perfectly level each time an observation is made.

For map scales smaller than perhaps 1 in. = 50 ft the plane table is set up over the station without any attempt to place the plotted point vertically above the station point. For maps of larger scale, the table is set up roughly and oriented approximately, and then it is shifted bodily until the point on the paper is practically over the station point, as indicated by plumbing. In any case the aim is to set up with sufficient care so that the plotted position of lines drawn from the station will be shown correctly within the scale of the map.

The table may be oriented (1) by use of the *magnetic compass,* (2) by use of the *Baldwin solar chart,* (3) by *backsighting,* (4) by solving the *three-point problem* (Art. 13-14), or (5) by solving the *two-point problem,* described in more detailed textbooks. As soon as the table is oriented, it is clamped in position and all mapping at the station is carried on without disturbing the board.

1 Orientation by Compass. For rough mapping at small scale, often orientation by the magnetic compass is sufficiently precise. This method is susceptible to the errors of compass work (Chap. 8), but an error in the plotted direction of one line introduces no systematic errors in the lines plotted from succeeding stations.

If the compass is fixed to the drawing board, the board is oriented by rotating it about the vertical axis until the fixed bearing (usually magnetic north) is observed. If the compass is attached to the alidade or to a movable plate, the edge of the ruler or the plate is alined with a meridian previously established on the plane-table sheet, and the board is turned until the needle reads north.

2 Orientation by Solar Chart. In regions where the local conditions will not permit orientation by the magnetic compass, the table may be oriented by means of the Baldwin solar chart. This chart consists of a series of elliptical lines indicating the sun's path for various latitudes intersected by a series of straight lines indicating various sun times. When the chart is laid on the leveled plane table and is turned until the shadow cast by a plumb line passes through two designated points on the chart, an orientation arrow on the chart will point true north.

3 *Orientation by Backsighting.* For mapping at intermediate or large scale, usually the board is oriented by backsighting along an established line, the direction of which has been plotted previously but the length of which need not be known. The method is equivalent to that employed in azimuth traversing with the transit. Greater precision is obtainable than with the compass, but an error in direction of one line is transferred to succeeding lines.

The plane table is set up as at *B* on the line *AB* which has been plotted previously as *ab,* the straightedge of the alidade is placed along the line *ba,* and the board is oriented by rotating it until the line of sight falls at *A*. Preferably the longest line available should be used.

13-6 Sighting. When a station or object is sighted and a ray is drawn through the plotted location of the station occupied toward the station or object, the sight is called a *foresight.* When the straightedge of the alidade is placed along a previously plotted line passing through the plotted location of the station occupied and that of another station or object, and then the board is turned until the line of sight cuts the station or object, the sight is called a *backsight.* When a known station is sighted and a line is drawn through the plotted location of that station toward the station occupied, the sight is called a *resection.* Resection is also a general term applied to the process of determining the location of the station occupied (Art. 13-11).

13-7 Radiation. When the table has been oriented, the direction to any object in the landscape may be drawn on the map by pivoting the alidade about the plotted location of the plane-table station, pointing the alidade toward the distant object, and drawing a line along the straightedge.

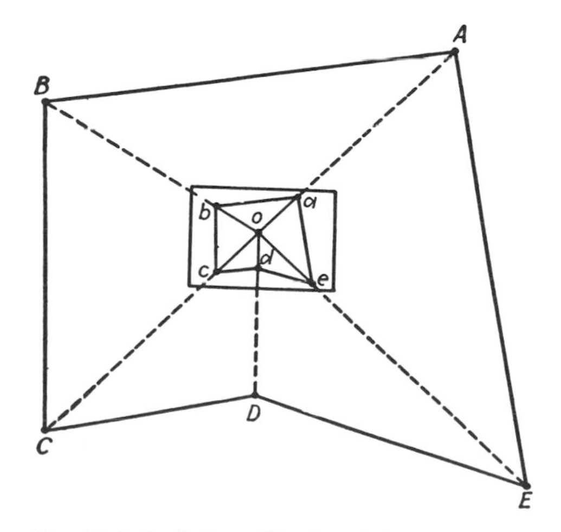

Fig. 13-2. Radiation with plane table.

Thus in Fig. 13-2 the plane table is shown in position over station *o* in the center of the field. The plotted location of the plane-table station is indicated at *o* on the plane-table sheet. The alidade is pivoted about this point; and as sights are taken to points as *A, B,* and *C,* rays are drawn along the edge of the ruler. The distances are measured and are then plotted to scale along the corresponding rays, thus locating the points *A, B,* and *C* on the map at *a, b,* and *c*. This procedure is called *radiation.*

13-8 Traversing. Traversing with the plane table involves the same principles as traversing with the transit. As each successive station is occupied, the table is oriented, sometimes with the compass but usually by backsighting on the preceding station. A foresight is then taken to the following station, and its location is plotted as in the radiation method just described. The distances between successive instrument stations must be measured.

Thus in Fig. 13-3 a series of traverse stations *A, B, C,* etc., is represented. The plane table is set up as at *A,* and *a* representing *A* is plotted

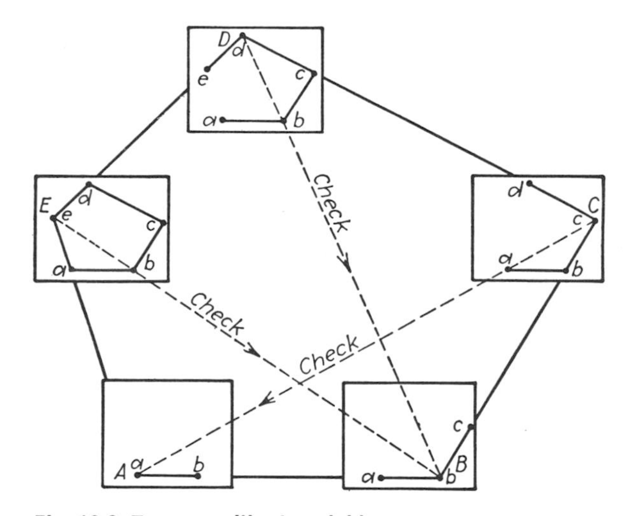

Fig. 13-3. Traverse with plane table.

in such location that the other stations will fall within the limits of the sheet. With the straightedge passing through *a,* a foresight is taken to station *B,* and *b,* its location on the map, is plotted by radiation as just described. The instrument is then set up at station *B* and is oriented by backsighting on station *A* as described in Art. 13-5. A foresight is taken to station *C,* and its location is plotted at *c*. By a similar procedure the locations of the remaining stations are plotted. If the traverse forms a closed figure, any error of closure will become apparent on the plane-table sheet when the initial station is again plotted at the end of the traverse. With reasonable care, the linear error of closure can be kept within 1/500 of the length of the traverse.

13-9 Intersection. This method is similar to that described for the transit in Art. 10-4. It is useful for locating objects otherwise not conveniently accessible. The location of an object is determined by sighting at the object from each of two plane-table stations (previously plotted) and by drawing rays as in the method of radiation; the intersection of the two rays thus drawn marks the plotted location of the object. No linear measurements are required except to determine the length of the line joining the two plane-table stations.

Thus the locations of the objects *A, B, C,* etc. (Fig. 13-4), may be plotted as follows: The plane table is set up at station *M,* a foresight to

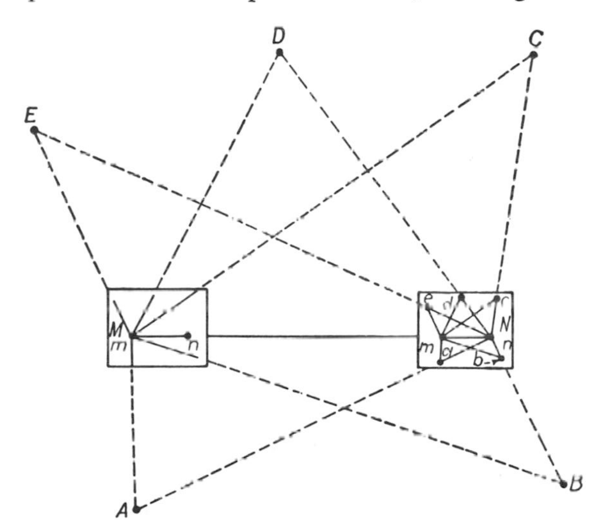

Fig. 13-4. Intersection with plane table.

station *N* is taken as in the method of traversing, and the line *mn* representing the line *MN* is drawn to scale. Rays of indefinite length are drawn from *m* toward the objects *A, B, C,* etc. The plane table is then set up at station *N* and is oriented by backsighting to station *M*. Rays are drawn from *n* toward the same objects. The intersections of these rays with the corresponding rays drawn from *m* mark the plotted locations of the objects at *a, b, c,* etc. Distances to the objects are not measured but may be scaled from the map. If the angle between the intersecting rays is small, the location will be indefinite.

13-10 Graphical Triangulation. Graphical triangulation achieves the same results as triangulation with the transit (Chap. 12), but the procedure differs in that the plotted locations of the distant signals are determined graphically on the plane-table sheet. The method involves both intersection and, where necessary, resection. It is advantageous where the terrain offers unobstructed sights, considerable relief, and many well-defined objects.

Two plane-table stations the locations of which are known (as *A* and *B* in Fig. 13-5) are marked by signals. Prior to the beginning of the

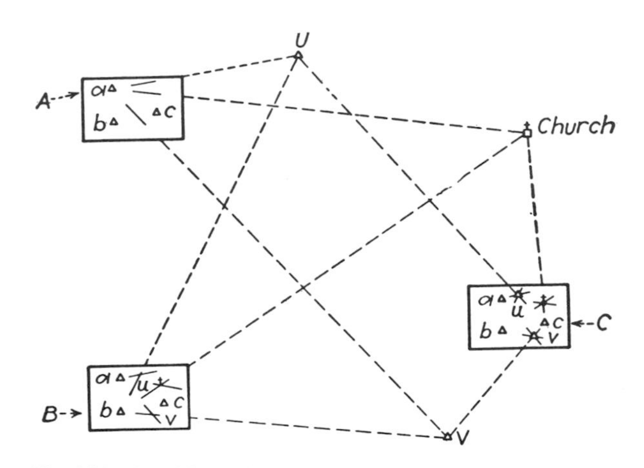

Fig. 13-5. Graphical triangulation.

work, the locations of these stations are plotted on the sheet at *a* and *b*. The field procedure is then as follows: The plane table is set up as at *A* and is oriented by sighting at *B;* rays are then drawn toward other stations. The table is then set up and oriented at station *B,* and the stations of unknown location are sighted again. The correct plotted location of each station is determined by the intersection of two rays, as shown at station *B* in Fig. 13-5. With the instrument at a third station of known location, as *C,* the plotted location of each station sighted from *A* and *B* will be verified if a third ray drawn toward the station from *c* passes through the intersection of the rays previously drawn from *a* and *b*.

RESECTION AND ORIENTATION

13-11 Resection. Resection is the process of determining the plotted location of a station occupied by the instrument by means of sights taken toward known points the locations of which have been plotted. Resection enables the topographer to select advantageous plane-table stations which have not been plotted previously.

With the plane table oriented over the desired station of unknown map location, two or more objects of known location are sighted; as each object is sighted, a line of indefinite length is drawn through the plotted location of that object on the map. The intersection of these lines marks the plotted location of the plane-table station. It is desirable to resect from nearby stations.

The table may be oriented by any of the methods stated in Art. 13-5. It is emphasized that, for the methods of orientation by magnetic compass and solar chart and by backsighting, resection can be accomplished only *after the board has been oriented*. If resection is by the three-point problem or the two-point problem, orientation and resection are accomplished in the same operation.

13-12 Resection after Orientation by Compass. If the plane table has been oriented by means of the compass, the method of resection is as follows: Let *P* be the station of unknown location occupied by the plane table, and let *A* and *B* be two visible stations which have been plotted on the sheet at *a* and *b*. Then the plotted location of *P* is determined by drawing a line, or resecting, through *a* in the direction of *A* and resecting through *b* in the direction of *B*. The point *p*, where the two (or more) lines cross, marks the plotted location of the instrument station *P*. The method is used only for small-scale or rough mapping for which the relatively large errors of orienting with the compass would not impair the usefulness of the map.

13-13 Resection after Orientation by Backsighting. If the table is oriented by backsighting along a line the direction of which has been plotted but the length of which may be unknown, the method of orientation and resection is as follows: Suppose that the topographer wishes to occupy station *C* (Fig. 13-6), the location of which has not been plotted but from which can be seen two points as *A* and *B* the locations of which have been plotted at *a* and *b*. With the plane table oriented at one of the known stations as *B*, he takes a foresight to *C* and draws through *b* a ray of indefinite length. He then sets up the plane table at station *C*, orients it by backsighting to *B*, and resects from *A* through *a*. The intersection *c* of the ray from *b* and the resection line from *a* is the plotted location of the plane-table station *C*, since the triangles *abc* and *ABC* are similar.

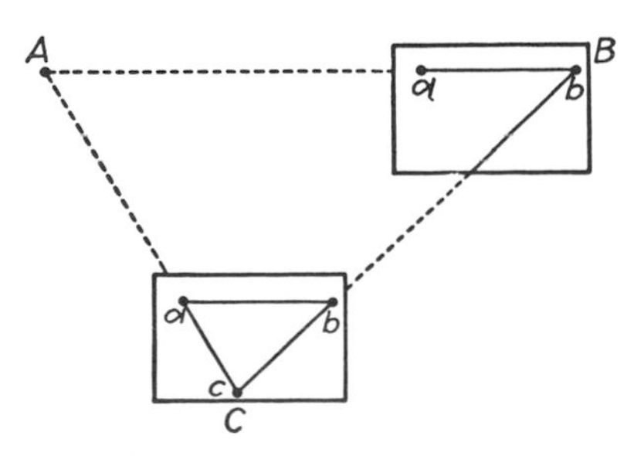

Fig. 13-6. Resection after orientation by backsighting.

If the angle between the ray and the resection line is small, the location will be indefinite; for strong location the acute angle between these lines should be greater than 30°.

13-14 Resection and Orientation: Three-point Problem.

Frequently the topographer wishes to occupy an advantageous station which has not been located on the map and toward which no ray from located stations has been drawn, and at the same time orientation by use of the compass is not sufficiently accurate. If three located stations are visible, the three-point problem offers a convenient method of orienting and resecting in the same operation. There are several solutions of the three-point problem. In the United States, experienced topographers commonly employ a method of direct trial, guided by rules (Art. 13-14*a*). The mechanical, or tracing-cloth, solution (Art. 13-14*b*) is simpler to understand but is not so satisfactory nor so expeditious under the usual field conditions.

13-14a Trial Method.

The plane table is set up over the station of unknown location and is oriented approximately either by compass or by estimation. Resection lines from the three stations of known location are drawn through the corresponding plotted points. These lines will not intersect at a common point unless the trial orientation happens to be correct. (An exception to this statement occurs when the station of unknown location happens to fall on the circumference of a circle passing through the three stations of known location, as discussed later in this article.) Usually a small triangle called the *triangle of error* is formed by the three lines.

Thus in Fig. 13-7, suppose that the plane table has been set up over a ground point *p* and oriented approximately. Resection lines are drawn from *A, B,* and *C* through the corresponding plotted points *a, b,* and *c,*

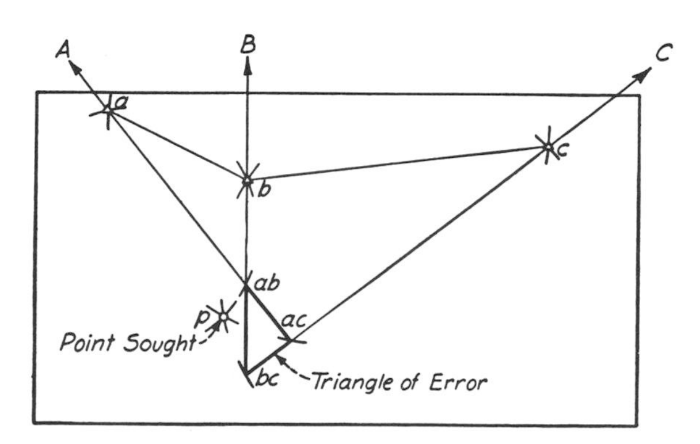

Fig. 13-7. Three-point problem with plane table.

respectively, forming a triangle of error. The correct plotted location p of the plane-table station, called the *point sought,* is then determined more closely. One method is to draw arcs of circles through the points as shown (through a, b, and point ab; b, c, and bc; and a, c, and ac); the circles will intersect at p, the point sought. Usually, however, the correct location of the point sought is estimated more conveniently by means of rules 1 and 2, given below.

The board is then reoriented by backsighting through the estimated location of p toward one of the known stations (preferably the most distant); and the orientation is checked by resecting from the other two known stations. If the three lines still do not meet at a point, the process is repeated until they do; the orientation is then correct, and the common intersection of the three lines is the correct plotted location of the plane-table station.

rule 1. *The point sought is on the same side of all resection lines.* That is, it lies either to the right of each line (as the observer faces the corresponding station) or to the left of each line.

rule 2. *The distance from each resection line to the point sought is proportional to the length of that line.* By "length" is meant either the actual distance from plane-table station to known station or the corresponding plotted distance.

Rule 2 can be proved by reference to Fig. 13-8, which shows the triangle of error as in Fig. 13-7. The triangle of error is actually small, and distances from a, b, and c to any point of it may be taken as the distance to p without error of consequence. Lines pp_a, pp_b, and pp_c are, respectively, perpendicular to the resection rays from a, b, and c. Then, by similar triangles,

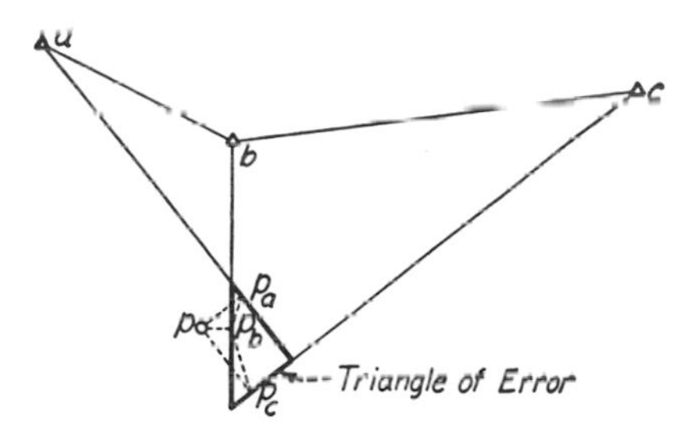

Fig. 13-8. Triangle of error.

$$\frac{pp_a}{pa} = \frac{pp_b}{pb} = \frac{pp_c}{pc} \tag{1}$$

Rules 1 and 2 are general and apply to any location of the plane-table station except on the *great circle* passing through the three known stations (Fig. 13-9). In this case, regardless of the orientation of the table, the lines will meet in a common point (see points 2 in the figure) which will not necessarily be the point sought. If it is suspected but not known that the plane-table station is on the great circle, either the great circle should be plotted on the plane-table sheet or the orientation of the board should be changed slightly and a second trial made. If it is found that the station is on the great circle, one (or more) of the three known stations must be replaced by a known station (or stations) suitably located.

Rules 1 and 2 are supplemented by several auxiliary rules (given in more detailed publications) which apply to particular locations of the plane-table station. For example, if the station happens to be within the great triangle, the point sought is within the triangle of error.

The strength of the determination varies with the location of the plane-table station. The strength of determination should be considered not only in selecting the most favorable of the available known stations but also in deciding whether the three-point problem can satisfactorily be used with the plane table at a given location.

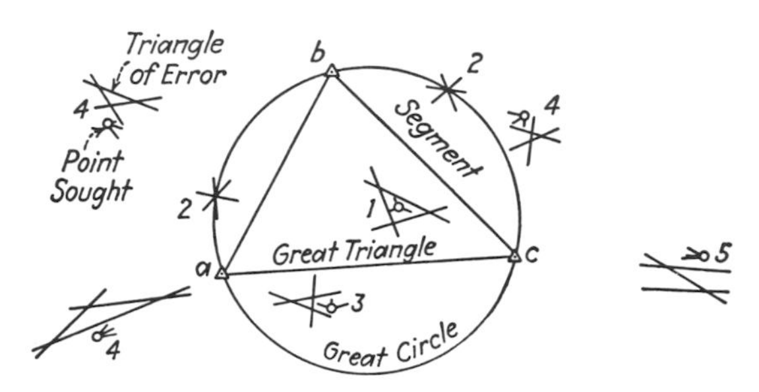

Fig. 13-9. Three-point problem; solution by trial.

13-14b Tracing-cloth Method. A piece of tracing cloth or tracing paper is fastened on the plane table over the map. Any convenient point on the tracing cloth is chosen to represent the unknown station over which the plane table is set, and from it rays are drawn toward the three known stations or objects. Then the cloth is loosened and is shifted over the map until the three rays pass through the corresponding plotted points. The intersection of the rays marks the plotted location of the plane-table station. It is pricked through onto the map, and the table is oriented by backsighting on one of the known stations (preferably the most distant).

USE AND ADJUSTMENT OF PLANE TABLE

13-15 Difference in Elevation. The methods of plotting the *horizontal projection* of ground points have been described. As the plane table is used principally in the work of topographic mapping, many *elevations* of ground points are determined by methods closely similar to those for leveling by means of the transit, namely, by direct leveling and trigonometric leveling including stadia leveling.

The plane table is not so precise nor so stable as the horizontal plate of the transit; accordingly in measuring vertical angles it is necessary either to determine the index error of each sight or preferably to employ a control level attached to the vernier arm (Art. 9-14).

The elevation of a distant point the location of which has been plotted by the method of intersection may be secured readily by the method of trigonometric leveling (Art. 4-5), in which the horizontal distance may be scaled directly from the map. The conditions as to distance and as to the accuracy required will determine whether or not corrections for curvature of the earth and atmospheric refraction should be applied.

The precision of plane-table leveling is not high; however, with reasonably careful work the maximum error in elevation, in feet, may be kept within 0.2 or 0.3 $\sqrt{\text{miles}}$.

13-16 Details. The operation of plotting details by radiation with the plane table is commonly known as taking *side shots*. Table 13-1 gives the sequence of operations for taking a side shot when the alidade is equipped with a stadia arc but not with a vertical control level. It is assumed that the party consists of a plane-table man (called a *topographer*), one or more rodmen, and a computer. In actual practice usually no record is kept of the side shots. In this article, however, it is assumed that a recorder is employed for purposes of training, checking, or record; and his operations are also listed. Each line of the table reads from left to right, and the operations in each line follow those of the preceding line. It is evident that the progress of the party depends largely upon systematic operation and upon the dispatch with which the topographer performs his duties.

If the alidade is equipped with a vertical vernier control level, it is not necessary for the topographer to level the telescope; instead he centers the control-level bubble. If the alidade is not equipped with a stadia arc or vertical control level, he levels the telescope, reads the index error, and then reads the vertical angle; the details of procedure are modified accordingly.

If field notes are taken, the form is similar to that for transit-stadia surveying (Chap. 11). Column headings will differ according to the type of vertical arc and type of telescope. With an external-focusing telescope not equipped with a stadia arc, pertinent column headings would include object, stadia interval, rod reading, vertical angle, stadia distance, horizontal distance, difference in elevation, elevation, and remarks. Such notes are often kept for the sights to control stations, whether or not the side shots are recorded.

Aerial photographs may be used in connection with the plotting of details on the plane table (Chap. 23). By examination of both photographs and terrain, the topographer can fill in details not visible from the air, can check against mistakes, and can expedite the work. If the

Table 13-1 Sequence of operations for side shot with plane table
Alidade equipped with stadia arc and fixed vertical vernier

Rodman	*Topographer*	*Computer*	*Recorder*
Sets rod on ground point	Sights on rod, with lower cross hair on a foot mark		
	Reads and calls stadia interval	Sets index of slide rule on stadia distance	Records stadia interval (or distance)
	Draws ray		
	Levels telescope, and sets stadia arc to zero (or 50)[a]		
	Sights on rod, with stadia index set on some mark of *V* arc		
	Calls *V* reading, *H* reading, and reading of middle cross hair of rod	. . . .	Records *V*, *H*, and rod reading
	Waves rodman ahead	Computes and calls horizontal distance	Records horizontal distance
Starts to new ground point	Plots point	Computes and calls vertical distance	Records vertical distance
		Computes and calls elevation	Records elevation
	Records elevation on sheet		
	Interpolates and sketches contours and features		

[a] If the alidade is equipped with a vertical vernier control level, the topographer centers the control-level bubble instead.

scale of the photograph is at or near the scale of the map, an overlay sheet of transparent cellulose acetate may be used on which to plot certain details, keeping in mind that the photograph is not an orthographic projection. Another technique is to print the compilation sheet obtained by photogrammetry onto a plane-table sheet sensitized with blueprint emulsion, then in the field to plot the contours on this sheet.

13-17 Adjustment of Conventional Alidade. The adjustments of the conventional tube-in-sleeve type of telescopic alidade are similar to those previously described for the wye level and the transit. In general, the adjustments need not be so refined, and in one or two cases they are omitted. As the telescope is never inverted, no large error is introduced through any lack of perpendicularity between the line of sight and the horizontal axis, nor any error through lack of parallelism between the line of sight and the straightedge. It may be assumed that the horizontal axis is parallel to the plane of the ruler. The edge of the ruler is not in a vertical plane through the line of sight, but the error from this source is negligible because, in plotting, the offset is multiplied by the scale fraction.

1 To Make the Axis of Each Plate Level Parallel to the Plate. Center the bubble (or bubbles) of the plate level (or levels) by manipulating the board. On the plane-table sheet mark a guide line along one edge of the straightedge. Turn the alidade end for end, and again place the straightedge along the guide line. If the bubble is off center, bring it back *halfway* by means of the adjusting screws. Again center the bubble by manipulating the board, and repeat the test. For a circular level the procedure is similar.

2 To Make the Vertical Cross Hair Lie in a Plane Perpendicular to the Horizontal Axis. Sight the vertical cross hair on a well-defined point not less than about 200 ft away, and swing the telescope through a small vertical angle. If the point appears to depart from the vertical cross hair, loosen two adjacent screws of the cross-hair ring, and rotate the ring in the telescope tube until by further trial the point sighted traverses the entire length of the hair. Tighten the same two screws.

3 To Make the Line of Sight Coincide with the Axis of the Telescope Sleeve. Sight the intersection of the cross hairs on some well-defined point. Rotate the telescope in the sleeve through 180° (usually the limits of rotation are fixed by a shoulder and a lug). If the cross hairs have apparently moved away from the point, bring each hair halfway back to its original position by means of the capstan screws holding the cross-hair ring. The adjustment is made by manipulating opposite screws, bringing first one cross hair and then the other to its estimated correct position. Again sight on the point, and repeat the test.

4 To Make the Axis of the Striding Level Parallel to the Axis of the Telescope Sleeve (and Hence Parallel to the Line of Sight). Place the striding level on the telescope, and center the bubble. Remove the level, turn it end for end, and replace it on the telescope tube. If the bubble is off center, bring it back halfway by means of the adjusting screw at one end of the level tube. Again center the bubble (by means of the tangent screw), and repeat the test.

5 (For Alidade Having a Fixed Vertical Vernier) *To Make the Vertical Vernier Read Zero When the Line of Sight Is Horizontal.* With the board level, center the bubble of the telescope level. If the vertical vernier does not read zero, loosen it and move it until it reads zero.

5a (For Alidade Having a Movable Vertical Vernier with Control Level) *To Make the Axis of the Vernier Control Level Parallel to the Axis of the Telescope Level When the Vertical Vernier Reads Zero.* Center the bubble of the telescope level, and move the vernier by means of its tangent screw until it reads zero. If the control-level bubble is off center, bring it to center by means of the capstan screws at the end of the control-level tube.

13-18 Sources of Error. In the main, the sources of error in plane-table work are the same as those which affect transit work and plotting, and the discussion relating to those subjects need not be repeated here. However, the following three sources of error should be considered:

1 *Setting over a Point.* Because plotted results only are required, it is not necessary to set the plotted location of the plane table over the corresponding ground point with any greater precision than is required by the scale of the map.

2 *Drawing Rays.* The accuracy of plane-table mapping depends largely upon the precision with which the rays are drawn; consequently the rays should be of considerable length. To avoid confusion, however, only enough of each ray is drawn to ensure that the plotted point will fall upon it, with one or two additional dashes drawn near the end of the alidade straightedge to mark its direction. Fine lines are desirable both for precision and for legibility.

3 *Instability of the Table.* The plane table is subject to continual disturbance by the topographer while he is working. Errors from this source can be kept within reasonable limits (*a*) by planting the tripod firmly in the ground, (*b*) by setting the table approximately waist-high so that the topographer can bend over it without leaning against it, (*c*) by avoiding undue pressure upon or against the table, and (*d*) by testing the orientation of the board occasionally and correcting its position if necessary. This test is always applied before a new instrument station is plotted.

13-19 Field Checks. The previously plotted location of any visible object is verified during a subsequent setup if a ray drawn toward the object passes through the plotted point. The plotted location of the plane table itself may be verified by resecting from distant visible objects the plotted locations of which are known to be correct. Checks of this sort should be applied at each station.

13-20 Advantages and Disadvantages. As compared with other methods of mapping, the plane-table method has these advantages: (1) relatively few points need be located because the map is drawn as the survey proceeds; (2) contours and irregular objects can be represented accurately because the terrain is in view as the outlines are plotted; (3) as numerical values of angles are not observed, the consequent errors and mistakes due to reading, recording, and plotting are avoided; (4) as all plotting is done in the field, omissions in the field data are avoided; (5) the useful principles of intersection and resection are made convenient; (6) checks on the location of plotted points are obtained readily; and (7) the amount of office work is relatively small.

The disadvantages are (1) the plane table and accessories are cumbersome; (2) considerable time is required for the topographer to gain proficiency; (3) the time required in the field is relatively large; and (4) the usefulness of the method is limited to relatively open country (where visibility is fair) and to favorable weather conditions.

With a long-legged tripod the table may be raised above low brush or cornstalks. In wet weather a sheet of celluloid may be used instead of drawing paper. The plane table may be both raised and protected from rain by being mounted in a covered automobile truck having sides which can be opened as desired; during a setup the truck is braced to keep the table stationary.

Comparing the plane table with the transit: If a large number of *points* are to be plotted on a map, the transit method is superior; but if *irregular lines and areas* are to be represented, the plane-table method is superior in accuracy, speed, and economy.

13-21 Office Problems

1 Verify the solution of the three-point problem in the drafting room, using a large (say, 24 by 30-in.) sheet of paper to represent the field and a small (say, 3 by 5-in.) card to represent the plane table. A long straightedge will serve as a line of sight to connect stations.

2 For each of the plane-table setups sketched in Fig. 13-10, indicate by estimation (in accordance with Rules 1 and 2) the location of the point sought, and state the direction in which the table should be rotated in order to orient it properly. In the sketches, the triangle of error is exaggerated for clearness.

13-22 Field Problems

Problem 34. Adjustment of Plane-table Alidade

Object. To make the field adjustments of the conventional tube-in-sleeve telescopic alidade.

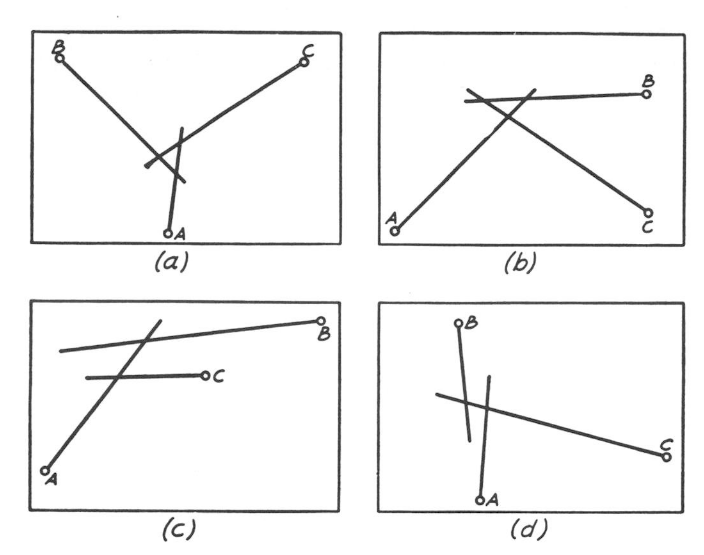

Fig. 13-10. Problems in orientation of plane table.

Procedure. As outlined in Art. 13-17.

Problem 35. Plane-table Survey of Field

Object. To make a plane-table survey of an assigned portion of the campus by a combination of the methods of radiation, intersection, and traversing. Either a plain map or a topographic map may be made.

Procedure. (1) Stake out an irregular field having perhaps five sides. (2) Set up the plane table at one corner of the field, and so locate the plotted position of the station that the area to be covered will fall within the limits of the plane-table sheet. (3) Draw a magnetic meridian on the plane-table sheet. (4) Locate and sketch all conveniently accessible objects by the method of radiation (Art. 13-7), employing the stadia. Also, draw and identify rays toward relatively inaccessible objects which can also be seen from another point on the traverse. (5) Move to the next corner of the field, orient the board by backsighting, check the orientation by the magnetic compass, and similarly locate objects by radiation and intersection. (6) Continue in this manner around the field to form a closed traverse, and plot sufficient data to construct a complete map. At each station, check the location of the plane table by resection (Art. 13-19). (7) If a marked error becomes evident, the orientation of the board should be checked, and, if necessary, one or more of the previous stations should be reoccupied. (8) If it is convenient to occupy some station other than a traverse station, locate the station occupied by resection, employing the method of backsighting (Art. 13-13). A ray must have been drawn previously toward the station to be occupied. (9) If a topographic map is to be made, locate a sufficient number of controlling ground points to enable contours to be drawn in the field. (10) Note the error of closure of the traverse; if the error is considerable, the work should be repeated. (11) Finish the map.

Problem 36. Orientation and Resection by Three-point Problem

Object. With three stations (or objects) of known location assigned, to plot the location of a fourth station occupied by the plane table and simultaneously to orient the board.

Procedure. (1) Plot the three known points on the plane-table sheet. (2) Set up the plane table at another assigned station and solve the three-point problem by the trial method as described in Art. 13-14*a*. (3) If other known points are visible, plot their locations and check the location and orientation of the plane table by sighting on these points. (4) Change the orientation of the board slightly, and again orient the board by the tracing-cloth method (Art. 13-14*b*). (5) Compare the results obtained by the two methods.

Hints and Precautions. If the station occupied is at or near a great circle passing through the three known points, the solution is indeterminate.

14
MAP DRAFTING

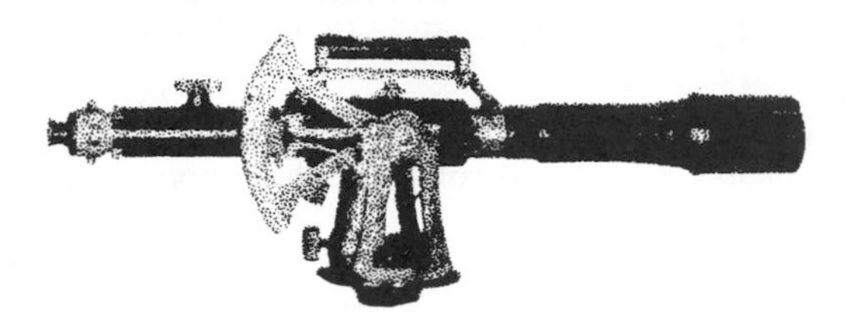

14-1 The Drawings of Surveying. It is assumed that the student is familiar with the use of the ordinary drafting instruments and with the elements of mechanical drawing. The drawings of surveying consist of maps, profiles, cross-sections, and (to some extent) graphical calculations; the usefulness of these drawings is largely dependent upon the accuracy with which points and lines are projected on paper. For the most part, few dimensions are shown, and the person who makes use of the drawings must rely upon distances as measured with a scale and upon angles as measured with a protractor. A consistent relation between the field measurements and the map requires great care in plotting.

Profiles and cross-sections are described in Chap. 7.

14-2 Map Projection. Since the surface of the earth is curved and the surface of the map is a plane, no map can be made to represent a given territory without some distortion. If the area is small, the earth's surface may be regarded as plane and a map constructed by orthographic projection, as in mechanical drawing, will represent the relative location of objects without measurable distortion. The maps of plane surveying are constructed in this manner, points being plotted either by rectangular coordinates or by horizontal angles and distances.

As the size of the territory increases, this method becomes inadequate, and various forms of projection are employed to minimize the effect of map distortion. (See Arts. 15-22 to 15-25.) The points of control for the maps of states and countries are plotted by spherical coordinates through the use of geographic tables. Recently state plane-coordinate systems have been devised whereby, even over large areas, points can be mapped accurately without the direct use of spherical coordinates (Art. 12-17).

14-3 Information Shown on Maps. Maps are classified variously according to their specific use or type, but in general either they become a part of public records of land division or they form the basis of a study for the works of man. In general, the following information should appear on any map:

a The direction of the meridian.

b A graphical scale of the map with a corresponding note stating the scale at which the map was drawn.

c A legend or key to symbols other than the common conventional symbols (Fig. 14-5*a* to *d*).

d An appropriate title.

e On topographic maps, a statement of the contour interval (Chap. 19).

In addition, a map that is to become a part of a public record of land division should contain the following information:

1 The length of each line.

2 The bearing of each line or the angle between intersecting lines.

3 The location of the tract with reference to established coordinate axes.

4 The number of each formal subdivision such as a section, block, or lot.

5 The location and kind of each monument set, with distances to reference marks.

6 The location and name of each road, stream, landmark, etc.

7 The names of all property owners, including owners of property adjacent to the tract mapped.

8 A full and continuous description of the boundaries of the tract by bearing and length of sides; and the area of the tract.

9 The witnessed signatures of those possessing title to the tract mapped; and if the tract is to be an addition to a town or city, a dedication of all streets and alleys to the use of the public.

10 A certification by the surveyor that the plat is correct to the best of his knowledge.

Explanatory notes may be employed to give such information as the sources of data for the map, the precision of the survey, or the datum to which the elevations are referred.

Maps made the basis of studies may or may not show the relief of the ground. In general, such maps show very few dimensions (often not any), the value of the map depending upon the correct representation of the location of features of the land rather than directly upon field measurements or computed values. In addition to items *a* to *e* above, usually such a map shows streams and shore lines, roads and railroads, political boundaries, significant property lines and structures, and perhaps the condition and culture of the land. If the relief of the ground surface is not to be indicated, the map is called a *plan,* a *plat,* or a *planimetric map.* If the relief is to be shown, the map is called a *topographic map.* Relief is usually indicated by *contour lines* drawn for stated equal intervals of elevation, each contour line joining points of equal elevation (Chap. 19).

In connection with lawsuits regarding automobile or train collisions, falls, injuries during construction, and other accidents, the surveyor is sometimes called on to prepare a large-scale map for exhibit in the courtroom. While the map for this purpose should be extremely simple, it should include all details that might have a bearing on the accident. Some of these details may be the grade and crown of the roadway; height of curb; depressions; location (at time of accident) of obstructions to traffic or to view such as trees, poles, signs, and parked automobiles; sources of light (if at night); and location of points (in plan and elevation) from which it is stated that the accident has been witnessed. Colors are sometimes employed to make the various features more intelligible to the layman.

14-4 Scales. The scale of a map is the fixed relation that every distance on the map bears to the corresponding distance on the ground. It may be stated either by numerical relations or graphically, as follows:

1 One inch on the drawing represents some whole number of tens, hundreds, or thousands of feet on the ground, as 1 in. = 200 ft. This type is called the *engineer's scale.* It is used for most maps for construction purposes. In another form of this scale often used on land maps, a whole number of inches on the drawing represents 1 mile on the ground, as 6 in. = 1 mile.

2 One unit of length on the drawing represents a stated number of the *same* units of length on the ground, as 1/62,500. This ratio of map distance to corresponding ground distance is called the *representative fraction.* The scale is independent of the units of measurement. It is used extensively for geographic and military maps.

3 A *graphical scale* is a line subdivided into map distances corresponding to convenient units of length on the ground. Various forms of subdivision are shown in Fig. 14-1, in which the top line represents a scale of 1 in. = 40 ft, and the two lower scales represent 1 in. = 200 ft. On a graphical scale the units of measurement should always be stated.

The numerical scales described above are subject to error if the drawing paper shrinks or swells, as often happens, but this error is not of consequence for many uses of the map. An important objection to the use of numerical scales alone, however, is that often maps are reproduced in other sizes by photographic means. If distances are to be determined accurately from the map, *a graphical scale should always be shown.* If, for convenience, a numerical scale is stated, it should be made clear that this is the scale at which the map was drawn or published; for example, "Original scale 1 in. = 200 ft." When a published map is a considerable enlargement, that fact should be stated.

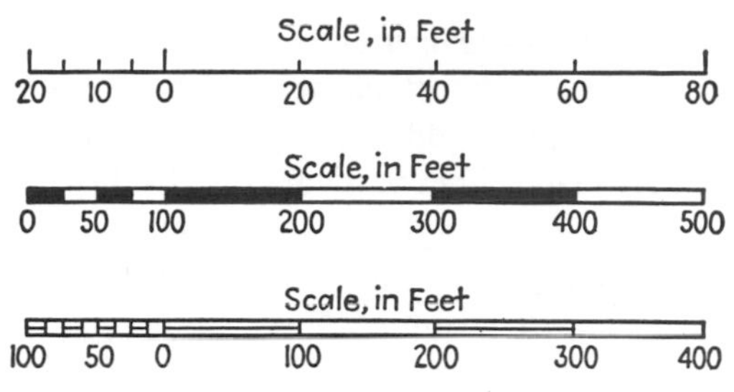

Fig. 14-1. Graphical scales.

As a general rule, the scale of a map should be no larger than is necessary to represent the location of details with the required precision. Maps for engineering projects have scales usually ranging from 1 in. = 20 ft to 1 in. = 800 ft. Maps of land subdivision have scales ranging from 6 in. = 1 mile to 1 in. = 1 mile. For convenience in discussion, maps are herein divided arbitrarily into those of

Large scale: 1 in. = 100 ft or less
Intermediate scale: 1 in. = 100 ft to 1 in. = 1,000 ft
Small scale: 1 in. = 1,000 ft or more

14-5 Meridian Arrows. The direction of the meridian is indicated by a needle or feathered arrow pointing north, of sufficient length to be transferred with reasonable accuracy to any part of the map. The true meridian is usually represented by an arrow with *full* head; the magnetic meridian by an arrow with *half* head. When both are shown, the angle between them should be indicated. The general tendency is to make the arrows too large, blunt, and heavy. A simple design is shown in Fig. 14-2.

Preferably the top of the map should represent north, although the shape of the area covered or the direction of some principal feature of a project may make another orientation preferable.

14-6 Lettering. In general, lettering should be freehand, of a style in keeping with the purpose of the drawing. Guide lines and slope lines should be employed. For drawings that are not to be used by the general public, the *Reinhardt* style of single-stroke lettering is employed almost entirely. The letters are constructed rapidly and are easy to read. Reinhardt letters are made either vertical (Fig. 14-3*b*) or inclined (Fig. 14-3*a*). Frequently vertical and inclined lettering including the extended and compressd forms may be combined advantageously on a single drawing.

On drawings that are to be used by the general public, *gothic, roman,* and *italic* letters may be employed. These several styles are frequently used on a single map (for example, see the maps of the U.S. Geological Survey).

Mechanical lettering guides are often used because letters of regular form can be made quickly. If the letters are spaced properly, satisfactory lettering can be secured with these guides. Freehand and mechanical lettering should not be used on the same drawing.

Detailed information concerning lettering is to be found in textbooks on drawing.

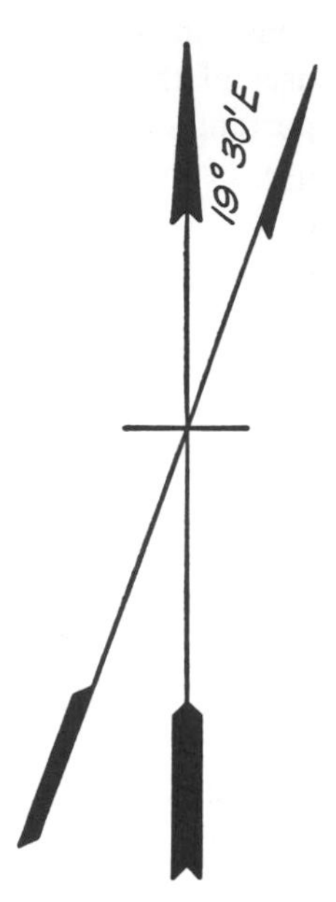

Fig. 14-2. Meridian arrows.

ABCDEFGHIJ
KLMNOPQ
RSTUVWXYZ
abcdefghijklmn
opqrstuvwxyz
1234567890
NORMAL
Hickory Tree 10 ft.
COMPRESSED
WASHINGTON Sta. 71+43.8
EXTENDED
NEVADA

Fig. 14-3*a*. Reinhardt letters, slope form.

ABCDEFGHI
JKLMNOPQR
STUVWXYZ
abcdefghijklmn
opqrstuvwxyz
1234567890
NORMAL
Excavation 23 cu. yd.
COMPRESSED
RICHARDSON ESTATE 300 Ac.
EXTENDED
RED RIVER

Fig. 14-3*b*. Reinhardt letters, vertical form.

14-7 Titles. The best position for the title is the lower right-hand corner of the sheet. Each line should be centered, and the title as a whole balanced. Only the common styles of letters should be used. A change in the style of lettering is sometimes permissible to accentuate the important items, but slope letters and vertical letters should not be included in the same title. The parts should be weighted in order of their importance, beginning with the principal object of the drawing or with the name of the area. Usually the title should state the purpose or kind of map, the name and location of the tract or project, the name of the owner, the scale of the drawing (unless this is shown elsewhere), the contour interval, the name of the engineer and the draftsman, and the date. A simple form of title is shown in Fig. 14-4. Revisions of the map should be shown by dated notes at the left of the title.

DEPARTMENT OF THE INTERIOR
BUREAU OF RECLAMATION

GRAND COULEE DAM – WASHINGTON

TOPOGRAPHIC MAP
OF
EAST SIDE GRAVEL PIT

DRAWN:________________ TRACED:________
CHECKED:________ APPROVED:____________

OCT. 10, 1967	DENVER, COLO.	222-D-539

Fig. 14-4. Title for map.

14-8 Symbols. Objects are represented on a map by *symbols,* many of which are conventional. A symbol is a diagram, design, letter, or abbreviation which by convention or reference to a legend is understood to represent a specific characteristic or object. Standard symbols have been adopted for the United States, and these symbols are commonly used by other organizations. Some topographic map symbols published by the U.S. Geological Survey are shown in Fig. 14-5*a* to *d*. A chart showing these symbols is available from the Survey without charge.

Where feasible, the information shown on topographic maps is distinguished by colors in which the symbols are printed. Black is used for man-made or cultural features such as roads, buildings, names, and boundaries. Blue is used for water or hydrographic features such as lakes, rivers, canals, and glaciers. Brown is used to show the relief or configuration of the ground surface as portrayed by contours or hachures. Green is used for wooded or other vegetative cover, with typical patterns to show such features as scrub, vineyards, or orchards. Red emphasizes important roads and public-land subdivision lines and shows built-up urban areas. Where colors are not available or where the map is to be

Hard surface, heavy duty road, four or more lanes

Hard surface, heavy duty road, two or three lanes

Hard surface, medium duty road, four or more lanes

Hard surface, medium duty road, two or three lanes

Improved light duty road

Unimproved dirt road—Trail

Dual highway, dividing strip 25 feet or less

Dual highway, dividing strip exceeding 25 feet

Road under construction

Railroad: single track—multiple track

Railroads in juxtaposition

Narrow gage: single track—multiple track

Railroad in street—Carline

Bridge: road—railroad

Drawbridge: road—railroad

Footbridge

Tunnel: road—railroad

Overpass—Underpass

Important small masonry or earth dam

Dam with lock

Dam with road

Canal with lock

Fig. 14-5*a*. Symbols for roads, railroads, and dams.

reproduced by contact or photographic printing and will therefore be in one color, black is used throughout; the conventional shapes of the symbols are the same as where color is used.

For many purposes, it would lessen the usefulness of the map if all the objects were shown for which symbols are available. The size of the symbols should be proportioned somewhat to the scale of the map.

In drawing the symbols for culture, marsh, etc., the separate symbols are distributed evenly over the area but are irregularly spaced so as not to give the appearance of rows. The water-surface lines for marsh are drawn with a ruling pen, but all other parts of the symbols are drawn freehand.

Buildings (dwelling, place of employment, etc.)

School—Church—Cemeteries ... Cem

Buildings (barn, warehouse, etc.)

Power transmission line

Telephone line, pipeline, etc. (labeled as to type)

Wells other than water (labeled as to type) ... Oil ... Gas

Tanks; oil, water, etc. (labeled as to type) ... Water

Located or landmark object—Windmill

Open pit, mine, or quarry—Prospect ... X

Shaft—Tunnel entrance

Horizontal and vertical control station:

tablet, spirit level elevation ... BM Δ3899

other recoverable mark, spirit level elevation ... Δ3938

Horizontal control station: tablet, vertical angle elevation ... VABM Δ2914

any recoverable mark, vertical angle or checked elevation ... Δ5675

Vertical control station: tablet, spirit level elevation ... BM ×945

other recoverable mark, spirit level elevation ... ×[illegible]

Checked spot elevation ... ×5923

Unchecked spot elevation—Water elevation ... × 5657 ... 870

Fig. 14-5*b*. Symbols for structures and stations.

Tree symbols are usually drawn in plan. It is common practice to differentiate between deciduous and evergreen trees.

To draw the water-line symbol, the draftsman begins by sketching the shore line as a heavy line and then drawing a fine line as close as possible to it. Each succeeding fine line is drawn in close conformity with the preceding line, and the spacing between lines is increased uniformly outward from the shore. Each fine line is composed of a series of smooth, intersecting arcs, each concave toward the shore. If the space is large, it may be left without lines except near the shore line.

Contour lines are drawn as fine, smooth, freehand lines of uniform width (Chap. 19). Preferably a contour pen (Fig. 14-11) is used.

A sand bar or flat is represented by dots spaced evenly over the area. The boundary of the area is formed by two or three rows of closely and evenly spaced dots somewhat larger than the rest.

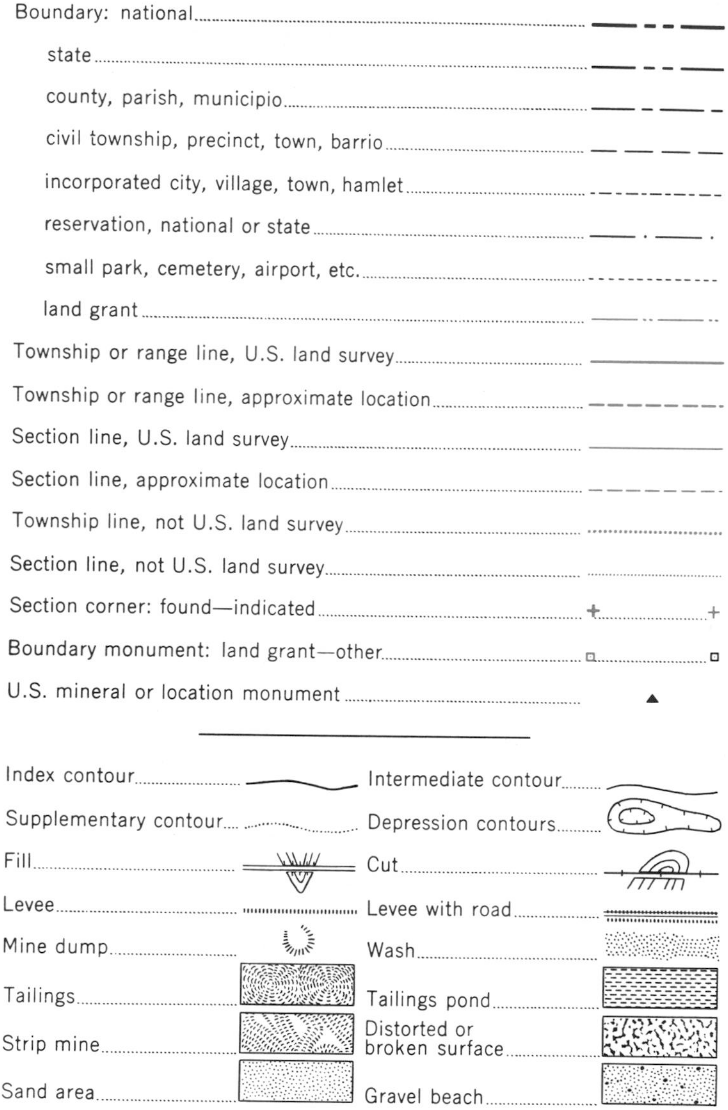

Boundary: national

state

county, parish, municipio

civil township, precinct, town, barrio

incorporated city, village, town, hamlet

reservation, national or state

small park, cemetery, airport, etc.

land grant

Township or range line, U.S. land survey

Township or range line, approximate location

Section line, U.S. land survey

Section line, approximate location

Township line, not U.S. land survey

Section line, not U.S. land survey

Section corner: found—indicated

Boundary monument: land grant—other

U.S. mineral or location monument

Index contour

Intermediate contour

Supplementary contour

Depression contours

Fill

Cut

Levee

Levee with road

Mine dump

Wash

Tailings

Tailings pond

Strip mine

Distorted or broken surface

Sand area

Gravel beach

Fig. 14-5*c*. Symbols for boundaries, contours, and excavation.

14-9 Drafting Papers. Pencil drawings and temporary drawings are often made on a smooth manila *detail paper*. For general map work a fairly smooth, tough *drawing paper* of uniform texture is desirable. The paper should take ink well and should stand erasures without its surface becoming fibrous. Drawings to be subjected to hard usage may be constructed on paper that is mounted on muslin. Plane-table sheets may be mounted on aluminum in order to prevent shrinkage.

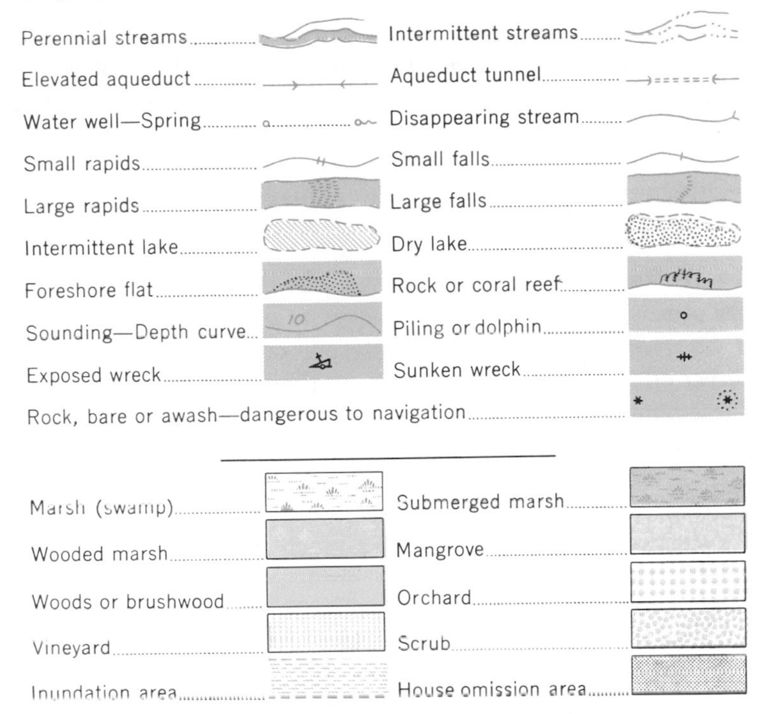

Fig. 14-5d. Symbols for hydrography and land classification.

A *tracing* is a drawing in ink or pencil on a transparent sheet of paper or cloth, for the purpose of reproduction.

Tracing paper is economical and is satisfactory for many purposes, but it will not stand repeated erasures well, and it will become torn and cracked unless it is handled carefully. Pencil tracings on paper have recently come into common use, including reproduction by photography.

Tracing cloth is made from fine starch-filled cotton or linen cloth specially treated to render it firm, transparent, and smooth. The unglazed side is preferred by most draftsmen. The conventional tracing cloth turns white and wrinkles when touched by water. Waterproofed tracing cloth is available.

Preparatory to tracing in ink on cloth, the surface is dusted with powdered talc or chalk and is rubbed with a dry cloth.

Erasures of ink on cloth are made with least damage to the surface by rubbing lightly with a soft pencil eraser, using an erasing shield. Ordinary ink erasers are too abrasive and produce a fibrous surface which does not take ink well. Pencil lines are removed and the tracing is cleaned by rubbing either with artgum or with a cloth saturated in cleaning fluid.

Recently polyester *film* has come into use for drawings and tracings because of its high transparency, dimensional stability, tearing strength, and resistance to heat and age. It is insoluble and waterproof. It takes either pencil or ink work, and either can be erased easily and clearly. It can be coated with semi-opaque material or in various colors for scribing, a process of forming lines with a sharp-pointed instrument which scratches through the coating. The points used in scribing are similar to phonograph needles, with steel or sapphire tips, and are available in various widths; they are fastened to handles similar to penholders.

14-10 Reproduction of Drawings. A drawing may be reproduced at the same scale by making a contact print on processed paper, resulting in a *blueprint,* a *brownprint* (vandyke), or a *direct blackline print.* By direct printing or by reprinting from a vandyke contact negative, any of these prints may be made with white lines on a dark background or with dark lines on a white background.

A drawing may be reproduced in black either at the same scale or at different scales (enlarged or reduced), from either drawings or tracings, by various photographic processes such as the *photostat* process, the *photo-offset* methods, *xerography,* or other methods.

Usually maps and other drawings for general distribution are either lithographed or printed from etchings.

1 Blueprints. A blueprint is made by placing the inked side of a tracing next to a sheet of glass, placing the sensitized side of processed paper or cloth next to the tracing, exposing this side to light, and developing the exposed sheet in a bath of water.

Unexposed blueprint paper should be protected from atmospheric moisture and from light. The proper time of exposure for printing is best determined by trial with small pieces of the paper. If underexposed, the body of the print when washed will be a pale blue; if overexposed or "burned," it will be a dark mottled blue and the fine lines of the drawing will be indistinct or missing. The exposed print is submerged quickly in water, agitated, and then allowed to soak for 10 to 30 min. The print is then hung to dry in a subdued light.

Alterations on blueprints can be made with a weak solution of caustic soda, which is used as an ink to produce white lines. If a colored line is desired, the solution may be mixed with ink. Several solutions of this nature are on the market in bottled form and are known as *erasing fluids.* Special inks for alteration of blueprints are available.

2 Brownprints. A print made from a tracing onto processed *vandyke* paper has white lines on a dark-brown background. The paper is exposed in the same manner as in blueprinting. After the exposed paper is washed for about 5 min in water (in subdued light), it is transferred to a fixing bath consisting of 2 oz of hyposulfate of soda to 1 gal of water. When the print takes on a deep brown, it is again washed thoroughly in clear water and is left to soak for 20 to 30 min. The principal use of the vandyke process is in the making of a negative from which blueline positive prints or other contact prints may be made.

3 *Blackline Prints.* A *direct blackline print* is made from the tracing by contact printing in sunlight or artificial light, using a special sensitized paper. The print is developed by applying a chemical furnished by the manufacturer, after which it is thoroughly washed in water and is dried. This type of print is increasing in use, as it has the advantages of printing without a negative, a white background, freedom from excessive shrinkage, and difficulty of alteration without detection.

4 *Ozalid Prints.* An *ozalid* or *diazo* print with red, blue, brown, or black lines is made on a special sensitized paper by direct contact printing in the usual manner. In the dry process, it is developed by being placed in a tight container, which is then filled with ammonia vapor. In the wet process, the exposed sensitized sheet is passed through a liquid ammonia solution.

5 *Photostats.* A drawing on any kind of paper or cloth may be reproduced to any scale by the photostat process, provided the lines are of a color which photographs well. The process is widely used, especially in the reproduction of pages from books.

The photostat machine is a modified form of camera. The drawing is strongly illuminated by artificial light, and a negative to the desired scale is made, in which black lines of the original appear white. By rephotographing, a positive is produced in which black lines of the original appear black on a white or gray background.

6 *Photo-offset Prints.* The photo-offset process, known by various trade names, is useful when many prints of a drawing are required. This process consists in making a negative to the desired scale by photographing the original; from this negative a plate is prepared and mounted for use in an offset type of printing press. The prints may be made on any good grade of bond paper; they have distinct black lines. For large quantities the cost is low.

14-11 Map-drafting Instruments. *Engineer's scales* are divided into 10, 20, 30, 40, 50, 60, 80, or 100 parts to the inch. The 12-in. triangular boxwood rule (Fig. 14-6) is most commonly used in mapping. A flat rule with two scales on edges of opposite bevel is also used.

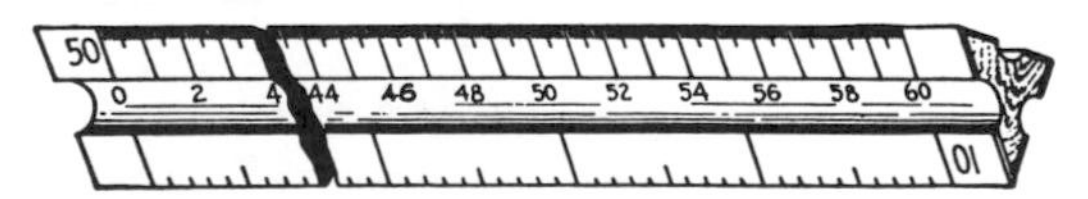

Fig. 14-6. Engineer's scale, triangular.

Probably $\frac{1}{100}$ in. is as close as distances can be plotted by the ordinary methods of drafting. For precise drafting, points should be pricked with a fine needle, a reading glass should be employed, and distances should be plotted with the eye directly above the graduation to which the distance is measured. Under these conditions, points can be plotted within perhaps $1/200$ in.

The *protractor* is a device for laying off and measuring angles on drawings. The usual form for mapping consists of a full circle or a semicircular arc of metal, plastic, or paper graduated in degrees or fractions of a degree. Figure 14-7 shows part of a full-circle paper protractor which before use is cut on a circle passing through the graduations; either the inner or the outer portion may be used. Figure 14-8 shows one type of semicircular protractor commonly used for plotting survey details; an added feature of this protractor is the linear scale for measuring distances at the time the corresponding angles are laid off from the meridian.

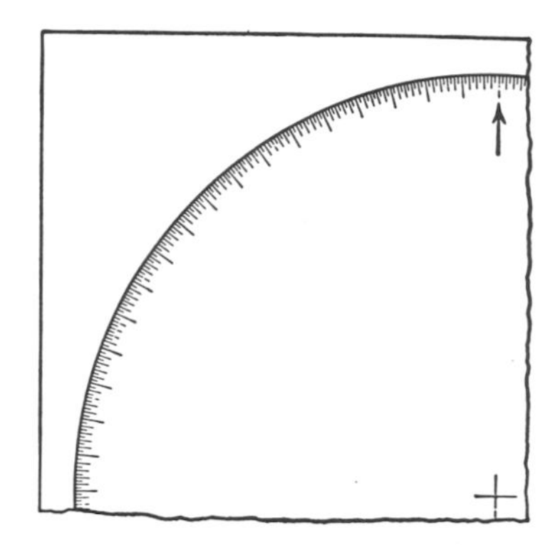

Fig. 14-7. Full-circle paper protractor.

To lay off an angle, the center of the protractor is placed at the vertex, and the protractor is oriented with reference to the established line. A mark is then made on the drawing at the proper graduation of the protractor arc, and a line is drawn joining this mark with the vertex. An angle is measured in a similar manner.

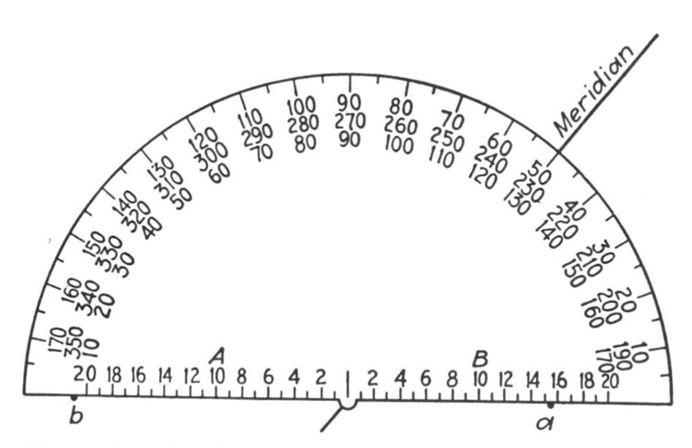

Fig. 14-8. Protractor for plotting details.

The *beam compass,* illustrated in Fig. 14-9, is used for drawing the arcs of large circles, say of radius greater than 6 in. A beam compass may easily be improvised from any thin strip of wood, or a stretched line or wire may be used.

Railroad curves are thin strips of cardboard, wood, metal, hard rubber, or plastic, the edges of which are arcs of circles. A number on each curve indicates its radius in inches, and sometimes also an additional number indicates the degree of curvature for a given scale. With these curves, arcs of circles can be drawn with much larger radius than could be used with a beam compass.

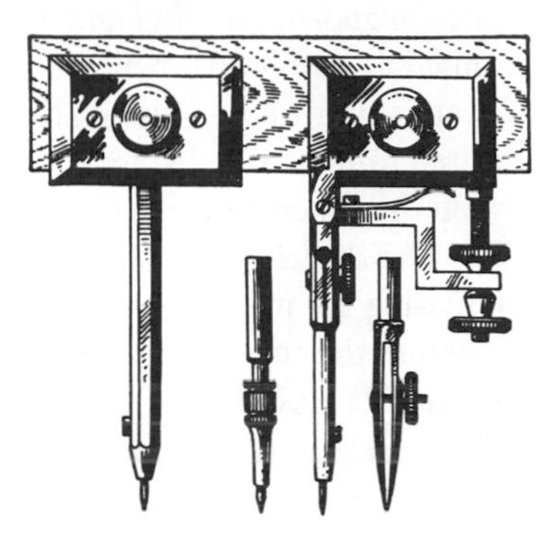
Fig. 14-9. Beam compass.

The *road pen* or *railroad pen* (Fig. 14-10) is used principally for drawing two parallel lines either freehand or by means of a straightedge or curve. It consists of two ruling pens with spring shanks attached to a handle, the distance between the two pens being controlled by a screw passing through the shanks.

The *contour pen* (Fig. 14-11) is useful for drawing contours or other freehand curves. The pen is connected rigidly to a shaft which turns freely in the handle. The point of the pen is eccentric with the axis of the shaft so that the pen will turn in whatever direction it is being drawn on the paper. In use, the handle is held vertical, and the line is preferably drawn toward the draftsman.

The most satisfactory form of *straightedge* for general office use is

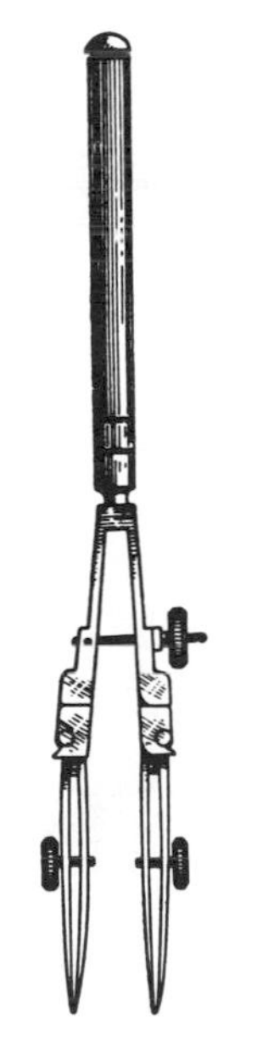

Fig. 14-10. Road pen.

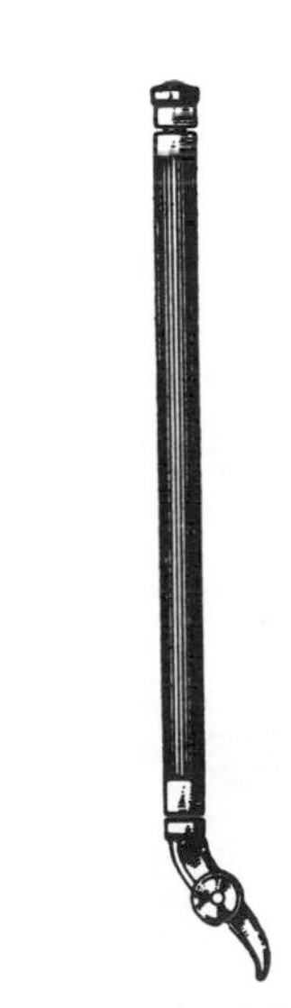
Fig. 14-11. Contour pen.

of nickel-plated or stainless steel, perhaps 42 in. long, with one edge beveled.

For transferring distances from one map to another at a different scale, *proportional dividers* are useful. They consist of two legs, each pointed at both ends, which are held together by means of a central pivot. The position of the pivot along the legs can be varied to produce any desired ratio of the distance between one pair of points to the distance between the other pair of points.

The *drafting machine,* one form of which is shown in Fig. 14-12, is increasing in use for map drafting, particularly for plotting details. It combines the functions of the T square, straightedge, triangles, scales, and protractor. Essentially, the drafting machine consists of a mechanical linkage bearing a pivoted *protractor head* to which are attached two mutually perpendicular graduated *arms*. The arrangement is such that each arm remains parallel to its original direction as the protractor head is moved about on the drafting table; thus, for example, a reference system of rectangular coordinates can be established on the drafting sheet by drawing lines along the edges of the arms. Further, the protractor head can be oriented quickly in any desired direction and clamped, thus enabling lines to be drawn or measuremennts to be made in oblique directions. By means of a vernier the protractor can be set to 05′ or, on some machines, to 01′. The arms are removable in order that engineer's scales of various graduations and lengths may be employed.

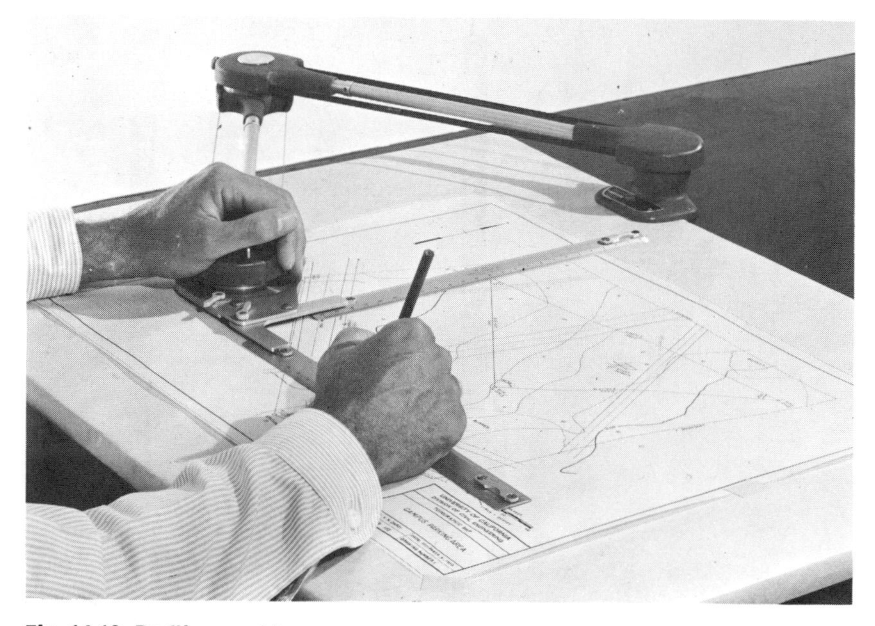

Fig. 14-12. Drafting machine.

15

MAP PLOTTING

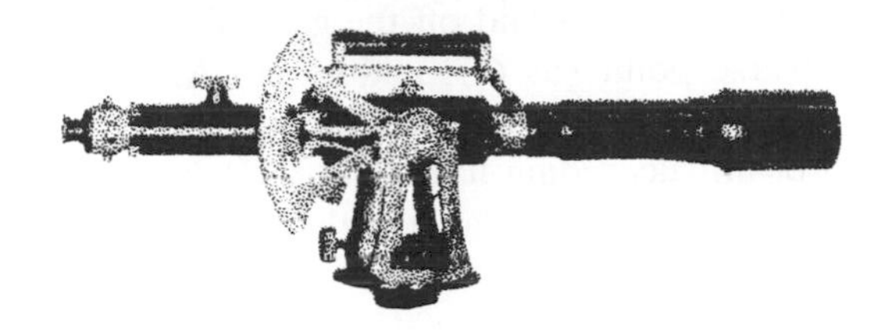

15-1 Process of Making a Map. The operations of plotting are in a sense the reverse of the operations of surveying. The general aim is to plot the more definite features within proper limits of error. The process of mapping involves (1) the plotting, by more precise methods, of points of horizontal *control* which are generally transit stations and which may be traverse points, triangulation points, or both; and (2) the plotting, by less precise methods, of the map *details*, employing angular and linear measurements from the lines and points in the horizontal control system.

Most maps are plotted wholly in the office; but where the objects to be shown are numerous, often the primary control only is plotted in the office and the secondary control and details are plotted in the field.

15-2 Plotting Control. Horizontal control may be plotted (1) by use of the protractor, (2) by the tangent method, (3) by the chord method, or (4) by the coordinate method. In any case, distances are measured with the engineer's scale; for precise work the points are pricked with a needle and a reading glass is employed. Traverse and triangulation stations are indicated by appropriate symbols (see Fig. 14-5*b*), and control lines are drawn carefully with a hard pencil having a fine point. Usually the control is not shown on the finished map.

For methods of plotting details, see Art. 15-16. For methods of plotting profiles and cross-sections, see Chap. 7.

15-3 Protractor Method. Where the control system is not extensive and the map is small, a fairly large protractor provides a sufficiently precise means of plotting angular values (see Art. 14-11). However, for the 6-in. protractor (a size in common use) the accidental error of plotting may amount to 15′. The protractor is useful in checking angles plotted by other methods.

When a deflection-angle traverse is to be plotted by protractor, the position of the first line is fixed by estimation and its length (as AB) is laid off by measurement. The protractor is oriented at the forward point (as B), the deflection angle to the succeeding line is laid off, and a light line of indefinite length is drawn. Along this line is laid off the given distance (as BC) to the succeeding traverse point (as C), and so on. An objection to this procedure is that any error in the direction of one line affects to a like degree the direction of all succeeding lines, and thus the linear error in the location of succeeding points increases with the distance.

Use of Meridians. If azimuths or bearings of the lines of a traverse are given or computed, a meridian line is drawn through each station and the direction of the succeeding line is laid off with respect to the meridian. (An alternative method is to erect a meridian near the middle of the sheet and, with any desired point on the meridian as an origin, to lay off the directions of all traverse lines, later transferring these directions to the appropriate positions on the sheet.) When angles are laid off from a meridian, any angular error made in plotting a given line does not affect the plotted *directions* of succeeding lines, although it results in a constant linear error in the *locations* of succeeding traverse points.

15-4 Tangent Method. In principle, the tangent method is similar to the protractor method, with the difference that an angle is laid off by a linear measurement which is a constant times the natural tangent of the angle. In Fig. 15-1, AB represents the plotted position of the initial line of a traverse, and α is the deflection angle at B which it is desired to lay off in order to determine the direction of BC. By the tangent method, the line last established (in this case AB) is prolonged some convenient distance, usually 10 in., to form a base line Bb, at the end of which a per-

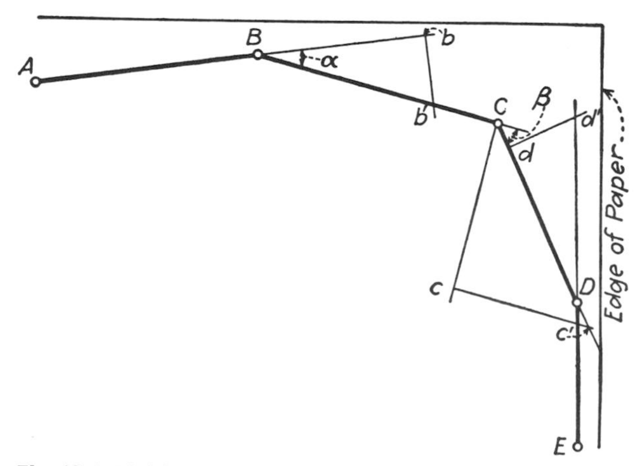

Fig. 15-1. Plotting by tangent offsets.

pendicular bb' is erected; and the distance bb' is laid off equal to the length of the base line Bb multiplied by the natural tangent of α. A line drawn from B through b' defines the direction of BC. If the traverse approaches the edge of the sheet, the base line may be measured *back* along the preceding line, as at Dd in Fig. 15-1.

If the deflection angle is greater than 45°, usually the base line is established not as a prolongation of the preceding line but as a perpendicular to the preceding line at the point last plotted. Thus in the figure Cc is the base line perpendicular to BC; and the perpendicular offset cc' is the length of the base line Cc times the natural cotangent of the angle β.

When a traverse is to be plotted by the method of tangent offsets, the deflection angles and distances are tabulated, and the tangents of angles less than 45° and cotangents of angles greater than 45° are recorded. The traverse is plotted roughly to small scale, the protractor being used for laying off the angles; and from the small-scale sketch a suitable position for the first line of the traverse is estimated.

The error in laying off an angle from a 10-in. base line need not exceed 03′.

Use of Meridians. In plotting deflection angles by tangents, any error in the direction of one line affects to a like degree the directions of all succeeding lines. If azimuths or bearings of lines are given or computed, a meridian and base line may be established at each transit point, and the succeeding line may be plotted by methods similar to those described for deflection angles. In this way any error in the direction of one line does not affect the direction of succeeding lines. A common central meridian may be employed, as described for the protractor method.

15-5 Chord Method. This method is much like that of plotting by tangents as just described, except that instead of erecting a perpendicular at the end of a 10-in. base line, an *arc* of 10-in. radius is struck. The chord distance for the given angle is then scaled from the point of intersection between arc and base line to a point on the arc. If a table of chords is available, the method is rapid; however, the chord method is not in as general use as the tangent method.

15-6 Checking. The methods of checking are essentially the same for the protractor, tangent, and chord methods of plotting just described. A traverse that has been brought to closure in the field should likewise close on paper. If the field work is assumed as correct, any error of closure on paper is because of inaccuracies in laying off angles and distances.

If the error of closure is small, it is usually assumed to have accumulated gradually, and the traverse is made to close by an arbitrary progressive change in the position of the plotted lines, as shown in Fig. 15-2. Another method of closure is illustrated in Fig. 15-3; each traverse point is moved in a direction parallel to the side of closure, by an amount proportional to the distance along the traverse from the initial point to the given point. For example, $C'C = AB'C'/AB'C'D'E'A' \times A'A$. Essentially this method is an application of the compass rule (Art. 15-12).

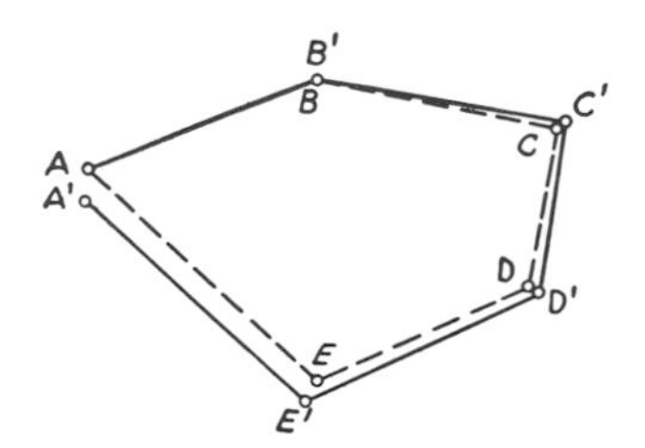

Fig. 15-2. Closure of traverse, assuming gradual accumulation of error.

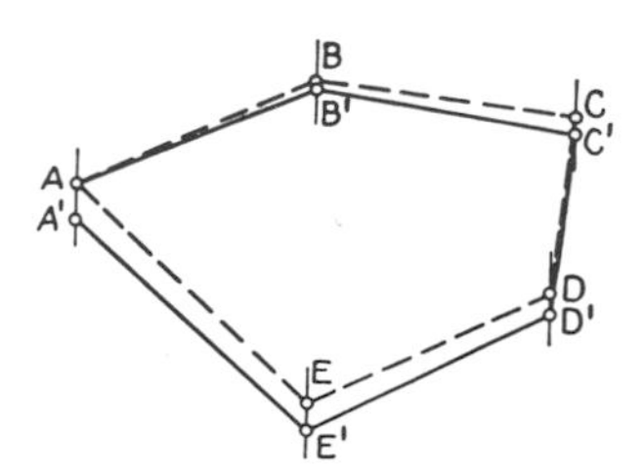

Fig. 15-3. Closure of traverse, based on compass rule.

If the error of closure is large, a mistake in laying off an angle or a distance has occurred; sometimes the mistake can be located quickly by observing the simplest way of bringing the traverse to a closure. Thus in Fig. 15-3, with the error of closure $A'A$ as shown, one might expect that a mistake had been made in the length of the side $C'D'$ which is nearly parallel to the side of closure. If the cause of the error of closure cannot be detected readily by measurement of the plotted angles and distances, the traverse may be plotted in the reverse direction, starting from the same initial point, until a point coincides with the corresponding point of the traverse as originally plotted.

For an *open* or continuous traverse, there is no absolute check on the precision of plotting. As a check on distances, the full length of the traverse may be marked off along the edge of a strip of paper, and the total length scaled and compared with the sum of the lengths in the notes used for plotting. If deflection angles are used in plotting, the direction of any line can be checked by computing its azimuth or bearing and observing whether the line makes this azimuth or bearing with an established meridian. If bearings or azimuths are used in plotting, the deflection angles at the several points can be computed, and the direction of the courses can be checked either with the protractor or by tangent offsets or chord distances.

15-7 Method of Rectangular Coordinates. This method, also known as the *method of total latitudes and departures,* is recognized as the most reliable method of plotting. It is the only practical method for plotting extensive systems of horizontal control, whether they be established by traversing or by triangulation. Rectangular coordinates are employed not only for plotting maps but also frequently for calculating areas, as described in Art. 16-2.

The coordinate axes are a *reference meridian* (true, magnetic, or assumed) and a line at right angles thereto called a *reference parallel.* The intersection of these lines, marking the origin, may be any point in the survey or may be outside the survey. With the direction of each survey line determined and its length known, computations are made to determine its *latitude* and *departure,* or the lengths of its orthographic projection upon the meridian and the parallel, respectively. The origin having been chosen, the coordinates for the several control points are computed by using the latitudes and departures. The coordinate of a point measured normal to the $\left\{\begin{matrix}\text{parallel}\\ \text{meridian}\end{matrix}\right\}$ is called the $\left\{\begin{matrix}\textit{total latitude}\\ \textit{total departure}\end{matrix}\right\}$ or the $\left\{\begin{matrix}\textit{parallel distance}\\ \textit{meridian distance}\end{matrix}\right\}$ of the point. $\left\{\begin{matrix}\text{Total latitudes}\\ \text{Total departures}\end{matrix}\right\}$ are positive or negative according to whether the corresponding points lie $\left\{\begin{matrix}\text{north or south of the reference parallel}\\ \text{east or west of the reference meridian}\end{matrix}\right\}$. With the coordinate axes established on paper, a point is plotted by laying off its total latitude and its total departure to the required scale. Articles 15-8 to 15-15 are devoted to a more detailed discussion of the processes involved.

Before the latitudes and departures of the traverse lines are computed, the angular error of closure of the traverse is determined by the known geometrical conditions, and the angles or bearings are so adjusted or corrected that the known geometrical conditions will be fulfilled (see Art. 15-11).

15-8 Latitudes and Departures. In Fig. 15-4, *AB* represents any line the latitude and departure of which it is desired to determine, and the lines NS and EW represent any meridian and any parallel. The line *AB* makes the angle α with the meridian.

As the latitude of a line is the orthographic projection of the line upon a meridian, the latitude of *AB* is $A'B' = Ab = AB \cos \alpha$. And as the departure of a line is its orthographic projection upon a parallel, the departure of *AB* is $A''B'' = Aa = AB \sin \alpha$. Stated in the form of a general rule applicable to any line:

$$\text{Latitude} = \text{length} \times \text{cosine bearing angle} \tag{1}$$
$$\text{Departure} = \text{length} \times \text{sine bearing angle} \tag{2}$$

Latitudes are designated as *North* or *positive* for all lines having a northerly bearing, and *South* or *negative* for all lines having a southerly bearing. Departures are designated as *East* or *positive* for lines having an easterly bearing, and *West* or *negative* for lines having a westerly bearing. Thus, in Fig. 15-5, for the line *AB* the latitude is North or + and the departure is East or +. For *BC* the latitude is South or − and the departure is East or +. For *CD* the latitude is South or − and the departure is West or −. For *DE* the latitude is North or + and the departure is West or −.

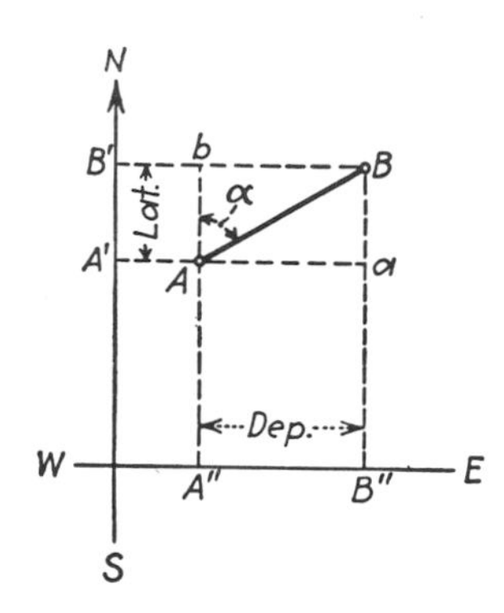

Fig. 15-4. Latitude and departure of line *AB*.

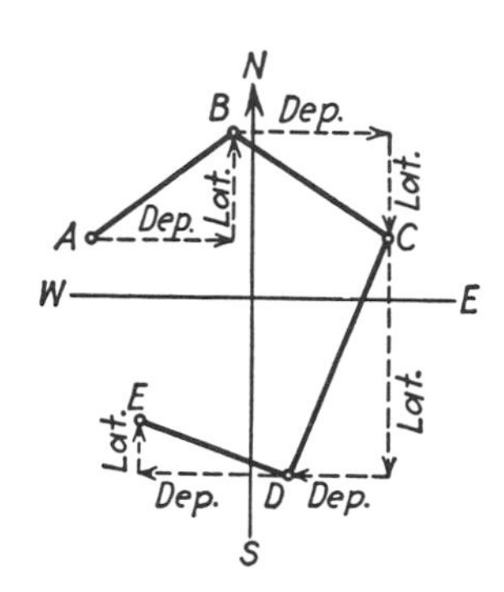

Fig. 15-5. Directions of latitudes and departures.

If the latitudes and departures are determined solely for purposes of map construction, the number of places to be used in computations should be such that points may be plotted correctly within the scale of the map. Computations are made either with a computing machine or by logarithms. "Traverse tables" found in various publications, such as Ref. 43, give for various bearing angles the latitudes and departures for various lengths of line.

If the computing machine is employed, the data may be kept in the following form:

Line	Bearing	Length, ft	Cos bearing	Sin bearing	Latitude, ft		Departure, ft	
					N	S	E	W

First the bearings and lengths of the lines are tabulated. Then the natural cosines and sines of the bearing angles are recorded. And finally for each line the latitude and departure are computed, the length being set on the machine as the multiplicand.

If logarithms are used, the form of the following example is convenient.

example. For a given course in a traverse, the computed bearing is N34°21′W and the observed length is 1,215.3 ft. The traverse is to be plotted to the scale of 1 in. = 100 ft. It is desired to compute the latitude and departure.

At the given scale, distances may be plotted within 1 or 2 ft, and therefore latitudes and departures should be correct to perhaps the nearest quarter foot if the traverse has many sides. Five-place logarithms will be used. The computations are given in the accompanying tabulation. Beginning with "log distance" in the tabulation, computations for latitude are made reading upward, and computations for departure are made reading downward.

Latitude	1,003.3 ft	
log latitude		3.00145
log cos bearing		9.91677
log distance		**3.08468**
log sin bearing		9.75147
log departure		2.83615
Departure	685.7 ft	

15-9 Error of Closure.

In any closed traverse it is obvious that the sum of the north latitudes should equal the sum of the south latitudes, and that the sum of the east departures should equal the sum of the west departures. In other words, for any closed traverse the algebraic sum of the latitudes (ΣL) should be equal to zero, and the algebraic sum of the departures (ΣD) should be equal to zero. However, owing to errors in field measurements of both angles and distances, in general an unadjusted traverse will not close on paper, even though the plotting be without error. The conditions stated above make it possible to determine the error of closure by means of the computed latitudes and departures.

Figure 15-6 shows a traverse that does not close, the line $A'A = e$ being the side of error. In order that the algebraic sum of the latitudes and the algebraic sum of the departures shall each be equal to zero, the latitude of the side of error must be $-\Sigma L$ and its departure must be $-\Sigma D$, considered algebraically. These two quantities form the base and altitude of the right-angle triangle of which the side of error is the hypotenuse; hence the linear error of closure is

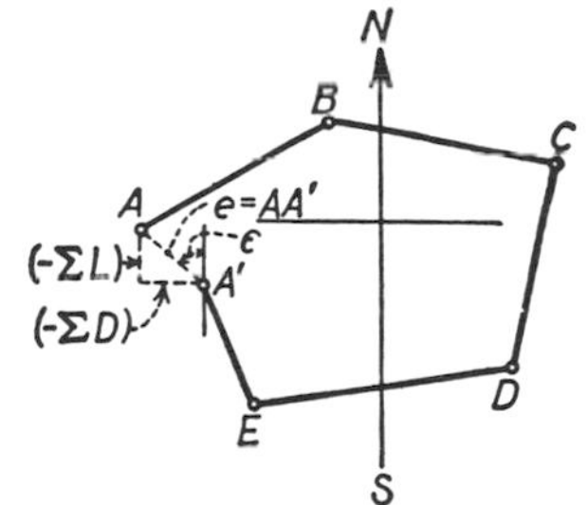

Fig. 15-6. Error of closure of traverse.

$$e = \sqrt{(\Sigma L)^2 + (\Sigma D)^2} \tag{3}$$

The direction of the side of error is given by the relation

$$\tan \epsilon = \frac{-\Sigma D}{-\Sigma L} \tag{4}$$

with due regard to sign. The data will make it apparent in which quadrant the bearing lies.

15-10 Balancing the Survey. When the error of closure has been determined as described in the preceding article, usually corrections are made so that the traverse will form a mathematically closed figure, and the corrections are applied to the latitudes and departures in such manner as to make their algebraic sum equal zero. This operation is called *balancing the survey* or *balancing the traverse*.

Electronic computers are used increasingly for computation of latitudes and departures from the lengths and directions observed in the field and for balancing the traverses and computing the coordinates of the stations. Standard computer programs are available in many localities.

If the error of closure is excessive, it indicates that a mistake in field work or in plotting has been made, and the work should be checked (Art. 15-6).

There are several rules for distribution of errors, each of which will produce a mathematically closed figure and each of which is assumed to be adapted to certain conditions as regards measurements. However, many surveyors rely on their own judgment with little or no regard for any established rule, and distribute the error arbitrarily in accordance with their estimation of the field conditions.

The principles of adjusting field observations are discussed in Chap. 2.

15-11 Adjustment of Angular Error. The total angular error determined from the known geometrical conditions is distributed among the angles or computed bearings of the traverse *before computations of the latitudes and departures are made.* Often this adjustment is arbitrary, based upon a knowledge of the field conditions; but if all angles have been measured under like conditions, the error is distributed equally to each angle in the traverse.

The angular adjustment does not yield true values, but it meets the known geometrical conditions; and the adjusted values are the most probable values that can be assigned.

15-12 Compass Rule for Balancing a Survey. The *compass rule* to balance a traverse is the one most commonly employed. In the light of the principles of least squares, it is logical for the assumptions made. The *transit rule* cannot be properly used in most cases, and it is not discussed herein. The Crandall method, which is well suited for electronic computation and which applies the method of least squares, is valid when the accidental errors in linear measurements are assumed to be much larger than those in angles.

The compass rule states that the correction to be applied to the $\left\{\begin{matrix}\text{latitude}\\\text{departure}\end{matrix}\right\}$ of any course is to the total correction in $\left\{\begin{matrix}\text{latitude}\\\text{departure}\end{matrix}\right\}$ as the *length* of the course is to the length of the traverse. It is based upon the assumptions (1) that the errors in traversing are accidental and therefore vary with the square root of the lengths of the sides, thus making the correction to each side proportional to its length (Chap. 2); and (2) that the effects of the errors in angular measurements are equal to the effects of errors in linear measurements.

example. In the accompanying tabulation are given the bearings, lengths, latitudes, and departures for a closed traverse of six sides. The survey is to be balanced by the compass rule.

			Latitude, ft		*Departure, ft*		*Correction, ft*	
Line	*Bearing*	*Length, ft*	*N*	*S*	*E*	*W*	*Latitude*	*Departure*
AB	N	500.0	500.0	......	0.0	......	+1.5	+3.2
BC	N45°00′E	848.6	600.0	......	600.0	......	+2.6	+3.4
CD	S69°27′E	854.4	......	300.0	800.0	......	−2.6	+5.4
DE	S11°19′E	1,019.8	......	1,000.0	200.0	......	−3.1	+6.6
EF	S79°42′W	1,118.0	......	200.0	......	1,100.0	−3.3	−7.2
FA	N54°06′W	656.8	385.0	......	......	532.0	+1.9	+4.2
Sum	....	4,997.6	1,485.0	1,500.0	1,600.0	1,632.0	15.0	32.0

The error of closure in latitude is 15.0 ft, and in departure is 32.0 ft. The sum of the lengths of the sides is 4,997.6 ft, or practically 5,000 ft. The length of the side of closure (linear error of closure) is $\sqrt{15^2 + 32^2} = 35.3$, and the relative linear error of closure is $35.3/5{,}000 = 1/142$. (This error is too large to be permitted in practice but was made large for the purpose of this example.)

For example, the correction in latitude of the course *CD* is computed to be $(15/5{,}000) \times 854.4$ ft $= 2.6$ ft. As the south latitudes are too large in this example, the correction is to be subtracted, that is, the correction is −2.6 ft. The correction to the departure of *CD* is $(32/5{,}000) \times 854.4 = 5.4$ ft; as the east departures are too small, the correction is to be added. The corrections to the other courses are computed in a similar manner, with due regard to sign.

In the eighth and ninth columns of the tabulation are given the corrections in feet. As a check, the arithmetical sum of the corrections in latitude or departure (given in the last line of each column) should equal the total error in latitude or departure. When the corrections have been applied, the algebraic sum of the latitudes and of the departures must be zero, and hence the survey is balanced.

15-13 Computation of Coordinates. When it is desired to plot points of horizontal control by the method of coordinates, the latitudes and departures of the control lines are computed, and in the case of a closed traverse the survey is balanced. The origin is chosen, and the coordinates, or total latitudes and departures, of the several points in the survey are computed by summing algebraically the latitudes and departures, respectively, of lines between that point and the origin.

In the accompanying tabulation are given (1) the adjusted latitudes and departures of the lines in the traverse of the preceding example, (2) the computed total latitudes and departures, or coordinates, referred to station *A* as the origin, and (3) the adjusted bearings and lengths of the traverse lines.

Sta.	*Adjusted lat.*		*Adjusted dep.*		*Total lat.*		*Total dep.*		*Adjusted bearing*	*Adjusted length, ft*
	N	*S*	*E*	*W*	*N*	*S*	*E*	*W*		
A					0	0	0	0		
	501.5	......	3.2	......					N0°22′E	501.5
B					501.5	....	3.2	..		
	602.6	......	605.4	......					N45°08′E	854.2
C					1,104.1	....	608.6	..		
	......	297.4	805.4	......					S69°44′E	858.6
D					806.7	....	1,414.0	..		
	......	996.9	206.6	......					S11°42′E	1,018.1
E					......	190.2	1,620.6	..		
	......	196.7	......	1,092.8					S79°48′W	1,110.4
F					......	386.9	527.8	..		
	386.9	......	......	527.8					N53°45′W	654.4
A					0	0	0	0		
Sum	1,491.0	1,491.0	1,620.6	1,620.6	......	....	......	..	....	4,997.2

In a closed traverse the additions are verified if on making the complete circuit the total latitude and total departure of the initial point check these quantities as assumed at the beginning. In the tabulation this check has been performed.

15-14 Plotting Control by Coordinates. When a system of horizontal control is to be plotted by rectangular coordinates, the size of the enclosing rectangle is determined from an examination of the total latitudes and departures. In order to ensure the proper location of points on the map sheet, usually the enclosing rectangle and the principal points of control are first plotted roughly to small scale. On the drawing paper the enclosing rectangle is drawn to the required scale, its position being fixed by estimation from the small-scale sketch.

Let *HJKL* (Fig. 15-7) be the enclosing rectangle for the traverse of the preceding article. The rectangle should be drawn with great care and checked by scaling the lengths of the diagonals. Perpendiculars con-

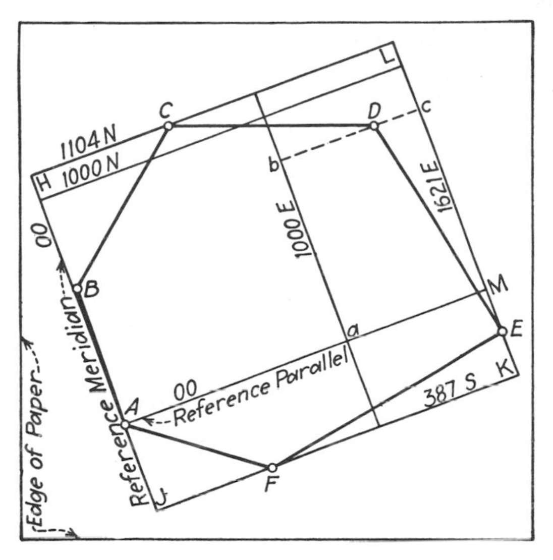

Fig. 15-7. Plotting control by coordinates.

structed by use of the ordinary triangle are not necessarily accurate; they should be checked by reversing the triangle, or the perpendicular should be erected by a 3:4:5 method or by the use of a drafting compass. The engineer's drafting machine (Art. 14-11) is useful for this work, or a recently developed coordinate-plotting machine may be available. The locations of the reference meridian and reference parallel are determined by scaling along the sides of the enclosing rectangle. For example, in Fig. 15-7 the reference parallel is located by scaling *JA* and *KM* equal to 387 ft, the total latitude of the most southerly point of the survey. The location is checked by scaling *AH* and *ML* equal to 1,104 ft, the total latitude of the most northerly point of the survey.

If the drawing is large and many control points are to be plotted, other meridians and parallels are constructed to divide the area into squares which represent some whole number of hundreds of feet at the given scale. Thus in the figure a meridian is drawn 1,000 ft east of the reference meridian, and a parallel is drawn 1,000 ft north of the reference parallel. If the scale of the map is 1 in. = 100 ft, the actual length of the sides of the resulting squares is 10 in.

Each point of horizontal control is located on the sheet by plotting its total latitude and departure, and its location is verified by scaling the length of the preceding course. For example, point *D* (Fig. 15-7) has a north total latitude of 807 ft and an east total departure of 1,414 ft. The point *D* is plotted by laying off to scale above the reference parallel the distances *ab* and *Mc* each equal to 807 ft, then drawing the line *bc,* and finally laying off to scale the distance $bD = 1{,}414 - 1{,}000 = 414$ ft. As a check, the distance *CD* as scaled from the drawing should agree with the length of this line as computed from the adjusted latitude and departure ($CD = 858.6$ ft by the tabulation in Art. 15-13).

For long open traverses the enclosing rectangle is not constructed. The reference meridian and reference parallel are drawn as accurately as possible, and with these as a basis other meridians and parallels are drawn as necessary for plotting the traverse points.

15-15 Advantages and Disadvantages of Coordinate Method. The coordinate method is recognized as the most reliable method of plotting control. Its principal advantages are as follows: (1) the size and shape of the drawing can be determined accurately beforehand; (2) the accuracy of the location of any point does not depend upon the accuracy with which previous lines in the control system are plotted; (3) the method of checking is simple and is unlike the method of plotting; (4) for closed traverses the field measurements are checked and the survey is balanced before plotting is begun; and (5) shrinkage or expansion of the paper is easily detected (by measuring the side of a control square) and proportionate corrections made.

The single disadvantage of the coordinate method as compared with the method of tangents or the method of chords is the greater amount of computation required preliminary to plotting. However, often in the case of a closed traverse the latitudes and departures are necessarily computed for the purpose of determining the area (Arts. 16-2 and 16-3); and with these values determined the labor of computing coordinates is not great.

15-16 Plotting Details. The processes of plotting details on the map are similar to those employed in making the field measurements but are in the reverse order. The work of plotting details is less refined than that of plotting the horizontal control, but the aim is still to plot objects of definite size and shape within the allowable limit of error. The angles are laid off with a protractor, or the drafting machine may be used.

If an object has been located in the field by the method of radiation, on the map a line is drawn from the plotted location of the instrument station in a direction corresponding to that on the ground; the distance from station to object is then scaled off along this line. If an object has been located by intersection, the corresponding angles are laid off from the two control stations, and the intersection of the two lines thus defined marks the plotted location of the object. If an object has been located by linear measurements from two stations, its location is plotted by intersecting arcs of the given radii drawn with the drafting compass. If objects of somewhat indefinite form have been located by perpendicular offsets from a traverse line, usually these offsets may be plotted by estimating the perpendiculars with the eye.

The location of all important details should be checked by measurement, and that of less important objects by inspection. If it happens that an object has more than one location (owing to a mistake in the field work) or that a part of the field notes is confusing, it is advisable to proceed with the plotting of adjacent portions, for these when mapped may help in clearing up the doubtful points.

OMITTED MEASUREMENTS

15-17 General. When for any reason it is impossible or impractical to determine by field observations the length and bearing of every side of a closed traverse, the missing data may generally be computed, provided that not more than two quantities (lengths and/or bearings) are omitted. (If only one measurement is omitted, a partial check is obtained on the work.) It must be assumed that the observed values are without error, and hence all errors of measurement are thrown into the calculated lengths or bearings.

There are three general cases: (1) length and bearing of one side unknown, (2) omitted measurements in *adjoining* sides of the traverse, and (3) omitted measurements in *nonadjoining* sides. In case 3, the solution involves changing the order of sides in the traverse in such a way as to make the two partly unknown sides adjoin.

When one of the sides is known in direction but unknown in length, the solution can be facilitated by assuming that side to lie on the reference meridian.

Methods of parting land, which involve the computation of lengths and bearings of unknown sides of a traverse, are described in Arts. 16-9 to 16-13.

15-18 Length and Bearing of One Side Unknown. The length S of the unknown side is

$$S = \sqrt{(\Sigma L)^2 + (\Sigma D)^2} \qquad (5)$$

in which ΣL and ΣD represent, respectively, the algebraic sum of the latitudes of known sides and the algebraic sum of the departures of known sides. The tangent of the bearing angle α is

$$\tan \alpha = \frac{-\Sigma D}{-\Sigma L} \qquad (6)$$

with due regard to sign.

15-19 Length of One Side and Bearing of Another Side Unknown. Figure 15-8 represents a closed traverse for which the direction of the line $DE = d$ and the length of the line $EA = e$ are not determined by field measurements. Let an imaginary line extend from D to A, cutting off the unknown sides from the remainder of the traverse. Then $ABCDA$

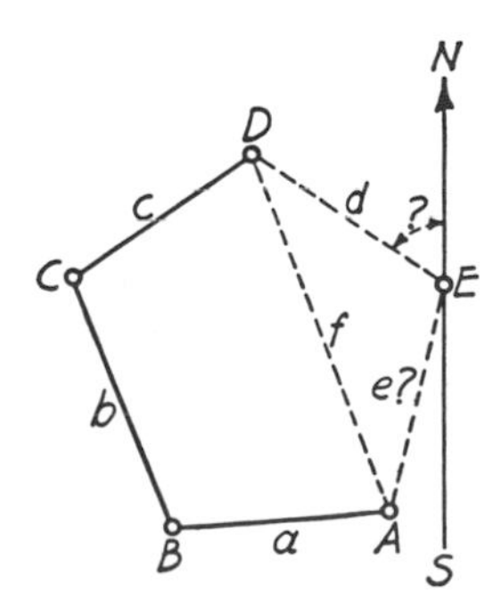

Fig. 15-8.

forms a closed traverse for which the side $DA = f$ is unknown in both direction and length. By the method of the preceding article,

$$\text{tan brg. angle of } f = \frac{\text{dep. } f}{\text{lat. } f} = \frac{-(\text{dep. } a + \text{dep. } b + \text{dep. } c)}{-(\text{lat. } a + \text{lat. } b + \text{lat. } c)} \tag{7}$$

and

$$\text{Length of } f = \frac{\text{lat. } f}{\text{cos brg. angle of } f} = \frac{\text{dep. } f}{\text{sin brg. angle of } f} \tag{8}$$

In computing the length of f by Eq. (8) it is desirable to use the larger of the two quantities, latitude or departure.

The angle between the lines e and f in triangle ADE is

$$\angle DAE = \text{azimuth of } AE - \text{azimuth of } AD \tag{9}$$

In the triangle ADE the length of the two sides d and f and one angle DAE are known. By the relation that sines of angles are proportional to sides opposite,

$$\sin DEA = \sin DAE \frac{f}{d} \tag{10}$$

With angle DEA known, angle ADE can be computed, and the remaining unknown length is given by the equation

$$e = f \frac{\sin ADE}{\sin DEA} = d \frac{\sin ADE}{\sin DAE} \tag{11}$$

Also,

$$\text{Azimuth of } DE = \text{azimuth of } DA - \angle ADE \tag{12}$$

The preceding method of solution is generally applicable even though two partly unknown courses are not adjoining. Obviously the latitude and departure of any line of fixed direction and length are the same for one location of the line as for any other. Regardless of the order in which the lines of a closed figure are placed, the algebraic sum of the latitudes and the algebraic sum of the departures must be zero. Hence, when two partly unknown sides of a closed traverse are not adjoining, one of the sides is considered as moved from its location to a second location parallel to the first, such that the two partly unknown sides adjoin; the solution then becomes identical with that just described. To simplify the problem, the data are usually plotted roughly to small scale.

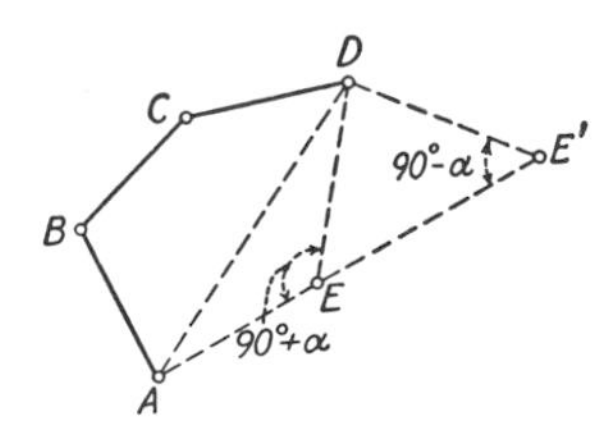

Fig. 15-9.

When the length of one side and the bearing of another are unknown, the solution described in this article will generally render two values of each of the unknowns, as illustrated by Fig. 15-9. Often it is impossible to tell which are the correct values unless the general direction of the side of unknown bearing is observed.

As the angle between the partly unknown lines approaches 90°, the solution becomes weak; and as the angle between these lines becomes small, the solution becomes strong.

15-20 Length of Two Sides Unknown. The solution is nearly identical with that of the preceding article. In Fig. 15-10 the solid lines represent courses for which the direction and length are known, and the lines *DE* and *EA* represent courses for which the direction is known but the length is unknown. From the latitudes and departures of the known sides, the length and bearing of the closing line *DA* are computed; and in the triangle *ADE* the angles *A, D,* and *E* are computed from the known directions of the sides. The lengths *DE* and *EA* are determined through the relation

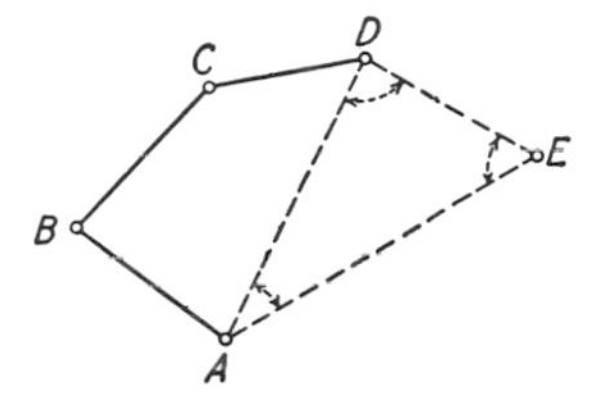

Fig. 15-10.

$$\frac{DE}{\sin A} = \frac{EA}{\sin D} = \frac{DA}{\sin E} \tag{13}$$

If the two lines are not adjoining, the problem may be solved as though they were, as explained in the preceding article. As the angle between the partly unknown lines approaches 90°, the solution becomes strong; and as the angle approaches 0 or 180°, the solution becomes weak, the problem being indeterminate when the lines are parallel.

15-21 Direction of Two Sides Unknown. The solution is similar to that of the preceding article. In Fig. 15-10, if *DA* is the closing side of the known portion of the traverse, its direction and length are computed; then the lengths of the three sides of the triangle *ADE* are known, and the angles *A, D,* and *E* can be computed.

The general direction of at least one of the partly unknown lines must be observed, as the values of the trigonometric functions merely determine the shape of the triangle but do not fix its position.

MAP PROJECTION

15-22 General. In maps of large areas where curvature becomes important, it is necessary to locate points by coordinates which are the geographical latitudes and longitudes expressed in angular units; for example, New York is at a latitude 40°45′ north of the equator and at a longitude 74°00′ west of Greenwich. Points are plotted with respect to a series of lines representing the earth's parallels and meridians. Any system of representing these parallels and meridians on a plane surface is called a *map projection.*

On a theoretically perfect map, without distortion, the following conditions will be satisfied: (1) All distances and areas would have correct relative magnitudes, (2) all azimuths and angles would be correctly shown, (3) all great circles would appear as straight lines, and (4) geographic latitudes and longitudes of all points would be correctly shown. Although in a plane map not all of these requirements can be satisfied at the same time, one or more conditions may be satisfied, as follows:

1 An *equal-area* projection results in a map showing all areas in proper relative *size,* although these areas may be much out of shape and the map may have other defects.

2 A *conformal* or *orthomorphic* projection results in a map showing the correct angle between any pair of short intersecting lines, thus making small areas appear in correct *shape.* As the scale varies from point to point, the shapes of larger areas are incorrect.

3 An *azimuthal* projection results in a map showing the correct *direction* or azimuth of any point from one central point.

15-23 Lambert Conformal Conic Projection. Attention was called to this projection by its use for the French battle maps during the First World War. Tables for its construction have been published by the U.S. Coast and Geodetic Survey. It is used for the state plane-coordinate systems of states (or zones thereof) of greater east-west than north-south extent (Art. 12-17).

This is a simple conic projection, the cone used being imagined to cut the surface of the earth along two parallels of latitude, called *standard parallels* (Fig. 15-11). When points on the earth's surface are projected to such a cone, there is a slight compression or decrease of scale between the standard parallels, and a stretching or increase of scale outside the standard parallels. Only a slight adjustment of scales is necessary to make the map conformal. It has been shown that, for a map of the United States, scale errors need not exceed 2 per cent at any point. For details the reader is referred to publications of the U.S. Coast and Geodetic Survey dealing with this projection.

15-24 Mercator Projection. This projection is cylindrical, but it cannot be constructed as a geometrical projection.

In a cylindrical projection, formed by means of a cylinder touching the earth along the equator, all meridians appear as straight parallel lines. But on the sphere any two such meridians are a maximum distance apart at the equator and converge toward the poles. Showing them parallel, therefore, results in a systematically increasing scale along the parallels of latitude as we pass from the equator toward the pole, with resulting dis-

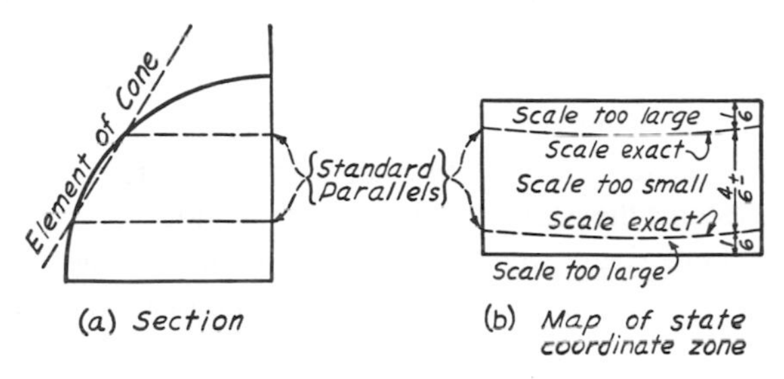

Fig. 15-11. Lambert conformal conic projection.

tortion of all areas shown on the map. The change of scale along the parallel, varying with the latitude, is readily computed (still assuming the earth to be spherical) by means of the formula

$$S' = S \cos \phi \tag{14}$$

in which S is the scale at the equator and S' is the scale at any latitude ϕ. For example, if the equatorial scale of the map is 1,000 miles per inch, then at latitude 60° (since the cosine of 60° is $\frac{1}{2}$) the scale is 500 miles to the inch.

The particular feature of the Mercator projection is that the scale along the meridian is varied to agree with the scale along the parallel, so that although the scale varies from point to point on the map at any given point the scale is the same in all directions. The map is therefore conformal. It has also the important property that a line of constant true bearing, or *rhumb line*, appears straight, which property renders it invaluable for purposes of navigation. The shortest course between two points is determined by drawing on a gnomonic chart a great circle, which there appears as a straight line. Selected points, at convenient distances apart, of this great circle are then plotted on the Mercator chart, after making any necessary corrections on account of shoals, wind, currents, etc. The rhumb line connecting any two adjacent points indicates the true bearing of the course, which is read by means of a protractor. This true bearing, corrected for magnetic declination, gives the compass bearing to be used in steering.

Owing to the rapid variation of scale, maps constructed on the Mercator projection give very inaccurate information as to relative sizes of areas in widely different latitudes. For example, on the map Greenland appears larger than South America, whereas in fact South America is nine times as large as Greenland. Consequently, such a map is not suited to general use, although because of its many other advantages it is widely published.

15-25 Transverse Mercator Projection. A transverse Mercator projection is the ordinary Mercator projection turned through an angle of 90° so that it is related to a central meridian in the same way that the ordinary Mercator projection is related to the equator. This projection is used for the state plane-coordinate systems of states (or zones thereof) of greater north-south than east-west extent (Art. 12-17). For the state systems the Mercator projection cylinder is made to cut the surface of the sphere along two standard lines parallel to the central meridian instead of being tangent to the sphere as in the ordinary Mercator projection.

15-26 Numerical Problems

1 With the protractor, plot to the scale of 1 in. = 200 ft the closed deflection-angle traverse for which the following notes are given. Measure the linear error of closure, and record it on the sheet. Distribute the error as suggested in Art. 15-6.

Station	*Deflection angle*	*Length, ft*
A	. . .	
		388
B	9°00′L	
		307
C	76°45′R	
		792
D	74°45′R	
		822
E	102°00′R	
		624
F	23°30′R	
		620
A	92°00′R	

2 Plot the following open deflection-angle traverse to a scale of 1 in. = 400 ft by the method of tangents, using a 10-in. base. Lay off successive lines by deflection angles, as described in Art. 15-4. Assume the direction of the first course to be north, and compute the bearings of the other courses. Check the accuracy of the plotting by methods described for open traverses in Art. 15-6. (See also office problem 37.)

Station	*Deflection angle*
118 + 75.0	. . .
98 + 95.6	39°47′L
73 + 01.4	17°28′L
70 + 13.5	14°08′L
49 + 41.3	3°11′L
40 + 00	49°59′L
37 + 18.8	32°18′R
18 + 26.0	18°44′R
5 + 03.2	7°31′L
0 + 00	. . .

3 Given the notes of problem 1. Assume the original direction of *AB* to be north, and compute the bearings of the several courses. Compute the latitudes and departures to tenths of feet, assuming that the given lengths are correct to tenths of feet, and using four-place logarithms. Compute the error of closure, and balance the survey by the compass rule (Art. 15-12).

4 Solve the preceding problem approximately, using the slide rule.

5 Given the following data of an open azimuth traverse. Compute the coordinates of the several points in the survey, assuming that the origin is at *A*. Plot the traverse by the coordinate method, using a scale of 1 in. = 400 ft. (See also office problem 38.)

Course	*Azimuth*	*Distance, ft*
AB	142°08′	815.3
BC	181°37′	1,146.0
CD	296°13′	520.8
DE	323°46′	816.5
EF	249°51′	726.4
FG	214°03′	1,862.0
GH	195°45′	2,795.5
HJ	191°28′	2,463.7
JK	138°42′	586.4

6 Given the following data for a closed traverse. Compute the length and bearing of the unknown side, using the slide rule.

Course	*Bearing*	*Distance, ft*
AB	N82°W	461
BC	unknown	unknown
CD	N68°15′E	829
DA	N80°45′E	441

7 Given the following data for a closed traverse, for which the length *DE* and the azimuth of *EA* have not been observed in the field. Determine the unknown quantities, using five-place logarithms. The general direction of *EA* is easterly.

Course	*Azimuth (from north)*	*Distance, ft*
AB	106°13′	1,081.3
BC	195°14′	1,589.5
CD	247°07′	1,293.7
DE	332°22′	unknown
EA	unknown	1,737.9

8 Solve the preceding problem by use of the slide rule, with corresponding precision (to 1 ft and to 05′).

9 Given the data on page 290 for a closed traverse, for which the lengths of *BC* and *DE* have not been measured in the field, compute the unknown lengths, using five-place logarithms.

10 Solve the preceding problem by use of the slide rule, with corresponding precision (to 1 ft and to 05′).

Course	Bearing	Distance, ft
AB	N9°30′W	689.32
BC	N56°55′W	unknown
CD	S56°13′W	678.68
DE	S2°02′E	unknown
EA	S89°31′E	1,082.71

15-27 Office Problems

Problem 37. Plotting by Tangents; Map Construction

Object. To plot a given traverse by the method of tangents. The data of field problems of Chaps. 8 and 10 may be used.

Procedure. (1) Tabulate angles and distances of the given traverse in the computation book. If angles are given as azimuths or bearings, change them to deflection angles. (2) Tabulate and check the natural tangent of each deflection angle less than 45° and the cotangent of each angle greater than 45°. (3) Plot the traverse roughly to small scale, using the protractor for angles; note its general form. (4) Carefully plot the first line of the traverse on drawing paper to the required scale, estimating the position of the line by means of the small-scale sketch. The line should lie so that the drawing, when finished, will be symmetrical with the sheet. (5) Plot the remaining lines of the traverse by the method described in Art. 15-4. Verify all measurements as soon as they have been plotted. (6) Check the traverse by the methods of Art. 15-6. If it is a closed traverse, distribute the error of closure as indicated by the conditions of the problem and perhaps by the direction of the side of closure. (7) Plot the details by methods corresponding to those used in the field. Use symbols wherever applicable. Show the meridian. Make a title.

Hints and Precautions. (1) Particular care should be taken in scaling the lengths of the lines of the traverse and the tangent distances. Points should be pricked with a needle, a reading glass should be used, and the eye should be above each point as it is plotted. (2) Lengths of lines and tangent distances plotted by estimation using the 10 scale are not sufficiently accurate; the 50 scale should be used. (3) A perpendicular should not be erected with the corner of the triangle at the point of intersection of base line and perpendicular, as the triangle corner may be rounded. The hypotenuse of the right triangle should be fitted to the base line, and a straightedge (or another triangle) placed in contact with the base of the triangle. Then the triangle should be turned through 90°, and its third edge placed in contact with the straightedge and moved along it until the hypotenuse passes through the point of intersection. (4) It is well to test each perpendicular by measuring the hypotenuse of a 45° right triangle having sides perhaps 8 in. long; in this case the length of the hypotenuse is 11.31 in. (5) If the construction lines would otherwise fall off the edge of the paper, measure back along the line as shown for the plotting of *DE* from *CD* in Fig. 15-1.

Problem 38. Plotting by Coordinates; Map Construction

Object. To plot a given traverse by the method of coordinates. The data of field problems of Chaps. 8 and 10 may be used.

Procedure. (1) Transcribe the given data to a computation book in the form shown by the first three columns of the tabulation in Art. 15-12. If directions of lines are given as azimuths or deflection angles, compute either the true or the assumed bearings. (2) Compute the latitudes and departures of the traverse lines; and if it is a closed traverse, balance the survey as described in Art. 15-12. (3) Assume one of the traverse points as the origin of coordinates, and compute the total latitudes and total departures as described in Art. 15-13. (4) Check all computations. (5) To small scale, plot roughly the traverse and enclosing rectangle (Art. 15-14), and note their relative positions. (6) On drawing paper plot the enclosing rectangle to the required scale, estimating its position by means of the small-scale sketch. The traverse, not the rectangle, should be symmetrical with the sheet. (7) Test the accuracy of the plotting by scaling the length of the diagonals. (8) Plot the reference meridian, reference parallel, and any supplementary meridians and parallels as described in Art. 15-14. Check the plotting. (9) Locate each traverse point by plotting its total latitude and total departure. Check by scaling the length of the preceding traverse line. (10) Plot the details by methods corresponding to those used in the field. Use conventional symbols wherever applicable. Finish the map without erasing construction lines. Label each traverse line with its corrected length *inside* the traverse and its corrected bearing *outside*. Show the meridian. Make a title.

Hints and Precautions. For details of plotting, see Hints and Precautions 1 and 2 of the preceding office problem.

16

CALCULATION OF AREAS OF LAND

16-1 Methods of Determining Area. One of the primary objects of most land surveys is to determine the area of the tract. In ordinary land surveying, the area is taken as the projection of the tract upon a horizontal plane. The area may be determined by any of the following methods:

1 By plotting the boundaries to scale, as described in Chap. 15; the area of the tract may then be found by use of the planimeter as described in Art. 7-6, or it may be calculated by dividing the tract into triangles and rectangles, scaling the dimensions of these figures, and computing their areas mathematically. This method is useful in roughly determining areas or in checking those that have been calculated by more exact methods.

2 By mathematically computing the areas of individual triangles into which the tract may be divided. Table XII gives the relations between the area of a triangle and its angles and lengths of sides.

3 By calculating the area from the coordinates, or meridian distances and parallel distances, of the *corners* of the tract (Art. 16-2).

4 By calculating the area from the double meridian distances and the latitudes of the *sides* of the tract (Art. 16-3).

5 For tracts having irregular or curved boundaries, the methods of Arts. 16-4 to 16-8 are employed.

For computation of areas of cross-sections, see Arts. 7-4 and 7-5.

16-2 Coordinate Method. If the coordinates of the corners of a tract have been found as described in Art. 15-13, the area can be calculated expeditiously. Essentially the calculation is that of finding the areas of trapezoids formed by projecting the lines upon one of a pair of coordinate axes.

In Fig. 16-1, *ABCDF* represents a tract the area of which is to be determined, *SN* a reference meridian, and *WE* a reference parallel. The coordinates of *A, B, . . . , F* are known; for any point (as *A*) the abscissa is the perpendicular distance from the reference meridian (as $aA = m_1$), defined as the *total departure* or the *meridian distance;* and the ordinate is the perpendicular distance from the reference parallel (as $a'A = p_1$), defined as the *total latitude* or the *parallel distance*. Meridian distances are regarded as positive or negative according to whether they lie east or west of the reference meridian; parallel distances are regarded as positive or negative according to whether they lie north or south of the reference parallel. In the figure all meridian and parallel distances are positive.

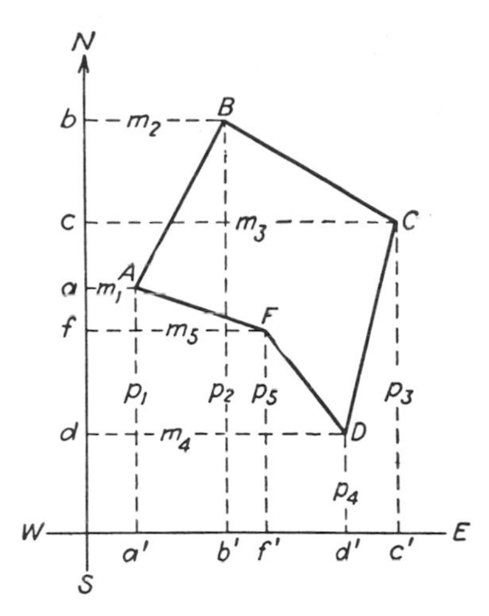

Fig. 16-1. Area by coordinates.

The area can be computed by summing algebraically the areas of the trapezoids formed by projecting the lines upon the reference meridian; thus

$$\text{Area } ABCDF = \text{area } BCcb + \text{area } CDdc - \text{area } DFfd - \text{area } FAaf - \text{area } ABba \quad (1)$$

or

$$\text{Area} = \tfrac{1}{2}(m_2 + m_3)(p_2 - p_3) + \tfrac{1}{2}(m_3 + m_4)(p_3 - p_4) - \tfrac{1}{2}(m_4 + m_5)(p_5 - p_4) - \tfrac{1}{2}(m_5 + m_1)(p_1 - p_5) - \tfrac{1}{2}(m_1 + m_2)(p_2 - p_1) \quad (2)$$

By multiplication and a rearrangement of terms in Eq. (2), there is obtained

$$2 \cdot \text{area} = -[p_1(m_5 - m_2) + p_2(m_1 - m_3) + p_3(m_2 - m_4) + p_4(m_3 - m_5) + p_5(m_4 - m_1)] \quad (3)$$

rule. To determine the area of a tract of land when the coordinates of its corners are known, multiply the parallel distance, or ordinate, of each corner by the difference between the meridian distances, or abscissas, of the following and the preceding corners, aways algebraically subtracting the following from the preceding. One half of the algebraic sum of the resulting products is the required area.

A result identical except for sign would be obtained by always subtracting the preceding from the following. The sign of the area is not significant.

example. Given the following data. Find the required area by applying the foregoing rule:

Corner	*1*	*2*	*3*	*4*	*5*
Meridian distance, ft	300	400	600	1,000	1,200
Parallel distance, ft	300	800	1,200	1,000	400

$$\begin{aligned} 2 \cdot \text{area} &= -[300(800) + 800(-300) + 1{,}200(-600) \\ &\qquad + 1{,}000(-600) + 400(700)] \\ &= -240{,}000 + 240{,}000 + 720{,}000 + 600{,}000 - 280{,}000 \\ &= 1{,}040{,}000 \text{ sq ft} \\ \text{Area} &= \frac{1{,}040{,}000}{2} = 520{,}000 \text{ sq ft} \end{aligned}$$

Equation (2) can also be expressed in the form

$$\begin{aligned} 2 \cdot \text{area} = m_2p_1 + m_3p_2 + m_4p_3 + m_5p_4 + m_1p_5 \\ -m_1p_2 - m_2p_3 - m_3p_4 - m_4p_5 - m_5p_1 \qquad (4) \end{aligned}$$

When this form is employed, computations can be made conveniently by tabulating each parallel distance below the corresponding meridian distance as follows:

$$\frac{m_1}{p_1} \times \frac{m_2}{p_2} \times \frac{m_3}{p_3} \times \frac{m_4}{p_4} \times \frac{m_5}{p_5} \times \frac{m_1}{p_1} \qquad (5)$$

Then in Eq. (5) the difference between the sum of the products of the coordinates joined by full lines and the sum of the products of the coordinates joined by dotted lines is equal to twice the area of the tract. The foregoing example may be quickly checked by this method.

16-3 Double-meridian-distance Method.

The D.M.D. method of calculating area is a convenient form of the method of coordinates just described, but the computations do not involve the direct use of coordinates. The latitudes and departures of all the courses are determined, and the survey is balanced. A reference meridian is then assumed to pass through some corner of the tract, usually for convenience the most westerly point of the survey; the double meridian distances of the lines are computed as described herein; and double the areas of the trapezoids or triangles formed by orthographically projecting the several traverse lines upon the meridian are computed. The algebraic sum of these double areas is double the area within the traverse.

The meridian distance of a point has been defined; thus in Fig. 16-2 the meridian distance of B is Bb and is positive. The meridian distance of a straight line is the meridian distance of its mid-point. The *double meridian distance* of a straight line is the sum of the meridian distances of the two extremities; thus the double meridian distance of *BC is Bb* $+Cc$.

From the figure it is seen that each projection trapezoid or triangle, for which a course in the traverse is one side, is bounded on the north and south by meridian distances and on the west by the latitude of that course. Thus the projection trapezoid for *BC* is *BCcb*. Therefore the double area of any triangle or trapezoid formed by projecting a given course upon the meridian is the product of the double meridian distance of the course and the latitude of the course, or

$$\text{Double area} = \text{D.M.D.} \times \text{latitude} \tag{6}$$

In computing double areas, account is taken of signs. If the meridian extends through the most westerly point, all double meridian distances are positive; hence the sign of a double area is the same as that of the corresponding latitude. Thus in the figure the double areas of *AbB, DdfF,* and *FfA* are positive, the latitudes *Ab, df,* and *fA* being positive; while the double areas of *CcbB* and *DdcC* are negative, the latitudes *bc* and *cd* being negative. Since the projected areas *outside* the traverse are considered once as positive and once as negative, the algebraic sum of their double areas is zero. Therefore the algebraic sum of all double areas is equal to twice the area of the tract *within* the traverse. Whether this algebraic sum is a positive or negative quantity is determined solely by the order in which the lines of the traverse are considered; the sign of the area is not significant.

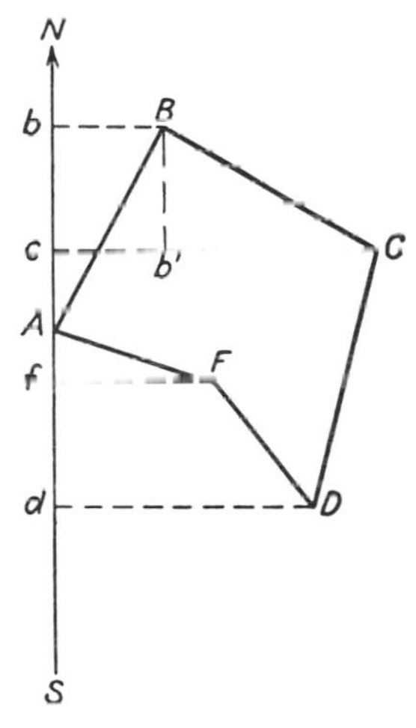

Fig. 16-2. Area by double meridian distances.

Following are three convenient rules for determining D.M.D.'s, which can be deduced from Fig. 16-2:

1 The D.M.D. of the first course (reckoned from the point through which the reference meridian passes) is equal to the departure of that course.

2 The D.M.D. of any other course is equal to the D.M.D. of the preceding course, plus the departure of the preceding course, plus the departure of the course itself.

3 The D.M.D. of the last course is numerically equal to the departure of the course but with opposite sign. This relation is useful as a check on the computations made in order around the traverse.

example. The latitudes and departures of the sides of the closed traverse shown in Fig. 16-2 are given in the accompanying tabulation. To calculate the area, first the D.M.D.'s are computed and checked by the three foregoing rules, then the double area between each line and the reference meridian is computed as the product of the latitude and the D.M.D., with due regard to sign; the computed values are shown in the tabulation.

Line	*AB*	*BC*	*CD*	*DF*	*FA*
Latitude	+ 190	− 110	− 230	+ 110	+ 40
Departure	+ 100	+ 190	− 60	− 90	− 140
D.M.D.	+ 100	+ 390	+ 520	+ 370	+ 140
Double area	+19,000	−42,900	−119,600	+40,700	+5,600

$$2 \cdot \text{area} = +19{,}000 + 40{,}700 + 5{,}600 - 42{,}900 - 119{,}600 = -97{,}200 \text{ sq ft}$$

$$\text{Area} = \frac{97{,}200}{2} = 48{,}600 \text{ sq ft}$$

Double parallel distances may be used in a manner similar to that just described for double meridian distances. In practice, however, they are little used except as an independent check on D.M.D.'s.

16-4 Area of Tract with Irregular or Curved Boundaries. If the boundary of a tract of land follows some irregular or curved line, such as a stream or road, it is customary to run a traverse in some convenient location near the boundary and to locate the boundary by offsets from the traverse line. Figure 16-3 represents a typical case, AB being one of the traverse lines. The offset distances are aa', bb', etc., and the corresponding distances along the traverse line are Aa, Ab, etc. Where the boundary is irregular, as from a' to f', it is necessary to take offsets at points of change and hence usually at irregular intervals. Where a segment of the boundary is straight, as from f' to g', offsets are taken only at the ends. Where the boundary is a gradual curve, as from g' to m', ordinarily the offsets are taken at regular intervals.

If the offsets are taken sufficiently close together, the error involved in considering the boundary as straight between offsets is small as compared with the inaccuracies of the measured offsets. The assumed boundary takes some such form as that illustrated by the dotted lines $g'h'$, $h'k'$, etc., in Fig. 16-3, and the areas between offsets are of trapezoidal shape. Under such an assumption, irregular areas are said to be calculated by the *trapezoidal rule* (Art. 16-5).

Where the curved boundaries are of such definite character as to make it justifiable, the area may be calculated somewhat more accurately by assuming that the boundary is made up of segments of parabolas as first suggested by Simpson. Under this assumption, irregular areas are said to be calculated by *Simpson's one-third rule* (Art. 16-6).

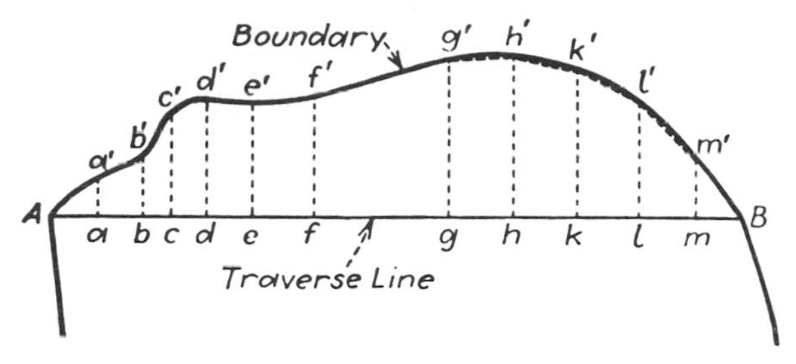

Fig. 16-3. Irregular boundary.

16-5 Offsets at Regular Intervals: Trapezoidal Rule. Let Fig. 16-4 represent a portion of a tract lying between a traverse line AB and an irregular boundary CD, offsets h_1, h_2, . . . , h_n having been taken at the regular intervals d. The summation of the areas of the trapezoids comprising the total area is

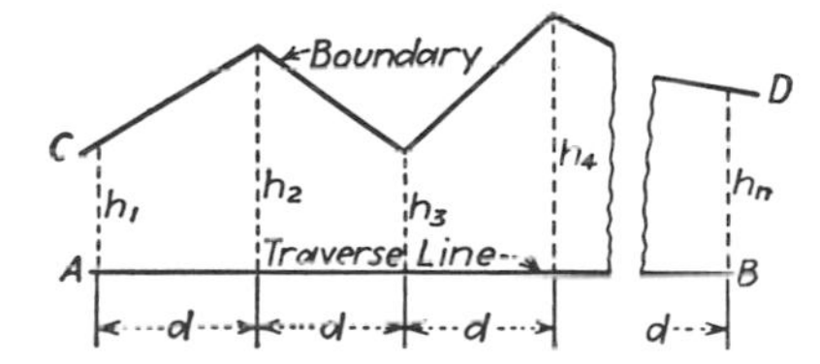

Fig. 16-4. Area by trapezoidal rule.

$$\text{Area} = \frac{h_1 + h_2}{2} \cdot d + \frac{h_2 + h_3}{2} \cdot d + \cdots + \frac{h_{(n-1)} + h_n}{2} \cdot d$$

$$\text{Area} = d\left(\frac{h_1 + h_n}{2} + h_2 + h_3 + \cdots + h_{(n-1)}\right) \tag{7}$$

Equation (7) may be expressed conveniently in the form of the following rule:

trapezoidal rule. Add the average of the end offsets to the sum of the intermediate offsets. The product of the quantity thus determined and the common interval between offsets is the required area.

example. By the trapezoidal rule find the area between a traverse line and a curved boundary, rectangular offsets being taken at intervals of 20 ft, and the values of the offsets in feet being $h_1 = 3.2$, $h_2 = 10.4$, $h_3 = 12.8$, $h_4 = 11.2$, and $h_5 = 4.4$. By the foregoing rule,

$$\text{Area} = 20\left(\frac{3.2 + 4.4}{2} + 10.4 + 12.8 + 11.2\right) = 764 \text{ sq ft}$$

16-6 Offsets at Regular Intervals: Simpson's One-third Rule. In Fig. 16-5 let AB be a portion of a traverse line, DFC a portion of the curved boundary assumed to be the arc of a parabola, and h_1, h_2, and h_3 any three consecutive rectangular offsets from traverse line to boundary taken at the regular interval d.

The area between traverse line and curve may be considered as composed of the trapezoid $ABCD$ plus the area of the segment between the parabolic arc DFC and the corresponding chord DC. One property of a parabola is that the area of a segment (as DFC) is equal to two-thirds the area of the enclosing parallelogram (as $CDEFG$). Then the area between the traverse line and curved boundary within the length of $2d$ is

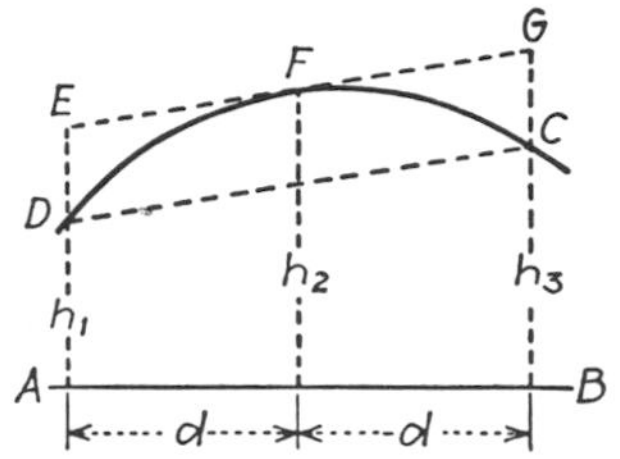

Fig. 16-5. Area by Simpson's rule.

$$A_{1,2} = \frac{h_1 + h_3}{2} 2d + \left(h_2 - \frac{h_1 + h_3}{2}\right) 2d \cdot \frac{2}{3}$$

$$= \frac{d}{3}(h_1 + 4h_2 + h_3)$$

Similarly for the next two intervals

$$A_{3,4} = \frac{d}{3}(h_3 + 4h_4 + h_5)$$

The summation of these partial areas for $(n - 1)$ intervals, n being an odd number and representing the number of offsets, is

$$\text{Area} = \frac{d}{3}[h_1 + h_n + 2(h_3 + h_5 + \cdots + h_{(n-2)}) + 4(h_2 + h_4 + \cdots + h_{(n-1)})] \quad (8)$$

Equation (8) may be expressed conveniently in the form of the following rule, which is applicable if the number of offsets is odd:

Simpson's one-third rule. Find the sum of the end offsets, plus twice the sum of the odd intermediate offsets, plus four times the sum of the even intermediate offsets. Multiply the quantity thus determined by one third of the common interval between offsets, and the result is the required area.

example. By Simpson's one-third rule find the area between the traverse line and the curved boundary of the example of the preceding article. By Simpson's rule,

$$\text{Area} = {}^{20}\!/_{3}[3.2 + 4.4 + 2(12.8) + 4(10.4 + 11.2)] = 797 \text{ sq ft}$$

If the total number of offsets is *even*, the partial area at either end of the series of offsets is computed separately, in order to make n for the remaining area an odd number and thus to make Simpson's rule applicable.

Simpson's rule is also useful in other applications, such as finding centers of areas. The prismoidal formula for computing volumes of earthwork (Art. 7-11) embodies Simpson's rule for area and a factor for the third dimension.

16-7 Rules Compared. Results obtained by using Simpson's rule are greater or smaller than those obtained by using the trapezoidal rule, according as the boundary curve is concave or convex toward the traverse line. The results obtained by using Simpson's rule are the more accurate, but the rule is not so easily applied as the trapezoidal rule.

16-8 Offsets at Irregular Intervals. The method of coordinates described in Art. 16-2 may be applied to this problem by assuming the origin as being on the traverse line and at the point where the first offset is taken. The coordinate axes are then the traverse line and a line at right angles thereto. The rule of Art. 16-2 may then be modified to the following:

rule. Multiply the distance (along the traverse) of each intermediate offset from the first by the difference between the two adjacent offsets, always subtracting the following from the preceding. Also multiply the distance of the last offset from the first by the sum of the last two offsets. The algebraic sum of these products, divided by two, is the required area.

PARTITION OF LAND

16-9 General. Where a given tract of land is to be divided into two or more parts, a resurvey is run, the latitudes and departures are computed, the survey is balanced, and the area of the entire tract is determined. The corrected latitudes and departures are further employed in the computations of subdivision.

16-10 Area Cut Off by a Line between Two Points. In Fig. 16-6 let *ABCDEFG* represent a tract of land to be divided into two parts by a line extending from *A* to *D*. It is desired to determine the length and direction

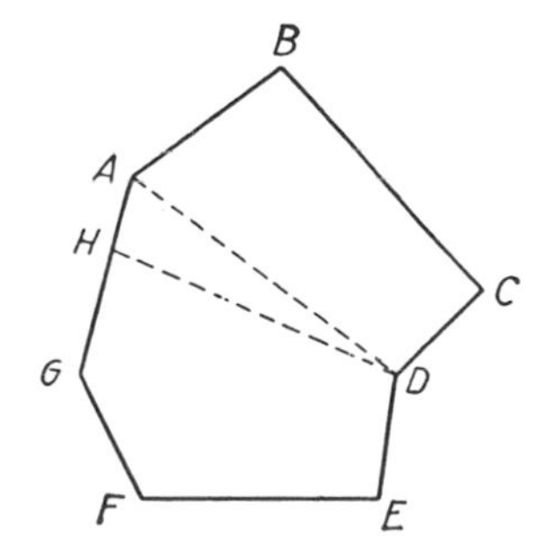

Fig. 16-6.

of the cut-off line *AD* without additional field measurements and to calculate the area of each of the two parts into which the tract is divided.

Either of the two parts may be considered as a closed traverse with the length and bearing of one side *DA* unknown. Considering the part *ABCDA,* the latitudes and departures of *AB, BC,* and *CD* are given; hence the latitude, departure, length, and bearing of *DA* can be determined as described in Art. 15-18. The area of either part can then be found by the D.M.D. method (Art. 16-3).

16-11 Area Cut Off by a Line Running in a Given Direction.

In Fig. 16-6, *ABCDEFG* represents a tract of known dimensions, for which the corrected latitudes and departures are given; and *DH* represents a line running in a given direction which passes through the point *D* and divides the tract into two parts. It is desired to calculate from the given data the lengths *DH* and *HA* and the area of each of the two parts into which the tract is divided.

Either of the two parts may be considered as a closed traverse for which the lengths of two sides are unknown; these lengths can be computed as described in Art. 15-20. Considering the part *ABCDHA,* the latitudes and departures of *AB, BC,* and *CD* are known; from these the length and bearing of *DA* are computed. In the triangle *ADH* the lengths of the sides *DH* and *HA* are found, and their latitudes and departures are computed. The area of *ABCDHA* is then calculated by the D.M.D. method.

In the field the length and direction of the side *DH* are laid off from *D,* and a check on field work and computations is obtained.

16-12 To Cut Off a Required Area by a Line through a Given Point.

In Fig. 16-7, *ABCDEF* represents a tract of known dimensions, for which the corrected latitudes and departures are given; and *G* represents a point on the boundary through which a line is to pass cutting off a required area from the tract. It is assumed that the area within the tract has been calculated by the D.M.D. method and that a sketch of the tract has been prepared.

To find the length and direction of the dividing line, the procedure is as follows: A line *GF* is drawn to that corner of the traverse which, from inspection of the sketch, will come nearest being on the required line of division. The latitude and departure of *CG* are

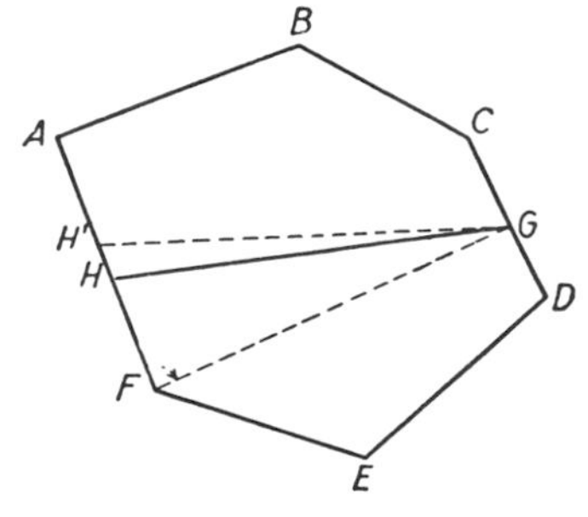

Fig. 16-7.

computed. Then in the traverse *ABCGFA* all sides are known except *GF*. By the methods of Art. 15-18 the latitude, departure, length, and bearing of *GF* are determined. By the D.M.D. method, the area of *FABCG,* the amount cut off by the line *FG,* is calculated. The difference between this area and that required is found.

In the figure it is assumed that *FABCG* has an area greater than the desired amount, *GH* being the correct position of the dividing line. Then the triangle *GFH* represents this excess area; and as the angle *F* may be computed from known bearings, there are given in this triangle one side *FG,* one angle *F,* and the area. The length *HF* is computed from the relation, area = ½*ab* sin *C,* given in Table XII; that is,

$$HF = \frac{2 \times \text{area } GFH}{FG \sin F} \tag{9}$$

The triangle is then solved for angle *G* and length *GH*. From the known direction of *GF* and the angle *G*, the bearing of *GH* is computed. The latitudes and departures of the lines *FH, GH,* and *HA* are then computed.

In the field the length *GH* is laid off in the required direction, and a check on field work and computations is obtained.

If desired, a trial line of subdivision such as *GH′* may be plotted near the estimated location of *GH*, and the scaled distance *AH′* used in the foregoing computations.

16-13 To Cut Off a Required Area by a Line Running in a Given Direction.

In Fig. 16-8, *ABCDEF* represents a tract of known dimensions and area, which is to be divided into two parts, each of a required area, by a line running in a given direction. The figure is assumed to be drawn at least roughly to scale, and the corrected latitudes and departures are known.

Through the corner that seems likely to be nearest the line cutting off the required area, a trial line *DG* is drawn in the given direction. Then in the closed traverse *GBCDG* the latitudes and departures of *BC* and *CD* and the bearings of *DG* and *GB* are known, and the lengths *DG* and *GB* are unknown. By the methods of Art. 15-20, these unknown quantities are found, and the latitudes and departures of the courses are determined. The area cut off by the trial line is calculated. The dif-

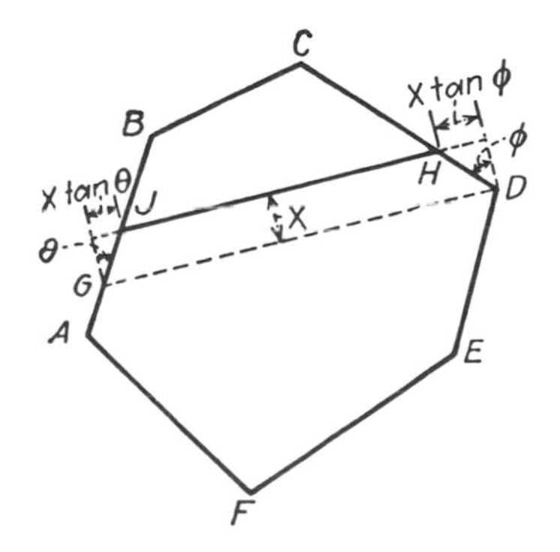

Fig. 16-8.

ference between this area and that required is represented in the figure by the trapezoid *DGJH* in which the side *DG* is known. The angles at *D* and *G* can be computed from the known bearings of adjacent sides, and in this way θ and ϕ are determined. Then

$$\text{Area of trapezoid} = DG \cdot x + \frac{x^2}{2} (\tan \theta + \tan \phi) \tag{10}$$

in which $\tan \theta$ or $\tan \phi$ is positive or negative according as θ or ϕ lies within or without the trapezoid and x is the altitude of the trapezoid. (In the figure both angles lie without the trapezoid and hence both tangents are negative.) The value of x is found by solving this equation. Then $GJ = x \sec \theta$; $DH = x \sec \phi$; and $JH = DG + x(\tan \theta + \tan \phi)$, in which the signs of $\tan \theta$ and $\tan \phi$ are as given above.

In the field the points *H* and *J* are established on the lines *CD* and *AB*. The side *JH* is then measured, and a check is obtained.

16-14 Numerical Problems

1 What is the area of a triangle having sides of length 219.0, 317.2, and 301.6 ft? Of a triangle having two sides of length 1,167.1 and 392.7 ft and an included angle of 39°46′?

2 In the following tabulation are given total latitudes and total departures of a closed traverse. Calculate the area by the coordinate method.

Corner	*A*	*B*	*C*	*D*
Total latitude, ft	+50.5	+203.4	−49.5	−75.0
Total departure, ft	−102.5	0	+100.3	0

3 Given the notes tabulated below, for a closed traverse. Compute the latitudes and departures, and balance the survey by the compass rule. Assume that the coordinates of *C* are 267.3N and 580.8E, and compute the coordinates of all other corners. Calculate the area by the coordinate method.

Course	*Bearing*	*Length, ft*
AB	N48°20′E	529.6
BC	N87°43′E	592.0
CD	S7°59′E	563.6
DE	S82°12′W	753.4
EA	N48°12′W	428.2

4 In the following tabulation are given the latitudes and departures of a balanced closed traverse. Calculate the area (*a*) by the D.M.D. method and (*b*) by the coordinate method, using five-place logarithms.

Course	*Latitude, ft*	*Departure, ft*
AB	S198.7	W213.6
BC	N181.1	W174.4
CD	N334.1	E 89.2
DE	N224.9	E110.7
EA	S541.4	E188.1

5 Find the error of closure of the following traverse, balance the survey by the compass rule, and calculate the area in acres by the D.M.D. method using four-place logarithms:

Course	*Bearing*	*Length, ft*
AB	S45°45′E	294.4
BC	N65°30′E	263.4
CD	N35°15′E	313.6
DE	N64°15′W	392.0
EF	S59°00′W	197.2
FA	S26°15′W	240.0

6 A traverse *ABCD* is established inside a four-sided field, and the corners of the field are located by angular and linear measurements from the traverse stations, all as indicated by the following data:

Course	*Bearing*	*Length, ft*
AB	S89°58′E	296.4
AE	N20°00′W	34.2
BC	S43°20′W	333.9
BF	N35°20′E	16.9
CD	S80°21′W	215.6
CG	S73°00′E	27.6
DA	N27°24′E	314.2
DH	S36°30′W	15.7

Compute the latitudes and departures, and balance the traverse by the compass rule. Compute the coordinates of each transit point and of each property corner, using *D* as an origin of coordinates. Compute the length and bearing of each side of the field *EFGH*, and tabulate results. Calculate the area of the field by the coordinate method.

7 Given the following offsets from traverse line to irregular boundary, measured at points 25 ft apart.

Distance, ft	*Offset, ft*	*Distance, ft*	*Offset, ft*
0	0.0	125	28.2
25	16.6	150	11.9
50	35.1	175	30.7
75	39.3	200	43.4
100	42.0	225	22.5

By the trapezoidal rule (Art. 16-5) calculate the area between traverse line and boundary.

8 Given the data of the preceding problem. Calculate the required area by Simpson's one-third rule. Note that the number of offsets is even.

9 Following are offsets taken at intervals of 50 ft, to the right and to the left of a traverse line:

Offset left, ft	*Distance, ft*	*Offset right, ft*
34.8	0	32.9
44.2	50	26.1
61.5	100	18.6
51.1	150	32.7
31.3	200	49.8
12.7	250	56.9
8.5	300	47.2

By the trapezoidal rule calculate the area between boundaries thus defined.

10 Given the data of the preceding problem. Calculate the required area by Simpson's one-third rule.

11 Following are offsets from a traverse line to an irregular boundary, taken at irregular intervals:

Distance, ft	*Offset, ft*	*Distance, ft*	*Offset, ft*
0	18.5	100	44.1
25	37.7	170	53.9
60	58.2	200	46.0
70	40.5	220	34.2

Calculate the area between traverse line and boundary by means of the rule of Art. 16-8.

16-15 Office Problem

Problem 39. Area of Field Surveyed with Tape

Object. To determine the area of a field surveyed with the tape. The data of field problem 3 (Chap. 3) may be used. For other methods of calculating areas, see the numerical problems of the preceding article.

Procedure. (1) Decide upon a convenient and systematic form of computation for each of the following methods, using four-place logarithms where possible; and transcribe the necessary data from field book to computation book. (2) By the protractor method, plot the boundaries of the field to a scale commensurate with the precision of the field measurements. (3) Determine the area of each part and of the entire field by use of the planimeter (Art. 7-6). (4) Calculate the area of the triangles and the total area in square feet and acres, following each method through before beginning another. Check the results with a slide rule. (5) Make the computations (*a*) by using the two sides and included angle of each triangle, (*b*) by using the three sides of the oblique triangles, and (*c*) by using the measured altitude and base of each triangle. (6) Calculate by the method of offsets the area of any portions of the field having an irregular boundary. (7) Compare the results obtained through the use of the various methods.

17

FIELD ASTRONOMY: LATITUDE AND AZIMUTH

PRINCIPLES

17-1 General. In plane surveying, often it is desired to determine the absolute direction of a line on the surface of the earth, as defined by the azimuth of the line from the true meridian. The azimuth of a line is established by angular observations on some celestial body, most commonly on the sun or on Polaris, the North Star or polestar. For the purpose of computing the azimuth from an astronomical observation, it is necessary that the latitude of the place be known. Ordinarily the latitude may be determined with sufficient precision by scaling from a map, but sometimes it is necessary to determine it by observation.

Geodetic surveying requires instruments and methods of high precision, but the requirements of plane surveying are met if the true azimuth of the survey lines is established with a degree of precision at least equal to that with which the angles between survey lines are measured. For plane surveying of ordinary precision, the use of the engineer's transit and the methods described herein will yield sufficiently precise results. Although these methods will enable the surveyor to determine the true azimuth of a line, for a thorough discussion of the principles involved reference is made to texts on geodesy and engineering astronomy.

The discussions herein are intended to apply in the Northern Hemisphere and for longitudes west of Greenwich.

17-2 Horizon System of Spherical Coordinates. The angular location of a celestial body at a given instant is determined by measuring its vertical angle (referred to the horizon plane) and its horizontal angle (referred to a given line on the ground). In defining the position of celestial bodies it is convenient to imagine their being attached to the inner surface of a hollow *celestial sphere* of infinite radius, of which the earth is the center. The radius of the earth is so small with relation to the distances to the stars that no account is taken of the fact that the observer is not at the center of the earth; but for observations on the sun a correction to the observed vertical angle is necessary, as will be explained later (Art. 17-19).

Figure 17-1 represents a portion of the celestial sphere in which O represents both the earth and the location of the observer, $NES'W$ the observer's horizon, and $S'ZN$ the meridian plane passing through his location. The point Z on the celestial sphere directly above the observer is called the *zenith*. The point S represents a celestial body, and BSZ is part of a great circle, called a *vertical circle,* through the body and the zenith. In this horizon system of spherical coordinates, the angular location of a celestial body is defined by its *azimuth* and *altitude;* thus in Fig. 17-1 the azimuth of S from south is the angle A, and the altitude is the angle h. In astronomical observations, azimuths are customarily reckoned from the true south, except that for circumpolar stars they are often reckoned from north. Also, for circumpolar stars west of north, often the azimuths are reckoned counterclockwise and are considered as negative values.

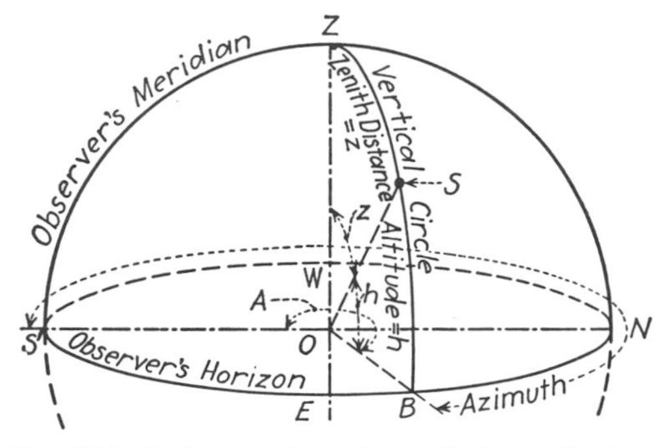

Fig. 17-1. Horizon system of spherical coordinates.

Altitudes are expressed in degrees and cannot exceed 90°; they correspond to the vertical angles of ordinary surveying. If the celestial body is above the horizon plane, the altitude is considered as being positive; if below, negative. The complement of the altitude is called the *zenith distance* or *coaltitude,* shown as the angle z in Fig. 17-1. Zenith distances are always positive.

17-3 Hour-angle Equator System of Spherical Coordinates.

The points where the earth's axis prolonged pierces the celestial sphere are called the *celestial poles,* and the great circle formed by the intersection of the earth's equatorial plane with the celestial sphere is called the *celestial equator*. Figure 17-2 represents a portion of the celestial sphere in which O represents both the earth and the location of the observer, P a celestial pole, and UQQ' the celestial equator.

It is customary to imagine the earth as being fixed and the celestial sphere as revolving about an axis through the poles. Thus to the naked eye the polestar appears to remain stationary, but the sun (and similarly the stars near the equator) appears above the horizon in the general direction of east, follows a curved path (convex southward) across the heavens, and disappears below the horizon in the general direction of west.

Just as the location of any place on the earth is conveniently fixed by its latitude and longitude, so may the location of any celestial body be similarly expressed by spherical coordinates referred to the celestial equator and a great circle normal thereto. (In the work described in this chapter it is necessary only to define the location with respect to the equator.) Comparable with the parallels of latitude of the earth are the *parallels of declination* of the celestial sphere; thus in Fig. 17-2, *R′SR* is the parallel of declination passing through the celestial body *S*. And comparable with the meridians of longitude of the earth are the *hour circles* of the celestial sphere; thus in Fig. 17-2, *PSU* is a portion of the hour circle passing through *S*.

The angular distance of a celestial body above or below the celestial equator, measured in a plane passing through the body and the axis of the celestial poles, is called the *declination*. Thus in Fig 17-2 the declination of *S* is the angle δ. Declinations are expressed in degrees and cannot exceed 90°. If the celestial body is above the equator, its declination is said to be north and is considered as positive; if below, south and negative. The complement of the declination is the *polar distance* or *codeclination,* shown as the angle *p* in Fig. 17-2. Polar distances are always positive.

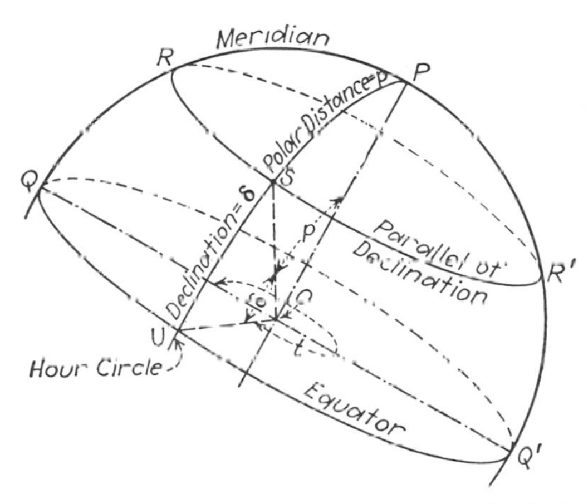

Fig. 17-2. Hour-angle equator system of spherical coordinates.

As the earth actually travels around the sun but not around the stars, the sun appears to move more slowly than do the stars, making in 1 year 365 apparent revolutions (approximately) while the stars make 366 apparent revolutions (approximately). Further, as the axis of rotation of the earth is not normal to the plane of the earth's orbit, the path apparently traced by the sun among the stars on the celestial sphere, called the *ecliptic,* is a continuous curved line; each year the sun crosses the equator northward on March 21, reaches a maximum positive declination (about N23½°) on June 21, crosses the equator southward on September 22, and reaches a maximum negative declination (about S23½°) on December 21. In making computations for latitude and azimuth based upon solar observations, the declination of the sun at the time of observation must be known; values of declination for each day are published in astronomical tables (Art. 17-6).

The *hour angle* of any celestial body is the angular distance measured westward along the equator from a fixed meridian of reference, say the observer's meridian, to the hour circle through the body. Thus in Fig. 17-2, t is the hour angle of S.

When any celestial body, real or imaginary, apparently crosses the upper branch of the meridian, above the axis of the celestial sphere, it is said to be at *upper transit* or *upper culmination;* if below, at lower culmination. Considering any celestial body that crosses the upper branch of the meridian between the pole and the zenith, when the body in its apparent travel is farthest east of the pole, it is said to be at *eastern elongation;* if west, at *western elongation.* In these positions its apparent path is tangent to the vertical circle through the zenith and the body appears to be traveling vertically for some time (Art. 17-27).

17-4 Horizon and Hour-angle Equator Systems Combined. In Fig. 17-3 are shown the horizon and hour-angle equator systems combined, the notation being the same as in Figs. 17-1 and 17-2. The meridians of the two systems coincide. As the latitude of a place on the earth is its angular distance from the equator measured along a meridian of longitude, the latitude of the observer is given by the angle ϕ between the equator and the zenith, measured in the plane of the meridian. The angle between the pole and the zenith is $90° - \phi = c$, which is called the *colatitude* of the place.

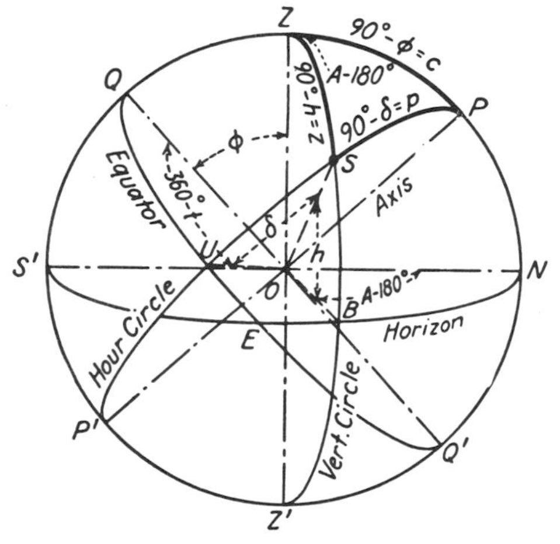

Fig. 17-3. Horizon and hour-angle equator systems combined.

In the horizon system the coordinates of S are A, the azimuth measured from the *south* point of the horizon ($A - 180°$, the azimuth from north, is shown in the figure), and h, the altitude. In the equator system the coordinate corresponding to latitude on the earth is δ, the declination measured above the equator. The colatitude ($90° - \phi = c$), the zenith distance ($90° - h = z$), and the polar distance ($90° - \delta = p$) define a spherical triangle the vertexes of which are the pole P, the zenith Z, and the celestial body S. This triangle is called the *PZS triangle* or the *astronomical triangle*. Most of the problems of field astronomy involve transposing from one system of spherical coordinates to the other and solving the *PZS* triangle for unknown coordinates, having certain coordinates in one or both systems known or observed.

In the figure the celestial body is shown as above the horizon and above the equator. If the body is below the horizon or below the equator, the sides of the *PZS* triangle are defined in a manner similar to that just described, but account is taken of the algebraic sign of the altitude or the declination.

In the figure the celestial body is shown as east of the observer's meridian; the angle Z of the spherical *PZS* triangle is therefore its azimuth from south minus 180° or $Z = A - 180°$. By means of a sketch it can be shown readily that when the body is west of the meridian,

$$Z = 180° - A.$$

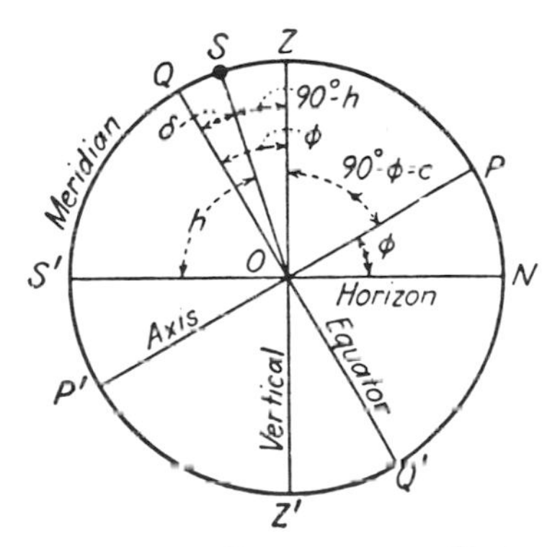

Fig. 17-4. Relation between latitude, altitude, and declination.

In Fig, 17-4, it is seen that the latitude ϕ is equal to the zenith distance ($90° - h$) plus the declination δ (north declinations considered as positive and south declinations as negative), or

$$\phi = (90° - h) + \delta \tag{1}$$

17-5 Solution of the PZS Triangle. In making determinations of azimuth, usually the altitude of the sun (or star) is measured, its declination at the instant of observation is determined from published tables, and the latitude of the place of observation is known or is determined by separate observation. Hence the three sides of the astronomical triangle are known (Fig. 17-5). The determination of azimuth A of the celestial body involves the computation of the angle at Z. By spherical trigonometry it can be shown that

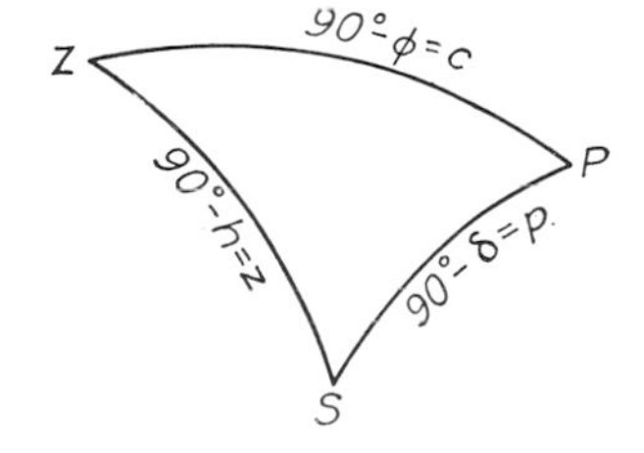

Fig. 17-5. PZS triangle.

$$\cos A = \tan h \tan \phi - \frac{\sin \delta}{\cos h \cos \phi} \tag{2}$$

which is a general expression for determining azimuth when the three sides of the astronomical triangle are known, A being the azimuth measured from *south,* clockwise if the celestial body is leaving the upper branch of the meridian, and counterclockwise if approaching. The azimuth A is less than 90° if the sign of cos A is found to be positive, and greater than 90° if cos A is negative.

17-6 Astronomical Tables Used by the Surveyor. By means of astronomical observations and calculations, the positions of many of the celestial bodies are predicted, and values of their right ascensions and declinations for various dates are available in various publications. The position of a celestial body at any time can be obtained by interpolation.

The publication most widely used by *astronomers* in the United States is "The American Ephemeris and Nautical Almanac" (about 500 pages); herein it is called the "American Ephemeris." It is published one or two years in advance for each year by the Nautical Almanac Office, U.S. Naval Observatory. Formerly its tables were based on "universal" (civil) time (Art. 17-8), but beginning in 1960 the tables are now based on "ephemeris" time; they are not convenient for use in surveying.

Most of the astronomical data used by surveyors are presented in the "Nautical Almanac" (about 300 pages), which is also published annually in advance by the Nautical Almanac Office, U.S. Naval Observatory, primarily for use in navigation.

In condensed form is the "Ephemeris of the Sun, Polaris, and Other Selected Stars" (about 30 pages), herein called the "Ephemeris of the Sun and Polaris." It is published annually in advance by the U.S. Bureau of Land Management. This ephemeris lists for each day of the current year the position of the sun and of Polaris, by means of which the surveyor can compute from his field observations the latitude and longitude of the point of observation, the time of observation, or the azimuth of a reference line. It is in a form useful for land surveying. The major points of difference between the arrangement of these tables and that of the tables published by the Nautical Almanac Office are explained in Arts. 17-12 and 17-14.

All the foregoing publications are sold by the Superintendent of Documents, Government Printing Office, Washington, D.C. 20402.

Useful condensed tables of data regarding the sun and Polaris are furnished to surveyors free of charge by various manufacturers of surveying instruments.

TIME

17-7 Solar Time. The interval of time occupied by one apparent revolution of the sun about the earth is called a *solar day,* the unit with which we are all familiar. The solar day is divided into 24 hr, for surveying purposes reckoned consecutively from 0 to 24. It is considered as beginning at the instant of lower transit of the sun (midnight).

17-8 Civil (Mean Solar) Time. On account of the elliptical shape of the earth's orbit, the apparent angular velocity of the sun that we see is not constant; hence the days as indicated by the apparent travel of the sun about the earth are not of uniform length. To make our solar days of uniform length, astronomers have invented the *mean sun,* a fictitious body which is imagined to move at a uniform rate along the celestial equator, making a complete circuit from west to east in one year. The time interval as measured by one daily revolution of the mean sun is called a *mean solar day,* which is the same as the civil day, beginning at midnight. The *mean solar time* at any place is given by the hour angle of the mean sun plus 12^h. With regard to time, the terms "mean" and "civil" are interchangeable.

Civil time has the same meaning as *mean solar time, mean time,* or *universal time* and, in the form of *standard time* (Art. 17-10), is the time in general use by the public. *Local civil time* is that for the meridian of the observer. Civil time for any other meridian is designated by name, for example, *Greenwich civil time.* Civil time for any meridian can be converted into terms of civil time for any other meridian by computations involving the longitude of the two meridians, as described in the following article.

17-9 Relation between Longitude and Time. As the sun apparently makes a complete revolution (360°) about the earth in one solar day (24 hr), and as the longitudes of the earth range from 0° to 360°, it follows that in 1 hr the sun apparently traverses $360/24 = 15°$ of longitude. This relation is used to determine the difference in time when the difference in longitude between two places is known, or *vice versa.*

Some solar ephemerides are for the meridian of Greenwich, and a problem of frequent occurrence is to find the local time corresponding to a given instant Greenwich time, or *vice versa.* The local time at a given instant of a place west of Greenwich is obtained by subtracting from the Greenwich time the difference in longitude (expressed in hours) between the two places.

example. It is desired to find the local civil time at longitude 122°38′15″W, at the instant of $18^h48^m15^s$ Greenwich civil time.

The difference in longitude, in hours, is equal to the difference in longitude, in degrees, divided by 15.

$$\text{Local civil time} = 18^h48^m15^s - \frac{122°38'15''}{15} = 10^h37^m42^s$$

Sketches are a valuable aid in the solution of problems involving longitude and time, as they enable the surveyor to visualize the relations. A simple "straight-line" type of sketch is shown in Fig. 17-6, for the instant of 9:00 A.M. Pacific standard time. For clarity, values are given only to 01^m; the actual computations of a surveying problem would be more precise.

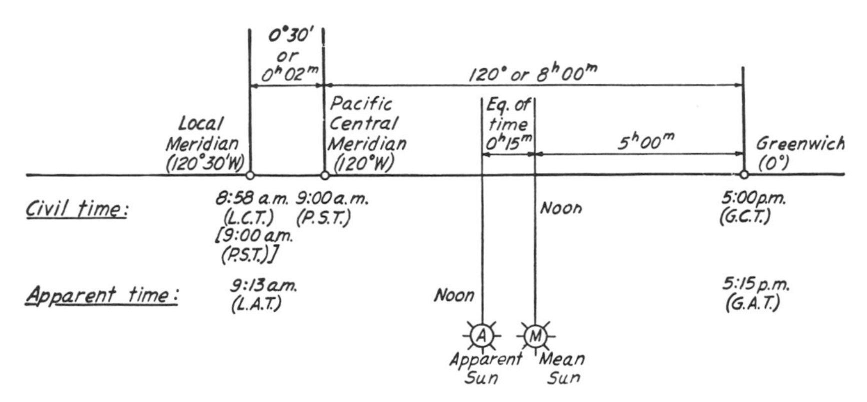

Fig. 17-6. Relations between longitude and time at a given instant.

17-10 Standard Time. In order to eliminate the industrial confusion attendant upon the use of local time by the public, the United States has been divided into belts, each of which occupies a width of approximately 15° or 1^h of longitude. In each belt the watches and clocks that control civil affairs all keep the same time, called *standard time,* which is the local civil time for a meridian near the center of the belt. The time in any belt is a whole number of hours slower than Greenwich civil time, as follows:

Standard time	*Abbreviation*	*Hours slower than Greenwich civil time*	*Central meridian*	*Where used*
Atlantic	A.S.T.	4	60°W	Maritime Provinces of Canada
Eastern	E.S.T.	5	75°W	Maine to central Ohio
Central	C.S.T.	6	90°W	Central Ohio to central Nebraska
Mountain	M.S.T.	7	105°W	Central Nebraska to western Utah
Pacific	P.S.T.	8	120°W	West of Utah to Pacific coast, excluding Alaska

Yukon	Y.S.T.	9	135°W	Eastern Alaska
Alaska	A.S.T.	10	150°W	Central Alaska, Hawaii
Bering Sea	B.S.T.	11	165°W	Western Alaska

The exact boundaries of the time belts are irregular and can be determined only from a map. Standard time, local civil time, and Greenwich civil time can be converted into terms of one another by computations similar to those of the preceding article.

Correct standard time can be obtained either from a clock known to be closely regulated or from radio signals, preferably those broadcast by the U.S. Naval Observatory.

In certain localities, "daylight saving time" is employed during the summer months. Daylight saving time is 1^h faster than standard time.

17-11 Apparent (True Solar) Time. For certain parts of the year, the sun we see, called the *apparent sun* or the *true sun,* travels somewhat faster than the mean sun, and for other parts it travels more slowly than the mean sun. The time interval as measured by one daily revolution of the apparent sun about the earth is called an *apparent solar day,* which begins at midnight. The *apparent solar time* at any place is given by the hour angle of the apparent sun plus 12^h. With regard to time, the terms "true" and "apparent" are interchangeable.

Apparent time has the same meaning as *apparent solar time. Local apparent time* is that for the meridian of the observer. Apparent time for any other meridian is designated by name, for example, *Greenwich apparent time.* Apparent time for any meridian can be converted into terms of apparent time for any other meridian by computations identical with those for civil time, as described in the previous articles; 1^h apparent time corresponds to 1^h or 15° of longitude.

17-12 Equation of Time. When the apparent (true) sun is ahead of the mean sun, apparent time is faster than mean (civil) time; when behind, slower. The difference between apparent time and civil time at any instant is called the *equation of time.* It is used to convert civil time at any instant into apparent time, and *vice versa.*

The maximum value of the equation of time is only about 16^m; hence for work in which its only use is for the determination of change in declination, it is sometimes neglected.

The equation of time may be obtained either from an ephemeris which gives values at given instants of civil time or from one which gives values for apparent time, as follows:

1 In the "Nautical Almanac," the equation of time, to 01^s, is given for each day at 0^h and 12^h Greenwich civil time. The civil time of meridian passage of the apparent (true) sun is also given. If the civil time of meridian passage is before noon, the equation of time should be added to civil time to obtain apparent time; if after noon, subtracted.

2 In the "Ephemeris of the Sun and Polaris" of the U.S. Bureau of Land Management, the equation of time, to 0.01^s, is given for each day at the instant of Greenwich apparent noon. The column headings state directly whether the equation of time is to be added to or subtracted from the apparent time when the civil time is desired.

To find the equation of time at any instant other than that for which a value is tabulated, it is necessary to interpolate, adding to or subtracting from the tabulated value of equation of time the change in the equation of time since the instant to which the tabulated value applies.

example 1. It is desired to determine by use of the "Nautical Almanac" the equation of time at the instant of $3^h30^m45^s$ P.M. Greenwich civil time on December 15, 1964. Greenwich civil time $= 12^h + 3^h30^m45^s = 15.51^h$.

From the "Nautical Almanac" the equation of time at 12^h G.C.T. is 04^m44^s. The change in the equation of time in 12 hr (to 0^h December 16) is

$$04^m29^s - 04^m44^s = -15^s$$

The change in the equation of time up to the given instant is

$$\frac{15.51 - 12}{12} \times (-15) = -4.4^s$$

The equation of time at the given instant is

$$04^m44^s - 4.4^s = 04^m39.6^s$$

The Greenwich civil (mean) time of meridian passage of the apparent (true) sun is given as 11^h55^m; therefore the equation of time is to be added to mean time to obtain apparent time or subtracted from apparent time to obtain mean time.

example 2. It is desired to determine by use of the "Ephemeris of the Sun and Polaris" the Greenwich civil time (G.C.T.) at the instant of $9^h00^m15^s$ Greenwich apparent time (G.A.T.) on October 10, 1964. The time that will elapse before G.A. noon is 3.00^h.

The equation of time at G.A. noon is $13^m00.8^s$, to be subtracted from apparent time. The change in one day is $13^m00.8^s - 12^m44.8^s = 16.0^s$.

The change before G.A. noon is $(3.00/24) \times 16.0 = 2.0^s$
Eq. time for $9^h00^m15^s$ G.A.T. $= 13^m00.8^s - 2.0^s = 12^m58.8^s$
G.A.T. $= 9^h00^m15^s$
Eq. time $= 12^m58.8^s$, to be subtracted from apparent time
G.C.T. $= 8^h47^m16.2^s$

SOLAR OBSERVATIONS

17-13 General. Measurements to the sun cannot be made with as high precision as to a star. However, the sun may be viewed at convenient times and with sufficient precision for most ordinary surveys to determine latitude and azimuth.

17-14 Declination of the Sun. For the determination of latitude or azimuth by solar observations, it is necessary that the declination of the sun at the instant of sighting be known. The declination at a given instant is obtained by interpolating between values given in a solar ephemeris for the current year. Either of the two common types of ephemeris may be used. One type, the "Nautical Almanac," gives the apparent declination for each hour of each day of the year and is especially adapted for use when the standard time or the Greenwich civil time is known. The other type, the "Ephemeris of the Sun and Polaris," gives the apparent declination for each day of the year at the instant of Greenwich apparent noon and is especially adapted for use when the longitude of the place and the local apparent time of the observation are known. It is widely used in land surveying. In the work described in this chapter the computations are based upon civil time; hence in order to use the second type of ephemeris the observed values of civil time must be converted into terms of apparent time.

The following examples illustrate the use of each of these types of ephemeris to determine declination:

example 1. An observation is taken on the sun at 10^h00^m A.M. Eastern standard time, on December 15, 1964. It is desired to determine the declination at the given instant.

The Greenwich civil time at the instant of observation is $10^h00^m + 5^h = 15^h00^m = 15.00^h$. By ephemeris for 0^h Greenwich civil time the declination for 0^h on December 15 is $-23°08.4'$ and the change per hour is $-0.16'$. The change in declination since 0^h G.C.T. is $-0.16 \times 15.00 = -2.4'$. The declination at the instant of observation is $-23°08.4' - 2.4' = -23°10.8'$. The "Nautical Almanac" gives $-23°10.9'$ directly for 15.00^h G.C.T.

example 2. An observation is taken on the sun as it crosses the meridian on November 16, 1964, at a place whose longitude is 87°49′30″ west of Greenwich. It is desired to determine the apparent declination at the given instant.

$$\text{G.A.T.} = \frac{87°49'30''}{15} = 5^h51^m18^s \text{ (after apparent noon)} = 5.86^h$$

From the "Ephemeris of the Sun and Polaris," the declination at Greenwich apparent noon is S18°48′56″. The average difference for $1^h = -37.2''$, the minus sign indicating that south declinations are increasing. The change in declination since Greenwich apparent noon is $37.2'' \times 5.86 = 03'38''$. The declination at local apparent noon at the place $= 18°48'56'' + 03'38'' = \text{S}18°52'34''$.

example 3. It is desired to determine the apparent declination of the sun at the instant of 1^h00^m P.M. Eastern standard time on November 18, 1964, from a solar ephemeris giving values for Greenwich apparent noon.

The difference between Eastern standard time and Greenwich mean time is 5^h; hence the instant of observation is 6^h00^m after Greenwich mean noon. At Greenwich apparent noon the equation of time as given in the ephemeris is $-14^m45.2^s$, to be subtracted from Greenwich apparent time to give Greenwich mean time. It follows that apparent time is faster than mean time, and the Greenwich apparent time is roughly $6^h00^m + 15^m = 6.25^h$. The daily rate of change in the equation of time is given by the difference between the equation of time for November 18 and that for November 19, or $14^m45.2^s - 14^m32.0^s = 13.2^s$. The change in the equation of time since Greenwich apparent noon is $(6.25 \times 13.2)/24 = 3.4^s$. The equation of time is decreasing, and hence the equation of time for the given instant is $14^m45.2^s - 3.4^s = 14^m41.8^s$. The interval since Greenwich apparent noon is $6^h00^m + 14^m41.8^s = 6^h14^m41.8^s = 6.245^h$. At Greenwich apparent noon the apparent declination is S19°18′02″; the average difference for 1^h is 35.5″. The change in apparent declination since Greenwich apparent noon is $35.5'' \times 6.245 = 3'42''$. South declinations are increasing; hence the apparent declination at the given instant is $19°18'02'' + 3'42'' = \text{S}19°21'44''$.

In Example 3 the equation of time has been determined for the given instant. For all practical purposes the equation of time for Greenwich apparent noon might have been employed, since the small error of 3.4^s in time would have no effect upon the computed change in the declination unless declinations were carried out to tenths of seconds.

If the equation of time were neglected entirely, the error introduced in the computed value of the apparent declination would not be of consequence in rough calculations. Similarly, an observation of time for the sole purpose of determining declination need not be exact.

17-15 Measurement of Angles. Whenever observations are made to determine azimuth, a part of the field work consists in measuring the horizontal angle between the celestial body and a reference mark on the earth's surface. As the sights to the celestial body are in general steeply inclined, it is highly important that the horizontal axis be in adjustment with respect to the vertical axis and that the transit be carefully leveled, preferably by means of the telescope level. Also, sights should be taken with the telescope in both the normal and the inverted positions.

Whenever altitudes are observed, the index error of the vertical circle should be determined at the time of the observation, since double-sighting does not eliminate any error due to inclination of the vertical axis (Art. 9-14). The transit should be leveled with great care, preferably by means of the telescope level.

The transit should be supported firmly; if the setup is on soft ground, pegs should be driven to support the tripod legs.

17-16 Observations on the Sun. To observe the sun directly through the telescopic eyepiece may result in serious injury to the eye. A piece of colored or smoked glass may be held between the eye and the eyepiece. Some transits are equipped with a colored *sun glass* that may be attached to the eyepiece.

Good observations can be made by bringing the sun's image to a focus on a white card held several inches in the rear of the telescopic eyepiece. A rough pointing on the sun is made by sighting over the telescope. The eyepiece is then drawn back, and the objective is focused until the sun's image and the cross hairs are clearly seen on the card. If the eyepiece of the telescope is erecting, the image on the card will be inverted; if the eyepiece is inverting, the image will be erect. The cross hairs are visible only on the image of the sun. As only two of the horizontal cross hairs are visible on the image of the sun at the same time, care must be taken not to mistake one of the stadia hairs for the middle horizontal cross hair. This mistake can be avoided by rotating the telescope slightly about the horizontal axis until all three hairs have been seen.

When the vertical angle of sighting is large, it is impossible to look directly through the transit telescope. The *prismatic eyepiece* is a device which, when attached to the telescopic eyepiece, reflects the rays through an angle of 90° with the axis of the telescope. The image appears upside down, but it is not reversed horizontally. The prismatic eyepiece is equipped with a sun glass. The sun may be sighted at high altitudes by means of a white card held in the rear of the eyepiece, as described in the preceding paragraph.

The *solar screen* is a device utilizing the principle of the card. It consists of a piece of ground white glass fixed to a metal arm which is screwed or clamped to the eyepiece end of the telescope. The sun and the cross hairs are brought to a focus on the ground glass, as previously described.

17-17 Semidiameter Correction. As the sun is large (apparent angular diameter about 32′), its center cannot be sighted precisely with the ordinary transit, and it is customary to bring the cross hairs tangent to the sun's image. When the horizontal cross hair is brought tangent to the lower edge of the sun, the sight is said to be taken to the sun's *lower limb,* and this is indicated in the notes by the symbol $\underline{\odot}$. Similarly the symbol $\overline{\odot}$ indicates a sight to the sun's *upper limb,* $\odot|$ a sight with the vertical cross hair to the sun's *right limb,* and $|\odot$ a sight to the sun's *left limb.*

When a vertical angle is measured to the sun's upper limb, it is necessary to subtract the sun's semidiameter in order to obtain the observed altitude of the sun's center; when to the lower limb, add the semidiameter. The "Ephemeris of the Sun and Polaris" and other solar ephemerides give values of the semidiameter of the sun for each day of the year. The semidiameter varies from about 15′46″ in July to about 16′18″ in January; for rough calculations it may be taken as 16′.

When a horizontal angle is measured to the sun's right or left limb, a correction equal to the sun's semidiameter times the secant of the altitude is applied. Thus if the altitude h is 60° and the semidiameter is 16′, the correction to a horizontal angle is $16' \sec h = 32'$. As the sun approaches the zenith, the correction becomes very large (approaching 90°), and readings should not be taken to one limb only.

For most solar observations, an equal number of sights are taken to opposite limbs of the sun; the mean of the horizontal angles and the mean of the vertical angles at the mean of the times of observation are taken, and no corrections for semidiameter are necessary.

17-18 Procedure of Sighting. The process of bringing both cross hairs tangent to the sun is illustrated by Fig. 17-7, which shows the position of the sun in the field of view of an erecting telescope just prior to tangency, in the morning. For the first of a pair of observations, the horizontal cross hair is sighted a short distance above the sun's lower limb as illustrated. As the altitude of the sun is increasing, the horizontal cross hair approaches tangency owing to the sun's apparent movement. At the same time, the vertical cross hair is kept continuously on the sun's western limb by means of the upper-motion tangent screw. At the instant when the vertical and horizontal cross hairs are simultaneously tangent to the sun's disk, the motion of the telescope is stopped, the time is observed, and the horizontal and vertical circles are read. The telescope is then plunged, and the second observation is taken with the sun in the lower right-hand quadrant as shown in Fig. 17-7, as follows: The vertical cross hair is set a short distance to the right of the sun's eastern limb. As the sun is traveling westward, the vertical cross hair approaches tangency owing to the sun's apparent movement. At the same time, the horizontal cross hair is kept continuously on the sun's upper limb by means of the telescope tangent screw. As before, observations are taken for the instant when the horizontal and vertical cross hairs are simultaneously tangent to the sun's disk. The procedure is such that the final setting for either observation requires the manipulation of only one tangent screw and that the cross hair which is approaching tangency is visible on the sun's disk. Also the procedure of double-sighting eliminates certain instrumental errors.

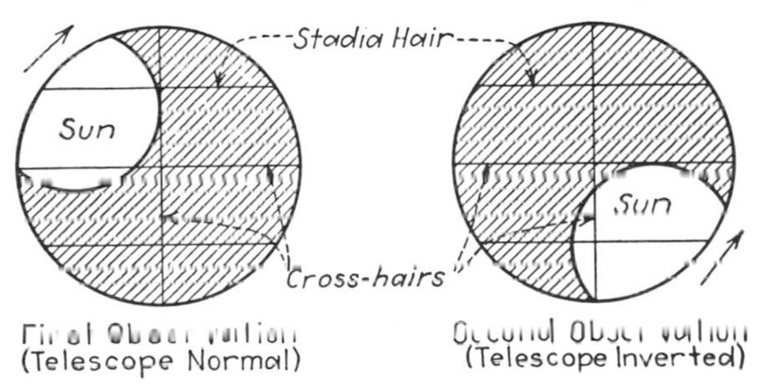

Fig. 17-7. Position of sun in field of view just prior to tangency in morning.

For afternoon observations the procedure is the same as that just described, except that the sun is sighted first in the upper right-hand quadrant and then in the lower left-hand quadrant.

If the transit is not equipped with a full vertical circle, it cannot be plunged between sights, but otherwise the procedure may be as just described, and the means for the two observations taken as the observed angles.

Special devices for observing the sun without correction for semi-diameter are the *Simplex solar shield* and the *solar prism*.

17-19 Parallax Correction. In the previous discussions, it has been assumed that the celestial sphere is of infinite radius and that a vertical angle measured from a station on the surface of the earth is the same as it would be if measured from a station at the center of the earth. For the fixed stars this assumption yields results that are sufficiently accurate for the work described herein; but the distance between the sun and the earth is relatively small, and for solar observations a *parallax correction* is added to the observed altitude to obtain the altitude of the sun from the center of the earth.

In Fig. 17-8*a*, h' is the altitude of the sun above the horizon of an observer at A, and h is the altitude of the sun above the celestial horizon. The parallax correction is equal to the difference between these two angles. As h is always larger than h', the correction must be added to the observed altitude.

The parallax correction C_p for any observed altitude h' is equal to $C_h \cos h'$, in which C_h is the *horizontal parallax,* or correction when the sun is on the observer's horizon. The horizontal parallax is always slightly less than 09″; hence the parallax correction at any altitude cannot exceed 09″. Values of the sun's parallax correction at various altitudes are given in the "Ephemeris of the Sun and Polaris."

Corrections for parallax and refraction are usually made together (see Art. 17-21).

17-20 Refraction Correction. When a ray of light emanating from a celestial body passes through the atmosphere of the earth, the ray is bent downward, as illustrated in Fig. 17-8*b*. Hence the sun and stars appear to be higher above the observer's horizon than they actually are. The angle of deviation of the ray from its direction on entering the earth's atmosphere to its direction at the surface of the earth is called the *refraction* of the ray. A *refraction correction* C_r of amount equal to the refraction is subtracted from the observed altitude to determine the actual altitude h' above the observer's horizon.

The magnitude of the refraction correction depends upon the temperature and barometric pressure of the atmosphere and upon the altitude of the ray, varying as the cotangent of the altitude. Table II herein

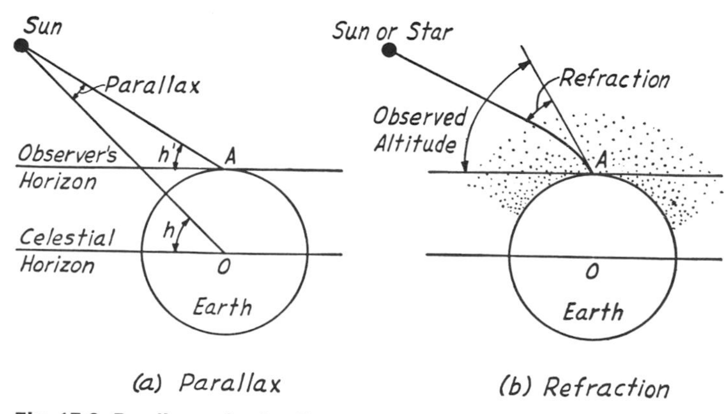

Fig. 17-8. Parallax and refraction.

gives values of refraction corrections for a barometric pressure of 29.5 in. (which may be assumed with sufficient precision for practical purposes), for various temperatures, and for altitudes between 10° and 90°. Owing to the uncertainties of the refraction correction for low altitudes, observations for precise determinations are never taken on a celestial body which is near the horizon.

17-21 Combined Correction. For solar observations, refraction and parallax corrections are usually made together. The refraction correction, which is subtractive, is many times larger than the parallax correction, which is additive; hence the combined correction is of the same sign as the refraction correction. Table I herein gives corrections for the combined effect of refraction and parallax, to be subtracted from observed altitudes of the sun to determine the true altitudes above the celestial horizon.

17-22 Latitude by Observation on Sun at Noon. The latitude of a given station can be determined with a fair degree of precision, ordinarily within 01′, by observing with the engineer's transit the altitude of the sun at local apparent noon, when the sun crosses the meridian. As explained in Art. 17-4, the latitude ϕ is equal to the zenith distance $(90° - h)$ plus the declination δ. The true altitude h of the center of the sun is determined by applying to the observed altitude the appropriate corrections for index error, semidiameter, and parallax and refraction. The sun's declination δ at the instant of sighting is computed from values given in an ephemeris (Art. 17-14), either the longitude or the standard time of the observation being roughly known. It is not necessary that the direction of the meridian be known.

The usual procedure is as follows: The transit is set up and very carefully leveled. The horizontal cross hair is sighted continuously on either the lower limb or the upper limb of the sun until the sun reaches its maximum altitude and begins its apparent descent. If the longitude of the place is unknown, the watch time is observed. The maximum altitude and the watch time are recorded. The telescope may be plunged, and a second sight taken on the opposite limb of the sun; if the time between sights does not exceed 3 or 4 min, the mean altitude may be considered as the altitude at apparent noon. With the telescope still approximately in the plane of the meridian, the index error (Art. 9-14) is determined. The watch is compared with a timepiece keeping correct standard time, and the error is noted. The observed values are corrected, and the latitude is computed as previously described.

If the longitude of the place is known, it is assumed that the instant of observation is local apparent noon. The approximate standard time of local apparent noon is computed in advance.

The field notes and computations are made in a form similar to that shown in Fig. 17-9. For these observations the longitude of the place was unknown, and the standard time was recorded. Only the sun's lower limb was sighted. The index error was determined by double-sighting at a

LATITUDE OF TOWN HALL 14

Circle	Obj.	Time	V. Circle	Index E.	
	(Observation on Sun at Apparent Noon)				
		Field Work			
Circle	Obj.	Time	V. Circle	Index E.	
L	A		+2°58'30"		h_1
R	A		+2°55'30"		h_2
				+0°01'30"	
L	☉	11ʰ34ᵐ01ˢ	47°51'00"		h'
		Computations			
Watch time					11ʰ34ᵐ01ˢ
" slow					29ˢ
G.C.T. (E.S.T.+5ʰ)					16ʰ34ᵐ30ˢ
Eq. of time (from Ephemeris)				+	4ᵐ51ˢ
G.A.T.					16ʰ39ᵐ21ˢ
δ_0 (decl. at G.A. Noon)				+	2°54'48"
$\Delta\delta$ (57.8" x 4.66ʰ)				−	4'29"
δ				+	2°50'19"
Obs. h' on ☉					47°51.0'
Index correction				−	01.5'
Refr. & Parallax (Table I)				−	00.7'
Sun's semidiameter				+	15.9'
h (corrected altitude)					48°04.7'
δ					2°50.3'
$\phi=90°-h+\delta$					44°45.6'

Wm. Bolton, H. L. Brown — Observers
Sept. 15, 1964
Remarks — Fair, Warm, Calm
"A" is mark on barn 400 ft. south
See Note — B. & B. Transit No. 142
By Watch — Waltham Watch
29ˢ slow E.S. Time

16.66ʰ

Note: - Index error found by reversal on point "A"

$$\text{Index E.} = \frac{+(2°58'30")-(2°55'30")}{2}$$

" " = +0°01'30" with circle left

Ephemeris gave values for G. A. Noon

Latitude of Town Hall

Fig. 17-9. Latitude by observation on sun at noon.

mark; the letters "L" and "R" in the column headed "Circle" indicate whether the vertical circle was left or right and therefore whether the telescope was normal or inverted. As the available ephemeris gave values of declination at Greenwich apparent noon, the watch time was converted into Greenwich apparent time. In the line beginning "$\Delta\delta$," the change in declination during the 4.66 hr that elapsed since Greenwich apparent noon was computed by multiplying the elapsed time by the variation per hour (57.8″) taken from the ephemeris.

17-23 Azimuth by Direct Solar Observation.

The azimuth of a line can be determined by a single observation of the sun at any time when it is visible, provided the latitude of the place is known.

At a known instant of time the sun is observed, and the altitude of the sun and the horizontal angle from the sun to a given reference line are measured. The declination of the sun at the given instant is found from a solar ephemeris. The *PZS* triangle can then be solved as described in Art. 17-5, the azimuth of the sun from south being determined by Eq. (2). The azimuth of the line is readily computed from the azimuth of the sun and the observed horizontal angle.

The usual procedure is as follows: The transit is set up and very carefully leveled over one end of the line. The *A* vernier is set at zero on the horizontal circle, and the reading of the *B* vernier is noted. A sight is taken along the given line with the telescope in, say the normal position, and the lower motion is clamped. The upper motion is loosened, and sights are taken on the sun as described in Art. 17-18. The field work is completed by again sighting along the line and reading the horizontal circle, this time with the telescope still inverted. The watch is compared with a timepiece keeping correct standard time, and the error is noted. The mean of each pair of observed angles is taken as the angle to the sun's center at the mean of the observed times. The observed values are corrected, and the azimuth of the sun is computed as previously described, five-place logarithms being used. The azimuth of the line is computed by subtracting algebraically the mean of the horizontal angles from the azimuth of the sun, angles taken in a clockwise direction from line to sun being considered positive. The field notes and computations are made in a form similar to that shown in Fig. 17-10.

AZIMUTH OF LINE Δ46—Δ63 15

(Direct Solar Observation)

Circle	Object	Time	V. Circle	Horiz. Circle Ver. A	Ver. B
R	Δ63			0°00′00″	180°00′00″
R	☉	8h42m48s	34°46′	331°37′40″	151°38′20″
L	☉	8h47m23s	34°55′	152°05′40″	332°05′20″
I	Δ63			179°59′40″	359°59′20″

			Logs	
Watch time		8h45m05.5s		
Watch slow	+	33s	tan h	9.84235
C.S.T.		8h45m38.5s	tan φ	9.99672
G.C.T.		14h45m38.5s		9.83907
			sin δ	8.69932
δ_0	+	3°06.3′	cos h	9.91431
Δδ	−	14.2′	cos φ	9.85112
δ	+	2°52.1′		8.93389
			Numbers	
h′		34°50.5′	tan h tan φ	0.69035
Ref. & Par.	−	01.2′	$-\dfrac{\sin\delta}{\cos h \cos\phi}$	−0.08588
h		34°49.3′	cos A	0.60447
			A	52°49′
φ		44°47′	Z	127°11′
			H	28°08′
				155°19′

Dietzgen Transit — J. C. Mac Dougal
Watch 33s slow — W. W. Wilson
Central Std. Time — Sept. 15, 1964, A. M.
Hot, Windy

Hor. angles on azimuth circle

$$\cos A = \tan h \tan \phi - \frac{\sin \delta}{\cos h \cos \phi}$$

Ephemeris gave values for 0h G.C.T.

N
46
155°19′
28°08′
A = 52°49′
S
63
Sun

Angle of sun from South
Azimuth of sun from North
Azimuth of line from North

Fig. 17-10. Azimuth by observation on sun.

The precision with which azimuths can be determined depends not only upon the precision of field observations and the exactitude of the corrections in altitude but also upon the shape of the astronomical triangle. In order to avoid either large uncertainties in the refraction correction (when the sun is low) or weak *PZS* triangles (when the sun is high), observations are usually taken between the hours of 8 and 10 A.M. or 2 and 4 P.M. For observations taken at these times, roughly an error of 01′ in latitude, declination, or altitude produces an error of 01′ to 02′ in the azimuth. Under ordinary conditions the azimuth of a line may be determined within about 01′.

POLARIS OBSERVATIONS

17-24 Observations on Stars. The methods of determining latitude and azimuth by direct solar observations are, with slight modifications, applicable to observations on the stars. If a high degree of precision is not required, the same procedure may be followed for stellar observations as for solar observations. Usually, however, it is expected that a higher degree of precision will be obtained; consequently a corresponding degree of refinement is necessary, and special care is taken to eliminate systematic errors. Usually observations to determine latitude and azimuth are made on Polaris, the North Star; and the methods described hereinafter apply to Polaris observations.

The measurement of horizontal and vertical angles to celestial bodies is discussed in Art. 17-15. The refraction correction is described in Art. 17-20. For observations on stars, no correction is required for parallax or semidiameter.

In sighting on a star, the objective should be focused until the star appears as a fine, brilliant point of light. The approximate position of the objective slide may be determined by focusing on a distant object in the landscape. When the atmosphere is clear, observations may be taken on stars during daylight hours near the hours of darkness, even when the stars are invisible to the naked eye.

During the hours of darkness, artificial illumination is required to make visible the cross hairs of the transit. The cross hairs can be illuminated by holding a flashlight several inches in front of the objective and a little to one side of the telescope barrel, thus causing the rays to enter the telescope diagonally. The location of any terrestrial mark is indicated by a light; usually the mark is a slit in the side of a box in which there is a lamp.

17-25 Polaris. Polaris (α Ursa Minor) is a second-magnitude star the position of which is readily identified by the neighboring constellations of Ursa Major and Cassiopeia (Fig. 17-11). Its polar distance is approximately 1°, its annual change in polar distance (or in declination) is less than 01′, and its maximum daily change in polar distance is less than ½″. Values giving the position of Polaris during each year are published in various ephemerides (Art. 17-6).

Observations to determine latitude are usually made when Polaris is at upper or lower *culmination,* when the star appears to be moving almost horizontally for some time; essentially the method is the same as that for solar observations. Observations to determine azimuth are usually made when Polaris is at eastern or western *elongation,* when the star is farthest east or farthest west of the pole, and when it appears to be traveling vertically for some time. Although it is not necessary that the time of culmination or elongation be known, this information is of convenience in scheduling the observations.

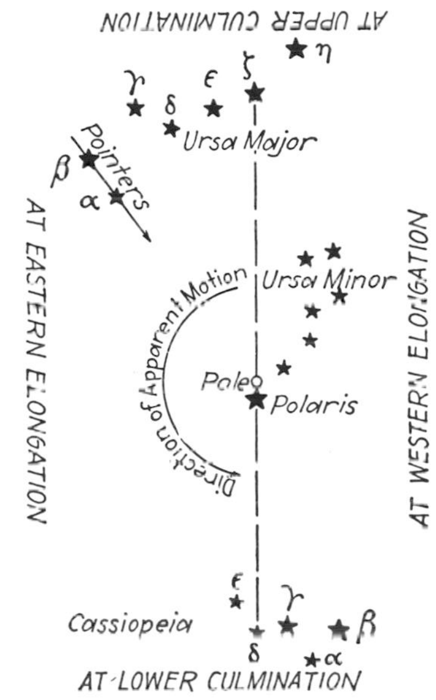

Fig. 17-11. Positions of constellations near the North Pole when Polaris is at culmination and elongation.

Data concerning Polaris can be found from a variety of sources. The "Ephemeris of the Sun and Polaris" gives the declination of Polaris and the Greenwich mean time of its culmination and elongation at the meridian of Greenwich for each day of the current year. The data in the "Nautical Almanac" are not in convenient form for use by the surveyor. Convenient tables are to be found in the ephemerides published by the manufacturers of surveying instruments.

Since Polaris, in common with other fixed stars, travels at an angular rate more rapid than that of the sun, it follows that at a given meridian it arrives at culmination a little earlier by mean solar (civil) time each day than it did the day before, the amount earlier being approximately equal to the gain of sidereal time on mean solar time for a 24^h interval, or approximately 3^m56^s per day or 9.86^s per hour. Furthermore, the time of culmination depends upon the longitude of the place. To determine the local mean time of upper culmination of Polaris at any given meridian on any given date, the value for the meridian of Greenwich is taken from the table and is reduced to the longitude of the place by means of the variation per hour.

example. It is desired to find from the "Ephemeris of the Sun and Polaris" the Eastern standard time of upper culmination of Polaris on December 9, 1964, at a place the longitude of which is $5^h15^m45^s$ west of Greenwich.

On December 9, 1964, U.C. at Greenwich occurs at	$20^h45.5^m$
Change in time for $\Delta\lambda$ is $-51.26 \times 9.86^s = -52^s =$	-0.9^m
Local civil time of U.C. at place	$20^h44.6^m$
$\Delta\lambda = 5^h15^m45^s - 5^h = +15^m45^s$	$= +15.8^m$
E.S.T. of upper culmination at place	$= 21^h00.4^m$
	$= 9^h00^m24^s$ P.M.

17-26 Latitude by Observation on Polaris at Culmination. In Fig. 17-4 it is seen that the latitude of the observer is equal to the altitude of the pole, as the two angles ϕ have mutually perpendicular sides. By a similar sketch it can be shown readily that if h is the true altitude of Polaris as it crosses the meridian and p its polar distance, then the latitude is

$$\phi = h \pm p \tag{3}$$

in which the sign preceding p is positive or negative according as the star is at lower or upper culmination.

The usual procedure of determining latitude by observation on Polaris at culmination is as follows: The standard time of culmination at the given station is determined within perhaps 5 min by use of an ephemeris, as illustrated in the preceding article. A few minutes before the estimated time of culmination the transit is set up and is leveled very carefully. The telescope is focused for a star, and the estimated latitude of the place is laid off on the vertical circle to facilitate finding Polaris. When Polaris is sighted, the cross hairs are illuminated if necessary, and the star is continuously bisected with the horizontal cross hair. When during a period of 3 or 4 min Polaris no longer appears to move away from the hair but moves horizontally along it, the star is practically at culmination. The vertical angle is read with dispatch, the transit is carefully releveled, the telescope is plunged, and a second observation on the star is taken with the telescope inverted. Usually the instrument is releveled and a second pair of observations is made. The mean of the altitudes, corrected for refraction (Table II) and index error, is taken as the true altitude of the star. The polar distance can be found from any ephemeris giving either declinations or polar distances of Polaris for the days of the current year. Finally the latitude is computed by applying to the true altitude the polar distance with proper sign. Under ordinary conditions the latitude can be determined within about 01′, or less if the mean of several observations is taken.

It is not essential that the altitude be observed at the exact instant

Polaris crosses the meridian. Within 12^{m} before to 12^{m} after culmination the maximum change in altitude is only 0.1′.

17-27 Azimuth by Observation on Polaris at Elongation.[1]

At the instant of elongation, as a circumpolar star appears to be traveling vertically, its apparent path in the celestial sphere is tangent to the vertical circle through the observer's zenith, as illustrated by Fig. 17-12, and the angle at S is a right angle. In the PZS triangle, angle $S = 90°$, side $PS = p = 90° - \delta$ is given in published tables of polar distance p or declination δ, and the side $ZP = 90° - \phi$ is known, provided the latitude ϕ of the observer is known. By spherical trigonometry it can be shown that

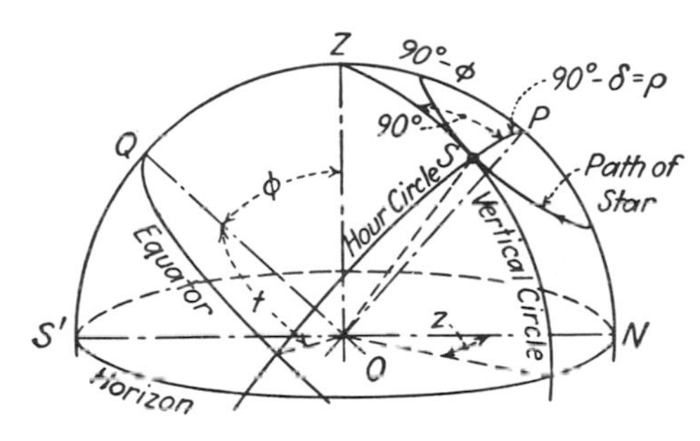

Fig. 17-12. Star at elongation.

$$\sin Z = \frac{\sin p}{\cos \phi} \tag{4}$$

which is the general expression employed for determining the azimuth of a circumpolar star when at elongation, Z being the azimuth reckoned east or west of north according as the star is at eastern or western elongation.

The azimuth of a line can be determined conveniently by an observation on Polaris at eastern or western elongation, provided the latitude of the place is known.

The direction of Polaris from the observer's station at the time of elongation is established by projecting a vertical plane from the star to the earth. The terrestrial line thus established has the same azimuth as the star at elongation. The angle between the line thus established and an established reference line is measured, and the azimuth of the reference line is computed.

The usual procedure is as follows: The standard time of elongation at the given station is determined by use of an ephemeris. A few minutes before the estimated time of elongation, the transit is set up over a given station and is leveled very carefully. The telescope is focused for a star, and the latitude of the place is laid off on the vertical circle to facilitate finding Polaris. When Polaris is sighted, the horizontal and vertical motions are clamped, the cross hairs are illuminated if necessary, and the

[1] For rough determination of meridian by ranging plumb lines on Polaris, see Art. 8-5.

star is continuously bisected with the vertical cross hair. When during a period of 2 or 3 min Polaris no longer appears to move away from the hair but moves vertically along it, the star is practically at elongation. The telescope is depressed, and a point on the line of sight is marked on a stake or other reference monument 300 ft or more away. The telescope is then plunged, and another sight is taken on Polaris. The line of sight is again depressed, and a second point is set on the stake beside the first. Usually the transit is releveled and a second pair of observations is made.

Later the mean of the points is found and marked on the stake, thus defining the direction from the observer's station to Polaris at elongation. The azimuth of Polaris at elongation either is computed by Eq. (4) or is found directly from an ephemeris. It is given to seconds in the "Ephemeris of the Sun and Polaris." The azimuth of any other line through the station can be determined by measuring the horizontal angle between the two lines, by the method of repetition (Art. 9-10) if necessary to secure the required precision. A true meridian can be established by a perpendicular offset from the established point on the stake.

example. In taking an observation on Polaris at western elongation, the reference point marking the azimuth of the star is 400 ft from the transit. The azimuth from north of the star at elongation is $-1°40'45''$. A point of the true meridian through the transit station is to be established by a perpendicular offset from the reference mark, using the relation that the perpendicular offset distance is equal to the distance from the instrument to the mark multiplied by the tangent of the offset angle.

$$\begin{aligned} \log 400 &= 2.60206 \\ \log \tan 1°40'45'' &= 8.46710 \\ \hline \log \text{offset} &= 1.06916 \\ \text{Offset} &= 11.725 \text{ ft} \end{aligned}$$

By the method just described, the azimuth of any line can be determined within perhaps $0.5'$, or less if the mean of several observations is taken. If an error of $01'$ is allowable, points need not be set beneath the star, and the angle may be measured directly.

It is not essential that the direction to Polaris be observed at the exact instant of elongation. Within 10^m before to 10^m after elongation, the maximum change in azimuth is only $0.1'$.

17-28 Numerical Problems

1 What is the Greenwich apparent time at a place the longitude of which is $96°15'10''$W, when the local apparent time is $8^h17^m12^s$?

2 When it is $15^h31^m12^s$ Greenwich civil time, it is $10^h16^m37^s$ local civil time at a given place. What is the longitude of the place?

3 What is the Greenwich civil time when it is 3^h15^m P.M. Central standard time?

4 If the local civil time at a place is $16^h23^m22^s$ and the longitude of the place is 78°36′20″W, what is the Eastern standard time?

5 From an ephemeris find the equation of time for the instant of $4^h15^m00^s$ P.M. Pacific standard time on April 21 of the current year. If the longitude of the place is $7^h46^m03^s$W, compute the local civil and local apparent times.

6 Find the apparent declination of the sun for the instant of 9^h00^m A.M. Central standard time on February 15 of the current year, using an ephemeris giving values of 0^h Greenwich civil time.

7 Find the apparent declination of the sun for the instant of local apparent noon at a place the longitude of which is $5^h52^m54^s$W for the date of July 21 of the current year, using an ephemeris giving values for Greenwich apparent noon.

8 In connection with a solar observation, sights to determine index error are taken on a mark in the general direction of the sun. The vertical angles to the mark are −3°17′00″ with telescope normal and −3°18′30″ with telescope inverted. The sun is sighted with telescope inverted, and the vertical angle is observed to be 36°02′30″. Correct the observed angle for index error of the vertical circle.

9 The observed altitude of a star is 23°15′20″. The temperature is 90°F. By Table II find the refraction correction, and compute the true altitude.

10 The observed altitude of the sun's center is 15°07′30″. The temperature is 15°F. By Table I find the correction for parallax and refraction, and compute the true altitude of the sun.

11 At a place the longitude of which is 113°W, the observed altitude of the upper limb of the sun as it crosses the meridian on March 4 of the current year is 52°13′. The temperature is 47°F, and the index error of the transit is +1′30″. What is the latitude of the place?

12 At a place the latitude of which is 41°58′30″N, the observed altitude of the sun's center (taken as the mean obtained by double-sighting) at 3^h12^m P.M. Central standard time on October 21 of the current year is 20°04′30″. The horizontal angle measured clockwise from a reference line to the sun is 81°32′20″. The temperature is 40°F. What is the azimuth (measured from south) of the sun at the given instant, and what is the azimuth of the reference line?

13 On August 1 of the current year the observed altitude of the sun at a given place is 30°51′45″ at $7^h42^m20^s$ A.M. local apparent time. The latitude of the place is 37°18′20″N, and the longitude is 102°17′30″W. The temperature is 75°F. The horizontal angle (measured clockwise) from reference line to sun is 89°39′15″. What is the azimuth of the sun measured from north, and what is the azimuth of the reference line?

14 At a given place on January 12 of the current year the observed altitude of Polaris at upper culmination is 44°36′25″. The temperature is 15°F. What is the latitude of the place?

15 Find the Central standard time of upper culmination of Polaris on December 7 of the current year at Des Moines, Iowa (longitude $6^h14^m30.6^sW$).

17-29 Field Problems

Problem 40. Latitude by Observation on Sun at Noon

Object. To determine the latitude of the place by an observation on the sun at local apparent noon, using the engineer's transit.

Procedure. (1) Follow the procedure outlined in Art. 17-22, assuming that the longitude of the place is unknown. If the transit has a full vertical circle, use the method of double-sighting to determine the mean vertical angle to the sun's upper and lower limbs. If the transit is not equipped with a full vertical circle, the altitude correction for semidiameter (which may be taken as 16′) must be applied.

Hints and Precautions. (1) See Art. 17-16. (2) Pay particular attention to the algebraic sign of each quantity and of each correction. (3) If the longitude of the place is known approximately, the approximate standard time of upper transit of the sun may be calculated in advance as a guide in observing.

Problem 41. Azimuth by Direct Solar Observation

Object. To determine the true azimuth of a line by an observation on the sun with the engineer's transit.

Procedure. (1) Follow the procedure outlined in Art. 17-23. If the transit is not equipped with a full vertical circle, the correction for semidiameter (taken as 16′) must be applied to the altitude; further, either sights must be taken to both right and left limbs of the sun, or the correction for semidiameter (taken at 16′ sec h) must be applied to the observed horizontal angle. (2) As a check, observe the magnetic bearing of the line.

Hints and Precautions. (1) See Art. 17-16. (2) Pay particular attention to algebraic signs.

Problem 42. Latitude by Observation on Polaris at Culmination

Object. To determine the latitude of the place by observing Polaris at upper or lower culmination.

Procedure. Follow the procedure outlined in Art. 17-26. If the transit is not equipped with a full vertical circle, make two observations with the telescope normal, releveling the instrument between observations.

Hints and Precautions. (1) See Art. 17-24. (2) Pay particular attention to algebraic signs. (3) As a check, the mean of the times of observation should agree (within a few minutes) with the computed time of culmination.

Problem 43. Azimuth by Observation on Polaris at Elongation

Object. To determine the azimuth of a line by observation on Polaris at eastern or western elongation.

Procedure. (1) Follow the procedure outlined in Art. 17-27. If the transit is not equipped with a full vertical circle, make two observations with the telescope normal, releveling the instrument between observations. (2) As a check, observe the magnetic bearing of the established line.

Hints and Precautions. (1) See Art. 17-24. (2) Pay particular attention to algebraic signs. (3) As a check, the mean of the times of observation should agree (within a few minutes) with the computed time of elongation.

18

LAND SURVEYING

18-1 General. Land surveying deals with the laying off or the measurement of the lengths and directions of lines forming the boundaries of real or landed property. Land surveys are made for one or more of the following purposes:

1 To secure the necessary data for writing the legal description and for finding the area of a designated tract of land, the boundaries of the property being defined by visible objects.
2 To reestablish the boundaries of a tract for which a survey has previously been made and for which the description as defined by the previous survey is known.
3 To subdivide a tract into two or more smaller units in accordance with a definite plan which predetermines the size, shape, and location of the units.

Whenever real estate is conveyed from one owner to another, it is important to know and state the location of the boundaries. Land surveys are of the following types:

1 *Original surveys,* to measure the unknown lengths and directions of boundaries already established by agreement, to find the area of the tract, and to write a legal description.
2 *Resurveys,* to reestablish the boundaries of a tract for which the description as defined by a previous survey is known.
3 *Subdivision surveys,* to subdivide land into smaller units in accordance with a definite plan. Examples are the United States public-land surveys and the subdivision of a city addition into blocks and lots.

The land surveyor makes the field surveys, calculates dimensions and areas, prepares maps, and writes legal descriptions for deeds. He must be familiar not only with technical procedures but also with the legal aspects of real property and boundaries. Usually he is required to be licensed by the state, either directly or as a civil engineer.

Methods of calculating and subdividing areas are described in Chap. 16.

18-2 Instruments and Methods. Nearly all land surveys are run with the transit and tape, as described in Chap. 10. Usually the directions of lines are expressed as true bearings. Ordinarily distances are measured to feet and decimals, but on United States public-land surveys the unit of length prescribed by law is the Gunter's (66-ft) chain.

Formerly the surveyor's compass and 66-ft link chain were used extensively, and the descriptions in many old deeds are given in terms of magnetic bearings and Gunter's chains. In retracing old surveys of this character, allowance must be made for change in magnetic bearing since the time of the original survey. Also, it must be kept in mind that the compass and link chain used on old surveys were relatively inaccurate instruments and that great precision was not regarded as necessary, since usually the land values were low. Further, for many years the United States public lands were surveyed under contract, at a low price. Many of the lines and corners established by old surveys are not where they theoretically should be; nevertheless these boundaries legally remain fixed as they were originally established.

Wherever possible, the lengths of boundary lines and the angles between boundaries are determined by direct measurement. Therefore the land survey is in general a traverse, the transit stations being at corners of the property and the traverse lines coinciding with property lines. Where obstacles render direct measurement of boundaries impossible, a traverse is run as near the property lines as practicable, and measurements are made from the traverse to property corners; the lengths and directions of the property lines are then calculated. In general, the required precision of land surveys depends upon the value of the land, being higher in urban than in rural areas.

18-3 Corners, Monuments, and Reference Marks. It is customary to mark the corners of landed property by visible monuments. The term *corner* is applied to a point established by a survey or by an agreement; the term *monument* is applied to an object placed to mark the corner point upon the surface of the earth. For early original surveys, many of the corners were marked by natural objects such as trees and large stones. In general, however, the corner monuments are established by the surveyor either to mark the intersections of boundaries already in existence or to define new boundaries. Unfortunately, many monuments (such as wooden stakes) are temporary, and many resurveys are made necessary by the obliteration of such temporary markers (Arts. 18-12 and 18-30).

Examples of markers of a more permanent character are an iron pipe or bar driven in the ground; a concrete or stone monument with drill hole, cross, or metal plug marking the exact corner; a stone with identifying mark, placed below the ground surface; charcoal placed below the surface; a mound of stones; a mound of earth above a buried stone; and a metal marker set in concrete below the surface, reached through a covered shaft. On many old governmental surveys, through wooded country where stones were not available, corners were established by building up a mound of earth over a quart of charcoal or a charred stake, or by building a mound around a tree at which the corner fell. The U.S. Bureau of Land Management has adopted as the standard for the monumenting of the public-land surveys a post made of iron pipe filled with concrete, the lower end of the pipe being split and spread to form a base, and the upper end being fitted with a brass cap with identifying marks (Art. 18-29). Metal caps for use as property markers are commercially available.

Damage to public and private survey monuments, or interference with the proper use of such monuments, is usually prohibited by law.

If there is a possibility that a corner monument will become displaced, the corner should be *referenced* (Art. 10-12), or connected, to nearby objects of more or less permanent character. Usually the recorded measurement is called a *connection,* and the object is called a *reference mark* or a *corner accessory*. Examples of corner accessories are trees, large stones, and buildings. On public-land surveys it is specified that every corner be referenced by bearing and distance to one or more objects, preferably to a tree, in this case called a *bearing tree*.

If the location of a corner within reasonable accuracy can be determined beyond reasonable doubt, the corner is said to *exist;* otherwise it is said to be *lost*. If the monument marking an existing corner cannot be found, the corner is said to be *obliterated,* but it is not necessarily lost.

Where a corner falls in such location as to make it impossible or impracticable to establish a monument in its true location, it is customary to establish a *witness corner* on one or more of the boundary lines leading to the corner, as near to the true corner as practicable. Witness corners are necessary where the true corner falls in a road, body of water, building, or precipitous slope.

The field notes should give detailed information concerning the character, size, and location of all monuments and reference marks. So far as possible, such objects should be clearly marked in the field.

18-4 Meander Lines. In surveying along the shore of a lake or the bank of a stream a traverse called a *meander line* is run, roughly following the shore or bank line. The process of establishing such a line is called *meandering.* Meander lines are for surveying and mapping purposes only and are not property lines except in the rare case where they are specifically stated as property lines in a deed.

At each intersection of a major boundary line with the shore line of a lake or with the mean high-water mark of a stream that marks a boundary, a corner called a *meander corner* is established. Usually the mean high-water mark is taken as the line along which vegetation ceases or changes. The surveyor begins at a meander corner and traverses by a succession of straight lines to the next meander corner, proceeding as close to the shore or bank as convenient. The meander line permits the traverse of the tract to be closed; also usually it defines the sinuosities of the shore or high-water line with sufficient accuracy. If more accurate location of the shore line is desired, offsets are taken from the meander line.

18-5 Boundary Records. Descriptions of the boundaries of real property may be found from deeds, official plats or maps, or notes of original surveys. Typical descriptions are given in Arts. 18-10, 18-14, and 18-28. Unfortunately, many descriptions are inadequate or incorrect.

Records of the transfer of land from one owner to another are usually kept in the county registry of deeds, exact copies of all deeds of transfer being filed in deed books. These files are open to the public and are a frequent source of information for the land surveyor. An alphabetic index is kept, usually by years, giving in one part the names of *grantors,* or persons selling property, and in the other part the names of *grantees,* or persons buying property. Usually the preceding transfer of the same property is noted on the margin of the deed.

Many states have special "land courts" where title to land can be confirmed by simple procedure and at nominal cost. Land courts, together with the state systems of plane coordinates (which in some states may be used as the legal basis for description of land), are gradually simplifying and rendering more certain the registration and transfer of land titles.

In most cases the deeds of transfer of city lots give only the lot or block number and the name of the addition or the subdivision. The official plat or map showing the dimensions of all lots and the character and location of permanent monuments is on file either in the office of the city clerk or in the county registry of deeds; copies are also on file in the offices of city and county assessors.

Some organizations, usually called *title companies,* for a fee will search the records for boundary descriptions and will guarantee the title against possible defects in description, legal transfer, and certain types of claims such as those for right of way. Title insurance does not necessarily mean that the property corners are correctly marked on the ground; and if assurance is desired, a survey should be made.

As the United States public lands are subdivided, official plats are prepared showing the dimensions of subdivisions and the character of monuments marking the corners. When the surveys within a state have been completed, the records are given to the state. An exception is Oklahoma, for which the United States survey records are filed with the Director of the Bureau of Land Management at Washington, D.C. States in possession of records have them on file at the state capitol. Usually information concerning these records can be obtained from the state secretary of state. Photographic copies of the official plats are obtainable at nominal cost.

18-6 Legal Interpretation of Deed Description. The descriptions of the boundaries of a tract include the objects that fix the corners, the lengths and directions of lines between the corners, and the area of the tract. A deed description may contain errors or mistakes of measurement or mistakes of calculation or record, thus introducing inconsistencies which cannot be reconciled completely when retracement becomes necessary. In such cases, where uncertainty has arisen as to the location of property lines, it is a universal principle of law that the endeavor is to make the deed effectual rather than void and to execute the intentions of the contracting parties. The following general rules are pursuant to this principle.

1 Monuments. It is presumed that the visible objects which marked the corners when a conveyance of ownership was made indicated best the intentions of the parties concerned; hence it is agreed that a corner is established by an existing material object or by conclusive evidence as to the previous location of the object. A corner thus established will prevail against all other conflicting evidence if there is reason to believe that it has not been disturbed.

2 Distance and Direction vs. Area. In the case of discord between the described courses and the calculated area of a tract, the deed-description requirements, or "calls," for distances or directions of courses will prevail against the call for area, again on the assumption that the boundary lines are more visible and actual evidence of the intentions of the parties than is the calculated area of the tract.

3 Mistakes. It is a well-established principle that a deed description which taken as a whole plainly indicates the intentions of the parties concerned will not be invalidated by evident mistakes or omissions. For example, such obvious mistakes as the omission of a full tape length in a dimension or the transposition of the words "northeast" for "northwest" will have no effect on the validity of a description, provided it is otherwise complete and consistent or provided its intention is manifest.

4 *Purchaser Favored.* In the case of a description that is capable of two or more interpretations, that one will prevail which favors the purchaser.

5 *Ownership of Highways.* Land described as being bounded by a highway or street conveys ownership to the center of the highway or street. Any variation from this interpretation must be explicitly stated in the description.

6 *Original Government Surveys Presumed Correct.* Errors found in original government surveys do not affect the location of the boundaries established under those surveys, and the boundaries remain fixed as originally established.

18-7 Riparian Rights. An owner of property that borders on a body of water is a riparian proprietor and has riparian rights (pertaining to the use of the shore or of the water) which may be valuable. Because of the difficulties arising from the irregularity of such boundaries, it is important that the land surveyor be familiar with the general principles relating to riparian rights and with the statutes and precedents established in his particular state. For a comprehensive treatment of the subject the reader is referred to detailed texts on land surveying.

18-8 Adverse Possession. Under conditions fixed by statute in the various states, property lines may be fixed by continued possession and use of the land (usually for 20 years) as against original survey boundaries. *Adverse possession* is the enjoyment of land under such circumstances as indicate that such enjoyment has been commenced and continued under an assertion of right on the part of the possessor. To become effective, adverse possession must be plainly evident to the owner, without his permission, to his exclusion, and hostile to his interests. Such possession may be evidenced by fencing, cultivation, erection of buildings, etc. The application of the principle of adverse possession is entirely a matter of intention and belief.

Right to title by adverse possession may be acquired by individuals, corporations, and even by the state. But the statute does not run *against* the state; that is, property in a street or highway cannot be acquired by adverse possession.

For a detailed treatment of the subject, the reader is referred to texts on land surveying.

18-9 Legal Authority and Liability of the Surveyor. A resurvey may be run to settle a controversy between owners of adjoining property. The surveyor should understand that, although he may act as an arbiter in such cases, it is not within his power legally to fix boundaries without the mutual consent and authority of all interested parties. A competent surveyor by wise counsel can usually prevent litigation; but if he cannot bring his clients to an agreement, the boundaries in dispute become valid and defined only by a decision of the court. In boundary disputes the surveyor is an expert witness, not a judge.

The right to enter upon property for the purpose of making public surveys is usually provided by law, but there is no similar provision regarding private surveys. The surveyor (or his employer, whether public or private) is liable for damage caused by cutting trees, destroying crops or fences, etc.

Permission should be obtained before entering private property in order to avoid a possible charge of trespassing. If permission is not granted, it may be necessary to traverse around the property or to resort to aerial photographs. In certain cases, as for example, where a boundary is in doubt or dispute, a court order may be obtained.

It has been held in court decisions that county surveyors and surveyors in private practice are members of a learned profession and may be held liable for incompetent services rendered. If the surveyor knows the purpose for which the survey is made, he is liable for damages resulting from incompetent work. The general principle is that the surveyor is bound to exhibit that degree of prudence, judgment, and skill which may reasonably be expected of a member of his profession. Ruling Case Law says, ". . . a person undertaking to make a survey does not insure the correctness of his work, nor is absolute correctness the test of the amount of skill the law requires. Reasonable care, honesty, and a reasonable amount of skill are all he is bound to bring to the discharge of his duties."

RURAL–LAND SURVEYS

18-10 Description of Rural Land. In the older portions of the United States nearly all of the original land grants were of irregular shape, many of the boundaries following stream and ridge lines. Also, in the process of subdivision the units were taken without much regard for regularity, and it was thought sufficient if lands were specified by natural or artificial features of the terrain and if the names of adjacent property owners were given.

1 By Metes and Bounds. As boundaries were obliterated and as land became more valuable, land litigations became numerous. It then became the general practice to determine the lengths and directions of the boundaries of land. In describing a tract not within the system of United States public-land surveys, the lengths and bearings of the several courses are given in order, and the objects marking the corners are described; if any boundary follows some prominent feature of the terrain, the fact is stated; and the calculated area of the tract is given. When the bearings and lengths of the sides are thus given, the tract is said to be described by *metes and bounds*. Later in this article is an illustrative description by metes and bounds.

2 *By Subdivisions of Public Land.* The type of description employed for lands which have been divided in accordance with the rectangular system of the Bureau of Land Management is described in detail in Arts. 18-21, 18-22, and 18-28.

3 *By Coordinates.* In some of the states the locations of land corners are legally described by their coordinates with respect to the state-wide plane-coordinate system. The following description by the Tennessee Valley Authority illustrates the description of land by metes and bounds and by coordinates, with further reference to corners and lines of the United States public-land survey. The public-land survey is referred to the Huntsville principal meridian. A map of the tract is shown in Fig. 18-1.

"A tract of land lying in Jackson County, State of Alabama, on the left side of the Tennessee River, in the South Half (S ½) of the Northwest Quarter (NW ¼) of section Three (3), Township Six (6) South, Range Five (5) East, and more particularly described as follows:

"Beginning at a fence corner at the southwest corner of the Northwest Quarter (NW ¼) of Section Three (3) (coordinates N 1,470,588; E 416,239), said corner being North six degrees twenty-four minutes West (N6°24′W) twenty-six hundred (2600) feet from the southwest corner of Section Three (3) (N 1,468,004; E 416,529), and a corner to the land of T. E. Morgan; thence with Morgan's line, the west line of Section Three (3), and a fence line, North five degrees thirty-three minutes West (N5°33′W) thirteen hundred four (1304) feet to a fence corner (N, 1,471,886; E 416,113), a corner of the lands of T. E. Morgan, and the G. T. Cabiness Estate . . . thence with Weeks's line, the south line of the Northwest Quarter (NW ¼) of Section Three (3), and a fence line, North eighty-nine degrees eleven minutes West (N89°11′W) two thousand five hundred fifty (2550) feet to a point on the ground shown by S. L. Cobler, a corner of the lands of H. O. Weeks and T. E. Morgan; thence with Morgan's line, the south line of the Northwest Quarter (NW ¼) of Section Three (3), and the fence line North eighty-nine degrees eleven minutes West (N89°11′W), one hundred twenty-five (125) feet to the point of beginning.

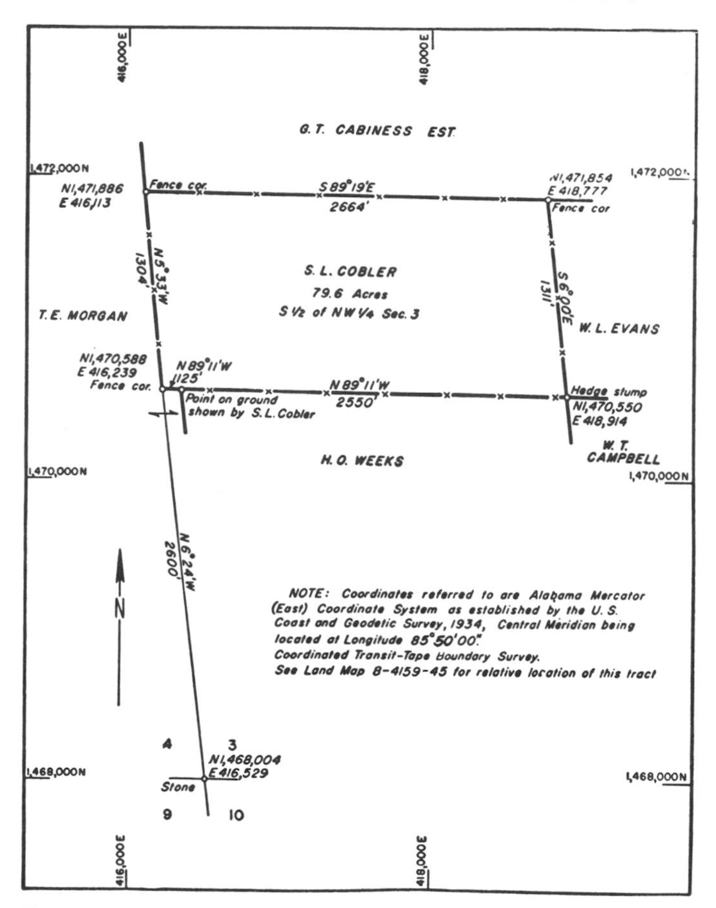

Fig. 18-1. Land map.

"The above described land contains seventy-nine and six-tenths (79.6) acres more or less, subject to the rights of a county road which affects approximately five-tenths (0.5) acre, and is known as Tract No. GR 275 as shown in Map No. 8-4159-45, prepared by the Engineers of the Tennessee Valley Authority.

"The coordinates referred to in the above description are for the Alabama Mercator (East) Coordinate System as established by the U.S. Coast and Geodetic Survey, 1934. The Central Meridian for this coordinate system is Longitude eighty-five degrees (85°) fifty minutes (50′) no seconds (00″)."

18-11 Original Survey. With the desired boundaries of the land given, the surveyor establishes and references monuments at the corners and runs a closed transit traverse around the property, measuring the lengths of lines and the angles between intersecting lines. Curved boundaries are located by offsets from traverse lines; and obstructed boundary lines are located by linear and angular measurements from nearby traverse lines. The direction of the true meridian is determined, usually by a solar observation (Art. 17-23). From the field data the true bearings of the boundaries and the area of the tract are calculated. A description of the tract, usually by metes and bounds, is prepared, and the area is calculated. Usually a plat is drawn. A copy of the description and a tracing and/or prints of the plat are submitted to the person for whom the survey is made.

18-12 Resurvey. The proper relocation of old lines calls for greater ingenuity and broader experience on the part of the surveyor than does any other kind of surveying.

The purpose of the resurvey is to reestablish boundaries in their original locations. To guide him the surveyor has available the description contained in the deed or obtained from old records, and descriptions of adjoining property.

As a first step the surveyor critically examines the descriptions for gross errors; he then computes the latitudes and departures of the several courses as given in the description, determines the error of closure, and plots the boundaries of the tract to scale.

If original bearings are magnetic, the magnetic declination at the time of the original survey is found, and true bearings are computed. If true bearings cannot be found in this manner (as when the date of the original survey is unknown) and if one or more boundaries can be positively identified, observations are made to determine the true bearings of these known lines; from these the true bearings of the other lines are computed.

1 One or More Boundaries Evident. If one or more boundary lines can be identified from the monuments or from reliable reference marks, a comparison is obtained between the length of the chain or tape used on the original survey and that to be used on the resurvey; the proportionate lengths of the other sides of the tract are then computed.

With the computed directions and proportionate lengths, the surveyor starts from a known corner and reruns the courses; at each estimated location of a corner he seeks physical evidence (such as monuments, the remains of monuments, or reference marks) of the location of the original corner. If such evidence is found and if the old monument is not in good condition, he sets a new monument. If the location of the corner does not agree with that in the description, he makes new measurements to the established monument.

At any point where physical evidence as to the original location of a corner is entirely lacking, the corner is located temporarily from the description. The survey is then continued until either positive evidence of the location of a succeeding corner is found or the traverse is brought to a closure at the initial point. If a succeeding corner is found, the traverse between the previously located lines and that corner is adjusted to meet the known conditions.

If no physical evidence except one boundary is found, the survey is run to the point of beginning, and the linear error of closure is measured. The survey is then balanced as described in Arts. 15-10 to 15-12, and the computed corrections are applied by moving the preceding temporary monuments and establishing them as permanent. Finally, the lengths and bearings of the adjusted courses are measured in the field.

2 One Corner Evident. Where only a single corner can be found, the true bearings are determined either directly from the description or by computation from magnetic bearings given in the description. If the date of the survey from which the description by magnetic bearings is derived is unknown, it is estimated as closely as possible, and the corresponding magnetic declination is found for computation. By use of the true bearings and the lengths of lines given in the description, the latitudes and departures of the boundaries are computed, and the linear error of closure of the original survey is determined. If this error is reasonably small (say not greater than $1/300$ if the old survey was run with a compass), it is indicated that there are no mistakes in the lengths and bearings given in the description, although this check does not detect systematic errors (which may be large) in chainage. About the only course open to the surveyor is to establish the true meridian and to rerun the survey in accordance with the old description, distributing the error of closure proportionately among the several courses.

3 No Corner Evident. If a description is available but all evidence of the location of original corners is lost, the surveyor will find it expedient to search the records for descriptions of adjoining property and by means of these descriptions to reestablish by measurement as many corners of the tract in question as seems feasible. It is possible that these locations may be considerably in error. A corner may be reestablished by measurements from several sources, each resulting in a different location; in such cases the surveyor is called upon to exercise his judgment as to the most probable location.

Sometimes it is possible to determine the location of an obliterated corner through evidences of previously existing lines such as fences and roads. Thus, if the surveyor has reason to believe that a fence once stood on the line, he may be able to find evidences of rotted posts in the ground.

Differences in the ground surface, or even differences in vegetation along a definite line, are valuable clues. Occasionally the surveyor may find it desirable to consult old settlers who were familiar with the original boundaries; but although such persons are usually very positive in their opinions, the information is seldom of much value and is frequently misleading.

Having thus tentatively fixed the locations of one or more corners, the surveyor attempts to reconcile these locations with the description of the given track. Readjustments are made to conform to the judgment of the surveyor in light of the information that he obtains as the survey progresses.

Report. When a resurvey has been completed, it is the duty of the surveyor to render a report to his client stating exactly what he found and what course of procedure he employed in attempting to reestablish missing corners. The report should be accompanied by a plat similar to that of an original survey (Art. 18-11). In addition, it should indicate which are original monuments and which are monuments established at the time of the resurvey. Mistakes in the original description should be pointed out, but the surveyor should clearly understand that it is his function to reestablish boundaries of a given tract in as nearly as possible their original location.

18-13 Subdivision Survey of Rural Land. The subdivision of the United States public lands is described in Arts. 18-27 and 18-28, and the subdivision of urban lands into lots in Art. 18-15. Surveys of irregular subdivisions for rural lands may be made for a variety of purposes, such as the establishing on the ground of a right of way for a highway or railroad or the division of a tract of land into parts stipulated by a will. In such cases a resurvey of the tract is run, new monuments are established on the new boundary lines, and a new plat and description are prepared as in the case of an original survey.

URBAN–LAND SURVEYS

18-14 Description of Urban Land. The manner of legally describing the boundaries of a tract of land within the corporate limits of a city depends upon conditions attached to the survey by which the boundaries of the tract were first established, as indicated by the following classifications:

1 By Lot and Block. If the boundaries of the tract coincide exactly with a lot which is part of a subdivision or addition for which there is recorded an official map, the tract may be legally described by a statement giving the lot and block numbers and the name and date of filing of the official map, as illustrated by the following:

"Lot 15 in Block 5 as said lots and blocks are delineated and so designated upon that certain map entitled *Map of Thousand Oaks, Alameda County, California,* filed August 23, 1909, in Liber 25 of Maps, page 2, in the office of the County Recorder of the said County of Alameda."

2 By Metes and Bounds from Points of Known Location. If the boundaries of a given tract within a subdivision for which there is a recorded map do not conform exactly to boundaries shown on the official map, the tract is described by metes and bounds (Art. 18-10) with the point of beginning referred to a corner shown on the official map. Also, the numbers of the lots of which the tract is composed are given.

If a system of reference monuments has been established in the city, the point of beginning of a boundary description may be fixed by stating its direction and distance from an official reference monument and by describing the monument that marks the corner. The boundaries of the tract may then be described by metes and bounds. The location of corners may also be defined by rectangular coordinates referred to the origin or initial point of the city system and/or the state system (Arts. 18-16 and 12-17).

If the tract is within an area not so monumented, the point of beginning of the boundary description may be referred by direction and distance to the intersection of the center lines of streets.

18-15 Subdivision Survey of Urban Land. As a city or town develops, unimproved lands are subdivided into lots which are placed on sale as residential or business property. In most instances such extensions are the result of the activities of real-estate operators who develop a plan of subdivision which is approved by the authorities of the municipality to which the tract is to be attached and cause surveys to be made to establish the boundaries of individual lots. A tract thus divided according to an acceptable plan is known as an *addition* or *subdivision.*

For large and important developments the work of originating the general plan is often carried out by persons specializing in city planning and landscape architecture, under whose direction the surveyor works. Such developments require a great deal of skill, and usually extensive surveys (particularly in hilly sections) are carried out before the actual plan of subdivision can be decided upon. Problems of this character can be adequately discussed only in treatises on city planning, to which the reader is referred, but it is appropriate to state here that the preliminary studies should consider the probable future character of the district; the probable location of business sections; the probable magnitude, direction, and character of future traffic; the topography of the land; the location, width, grade, and character of paving of streets; the size and shape of lots and blocks; the location and size of storm and sanitary sewers; and the disposition of electric and telephone wires and cables.

For the ordinary real-estate development the owner usually calls for the services of an engineer or surveyor who has had experience in such work. The surveyor confers with the owner, and they discuss a general plan. The surveyor makes a resurvey of the entire property; and if the character of the topography is irregular, he usually makes certain preliminary surveys for the purpose of finding the location and elevation of the governing features of the terrain. In some cases a complete topographic survey may be made. With the general plan fixed, the surveyor works out a detailed plan on paper, showing on the drawing the names of all streets and the numbers of all blocks and lots, the dimensions of all lots, the width of streets, the length and bearing of all street tangents, and the radius and length of all street curves. He also prepares a report which, in addition to a discussion of the plan of subdivision, may consider the cost of subdividing, including not only the establishing of boundaries but also the work of grading, paving, constructing sewers, and landscaping.

This detailed plan, when approved by the owner, is submitted to the governing body in the municipality. If it meets with the requirements of this body, it is approved.

Upon the authority of the owner, the surveyor then proceeds to execute the necessary subdivision surveys, including the laying out of roads, walks, blocks, and lots. Often the lot and block corners are marked with permanent monuments; but in many cases, contrary to what may be considered good practice, the lot corners are marked by wooden stakes. When the surveys are completed, the map of the subdivision is revised to show minor changes made during the survey, together with the location and character of permanent monuments. A tracing is submitted to the municipality, and this, when duly signed by those in authority, becomes the official map of the subdivision. It then becomes a part of the public records and is usually filed in the registry of deeds of the county in which the municipality lies. Upon this approval, if the subdivision is outside the corporate limits of the municipality, they are extended to include it.

18-16 City Surveying. The term *city surveying* is frequently applied to the surveying operations within a municipality with regard to mapping its area, laying out new streets and lots, and constructing streets, sewers and other public utilities, and buildings. Although the principles of city surveying are not different from those of ordinary surveying, there are some differences in the details of the methods employed. Some features pertinent to city surveying are as follows:

1 Measurements are made with a greater degree of refinement than for land of less value.

2 Some cities maintain a standard of length with which tapes may be compared.

3 Usually the horizontal control of the survey for the map of a city is by triangulation rather than by traversing, which would be employed for an equal area outside the city.

4 A system of reference points and bench marks is established, usually by traversing, at points a few block apart, usually at street intersections. Preferably this system is tied in with the United States precise surveys and with the state plane-coordinate system. Points are located in the street, at the curb, or on the sidewalk, one such point being sufficient for each chosen intersection. On subsequent surveys, it is good practice to tie in to more than one of these established points, as monuments may have been moved. (For a description of monuments and reference marks, see Art. 18-3.)

5 The established points are well referenced (see Art. 10-12) to more or less permanent objects such as building corners, curb or walk lines, centers of street intersections, and manhole covers. In undeveloped districts, these points are referenced to stakes.

6 Maps showing the location of proposed sewers, street extensions, and other improvements usually show, to scale and in figures, the exact dimensions of adjacent lots and of all other lots that will be benefited by, or assessed for, the proposed improvement.

7 Sometimes separate maps are made of surface and underground utilities such as car lines, sewers, water lines, gas lines, electric power and telephone lines and conduits, tunnels, etc., both for convenient reference and in order to avoid interference in the location of new projects.

The subdivision of urban lands is discussed in Art. 18-15, and typical descriptions of urban lands are given in Art. 18-14. The usual methods of keeping records are described in Art. 18-5.

The operations of surveying in connection with the construction of buildings, bridges, sewers, pipelines, pavements, and railroads are described in Chap. 22. Details regarding the width of streets, size of blocks and lots, location of utilities, etc., are to be found in texts and manuals on city planning, highway engineering, and sanitary engineering.

18-16a City Survey. The term *city survey* has come to mean an extensive coordinated survey of the area in and near a city for the purposes of fixing reference monuments, locating property lines and improvements, and determining the configuration and physical features of the land. Such a survey is of value for a wide variety of purposes, particularly for planning city improvements. The technical procedure for a city survey of this type is described in detail in Ref. 5. Briefly, the work consists in:

1 Establishing *horizontal and vertical control,* as described for topographic surveying. The primary horizontal control is usually by triangulation, supplemented as desired by precise traversing. Secondary horizontal control is by traversing of appropriate precision. Primary vertical control is by precise leveling.

2 Making a *topographic survey* and *topographic map.* Usually the scale of the topographic map is 1 in. = 200 ft. The map is divided into sheets which cover usually 5,000 ft of longitude and 4,000 ft of latitude. Points are plotted by rectangular plane coordinates.

3 *Monumenting* a system of selected points at suitable locations such as street corners, for reference in subsequent surveys. These monuments are referred to the plane-coordinate system and to the city datum.

4 Making a *property map.* The survey for the map consists in (*a*) collecting recorded information regarding property; (*b*) determining the location on the ground of street intersections, angle points, and curve points; (*c*) monumenting the points so located; and (*d*) traversing to determine the coordinates of the monuments. Usually the scale of the property map is 1 in. = 50 ft. The map is divided into sheets which cover usually 1,250 ft of longitude and 1,000 ft of latitude, thus bearing a convenient relation to the sheets of the topographic map. The property map shows the length and bearing of all street lines and boundaries of public property, coordinates of governing points, control, monuments, important structures, natural features of the terrain, etc., all with appropriate legends and notes.

5 Making a *wall map* which shows essentially the same information as the topographic map but which is drawn to a smaller scale; the scale should be not less than 1 in. = 2,000 ft. The wall map is reproduced in the usual colors—culture in black, drainage in blue, wooded areas in green, and contours in brown.

6 Making an *underground map.* Usually the scale and the size of the map sheets are the same as those for the property map. The underground map shows street and easement lines, monuments, surface structures and natural features affecting underground construction, and underground structures and utilities (with dimensions), all with appropriate legends and notes.

18-17 Cadastral Surveying. Cadastral surveying is a general term referring to extensive surveys relating to land boundaries and subdivisions, whether they are city surveys as described in the preceding article or surveys of rural land. The term is applied to the United States public-land surveys by the U.S. Bureau of Land Management. A cadastral map shows individual tracts of land with corners, length and bearing of boundaries, acreage, ownership, and sometimes the cultural and drainage features. The surveying methods are the same as those described for topographic surveying for maps of intermediate and large scale (Chap. 20).

In Ref. 5, *Manual* 34, of the American Society of Civil Engineers, a cadastral survey is defined as follows:

"A survey relating to land boundaries and subdivisions, made to create units suitable for transfer or to define the limitations of title. Derived from 'cadastre' meaning register of the real property of a political subdivision with details of area, ownership, and value. The term cadastral survey is now (1954) used to designate the surveys of the public lands of the United States, including retracement surveys for the identification and resurveys for the restoration of property lines; the term can also be applied properly to corresponding surveys outside the public lands, although such surveys are usually termed *land surveys* through preference."

U.S. PUBLIC–LAND SURVEYS

18-18 General. This section deals with the methods of subdividing the public lands of the United States in accordance with regulations imposed by law. The public lands are subdivided into townships, sections, and quarter sections—in early years by private surveyors under contract, later by the Field Surveying Service of the General Land Office, and currently by civil-service employees of the Bureau of Land Management which succeeded the General Land Office in 1946. Further subdivision of such lands is made after the lands have passed into the hands of private owners, the work being carried out by surveyors in private practice. Under certain conditions, resurveys of public land which has passed into private ownership may be made by the government.

The methods described herein are those now in force, but with minor differences they have been followed in principle since 1785, when the rectangular system of subdivision was inaugurated. Under this system, surveys of the public lands either have been completed or are in progress in the following states: Alabama, Alaska, Arizona, Arkansas, California, Colorado, Florida, Idaho, Illinois, Indiana, Iowa, Kansas, Louisiana, Michigan, Minnesota, Mississippi, Missouri, Montana, Nebraska, Nevada, New Mexico, North Dakota, Ohio, Oklahoma, Oregon, South Dakota, Utah, Washington, Wisconsin, and Wyoming. In general, these methods of subdividing public land do not apply in the thirteen original states and in Hawaii, Kentucky, Tennessee, Texas, and West Virginia. As the progress of the public-land surveys has been from east to west, the details in states east of the Mississippi River differ somewhat from those of present practice.

The laws regulating the subdivision of public lands and the surveying methods employed are fully described in the "Manual of Instructions for the Survey of the Public Lands of the United States," published by the Bureau of Land Management (Ref. 41). From the manual is drawn much of the material for the succeeding articles of this chapter.

Field notes and plats of the public-land surveys may be examined in the regional offices of the Bureau, and copies may be procured for a nominal fee.

Prior to 1910, public-land surveys were made by contract. The field work is now performed by a permanent corps of engineers under civil-service regulations. It must be kept in mind that the early surveys made under contract were made with relatively crude instruments and often under unfavorable field conditions; some were incompletely or even fraudulently executed. Hence, often the lines and corners will be found in other than their theoretical locations. However, the original corners as established legally stand as the true corners, and the surveyor must be guided by them in making resurveys or subdivisions, regardless of irregularities in the original survey. The primary purpose of the public-land surveys is to *mark the boundaries on the ground;* the field notes and plats are subordinate.

The unit of length in public-land surveys is the Gunter's (66-ft) chain divided into 100 links each 7.92 in. long.

18-19 Scheme of Subdivision. The primary unit of subdivision is the *township,* bounded by meridional and latitudinal lines and as nearly as may be 6 miles square. The township is divided into 36 secondary units called *sections,* as nearly as may be 1 mile square. Because the meridians converge (Art. 18-24), it is impossible to lay out a square township by such lines; and because the township is not square, not all of the 36 sections can be square.

18-20 Standard Lines. Since the time of the earliest surveys, the townships and sections have been located with respect to principal axes passing through an origin called an *initial point;* the north-south axis is a true meridian called the *principal meridian,* and the east-west axis is a true parallel of latitude called the *base line.*

The principal meridian is given a name to which all subdivisions are referred. Thus the principal meridian which governs the rectangular surveys (wholly or in part) of the states of Ohio and Indiana is called the First Principal Meridian; its longitude is 84°48′50″W, and the latitude of the base line is 41°00′00″N. The extent of the surveys which are referred to a given initial point may be found by consulting a map, published by the Bureau of Land Management, entitled "United States, Showing Principal Meridians, Base Lines, and Areas Governed Thereby," or from Ref. 41.

Secondary axes are established at intervals of 24 miles east or west of the principal meridian and at intervals of 24 miles north or south of the base line, thus dividing the tract being surveyed into quadrangles bounded by true meridians 24 miles long and by true parallels, the south boundary of each quadrangle being 24 miles long and the north boundary being 24 miles long less the convergency of the meridians in that distance. (In some early surveys, these distances were 30 or 36 miles.) The secondary parallels are called *standard parallels* or *correction lines,* and each is continuous throughout its length. The secondary meridians are called *guide meridians,* and each is broken at the base line and at each standard parallel.

The principal meridian, base line, standard parallels, and guide meridians are called *standard lines.*

A typical system of principal and secondary axes is shown in Fig. 18-2. The base line and standard parallels, being everywhere perpendicular to the direction of the meridian, are laid out on the ground as curved lines, the rate of curvature depending upon the latitude. The

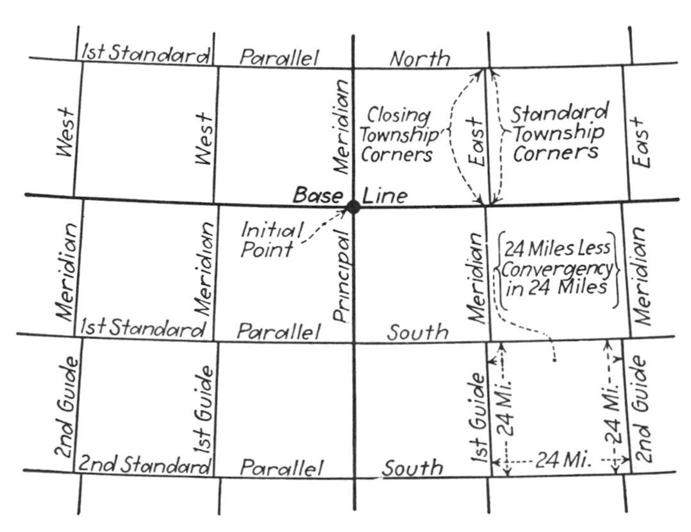

Fig. 18-2. Standard lines.

principal meridian and guide meridians, being true north-and-south lines, are laid out as straight lines but converge toward the north, the rate of convergency depending upon the latitude.

Standard parallels are counted north or south of the base line; thus the *second standard parallel south* indicates a parallel 48 miles south of the base line. Guide meridians are counted east or west of the principal meridian; thus the *third guide meridian west* is 72 miles west of the principal meridian.

18-21 Townships. The division of the 24-mile quadrangles into townships is accomplished by laying off true meridional lines called *range lines* at intervals of 6 miles along each standard parallel, the range line extending north 24 miles to the next standard parallel; and by joining the township corners established at intervals of 6 miles on the range lines, guide meridians, and principal meridian with latitudinal lines called *township lines*.

The plan of subdivision is illustrated by Fig. 18-3. A row of townships extending north and south is called a *range;* and a row extending

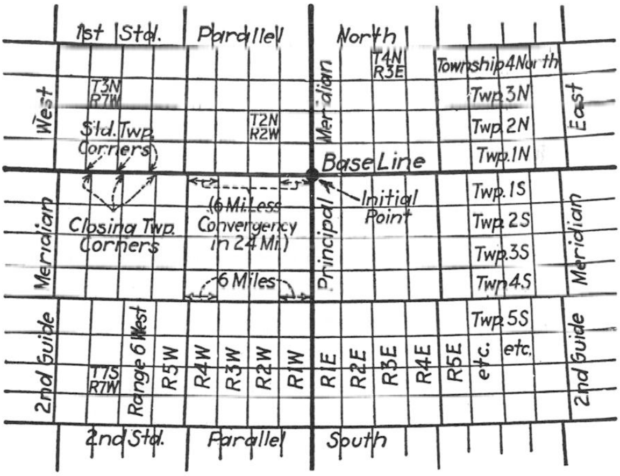

Fig. 18-3. Township and range lines.

east and west is called a *tier*. Ranges are counted east or west of the principal meridian, and tiers are counted north or south of the base line. Usually for purposes of description the word "township" is substituted for "tier." A township is designated by the number of its tier and range and the name of the principal meridian. For example, the location of a township is defined by the description T7S, R7W (read *township seven south, range seven west*), of the Third Principal Meridian.

18-22 Sections. The division of townships into sections is performed by establishing, at intervals of 1 mile, "meridional" lines parallel to the east boundary of the township and by joining the section corners at intervals of 1 mile with straight latitudinal lines. (The north-south lines are not exactly meridional but are parallel to the east boundary of the township, which is a meridional line.) These lines, called *section lines,* divide each township into 36 sections, as shown in Fig. 18-4. The sections are numbered consecutively from east to west and from west to east, beginning with No. 1 in the northeast corner of the township and ending with No. 36 in the southeast corner.

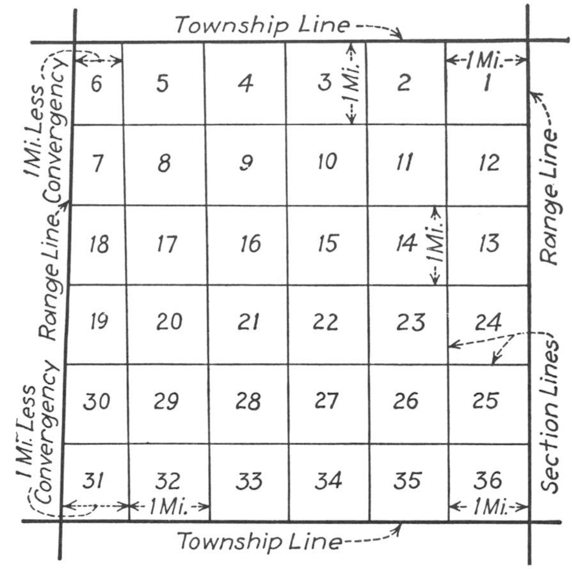

Fig. 18-4. Numbering of sections.

A section is legally described by giving its number, the tier and range of the township, and the name of the principal meridian; for example, Section 16, T7S, R7W, of the Third Principal Meridian.

On account of the convergency of the range lines (true meridians) forming the east and west boundaries of townships, the latitudinal lines forming the north and south boundaries of townships are less than 6 miles in length, except for the south boundary of townships that lie just north of a standard parallel. As the north-south section lines are run parallel to the *east* boundary of the township, it follows that all sections except those adjacent to the west boundary will be 1 mile square but those adjacent to the west boundary will have a latitudinal dimension less than 1 mile by an amount equal to the convergency of the range lines.

The subdivision of sections is described in Art. 18-28.

18-23 Establishing the Standard Lines. The principal meridian is established as a true meridian through the initial point, either north or south, or in both directions, as conditions require. Permanent quarter-section and section corners, called *standard corners,* are established alternately at intervals of 40 chains (½ mile), and regular township corners are placed at intervals of 480 chains (6 miles).

From the initial point the base line is extended east and west on a true parallel of latitude, standard quarter-section and section corners being established alternately at intervals of 40 chains (½ mile) and standard township corners being placed at intervals of 480 chains (6 miles). At intervals of 24 miles north and south of the base line, true parallels of latitude called *standard parallels* or *correction lines* are run east and west from the principal meridian.

The guide meridians are extended north from the base line and standard parallels at intervals of 24 miles east and west of the principal meridian. Each guide meridian is established as a true meridian. The guide meridians terminate at the points of their intersection with the standard parallels, and hence are broken lines, each segment being theoretically 24 miles long (Fig. 18-3). Errors of measurement are placed in the most northerly half mile of each 24-mile segment. At the point of intersection of the guide meridian and standard parallel a township corner, called a *correction corner* or *closing corner* (Fig. 18-3), is established by retracing the standard parallel between the first standard corners to the east and to the west of the point for the closing corner; and the distance from the closing corner to the nearest standard corner on the standard parallel is measured.

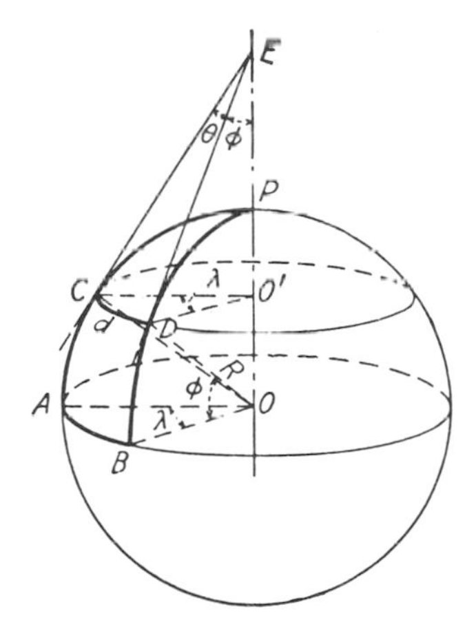

Fig. 18-5. Convergency of meridians.

18-24 Convergency of Meridians. In Fig. 18-5, let ACP and BDP represent two meridians, P being the north pole of the earth, O the center of the earth, and AB an arc of the equator intercepted by the two meridians; and let CD be the arc of a parallel of latitude at a latitude $\phi = COA = DOB$ at which it is desired to determine the angular and linear convergency of the meridians. Consider the earth as a perfect sphere.

The difference in longitude between the two meridians is

$$\lambda = \frac{CD}{CO'} \qquad \text{or} \qquad CD = CO' \cdot \lambda$$

The latitude of the arc CD is

$$\phi = DOB = DEO'$$

Then

$$\sin \phi = \frac{DO'}{DE} \qquad \text{or} \qquad DE = \frac{DO'}{\sin \phi}$$

With negligible error the angle of convergency is

$$\theta = \frac{CD}{DE}$$

By substitution of the values for CD and DE obtained above

$$\theta = \lambda \sin \phi \tag{1}$$

Let the distance between two meridians measured along a parallel be $d = CD$, and let the radius of the earth at the parallel be R. Then, from the figure,

$$\lambda = \frac{CD}{CO'} = \frac{d}{R \cos \phi}$$

By substitution of this value of λ in Eq. (1), there results

$$\theta = \frac{d \sin \phi}{R \cos \phi} = \frac{d \tan \phi}{R} \tag{2}$$

in which θ is in radians.

If d is in miles and $R = 20{,}890{,}000$ ft, the approximate mean radius of the earth, then θ in seconds is, from Eq. (2),

$$\theta'' = 52.13d \tan \phi \tag{3}$$

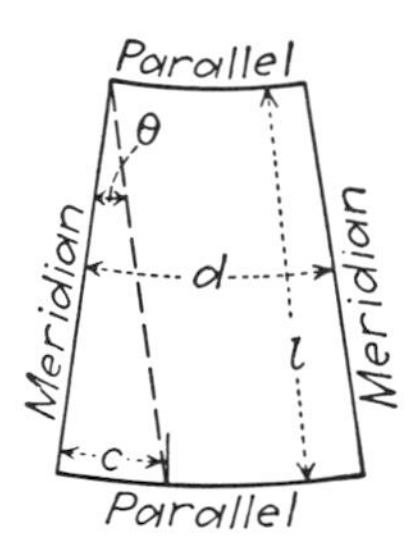

Fig. 18-6. Linear convergency of meridians.

In Fig. 18-6 let l be the length of the meridian between two parallels, and let θ be the mean angle of convergency of two meridians whose mean latitude is ϕ and whose mean distance apart measured on a parallel is d. Also let the linear convergency of the two meridians, measured along a parallel, be c. Then, with small approximation, $\theta = c/l$.

By substitution of this value in Eq. (2), there results

$$c = \frac{dl \tan \phi}{R} \tag{4}$$

which gives with sufficient precision for land surveying the linear convergency between two meridians. If d and l are in miles and R is the mean radius of the earth, then c in feet is given approximately by the expression

$$c_f = \tfrac{4}{3} dl \tan \phi \text{ (approx.)} \tag{5}$$

example. Find the convergency in feet of two range lines 6 miles apart and 6 miles long at a mean latitude of 43°20′. By Eq. (5),

$$c_f = \tfrac{4}{3} \times 6 \times 6 \tan 43°20'$$
$$= 45.29 \text{ ft}$$

In Table IV are given, for each degree of latitude, the linear and angular convergency of meridians 6 miles long and 6 miles apart. The table also gives the difference in longitude for 6 miles in both angle and time, and the difference in latitude for both 1 and 6 miles in angular measure.

Meridional Section Lines. In the subdivision of townships into sections, the establishment of section lines parallel to the east boundary of the township necessitates a correction in azimuth of these section lines on account of the angular convergency of the meridians. While meridional section lines are being run north, they are made to deflect to the left or west of the true meridian by an angle equal to the convergency in the distance to the section line from the east boundary. Hence, 1/6, 1/3, 1/2, 2/3, and 5/6 of the angles of convergency given in Table IV represent, respectively, the deflections from the true meridian for section lines, respectively, 1, 2, 3, 4, and 5 miles west of the east boundary of the township.

18-25 Secant Method of Laying Off a Parallel of Latitude.

As the base line, standard parallels, and latitudinal township lines are true parallels of latitude, they are curved lines when established on the surface of the earth. Ordinarily a latitudinal line is laid off on the ground by the method of offsets from a straight line either tangent to or intersecting the parallel. (It may also be laid off through the use of a solar attachment whereby the transit is oriented with respect to the true meridian at each transit station.) The amounts of the offsets either may be calculated from the convergency of the meridians or may be obtained from published tables.

The *secant method* of laying off a parallel of latitude is recommended by the Bureau of Land Management for its simplicity of execution and for the proximity of the straight line (secant) to the true latitude curve. It is the method most commonly employed. All measurements and all cutting to clear the line are substantially on the true parallel. In Fig. 18-7 the secant is a straight line 6 miles in length, which intersects the true parallel at the end of the first and fifth miles from the point of beginning. For the latitude of the given parallel, the offsets (in links) from secant to parallel are given in the figure, at intervals of 1/2 mile. In Table V are given, for various latitudes, the azimuths (measured in either direction from true north) of the secant at intervals of 1 mile. In Table VI are tabulated the offsets from the secant to the parallel at intervals of ½ mile.

The procedure employed in establishing a true parallel 6 miles long by the secant method is as follows: The initial point on the secant is located by measuring south of the beginning corner a distance equal to the secant offset for 0 miles given in Table VI (5 links in Fig. 18-7). The transit is set up at this point, and the direction of the secant line is established by laying off from true north the azimuth, either calculated or taken from published tables such as Table V; for the conditions illustrated by Fig. 18-7 the bearing of the secant which extends east from the

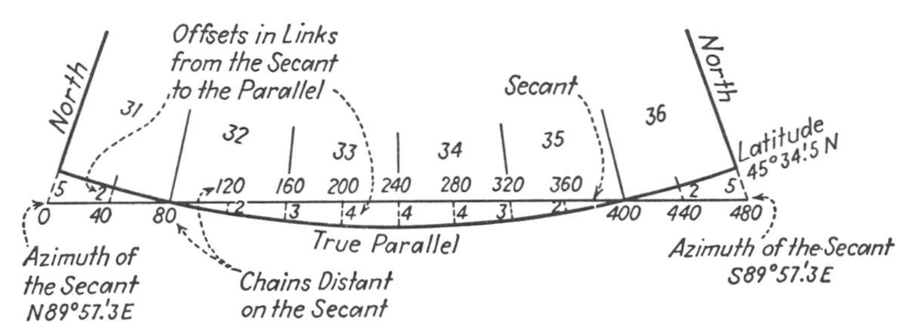

Fig. 18-7. Parallel of latitude by secant method.

point of beginning is N89°57.3′E. (Owing to the convergency of meridians, the azimuth of the secant—a straight line in plan—varies along its length.) The secant is then projected in a straight line for 6 miles; and as each 40 chains (½ mile) is laid off along the secant, the proper offset is taken to establish the corresponding section or quarter-section corner on the true parallel.

At the end of 6 miles, if it is not convenient to determine the true meridian, the succeeding secant may be established by laying off the deflection angle given in the last column of Table V.

18-26 Allowable Error of Closure. The maximum allowable error of closure prescribed for the United States rectangular surveys is 1/452, provided the error of closure in either latitude or departure does not exceed 1/640. Whenever a closure is effected, the latitudes, departures, and error of closure of the lines composing the figure (quadrangle, township, section, meander, etc.) must be calculated, and corrective steps must be taken whenever the test discloses an error in excess of the allowable value.

With regard to the allowable limits of precision, the Manual of surveying instructions (Ref. 41) of the Bureau of Land Management is quoted as follows:

"In the administration of surveying laws it has been necessary to establish a definite relation between rectangularity (square miles of 640 acres, or aliquot parts thereof), as contemplated by law, and the resulting unit of subdivision consequent upon the practical application of surveying theory to the marking out of the lines on the earth's surface, wherein the ideal section is allowed to give way to one which may be termed 'regular.' Such relation, as applied to the boundaries of a section, has been placed at the following limits: (*a*) For alinement, not to exceed 21′ from cardinal in any part; (*b*) for measurement, the distance between regular corners to be normal according to the plan of the survey, with certain allowable adjustments not to exceed 25 links in 40 chains; and (*c*) for closure, not to exceed 50 links in either latitude or departure.

"Township exteriors, or portions thereof, will be considered defective when they do not qualify within the above limits. It is also necessary, in order to subdivide a township regularly, to consider a fourth limit, as follows: (*d*) For position, the corresponding section corners upon the opposite boundaries of the township to be so located that they may be connected by true lines which will not deviate more than 21′ from cardinal."

18-27 Subdivision of Townships. In the normal subdivision of townships with regular boundaries, the south and east boundaries are the governing lines. Briefly, the order of establishing sections is as follows: The subdivisional survey is begun on the south boundary of the township, at the section corner between sections 35 and 36 (see Fig. 18-4). The line is run in a northerly direction parallel to the east boundary of the township; at each section corner a random line is run eastward 1 mile, to connect with the corresponding corner on the east boundary; this line is then corrected if necessary. When the north boundary of the township is reached, the process is repeated, beginning on the south boundary at the section corner between sections 34 and 35. When running the last north-south line, starting between sections 31 and 32, at each section corner an additional random line is run 1 mile west, to connect with the corresponding corner on the west boundary.

The method of establishing section corners is such that all sections except the eleven that lie next to the north and west boundaries of the township are as nearly as may be 1 mile square, containing only the errors of measurement. The error of closure in latitude and departure is determined for each section established during the survey, and if necessary the survey is rerun to bring the error within the permissible limit. Any excess or deficiency in the measured distance between the south and north boundaries of the township is placed in the most northerly half mile. Any deficiency due to the convergency of meridians and any excess or deficiency in the measured distance between the east and west boundaries of the township are placed in the most westerly half mile. (If the section is further subdivided, all of the excess or deficiency is placed in the most northerly quarter mile and the most westerly quarter mile.) Typical dimensions, in chains, are shown in Fig. 18-8 for four sections which lie in the northwest corner of a township.

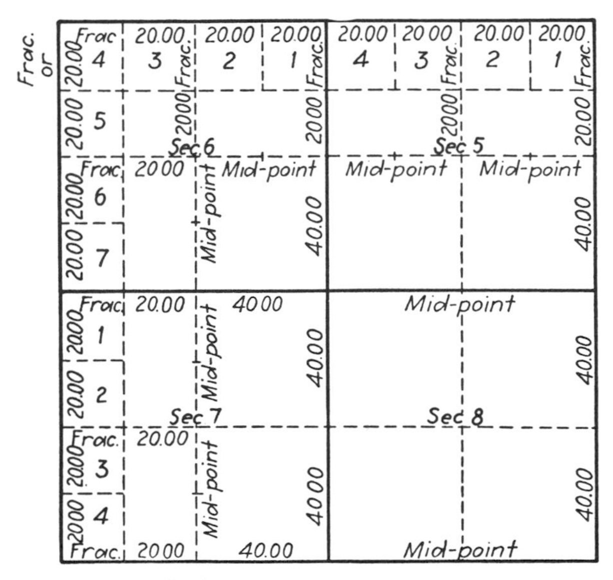

Fig. 18-8. Subdivision of sections.

18-28 Subdivision of Sections. Although the public-land laws provide for the subdivision of sections into quarter sections and quarter-quarter sections, only rarely are sections subdivided by United States surveyors. However, certain lines of subdivision are shown on the official plats, and the surveyor in private practice is compelled to correlate conditions found on the ground with those on the approved plat.

The *regular* subdivisions of a section are the quarter section (½ mile square), the half-quarter section (¼ by ½ mile), and the quarter-quarter section (¼ mile square); the last contains 40 acres and is the legal minimum for purposes of disposal under the general land laws. An example of a complete legal description of a normal quarter-quarter section is as follows: "The northeast quarter of the southwest quarter of section ten (10), Township four (4) south, Range six (6) east, of the Mount Diablo Meridian, containing forty (40) acres, more or less, according to the United States Survey."

Of the 36 sections in each normal township (Fig. 18-4), 25 are returned as containing 640 acres each; 10 sections adjacent to the north and west boundaries (comprising sections 1–5, 7, 18, 19, 30, and 31) each contain regular subdivisions totaling 480 acres and in addition 4 *fractional lots* each containing 40 acres plus or minus definite differences to be determined in the survey (see Fig. 18-8); and 1 section (section 6) in the northwest corner contains regular subdivisions totaling 360 acres and in addition 7 fractional lots (Fig. 18-8). The system of numbering fractional lots is shown in the figure. Sections or lots may also be made fractional on account of meanderable bodies of water (Fig. 18-9), mining claims, and other segregated areas within their limits.

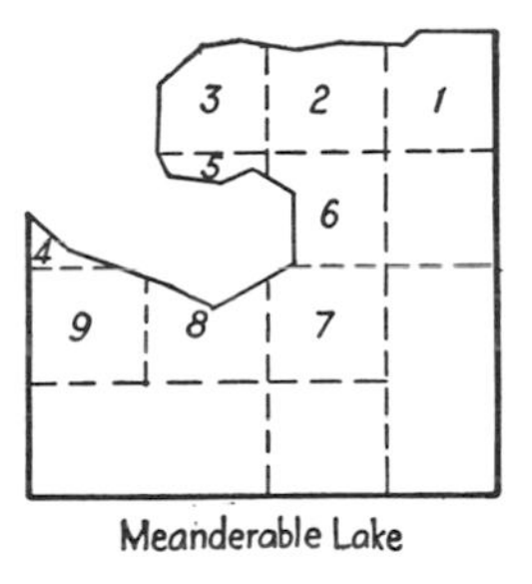

Meanderable Lake

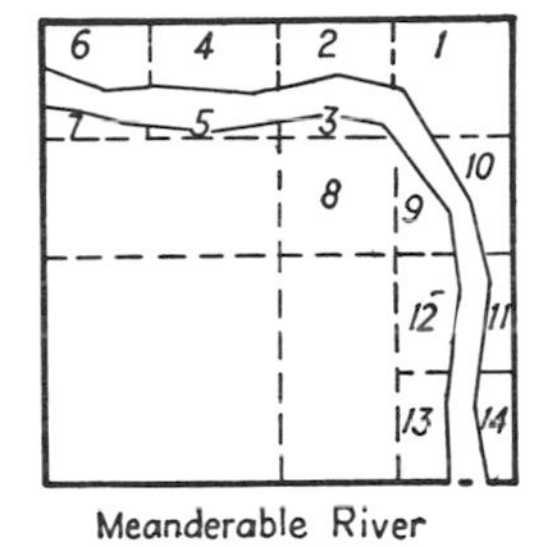

Meanderable River

Fig. 18-9. Fractional lots.

On all section lines, monuments are placed at intervals of 40 chains, thus marking all section corners and all exterior quarter-section corners. Further subdivision of sections must be done by surveyors in private practice, but such subdivision must conform to the established dimensions of the official plat, including certain lines of subdivision which are shown on the plat but are not laid off on the ground. For example, in Fig. 18-8, the dash lines are established on the plat by protraction from definite points on the opposite boundaries (represented by solid lines) which have been established in the field.

A section cannot be legally subdivided until the section and exterior quarter-section corners have been either found or restored and until the resulting courses and distances have been redetermined in the field. When the opposite quarter-section corners have been located, the interior quarter-section corner may be established. If the boundaries of quarter-quarter sections or of fractional lots are to be established on the ground, it is necessary to measure the boundaries of the quarter section and to fix thereon the quarter-quarter-section corners at distances in proportion to those given on the official plat; then the interior quarter-quarter-section corner may be placed. The subdivisional lines of fractional quarter sections are run from properly established quarter-quarter-section corners, with courses governed by the conditions represented on the official plat.

18-29 Marking Corners and Lines. As a final step in the government's survey of the public lands, permanent monuments (Art. 18-3) are set at the section and exterior quarter-section corners, and the locations of these lines through living timber are indicated by blazing trees and by hack marks on trees. Each corner is referenced to at least one corner accessory (Art. 18-3). The monument may be an iron pipe filled with concrete, a durable stone on which the markings are cut with a chisel, or a tree on which the markings are scribed in a vertical blaze. For further details, the reader is referred to the "Manual of Surveying Instructions" of the Bureau of Land Management (Ref. 41).

All monuments are marked in accordance with a system which provides ready identification. Iron posts and tree corners are marked with capital letters which are themselves keys to the character of the monument and with arabic figures giving the section and township and range numbers of the adjacent subdivisions and the year in which the survey was made. Marks in the shape of *notches* and *grooves* are placed on the vertical edges or faces of stone monuments.

The ordinary markings common to all classes of corners are as shown in the accompanying table, and typical markings on iron monuments are arranged as indicated in Fig. 18-10. All standard township, section, and quarter-section corners on base lines and standard parallels are marked "SC." All closing township and section corners on these lines are marked "CC." A witness corner and its accessories are constructed and marked similarly to a regular corner for which it stands, with the additional letters "WC" and with an arrow pointing to the true corner.

SC
T25N
R17E | R18E
S36 | S31
1916

(*a*) Standard township corner

T27N | R17W
S31 | S32
T26N R17W
S6
1916

(*b*) Section corner

Fig. 18-10. Typical markings on iron monuments.

Mark	*Meaning*	*Mark*	*Meaning*
AMC	Auxiliary meander corner	S	Section
BO	Bearing object	S	South
BT	Bearing tree	SC	Standard corner
C	Center	SMC	Special meander corner
CC	Closing corner	T	Township
E	East	W	West
MC	Meander corner	WC	Witness corner
N	North	WP	Witness point
R	Range	¼	Quarter section
RM	Reference monument	1/16	Quarter-quarter section

18-30 Restoration of Lost Corners. Many corner marks become obliterated with the progress of time. It is one of the important duties of the local or county surveyor, in the relocation of property lines or in the further subdivision of lands, to examine all available evidence and to identify the official corners if they exist. Should a search of this kind result in failure, it is the duty of the surveyor to employ a process of field measurement which will result in the obliterated corner's being restored to its most probable original location (see Art. 18-12). If the original location of a corner cannot be determined beyond reasonable doubt, the corner is said to be *lost;* and it is then restored to its original location, as nearly as possible, by processes of surveying that involve the retracement of lines leading to the corner.

Where linear measurements are necessary to the restoration of a lost corner, the principle of *proportionate measurement* must be employed. Proportionate measurement distributes an excess or deficiency in an overall remeasured distance so that each of the remeasured parts will have the same ratio to the remeasured distance as the corresponding original parts had to the originally measured distance. Single proportionate measurement consists in first comparing the resurvey measurement with the original measurement between two existing corners on opposite sides of the lost corner, and then laying off a proportionate distance from one of the existing corners to the lost corner (see numerical problem 3, Art. 18-31). Double proportionate measurement consists in single proportionate measurement on each of two such lines perpendicular and intersecting at the lost corner. The restorative process must be in harmony with the methods employed in originally establishing the lines involved, and the preponderant lines must be given the greater weight in determining whether a corner should be relocated by single or double proportionate measurement or by some other method. Detailed instructions for the relocation of lost corners are given in the manual of the Bureau of Land Management.

18-31 Numerical Problems

1 Find the angle of convergency between two meridians 24 miles apart at a mean latitude of 45°. Compute the linear convergency, measured along a parallel of latitude, in a distance of 24 miles.

2 Show the dimensions and areas of the protracted subdivisions of Section 6, as required by law to be shown on the official plat, when the north, east, south, and west boundaries are, respectively, 76.36, 80.44, 76.60, and 80.00 chains.

3 A lost section corner on a range line is to be restored by a resurvey. One mile to the south the township corner is identified, and 2½ miles to the north the quarter-section corner is found. According to the records the corresponding distances measured at the time of the original survey were 80.00 and 200.00 chains. The resurvey distance between the existing corners is 279.64 chains. State the procedure to be followed in restoring the lost corner, and calculate the proportionate distances to be employed.

18-32 Field and Office Problem

Problem 44. Survey of Tract for Deed Description

Object. To obtain sufficient data for a proper legal description of a tract and to prepare such a description.

Procedure. (1) Around the assigned field run an azimuth traverse with the transit, measuring the sides with a steel tape and setting hubs at the corners. The angular error of closure in minutes should not exceed $\frac{1}{2} \times \sqrt{\text{number of sides}}$. Distribute the error of closure among the angles of the traverse. Refer azimuths to the true meridian. (2) Compute the latitudes and departures and the linear error of closure; the linear error of closure should not exceed 1/5,000. (3) Balance the survey, and calculate the area of the tract by the coordinate method. (4) Determine the location of one corner of the traverse from an established reference point (Art. 18-14) and, beginning at this corner, write a description of the tract by metes and bounds.

19
TOPOGRAPHIC MAPS

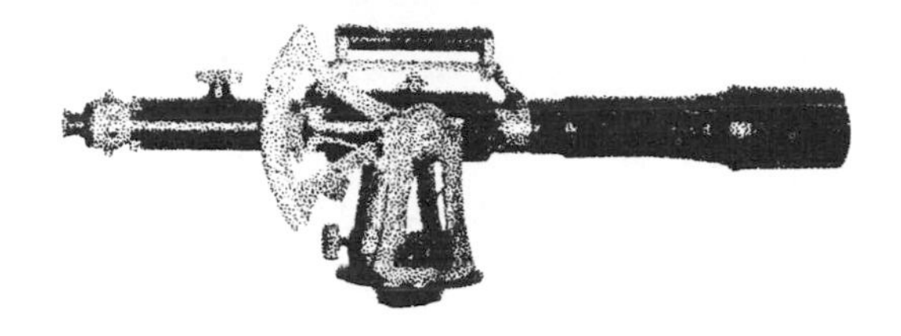

19-1 Relief. A topographic map shows not only the natural and artificial features of the terrain but also the *relief,* or configuration of the earth's surface. Topographic maps are a necessary aid in the design of engineering projects which require a consideration of land forms, elevations, or gradients. Small-scale topographic maps supplying information for general use are published by various governmental organizations.

As an aid to any survey, the surveyor should obtain available maps and/or aerial photographs of the region, even though they may not be of the particular nature or scale desired for his purpose. The central source of information regarding all Federal maps and aerial photographs is the Map Information Office, U.S. Geological Survey, Washington, D.C. Likewise many maps are available from state, county, and city agencies.

19-2 Representation of Relief. Relief may be represented by *relief models, shading, hachures, form lines,* or *contour lines.* Of the symbols used on maps, only contour lines indicate elevations directly and quantitatively; they have by far the widest use. They are the principal topic of this chapter. Form lines are similar to contour lines but are not true to scale.

1 Relief Model. A relief model is a representation of ground forms done in three dimensions to suitable horizontal and vertical scales. Plastic materials such as wax or clay are used; also laminated models are made by cutting cardboard sheets to the shape of successive contours and then assembling the sheets. The U.S. Army Map Service has developed a process for producing three-dimensional contour maps in the form of sheets of plastic which are flat-printed and then molded over a relief model. Cardboard relief maps of certain areas are commercially available.

2 Shading. Shading is a method of showing roughly in plan the terrestrial relief as it would appear from a point vertically above and with parallel rays of light flooding the landscape from a given angle, causing shadows to lie upon the less-illuminated areas. Shading is useful where the relief is high and the slopes are steep; it is sometimes used in combination with hachures or contour lines to render the map more legible.

3 Hachures. Hachures show relief more definitely but less legibly than does shading. The symbol consists of rows of short, nearly parallel lines drawn parallel to the steepest slopes. In the best practice a standard scale of lengths and weights of lines is used to represent the various degrees of inclination of slopes.

19-3 Contours and Contour Lines.

A *contour* is an imaginary line of constant elevation on the ground surface. It may be thought of as the trace formed by the intersection of a level surface with the ground surface, for example, the shore line of a still body of water. Contours on the ground are represented on the map by *contour lines;* loosely, the terms *contour* and *contour line* are often used interchangeably. On a given map, successive contour lines represent elevations differing by a fixed vertical distance called the *contour interval.*

The principal characteristics of contour lines may be illustrated by reference to Fig. 19-1. For the purpose of this discussion the slope of the river surface is disregarded, and it is considered that the shore line on the map marks the position of the 510-ft contour line. For this map the contour interval is 5 ft.

1 The horizontal distance between contour lines is inversely proportional to the slope. Hence on steep slopes (as at the railroad and at the river banks in Fig. 19-1) the contour lines are spaced closely.

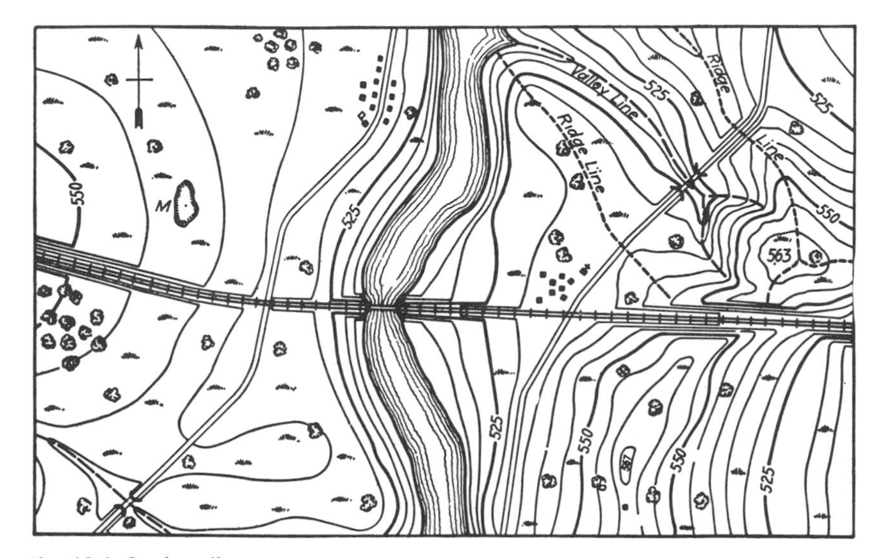

Fig. 19-1. Contour lines.

2 On uniform slopes the contour lines are spaced uniformly.

3 Along plane surfaces (such as those of the railroad cuts and fills in Fig. 19-1) the contour lines are straight and parallel to one another.

4 As contour lines represent level lines, they are perpendicular to the lines of steepest slope. They are perpendicular to ridge and valley lines where they cross such lines.

5 As all land areas may be regarded as summits or islands above sea level, all contour lines must close upon themselves either within or without the borders of the map. It follows that a closed contour line on a map always indicates either a summit or a depression. If water lines or the elevations of adjacent contour lines do not indicate which condition is represented, a depression is shown by a hachured contour line, called a *depression contour,* as shown at *M* in Fig. 19-1.

6 As contour lines represent contours of different elevation on the ground, they cannot merge or cross one another on the map, except in the rare cases of vertical surfaces (see bridge abutments of Fig. 19-1) or overhanging ground surfaces as at a cliff or a cave.

7 A single contour line cannot lie between two contour lines of higher or lower elevation.

19-4 Contour Interval. The appropriate contour interval, or vertical distance between contours, depends upon the purpose and scale of the map and upon the character of terrain represented. For small-scale maps of rough country the interval may be 50 ft, 100 ft, or more; for large-scale maps of flat country the interval may be as small as ½ ft. For maps of intermediate scale, such as are used for many engineering studies, the interval is usually 2 or 5 ft (see also Art. 19-11).

19-5 Contour-map Construction. Normally the construction of a topographic map consists of three operations: (1) the plotting of the horizontal control, or skeleton (Chap. 15); (2) the plotting of details, including the map location of points of known ground elevation, called *ground points,* by means of which the relief is to be indicated; and (3) the construction of contour lines at a given contour interval, the ground points being employed as guides in the proper location of the contour lines. A ground point on a contour is called a *contour point.*

Ridge and valley lines are important aids to the correct drawing of contour lines (Fig. 19-2). Frequently they are drawn first, and the contour crossings are spaced along them before the contour lines are drawn.

Contour lines are shown for elevations which are multiples of the contour interval. They are drawn as fine smooth freehand lines of uniform width, preferably by means of a contour pen (Fig. 14-11). Usually each fifth line is made heavier than the rest. Elevations of contours are indicated by numbers placed at appropriate intervals; usually only the fifth or heavier contour lines are numbered. The line is broken to leave a space for the number. So far as possible the numbers are faced so as to be read from one or two sides of the map; but on some maps the numbers are faced so that the top of the number is uphill. *Spot elevations* are shown by numbers at significant points such as road intersections, bridges, water surfaces, summits, and depressions.

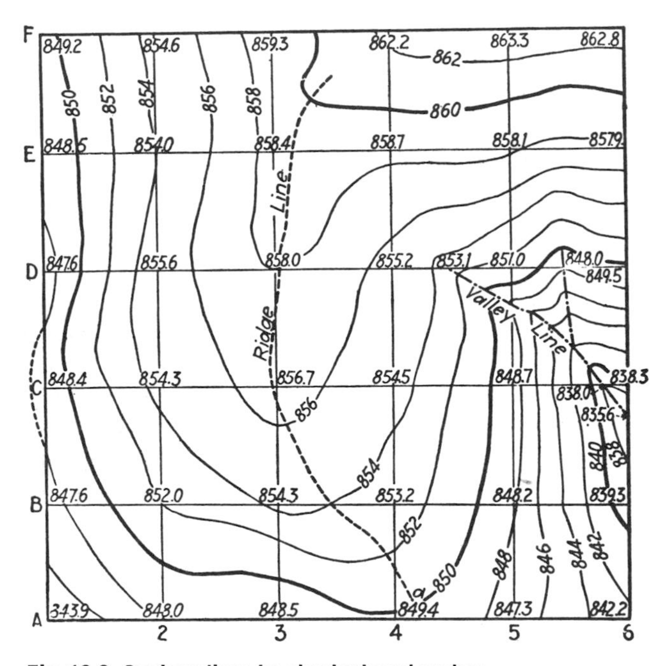

Fig. 19-2. Contour lines by checkerboard system.

Any contour line must be drawn, to some degree, by estimation. Skill and judgment are required to the end that the contour lines may best represent the actual configuration of the ground surface.

19-6 Interpolation. The process of spacing the contour lines proportionately between plotted points is called *interpolation*. For example, in Fig. 19-2 the elevations of points *A*-2 and *B*-2 are, respectively, 848.0 and 852.0 ft, and the 848 and 852-ft contour lines pass through these points. Under the assumption that the slope is uniform, the 850-ft contour line passes through a point midway between *A*-2 and *B*-2. Interpolations are made by estimation, by computation, or by graphical means, as follows:

1 Estimation. On intermediate-scale and small-scale maps, often the desired precision can be obtained if the interpolation is made by careful estimation supplemented by approximate mental computations.

2 Computation. Where considerable precision is desired in the map, the computations for interpolation may be made with the aid of a slide rule. For example, the elevations of points *E*-6 and *F*-6 (Fig. 19-2) are 857.9 and 862.8 ft, respectively. The difference in elevation between *E*-6 and the 858-ft contour is 0.1 ft. Then, since the total difference in elevation is 4.9 ft, the proportional part of the distance from *E*-6 to *F*-6 to locate the 858-ft contour line is 0.1/4.9 of the map distance between these points.

3 Graphical Means. One method of graphical interpolation is illustrated by Fig. 19-3. Parallel lines are drawn at equal intervals on tracing

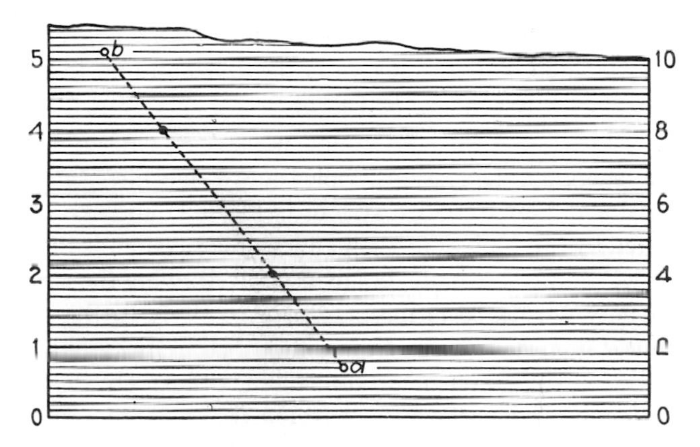

Fig. 19-3. Graphical interpolation of contour lines.

cloth, each fifth or tenth line being made heavier than, or of a different color from, the rest and being numbered as shown. If it is desired to interpolate the position of, say the 52 and 54-ft contours between *a* with elevation of 50.7 and *b* with elevation of 55.1, the line on the tracing cloth corresponding to 0.7 ft (scale at left end) is placed over *a*, and the tracing is turned about *a* as a center until the line corresponding to 5.1 ft (scale at same end) covers *b*. The interpolated points are at the intersections of lines 2.0 and 4.0 (representing elevations 52 and 54) and the line *ab* and may be pricked through the tracing cloth. By assigning different values to the spaces, as by using the scale at the right end of Fig. 19-3, a single piece of tracing cloth can be made to suit a variety of conditions.

Another convenient graphical means of interpolation is by the use of a rubber band graduated at equal intervals with lines which form a scale similar to that just described for the tracing cloth. The band is stretched between two plotted points so that these points fall at scale divisions corresponding to their elevations. The intermediate contour points are then marked on the map.

19-7 Systems of Ground Points. The typical systems of ground points commonly used in topographic surveying (Chap. 20) and in topographic mapping are the *trace-contour, checkerboard, controlling-point,* and *cross-profile* systems.

1 Tracing Contours. A number of points on a given contour are located on the ground, and their corresponding locations are plotted on the map. The contour line is then drawn through these plotted points.

2 Checkerboard. A system of squares or rectangles is plotted as in Fig. 19-2, and near each corner is written its elevation. Also, the locations of valley and ridge lines are shown. The contour crossings are interpolated on the valley and ridge lines and on the sides of the squares, and the contour lines are drawn.

3 Controlling Points. Ground points are chosen as necessary to define the summits, depressions, valley and ridge lines, and important changes in slope. This irregular system is plotted, and the contour lines are located by interpolation. Such an irregular system is illustrated in Fig. 19-4, in which are shown the points used in drawing the summit and immediate vicinity illustrated in the map of Fig. 19-1. In this sketch the ridge and valley lines have been drawn, the contour lines have been spaced along them and between other controlling points (by interpolation), and the fifth contour lines have been sketched.

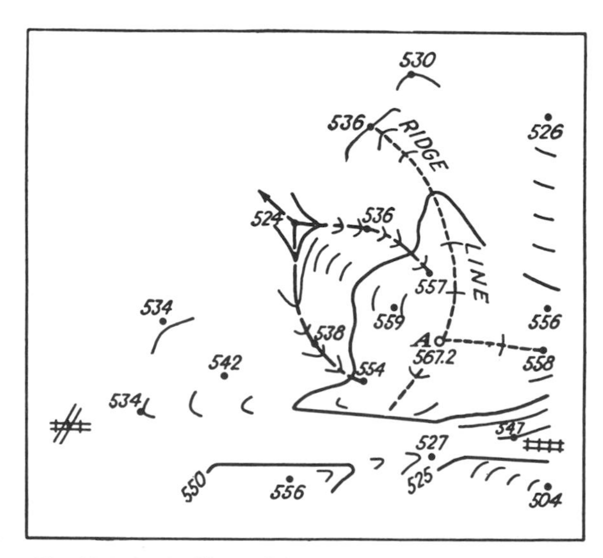

Fig. 19-4. Controlling points.

4 Cross Profiles. The cross-profile method is most frequently used in connection with route surveys. The field surveys determine the location either of all contour points or of all points of change in slope, along selected lines normal to the route traverse line. The traverse, cross-profile lines, and ground points are plotted, and the contour lines are drawn. In Fig. 19-5 a transit traverse is represented as a straight line along which the 100-ft stations are shown. The cross-profile lines are dashed, and the contour points are shown as dots.

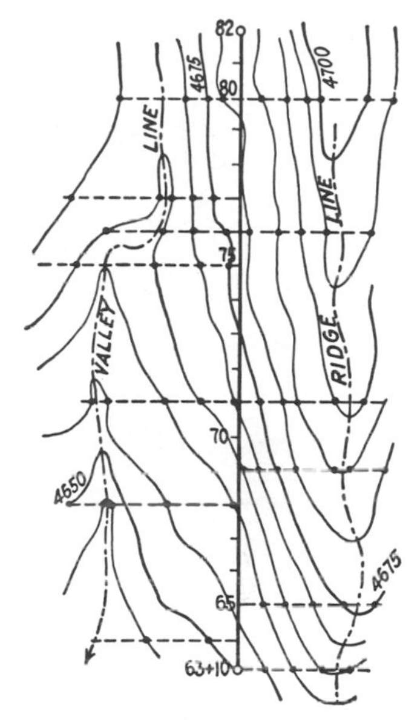

Fig. 19-5. Cross profiles.

19-8 Finishing the Map. In Chap. 14 the general subject of map drafting is discussed and map symbols are given. Methods of plotting are explained in Chap. 15. Modern map drafting tends toward restraint in the use of titles and symbols. The standards most generally used for topographic map drafting and for any colors used in finishing the map are those employed by the U.S. Geological Survey (see Fig. 19-6) and the U.S. Coast and Geodetic Survey.

19-9 Tests for Accuracy. A topographic map can be tested for accuracy, both in plan and in elevation. In this discussion it is assumed that the errors in field measurement may be disregarded and that a graphical scale is provided on the map to render negligible any effect of shrinkage of the paper.

The test for *horizontal dimensions* consists in comparing distances scaled from the map and distances measured on the ground between the corresponding points. The precision with which distances may be scaled from a map depends upon the scale of the map and upon the size of the plotting errors. Thus, if for a map scale of 1 in. = 100 ft it is known that the error in location of any one point with respect to any other on the map is $\frac{1}{40}$ in., then the error represents 2.5 ft on the ground.

One test for *elevations* consists in comparing, for selected points, the elevations determined by field levels and the corresponding elevations taken from the map. Usually the points are taken at 100-ft stations along traverse lines crossing typical features of the terrain. A more searching test is to plot selected profiles of the ground surface as determined by the field levels and the corresponding profiles taken from the

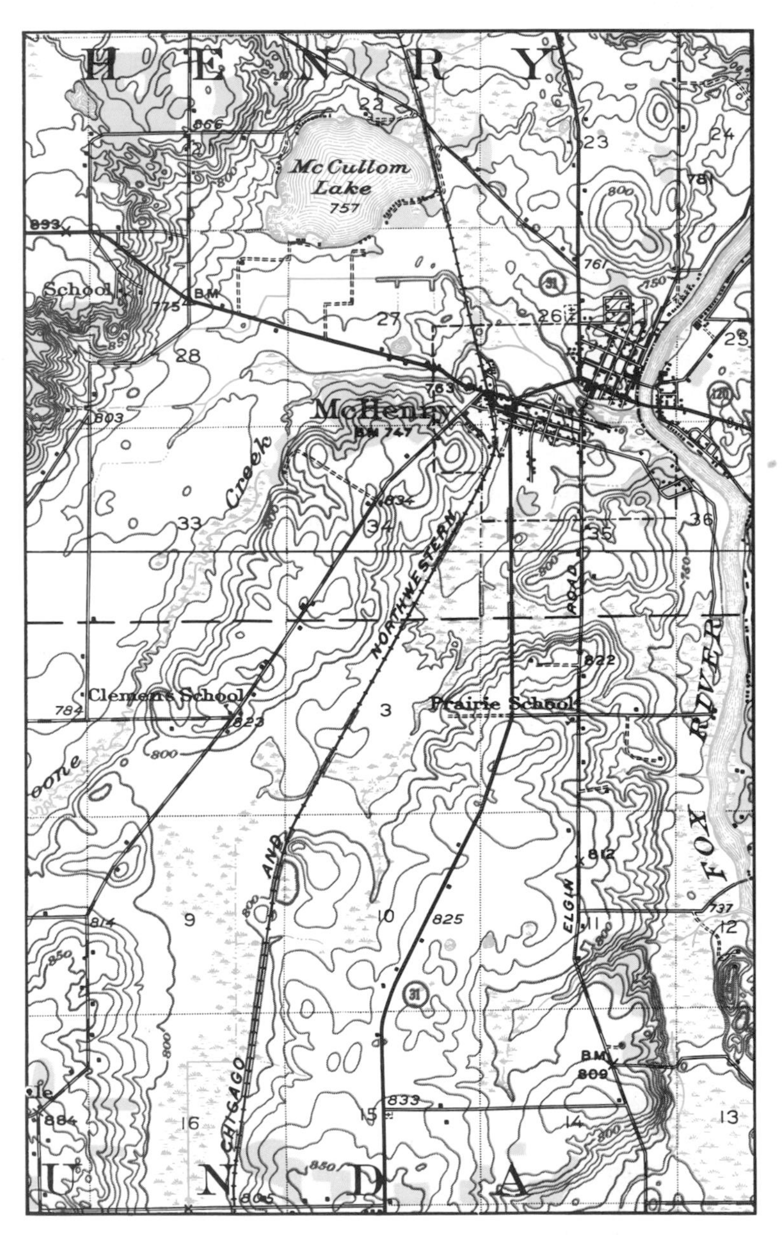

Fig. 19-6. Typical contour map of U.S. Geological Survey. Scale approximately 1 in. = 1 mile (representative fraction 1/62,500). Contour interval 10 ft.

map. The presence of systematic errors will be evidenced if the map profile is above or below the ground profile for an undue proportion of its length.

19-10 Choice of Map Scale. From a consideration of the tests described in the preceding article it is possible to choose a map scale consistent with the purpose of the survey if the approximate size of the plotting error is known. For example, if it is known that (with reasonable care in plotting) the average error in distance between any two definite points on the map is $\frac{1}{40}$ in., and if it is known that the purpose of the survey will be met if the average error in scaled distances is 10 ft, these conditions are satisfied by a map scale of 1 in. = 400 ft. By a similar course of reasoning, a map scale may be chosen that will represent a given *area* within desired limits of plotting error.

In addition to accuracy, considerations in the choice of a map scale are (1) the clarity with which features can be shown, (2) cost (the larger the scale the greater the cost), (3) correlation of data with related maps, and (4) physical factors such as the number and character of features to be shown, the nature of the terrain, and the necessary contour interval.

19-11 Choice of Contour Interval. The contour interval may be thought of as the scale by which the vertical distances or elevations are measured on a map. The choice of a proper contour interval for a topographic survey and map is based upon four principal considerations, as follows:

1 *Accuracy*. The greater the desired accuracy of elevations read from the map, the smaller should be the contour interval. In general, the average map error in elevation of points chosen at random should not exceed one half of one contour interval.

2 *Features*. The contour interval should generally be smaller for a map of flat country than for a map of hilly country. If a given kind of terrain is irregular rather than smooth, the contour interval should be relatively small in order to show the complexity of configuration.

3 *Legibility*. To ensure legibility of the map, contour lines should not be spaced more closely than 20 or 30 to the inch. Under this limitation the smaller the scale of the map, the larger should be the contour interval.

4 *Cost*. The smaller the contour interval, the higher the cost, especially if the usual accuracy of one-half contour interval is maintained.

The contour interval and the scale of the map are interrelated; in general, the smaller the scale, the larger the contour interval. One rough rule is that the scale, in feet per inch, should be about 40 times the contour interval. Table 19-1 represents good practice in the choice of a contour interval for general-purpose maps under usual conditions. Maps for special purposes may depart considerably from these values.

Table 19-1 Relation between scale of map, slope of ground, and contour interval

Scale of map	*Slope of ground*	*Interval, ft*
Large (1 in. = 100 ft or less)	Flat	0.5 or 1
	Rolling	1 or 2
	Hilly	2 or 5
Intermediate (1 in. = 100 ft to 1,000 ft)	Flat	1, 2, or 5
	Rolling	2 or 5
	Hilly	5 or 10
Small (1 in. = 1,000 ft or more)	Flat	2, 5, or 10
	Rolling	10 or 20
	Hilly	20 or 50
	Mountainous	50, 100, or 200

19-12 Accuracy of Topographic Maps. The principles stated in the preceding articles provide the criteria by which the accuracy of topographic maps may be specified. Thus, the average errors in scaled horizontal *dimensions* between definite points chosen at random, or in *areas* scaled from the map, may be limited to a stated value. The accuracy of contour lines may be specified by assigning maximum values to (1) the average error in elevations taken from the map, (2) the maximum error indicated by random test profiles, and (3) the ratio of the length of the map test profile that lies above the ground test profile to the length that lies below the ground test profile. For example, the accuracy of a given topographic map might be specified as follows: The average error in distances between definite points as scaled from the map shall not exceed 8 ft; the average error in elevations read from the map shall not exceed 1 ft; the maximum error indicated by random test profiles shall not exceed 4 ft; and the ratio of the length of the map test profile that lies above the ground test profile to the length that lies below the ground test profile shall lie between $\frac{1}{3}$ and 3.

Federal agencies engaged in mapping have agreed on minimum requirements which entitle the following statement to be printed on the map, "This map complies with the National Standards of Map Accuracy requirements." With regard to horizontal accuracy, it is required that for maps on publication scales larger than 1:20,000 not more than 10 per cent of the well-defined points tested shall be in error by more than $\frac{1}{30}$ in., measured on the publication scale; and for maps on publication scales of 1:20,000 or smaller, $\frac{1}{50}$ in. Well-defined points are those that are easily visible or recoverable on the ground—in general, those which are plottable on the scale of the map within $\frac{1}{100}$ in. With regard to vertical accuracy, it is required that not more than 10 per cent of the elevations tested shall be in error more than one half the contour interval; the apparent vertical error may be decreased by assuming a horizontal displacement within the permissible horizontal error. The accuracy of the map may be tested by comparing points whose locations or elevations are shown on it with corresponding points as determined by surveys of a higher accuracy.

USES OF TOPOGRAPHIC MAPS

19-13 Cross-sections and Profiles. Figure 19-7 shows (below) a portion of a contour map for use in connection with roadway construction and (above) a vertical section on which the profile of the ground line and that of the grade line are plotted. In the case illustrated the center line of the roadway is straight, and the profile of the ground along this center line is constructed by projecting points (either full stations or contour points) to corresponding points on a base line of the vertical section and then by plotting the elevations of these points to scale on the vertical section. If the horizontal alinement of the roadway either is curved or is not parallel to the base line of the vertical section, usually

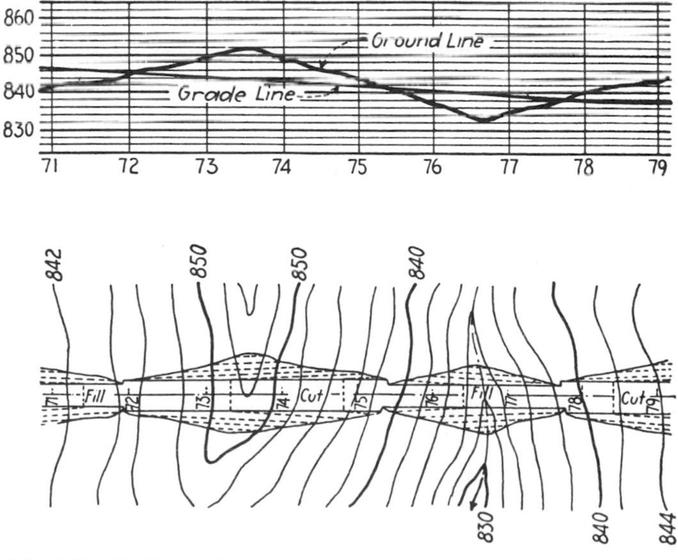

Fig. 19-7. Earthwork for roadway.

the distances from full stations to contour crossings are scaled from the map, and then these distances are plotted on profile paper as if the stations and elevations were being taken from profile notes (Art. 7-1).

Vertical cross-sections for estimating earthwork for other purposes are constructed in a manner similar to that just described for the profile of a roadway. In practice, parallel lines along which the cross-sections are to be taken are drawn on the map, and the distance between them is scaled. One person scales horizontal distances to contour crossings and reads contour elevations, while a second person plots the data on regular cross-section paper. Frequently the horizontal scale to which the profile is plotted is not the same as that used on the map.

19-14 Earthwork for Grading Areas. Quantities of earthwork for grading areas can be estimated from contour maps by (1) vertical cross-sections, (2) horizontal planes, or (3) equal-depth (or equal-height) contours.

1 Cross-sections. A cross-section is plotted for each of a number of selected parallel lines on the contour map. On each cross-section is shown the profile of (*a*) the original ground line and (*b*) the ground line after the proposed grading. To obtain the volume of a given cut or fill between adjacent cross-sections, the area between these ground lines is averaged for the two cross-sections and is multiplied by the distance between them.

2 Horizontal Planes. Where the graded surface itself is irregular, the common practice is to use the topographic map directly as a basis for calculations of volume. On the map are shown by full lines the contours for the original ground surface and by dash lines the contours for the proposed graded surface. The volume of earth to be moved, between the two horizontal planes at the elevations of successive contours, is a solid the altitude of which is the contour interval and the top and bottom bases of which are the horizontal projections of the cut or fill at the contour elevations, as indicated by the horizontal area between the corresponding full and dash contour lines. The end volumes may be considered as pyramids, and their limits estimated by proportion.

example 1. By planimeter observation on the map, the base area between the 96-ft solid and dash contour lines is 328 sq ft, and the base area between the 94-ft solid and dash contour lines is 564 sq ft. The volume of earthwork between these horizontal planes is

$$(96 - 94) \times \frac{328 + 564}{2} = 892 \text{ cu ft, or } 33.0 \text{ cu yd}$$

3 Equal-depth Contours. This method consists in determining volumes between irregularly inclined upper and lower surfaces bounding certain increments of cut or fill. On the map are shown by light full lines the contours for the original ground surface and by dash lines the contours for the proposed graded surface. At the intersection of each light full line with each of the dash lines the depth of fill (or cut) is recorded. Heavy full lines similar to contour lines are drawn through points of equal fill. The heavy outer line, through points of zero fill, marks the limit of the fill; the next heavy line encloses the area over which the fill is a minimum of one contour interval; and so on.

The fill between the graded surface and the surface one contour interval below is represented by the solid the altitude of which is the contour interval and the upper and lower surfaces of which are shown in horizontal projection by the line of zero fill and the line of one-contour-interval fill, respectively. The areas of these two surfaces are determined, and the volume is calculated by the methods of Arts. 7-10 and 7-11; and so on for the remaining increments of fill. The end volumes may be considered as pyramids, and their limits estimated by proportion.

example 2. By planimeter observation on the map, the projected area of the surface of 6-ft fill is 5,000 sq ft, and the projected area of the surface of 4-ft fill is 17,000 sq ft. The volume of earthwork between these surfaces is

$$(6-4) \times \frac{5{,}000 + 17{,}000}{2} = 22{,}000 \text{ cu ft, or } 815 \text{ cu yd}$$

19-15 Earthwork for Roadway. Figure 19-7 shows (below) the contour lines for a proposed roadway (the grade line of which has been fixed on the profile) drawn dotted over the existing contour map. The contour crossings of the roadway are located on the map by projecting the contour crossings of the grade-line profile from the vertical section (above) to the map. The side slopes of the earthwork are $1\frac{1}{2}$ to 1. The width of the roadway is 36 ft in cut and 30 ft in fill. From a study of the figure the following observations may be made:

1 On the side slopes of the earthwork at any station the distance out from the edge of the roadway to a contour line is given by the difference in elevation multiplied by the side-slope ratio. Thus at station 76 + 40 the elevation of grade is 840.0 ft and the elevation of the first contour line out from the edge of the fill is 838.0 ft; hence the distance out is $2 \times 1\frac{1}{2} = 3$ ft. (For clearness, in the illustration the lateral scale is exaggerated.)

2 As the grade line is not level, the contour lines on the earthwork slopes are not parallel to the roadway. Thus the 844-ft dotted contour line which crosses the roadway at station 73 + 30 is so inclined in direction that at station 74 + 80 where the elevation of grade is 842 ft, the 844-ft contour line is out from the edge of the roadway a distance of $2 \times 1\frac{1}{2} = 3$ ft.

3 The toe of a slope is drawn on the contour map by connecting the points where the dotted lines intersect the corresponding full lines.

If a line is drawn across a plotted roadway normal to the center line, a cross-section of the proposed roadway (Art. 7-3) can be plotted from the data of the contour map. Between adjacent cross-sections, the quantity of earthwork can be calculated as explained in Arts. 7-10 and 7-11. The volume of earthwork may also be estimated by the methods explained in the preceding article.

19-16 Reservoir Areas and Volumes. A contour map can be employed to determine the capacity of a reservoir, the location of the flow line, the area of the reservoir, and the area of the drainage basin. The procedure may be illustrated by reference to the fill across the valley in Fig. 19-7, the fill being considered as a dam. If water is imagined to stand at the elevation of 834 ft, the water surface is represented by that within the full and the dotted 834-ft contour line. If the water were to rise through a 2-ft stage to the elevation of 836 ft, the water surface would be represented by that within the full and the dotted 836-ft contour line (the full line being continued until the two parts meet). The volume of water that caused the 2-ft rise is given by the average of the two surface areas multiplied by the vertical distance of 2 ft. By a similar procedure the volume of successive layers of the reservoir may be estimated.

The outline of the submerged area of a proposed reservoir is given by the contour line representing the maximum stage of the impounded water. The drainage area may be estimated by sketching on the map the watershed line and measuring the extent of the watershed with a planimeter.

19-17 Route Location. A contour map is useful in locating a proposed route for such projects as highways, railroads, and canals.

In roadway location it is desired to fix the center line of the proposed construction so that the subgrade will conform as nearly as practicable to the original ground surface (Chap. 21). Suppose that a proposed highway is to be located in the valley shown in Fig. 19-8, joining the existing highways at opposite corners of the map; and let the maximum permissible gradient be 4 per cent. Beginning at the lower highway, the proposed route is projected on the map up the valley until the steep slopes require a careful study of the ground, say to the 1,130-ft contour line. Then a pair of dividers is set at a map distance equal to the contour interval divided by the desired gradient, in this case 10 ft÷0.04=250 ft. One foot of the dividers is placed on the intersection of the proposed roadway and the 1,130-ft contour line, and the other foot of the dividers is placed on the 1,140-ft contour line, as indicated in the figure by a heavy dot; and so on for successive contour lines. The series of dots marks on the plotted ground surface the location of a 4 per cent grade line up the valley. Hence, the route is made to follow this line as closely as other limitations, such as the radius of curves, will permit.

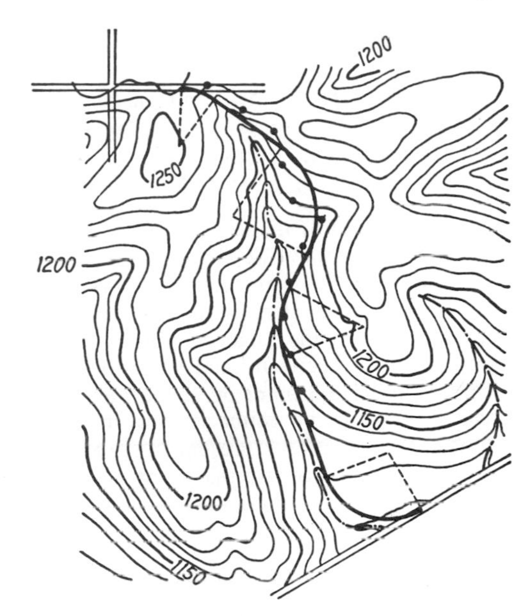

Fig. 19-8. Route location.

19-18 Numerical Problems

1 On a map of scale 1 in. = 400 ft with a contour interval of 5 ft, two adjacent contour lines are 0.54 in. apart. What is the slope of the ground in per cent?

2 In Fig. 19-2, assume that the corners of the squares are 100 ft apart. Plot the ground profile along line *C*-1 to *C*-6, using a horizontal scale of 1 in. = 50 ft and a vertical scale of 1 in. = 10 ft.

3 The accompanying tabulation gives elevations of points over the area of a 60 by 100-ft city lot. The elevations were obtained by the checkerboard method, using 20-ft squares. Point *A*-1 is at the northwest corner of the lot, and point *F*-1 is at the southwest corner. Plot the contours, using a horizontal scale of 1 in. = 10 ft and a contour interval of 2 ft.

	Elevation, Ft			
Point	*1*	*2*	*3*	*4*
A	322.9	327.0	327.5	328.4
B	326.6	331.0	333.3	332.2
C	327.4	333.3	335.7	333.5
D	326.6	334.6	337.0	334.2
E	327.5	333.0	337.4	337.7
F	328.2	333.6	338.3	341.2

19-19 Office Problems

Problem 45. Topographic-map Construction

Object. To construct a complete topographic map from field notes, relief being represented by contours. The data of field problem 31 or 32 (Chap. 11), 48 or 49 (Chap. 20), or 50 (Chap. 21) may be used.

Procedure. (1) If the skeleton of the survey is a traverse, plot the traverse either by the method of tangents (Art. 15-4) or by the method of coordinates (Art. 15-7); if the horizontal control is in the form of rectangles or squares, plot the control by the method of coordinates. (2) Plot the details of the map by methods corresponding to those used in the field, as described in Art. 15-16. Use conventional symbols wherever applicable (Art. 14-8). (3) Mark each ground point by a dot, and mark each elevation in such location that there will be no doubt as to which point it refers, by letting the decimal point of the elevation represent the ground point. (4) Interpolate the contour crossings. (5) Place necessary notes so that they will not interfere with the map. (6) Draw a meridian arrow, and make an appropriate title. (7) Ink the map.

Problem 46. Profile from Topographic Map

Object. To plot the profile for a proposed highway or similar route from data of a contour map. It is assumed that the governing points are given and that the maximum rates of grade, the width of roadbed, and the side slopes are fixed.

Procedure. (1) Sketch in pencil a route between governing points that appears favorable. (2) Set bow dividers to measure 100 ft or some multiple thereof at the scale of the map. From the point of beginning of the route, step off distances and read elevations as indicated by the contours. (3) Plot the corresponding profile. (4) Fix the grade line, making such readjustments of the proposed route as seem necessary to secure the most favorable location. (5) Compute the volumes of cuts and fills by the second method of Art. 7-12; check the computations by the first method of that article. (6) On each of the cuts and fills of the profile show the volume in cubic yards.

Problem 47. Volume of Earthwork from Contours

Object. To determine volumes of earthwork from a topographic map showing contours before and after grading. It is assumed (1) that a map showing contours of the original ground is assigned and (2) that other conditions attached to the problem, such as the area to be graded and the slopes of the finished surface, are given.

Procedure. (1) On the assigned map draw dash contour lines of the proposed ground surface. Draw heavy lines of no cut and fill. Ink all the foregoing lines in black. (2) With the planimeter measure the horizontal sections of earth cut and filled at each contour elevation. By Method 2 of Art. 19-14, determine the volume between successive contour planes and the total volume for each cut and fill. (3) Draw all lines of equal cut and fill, and ink these lines in red. With the planimeter measure the horizontal projections of the areas enclosed by successive lines of equal cut and fill. By Method 3 of Art. 19-14, determine the volume between successive surfaces of equal cut and fill. (4) Compare the total volumes given by the two methods, and show these total volumes on the drawing.

20
TOPOGRAPHIC SURVEYING

20-1 General. The distinguishing feature of a topographic survey is the determination of the location, both in plan and in elevation, of selected ground points which are necessary to the plotting of the contour lines and to the construction of the topographic map. The topographic survey of a tract consists in (1) establishing over the area a system called the horizontal and vertical *control,* which consists of key stations connected by measurements of relatively high precision, and (2) locating the *details* (Arts. 10-13 and 10-14), including the selected ground points, by measurements of lower precision from the control stations.

Topographic surveys fall roughly into three classes, according to the map scale to be employed, as follows:

Large scale: 1 in. = 100 ft or less
Intermediate scale. 1 in. = 100 ft to 1 in. = 1,000 ft
Small scale: 1 in. = 1,000 ft or more

This chapter deals primarily with ground surveys for areas of moderate size and for maps of intermediate and large scale. The choice of a suitable map scale is discussed in Art. 19-10, and the choice of a suitable contour interval in Art. 19-11. Aerial surveys are discussed briefly in Chap. 23; even for ground surveys the surveyor should secure and study aerial photographs of the area whenever they are available; examination of overlapping pairs of photographs by means of a simple stereoscope affords vision as in three dimensions and is of great aid. Other sources of information which are helpful in planning and conducting the survey are existing maps, land and survey records, and reports on geological and other features of the area.

20-2 General Field Methods. The choice of field methods for topographic surveying is governed by the following items:

1 Intended Use of Map. Surveys for detailed maps should be made by more refined methods than surveys for maps of a general character.

2 Area of Tract. Control measurements for a large area should be more precise than those for a small area.

3 Scale of Map. The probable errors in the field measurements should be considerably less than the probable errors in plotting at the given scale. Precision may be increased more easily and at less cost in plotting than in field measurements.

4 Contour Interval. The smaller the contour interval, the more refined should be the field methods.

The principal instruments used are the engineer's transit, the plane table, the engineer's level, the hand level, and the clinometer. The use of the transit has advantages over the use of the plane table where there are many definite points to be located or where the ground cover limits the visibility and requires many setups. Conditions favorable to the use of the plane table are open country and many irregular lines to be mapped; the plane table is also advantageous for small-scale mapping. Sometimes the transit and the plane table, or the transit and the engineer's level, may be used together to advantage. Through dense woods, elevations of details are determined most advantageously by means of the hand level or the clinometer, and distances are usually determined by taping.

The horizontal control (Art. 20-4) is established by triangulation or by traversing, and the vertical control (Art. 20-5) is established by leveling, usually by direct leveling.

The details are located by methods described in Chaps. 3, 10, 13, and 19 and Arts. 20-6 to 20-11. The selected ground points may or may not be contour points. The horizontal locations of the ground points are determined in the same manner as for definite details, usually by radiation. The elevations of ground points are determined usually by trigonometric leveling or, where the terrain is flat, by direct leveling. The stadia is used extensively except on large-scale surveys, for which the errors in stadia distances are large compared with the errors of plotting; on large-scale surveys the distances to definite details are usually measured with the tape. The details may be located either at the time of establishing the control or later.

For the four systems of ground points commonly employed in locating details (Art. 19-7), the general field methods are as follows:

1 Where the *controlling-point* system is used (Art. 20-8), the ground points form an irregular system along ridge and valley lines and at other critical features of the terrain. The ground points are located in plan by radiation or intersection with transit or plane table, and their elevations are determined commonly by trigonometric leveling or sometimes by direct leveling.

2 Where the *cross-profile* system is used (Art. 20-9), as on route surveys, the ground points are on relatively short lines transverse to the main traverse. The distances from traverse to ground points are measured with the tape, and the elevations of ground points are determined by direct leveling, often with the hand level.

3 Where the *checkerboard* system is used (Art. 20-10), as where the scale is large and the tract is wooded or the topography smooth, the tract is divided into squares or rectangles with stakes set at the corners. The elevation of the ground is determined at these corners and at intermediate critical points where changes in slope occur, usually by direct leveling.

4 Where the *trace-contour* system is used (Art. 20-11), the contours are traced out on the ground. The various contour points occupied by the rod are located by radiation with transit or plane table. Frequently the engineer's level is employed as an auxiliary instrument.

Summary. The following statements summarize the use of the various systems of ground points employed in locating details:

1 *Intermediate-scale surveys.* Usually the controlling-point system is used on hilly or rolling ground, and the cross-profile system is used on flat ground or for route surveys.

2 *Large-scale surveys.* Usually the trace-contour system is used if the required accuracy is high and the ground is somewhat irregular in form, and the checkerboard system is used if the ground is smooth and the contour lines may be generalized to some extent.

3 *Small-scale surveys.* The controlling-point system is used almost universally. A relatively small number of ground points are located, often by triangulation with the plane table; their elevations are determined by trigonometric leveling, the horizontal distances used in computing the differences in elevation often being scaled from the map.

CONTROL

20-3 General. Control consists of (1) *horizontal control,* for which by triangulation and/or traversing the control stations are located in plan; and (2) *vertical control,* for which by leveling the bench marks are established and the control stations are located in elevation.

On surveys of wide extent a relatively few stations distributed over the tract are connected by more precise measurements, forming the *primary control;* within this control system other control stations are located by less precise measurements, forming the *secondary control.* For small areas only one control system is necessary, corresponding in precision to the secondary control for larger areas. The terms "primary" and "secondary" are purely relative. Table 20-2 gives approximate values of the limits of permissible error for control measurements suitable to the different map scales.

Another classification of control—either triangulation, traversing, or leveling—with regard to precision is by *orders*. The various orders are absolute, not relative. The extensive surveys executed by the Federal agencies include *first-order, second-order, third-order,* and *fourth-order* control; roughly these correspond respectively to primary, secondary, tertiary, and quaternary control for small-scale maps. Selected items of the standards of accuracy of the Federal Board of Surveys and Maps for third-order control surveys are given in Table 20-1.

Table 20-1 Standards of accuracy of third-order control surveys

(*Federal Board of Surveys and Maps*)
Selected items

Triangulation	Strength of figure:	
	Best chain: ΣR not to exceed	175
	Single figure: R not to exceed	50
	Base measurement: Probable error not to exceed	1/250,000
	Triangle closure: Average not to exceed	5″
	Astronomical azimuth: Probable error not to exceed	2.0″
	Closure in length after side and angle conditions have been satisfied should not exceed	1/5,000
Traverse	Astronomical azimuth: Probable error not to exceed	5.0″
	Distance measurements accurate within	1/7,500
	After azimuth adjustment, closing error in position not to exceed	3.34 ft $\sqrt{\text{miles}}$ or 1/5,000 (whichever is smaller)
Leveling	Average spacing of permanently marked bench marks along lines not to exceed	3 miles
	Check between forward and backward running, between fixed elevations or loop closures, not to exceed	0.050 ft $\sqrt{\text{miles}}$

20-4 Horizontal Control. The horizontal control may consist of a traverse system, a triangulation system, or a combination of the two. The required precision of horizontal control depends upon the scale of the map and upon the size and shape of the tract. In Table 20-2 are given approximate values of permissible error on ordinary surveys; for small areas the tabulated values for secondary or quaternary control may be used for primary or tertiary control, respectively.

Table 20-2 Topographic survey control data (approximate values)

Scale of map	Kind of control, for given scale	Triangulation: Length of sight, miles	Triangulation: Average error of closure in triangles	Triangulation: Distance between bases, miles	Triangulation: Probable error in base measure	Triangulation: Maximum discrepancy between bases	Traverse: Length of traverse, miles	Traverse: Maximum error of angles	Traverse: Maximum linear error of closure	Levels: Length of circuit, miles	Levels: Maximum error of closure
Small	Tertiary	1 to 10	6″	10 to 100	$\frac{1}{250,000}$	$\frac{1}{5,000}$	10 to 100	30″	$\frac{1}{5,000}$	10 to 100	0.05 ft $\times \sqrt{\text{miles}}$
	Quaternary	½ to 2	1′ or graphical	2 to 10	...	...	[illegible] to 2	2′ or compass	$\frac{1}{1,000}$	1 to 10	0.1 to 0.5 ft $\times \sqrt{\text{miles}}$
Intermediate	Primary	1 to 5	10″ to 20″	5 to 50	$\frac{1}{10,000}$ to $\frac{1}{40,000}$	$\frac{1}{1,000}$ to $\frac{1}{4,000}$	1 to 20	10″ to 1′	$\frac{1}{1,000}$ to $\frac{1}{5,000}$	1 to 25	0.05 to 0.3 ft $\times \sqrt{\text{miles}}$
	Secondary	½ to 2	Graphical	1 to 5	...	...	1 to 5	30″ to 3′	$\frac{1}{500}$ to $\frac{1}{2,500}$	1 to 5	0.1 to 0.5 ft $\times \sqrt{\text{miles}}$
Large	Primary	1 to 5	2″ to 10″	2 to 20	$\frac{1}{20,000}$ to $\frac{1}{80,000}$	$\frac{1}{2,000}$ to $\frac{1}{5,000}$	1 to 5	30″ to 1′	$\frac{1}{5,000}$ to $\frac{1}{20,000}$	1 to 10	0.05 to 0.1 ft $\times \sqrt{\text{miles}}$
	Secondary	¼ to 1	5″ to 20″	1 to 5	...	...	½ to 3	30″ to 2′	$\frac{1}{1,000}$ to $\frac{1}{5,000}$	1 to 3	0.05 to 0.1 ft $\times \sqrt{\text{miles}}$

1 Traversing. The *primary traverse* for an intermediate-scale map is usually run with transit and tape. Convenient routes are chosen which will result in the advantageous location of stations. Because of the cumulative effects of the errors in transit-tape traversing, in primary traversing it is desirable as a check to arrange closed circuits of length not to exceed perhaps 10 miles, dividing the tract into roughly equal areas. If closed circuits cannot be secured conveniently, checks for distance should be applied as the work proceeds, and checks for azimuth by astronomical observations (Chap. 17) should be applied at intervals of perhaps 10 miles. In applying the azimuth checks, it is of course necessary to take the convergency of meridians (Art. 18-24) into account.

Wherever *secondary traverses* are required to establish the instrument stations from which the details are located, an area within a closed primary traverse may be divided by means of the secondary traverses into a series of roughly parallel strips. Secondary traverses are usually run with the transit but are sometimes run with the plane table and occasionally with the surveyor's compass. The lengths of the traverse lines are determined commonly by stadia or, if greater precision is required, by means of the tape.

2 Triangulation. The field conditions favorable to the use of triangulation (Chap. 12 and Art. 13-10) to establish the horizontal control are (1) a fairly extended area in an open hilly region, (2) a city where traversing is difficult because of street traffic, or (3) a rugged mountainous region where traversing would be slow and laborious.

A general layout of the scheme of *primary* triangulation is planned on an existing small-scale map; the field stations are established on summits where visibility is good; and signals are erected. One or more base lines are established and measured, and their true azimuths are determined by astronomical observations (Chap. 17). Observations of angles are made on (1) major stations, which are marked by signals and which are to be occupied by the instrument, and (2) minor stations marked by such objects as trees, spires, and chimneys. When the field measurements have been completed, the necessary computations and adjustments are made (Chap. 12); then the coordinates of each station are determined for use in plotting.

Although *secondary* control is commonly established by traversing, triangulation is employed where instrument stations can be advantageously located by this method, particularly in open rough country where taping would be difficult. The secondary triangulation may be established with either the transit or the plane table. For small-scale maps, usually it is possible to obtain the required precision by the method of graphical triangulation employing the plane table (Art. 13-10); this method has the advantage that no computations are required.

20-5 Vertical Control. Bench marks are established at convenient intervals over the area, to serve as points of departure and closure for the leveling operations of the topographic parties when locating details.

Primary and secondary level routes are required in about the same amount, and bear about the same relation to each other, as do the primary and secondary traverses or triangulation systems. Often the level routes follow the traverse lines, the traverse stations being used as bench marks. In Table 20-2 are given approximate values of permissible error on ordinary surveys.

The methods of leveling are described in Chaps. 4 and 5. Vertical control is usually accomplished by direct leveling, but for small areas or in rough country frequently the vertical control is established by trigonometric leveling.

LOCATION OF DETAILS

20-6 General. The instruments used in locating details, and the four typical systems of ground points used in map construction, are discussed briefly in Art. 19-7 and more thoroughly in Arts. 20-8 to 20-11. A combination of methods may be used; the aim is to locate the details with a minimum of time and effort. Aerial photographs, or even ground views with the ordinary camera, may be of value in locating and plotting details.

20-7 Precision. The precision required in locating such definite objects as buildings, bridges, and boundary lines should be consistent with the precision of plotting, which may be assumed to be a map distance of about $\frac{1}{50}$ in. Such less definite objects as shore lines, streams, and edges of woods are located with a precision corresponding to a map distance of perhaps $\frac{1}{30}$ or $\frac{1}{20}$ in. For use in maps of the same relative precision and for a given area, more located points are required on large-scale surveys than on intermediate-scale surveys; hence the location of details is relatively more important on large-scale surveys.

The veracity with which contour lines represent the terrain depends upon (1) the accuracy and precision of the observations, (2) the number of observations, and (3) the distribution of the points located. Ground points are definite, but as the contour lines must necessarily be generalized to some extent it would be inappropriate to locate the ground points with refined measurements. The error of field measurement in plan should be consistent with the error in elevation, which in general should not exceed one fifth of a contour interval; thus generally the error in plan should not exceed one fifth of the horizontal distance between contours. The purpose of a topographic survey will be better served by locating a greater number of points with less precision, within reasonable limits, than by locating fewer points with greater precision.

The precision needed in the field measurements of angles to details should be consistent with the required precision of corresponding distances.

20-8 Details by Controlling-point Method. Details may be located by the controlling-point method employing the transit and stadia, the plane table, or the transit and plane table together. The distances are usually measured by stadia, but on large-scale surveys distances to definite details may be measured with the tape (see also Para. 3 of Art. 19-7).

Transit and Stadia. In locating ground points, usually the vertical angles are observed more precisely than the horizontal angles.

The transit is set up and oriented at a control station. Details in the vicinity of the station are located by angle and distance measurements (Art. 11-9). Notes may be kept in the form shown in Fig. 11-5, supplemented by sketches if necessary to make the location of all points clear to the map draftsman.

The rodmen choose ground points along valley and ridge lines and at summits, depressions, important changes in slope, and definite details. The selection of points is important, and the rodmen should be instructed and trained for their work. They should follow a systematic arrangement of routes such that the entire area is covered. They should observe the terrain carefully and report important features which cannot be seen from the transit station.

Plane Table. Before the plane table is taken into the field, the horizontal control is adjusted and plotted on the plane-table sheet. The elevations of all bench marks either are recorded on the sheet or are in the hands of the computer.

The instrumentman sets up and orients the plane table at a control station. He then directs the rodmen to the controlling points of the terrain, as just described for the transit. When a rodman holds the rod on a ground point, the instrumentman sights on the rod, reads the stadia intercept, draws a short portion of the ray near the end of the alidade farthest from the station point, sets the cross hair preferably on the H.I. point, and motions the rodman forward. He next centers the control bubble on the vernier arm and reads the vertical angle. He then plots the point by scaling the horizontal distance (corrected for slope, if necessary, by the computer). The computer has now calculated the elevation of the point, and the instrumentman records it on the map near the plotted point. As rapidly as sufficient data are secured, the instrumentman sketches the contour lines. Other objects of the terrain are located and are drawn either in finished form or with sufficient detail so that they may be completed in the office. A more detailed account of the procedure of taking side shots, with an alidade equipped with stadia arc, is given in Art. 13-16.

Transit and Plane Table. For large-scale maps and where many details are to be sighted, sometimes it is advantageous to use both the transit and the plane table. This method saves time in the field, but it may not reduce the total cost, as a larger party is required than for the plane table alone.

The transit is set up and oriented at the control station, the location of which is plotted on the plane-table sheet. The plane table is set up and oriented nearby, and its location is plotted on the map in its correct relation to the transit station. When a rodman has selected a ground point, the transitman observes the stadia distance and vertical angle to it; the plane-table man sights in the direction of the point, draws a ray toward it from the plotted location of the plane-table station, plots the point at the correct distance scaled from the plotted location of the transit station, and records on the map the elevation (computed by the transitman or the computer) of the plotted point.

20-9 Details by Cross-profile Method. For the cross-profile method of locating details, the party consists of a topographer and usually two men, herein called "chainmen," who act either as chainmen or as rodmen. Sometimes only one chainman is employed, and the topographer assists in chaining. The equipment consists of a leveling rod (usually a topographer's rod, Art. 4-17), a steel or a metallic tape, a hand level or a clinometer, and usually a cross-ruled wide-page sketchbook. Sometimes a Jacob's staff or other rod about 5 ft long is used as a support for the hand level or clinometer while sights are being taken (see also Para. 4 of Art. 19-7).

The control points are the 100-ft stations of the transit traverse; these points have been marked on the ground by stakes, and their elevations have been determined by profile leveling and have been furnished to the topography party.

The ground points are on relatively short crosslines transverse to the traverse line. They are either contour points or more commonly points of change in slope; in the latter case the intermediate contour points are located by interpolation.

The party proceeds from station to station along the traverse. At each station the topographer notifies the chainmen of the elevation of the station. The head chainman carrying the rod moves out on a line estimated to be at right angles with the traverse line until the rod is on the next contour (either higher or lower) from the station, as determined by the rear chainman employing the hand level; the distance out to the contour is then measured with the tape. The rear chainman then goes out to the point occupied by the rod, and the head chainman again moves out until the next contour is reached; and so the process is repeated until all contour points are located out to the edge of the strip being surveyed. A similar procedure is followed on the other side of the traverse line. Usually the trends or directions of the contours are sketched at each crossline and along ridge and valley lines, but on the field sheets the contour lines are not sketched for their full length. Definite details are located with relation to the transit line by tape measurements. If the topography is regular, sometimes the sketches are omitted, and the distances from traverse to contour points are recorded numerically.

example. Suppose that the contour interval is 5 ft, that the topography party has reached a station the elevation of which is 821.1 ft, and that on the left side of the traverse at this station the ground slopes downward. To locate the 820-ft contour, the head chainman carrying the rod (which is graduated from the bottom) and the zero end of the tape moves out to the left until the rear chainman by the use of the hand level supported, say on a 5.0-ft staff, reads 6.1 on the rod ($821.1 + 5.0 - 6.1 = 820.0$). The horizontal distance out from the station is read on the tape and called to the topographer, who plots its location. In order to locate the 815-ft contour, the rear chainman moves out to the 820-ft contour, and the head chainman moves out until the rear chainman reads 10.0 on the rod ($820.0 + 5.0 - 10.0 = 815.0$); the horizontal distance between the two contours is read from the tape, and the second point is plotted; and so on.

For relatively small-scale maps, sometimes the clinometer is employed to determine the elevations of controlling points of change in slope on the crosslines by rough indirect leveling, and the distances are measured by pacing. The method is considerably faster than that just described for the hand level and tape, but is of lower precision.

The maximum lengths of hand-level or clinometer sights should be limited under ordinary conditions to about 100 ft, for small-scale maps or large contour intervals to perhaps 400 ft, and for contour intervals of 2 ft or less to about 50 ft.

For large-scale surveys or in flat country where the contour interval is small, usually elevations are determined either by direct leveling with the engineer's level or by indirect stadia leveling with the transit or the plane table.

The cross-profile method is primarily suitable for route surveys. It is also sometimes used for area surveys if the ground cover is dense, because hand-level or clinometer sights can be taken through very small openings in the underbrush; the area is surveyed by means of a series of overlapping strips.

20-10 Details by Checkerboard Method.

The checkerboard method of locating details is well adapted for large-scale surveys, also where the tract is wooded, where the topography is smooth, and on urban surveys. The tract is staked off into squares or rectangles—usually 50- or 100-ft squares. The ground points and other details are then located with reference to the stakes and connecting lines.

The usual procedure is first to run a rectangular transit-tape traverse near the perimeter of the tract, with stakes set at each 100-ft station. The error of closure becomes apparent in the field; if this is greater than the permissible error, the stakes are reset. The interior of the figure bounded by this rectangular traverse is then filled in with stakes set at the corners of the 100-ft squares; each stake is marked usually with a letter and a number indicating its position with respect to a pair of coordinate axes, as illustrated in Fig. 20-1. The ground elevation at each stake is determined by direct leveling. Sketch sheets are prepared, on which are shown the elevations of the corners of the squares. The location of ground irregularities or other details inside the squares is determined by measurements either from adjacent coordinate points or from the sides of the squares. The elevation of such details is determined by use of the hand level. The map is constructed in the office.

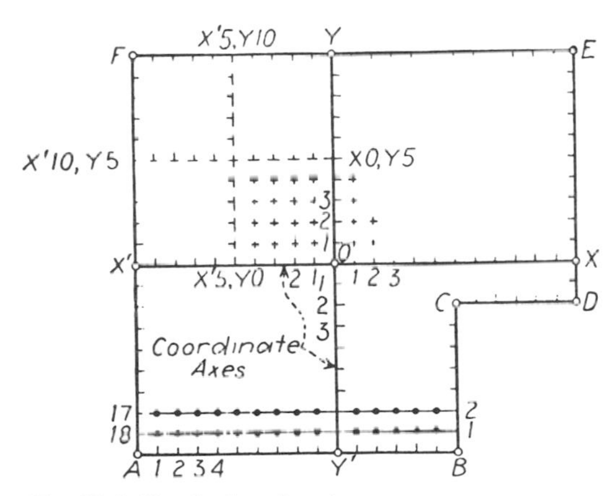

Fig. 20-1. Checkerboard system.

If many irregular features are to be mapped, the plane table may be used advantageously. Before the plane table is taken into the field, the corners of the squares are established on the ground with transit and tape, their elevations are determined by direct leveling, and the plane-table sheet is prepared showing the elevations of the corners of the squares, all as described earlier in this article. The plane table is then set up over the corner of a square and is oriented by backsighting along one of the control lines marked by stakes. Directions to details inside the squares are determined usually with a peepsight alidade, and distances to these details are determined usually by taping either from the instrument station or from a convenient corner or line. Only as many stations are occupied by the plane table as are necessary to cover the area.

20-11 Details by Trace-contour Method. The trace-contour method of locating contour points on the ground is commonly used on large-scale surveys and sometimes on intermediate-scale surveys where the ground is irregular. Under these conditions, if visibility is good, the trace-contour method is more rapid and more accurate than the checkerboard method.

Although the transit may be used in this work, either alone or with the engineer's level, the plane table is commonly used. Often the plane table and the engineer's level are used together. In this case the levelman sets up the level at a convenient location and directs the rodman up or down the slope until a point on a given contour is located. This point is immediately sighted by the plane-table man and is plotted on the plane-table sheet. The rodman then moves to another contour point, usually on the same contour. The distances from plane-table station to contour points are measured by stadia; if the scale is large, definite objects may be located by taped distances.

SPECIAL SURVEYS

20-12 Mine Surveying. Mine surveying includes *underground surveying* as practiced in mining and tunnel operations and *mineral-land surveying* which involves location and patent surveys.

1 Underground Surveying. Underground surveying differs from surface work in the following ways: the transit station is usually in the roof instead of the floor; the object to be sighted and the cross hairs of the telescope must be illuminated; distances are usually measured on the slope instead of along horizontal lines; and the transit tripod has adjustable legs to adapt its use to low workings or to very irregular or steeply inclined surfaces.

When the station is in the roof, the transit may be centered either by first plumbing from the station mark to a point on the floor or by centering the transit beneath a plumb bob suspended from the roof station.

Illumination of the cross hairs may be accomplished by slipping a rolled piece of paper into the sunshade, then holding a light in front of and a little to one side of the objective. The signal or target, usually a plumb bob hung from the roof station, is illuminated either by directing a light from one side or by placing a piece of thin paper behind the plumb bob and illuminating the paper from beyond.

Desirable features of the mining transit for underground work are an extension-leg tripod and/or a trivet, a striding level for the horizontal axis, a vertical circle graduated on the edge instead of the side, water-excluding construction (so far as possible), a "shifting center" to permit moving the transit head laterally for short distances, a center point marked on top of the telescope, and a prismatic eyepiece. For taking steeply inclined sights, an auxiliary telescope may be attached either at one end of the horizontal axis or above the main telescope, with the line of sight of the auxiliary telescope parallel to that of the main telescope. When an auxiliary telescope on the side is used to measure a horizontal angle, or an auxiliary telescope on top is used to measure a vertical angle, the correct angle (from the main telescope) is computed from the observed value, or *reduced to center*.

In order to connect an underground survey with a surface survey, no special methods are necessary unless the angle of inclination is greater than about 60°. A point of known horizontal location at the surface can be projected down a vertical shaft by plumbing either with a plumb bob or by means of an optical plummet. In the case of a single vertical shaft, underground directions are referred to a vertical plane defined by two plumb lines in the shaft, the azimuth of the plane being established by the surface survey. If there are two vertical shafts, a plumb line is established in each, and surface and underground traverses are run from one plumb point to the other. Differences in elevation down a vertical shaft are best measured by means of the steel tape.

2 *Mineral-land Surveying.* Ordinarily the ownership of land implies ownership within vertical planes through the boundaries. To encourage the development of mining, however, the United States Government has passed laws modifying in certain cases the usual rule of vertical planes and specifying the manner in which the person discovering a mineral vein or lode on government land may acquire title. The laws are based upon the concept of a relatively thin vein, or lode, limited between surfaces which are essentially plane. It provides that a mining claim of maximum length 1,500 ft and maximum width 600 ft may be located on the surface along the outcrop of the vein and that after the serious intent of the claimant has been proved the government will grant a patent carrying a clear title to the land with ownership of the vein anywhere between vertical planes through the end lines. The lode claim need not be rectangular, but the end lines must be parallel straight lines. The length of the claim must be measured along the center line. The outcrop must cross the end lines but not the side lines.

The location survey may be made by the claimant or by someone employed by him. The final survey for patent must be made by a mineral surveyor commissioned by the United States to do that work.

20-13 Hydrographic Surveying. Hydrographic surveys are those which are made in relation to any considerable body of water such as a bay, harbor, lake, or river. These surveys are made for the purposes of (1) determination of channel depths for navigation; (2) determination of quantities of subaqueous excavation; (3) location of rocks, sand bars, lights, and buoys for navigation purposes; and (4) measurement of areas subject to scour or silting. In the case of rivers, surveys are made for flood control, power development, navigation, water supply, and water storage.

Since a certain amount of shore location is included in most hydrographic surveys, a single control survey is located on shore to serve both for soundings and for shore details.

As in topographic surveying, the horizontal control is a series of connected lines whose directions and lengths have been determined. The control may be established either by traversing or by triangulation of appropriate precision. Long narrow inlets or rivers are usually surveyed from a single traverse on one shore. If the shore line is obscured by woods, a system of triangulation is used. The control for large lakes and ocean shore lines consists of a chain or network of triangles onshore.

A chain of bench marks is established to serve as a vertical control. These bench marks are near the shore line and are located at frequent intervals so that water-stage gages may be set conveniently.

Most hydrographic surveys require the location of all irregularities in shore line, all prominent features of topography and culture, and all lighthouses, buoys, etc., in order that these points may be used for references in sounding work.

The determination of the relief of the bottom of a body of water is made by soundings. The depth of the sounding is referred to water level at the time it is made, and later is corrected to the datum water level through the use of gage readings. Before the corrected soundings can be plotted on the map, their location with reference to the shore traverse is determined by one of the following methods:

1 By taking soundings on a known range line and reading one angle either from a boat or from a fixed point on shore.

2 By propelling a boat at a uniform rate along a known range line and taking soundings at equal intervals of time.

3 By taking soundings from a boat at the intersections of known range lines.

4 By reading two angles simultaneously from two fixed points on shore, that is, by intersection.

5 By taking readings with the transit and stadia.

6 By taking soundings at known distances along a wire stretched between stations.

7 By reading two angles from a boat to three fixed points on shore, by means of the sextant.

8 By electronic means.

The transit and other instruments used in land surveys are not adapted for use in a boat, where the support is unstable. The sextant (Fig. 20-2) is well suited to hydrographic work and has the added advantage of measuring angles in any plane. It is called a "sextant" because its limb includes but one sixth of a circle. Although the arc is limited to 60°, the instrument will measure angles to 120°. It is used principally by navigators and surveyors for measuring angles from a boat, but it is also employed on exploratory, reconnaissance, and preliminary surveys on land. The handle of the sextant is held in the right hand, and the plane of the arc is made to coincide with the plane of the two objects between which the angle is to be measured. The sextant is turned in the plane of the objects until the left-hand object can be viewed through the telescope and the clear portion of the horizon glass. With the instrument held in this position, the index arm is moved with the left hand until the images

Fig. 20-2. Sextant.

of the two objects coincide. The final setting is made with the tangent screw, and a test for coincidence is made by twisting the sextant slightly in the hand to make the reflected image move back and forth across the position of coincidence. When the setting is thus verified, the vernier arm is clamped and the angle is read.

Up to depths of about 12 ft and with low current velocities, sounding rods can be used to advantage. For greater depths, weights called "sounding leads" are suspended on graduated sounding lines. Continuous-recording echo sounders are widely used.

The methods used in plotting the soundings are the inverse of the field methods used in locating the soundings. They are closely similar to the methods of plotting details in topographic surveying. Various special types of protractors or charts may be employed.

A hydrographic map is similar to the ordinary topographic map but has its own particular symbols. These may be found in almost any book on topographic drawing or in the manual issued by the U.S. Coast and Geodetic Survey (Ref. 22; see also Art. 14-8).

20-14 Field Problems

Field problems 32 (Chap. 11) and 35 (Chap. 13) provide exercises in the controlling-point method of locating details, and field problem 50 (Chap. 21) provides an exercise in the cross-profile method. Elementary field problems in topographic surveying are given at the end of Chaps. 11 and 13.

Problem 48. Topographic Survey by Checkerboard Method

Object. To obtain sufficient data for an accurate topographic map of large scale and small contour interval. The area to be mapped is small and possesses few details and the topography is smooth. The data may be used in office problem 45 of Chap. 19.

Procedure. (1) With transit and tape divide the tract into 100-ft squares, setting stakes at the corners. Letter and number each stake to conform to a coordinate system. (2) With the engineer's level run levels over the area, taking rod readings at summits, depressions, corner stakes, and points of change in slope along the sides of the squares. (3) Locate the details (either definite details or ground points) inside the squares by taking offsets, ties, etc. (Arts. 10-13 and 10-14). (4) Keep notes in a form similar to that of Fig. 6-3; identify each point by its coordinates (a letter and a number).

Problem 49. Topographic Survey by Trace-contour Method Using Plane Table and Engineer's Level

Object. To obtain sufficient data for an accurate topographic map of intermediate or large scale and small contour interval. The area to be mapped is small and the topography is irregular. It is assumed that control points have been established at advantageous locations within the tract and that the control has been plotted (office problem 45, Chap. 19) on the plane-table sheet.

Procedure. (1) The plane-table man sets up and orients the table at one or more control stations such that the entire area may be mapped. (2) The levelman sets up the engineer's level, takes a backsight on a bench mark, and computes a rod reading such that the foot of the rod will be at the elevation of a given contour. (3) The rodman moves about as directed by the levelman, locating critical contour points. When a contour point has been located, the plane-table man sights on the rod, determines the distance by stadia, and plots the contour point. (4) Definite details are located either by radiation or by intersection; distances to such details are determined either by stadia or by tape measurements, depending upon the scale of the map.

21
ROUTE SURVEYING

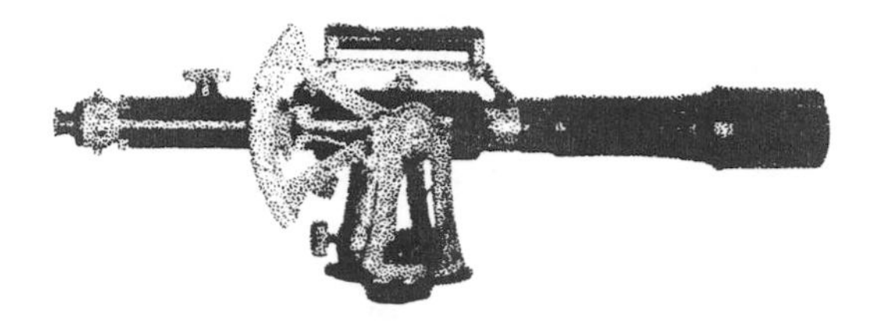

GENERAL

21-1 General. Route surveys are made for the purpose of locating and building cross-country utilities such as highways, railways, canals, transmission lines, and pipelines. Route surveying consists in determining the ground configuration and the location of objects along the proposed route, establishing the line on the ground, and computing volumes of earthwork.

The location of any route involves a study to determine the manner in which certain definite requirements of the enterprise may be met at the least expense, including the cost of construction and the cost of maintenance and operation, and taking into consideration the benefits to traffic and social benefits to the community. It is therefore a problem in economics. Related considerations are right of way, geology, soils and foundations, and drainage.

The details of the surveying methods employed naturally vary with the character of the project, but certain general field methods are widely applicable. The methods described in this brief chapter apply primarily to ground surveys for railways and for secondary highways, excluding the specialized methods used for major highways and freeways. Some special considerations pertaining to various structures are stated in Arts. 21-8 to 21-13. Construction surveys are discussed in Chap. 22.

Aerial mapping (Chap. 23) and electronic computers have made possible certain alternatives to the procedures described in this chapter for ground surveys. In any case, however, the nature and sequence of the various operations are essentially the same as those described herein.

In planning a location survey, usually the points to be connected and certain points to be passed or avoided are first defined in general terms. Also, certain governing standards are set with regard to such items as maximum grade, minimum radius of curvature, width of roadway and right of way, and accuracy of the survey. Aerial photographs, which for many areas are available from government agencies, are of great aid in planning even if the survey is not to be made by photogrammetric methods.

The general procedure for the location of a new route is as follows: First a general study called a *reconnaissance* is made of the whole area under consideration, and one or more general routes are selected for further investigation. A *preliminary survey* is then run over each selected route, and a topographic map is prepared. A center line of the proposed route is tentatively established on the topographic map, and the ground profile is plotted. The final line is then located and staked on the ground by transit and tape, the tentative location being modified as necessary along the way; this is called the *field location* or *final location.* Subsequently *construction surveys* may be necessary to establish lines and grades for special structures or to measure the amount of earthwork. Land boundaries near the line are surveyed and monumented, and right-of-way maps are drawn. Special surveys and plans are required for such structures as overcrossings, bridges, and culverts. The general sequence of operations is somewhat as listed in Table 21-1.

21-2 Reconnaissance. Usually a reconnaissance involves field exploration of the area under consideration, supplemented by a study of available maps and photographs. It may be desirable to map the area roughly to small scale. Particular attention is given to streams and divides, and decision is made about certain controlling points such as communities, passes, and bridge sites. With these controlling points as a guide, one or more general routes are selected for further investigation. Important factors affecting the choice of highway location are traffic service, land use, and terrain.

The information obtained by reconnaissance should include the general rise and fall of the country, possible ruling and maximum grades, general slope of the sidehills, classification of material, drainage, snow conditions, character of clearing, development of country, service to existing communities, etc. The reconnaissance report should include the alternative routes with the advantages and disadvantages of each, the more important controls, comparative cost estimates, an economic analysis, and recommendations.

21-3 Preliminary Survey. A narrow strip of country along each proposed general route is surveyed and mapped, the strip being of sufficient width to contain the final location. The preliminary survey may be run by use of (1) the transit, tape, and level; (2) the transit and stadia; or (3) the plane table.

1 Transit-tape-level Method. The survey corps usually consists of a transit party, a level party, and a topography party. A field draftsman may also be included. The chief of the transit party, following the instructions

Table 21-1 Typical sequence of operations in route surveying

Survey	*Party*	*Operation*	*Maps and reports*
Reconnaissance	Locating engineer	Select general routes; establish controls	Reconnaissance report
Preliminary survey [1]	Transit-tape	Traverse	Preliminary map (contours)
	Level	Profile; set bench marks	Preliminary profile
	Topography	Cross-section to locate contours and principal details	Tentative location (drawn on preliminary map) Preliminary cost estimate (optional)
Location survey	Transit-tape	Stake final location, with circular curves	Location map
	Level	Profile; check bench marks	Location profile
	Cross-section	Cross-section; if line is fixed, also set slope stakes	Cross-sections Earthwork estimates
	Land-line	Property lines and details	Right-of-way map
	Special	Special surveys for structures	Structure maps and plans
Construction surveys	Various	Set slope stakes; set finishing stakes; stake borrow pits; stake spirals; give line and grade for track or pavement and for culverts and structures; set monuments; estimate quantities of earthwork moved	
			Final plans (include location map and profile as revised during construction, cross-sections, etc.)

[1] Alternative method is to traverse, profile, and take topography all in one operation. In flat country or where the route is fixed, the preliminary survey may be omitted.

and markers of the locating engineer, runs an open traverse at random approximately along the middle of the strip through which it appears that the final line will lie. Stakes are set at all full stations, and hubs are set at all transit stations. No curves are run in at this time.

The level party (levelman and rodman) follows the transit party, takes profile levels along the traverse, and establishes bench marks at intervals of a mile or less. Ground elevations are determined on the traverse line at all stakes set by the transit party, at changes in slope, and at crossings of roads and streams. Where there is a possibility that the maximum grade may be exceeded, the preliminary profile (Art. 21-4) is brought up to date each day.

The topography party (topographer, rodman, and frequently a tapeman) follows the level party and takes preliminary cross-sections, as described in Art. 20-9. Normally the cross-sections are taken at each 100-ft station, but in very irregular country they may be as close together as 25 ft and in very smooth country they may be as far apart as 500 ft. The topography party also determines the location of pertinent planimetric details and notes the character of cultivation, quality of the land, and probable character of excavation.

2 *Transit-stadia Method.* Where the country is fairly open, often the transit-stadia method of running the preliminary survey is satisfactory. The usual procedure is to run the traverse, measuring the vertical angles and stadia distances, and to take side shots at the same time, as described in Arts. 11-9 and 20-8; thus the horizontal control and vertical control are established and the details are observed in one operation. Usually the party consists of a chief of party who may also act as recorder, a transitman, two or more rodmen, and sometimes a recorder. Hubs are set at the transit stations, but no intermediate stakes are set. On extensive surveys, vertical control may be established by direct leveling, and distances between transit stations may be measured with the tape.

3 *Plane-table Method.* The field procedure employing the plane table is much the same as that just described for the transit-stadia method, except that the map of the strip of country is constructed in the field as the work progresses. The use of the plane table for such work is described in Arts. 13-8, 13-9, and 20-8. The use of the plane table is advantageous where the topography is irregular and the country is open. Often the plane table is employed for locating details only, the control being established by a transit-tape traverse and direct leveling.

21-4 Preliminary Profile and Map. A profile of the ground along the traverse line is prepared as described in Art. 7-1. For highway surveys, a common horizontal scale is 1 in. = 100 ft and a vertical scale is 1 in. = 10 ft.

From the notes of the preliminary survey there is also prepared a *preliminary map* showing the topography and details along the selected strip of country. Usually the contour interval is 5 ft, but in level country it may be 2 ft or even 1 ft, and in rough country it may be 10 ft or greater.

Both the map and the profile are employed by the locating engineer as a guide during the progress of the preliminary survey, and hence each day's work is preferably plotted before the next day's work is begun.

21-5 Location Survey. Based upon a study of the preliminary map and profile and upon further detailed study of the ground surface, the tentative alinement of a route (including curves) is chosen and this trial projection, or *paper location,* is drawn on the map. Usually a profile of the paper location is drawn by use of elevations taken from the contour lines, a grade line is fixed on the profile (Art. 7-2), and the cost of construction is roughly estimated. The paper location is then used as a guide in locating the line in the field.

In fixing the location and grade of a roadway, some of the primary considerations are (1) to keep changes in alinement at a minimum; (2) to keep grades at a minimum; (3) to make the sum of the volumes in cut and borrow as small as possible consistent with suitable alinement and grades, by making the volume of earthwork in fills as nearly as practicable equal to that in adjacent cuts; (4) to keep at a minimum the amount of *haul* (Art. 21-7) that will be necessary to transport excavated material from the cuts or borrow pits to the adjacent fills; and (5) to provide for drainage.

For an elementary example of the use of a contour map in the location of a route for a highway, see Art. 19-17.

The field work of final location starts with staking the projected center line on the ground so that, as closely as practicable, it bears the same relation to the preliminary traverse on the ground that the paper location bears to the preliminary traverse on the map. This relation may be determined either by intersections between paper location and preliminary traverse or by scaling from the map the offsets from stations on the preliminary line to the tangents of the paper location. As the field location is run (see following paragraph), the center-line tangents of the field location are established on the ground either by intersections with the preliminary traverse or by taping the scaled offsets from various stations on the preliminary traverse. Where practicable, adjoining tangents are run to an intersection, the intersection angles are measured, and the curve notes are computed as explained in Art. 10-18. Usually the degree of curve for a given curve is the same as that assumed in the paper location.

Beginning at some point where the stationing has been established, the field location is extended in the manner described in the preceding paragraph. Usually the located line is the center line of the roadway, but for highway surveys sometimes the line is located offset from one edge of the roadway. Stakes are set on tangents and circular curves at all full stations and in some cases at 50-ft or even 25-ft stations; and hubs are set and referenced at all P.I.'s, P.C.'s, P.T.'s, and intermediate transit stations. Transit notes are kept on the page as illustrated by Fig. 10-16. All important features such as roads, streams, and property lines are sketched on the right-hand page in their proper relation to the located line, the center line of the page being considered as the located line.

Profile levels are then run over the located line in the same manner as for the preliminary line, and from the data thus obtained a location profile is prepared showing the ground line and proposed grade line. For railway location usually an alinement diagram (Fig. 7-1) is drawn on the location profile, whereas for highway location usually the plan (without contours) and profile are drawn on the same sheet of standard form called a "Federal Aid plan-profile sheet" (Fig. 21-1).

In the light of a study of this profile and of the preliminary map, the grades are adjusted so that the line will better fit the ground. If minor modifications in alinement appear desirable, parts of the location are revised in the field. The final location of the line as located on the ground is plotted both in plan and in profile. On the final-location map are shown all features of importance, including the location of bench marks and of the objects to which hubs are referenced.

Cross-sections of the located line are plotted (Art. 7-3) in order to estimate earthwork quantities and for purposes of letting the construction contract. For approximate estimates, the cross-sections may be plotted from the data of the preliminary contour map; but usually the final cross-sections are taken while the slope stakes are being set (Art. 6-9), for which case the form of notes is shown in Fig. 6-6. The cross-sections are plotted on cross-ruled paper, which can be obtained either in rolls 20 in. wide or in standard sheets called "Federal Aid cross-section sheets" to match the Federal Aid plan-profile sheets. The usual scale is 1 in. = 10 ft.

When the line is located definitely, a survey is run to determine and monument the boundaries of property which will be needed for the project, in order to secure rights of way. The results of the survey are plotted on a *property-line map,* or *right-of-way map,* in the usual form for a land map (Chap. 18); and legal descriptions of the property are prepared.

21-6 Construction Surveys. The construction surveys for a roadway consist essentially in (1) staking out earthwork and structures preparatory to, and during the process of, grading and construction; and (2) making the measurements necessary to determine the volume of work actually performed up to a given date, as a basis for payment to the contractor. With regard to the final cross-sectioning, setting of slope

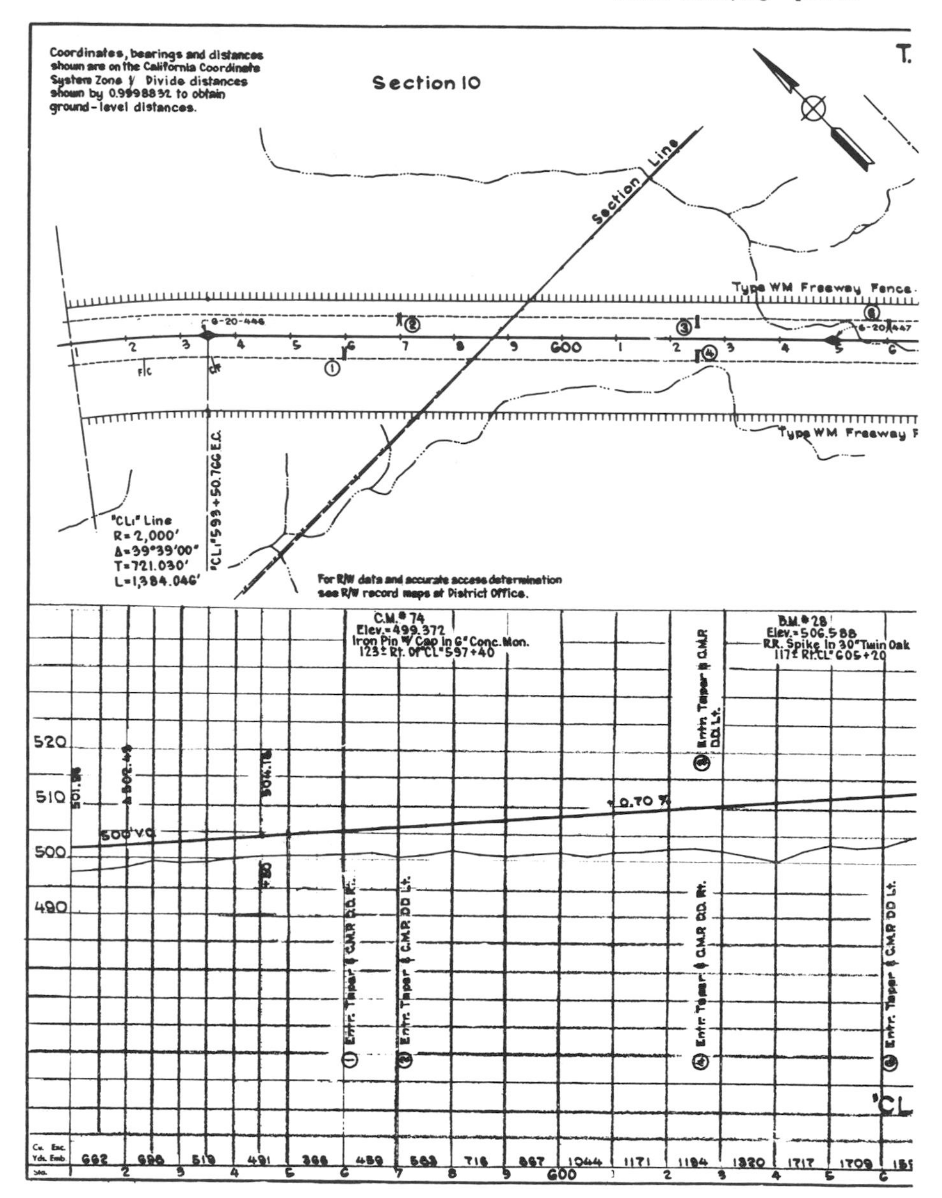

Fig. 21-1. **Typical location plan and profile.** (*California Division of Highways.*)

stakes, and staking out of curves, the dividing line between the location survey and the construction survey is not definite; the practice varies according to the organization.

Some general considerations relating to construction surveys are given in Chap. 22, and special features pertaining to route structures in Arts. 22-6 to 22-14.

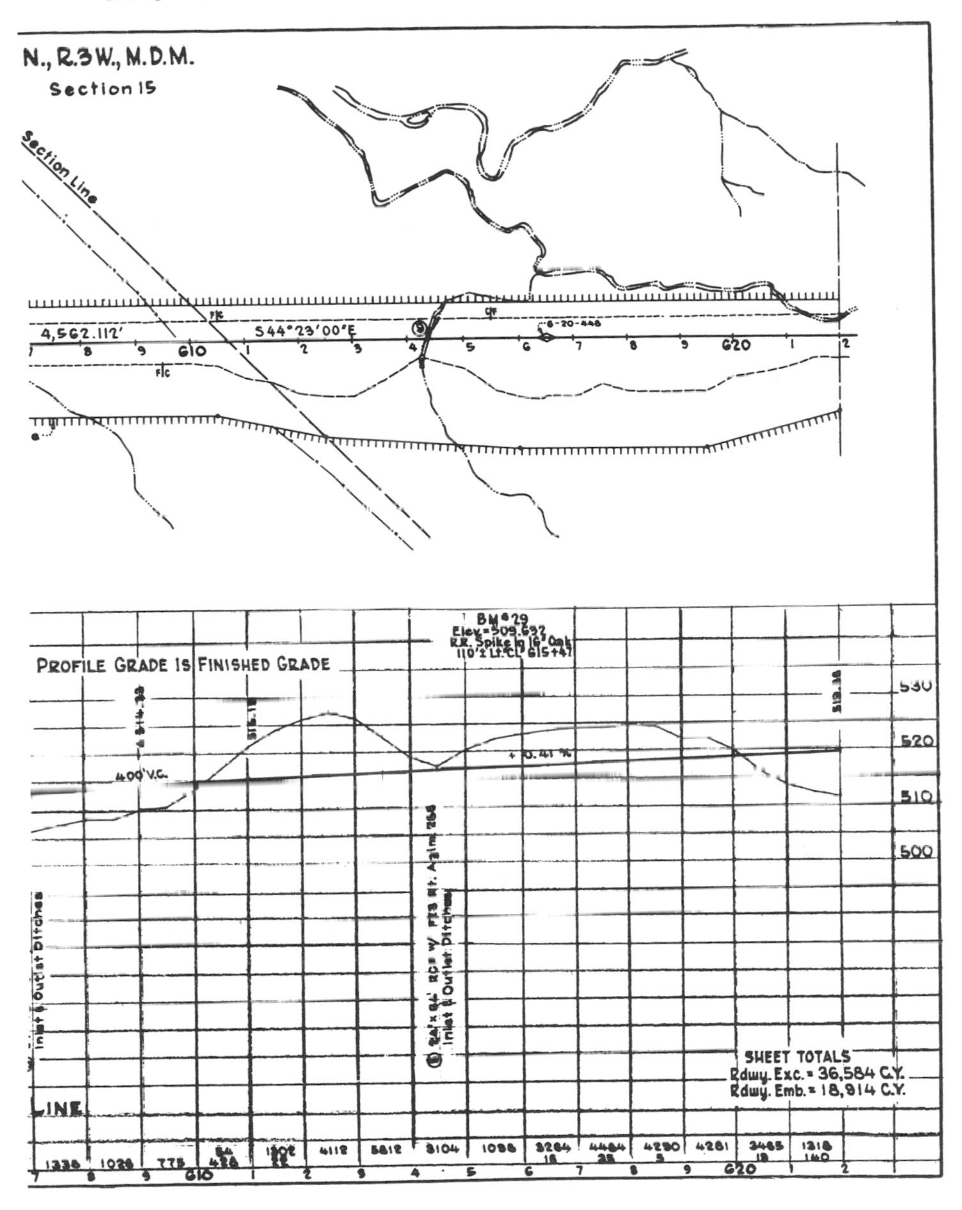

21-7 Haul. One primary consideration in fixing the location and grade of a roadway is the amount of *haul* that will be necessary to transport excavated material from the cuts or borrow pits to the adjacent fills or to waste. The construction contract usually names a price per cubic yard to be paid for excavation either unclassified or for each class of material (earth, loose rock, solid rock, etc.) and for transporting this material for any distance up to a limit of *free haul*. Transportation of material beyond this distance is termed *overhaul* and is paid for at a rate fixed by the contract. The unit of measurement for overhaul is usually the *station yard,* 1 station yard being 1 cu yd of material transported 100 ft.

The limits of free haul are determined by fixing (on the profile) one point in cut and one point in the adjacent fill, at the specified free-haul distance apart, such that the included quantities of excavation and embankment balance. The overhaul distance is computed as the distance between the center of gravity of the remaining mass of excavation and the center of gravity of the resulting fill, less the limit of free haul.

In order to determine in advance the proper distribution of excavated material and the amount of waste and borrow, and as a basis for estimate of cost, a *mass diagram* is usually prepared. The abscissas in the mass diagram are the distances along the survey line, and the ordinates are the algebraic sums of earthwork quantities from the beginning to each ordinate, considering cut volumes positive and fill volumes negative, and making due allowance for shrinkage, swell, and subsidence. The use of the mass diagram is discussed in detail in texts on railway location and earthwork.

LOCATION OF VARIOUS ROUTES

21-8 Highways. Many highway surveys are made along established roads as a basis for improvement; only portions are relocated as, for example, at sharp curves. As the general route is fixed beforehand, little or no reconnaissance is necessary. No preliminary survey may be required, and the location survey may be run at once, subject to small changes and adjustments after further study.

In new location for a highway, the complete procedure previously described is applicable. Usually the measurements for the preliminary survey for a highway are taken more completely and more precisely than those for a railway.

In highway practice, curvature is expressed on the arc basis (Art. 10-15). Main roads are designed with grades flat enough to be climbed by automobiles without shifting gears. Should such a rate of grade prove too expensive to construct, steeper grades are used, the rate of grade selected depending upon the topography and the density of traffic to be handled. On very steep grades, safety of descent is probably the controlling factor. Grades should be made slightly flatter, or *compensated,* on curves.

In improving an existing road it is particularly important to balance the earthwork quantities so that the excavated material will make all the fills, with no excess or deficiency, by reason of the fact that frequently there is no opportunity for waste or borrow of material along the line.

The planning of a suitable foundation (subgrade) and suitable drainage for a highway are greatly aided if the results of a soil survey are available. Also, as part of the location survey the earthwork along the route should be classified in order to determine unit costs; proper subbase and/or surfacing materials; proper slopes for the cuts and fills; method of compacting fills; probable shrinkage and swell; probable settlement; and probability of slides, frost heaves, or erosion.

21-9 Photogrammetry in Highway Location. Methods of producing maps from aerial photographs have demonstrated a considerable direct economic advantage over ground-survey methods and shorten materially the time necessary to produce maps on which location studies can be based. When they are employed, the ground surveying is reduced largely to the obtaining of control data for the aerial photography.

Photogrammetric surveying and mapping in general are described briefly in Chap. 23. Usually the actual photography and sometimes the mapping are done under contract by organizations specializing in that work. For many regions, aerial photographs are available from governmental agencies.

For highway-location studies, the most important aerial photographic products are the mosaic, paired stereoscopic prints, and the contour map. The mosaic, an assemblage of matched photographs which covers an extended area, is used primarily for aerial studies such as reconnaissance. Paired stereoscopic prints serve the purpose of detailed studies. The contour map, used for quantitative analyses of routes being studied, is produced by stereoscopic plotting instruments.

Application of aerial photography to highway location is effectively accomplished in four stages, as follows:

1 Reconnaissance of a Wide Area. The mosaic is used as a plotting sheet on which are drawn all possible routes between the established termini. The selection of the scale will depend upon the necessary width of ground coverage, the topography, and the land use.

2 Comparison of Alternative Routes and Selection of Best Route. Larger-scale photographs are necessary to permit rating for directness of route between controlling points, applicability of design standards (grades, curvature, etc.), economics of construction and operation, and esthetics.

3 Preliminary Survey and Design. The selected route is analyzed to determine the exact location of the line on that route. The preliminary survey may be either a ground or an aerial survey, depending upon administrative and physical factors. In either case, the line is laid out on a contour map. Paired stereoscopic photographs are used to supplement the contour-map data. The selected line is computed for stationing, including alinement on curves.

4 *Location Survey and Construction Survey.* These surveys consist in the actual staking of the highway alinement, profile grade line, cross-sections, and structures on the ground in readiness for construction.

Aerial photographs are also suitable for application to related phases of highway engineering. A mosaic, on which is shown the selected route location, is especially suited for display to property owners or to agencies concerned with the route (such as local planning groups), or as an exhibit at public meetings. Paired stereoscopic prints may be used in the study of drainage areas, the discharge from which is to be handled by culverts or bridges along the route. Sources of suitable construction materials may be found by a study of the mosaic and stereoscopic prints. Right-of-way agents are frequently able to use large-scale enlargements of aerial photographs to determine the fair value of property to be acquired and in the negotiation for its purchase.

21-10 Railways. The procedures described in the preceding articles in this chapter apply for the most part to surveys for railway as well as highway location. The width of roadbed and right of way for railways is usually much less than that for major highways, and the grade rates, particularly the maximum grade, are more critical. As in the case of highways, the justification for the location or relocation of a railway is based upon the added benefits to traffic as compared with the initial cost. For railways the benefits may be lower operating costs over a period of years for the estimated volumes of freight and passenger traffic, or they may be the result of new service which will be provided.

Owing primarily to the much lower rate of maximum grade usually allowable for railways, tunnels are used more frequently in railway than in highway location.

The economics of railway location, which depends to a great extent upon the maximum grade, is beyond the scope of this text. However, it is appropriate to state that for standard lines the grade seldom exceeds 1 or 2 per cent and generally is less than 0.5 per cent. Maximum grades are compensated on curves to allow for curve resistance. On a heavy-traffic high-speed railway, every effort is made to have no curve sharper than 5 or 6° (Art. 10-15).

Railway construction surveys are discussed in Chap. 22.

21-11 Canals. The survey for location of a major canal is somewhat similar to that previously described for a roadway except that the grades used are relatively flat and small differences in elevation are relatively more important.

Owing to lack of flexibility in permissible grades, the number and spread of the alternatives to be investigated during reconnaissance are usually much less than in highway location. On the reconnaissance survey the engineer's level is usually used, with hubs being set every few hundred feet at the required grade elevation and the distances being measured by pacing or by stadia. The reconnaissance survey is run from a controlling point at one end of the line. The grade to be used is selected so as to result in the desired velocity of flow with the chosen cross-section.

As in the case of highways, the preliminary survey can be made either exclusively by field methods or by a combination of field and aerial survey methods. Where field methods are employed, the level party usually works ahead, setting stakes at grade as a guide for proper placing of the line. The transit or plane-table party then runs a tape or stadia traverse along the staked line and obtains sufficient topographic and planimetric detail that the final-location line can be laid out in its proper location.

In general the location and construction surveys for a canal are the same as those for a highway or railway. There are, however, some differences in the office projection of the center line due primarily to the shape of the cross-section. In shallow cuts, the canal cross-section is in the form of an excavated channel having an embankment constructed of the excavated material on each side. In sidehill work the excavated material is used to form an embankment on the downhill side of the channel. Instead of a fill across low areas, as might be used in highway or railway construction, either a flume or an inverted siphon is commonly used. The water section is placed entirely in cut wherever possible.

21-12 Transmission Lines. The required precision of surveys for transmission lines is generally much lower than that for a highway or a railroad. Reconnaissance can be carried out either by thorough field study or by use of aerial photographs supplemented by brief field inspections. If aerial photographs are available, tentative tower locations can be marked on the photographs and checked by field inspection if necessary.

The controlling considerations are usually economy of tower and insulator design, at the same time obtaining the shortest practicable location between termini in order to minimize power loss. Where there is no change of direction at a tower, the only loads to be considered in the design are the vertical load due to the weight of the cables, probable ice and wind loads, and possible occasional loading caused by breaking of a cable. At the end of a line of towers or where a change of directions causes similar conditions, there is a large horizontal force applied which requires special design. Therefore, the line is made as straight as possible, changes in direction being avoided wherever it is practicable to do so.

Although construction is less costly in level country, fairly heavy grades may be adopted to avoid changes in direction or unnecessarily high cost of right of way. Where possible, it is desirable to follow section lines or property lines in order to reduce the cost of right of way. If the line can be located near a highway or railway, construction cost as well as costs of patrolling and maintenance may sometimes be reduced. Rights of way are frequently obtained in the form of easements based upon a center-line description.

After the best location has been selected, a traverse is run following the proposed line as closely as practicable. Where distances between towers are relatively long, particularly in rugged terrain, electronic distance-measuring instruments (Art. 3-1) can be employed effectively. Stakes are usually necessary only at tentative tower locations, at any high points where cable clearance might be a factor, and at locations where buildings or other improvements might be affected. No curves are used in the alinement, changes in directions being made by angles at towers.

Elevations at staked points are obtained either by field leveling or by photogrammetric methods. A profile of these points is then plotted, and any necessary adjustments are made in the tentative tower locations. The selected locations are then marked on the ground, and the necessary stakes are set as a guide to placing the poles or the tower foundations.

21-13 Pipelines. Procedures for high-pressure pipeline surveys are more on the order of those required for transmission lines than those for highways or railways. In addition to right-of-way costs, controlling considerations are total length and avoidance of rock excavation and expensive stream crossings. As in the case of transmission-line surveys, reconnaissance work can be minimized by the use of aerial photographs. In some cases their use will permit deferring of practically all field-survey work until the time of construction.

General practice is to run a field-survey traverse after the route is selected and to obtain elevations at summits and sags in the proposed line and at crossings of major streams where structures may be required. As an actual design of the grade line is unimportant, a complete profile is usually unnecessary.

21-14 Field Problem

Problem 50. Preliminary Survey for Road (Topography by Cross-profile Method)

Object. To obtain data for a topographic map along the route of a proposed highway or railroad, locating the contours directly by means of the hand level and metallic tape. The data may be used in office problem 45 (chap. 19).

Steps 1 and 2 of the following procedure may have been accomplished in field problems 26 of Chap. 10 and 8 of Chap. 6. For a preliminary survey run by an alternative method, employing transit and stadia, see field problem 31 of Chap. 11. For an exercise in setting slope stakes and taking final cross-sections, see field problem 10 of Chap. 6.

Procedure. (1) Over the assigned route run an open deflection-angle traverse with transit and tape. (2) Establish vertical control for the route by direct profile leveling. (3) At each full station and at any necessary plus stations, locate the 5-ft contours on a crossline extending 300 to 800 ft on either side of the line; also locate the points where the 5-ft contours cross the traverse line. Employ the hand level, topographer's rod, and metallic tape (see Art. 20-9). (4) Measure distances from the traverse line to other topographic features such as land lines, streams, roads, and buildings. (5) Note the quality of the land, as to whether it is clay, rock, or sand. Note the condition of the land, as to whether it is cleared land, pasture land, or woods; note the kind of trees and density of growth in woods. (6) Keep notes as shown in Fig. 10-3.

22
CONSTRUCTION SURVEYING

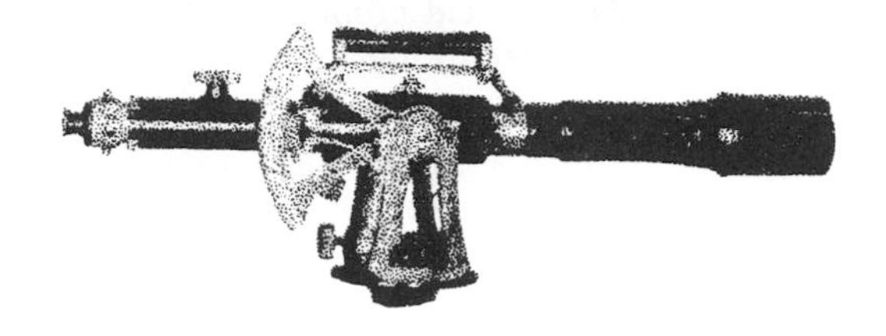

22-1 General. Surveys for construction usually involve (1) a topographic survey of the site, to be used in the preparation of plans for the structure; (2) establishment on the ground of a system of stakes or other markers, both in plan and in elevation, from which measurement of earthwork and structures can be taken conveniently by the construction force; (3) giving of line and grade as needed either to replace stakes disturbed by construction or to reach additional points on the structure itself; and (4) making of measurements necessary to verify the location of completed parts of the structure and to determine the volume of work actually performed up to a given date (usually each month), as a basis of payment to the contractor.

In connection with construction, often it is necessary to make property-line surveys (Chap. 18) as a basis for the acquisition of lands or rights of way (Chap. 21).

The detailed methods employed on construction surveys vary greatly with the type, location, and size of structure and with the preference of the engineering and construction organizations. Much depends on the ingenuity of the surveyor to the end that the correct information is given without confusion or needless effort.

The topographic survey of the structure site should include adjacent areas that are likely to be used for construction plant, roads, or auxiliary structures. Aerial photographs are useful aids for planning the construction.

22-2 Alinement. Temporary stakes or other markers are usually set at the corners of the proposed structure, as a rough guide for beginning the excavation. Outside the limits of excavation or probable disturbance but close enough to be convenient, permanent stations are set and well referenced. Permanent targets or marks called *foresights* may be erected as convenient means of orienting the transit on the principal lines of the structure and for sighting along these lines by eye. Stakes or other markers are set on all important lines in order to mark clearly the limits of the work.

In many cases, line and grade are given more conveniently by means of *batter boards* than by means of stakes. A batter board is a board (usually 1 by 6 in.) nailed to two substantial posts (usually 2 by 4 in.) with the board horizontal and its top edge preferably either at grade or at some whole number of feet above or below grade. The alinement is fixed by a nail driven in the top edge of the board. Between two such batter boards a stout cord or wire is stretched to define the line and grade.

Often it is impracticable to establish permanent markers on the line of the structure. In such cases the survey line is established parallel to the structure line, as close as practicable and with the offset distance some whole number of feet.

22-3 Grade. A system of bench marks is established near the structure in convenient locations that will probably not be subject to disturbance. From time to time these bench marks should be checked against one another to detect any disturbance. Every care should be taken to preserve existing bench marks of state and Federal surveys; if construction necessitates the removal of such marks, the proper organization should be notified and the marks transferred in accordance with its instructions.

The various grades and elevations are defined on the ground by means of pegs and/or batter boards, as a guide to the workmen. The grade pegs may or may not be the same as the stakes used in giving line. When stakes are used, the vertical measurements may be taken from the top of the stake, from a keel mark or a nail on the side of the stake, or (for excavation) from the ground surface at the stake; in order to avoid mistakes, only one of these bases for measurement should be used for a given kind of work, and the basis should be made clear at the beginning of construction. When batter boards are used, the vertical measurements are taken from the top edge of the board, which is horizontal. The stake or the batter board may be set either at grade or at a fixed whole number of feet above or below grade.

When a stake is to be driven with its top at a given elevation, the rodman starts the stake and then holds the rod on the stake. The levelman reads the rod and calls out the approximate distance the stake must be driven to reach grade. The rodman drives the stake nearly the desired amount, and a second rod reading is taken; and so the process is continued until the rod reading is made equal to the difference between the height of instrument and the desired elevation. A mark or nail on the side of the stake may be used instead of the top of the stake. In some cases, the stake is sawed off at the desired elevation. If the grade elevation is only a short distance below the ground elevation, a hole may be dug in order that the stake may be driven to grade.

22-4 Precision. For purposes of excavation only, usually elevations are given to the nearest 0.1 ft. For points on the structure, usually elevations to 0.01 ft are sufficiently precise. Alinement to the nearest 0.01 ft will serve the purposes of most construction, but greater precision may be required for prefabricated steel structures or members.

It is desirable to give dimensions to the workmen in feet, inches, and fractions of an inch. Ordinarily measurements to the nearest $\frac{1}{4}$ or $\frac{1}{8}$ in. are sufficiently precise, but certain of the measurements for the construction of buildings and bridges should be given to the nearest $\frac{1}{16}$ in. Often it is convenient to use the relation that $\frac{1}{8}$ in. equals approximately 0.01 ft.

22-5 Establishing Points by Intersection. Where conditions render the use of the tape difficult or impossible, often points are established at the intersection of two transit lines by simultaneous sighting with two transits in known locations. The process is the inverse of that in which the location of a ground point is determined by the method of intersection (Art. 10-4). By this method, points may be located in elevation as well as in plan. The precision of measurement is made commensurate with the requirements of construction.

22-6 Highways. Usually just prior to the beginning of construction of a section of highway, the located line is rerun, missing stakes are replaced, and hubs are referenced. Borrow pits (if necessary) are staked out and cross-sectioned (Art. 6-3). Lines and grades are staked out for bridges, culverts, and other structures. If slope stakes have not already been set during the location survey, they are set except where clearing is necessary; in that case they are set when the right of way has been cleared. For purposes of clearing, only rough measurements from the center-line stakes are necessary.

The method of setting and marking slope stakes is described in Art. 6-9. Additional stakes may be offset a uniform distance away from the work, with appropriate marking to indicate the offset. If intercepting ditches are to be placed along the cuts, these are staked out also.

Where the depth of cuts and fills does not average more than about 3 ft, the slope stakes may be omitted; in this case the line and grade for earthwork may be indicated by a line of pegs (with guard stakes) along one side of the road and offset a uniform distance such that they will not be disturbed by the grading operations. Pegs are usually placed on both sides of the road at curves, and may be so placed on tangents; when this is done, measurements for grading purposes may be taken conveniently by sighting across the two pegs or by stretching a line or tape between them.

When rough grading has been completed, a line of finishing stakes is set on both sides of the roadway at the edge of the shoulder. It should be understood whether these are to subgrade or to finish grade (final grade).

For concrete highways, to give line and grade for the pavement, a line of stakes is set along each side, offset a uniform distance (usually 2 ft) from the edge of the pavement. The grade of the top of the pavement, at the edge, is indicated either by the top of the stake or by a nail or line on the side of the stake. The alinement is indicated on one side of the roadway only, by means of a tack in the top of each stake. In another method, pegs are set so that the side forms may be placed directly upon them, and a line of stakes is set near one edge to give line for the forms. The distance between stakes in a given line is usually 100 or 50 ft on tangents at uniform grade and half the normal distance on horizontal or vertical curves. The dimensions of the finished subgrade and of the finished pavement are checked by the construction inspector, usually by means of a templet. For asphalt pavements, the line of finishing stakes for grading is usually sufficient.

As construction proceeds, monthly estimates are made of the work completed to date. A quantity survey is made near the close of each month, and the volumes of earthwork, etc., are classified and summarized as a basis for payment.

22-7 Streets. For street construction the procedure of surveying is similar to that just described for highways. Ordinarily the curb is built first. The line and grade for the top of each curb are indicated by pegs driven just outside the curb line, usually at 50-ft intervals. The grade for the edge of the pavement is then marked on the face of the completed curb; or for a combined curb and gutter it is indicated by the completed gutter. Ground pegs are set on the center line of the pavement, either at the grade of the finished subgrade (in which case holes are dug when necessary to place pegs below the ground surface) or with the cut or fill indicated on the peg or on an adjacent stake. Where the street is wide, an intermediate row of pegs may be set between center line and each curb. It is usually necessary to reset the pegs after the street is graded. Where driving stakes is impractical because of hard or paved ground, nails or spikes may be driven or marks may be cut or painted on the surface.

The surveys for street location and construction should determine the location of all surface and underground utilities that may affect the project, and notification of necessary changes should be given well in advance. Information regarding the desirable location of underground utilities, together with methods of surveying and mapping, is given in

Manual 14, "Location of Underground Utilities," of the American Society of Civil Engineers.

22-8 Railroads. Surveys for railroad earthwork are similar to those described in Art. 22-6 for highways. Prior to construction the located line is rerun, missing stakes are replaced, hubs are referenced, borrow pits are staked out, slope stakes are set, and lines and grades for structures are established on the ground. When rough grading is completed, finishing stakes are set to grade at the outer edges of the roadbed.

When the roadbed has been graded, alinement is established precisely by setting tacked stakes along the center line at full stations on tangents and usually at fractional stations on horizontal and vertical curves. Spiral curves are staked out at this time. An additional line of pegs is set on one side of the track and perhaps 3 ft from the proposed line of the rail, with the top of the peg usually at the elevation of the top of the rail. Track is usually laid on the subgrade and is lifted into position after the ballast has been dumped.

22-9 Sewers and Pipelines. The center line for a proposed sewer is located on the ground with stakes or other marks set usually at 50-ft intervals where the grade is uniform and as close as 10 ft on vertical curves. At one side of this line, just far enough from it to prevent being disturbed by the excavation, a parallel line of ground pegs is set. A guard stake is driven beside each peg, with the side to the line; on the side of the guard stake farthest from the line is marked the station number and offset, and on the side nearest the line is marked the cut (to the nearest ⅛ in.). In paved streets or hard roads where it is impossible to drive stakes and pegs, the line and grade are marked with spikes (driven flush), chisel marks, or paint marks.

When the trench has been excavated, batter boards are set across the trench at the intervals used for stationing. The top of the board is set at a fixed whole number of feet above the sewer invert (inside surface of bottom of sewer); and a measuring stick of the same length is prepared. A nail is driven in the top edge of each batter board to define the line. As the sewer is being laid, a cord is stretched tightly between these nails, and the free end of each section of pipe is set at the proper distance below the cord as determined by measuring with the stick.

If the trench is to be excavated by hand, the side pegs may be omitted and the batter boards set at the beginning of excavation.

For pipelines, the procedure is similar to that for sewers, but the interval between grade pegs or batter boards may be greater, and less care need be taken to lay the pipe at the exact grade.

For both sewers and pipelines, the extent of excavation in earth and in rock is measured in the trench, and the volumes of each class of excavation are computed as a basis of payment to the contractor.

The records of the survey should include the location of underground utilities crossed by, or adjacent to, the trench.

22-10 Canals. The location survey for a canal is described in Art. 21-11. Slope stakes for each bank are set as described in Art. 6-9. So far as possible, the cross-section is balanced, that is, the excavated material forms the fill at the same station, and little or no material needs to be moved longitudinally.

22-11 Tunnels. The methods employed in tunnel surveys vary with the purpose of the tunnel and the magnitude of the work. A coordinate system is particularly appropriate for tunnel work.

For a short tunnel such as a highway tunnel through a ridge, a traverse and a line of levels are run between the terminal points; and the length, direction, and grade of the connecting line are computed. Where practicable, the surface traverse between the terminals takes the form of a straight line. Outside the tunnel, on the center line at both ends, permanent monuments are established. Additional points are established in convenient surface locations on the center line, to fix the direction of the tunnel on each side of the ridge. As construction proceeds, the line at either end is given by setting up at the permanent monument outside the portal, taking a sight at the fixed point on line, and then setting points along the tunnel, usually in the roof. Grade is given by direct levels taken to points in either the roof or the floor, and distances are measured from the permanent monuments to stations along the tunnel (see Art. 20-12). If the survey line is on the floor of the tunnel, it is usually offset from the center line to a location relatively free from construction traffic and disturbance; from this line a rough temporary line is given as needed by the construction force.

The dimensions of the tunnel are usually checked by some form of templet transverse to the line of the tunnel, but may be checked by direct measurement with the tape.

Railroad and aqueduct tunnels in mountainous country are often several miles in length and are not uniform in either slope or direction. Tunnels of this character are usually driven not only from the ends but also from several intermediate points where shafts are sunk or adits are driven to intersect the center line of the tunnel. The surface surveys for the control of the tunnel work usually consist of a precise triangulation system tied to monuments at the portals of the main tunnel and at the entrances of shafts and adits, and a precise system of differential levels connecting the same points. With these data as a basis the length, direction, and slope of each of the several sections of the tunnel are calculated; and construction is controlled by establishing these lines and grades as the work progresses.

22-12 Bridge Sites. Normally the location survey will provide sufficient information for use in the design of culverts and small bridges; but for long bridges and for grade-separation structures usually a special topographic survey of the site is necessary. This survey should be made as early as possible in order to allow time for design and—in the case of grade crossings or navigable streams—to permit approval of the appropriate governmental agency to be secured. The site map should show all the data of the location survey, including the line and grade of the roadway and the marking and referencing of all survey stations. The usual map scale is 1 in. = 100 ft, and the usual contour interval is 5 ft on steep slopes and 2 ft over flat areas.

The preliminary report submitted with the site map should give all available information necessary for economic design, such as the character of the watershed and the stream bed, the elevation of the highest water, the character of the foundation, and local sources of materials. For a grade-separation structure, the required information is similar except that it relates to the intersecting roadway and its traffic instead of the intersecting stream and its flow. Photographs are useful adjuncts to the report.

22-13 Bridges. For a short bridge with no offshore piers, first the center line of the roadway is established, the stationing of some governing line such as the abutment face is established on the located line, and the angle of intersection of the face with the located line is turned off. This governing crossline may be established by two well-referenced transit stations at each end of the crossline beyond the limits of excavation or, if the face of the abutment is in the stream, by a similar transit line offset on the shore. Similarly, governing lines for each of the wing walls are established on shore beyond the limits of excavation, with two transit stations on the line prolonged at one, or preferably both, ends of the wing-wall line. If the faces are battered, usually one line is established for the bottom of the batter and another for the top. Stakes are set as a guide to the excavation and are replaced as necessary. When the foundation concrete is cast, line is given on the footings for the setting of forms and then by sighting with the transit for the top of the forms. As the structure is built up, grades are carried up by leveling, with marks on the forms or on the hardened portions of the concrete. Also, the alinement is established on completed portions of the structure. The data are recorded in field books kept especially for the structure, principally by means of sketches.

For long sights or for work of high precision, as in the case of offshore piers, various transit stations are established on shore and/or on specially built towers by a system of triangulation, such that favorable intersection angles and checks will be obtained for all parts of the work. To establish the offshore piers, simultaneous sights are taken from the ends of a line of known length.

For long bridges it is necessary that distances be measured with precision even higher than that required for first-order work in order that the parts of the superstructure can be fitted together with close tolerances. Also the coordinate system adopted for horizontal control should be based upon the average elevation of the structure. The survey is accomplished in stages, the pier locations being established with somewhat lower precision than those of the fittings subsequently established on the top of the pier.

Where cofferdams are used, reference points are established on the cofferdams for measurements to the pier.

When the structure has been completed, permanent survey points are established and referenced for use in future surveys to determine the direction and extent of any movement.

22-14 Culverts. At the intersection of the center line of the culvert with the located line, the angle of intersection is turned off, and a survey line defining the direction of the culvert is projected for a short distance beyond its ends and is well referenced. At (or offset from) each end of the culvert, a line defining the face is turned off and referenced. If excavation is necessary for the channel to and from the culvert, it is staked out in a manner similar to that for a roadway cut. Bench marks are established nearby, and pegs are set for convenient leveling to the culvert. Line and grade are given as required for the particular type of structure.

22-15 Building Sites. In the preparation of his plans for a building, the architect requires a large-scale map of the site to show the information necessary to the proper location of the building both in plan and in elevation. Such maps are usually drawn to the scale of 1 in. = 10 ft or 1 in. = 20 ft.

The survey party consists of an instrumentman and one or two chainmen, and the equipment includes that of a transit party. Because of the large number of elevations to be determined, frequently an engineer's level is also used. The notes may be kept in a transit or topography notebook; or a sketch board may be used to good advantage, as it provides more space for sketching and recording details than does the single page of a notebook.

First the lot corners are located, and permanent markers are set at these points. Then, with the property lines being used as reference lines, all objects are located, usually by tape measurements only, the instrumentman recording the data and drawing such sketches as are necessary. On extensive sites, or where it is not convenient for the chainmen to locate objects by coordinate measurements, transit angles and taped distances may be used.

The details should be shown and described as follows: (1) lot corners, state kind; (2) property lines, give dimensions and the distances from the walks; (3) street lines, show widths; (4) sidewalks and drives, give kind and widths; (5) pavements, give kind and widths; (6) gas and water mains, state size and show exact location; (7) manholes and storm and sanitary sewers, give size and kind of pipe; (8) trees, state kind and size; (9) poles of all kinds; (10) fire hydrants; and (11) existing structures on or near the site, state materials of construction. Also give the elevations of: (*a*) inverts of sewer outlets from manholes and the gradients of the sewers; (*b*) the reference bench mark, with description; (*c*) points along the sidewalks, curbs, and lot lines at intervals at 50 ft; and (*d*) ground points at the corners of 50-ft squares. The elevations of the sewer inverts, sidewalks, and curbs should be taken to hundredths of feet and all ground points to tenths of feet; also contour lines are shown if the ground is irregular.

The map should give the legal description of the tract and the other information ordinarily shown on the plot of an urban land survey. The drawing is made on tracing cloth, the size being the same as the other sheets of the architect's plans so that a print may be bound with each set of plans. A typical map of this kind is shown in Fig. 22-1.

22-16 Buildings. At the beginning of excavation, the corners of the building are marked by stakes, which will of course be lost as excavation proceeds. Sighting lines are established and referenced on each outside building line and line of columns, preferably on the center line of wall or column. A batter board is set at each end of each outside building line, about 3 ft outside the excavation. If the ground permits, the tops of all boards are set at the same elevation; in any event the boards at opposite ends of a given line (or portion thereof) are set at the same elevation so that the cord stretched between them will be level. The elevations are chosen at some whole number of feet above the bottom of the excavation, usually that for the floor rather than that for the footings.

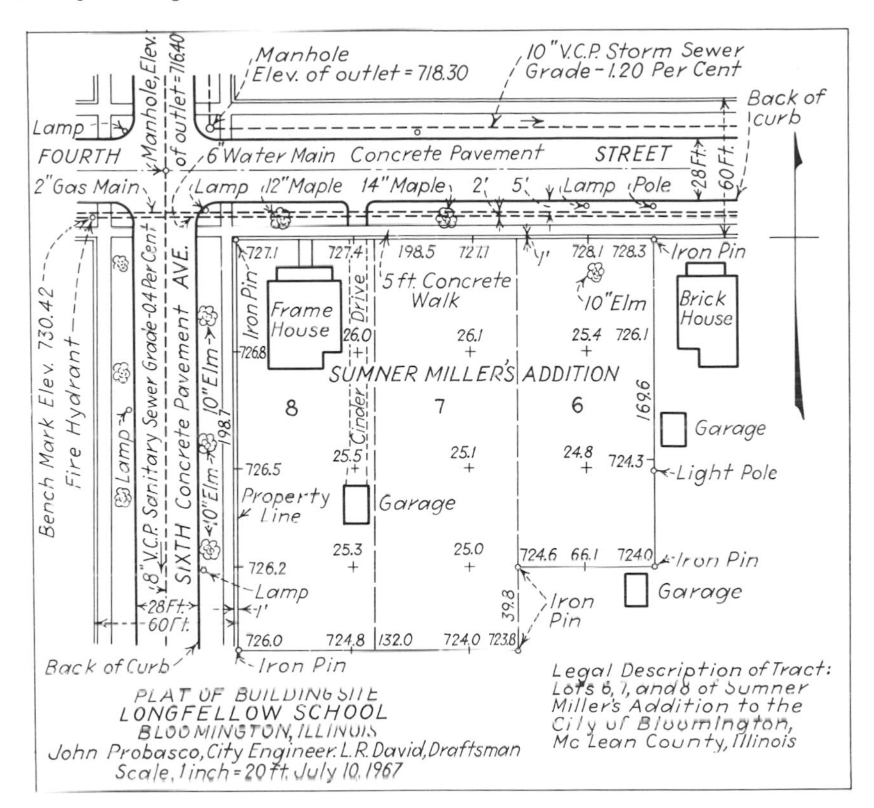

Fig. 22-1. Map of building site.

When the board has been nailed on the posts, a nail is driven in the top edge of the board on the building line, which is given by the transit. Carpenter's lines stretched between opposite batter boards define both the line and the grade, and measurements can be made conveniently by the workmen for excavation, setting forms, and alining masonry and framing.

If the space around the building is obstructed so that batter boards cannot be set, other means of marking the line and grade are substituted to meet the requirements of the situation.

When excavation is completed, grades for column and wall footings are given by ground pegs driven to the elevation of either the top of the footing or the top of the floor. Lines for footings are given by batter boards set in the bottom of the excavation. Column bases and wall plates are set to grade directly by the leveler. The position of each column or wall is marked in advance on the footing; and when a concrete form, a steel member, or a first course of masonry has been placed on the footing, its alinement and grade are checked directly.

In setting the form for a concrete wall, the bottom is alined and fixed in place before the top is alined.

Similarly, at each floor level the governing lines and grade are set and checked, except that for prefabricated steel framing the structure as a whole is plumbed by means of the transit at every second- or third-story level. Notes are kept in a field book used especially for the purpose, principally by means of sketches.

Whenever the elevation of a floor is given, it should be clearly understood whether the value refers to the bottom of the base course, the top of the base course, or the top of the finished floor.

Throughout the construction of large buildings, selected key points are checked by means of stretched wires, plumb lines, optical plummet, transit, or level in order to detect settlement, excessive deflection of forms or members, or mistakes. Bench marks are checked to detect any disturbance.

22-17 Dams. Prior to the design of a dam, a topographic survey is made to determine the feasibility of the project, the approximate size of the reservoir, and the optimum location and height of the dam. To provide information for the design, a topographic survey of the site similar in many respects to that for a bridge (Art. 22-12) is made by field and/or photogrammetric methods. Extensive soundings and borings are made, and topography is taken in detail sufficient to define not only the dam itself but also the appurtenant structures, necessary construction plant, roads, and perhaps a railroad branch. A property-line survey is made of the area to be covered by, or directly affected by, the proposed reservoir.

Prior to construction a number of transit stations, sighting points, and bench marks are permanently established and referenced upstream and downstream from the dam, at advantageous locations and elevations for sighting on the various parts of the structure as work proceeds. These reference points are usually established by triangulation from a measured base line on one side of the valley, and all points are referred to a system of rectangular coordinates, both in plan and in elevation. To establish the horizontal location of a point on the dam, as for the purpose of setting concrete forms or of checking the alinement of the dam, simultaneous sights are taken from two transits set up at reference stations, each transit being sighted in a direction previously computed from the coordinates of the reference station and of the point to be established. The elevation of the point is usually established by direct leveling. However, it may be established by setting off on one (or, as a check, both) of the transits the computed vertical angle, the height of instrument being known.

A traverse is run around the reservoir, above the proposed shore line, and monuments are set for use in connection with property-line surveys and for future reference. Similarly, bench marks are established at points above the shore line. The shore line may be marked with stakes set at intervals. The area to be cleared is defined with reference to these stakes. The area and volume of the reservoir may be computed as described in Art. 19-16.

22-18 Industrial Surveying. The manufacture and assembly of large machines, aircraft, missiles, and guiding and tracking systems require extremely close tolerances of measurement for location, orientation, dimensioning, and alinement. For reasons of size as well as precision, the ordinary tools of the mechanic and the jigs and techniques of conventional tool engineering are not adequate, and it is necessary to use apparatus and methods similar to those of surveying. In recent years many optical and related devices have been developed such as special transits, telescopes, targets, linear scales, and micrometers together with equipment and systems for supporting these instruments. A precision of up to 1/200,000 can be attained, the linear measurements being made to 0.001 in. and the angular measurements and alinements to 1 second of arc.

The principles of industrial surveying are the same as those of surveying in general, but the apparatus and methods are highly specialized and are beyond the scope of this book. Herein are discussed briefly some of the distinctive features.

Optical tooling is the term commonly applied to precise measurement and alinement of large machines and industrial assemblies by optical and closely related means.

Apparatus. A *jig transit,* or *jig collimator,* is used to establish vertical planes. One form is shown in Fig. 22-2. It is essentially a precise transit, similar to a theodolite, but has no horizontal or vertical graduated circles and no compass. Its range of focusing is from several inches to infinity. Its level tubes are sensitive. Through its hollow spindle, vertical sights can be taken by means of a right-angle eyepiece assembly. On the telescope is an optical micrometer for measuring short vertical or horizontal distances from the line of sight without moving the telescope. At either end of the horizontal axis is mounted an optically flat mirror parallel to the line of sight for use in establishing perpendiculars to the line of sight.

An *alinement telescope* has a cylindrical telescope barrel fitted with an objective lens, eyepiece, and reticule for defining the line of sight. It is provided with mutually perpendicular optical micrometers to measure short transverse distances. The barrel is precisely ground, and it may be mounted interchangeably in fixtures at various locations.

The *level* used in optical tooling is a precise engineer's level such as that used in conventional surveying and equipped with an optical micrometer.

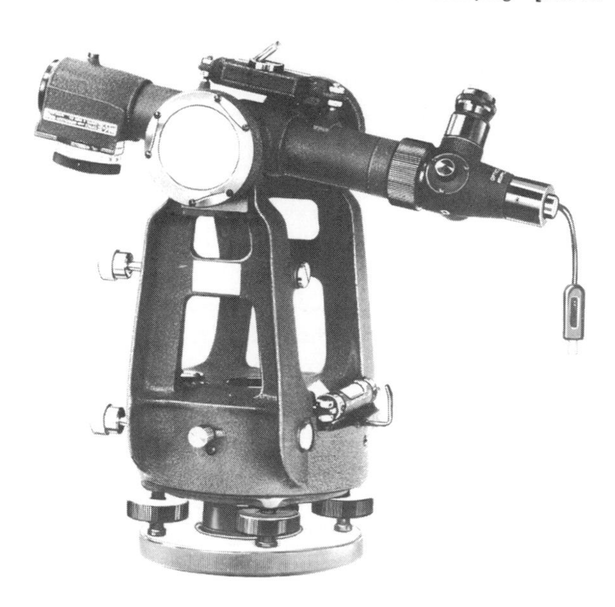

Fig. 22-2. Jig transit. (*Keuffel & Esser Co.*)

Specially designed *targets* are used for ease and precision in sighting. *Measuring bars* and optical or screw-thread *micrometers* can be coupled to form a precise measuring rod over a range of lengths; precise *tapes* are also used. Specially graduated flat linear *scales* are provided with devices for temporary or continued attachment to points on the structure; the scales are read directly through the transit, level, or alinement telescope, the distance of the line of sight from the nearest graduation of the scale being read with an optical micrometer. Auxiliary apparatus includes special stands, frames, carriages, tracks, and brackets for supporting the instruments and targets. According to the particular needs, instruments may be attached to the structure or supported separately from it or some attached and some supported separately.

Some Methods. A base line, or *master line,* is established in a principal plane of the structure, to which line directions and transverse measurements (usually horizontal and vertical) are referred. Also a base point, or *master button,* is established near one end of the base line as a reference for linear measurements. Other reference lines and points may be established for convenience.

To erect a perpendicular to a base line, a jig transit is set up and lined in on the base line (or parallel to it) at the desired longitudinal distance from the master button and is sighted along the line. A suitably supported alinement telescope is directed toward the mirror on the end of the horizontal axis of the jig transit and is moved about until its line of sight is reflected back on itself; the line of sight is then perpendicular to the base line. A precise pentagonal sighting prism may be attached to a telescope to project a perpendicular to its line of sight.

To check a level surface, the engineer's level is used in connection with a graduated linear scale which is held vertical and is moved about on the surface, as in leveling with a leveling rod. To level a given surface, a similar procedure is followed and the supports of the surface are adjusted until the vertical scale reading is the same over all the surface.

To aline a wall or other vertical plane of a structure, a line of sight is established in the desired direction at a short offset distance from the desired location of the surface. A graduated linear scale is held or fastened horizontally at two or more points on the surface, and the distance from surface to line of sight is read. The wall is moved as necessary until the horizontal scale reading is the same at all points. This process is similar to direct leveling but is in vertical instead of horizontal planes.

Plumbing is accomplished by means of an optical plummet, one form of which is an alinement telescope in a vertical mounting.

22-19 Ship Surveying. Ships not built in dry docks are built on inclined ways down which they will slide at launching. Assembly of the major parts of a ship is controlled by surveying methods similar to those for the construction of a building, except that alinements and measurements are either made in or based upon planes parallel and perpendicular to the plane of the ways. Account is taken as necessary of the ratio of horizontal or vertical distances to inclined distances, which are those parallel or perpendicular to the ship's structure as shown on the plans. For measurements in longitudinal planes, this ratio is equal to the cosine of the slope (usually about 3°) of the ways.

23
PHOTOGRAMMETRIC SURVEYING

23-1 General. *Photogrammetry* is the science of measurement by means of photographs. The scale and position of objects in photographs vary according to the distance and position of the corresponding actual objects relative to the camera. *Photogrammetric surveying* is accomplished by the measurement of these differences in scale and displacements in position in single or overlapping photographs, in order to determine the location and elevation of ground points and other details.

Although oblique photographs taken from ground stations or from airplanes are used to some extent, the principal development has been in the field of photographs taken from airplanes with the axis of the camera vertical (as nearly as may be). In connection with limited ground surveys made for the purpose of accurately establishing visible control points, vertical aerial photogrammetry has become the most rapid and accurate method of topographic surveying except perhaps where the ground is relatively flat, where elevations must be determined within less than 5 ft, or where the area is small. Its advantages are the speed with which the work is accomplished, the wealth of detail secured, and its application in locations otherwise difficult or impossible of access.

Since the First World War, aerial photogrammetry has come into general use for most surveying applications. Prominent in the application of aerial photogrammetry in the United States have been such organizations as the U.S. Geological Survey, the U.S. Coast and Geodetic Survey, the U.S. Army, the U.S. Navy, the U.S. Air Force, and agencies of the U.S. Department of Agriculture as well as numerous commercial companies concerned with the development of aerial cameras and related equipment and the training of photogrammetrists in the field. Technical societies within the photogrammetric field consist of one of the technical divisions of the American Society of Civil Engineers, the American Congress on Surveying and Mapping, the International Society for Photogrammetry, and the American Society of Photogrammetry. In addition to the extensive photogrammetric operations involved in the national military photogrammetric mapping, the use of photogrammetric techniques in the design and planning of civil-engineering projects is of increasing significance.

Most of the United States has been photographed from the air, at one scale or another, by agencies of Federal, state, or local government. Prints of most of these photographs can be obtained at low cost. Detailed information concerning governmental photographic coverage of a given area can be obtained from the Map Information Office of the U.S. Geological Survey, Washington, D.C. Pertinent information includes the date of photography, extent of area covered, type of camera used (precision or reconnaissance), flight height, focal length, approximate scale, and format. The Map Information Office can also furnish information on the additional coverage available in the possession of private organizations.

Maps are compiled in conventional symbols (Art. 14-8) and the space required to represent cartographic features on the map limits the number of features that can be shown to fewer than those on photographs of the same scale. In small-scale maps the conventional symbols are not true to scale, and it is not essential that the physical features on the photograph be at the same or an easily measurable scale. It is only necessary that each image be of sufficient size and clarity to permit correct interpretation of the photograph.

23-2 Definitions. Some of the special terms used in aerial photogrammetry are as follows:

Aerotriangulation. Any of several methods of extending horizontal control from ground control by means of aerial photographs.

Altitude. Vertical distance to points or objects *above* the earth's surface; for example, an airplane in flight. Not to be confused with "elevation," which is the vertical distance from a datum surface to points or objects *on* the earth's surface.

Anaglyph. A picture printed or projected in complementary colors combining the two images of a stereoscopic pair and giving a spatial model when viewed through spectacles having filters of corresponding complementary colors (usually red and blue-green). See also *vectograph.*

Mosaic. A matched assemblage of separate photographs to show an area. Not a true map, but useful for some purposes such as regional planning. A *controlled mosaic* is oriented and scaled to horizontal ground control. Contour lines may be drawn on a mosaic.

Parallax. The displacement of the images of objects with respect to other objects due to the difference in their respective distances from the observer. In overlapping vertical aerial photographs, the difference in parallax between two points is a measure of the difference between their elevations.

Principal Point. The point of intersection of the optical axis of the camera with the plane of the photograph. For vertical photographs, it is usually the central point.

Spatial Model. A three-dimensional image formed in the mind of the observer while stereoscopically observing two views of the same object.

Stereoscope. Any mechanical device to facilitate seeing as in three dimensions. It is usually formed of a combination of mirrors, lenses, and/or prisms; but stereoscopic vision may be accomplished by means of dichromatic or polaroid projection and printing. The observer views two overlapping photographs of the same object—one with each eye—which is three-dimensional. See *anaglyph* and *vectograph.*

Vectograph. A stereoscopic photograph composed of two superimposed images which polarize light in planes 90° apart to yield a spatial model when viewed through polaroid spectacles. See also *anaglyph.*

23-3 Terrestrial Photogrammetry. Photographs for terrestrial (ground) photogrammetry are taken with a *phototheodolite,* which is a camera mounted on a transit tripod and having a transit telescope with its optical axis parallel to that of the camera. The photographs are later inserted in an automatic plotting machine for the compilation of the map.

Terrestrial photographs are taken in pairs from the ends of a measured base line. The pairs are usually made with the two positions of the camera axis parallel to each other. The camera stations should be at the higher points in the area, in order that the direction of pointing may be as nearly normal to the slope as possible.

AERIAL SURVEYING

23-4 General Procedure of Aerial Surveying. Aerial photographs of the area to be surveyed are taken with specially designed automatic cameras with the axis of the camera vertical (as nearly as possible). The airplane is flown at a constant altitude (as nearly as possible) above the mean terrain. The area is covered in a series of parallel strips, with photographs of the same strip overlapping usually about 60 per cent and photographs of adjacent strips overlapping about 30 per cent.

Ground control consists of a relatively small number of points the location and elevation of which either are known or are to be determined on the ground. With adjacent ground-control points identified on a photograph or series of photographs and with the flight altitude and the focal length of the camera known, the scale of the photographs can be computed. A series of photographs can be brought to the same scale by enlargement or reduction of the individual photographs.

Secondary control, or photogrammetric control, consists of selected image points on photographs; these points are used to bring a series of photographs between adjacent ground-control stations into the desired relations for plotting to scale. With the control plotted on a transparent sheet and with the photographs brought to the same scale and placed successively under the sheet, the map details surrounding each control point are traced onto the sheet. Conventional symbols are employed as desired.

For purposes of representing relief, or *contouring,* use is made of the difference in parallax displacement between selected ground points of unknown elevation and control points of known elevation. To compute the displacement of all the image points of controlling points of the terrain would be unduly laborious, and in practice it is customary to employ automatic plotting machines based upon the principle of stereoscopic observation. Contouring may be accomplished in the field with the plane table, a copy of the planimetric map being used as the plane-table sheet.

Aerial surveying for topographic mapping, as accomplished by the Topographic Division of the U.S. Geological Survey, consists of eight operations: advance planning, aerial photography, ground-control surveying, manuscript preparation, aerotriangulation, stereocompilation of detail from photographs, field completion, and reproduction (see also Fig. 23-1). The general procedure is briefly as follows:

1 Advance planning includes the selection and specification of the map scale and contour interval; of the instrumentation, procedure, and methods; and of the scheduling of the various phases.

2 Aerial photography is planned to accommodate the map requirements, type of photography and aerial camera, and stereoscopic plotting instrument. Complete stereoscopic coverage must be obtained for the area to be mapped. (The area to be photographed must be somewhat larger.) The flight design specifies the height of flight, distance between flight lines, amount of forward overlap between exposures, and allowable tilts, with the allowable variations in each of these. Photomosaic indexes

Fig. 23-1. Sequence of topographic-mapping operations. (*U.S. Geological Survey.*)

are usually furnished by the contractor for inspection purposes as areas are completed.

3 Necessary *ground control,* both horizontal and vertical, is then planned on the aerial photographs and is obtained in the field by ground-survey methods. The location of each horizontal and vertical control point, either existing or planned, is carefully plotted and identified on the photographs.

4 *Manuscript preparation* consists in plotting the coordinate grid, ground control, and related information by scribing them on a dimensionally stable transparent plastic sheet coated with a thin opaque painted surface layer.

5 *Aerotriangulation,* or aerial triangulation, is any type of control extension accomplished by means of aerial photographs. For surveys of medium and small scale the ground control need not be in every photograph, but may appear in every third, fifth, or in some cases as much as every fifteenth photograph; the intervening distances are bridged by aerial triangulation.

6 Compilation of detail from photographs consists in scribing the details observed in the stereoscopic model onto the previously prepared base manuscript. For color reproduction, additional manuscript components are provided as necessary for the various colors, and these manuscript components are later overprinted to obtain the finished map.

7 Field-completion surveys are usually accomplished by ground inspection and mapping. Details not obtainable from the photographs are obtained as well as place names, public-land subdivisions, and road and stream classifications. Deficiencies and minor mistakes in the photogrammetric compilation are corrected, and a vertical and horizontal test of accuracy is also accomplished.

8 For *reproduction,* images of the map detail are printed photographically on sheets of dimensionally stable plastic material coated with a thin film of actinically opaque material. On such a sheet, the scriber traces the detail for a particular color by cutting through the coating with precision-ground needles or blades. The sheets are then exposed to sensitized metal pressplates, which subsequently print images of only the scribed symbols of that color.

23-5 Aerial Photographs. If the aerial photographs were truly vertical and if the terrain were flat, the photographs would be true maps to some scale. In general, however, the features are displaced and their size is varied because of longitudinal and transverse inclination (tilt) of the camera axis and because of the relief of the ground. The effect of tilt can be corrected either by photographic methods or (more laboriously) by computation; usually the average tilt is less than 3° and often it is neglected. The displacement, or parallax, of features due to ground relief makes necessary special methods of plotting to secure a true scale map; but it is useful, in fact necessary, as a means of determining the elevations of points by computation or by stereoscopic measurement.

The scale of a photograph can be computed from the relationship of a measured distance on the photograph to the corresponding distance on the ground. In many instances, a number of identifiable points of known position will appear on the photographs, and the known distances between these points offer means of determining scale. Scale may also be computed approximately as the ratio of the focal length of the aerial camera to the flight altitude, or height of the airplane above the average elevation of the terrain, by means of the following relationship:

$$S = \frac{f}{H - h} \tag{1}$$

in which H = height of camera lens above sea level
h = average elevation of ground points above sea level
f = focal length of the aerial camera

Conversely, the flight altitude to produce photographs of a desired scale can be computed from the same relationship.

An *index map* of the photographs is prepared, as follows: for large jobs, the position and coverage of each photograph is plotted on the flight map; for small jobs, the photographs are laid in the form of a mosaic and rephotographed.

23-6 Photographic Airplanes. Any airplane which is stable, has the required service ceiling, and has sufficient space for pilot, photographer, and aerial camera can be used for aerial photography. In addition, the airplane should allow excellent visibility for the pilot, permit operation from small and relatively unimproved fields, and have economical speed and performance characteristics. For flying at high altitudes, the airplane should afford protection from the cold and should have provision for oxygen for the crew.

23-7 Aerial Cameras. Aerial cameras may be of the *single-lens* or the *multiple-lens* type. Camera lenses are classified according to angular field of view as *normal-angle,* up to 75°; *wide-angle,* 75 to 100°; and *superwide-angle,* over 100°. The cameras are fully automatic, being operated by electricity; the operator has only to concern himself that the camera is level and fully alined with the line of flight. However, they can be manually controlled if desired.

The trimetrogon multiple-lens camera shown in Fig. 23-2 is used for the preparation of small-scale maps and charts of reconnaissance accuracy. Compilation of data is accomplished by using special plotting machines.

The U.S. Coast and Geodetic Survey multiple-lens aerial camera uses nine lenses, all of which point vertically downward; eight of them are mounted in a circle around one central lens which views the ground directly. Rays from the ground are reflected through the other eight from highly polished steel mirrors. The images are projected onto a single roll of film 24 in. wide. The photographs are printed onto a single paper by means of a rectifying printer to form a nine-lens composite photograph approximately 36 in. square.

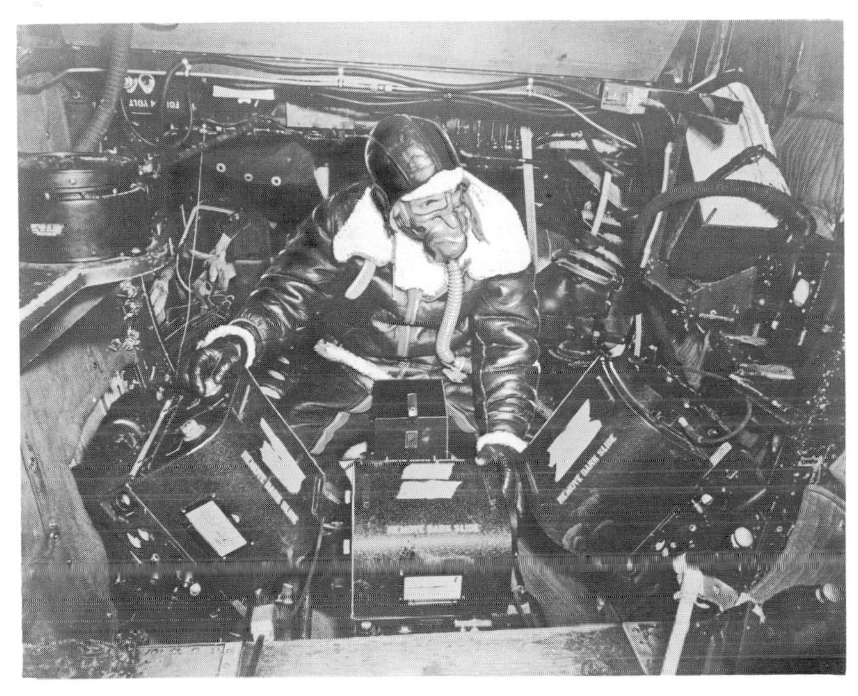

Fig. 23-2. Trimetrogon camera installation. (*U.S. Air Force.*)

23-8 Determination of Elevations by Measurement of Parallax.

The object of measuring the parallax difference between two ground points is to determine the difference in their elevations and thus to determine their respective heights above sea level. In Fig. 23-3 it is desired to measure the ground elevations above sea level of points A and B. Point A appears on photograph positive I at a and on positive II at a'. Line I_1a'' is constructed parallel to line I_2a', making a triangle $a''I_1a$ similar to triangle I_1AI_2. Then

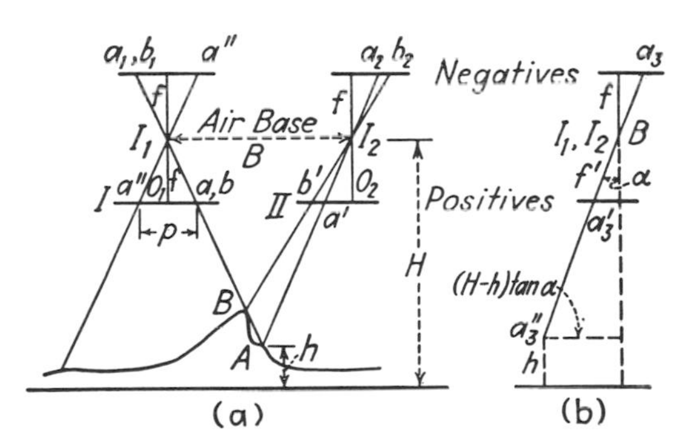

Fig. 23-3. Determination of elevation by measurement of parallax.

$$p = \frac{fB}{H - h} \tag{2}$$

in which H is the altitude of flight, h is the elevation of point A above sea level, B is the "air base," or distance the plane flew between exposures, and the distance p is the absolute stereoscopic parallax of point A.

Similarly the absolute parallax of point B can be determined. The difference between the absolute parallax of A and that of B is a measure of the difference between their elevations.

For purposes of contouring, use is made of Eq. (3) which is derived from Eq. (2) and which expresses the value of the change in parallax Δp (in millimeters) for a corresponding change in elevation Δh (in feet) between two points or between two contours on the photograph.

$$\Delta p = \frac{B_m \, \Delta h}{H - h} \tag{3}$$

in which H and h are in feet and B_m (in millimeters) is the measured stereoscopic base of the photographs $[(B_m = Bf/(H - h)]$.

In an aerial photograph, the displacement due to relief is a natural phenomenon and should not be considered an error. Were it not for the parallax displacement due to relief, stereoscopic measurement would be impossible.

23-9 Map Control. In addition to the necessary ground control, photogrammetric maps are based upon *secondary control* which is usually obtained graphically in a drafting room. Secondary control is said to tie the pictures to the ground. It may be established by the radial-line method (Art. 23-10), by the slotted-templet or the spider-templet method (Art. 23-11), or by aerial triangulation.

23-10 Radial-line Method of Control. The radial-line method of providing secondary map control from vertical aerial photographs is based upon the following perspective properties of such photographs: (1) that points near the center of a photograph are nearly free from errors of tilt; (2) that all errors due to small amounts of tilt and to differences of ground elevation are, within the limits of graphical measurement, radial from the principal point of each photograph; and (3) that objects included in properly overlapping photographs may be located by rays drawn to them from the principal points of photographs, the location of the objects being at the intersection of such lines. As in plane-table operations, the locations of points on aerial photographs are found at the intersections of rays drawn from two or more stations.

Marking of Photographs. In applying the radial-line method, a group of several consecutive photographs which include at least two ground-control points—one near each end of the group—is selected. Upon each photograph certain points (objects) are selected and marked and lines are drawn, as follows:

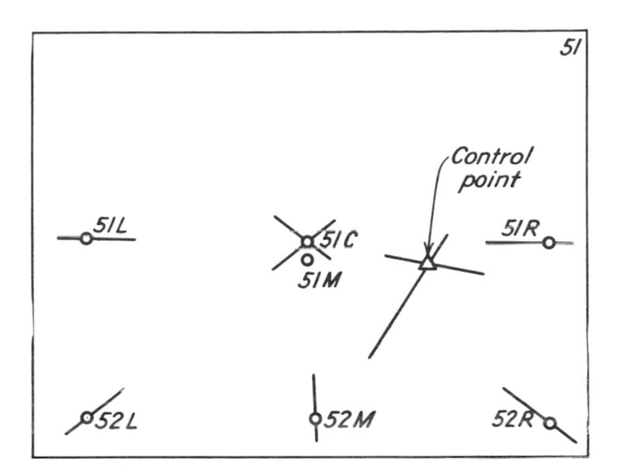

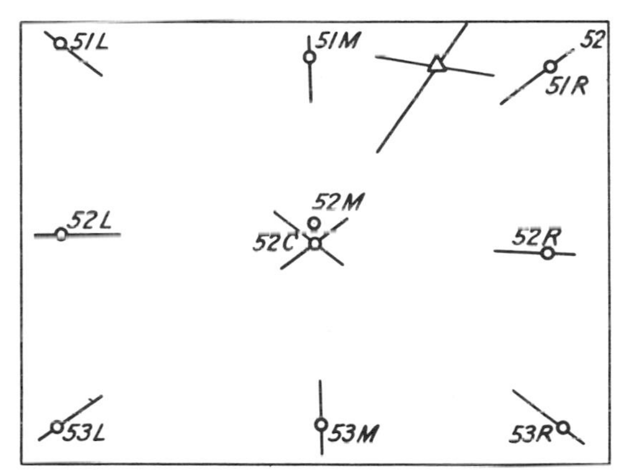

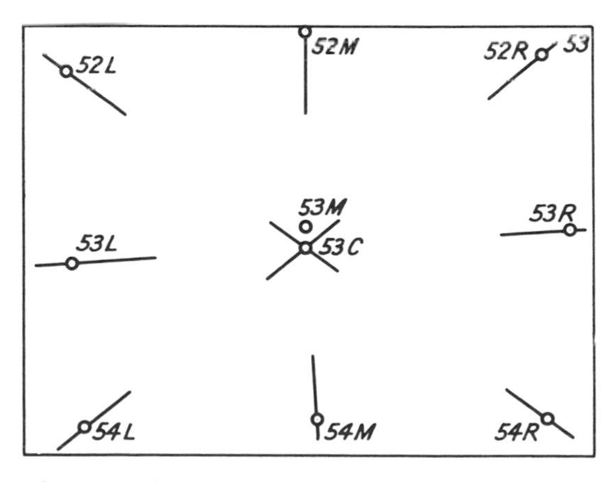

Fig. 23-4. Photographs marked for radial-line control.

1 The principal point of each photograph is plotted.

2 Beginning with the first photograph of the group, as, for example, 51, Fig. 23-4, a definite object called the "substitute center" (51*M*), which also appears on the adjoining photograph (52), is chosen near the principal point 51*C* and is marked on both photographs.

3 Points 51*R* and 51*L* are chosen at objects near the right and left edges, respectively, of the photograph, which objects also appear on photograph 52.

4 Similarly, points 52*R* and 52*L* are chosen at objects near the lower corners of photograph 51, and these two points must appear on the two succeeding photographs (52 and 53) opposite the center of 52 and near the upper corners of 53.

5 52*M* is chosen as a point which appears in photograph 52, near its center and also in photographs 51 and 53, near the lower edge of 51 and near the upper edge of 53.

This procedure of selecting and marking points having been carried through the group of photographs, it will be seen that each photograph of the group, except the first and last photographs, will have nine marked points in addition to the principal point. The first and last photographs will each have six marked points and the principal point as shown on 51 of Fig. 23-4, and the two photographs at each end of the group must also include at least one control point.

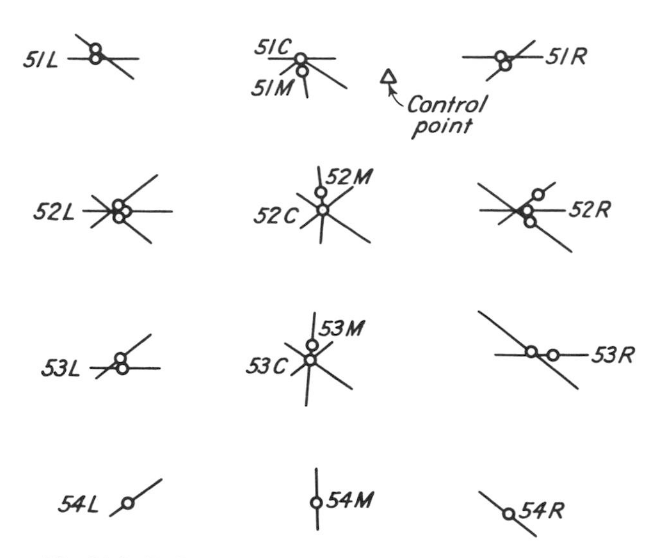

Fig. 23-5. Plotted map-control points.

Compilation of Control. The method of combining the data of the separate photographs into a map showing the correct relative locations of the selected points and the two control points is as follows:

A sheet of transparent film base (cellulose acetate) large enough to span the entire group of photographs when matched together is laid over the first photograph of the series, and it is so placed that each photograph in turn can be laid under the film base in correct relation. Three points, namely, the principal point 51*C* and the points 51*M* and 52*M* are traced. The radial line to each of the other points, including the control point, is also traced. Photograph 51 is removed, the film base is placed over photograph 52 so that the traced position of 52*M* falls on its position as it appears on photograph 52, and the film base is swung about this point until the traced position of 51*M* falls on its radial line on the photograph. The film base is now held with weights in this correctly oriented position. Radial lines to the other points are traced. These lines will give the location of all marked points that appear on photograph 51 and radial lines to three other points (53*L*, 53*M*, and 53*R*) which appear on photograph 52.

The location of a point thus determined may or may not coincide with its pictured location on either photograph, depending upon the tilt and relief displacements in the photograph. If the pictures are without errors due to tilt, etc., the pictured location of each point should lie at the intersection of the two lines drawn toward it but if displacement errors are present, the pictured location of any point will not coincide with the intersection of the radial lines. This condition is shown in Fig. 23-5, in which the pictured locations of the various points are shown by small circles and the correct locations are shown by the intersections of the corresponding radial lines.

When the location of points on the film base is begun, no attempt is made to use any particular scale, but a definite scale, as yet unknown, is established by the distance between the two central points, 51*M* and 52*M*, which have been traced as they appear on photograph 51. It is essential to accuracy that each central point, such as 51*M* or 52*M*, be selected as a point which lies close to the line connecting adjacent principal points and which lies close to its corresponding principal point and that the control point fall within the zone of overlap so that it can be intersected. The scale of the plotting in this case is the scale of the first photograph. As the procedure is continued it is evident that, as each succeeding photograph is treated, there have been located on the film base three points which appear along the upper border of the photograph and a direction line to the next central point. The film base is shifted until (1) the plotted locations of the three points appearing along the upper border of the photograph fall on their radial lines of the photograph and (2) the radial line on the film base to the principal point of the photograph now being treated falls on the point as marked on that photograph. When these conditions are satisfied, the film base is correctly placed with respect to the photograph; the principal point is then traced, and radial lines are drawn. This procedure is similar to the graphical solution of the three-point problem (Art. 13-14).

This process is carried through the group of photographs until the second control point is located. The scale of the data assembled on the film base can then be determined by measuring the distance between the two control points on the film base. Proper enlargement or reduction can be accomplished with the pantograph or graphically.

Topographical Features. The radial-line method is used even more extensively in the determination of the true locations of topographic features to be depicted on the final map. This application usually follows the adjustment of the radial-line strip in order to avoid shifting the locations of more than the minimum number of points. If control is in every photograph, the radial-line method of intersection may be used to "cut in" additional points, which operation can be done more expeditiously in the drafting room than in the field.

23-11 Slotted-templet Method of Control. A variation of the radial-line method of secondary map control is known as the *slotted-templet* method. Instead of the rays being drawn on the photographs as heretofore explained, the points are selected, marked on the photographs, and transferred to cardboard templets. Slots representing rays radiating from the nadir point to the selected photo points are cut into the templets by means of a mechanical slot cutter; the slotted templets are approximately oriented on the manuscript sheet; and movable metal studs are inserted through the slots which represent the rays to each selected photo point. Those studs which correspond to known control points are then fixed in position, and the system of slotted templets is shifted slightly about these fixed points until the arrangement having the least apparent residual strain is found. The movable studs are then in the most probable position for the corresponding photo points.

An alternative slotted-templet method of control involves the use of slotted strips of spring steel radiating about a center bolt in the form of "spiders," with each strip in the direction of a ray to a selected photo point. The procedure of orienting, connecting, and adjusting the spider-templet system is similar to that just described.

23-12 Compilation of Detail (Radial-line Method). In an aerial photograph, unimportant details are shown with the same degree of intensity as the important. The purpose of map compilation is to separate all these features and to represent them by conventional symbols according to their importance to the task at hand.

First the secondary control is transferred to the compilation manuscript. The manuscript is then oriented over each photograph in turn, and the detail is traced onto the sheet by the radial-line method, as follows:

1 Beginning at the center of the photograph, symbols are traced in for all the features in the central area, halfway to the nearest radial-line points.

2 The sheet is shifted slightly (if necessary) so that the location of one of the nearest radial-line points is exactly over the corresponding point on the photograph, the sheet is oriented, and the detail is traced in for the area surrounding the point, halfway to the adjacent radial-line points.

3 The detail between the two areas is connected, adjustments being made if necessary so that no feature is out of its relative location by more than about half the dimensions of the conventional sign.

4 The foregoing process is continued for all the other points in the area of the photograph, and for successive photographs in the series.

23-13 Contouring. There are two general methods of completing the contours on a map made from photographs: by field methods (usually plane-table mapping) and by application of the stereoscopic principle to the overlapping photographs.

The field method usually involves plane-table mapping wherein the plane-table board or worksheet is a copy of the radial plot reproduced on metal-mounted paper or dimensionally stable coated plastic. Aerial contact prints, enlargements, and precise mosaics are also used as plane-table sheets in reasonably flat country where the picture or mosaic, properly rectified, becomes a true map.

The other general method involves setting up the stereo pairs, oriented to ground or photogrammetric control, in stereoscopic plotting machines at proper scale and orientation. With this method there must be sufficient control to fix the datum of each model, either through photogrammetric extension or by ground-survey methods.

Contouring may also be accomplished on the stereocomparagraph (Art. 23-16) on a separate templet made for each photograph. Later the contours are compiled on the planimetric base sheet in the same manner as the detail of the photograph. Automatic plotting machines such as the Zeiss stereoplanigraph compile all the detail of the map in one operation; the radial-line technique is not used.

PLOTTING INSTRUMENTS

23-14 General. Photogrammetric plotting instruments are used to solve the equations of the loci of points on photographs automatically; they permit rapid and accurate compilation of the photographic detail in orthographic projection on a selected datum. Monocular or single-photo instruments and stereoscopic instruments are used.

Monocular and single-photo instruments are used mainly in the original compilation of small-scale planimetric charts and in the revision of existing map coverage where the planimetry is obsolete. Some are based upon the principle of the camera lucida, with which the eye receives two superimposed images, one from the photograph and the other from a map manuscript. Others employ optical projection.

Stereoscopic plotting instruments are used when it is desired to measure elevations and to trace contours and when relatively high precision is required. Automatic stereoscopic plotting instruments compensate for tilt of the photographs, measure elevations and delineate contours, and compile the perspective projections of the photographs onto the map manuscript in orthographic projection at one common scale.

The three main elements in all automatic stereoscopic plotting machines are a projection system, a viewing system, and a measuring system. Another element often classified is the tracing system. Projection systems are optical, mechanical, or combinations of optical and mechanical.

In one viewing system the image is viewed through a binocular microscope so arranged that the vision of one eye is directed to one photograph and the vision of the other eye is directed to the other photograph of a stereoscopic pair. In the other common viewing system the projections of the photograph are viewed on a clear white reflecting surface which also contains an index mark. In this system, image separation is obtained by the use of colored filters, polarizing filters, or a flicker system which alternately exposes the left and right photographs to view.

The measuring system in many instruments incorporates a simple device known as the Zeiss parallelogram, which enables either projector of the instrument to be used for the right or left photograph of a pair.

Anaglyphic projection stereoplotters have as main components (1) the supporting frame and projector bar, (2) the projectors, (3) the tracing table (stand) with platen and floating mark for the measuring of elevations and the tracing of planimetric details and of contours, and (4) the spectacles for viewing the projection stereoscopically. Examples are the Multiplex (Art. 23-15), the Balplex, and the Kelsh plotter. Precision stereoscopic plotting instruments of European manufacture include the Zeiss Stereoplanigraph and the Wild Autograph.

Analytical plotters make use of orthographic direct viewing; they are distinguished by the fact that the photographs are mounted in a common plane and that a parallax bar provides the basic means of measuring elevation. They are stereographs. Examples are the Zeiss Stereoscope and the Nistri-O.M.I. analytical stereoplotter.

Stereometer-type plotting instruments are relatively simple and portable. Differences in elevation are determined by measuring the differences in absolute stereoscopic parallax in the stereopair and converting these to differences in elevation. Two examples are the Fairchild Stereocomparagraph (Art. 23-16) and the Zeiss Stereopret.

In the following articles are briefly described one elaborate and one simple plotting machine.

23-15 Multiplex. The Multiplex is an anaglyphic projection stereoplotter. As shown in Fig. 23-6, it consists of a series of projectors which are small-scale reproductions of the taking cameras. A small diapositive, or positive transparency, is used in the projectors. A colored filter may be inserted in each projector in order to project with either red or blue-green light, as may be necessary.

In operation, the diapositives are inserted in the projectors; the projectors are then mutually adjusted through procedures called relative and absolute orientation so that their projected images will intersect in space above the plotting table and form a spatial model which is a true small-scale reproduction or image of the area common to two successive photographs of a strip. The spatial model is obtained by projecting one photograph of an overlapping pair in red light and the other in blue-green light, and by observing the combination of colors through spectacles containing one red and one blue-green lens. Only the red image is seen with

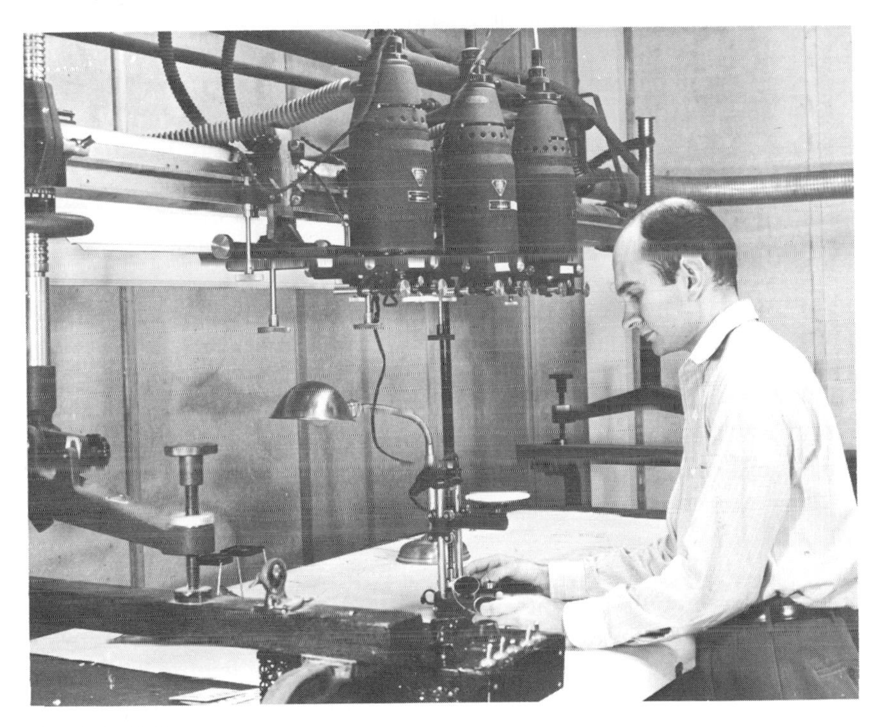

Fig. 23-6. Multiplex.

one eye, and only the blue-green image is seen with the other eye; thus the condition of stereoscopic vision is fulfilled. When the instrument is properly adjusted, the intersections of the bundles of rays from the projectors form a true image, in space, of the original landscape.

As the fusion of the images occurs in space above the drafting table, the image can be cut and measured at any desired height. An index mark, or "floating mark," is carried in the center of a circular disk on the tracing table. This disk is raised and lowered by means of a screw on the center post at the back of the tracing stand. On the left post of the tracing stand is a millimeter scale on which is read the height of the disk above the drafting table, which may be considered as the datum plane. Carried in the tracing stand directly below the floating mark is the drawing pencil which traces on the plotting sheet the horizontal movements of the floating mark. The height of any point is measured by bringing the floating mark into contact with the spatial model at the point, and reading the elevation in millimeters (at the plotting scale) directly on the millimeter scale on the tracing stand.

In contouring, the floating mark is set at the correct height of a contour (the value of the contour, in millimeters, multiplied by the representative fraction of the plotting scale), and the pencil is lowered onto the plotting sheet. The tracing stand is then moved about over the sheet while the floating mark is maintained in constant visual contact with the apparent terrain in the model. Although this process may seem difficult, it is easily accomplished on the instrument.

Planimetric features are delineated by following the feature with the floating mark and adjusting the elevation of the platen so that the floating mark is always at ground elevation as the plotting proceeds.

The projectors can be adjusted for tilt and in all directions so that they are located and oriented into the same relative positions as those of the aerial camera at the instant of exposure of the several photographs.

The advantages of the Multiplex are its ease of adjustment and the facility with which control can be extended from one photograph to the next. Control may be carried by aerial triangulation between successive bands of ground control several miles apart; it may even be extended from a single band of ground control. In order to increase production, control may be accomplished with the Multiplex and plotting with the stereocomparagraph.

23-16 Stereocomparagraph. The stereocomparagraph (Fig. 23-7) was invented by Benjamin B. Talley in 1935. It consists of a drawing attachment, a reflecting stereoscope, a parallax bar, a light fixture, and a parallel-motion device to keep the instrument in proper alinement. The stereoscope is equipped with matching lenses for the magnification of photographic detail. The measuring system consists of two index marks, one etched at the center of each of the two lenses attached to the base of the instrument; these marks may be fused when observed in the stereoscope. The right-hand mark is attached to a micrometer screw to change and measure the spacing between the marks. The micrometer is graduated from 0 to 25 mm and reads directly to 1/100 mm when clamped for reading. The right-hand mark may also be adjusted through a small range in a direction perpendicular to the micrometer screw. The left-hand mark is held in place by a locating pin and may be adjusted in increments of 5 mm through a range of 25 mm. Nominal spacing between the marks is 150 mm.

The lenses containing the index marks are in contact with the photographs. When the micrometer screw is turned, the two index marks fuse into a single image, or floating mark, which appears to rise or fall in space as the distance between the two lenses is decreased or increased respectively. As indicated by Eq. (3) (Art. 23-8), the movement of the micrometer (Δp), in millimeters, corresponding to an apparent change in height of the floating mark is equal to the mean stereoscopic base (B_m), in millimeters, of the photographs multiplied by the difference in elevation (Δh), in feet, between the two pictured points on which the floating mark is held, divided by the altitude of the airplane ($H - h$), in feet, above these points.

Fig. 23-7. Stereocomparagraph (Fairchild). (*Fairchild Space and Defense Systems.*)

In practice, the difference in elevation of points on aerial photographs is determined on the stereocomparagraph by measuring directly the difference in parallax of the points and by converting this difference into feet by reference to a parallax table which is constructed by solving Eq. (3) for an assumed stereoscopic base, for given increments of difference in elevation, and for variations in flight altitude throughout the range likely to be encountered.

For contouring, the micrometer is set at a value corresponding to the elevation of the contour to be drawn, and the floating mark is maintained in contact with the spatial model while the instrument is moved about over the photographs. Each photograph is contoured separately, and the contours are compiled onto the map sheet as previously described for the cultural detail of the photographs. Thus the scale is made uniform and the horizontal displacements of the contours due to relief are adjusted in the same operation. Control is required in every photograph.

APPLICATIONS OF PHOTOGRAMMETRY

23-17 General. Although the most common application of photogrammetry is in the preparation of topographic maps, photogrammetry is by no means limited to this field.

During recent years aerial photographs and photogrammetry have become widely used in the preparation of property-ownership maps for tax-equalization studies. Through their use the petroleum geologist, who formerly spent about 90 per cent of his time and effort in keeping himself located and oriented on the ground and 10 per cent of his time on geology is now able to reverse these percentages. Aerial photographs are used to study and catalogue geological formations and land classifications in mining and in aerial exploration. In forestry they are used for the classification of growing timber, for the determination of tree heights for the estimation of merchantable timber, and for other studies. The Federal government has used tremendous numbers of photographs for rural-rehabilitation studies and in connection with soil-conservation and erosion projects. Aerial photographs are increasingly used in the determination of correct land usages for agricultural purposes. Aerial maps find a wide application in city, county, and regional planning and development and in general engineering studies such as those for highway, pipeline, and transmission-line locations and for construction.

One of the more recent developments and one which is worldwide in its scope is the use of photogrammetry in connection with geophysical prospecting by means of the magnetometer. Color photography is often used in this work, since the varying tints of the geological formations are more readily discernible in natural color than in black-and-white photographs.

Military applications of aerial photography and photogrammetry are myriad in extent and usefulness.

REFERENCES

1. ADAMS, O. S., and C. N. CLAIRE, "Manual of Plane-coordinate Computation," *U.S. Coast and Geodetic Survey Spec. Pub.* 193, Government Printing Office, Washington, D.C., 1935.
2. ALLEN, C. F., "Railroad Curves and Earthwork: Field and Office Tables," 7th ed., McGraw-Hill Book Company, New York, 1931.
3. AMERICAN CONGRESS ON SURVEYING AND MAPPING, "Minimum Standard Detail Requirements for Land Title Surveys," *Surveying and Mapping*, pp. 339–341, June, 1962.
4. AMERICAN RAILWAY ENGINEERING ASSOCIATION, "Manual," current edition.
5. AMERICAN SOCIETY OF CIVIL ENGINEERS, Manuals of Engineering Practice, as follows:
   10. "Technical Procedure for City Surveys," 1963.
   14. "Location of Underground Utilities," 1937.
   16. "Land Subdivision," 1939.
   34. "Definitions of Surveying, Mapping, and Related Terms," 1954.
   44. "Report on Highway and Bridge Surveys," 1962 (Chap. 1, state plane coordinates; Chaps. 7 and 8, construction surveys).
6. AMERICAN SOCIETY OF PHOTOGRAMMETRY, "Manual of Photogrammetry," 2d ed., Washington, D.C., 1952.
7. AMERICAN SOCIETY OF PHOTOGRAMMETRY, "Manual of Photographic Interpretation," Washington, D.C., 1960.
8. BIRDSEYE, C. H., "Topographic Instructions of the United States Geological Survey," *U.S. Geological Survey Bull.* 788, Government Printing Office, Washington, D.C., 1928 (see also U.S. Geological Survey, hereinafter).
9. CLARK, FRANK E., "Law of Surveying and Boundaries," 3d ed., The Bobbs-Merrill Company, Inc., Indianapolis, 1959.
10. DAVIS, RAYMOND E., et al., "Surveying: Theory and Practice," 5th ed., McGraw-Hill Book Company, New York, 1966.
11. DEEL, SAMUEL A., "Magnetic Declination in the United States—1945," *U.S. Coast and Geodetic Survey Serial* 664, Government Printing Office, Washington, D.C., 1946 (see also U.S. Coast and Geodetic Survey, hereinafter).
12. DEETZ, C. H., and O. S. ADAMS, "Elements of Map Projection, with Applications to Map and Chart Construction," 5th ed., *U.S. Coast and Geodetic Survey Spec. Pub.* 68, Government Printing Office, Washington, D.C., 1944.
13. DURHAM, E. B., "Mine Surveying," McGraw-Hill Book Company, New York, 1927.

14. Franklin, W. S., "An Elementary Treatise on Precision of Measurements," Franklin and Charles, Lancaster, Pa., 1925.
15. Grover, N. C., and A. W. Harrington, "Stream Flow," John Wiley & Sons, Inc., New York, 1943.
16. Harrison, A. E., "Electronic Surveying: Electronic Distance Measurements," *Journal of the Surveying and Mapping Division,* pp. 97–116, American Society of Civil Engineers, October, 1963.
17. Hickerson, Thomas F., "Route Surveys and Design," 4th ed., McGraw-Hill Book Company, New York, 1959.
18. Hodgman, F., "Land Surveying," The F. Hodgman Company, Climax, Mich., 1907.
19. Hodgson, C. V., "Manual of Second and Third Order Triangulation and Traverse," *U.S. Coast and Geodetic Survey Spec. Pub.* 145, Government Printing Office, Washington, D.C., 1935.
20. Hosmer, G. L., and J. M. Robbins, "Practical Astronomy," 4th ed., John Wiley & Sons, Inc., New York, 1948.
21. Ives, H. C., and Philip Kissam, "Highway Curves," 4th ed., John Wiley & Sons, Inc., New York, 1952.
22. Jeffers, K. B., "Hydrographic Manual," *U.S. Coast and Geodetic Survey Pub.* 20–2, Government Printing Office, Washington, D.C., 1960.
23. Kissam, Philip, "Optical Tooling: For Precise Manufacture and Alignment," McGraw-Hill Book Company, New York, 1962.
24. Kissam, Philip, "Surveying for Civil Engineers," McGraw-Hill Book Company, New York, 1956.
25. Low, Julian W., "Plane Table Mapping," Harper & Row, Publishers, Incorporated, New York, 1952.
26. Meyer, Carl F., "Route Surveying," 3d ed., International Textbook Company, Scranton, Pa., 1962.
27. Mitchell, Hugh C., and Lansing G. Simmons, "The State Coordinate Systems: A Manual for Surveyors," *U.S. Coast and Geodetic Survey Spec. Pub.* 235, Government Printing Office, Washington, D.C., 1945.
28. Moffitt, Francis H., "Photogrammetry," International Textbook Company, Scranton, Pa., 1959.
29. Mussetter, William, "Tacheometric Surveying—Methods and Instruments," *Surveying and Mapping,* pp. 137–156, April–June 1956, and pp. 473–487, October–December, 1956.
30. Nassau, J. J., "Textbook of Practical Astronomy," 2d ed., McGraw-Hill Book Company, New York, 1948.
31. Pickels, G. W., and C. C. Wiley, "Route Surveying," 3d ed., John Wiley & Sons, Inc., New York, 1949.
32. Raisz, Erwin, "Principles of Cartography," McGraw-Hill Book Company, New York, 1962.
33. Reynolds, W. F., "Manual of Triangulation Computation and Adjustment," *U.S. Coast and Geodetic Survey Spec. Pub.* 138, Government Printing Office, Washington, D.C., 1934.
34. Rubey, Harry, "Route Surveys and Construction," 3d ed., The Macmillan Company, New York, 1956.
35. Searles, W. H., et al., "Field Engineering," 22d ed., John Wiley & Sons, Inc., New York, 1949.
36. Sloane, R. C., and J. M. Montz, "Elements of Topographic Drawing," 2d ed., McGraw-Hill Book Company, New York, 1943.
37. Swainson, O. W., "Topographic Manual," *U.S. Coast and Geodetic*

Survey Spec. Pub. 144, Government Printing Office, Washington, D.C., 1928.

38. Swanson, L. W., "Topographic Manual: Part II, Photogrammetry," *U.S. Coast and Geodetic Survey Spec. Pub.* 249, Government Printing Office, Washington, D.C., 1951.
39. Talley, B. B., and P. H. Robbins, "Photographic Surveying," Pitman Publishing Corporation, New York, 1945.
40. U.S. Bureau of Land Management, "Ephemeris of the Sun, Polaris, and Other Selected Stars for the Year ____," Government Printing Office, Washington, D.C., annual (in advance).
41. U.S. Bureau of Land Management, "Manual of Instructions for the Survey of the Public Lands of the United States," Government Printing Office, Washington, D.C., 1947.
42. U.S. Bureau of Land Management, "Restoration of Lost or Obliterated Corners and Subdivision of Sections," *Circ.* 1452, Government Printing Office, Washington, D.C., 1939.
43. U.S. Bureau of Land Management, "Standard Field Tables," 8th ed., Government Printing Office, Washington, D.C., 1956.
44. U.S. Coast and Geodetic Survey, "Control Leveling," *Spec. Pub.* 226, Government Printing Office, Washington, D.C., 1961.
45. U.S. Coast and Geodetic Survey, "United States Magnetic Tables for 1960," Pub. 40–2, Government Printing Office, Washington, D.C., 1962.
46. U.S. Geological Survey, "Topographic Instructions," Government Printing Office, Washington, D.C. (Published in separate volumes according to subject; list available from USGS.)
47. U.S. Government, "Transits, One-minute, and Transit Tripods," *Federal Specification* GG-T-621, Government Printing Office, Washington, D.C., 1948.
48. U.S. Naval Observatory, "The Nautical Almanac for the Year ____," Nautical Almanac Office, Washington, D.C., annual (in advance).
49. Urquhart, L. C., ed., "Civil Engineering Handbook," 4th ed., McGraw-Hill Book Company, New York, 1959.
50. Wright, T. W., and J. F. Hayward, "The Adjustment of Observations," D. Van Nostrand Company, Inc., Princeton, N.J., 1906.
51. Condensed tables regarding the sun and Polaris; various manufacturers of surveying instruments (annual, in advance).
52. Manuals of instruction for surveys and for construction of roads and bridges; various state highway departments.

GENERAL TABLES

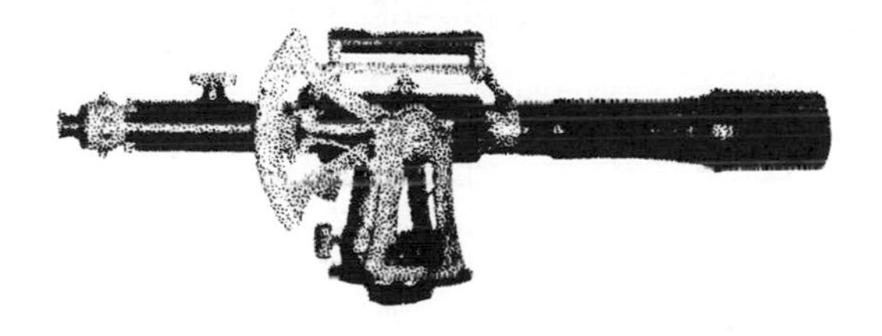

Table I Correction for refraction and parallax, to be subtracted from the observed altitude of the sun

Barometric pressure 29.5 in.

App't alt.	Temperature										App't alt.
	−10° C +14° F	−5° C +23° F	0° C +32° F	+5° C +41° F	+10° C +50° F	+15° C +59° F	+20° C +68° F	+25° C +77° F	+30° C +86° F	+35° C +95° F	
°	′	′	′	′	′	′	′	′	′	′	°
10	5.52	5.42	5.30	5.20	5.10	5.00	4.92	4.83	4.75	4.67	10
11	5.02	4.92	4.82	4.73	4.63	4.55	4.47	4.38	4.32	4.23	11
12	4.60	4.50	4.42	4.33	4.25	4.17	4.10	4.03	3.97	3.88	12
13	4.23	4.15	4.07	4.00	3.92	3.85	3.78	3.72	3.65	3.58	13
14	3.92	3.83	3.77	3.70	3.62	3.55	3.50	3.45	3.37	3.32	14
15	3.65	3.58	3.50	3.43	3.37	3.32	3.25	3.20	3.13	3.08	15
16	3.43	3.35	3.30	3.23	3.17	3.12	3.07	3.00	2.95	2.90	16
17	3.22	3.15	3.10	3.03	2.98	2.92	2.88	2.82	2.77	2.72	17
18	3.02	2.95	2.90	2.85	2.80	2.75	2.70	2.65	2.60	2.55	18
19	2.83	2.78	2.73	2.68	2.63	2.58	2.53	2.48	2.43	2.40	19
20	2.68	2.63	2.58	2.53	2.48	2.43	2.38	2.33	2.30	2.27	20
21	2.53	2.48	2.43	2.38	2.35	2.30	2.27	2.22	2.17	2.13	21
22	2.38	2.35	2.30	2.25	2.22	2.18	2.13	2.08	2.05	2.02	22
23	2.28	2.25	2.20	2.15	2.12	2.08	2.03	1.98	1.95	1.93	23
24	2.17	2.13	2.08	2.05	2.02	1.98	1.93	1.88	1.87	1.83	24
25	2.07	2.03	1.98	1.95	1.92	1.88	1.83	1.80	1.77	1.75	25
26	1.99	1.95	1.90	1.87	1.83	1.80	1.75	1.72	1.70	1.67	26
27	1.88	1.85	1.82	1.78	1.75	1.72	1.68	1.63	1.62	1.60	27
28	1.80	1.77	1.72	1.70	1.67	1.63	1.60	1.57	1.53	1.52	28
29	1.72	1.68	1.65	1.63	1.60	1.57	1.53	1.50	1.47	1.46	29
30	1.65	1.62	1.58	1.57	1.53	1.50	1.47	1.45	1.42	1.40	30
32	1.53	1.50	1.47	1.45	1.42	1.38	1.35	1.33	1.30	1.28	32
34	1.41	1.37	1.35	1.32	1.30	1.27	1.25	1.23	1.20	1.18	34
36	1.30	1.27	1.25	1.22	1.20	1.18	1.15	1.13	1.10	1.08	36
38	1.20	1.18	1.15	1.13	1.12	1.10	1.07	1.05	1.02	1.02	38
40	1.11	1.10	1.07	1.05	1.03	1.02	0.98	0.97	0.95	0.93	40
42	1.03	1.00	0.98	0.97	0.95	0.93	0.90	0.88	0.87	0.87	42
44	0.96	0.93	0.92	0.90	0.88	0.87	0.85	0.83	0.82	0.80	44
46	0.89	0.88	0.87	0.85	0.83	0.82	0.80	0.78	0.77	0.75	46
48	0.83	0.82	0.80	0.78	0.77	0.75	0.73	0.72	0.70	0.68	48
50	0.77	0.75	0.73	0.72	0.70	0.68	0.67	0.67	0.65	0.63	50
55	0.63	0.62	0.60	0.60	0.58	0.57	0.57	0.55	0.53	0.52	55
60	0.52	0.52	0.50	0.50	0.48	0.47	0.47	0.45	0.45	0.43	60
65	0.42	0.40	0.40	0.40	0.38	0.38	0.37	0.37	0.35	0.33	65
70	0.32	0.32	0.32	0.30	0.30	0.30	0.28	0.28	0.28	0.27	70
75	0.23	0.23	0.23	0.22	0.22	0.22	0.20	0.20	0.20	0.18	75
80	0.15	0.15	0.13	0.13	0.13	0.13	0.13	0.12	0.12	0.12	80
85	0.07	0.07	0.07	0.07	0.07	0.07	0.07	0.05	0.05	0.05	85
90	0.00	0.00	0.00	0.00	0.00	0.00	0.00	0.00	0.00	0.00	90

Table II Correction for refraction, to be subtracted from the observed altitude of a star

Barometric pressure 29.5 in.

App't alt.	Temperature										App't alt.
	−10° C +14° F	−5° C +23° F	0° C +32° F	+5° C +41° F	+10° C +50° F	+15° C +59° F	+20° C +68° F	+25° C +77° F	+30° C +86° F	+35° C +95° F	
°	′	′	′	′	′	′	′	′	′	′	°
10	5.67	5.57	5.45	5.35	5.25	5.15	5.07	4.98	4.90	4.82	10
11	5.17	5.07	4.97	4.88	4.78	4.70	4.62	4.53	4.47	4.38	11
12	4.75	4.65	4.57	4.48	4.40	4.32	4.25	4.18	4.12	4.03	12
13	4.38	4.30	4.22	4.15	4.07	4.00	3.93	3.87	3.80	3.73	13
14	4.06	3.97	3.91	3.84	3.76	3.69	3.64	3.59	3.51	3.46	14
15	3.79	3.72	3.64	3.57	3.51	3.46	3.39	3.34	3.27	3.22	15
16	3.57	3.49	3.44	3.37	3.31	3.26	3.21	3.14	3.09	3.04	16
17	3.36	3.29	3.24	3.17	3.12	3.06	3.02	2.96	2.91	2.86	17
18	3.16	3.09	3.04	2.99	2.94	2.89	2.84	2.79	2.74	2.69	18
19	2.97	2.92	2.87	2.82	2.77	2.72	2.67	2.62	2.57	2.54	19
20	2.82	2.77	2.72	2.67	2.62	2.57	2.52	2.47	2.44	2.41	20
21	2.67	2.62	2.57	2.52	2.49	2.44	2.41	2.36	2.31	2.27	21
22	2.52	2.49	2.44	2.39	2.36	2.32	2.27	2.22	2.19	2.16	22
23	2.42	2.39	2.34	2.29	2.26	2.22	2.17	2.12	2.09	2.07	23
24	2.31	2.27	2.22	2.19	2.16	2.12	2.07	2.02	2.01	1.97	24
25	2.21	2.17	2.12	2.09	2.06	2.02	1.97	1.94	1.91	1.89	25
26	2.12	2.08	2.03	2.00	1.96	1.93	1.88	1.85	1.83	1.80	26
27	2.01	1.98	1.95	1.91	1.88	1.85	1.81	1.76	1.75	1.73	27
28	1.93	1.90	1.85	1.83	1.80	1.76	1.73	1.70	1.66	1.65	28
29	1.85	1.81	1.78	1.76	1.73	1.70	1.66	1.63	1.60	1.59	29
30	1.78	1.75	1.71	1.70	1.66	1.63	1.60	1.58	1.55	1.53	30
32	1.65	1.62	1.59	1.57	1.54	1.50	1.47	1.45	1.42	1.40	32
34	1.53	1.49	1.47	1.44	1.42	1.39	1.35	1.35	1.32	1.30	34
36	1.42	1.39	1.37	1.34	1.32	1.30	1.27	1.25	1.22	1.20	36
38	1.32	1.30	1.27	1.25	1.24	1.22	1.19	1.17	1.14	1.14	38
40	1.22	1.21	1.18	1.16	1.14	1.13	1.09	1.08	1.06	1.04	40
42	1.14	1.11	1.09	1.08	1.06	1.04	1.01	0.99	0.98	0.98	42
44	1.07	1.04	1.03	1.01	0.99	0.98	0.96	0.94	0.93	0.91	44
46	0.99	0.98	0.97	0.95	0.93	0.92	0.90	0.88	0.87	0.85	46
48	0.93	0.92	0.90	0.88	0.87	0.85	0.83	0.82	0.80	0.78	48
50	0.86	0.84	0.82	0.81	0.79	0.77	0.76	0.76	0.74	0.72	50
55	0.72	0.71	0.69	0.69	0.67	0.66	0.66	0.64	0.62	0.61	55
60	0.59	0.59	0.57	0.57	0.55	0.54	0.54	0.52	0.52	0.50	60
65	0.48	0.46	0.46	0.46	0.44	0.44	0.43	0.43	0.41	0.39	65
70	0.37	0.37	0.37	0.35	0.35	0.35	0.33	0.33	0.33	0.32	70
75	0.27	0.27	0.27	0.26	0.26	0.26	0.24	0.24	0.24	0.22	75
80	0.18	0.18	0.16	0.16	0.16	0.16	0.16	0.15	0.15	0.15	80
85	0.08	0.08	0.08	0.08	0.08	0.08	0.08	0.06	0.06	0.06	85
90	0.00	0.00	0.00	0.00	0.00	0.00	0.00	0.00	0.00	0.00	90

Table III * Horizontal distances and elevations from stadia readings

Minutes	0° Hor. Dist.	0° Diff. Elev.	1° Hor. Dist.	1° Diff. Elev.	2° Hor. Dist.	2° Diff. Elev.	3° Hor. Dist.	3° Diff. Elev.
0	100.00	0.00	99.97	1.74	99.88	3.49	99.73	5.23
2	100.00	0.06	99.97	1.80	99.87	3.55	99.72	5.28
4	100.00	0.12	99.97	1.86	99.87	3.60	99.71	5.34
6	100.00	0.17	99.96	1.92	99.87	3.66	99.71	5.40
8	100.00	0.23	99.96	1.98	99.86	3.72	99.70	5.46
10	100.00	0.29	99.96	2.04	99.86	3.78	99.69	5.52
12	100.00	0.35	99.96	2.09	99.85	3.84	99.69	5.57
14	100.00	0.41	99.95	2.15	99.85	3.90	99.68	5.63
16	100.00	0.47	99.95	2.21	99.84	3.95	99.68	5.69
18	100.00	0.52	99.95	2.27	99.84	4.01	99.67	5.75
20	100.00	0.58	99.95	2.33	99.83	4.07	99.66	5.80
22	100.00	0.64	99.94	2.38	99.83	4.13	99.66	5.86
24	100.00	0.70	99.94	2.44	99.82	4.18	99.65	5.92
26	99.99	0.76	99.94	2.50	99.82	4.24	99.64	5.98
28	99.99	0.81	99.93	2.56	99.81	4.30	99.63	6.04
30	99.99	0.87	99.93	2.62	99.81	4.36	99.63	6.09
32	99.99	0.93	99.93	2.67	99.80	4.42	99.62	6.15
34	99.99	0.99	99.93	2.73	99.80	4.48	99.62	6.21
36	99.99	1.05	99.92	2.79	99.79	4.53	99.61	6.27
38	99.99	1.11	99.92	2.85	99.79	4.59	99.60	6.33
40	99.99	1.16	99.92	2.91	99.78	4.65	99.59	6.38
42	99.99	1.22	99.91	2.97	99.78	4.71	99.59	6.44
44	99.98	1.28	99.91	3.02	99.77	4.76	99.58	6.50
46	99.98	1.34	99.90	3.08	99.77	4.82	99.57	6.56
48	99.98	1.40	99.90	3.14	99.76	4.88	99.56	6.61
50	99.98	1.45	99.90	3.20	99.76	4.94	99.56	6.67
52	99.98	1.51	99.89	3.26	99.75	4.99	99.55	6.73
54	99.98	1.57	99.89	3.31	99.74	5.05	99.54	6.78
56	99.97	1.63	99.89	3.37	99.74	5.11	99.53	6.84
58	99.97	1.69	99.88	3.43	99.73	5.17	99.52	6.90
60	99.97	1.74	99.88	3.49	99.73	5.23	99.51	6.96
C = 0.75	0.75	0.01	0.75	0.02	0.75	0.03	0.75	0.05
C = 1.00	1.00	0.01	1.00	0.03	1.00	0.04	1.00	0.06
C = 1.25	1.25	0.02	1.25	0.03	1.25	0.05	1.25	0.08

* From "Theory and Practice of Surveying," by J. B. Johnson. By permission of the publishers, John Wiley & Sons, Inc., New York.

Table III Horizontal distances and elevations from stadia readings—Continued

	4°		5°		6°		7°	
Minutes	Hor. Dist.	Diff. Elev.	Hor. Dist.	Diff. Elev.	Hor. Dist.	Diff. Elev.	Hor. Dist.	Diff. Elev.
0	99.51	6.96	99.24	8.68	98.91	10.40	98.51	12.10
2	99.51	7.02	99.23	8.74	98.90	10.45	98.50	12.15
4	99.50	7.07	99.22	8.80	98.88	10.51	98.48	12.21
6	99.49	7.13	99.21	8.85	98.87	10.57	98.47	12.26
8	99.48	7.19	99.20	8.91	98.86	10.62	98.46	12.32
10	99.47	7.25	99.19	8.97	98.85	10.68	98.44	12.38
12	99.46	7.30	99.18	9.03	98.83	10.74	98.43	12.43
14	99.46	7.36	99.17	9.08	98.82	10.79	98.41	12.49
16	99.45	7.42	99.16	9.14	98.81	10.85	98.40	12.55
18	99.44	7.48	99.15	9.20	98.80	10.91	98.39	12.60
20	99.43	7.53	99.14	9.25	98.78	10.96	98.37	12.66
22	99.42	7.59	99.13	9.31	98.77	11.02	98.36	12.72
24	99.41	7.65	99.11	9.37	98.76	11.08	98.34	12.77
26	99.40	7.71	99.10	9.43	98.74	11.13	98.33	12.83
28	99.39	7.76	99.09	9.48	98.73	11.19	98.31	12.88
30	99.38	7.82	99.08	9.54	98.72	11.25	98.29	12.94
32	99.38	7.88	99.07	9.60	98.71	11.30	98.28	13.00
34	99.37	7.94	99.06	9.65	98.69	11.36	98.27	13.05
36	99.36	7.99	99.05	9.71	98.68	11.42	98.25	13.11
38	99.35	8.05	99.04	9.77	98.67	11.47	98.24	13.17
40	99.34	8.11	99.03	9.83	98.65	11.53	98.22	13.22
42	99.33	8.17	99.01	9.88	98.64	11.59	98.20	13.28
44	99.32	8.22	99.00	9.94	98.63	11.64	98.19	13.33
46	99.31	8.28	98.99	10.00	98.61	11.70	98.17	13.39
48	99.30	8.34	98.98	10.05	98.60	11.76	98.16	13.45
50	99.29	8.40	98.97	10.11	98.58	11.81	98.14	13.50
52	99.28	8.45	98.96	10.17	98.57	11.87	98.13	13.56
54	99.27	8.51	98.94	10.22	98.56	11.93	98.11	13.61
56	99.26	8.57	98.93	10.28	98.54	11.98	98.10	13.67
58	99.25	8.63	98.92	10.34	98.53	12.04	98.08	13.73
60	99.24	8.68	98.91	10.40	98.51	12.10	98.06	13.78
C = 0.75	0.75	0.06	0.75	0.07	0.75	0.08	0.74	0.10
C = 1.00	1.00	0.08	0.99	0.09	0.99	0.11	0.99	0.13
C = 1.25	1.25	0.10	1.24	0.11	1.24	0.14	1.24	0.16

Table III Horizontal distances and elevations from stadia readings—Continued

	8°		9°		10°		11°	
Minutes	Hor. Dist.	Diff. Elev.	Hor. Dist.	Diff. Elev.	Hor. Dist.	Diff. Elev.	Hor. Dist.	Diff. Elev.
0	98.06	13.78	97.55	15.45	96.98	17.10	96.36	18.73
2	98.05	13.84	97.53	15.51	96.96	17.16	96.34	18.78
4	98.03	13.89	97.52	15.56	96.94	17.21	96.32	18.84
6	98.01	13.95	97.50	15.62	96.92	17.26	96.29	18.89
8	98.00	14.01	97.48	15.67	96.90	17.32	96.27	18.95
10	97.98	14.06	97.46	15.73	96.88	17.37	96.25	19.00
12	97.97	14.12	97.44	15.78	96.86	17.43	96.23	19.05
14	97.95	14.17	97.43	15.84	96.84	17.48	96.21	19.11
16	97.93	14.23	97.41	15.89	96.82	17.54	96.18	19.16
18	97.92	14.28	97.39	15.95	96.80	17.59	96.16	19.21
20	97.90	14.34	97.37	16.00	96.78	17.65	96.14	19.27
22	97.88	14.40	97.35	16.06	96.76	17.70	96.12	19.32
24	97.87	14.45	97.33	16.11	96.74	17.76	96.09	19.38
26	97.85	14.51	97.31	16.17	96.72	17.81	96.07	19.43
28	97.83	14.56	97.29	16.22	96.70	17.86	96.05	19.48
30	97.82	14.62	97.28	16.28	96.68	17.92	96.03	19.54
32	97.80	14.67	97.26	16.33	96.66	17.97	96.00	19.59
34	97.78	14.73	97.24	16.39	96.64	18.03	95.98	19.64
36	97.76	14.79	97.22	16.44	96.62	18.08	95.96	19.70
38	97.75	14.84	97.20	16.50	96.60	18.14	95.93	19.75
40	97.73	14.90	97.18	16.55	96.57	18.19	95.91	19.80
42	97.71	14.95	97.16	16.61	96.55	18.24	95.89	19.86
44	97.69	15.01	97.14	16.66	96.53	18.30	95.86	19.91
46	97.68	15.06	97.12	16.72	96.51	18.35	95.84	19.96
48	97.66	15.12	97.10	16.77	96.49	18.41	95.82	20.02
50	97.64	15.17	97.08	16.83	96.47	18.46	95.79	20.07
52	97.62	15.23	97.06	16.88	96.45	18.51	95.77	20.12
54	97.61	15.28	97.04	16.94	96.42	18.57	95.75	20.18
56	97.59	15.34	97.02	16.99	96.40	18.62	95.72	20.23
58	97.57	15.40	97.00	17.05	96.38	18.68	95.70	20.28
60	97.55	15.45	96.98	17.10	96.36	18.73	95.68	20.34
C = 0.75	0.74	0.11	0.74	0.12	0.74	0.14	0.73	0.15
C = 1.00	0.99	0.15	0.99	0.16	0.98	0.18	0.98	0.20
C = 1.25	1.23	0.18	1.23	0.21	1.23	0.23	1.22	0.25

Table III Horizontal distances and elevations from stadia readings—Continued

	12°		13°		14°		15°	
Minutes	Hor. Dist.	Diff. Elev.	Hor. Dist.	Diff. Elev.	Hor. Dist.	Diff. Elev.	Hor. Dist.	Diff. Elev.
0	95.68	20.34	94.94	21.92	94.15	23.47	93.30	25.00
2	95.65	20.39	94.91	21.97	94.12	23.52	93.27	25.05
4	95.63	20.44	94.89	22.02	94.09	23.58	93.24	25.10
6	95.61	20.50	94.86	22.08	94.07	23.63	93.21	25.15
8	95.58	20.55	94.84	22.13	94.04	23.68	93.18	25.20
10	95.56	20.60	94.81	22.18	94.01	23.73	93.16	25.25
12	95.53	20.66	94.79	22.23	93.98	23.78	93.13	25.30
14	95.51	20.71	94.76	22.28	93.95	23.83	93.10	25.35
16	95.49	20.76	94.73	22.34	93.93	23.88	93.07	25.40
18	95.46	20.81	94.71	22.39	93.90	23.93	93.04	25.45
20	95.44	20.87	94.68	22.44	93.87	23.99	93.01	25.50
22	95.41	20.92	94.66	22.49	93.84	24.04	92.98	25.55
24	95.39	20.97	94.63	22.54	93.81	24.09	92.95	25.60
26	95.36	21.03	94.60	22.60	93.79	24.14	92.92	25.65
28	95.34	21.08	94.58	22.65	93.76	24.19	92.89	25.70
30	95.32	21.13	94.55	22.70	93.73	24.24	92.86	25.75
32	95.29	21.18	94.52	22.75	93.70	24.29	92.83	25.80
34	95.27	21.24	94.50	22.80	93.67	24.34	92.80	25.85
36	95.24	21.29	94.47	22.85	93.65	24.39	92.77	25.90
38	95.22	21.34	94.44	22.91	93.62	24.44	92.74	25.95
40	95.19	21.39	94.42	22.96	93.59	24.49	92.71	26.00
42	95.17	21.45	94.39	23.01	93.56	24.55	92.68	26.05
44	95.14	21.50	94.36	23.06	93.53	24.60	92.65	26.10
46	95.12	21.55	94.34	23.11	93.50	24.65	92.62	26.15
48	95.09	21.60	94.31	23.16	93.47	24.70	92.59	26.20
50	95.07	21.66	94.28	23.22	93.45	24.75	92.56	26.25
52	95.04	21.71	94.26	23.27	93.42	24.80	92.53	26.30
54	95.02	21.76	94.23	23.32	93.39	24.85	92.49	26.35
56	94.99	21.81	94.20	23.37	93.36	24.90	92.46	26.40
58	94.97	21.87	94.17	23.42	93.33	24.95	92.43	26.45
60	94.94	21.92	94.15	23.47	93.30	25.00	92.40	26.50
C = 0.75	0.73	0.16	0.73	0.17	0.73	0.19	0.72	0.20
C = 1.00	0.98	0.22	0.97	0.23	0.97	0.25	0.96	0.27
C = 1.25	1.22	0.27	1.21	0.29	1.21	0.31	1.20	0.34

Table III Horizontal distances and elevations from stadia readings—Continued

	16°		17°		18°		19°	
Minutes	Hor. Dist.	Diff. Elev.	Hor. Dist.	Diff. Elev.	Hor. Dist.	Diff. Elev.	Hor. Dist.	Diff. Elev.
0	92.40	26.50	91.45	27.96	90.45	29.39	89.40	30.78
2	92.37	26.55	91.42	28.01	90.42	29.44	89.36	30.83
4	92.34	26.59	91.39	28.06	90.38	29.48	89.33	30.87
6	92.31	26.64	91.35	28.10	90.35	29.53	89.29	30.92
8	92.28	26.69	91.32	28.15	90.31	29.58	89.26	30.97
10	92.25	26.74	91.29	28.20	90.28	29.62	89.22	31.01
12	92.22	26.79	91.26	28.25	90.24	29.67	89.18	31.06
14	92.19	26.84	91.22	28.30	90.21	29.72	89.15	31.10
16	92.15	26.89	91.19	28.34	90.18	29.76	89.11	31.15
18	92.12	26.94	91.16	28.39	90.14	29.81	89.08	31.19
20	92.09	26.99	91.12	28.44	90.11	29.86	89.04	31.24
22	92.06	27.04	91.09	28.49	90.07	29.90	89.00	31.28
24	92.03	27.09	91.06	28.54	90.04	29.95	88.96	31.33
26	92.00	27.13	91.02	28.58	90.00	30.00	88.93	31.38
28	91.97	27.18	90.99	28.63	89.97	30.04	88.89	31.42
30	91.93	27.23	90.96	28.68	89.93	30.09	88.86	31.47
32	91.90	27.28	90.92	28.73	89.90	30.14	88.82	31.51
34	91.87	27.33	90.89	28.77	89.86	30.19	88.78	31.56
36	91.84	27.38	90.86	28.82	89.83	30.23	88.75	31.60
38	91.81	27.43	90.82	28.87	89.79	30.28	88.71	31.65
40	91.77	27.48	90.79	28.92	89.76	30.32	88.67	31.69
42	91.74	27.52	90.76	28.96	89.72	30.37	88.64	31.74
44	91.71	27.57	90.72	29.01	89.69	30.41	88.60	31.78
46	91.68	27.62	90.69	29.06	89.65	30.46	88.56	31.83
48	91.65	27.67	90.66	29.11	89.61	30.51	88.53	31.87
50	91.61	27.72	90.62	29.15	89.58	30.55	88.49	31.92
52	91.58	27.77	90.59	29.20	89.54	30.60	88.45	31.96
54	91.55	27.81	90.55	29.25	89.51	30.65	88.41	32.01
56	91.52	27.86	90.52	29.30	89.47	30.69	88.38	32.05
58	91.48	27.91	90.48	29.34	89.44	30.74	88.34	32.09
60	91.45	27.96	90.45	29.39	89.40	30.78	88.30	32.14
C = 0.75	0.72	0.21	0.72	0.23	0.71	0.24	0.71	0.25
C = 1.00	0.96	0.28	0.95	0.30	0.95	0.32	0.94	0.33
C = 1.25	1.20	0.35	1.19	0.38	1.19	0.40	1.18	0.42

Table III Horizontal distances and elevations from stadia readings—Continued

	20°		21°		22°		23°	
Minutes.	Hor. Dist.	Diff. Elev.	Hor. Dist.	Diff. Elev.	Hor. Dist.	Diff. Elev.	Hor. Dist.	Diff. Elev.
0	88.30	32.14	87.16	33.46	85.97	34.73	84.73	35.97
2	88.26	32.18	87.12	33.50	85.93	34.77	84.69	36.01
4	88.23	32.23	87.08	33.54	85.89	34.82	84.65	36.05
6	88.19	32.27	87.04	33.59	85.85	34.86	84.61	36.09
8	88.15	32.32	87.00	33.63	85.80	34.90	84.57	36.13
10	88.11	32.36	86.96	33.67	85.76	34.94	84.52	36.17
12	88.08	32.41	86.92	33.72	85.72	34.98	84.48	36.21
14	88.04	32.45	86.88	33.76	85.68	35.02	84.44	36.25
16	88.00	32.49	86.84	33.80	85.64	35.07	84.40	36.29
18	87.96	32.54	86.80	33.84	85.60	35.11	84.35	36.33
20	87.93	32.58	86.77	33.89	85.56	35.15	84.31	36.37
22	87.89	32.63	86.73	33.93	85.52	35.19	84.27	36.41
24	87.85	32.67	86.69	33.97	85.48	35.23	84.23	36.45
26	87.81	32.72	86.65	34.01	85.44	35.27	84.18	36.49
28	87.77	32.76	86.61	34.06	85.40	35.31	84.14	36.53
30	87.74	32.80	86.57	34.10	85.36	35.36	84.10	36.57
32	87.70	32.85	86.53	34.14	85.31	35.40	84.06	36.61
34	87.66	32.89	86.49	34.18	85.27	35.44	84.01	36.65
36	87.62	32.93	86.45	34.23	85.23	35.48	83.97	36.69
38	87.58	32.98	86.41	34.27	85.19	35.52	83.93	36.73
40	87.54	33.02	86.37	34.31	85.15	35.56	83.89	36.77
42	87.51	33.07	86.33	34.35	85.11	35.60	83.84	36.80
44	87.47	33.11	86.29	34.40	85.07	35.64	83.80	36.84
46	87.43	33.15	86.25	34.44	85.02	35.68	83.76	36.88
48	87.39	33.20	86.21	34.48	84.98	35.72	83.72	36.92
50	87.35	33.24	86.17	34.52	84.94	35.76	83.67	36.96
52	87.31	33.28	86.13	34.57	84.90	35.80	83.63	37.00
54	87.27	33.33	86.09	34.61	84.86	35.85	83.59	37.04
56	87.24	33.37	86.05	34.65	84.82	35.89	83.54	37.08
58	87.20	33.41	86.01	34.69	84.77	35.93	83.50	37.12
60	87.16	33.46	85.97	34.73	84.73	35.97	83.46	37.16
C = 0.75	0.70	0.26	0.70	0.27	0.69	0.29	0.69	0.30
C = 1.00	0.94	0.35	0.93	0.37	0.92	0.38	0.92	0.40
C = 1.25	1.17	0.44	1.16	0.46	1.15	0.48	1.15	0.50

Table III Horizontal distances and elevations from stadia readings—Continued

	24°		25°		26°		27°	
Minutes	Hor. Dist.	Diff. Elev.	Hor. Dist.	Diff. Elev.	Hor. Dist.	Diff Elev.	Hor. Dist.	Diff. Elev.
0	83.46	37.16	82.14	38.30	80.78	39.40	79.39	40.45
2	83.41	37.20	82.09	38.34	80.74	39.44	79.34	40.49
4	83.37	37.23	82.05	38.38	80.69	39.47	79.30	40.52
6	83.33	37.27	82.01	38.41	80.65	39.51	79.25	40.55
8	83.28	37.31	81.96	38.45	80.60	39.54	79.20	40.59
10	83.24	37.35	81.92	38.49	80.55	39.58	79.15	40.62
12	83.20	37.39	81.87	38.53	80.51	39.61	79.11	40.66
14	83.15	37.43	81.83	38.56	80.46	39.65	79.06	40.69
16	83.11	37.47	81.78	38.60	80.41	39.69	79.01	40.72
18	83.07	37.51	81.74	38.64	80.37	39.72	78.96	40.76
20	83.02	37.54	81.69	38.67	80.32	39.76	78.92	40.79
22	82.98	37.58	81.65	38.71	80.28	39.79	78.87	40.82
24	82.93	37.62	81.60	38.75	80.23	39.83	78.82	40.86
26	82.89	37.66	81.56	38.78	80.18	39.86	78.77	40.89
28	82.85	37.70	81.51	38.82	80.14	39.90	78.73	40.92
30	82.80	37.74	81.47	38.86	80.09	39.93	78.68	40.96
32	82.76	37.77	81.42	38.89	80.04	39.97	78.63	40.99
34	82.72	37.81	81.38	38.93	80.00	40.00	78.58	41.02
36	82.67	37.85	81.33	38.97	79.95	40.04	78.54	41.06
38	82.63	37.89	81.28	39.00	79.90	40.07	78.49	41.09
40	82.58	37.93	81.24	39.04	79.86	40.11	78.44	41.12
42	82.54	37.96	81.19	39.08	79.81	40.14	78.39	41.16
44	82.49	38.00	81.15	39.11	79.76	40.18	78.34	41.19
46	82.45	38.04	81.10	39.15	79.72	40.21	78.30	41.22
48	82.41	38.08	81.06	39.18	79.67	40.24	78.25	41.26
50	82.36	38.11	81.01	39.22	79.62	40.28	78.20	41.29
52	82.32	38.15	80.97	39.26	79.58	40.31	78.15	41.32
54	82.27	38.19	80.92	39.29	79.53	40.35	78.10	41.35
56	82.23	38.23	80.87	39.33	79.48	40.38	78.06	41.39
58	82.18	38.26	80.83	39.36	79.44	40.42	78.01	41.42
60	82.14	38.30	80.78	39.40	79.39	40.45	77.96	41.45
C = 0.75	0.68	0.31	0.68	0.32	0.67	0.33	0.66	0.35
C = 1.00	0.91	0.41	0.90	0.43	0.89	0.45	0.89	0.46
C = 1.25	1.14	0.52	1.13	0.54	1.12	0.56	1.11	0.58

Table III Horizontal distances and elevations from stadia readings—Concluded

	28°		29°		30°	
Minutes	Hor. Dist.	Diff. Elev.	Hor. Dist.	Diff. Elev.	Hor. Dist.	Diff. Elev.
0	77.96	41.45	76.50	42.40	75.00	43.30
2	77.91	41.48	76.45	42.43	74.95	43.33
4	77.86	41.52	76.40	42.46	74.90	43.36
6	77.81	41.55	76.35	42.49	74.85	43.39
8	77.77	41.58	76.30	42.53	74.80	43.42
10	77.72	41.61	76.25	42.56	74.75	43.45
12	77.67	41.65	76.20	42.59	74.70	43.47
14	77.62	41.68	76.15	42.62	74.65	43.50
16	77.57	41.71	76.10	42.65	74.60	43.53
18	77.52	41.74	76.05	42.68	74.55	43.56
20	77.48	41.77	76.00	42.71	74.49	43.59
22	77.42	41.81	75.95	42.74	74.44	43.62
24	77.38	41.84	75.90	42.77	74.39	43.65
26	77.33	41.87	75.85	42.80	74.34	43.67
28	77.28	41.90	75.80	42.83	74.29	43.70
30	77.23	41.93	75.75	42.86	74.24	43.73
32	77.18	41.97	75.70	42.89	74.19	43.76
34	77.13	42.00	75.65	42.92	74.14	43.79
36	77.09	42.03	75.60	42.95	74.09	43.82
38	77.04	42.06	75.55	42.98	74.04	43.84
40	76.99	42.09	75.50	43.01	73.99	43.87
42	76.94	42.12	75.45	43.04	73.93	43.90
44	76.89	42.15	75.40	43.07	73.88	43.93
46	76.84	42.19	75.35	43.10	73.83	43.95
48	76.79	42.22	75.30	43.13	73.78	43.98
50	76.74	42.25	75.25	43.16	73.73	44.01
52	76.69	42.28	75.20	43.18	73.68	44.04
54	76.64	42.31	75.15	43.21	73.63	44.07
56	76.59	42.34	75.10	43.24	73.58	44.09
58	76.55	42.37	75.05	43.27	73.52	44.12
60	76.50	42.40	75.00	43.30	73.47	44.15
C = 0.75	0.66	0.36	0.65	0.37	0.65	0.38
C = 1.00	0.88	0.48	0.87	0.49	0.86	0.51
C = 1.25	1.10	0.60	1.09	0.62	1.08	0.64

Table IV Convergency of meridians, six miles long and six miles apart, and differences of latitude and longitude

Lat.	Convergency		Difference of longitude per range		Difference of latitude for—	
	On the parallel	Angle	In arc	In time	1 mi.	1 Tp.
°	*Lks.*	′ ″	′ ″	*Seconds*	′	′
25	33.9	2 25	5 44.34	22.96		
26	35.4	2 32	5 47.20	23.15		
27	37.0	2 39	5 50.22	23.35	0.871	5.229
28	38.6	2 46	5 53.40	23.56		
29	40.2	2 53	5 56.74	23.78		
30	41.9	3 0	6 0.26	24.02		
31	43.6	3 7	6 3.97	24.26		
32	45.4	3 15	6 7.87	24.52	0.871	5.225
33	47.2	3 23	6 11.96	24.80		
34	49.1	3 30	6 16.26	25.08		
35	50.9	3 38	6 20.78	25.39		
36	52.7	3 46	6 25.53	25.70		
37	54.7	3 55	6 30.52	26.03	0.870	5.221
38	56.8	4 4	6 35.76	26.38		
39	58.8	4 13	6 41.27	26.75		
40	60.9	4 22	6 47.06	27.14		
41	63.1	4 31	6 53.15	27.54		
42	65.4	4 41	6 59.56	27.97	0.869	5.216
43	67.7	4 51	7 6.29	28.42		
44	70.1	5 1	7 13.39	28.89		
45	72.6	5 12	7 20.86	29.39		
46	75.2	5 23	7 28.74	29.92		
47	77.8	5 34	7 37.04	30.47	0.869	5.211
48	80.6	5 46	7 45.80	31.05		
49	83.5	5 59	7 55.05	31.67		
50	86.4	6 12	8 4.83	32.32		
51	89.6	6 25	8 15.17	33.03		
52	92.8	6 39	8 26.13	33.74	0.868	5.207
53	96.2	6 54	8 37.75	34.52		
54	99.8	7 9	8 50.07	35.34		
55	103.5	7 25	9 3.18	36.22		
56	107.5	7 42	9 17.12	37.14		
57	111.6	8 0	9 31.97	38.13	0.867	5.202
58	116.0	8 19	9 47.83	39.19		
59	120.6	8 38	10 4.78	40.32		
60	125.5	8 59	10 22.94	41.52		
61	130.8	9 22	10 42.42	42.83		
62	136.3	9 46	11 3.38	44.22	0.866	5.198
63	142.2	10 11	11 25.97	45.73		
64	148.6	10 38	11 50.37	47.36		
65	155.0	11 8	12 16.82	49.12		
66	162.8	11 39	12 45.55	51.04		
67	170.7	12 13	13 16.88	53.12	0.866	5.195
68	179.3	12 51	13 51.15	55.41		
69	188.7	13 31	14 28.77	57.92		
70	199.1	14 15	15 10.26	60.68	0.866	5.193

Table V Azimuths of the secant

Lat.	0 mi.	1 mi.	2 mi.	3 mi.	Deflection angle 6 mi.
°	° ′	° ′	° ′		′ ″
25	89 58.8	89 59.2	89 59.6	90°	2 25
26	58.7	59.2	59.6	E or W.	2 32
27	58.7	59.1	59.6	" " "	2 39
28	58.6	59.1	59.5	" " "	2 46
29	58.6	59.0	59.5		2 53
30	58.5	59.0	59.5	" " "	3 0
31	58.4	59.0	59.5	" " "	3 7
32	58.4	58.9	59.5	" " "	3 15
33	58.3	58.9	59.4	" " "	3 23
34	58.2	58.8	59.4	" " "	3 30
35	58.2	58.8	59.4	" " "	3 38
36	58.1	58.7	59.4	" " "	3 46
37	58.0	58.7	59.3	" " "	3 55
38	58.0	58.6	59.3	" " "	4 4
39	57.9	58.6	59.3	" " "	4 13
40	57.8	58.5	59.3	" " "	4 22
41	57.7	58.5	59.2	" " "	4 31
42	57.7	58.4	59.2	" " "	4 41
43	57.6	58.4	59.2	" " "	4 51
44	57.5	58.3	59.2	" " "	5 1
45	57.4	58.3	59.1	" " "	5 12
46	57.3	58.2	59.1	" " "	5 23
47	57.2	58.1	59.1	" " "	5 34
48	57.1	58.1	59.0	" " "	5 46
49	57.0	58.0	59.0	" " "	5 59
50	56.9	57.9	59.0	" " "	6 12
51	56.8	57.9	58.9	" " "	6 25
52	56.7	57.8	58.9	" " "	6 39
53	56.6	57.7	58.8	" " "	6 54
54	56.4	57.6	58.8	" " "	7 9
55	56.3	57.5	58.8	" " "	7 25
56	56.2	57.4	58.7	" " "	7 42
57	56.0	57.3	58.7	" " "	8 0
58	55.8	57.2	58.6	" " "	8 19
59	55.7	57.1	58.6	" " "	8 38
60	55.5	57.0	58.5	" " "	8 59
61	55.3	56.9	58.4	" " "	9 22
62	55.1	56.7	58.4	" " "	9 46
63	54.9	56.6	58.3	" " "	10 11
64	54.7	56.5	58.2	" " "	10 38
65	54.4	56.3	58.1	" " "	11 8
66	54.2	56.1	58.1	" " "	11 39
67	53.9	55.9	58.0	" " "	12 13
68	53.6	55.7	57.9	" " "	12 51
69	53.2	55.5	57.8	" " "	13 31
70	89° 52′.9	89° 55′.3	89° 57′.6	" " "	14′ 15″
	6 mi.	5 mi.	4 mi.	3 mi.	

Table VI Offsets, in links, from the secant to the parallel

Lat.	0 mi.	½ mi.	1 mi.	1½ mi.	2 mi.	2½ mi.	3 mi.
°							
25	2 N.	1 N.	0	1 S.	1 S.	2 S.	2 S.
26	2	1	0	1	1	2	2
27	3	1	0	1	2	2	2
28	3	1	0	1	2	2	2
29	3	1	0	1	2	2	2
30	3	1	0	1	2	2	2
31	3	1	0	1	2	2	2
32	3	1	0	1	2	2	3
33	3	1	0	1	2	2	3
34	3	2	0	1	2	3	3
35	4	2	0	1	2	3	3
36	4	2	0	1	2	3	3
37	4	2	0	1	2	3	3
38	4	2	0	1	2	3	3
39	4	2	0	1	2	3	3
40	4	2	0	1	3	3	3
41	4	2	0	2	3	3	4
42	5	2	0	2	3	3	4
43	5	2	0	2	3	4	4
44	5	2	0	2	3	4	4
45	5	2	0	2	3	4	4
46	5	2	0	2	3	4	4
47	5	2	0	2	3	4	4
48	6	3	0	2	3	4	4
49	6	3	0	2	3	4	5
50	6	3	0	2	4	4	5
51	6	3	0	2	4	5	5
52	6	3	0	2	4	5	5
53	7	3	0	2	4	5	5
54	7	3	0	2	4	5	6
55	7	3	0	3	4	5	6
56	7	3	0	3	4	6	6
57	8	3	0	3	5	6	6
58	8	4	0	3	5	6	6
59	8	4	0	3	5	6	7
60	9	4	0	3	5	7	7
61	9	4	0	3	5	7	7
62	9	4	0	3	6	7	8
63	10	4	0	3	6	7	8
64	10	5	0	4	6	8	8
65	11	5	0	4	6	8	9
66	11	5	0	4	7	8	9
67	12	5	0	4	7	9	9
68	12	6	0	4	7	9	10
69	13	6	0	5	8	10	10
70	14 N.	6 N.	0	5 S.	8 S.	10 S.	11 S.
	6 mi.	5½ mi.	5 mi.	4½ mi.	4 mi.	3½ mi.	3 mi.

No. 100 **No. 109**
Log. 000 **Table VII Logarithms of numbers** **Log. 040**

N.	0	1	2	3	4	5	6	7	8	9	Diff.
100	00 0000	0434	0868	1301	1734	2166	2598	3029	3461	3891	432
1	4321	4751	5181	5609	6038	6466	6894	7321	7748	8174	428
2	8600	9026	9451	9876	0300	0724	1147	1570	1993	2415	424
3	01 2837	3259	3680	4100	4521	4940	5360	5779	6197	6616	420
4	7033	7451	7868	8284	8700	9116	9532	9947	0361	0775	416
105	02 1189	1603	2016	2428	2841	3252	3664	4075	4486	4896	412
6	5306	5715	6125	6533	6942	7350	7757	8164	8571	8978	408
7	9384	9789	0195	0600	1004	1408	1812	2216	2619	3021	404
8	03 3424	3826	4227	4628	5029	5430	5830	6230	6629	7028	400
9	7426	7825	8223	8620	9017	9414	9811	0207	0602	0998	397
	04										

PROPORTIONAL PARTS

Diff.	1	2	3	4	5	6	7	8	9
434	43.4	86.8	130.2	173.6	217.0	260.4	303.8	347.2	390.6
433	43.3	86.6	129.9	173.2	216.5	259.8	303.1	346.4	389.7
432	43.2	86.4	129.6	172.8	216.0	259.2	302.4	345.6	388.8
431	43.1	86.2	129.3	172.4	215.5	258.6	301.7	344.8	387.9
430	43.0	86.0	129.0	172.0	215.0	258.0	301.0	344.0	387.0
429	42.9	85.8	128.7	171.6	214.5	257.4	300.3	343.2	386.1
428	42.8	85.6	128.4	171.2	214.0	256.8	299.6	342.4	385.2
427	42.7	85.4	128.1	170.8	213.5	256.2	298.9	341.6	384.3
426	42.6	85.2	127.8	170.4	213.0	255.6	298.2	340.8	383.4
425	42.5	85.0	127.5	170.0	212.5	255.0	297.5	340.0	382.5
424	42.4	84.8	127.2	169.6	212.0	254.4	296.8	339.2	381.6
423	42.3	84.6	126.9	169.2	211.5	253.8	296.1	338.4	380.7
422	42.2	84.4	126.6	168.8	211.0	253.2	295.4	337.6	379.8
421	42.1	84.2	126.3	168.4	210.5	252.6	294.7	336.8	378.9
420	42.0	84.0	126.0	168.0	210.0	252.0	294.0	336.0	378.0
419	41.9	83.8	125.7	167.6	209.5	251.4	293.3	335.2	377.1
418	41.8	83.6	125.4	167.2	209.0	250.8	292.6	334.4	376.2
417	41.7	83.4	125.1	166.8	208.5	250.2	291.9	333.6	375.3
416	41.6	83.2	124.3	166.4	208.0	249.6	291.2	332.8	374.4
415	41.5	83.0	124.5	166.0	207.5	249.0	290.5	332.0	373.5
414	41.4	82.8	124.2	165.6	207.0	248.4	289.8	331.2	372.6
413	41.3	82.6	123.9	165.2	206.5	247.8	289.1	330.4	371.7
412	41.2	82.4	123.6	164.8	206.0	247.2	288.4	329.6	370.8
411	41.1	82.2	123.3	164.4	205.5	246.6	287.7	328.8	369.9
410	41.0	82.0	123.0	164.0	205.0	246.0	287.0	328.0	369.0
409	40.9	81.8	122.7	163.6	204.5	245.4	286.3	327.2	368.1
408	40.8	81.6	122.4	163.2	204.0	244.8	285.6	326.4	367.2
407	40.7	81.4	122.1	162.8	203.5	244.2	284.9	325.6	366.3
406	40.6	81.2	121.8	162.4	203.0	243.6	284.2	324.8	365.4
405	40.5	81.0	121.5	162.0	202.5	243.0	283.5	324.0	364.5
404	40.4	80.8	121.2	161.6	202.0	242.4	282.8	323.2	363.6
403	40.3	80.6	120.9	161.2	201.5	241.8	282.1	322.4	362.7
402	40.2	80.4	120.6	160.8	201.0	241.2	281.4	321.6	361.8
401	40.1	80.2	120.3	160.4	200.5	240.6	280.7	320.8	360.9
400	40.0	80.0	120.0	160.0	200.0	240.0	280.0	320.0	360.0
399	39.9	79.8	119.7	159.6	199.5	239.4	279.3	319.2	359.1
398	39.8	79.6	119.4	159.2	199.0	238.8	278.6	318.4	358.2
397	39.7	79.4	119.1	158.8	198.5	238.2	277.9	317.6	357.3
396	39.6	79.2	118.8	158.4	198.0	237.6	277.2	316.8	356.4
395	39.5	79.0	118.5	158.0	197.5	237.0	276.5	316.0	351.5

No. 110
Log. 041

Table VII—Continued

No. 119
Log. 078

N.	0	1	2	3	4	5	6	7	8	9	Diff.
110	04 1393	1787	2182	2576	2969	3362	3755	4148	4540	4932	393
1	5323	5714	6105	6495	6885	7275	7664	8053	8442	8830	390
2	9218	9606	9993	0380	0766	1153	1538	1924	2309	2694	386
3	05 3078	3463	3846	4230	4613	4996	5378	5760	6142	6524	383
4	6905	7286	7666	8046	8426	8805	9185	9563	9942	0320	379
115	06 0698	1075	1452	1829	2206	2582	2958	3333	3709	4083	376
6	4458	4832	5206	5580	5953	6326	6699	7071	7443	7815	373
7	8186	8557	8928	9298	9668	0038	0407	0776	1145	1514	370
8	07 1882	2250	2617	2985	3352	3718	4085	4451	4816	5182	366
9	5547	5912	6276	6640	7004	7368	7731	8094	8457	8819	363

PROPORTIONAL PARTS

Diff.	1	2	3	4	5	6	7	8	9
395	39.5	79.0	118.5	158.0	197.5	237.0	276.5	316.0	355.5
394	39.4	78.8	118.2	157.6	197.0	236.4	275.8	315.2	354.6
393	39.3	78.6	117.9	157.2	196.5	235.8	275.1	314.4	353.7
392	39.2	78.4	117.6	156.8	196.0	235.2	274.4	313.6	352.8
391	39.1	78.2	117.3	156.4	195.5	234.6	273.7	312.8	351.9
390	39.0	78.0	117.0	156.0	195.0	234.0	273.0	312.0	351.0
389	38.9	77.8	116.7	155.6	194.5	233.4	272.3	311.2	350.1
388	38.8	77.6	116.4	155.2	194.0	232.8	271.6	310.4	349.2
387	38.7	77.4	116.1	154.8	193.5	232.2	270.9	309.6	348.3
386	38.6	77.2	115.8	154.4	193.0	231.6	270.2	308.8	347.4
385	38.5	77.0	115.5	154.0	192.5	231.0	269.5	308.0	346.5
384	38.4	76.8	115.2	153.6	192.0	230.4	268.8	307.2	345.6
383	38.3	76.6	114.9	153.2	191.5	229.8	268.1	306.4	344.7
382	38.2	76.4	114.6	152.8	191.0	229.2	267.4	305.6	343.8
381	38.1	76.2	114.3	152.4	190.5	228.6	266.7	304.8	342.9
380	38.0	76.0	114.0	152.0	190.0	228.0	266.0	304.0	342.0
379	37.9	75.8	113.7	151.6	189.5	227.4	265.3	303.2	341.1
378	37.8	75.6	113.4	151.2	189.0	226.8	264.6	302.4	340.2
377	37.7	75.4	113.1	150.8	188.5	226.2	263.9	301.6	339.3
376	37.6	75.2	112.8	150.4	188.0	225.6	263.2	300.8	338.4
375	37.5	75.0	112.5	150.0	187.5	225.0	262.5	300.0	337.5
374	37.4	74.8	112.2	149.6	187.0	224.4	261.8	299.2	336.6
373	37.3	74.6	111.9	149.2	186.5	223.8	261.1	298.4	335.7
372	37.2	74.4	111.6	148.8	186.0	223.2	260.4	297.6	334.8
371	37.1	74.2	111.3	148.4	185.5	222.6	259.7	296.8	333.9
370	37.0	74.0	111.0	148.0	185.0	222.0	259.0	296.0	333.0
369	36.9	73.8	110.7	147.6	184.5	221.4	258.3	295.2	332.1
368	36.8	73.6	110.4	147.2	184.0	220.8	257.6	294.4	331.2
367	36.7	73.4	110.1	146.8	183.5	220.2	256.9	293.6	330.3
366	36.6	73.2	109.8	146.4	183.0	219.6	256.2	292.8	329.4
365	36.5	73.0	109.5	146.0	182.5	219.0	255.7	292.0	328.5
364	36.4	72.8	109.2	145.6	182.0	218.4	254.8	291.2	327.6
363	36.3	72.6	108.9	145.2	181.5	217.8	254.1	290.4	326.7
362	36.2	72.4	108.6	144.8	181.0	217.2	253.4	289.6	325.8
361	36.1	72.2	108.3	144.4	180.5	216.6	252.7	288.8	324.9
360	36.0	72.0	108.0	144.0	180.0	216.0	252.0	288.0	324.0
359	35.9	71.8	107.7	143.6	179.5	215.4	251.3	287.2	323.1
358	35.8	71.6	107.4	143.2	179.0	214.8	250.6	286.4	322.2
357	35.7	71.4	107.1	142.8	178.5	214.2	249.9	285.6	321.3
356	35.6	71.2	106.8	142.4	178.0	213.6	249.2	284.8	320.4

No. 120 **Log. 079** — **Table VII—Continued** — **No. 134** **Log. 130**

N.	0	1	2	3	4	5	6	7	8	9	Diff.
120	07 9181	9543	9904	0266	0626	0987	1347	1707	2067	2426	360
1	08 2785	3144	3503	3861	4219	4576	4934	5291	5647	6004	357
2	6360	6716	7071	7426	7781	8136	8490	8845	9198	9552	355
3	9905	0258	0611	0963	1315	1667	2018	2370	2721	3071	352
4	09 3422	3772	4122	4471	4820	5169	5518	5866	6215	6562	349
125	6910	7257	7604	7951	8298	8644	8990	9335	9681	0026	346
6	10 0371	0715	1059	1403	1747	2091	2434	2777	3119	3462	343
7	3804	4146	4487	4828	5169	5510	5851	6191	6531	6871	341
8	7210	7549	7888	8227	8565	8903	9241	9579	9916	0253	338
9	11 0590	0926	1263	1599	1934	2270	2605	2940	3275	3609	335
130	3943	4277	4611	4944	5278	5611	5943	6276	6608	6940	333
1	7271	7603	7934	8265	8595	8926	9256	9586	9915	0245	330
2	12 0574	0903	1231	1560	1888	2216	2544	2871	3198	3525	328
3	3852	4178	4504	4830	5156	5481	5806	6131	6456	6781	325
4	7105	7429	7753	8076	8399	8722	9045	9368	9690	0012	323
	13										

PROPORTIONAL PARTS

Diff.	1	2	3	4	5	6	7	8	9
355	35.5	71.0	106.5	142.0	177.5	213.0	248.5	284.0	319.5
354	35.4	70.8	106.2	141.6	177.0	212.4	247.8	283.2	318.6
353	35.3	70.6	105.9	141.2	176.5	211.8	247.1	282.4	317.7
352	35.2	70.4	105.6	140.8	176.0	211.2	246.4	281.6	316.8
351	35.1	70.2	105.3	140.4	175.5	210.6	245.7	280.8	315.9
350	35.0	70.0	105.0	140.0	175.0	210.0	245.0	280.0	315.0
349	34.9	69.8	104.7	139.6	174.5	209.4	244.3	279.2	314.1
348	34.8	69.6	104.4	139.2	174.0	208.8	243.6	278.4	313.2
347	34.7	69.4	104.1	138.8	173.5	208.2	242.9	277.6	312.3
346	34.6	69.2	103.8	138.4	173.0	207.6	242.2	276.8	311.4
345	34.5	69.0	103.5	138.0	172.5	207.0	241.5	276.0	310.5
344	34.4	68.8	103.2	137.6	172.0	206.4	240.8	275.2	309.6
343	34.3	68.6	102.9	137.2	171.5	205.8	240.1	274.4	308.7
342	34.2	68.4	102.6	136.8	171.0	205.2	239.4	273.6	307.8
341	34.1	68.2	102.3	136.4	170.5	204.6	238.7	272.8	306.9
340	34.0	68.0	102.0	136.0	170.0	204.0	238.0	272.0	306.0
339	33.9	67.8	101.7	135.6	169.5	203.4	237.3	271.2	305.1
338	33.8	67.6	101.4	135.2	169.0	202.8	236.6	270.4	304.2
337	33.7	67.4	101.1	134.8	168.5	202.2	235.9	269.6	303.3
336	33.6	67.2	100.8	134.4	168.0	201.6	235.2	268.8	302.4
335	33.5	67.0	100.5	134.0	167.5	201.0	234.5	268.0	301.5
334	33.4	66.8	100.2	133.6	167.0	200.4	233.8	267.2	300.6
333	33.3	66.6	99.9	133.2	166.5	199.8	233.1	266.4	299.7
332	33.2	66.4	99.6	132.8	166.0	199.2	232.4	265.6	298.8
331	33.1	66.2	99.3	132.4	165.5	198.6	231.7	264.8	297.9
330	33.0	66.0	99.0	132.0	165.0	198.0	231.0	264.0	297.0
329	32.9	65.8	98.7	131.6	164.5	197.4	230.3	263.2	296.1
328	32.8	65.6	98.4	131.2	164.0	196.8	229.6	262.4	295.2
327	32.7	65.4	98.1	130.8	163.5	196.2	228.9	261.6	294.3
326	32.6	65.2	97.8	130.4	163.0	195.6	228.2	260.8	293.4
325	32.5	65.0	97.5	130.0	162.5	195.0	227.5	260.0	292.5
324	32.4	64.8	97.2	129.6	162.0	194.4	226.8	259.2	291.6
323	32.3	64.6	96.9	129.2	161.5	193.8	226.1	258.4	290.7
322	32.2	64.4	96.6	128.8	161.0	193.2	225.4	257.6	289.8

No. 135 Log. 130 — **Table VII—Continued** — No. 149 Log. 175

N.	0	1	2	3	4	5	6	7	8	9	Diff.
135	13 0334	0655	0977	1298	1619	1939	2260	2580	2900	3219	321
6	3539	3858	4177	4496	4814	5133	5451	5769	6086	6403	318
7	6721	7037	7354	7671	7987	8303	8618	8934	9249	9564	316
8	9879	0194	0508	0822	1136	1450	1763	2076	2389	2702	314
9	14 3015	3327	3639	3951	4263	4574	4885	5196	5507	5818	311
140	6128	6438	6748	7058	7367	7676	7985	8294	8603	8911	309
1	9219	9527	9835	0142	0449	0756	1063	1370	1676	1982	307
2	15 2288	2594	2900	3205	3510	3815	4120	4424	4728	5032	305
3	5336	5640	5943	6246	6549	6852	7154	7457	7759	8061	303
4	8362	8664	8965	9266	9567	9868	0168	0469	0769	1068	301
145	16 1368	1667	1967	2266	2564	2863	3161	3460	3758	4055	299
6	4353	4650	4947	5244	5541	5838	6134	6430	6726	7022	297
7	7317	7613	7908	8203	8497	8792	9086	9380	9674	9968	295
8	17 0262	0555	0848	1141	1434	1726	2019	2311	2603	2895	293
9	3186	3478	3769	4060	4351	4641	4932	5222	5512	5802	291

PROPORTIONAL PARTS

Diff.	1	2	3	4	5	6	7	8	9
321	32.1	64.2	96.3	128.4	160.5	192.6	224.7	256.8	288.9
320	32.0	64.0	96.0	128.0	160.0	192.0	224.0	256.0	288.0
319	31.9	63.8	95.7	127.6	159.5	191.4	223.3	255.2	287.1
318	31.8	63.6	95.4	127.2	159.0	190.8	222.6	254.4	286.2
317	31.7	63.4	95.1	126.8	158.5	190.2	221.9	253.6	285.3
316	31.6	63.2	94.8	126.4	158.0	189.6	221.2	252.8	284.4
315	31.5	63.0	94.5	126.0	157.5	189.0	220.5	252.0	283.5
314	31.4	62.8	94.2	125.6	157.0	188.4	219.8	251.2	282.6
313	31.3	62.6	93.9	125.2	156.5	187.8	219.1	250.4	281.7
312	31.2	62.4	93.6	124.8	156.0	187.2	218.4	249.6	280.8
311	31.1	62.2	93.3	124.4	155.5	186.6	217.7	248.8	279.9
310	31.0	62.0	93.0	124.0	155.0	186.0	217.0	248.0	279.0
309	30.9	61.8	92.7	123.6	154.5	185.4	216.3	247.2	278.1
308	30.8	61.6	92.4	123.2	154.0	184.8	215.6	246.4	277.2
307	30.7	61.4	92.1	122.8	153.5	184.2	214.9	245.6	276.3
306	30.6	61.2	91.8	122.4	153.0	183.6	214.2	244.8	275.4
305	30.5	61.0	91.5	122.0	152.5	183.0	213.5	244.0	274.5
304	30.4	60.8	91.2	121.6	152.0	182.4	212.8	243.2	273.6
303	30.3	60.6	90.9	121.2	151.5	181.8	212.1	242.4	272.7
302	30.2	60.4	90.6	120.8	151.0	181.2	211.4	241.6	271.8
301	30.1	60.2	90.3	120.4	150.5	180.6	210.7	240.8	270.9
300	30.0	60.0	90.0	120.0	150.0	180.0	210.0	240.0	270.0
299	29.9	59.8	89.7	119.6	149.5	179.4	209.3	239.2	269.1
298	29.8	59.6	89.4	119.2	149.0	178.8	208.6	238.4	268.2
297	29.7	59.4	89.1	118.8	148.5	178.2	207.9	237.6	267.3
296	29.6	59.2	88.8	118.4	148.0	177.6	207.2	236.8	266.4
295	29.5	59.0	88.5	118.0	147.5	177.0	206.5	236.0	265.5
294	29.4	58.8	88.2	117.6	147.0	176.4	205.8	235.2	264.6
293	29.3	58.6	87.9	117.2	146.5	175.8	205.1	234.4	263.7
292	29.2	58.4	87.6	116.8	146.0	175.2	204.4	233.6	262.8
291	29.1	58.2	87.3	116.4	145.5	174.6	203.7	232.8	261.9
290	29.0	58.0	87.0	116.0	145.0	174.0	203.0	232.0	261.0
289	28.9	57.8	86.7	115.6	144.5	173.4	202.3	231.2	260.1
288	28.8	57.6	86.4	115.2	144.0	172.8	201.6	230.4	259.2
287	28.7	57.4	86.1	114.8	143.5	172.2	200.9	229.6	258.3
286	28.6	57.2	85.8	114.4	143.0	171.6	200.2	228.8	257.4

Table VII—Continued

N.	0	1	2	3	4	5	6	7	8	9	Diff.
150	17 6091	6381	6670	6959	7248	7536	7825	8113	8401	8689	289
1	8977	9264	9552	9839	0126	0413	0699	0986	1272	1558	287
2	18 1844	2129	2415	2700	2985	3270	3555	3839	4123	4407	285
3	4691	4975	5259	5542	5825	6108	6391	6674	6956	7239	283
4	7521	7803	8084	8366	8647	8928	9209	9490	9771	0051	281
155	19 0332	0612	0892	1171	1451	1730	2010	2289	2567	2846	279
6	3125	3403	3681	3959	4237	4514	4792	5069	5346	5623	278
7	5900	6176	6453	6729	7005	7281	7556	7832	8107	8382	276
8	8657	8932	9206	9481	9755	0029	0303	0577	0850	1124	274
9	20 1397	1670	1943	2216	2488	2761	3033	3305	3577	3848	272
160	4120	4391	4663	4934	5204	5475	5746	6016	6286	6556	271
1	6826	7096	7365	7634	7904	8173	8441	8710	8979	9247	269
2	9515	9783	0051	0319	0586	0853	1121	1388	1654	1921	267
3	21 2188	2454	2720	2986	3252	3518	3783	4049	4314	4579	266
4	4844	5109	5373	5638	5902	6166	6430	6694	6957	7221	264
165	7484	7747	8010	8273	8536	8798	9060	9323	9585	9846	262
6	22 0108	0370	0631	0892	1153	1414	1675	1936	2196	2456	261
7	2716	2976	3236	3496	3755	4015	4274	4533	4792	5051	259
8	5309	5568	5826	6084	6342	6600	6858	7115	7372	7630	258
9	7887	8144	8400	8657	8913	9170	9426	9682	9938	0193	256
	23										

PROPORTIONAL PARTS

Diff.	1	2	3	4	5	6	7	8	9
285	28.5	57.0	85.5	114.0	142.5	171.0	199.5	228.0	256.5
284	28.4	56.8	85.2	113.6	142.0	170.4	198.8	227.2	255.6
283	28.3	56.6	84.9	113.2	141.5	169.8	198.1	226.4	254.7
282	28.2	56.4	84.6	112.8	141.0	169.2	197.4	225.6	253.8
281	28.1	56.2	84.3	112.4	140.5	168.6	196.7	224.8	252.9
280	28.0	56.0	84.0	112.0	140.0	168.0	196.0	224.0	252.0
279	27.9	55.8	83.7	111.6	139.5	167.4	195.3	223.2	251.1
278	27.8	55.6	83.4	111.2	139.0	166.8	194.6	222.4	250.2
277	27.7	55.4	83.1	110.8	138.5	166.2	193.9	221.6	249.3
276	27.6	55.2	82.8	110.4	138.0	165.6	193.2	220.8	248.4
275	27.5	55.0	82.5	110.0	137.5	165.0	192.5	220.0	247.5
274	27.4	54.8	82.2	109.6	137.0	164.4	191.8	219.2	246.6
273	27.3	54.6	81.9	109.2	136.5	163.8	191.1	218.4	245.7
272	27.2	54.4	81.6	108.8	136.0	163.2	190.4	217.6	244.8
271	27.1	54.2	81.3	108.4	135.5	162.6	189.7	216.8	243.9
270	27.0	54.0	81.0	108.0	135.0	162.0	189.0	216.0	243.0
269	26.9	53.8	80.7	107.6	134.5	161.4	188.3	215.2	242.1
268	26.8	53.6	80.4	107.2	134.0	160.8	187.6	214.4	241.2
267	26.7	53.4	80.1	106.8	133.5	160.2	186.9	213.6	240.3
266	26.6	53.2	79.8	106.4	133.0	159.6	186.2	212.8	239.4
265	26.5	53.0	79.5	106.0	132.5	159.0	185.5	212.0	238.5
264	26.4	52.8	79.2	105.6	132.0	158.4	184.8	211.2	237.6
263	26.3	52.6	78.9	105.2	131.5	157.8	184.1	210.4	236.7
262	26.2	52.4	78.6	104.8	131.0	157.2	183.4	209.6	235.8
261	26.1	52.2	78.3	104.4	130.5	156.6	182.7	208.8	234.9
260	26.0	52.0	78.0	104.0	130.0	156.0	182.0	208.0	234.0
259	25.9	51.8	77.7	103.6	129.5	155.4	181.3	207.2	233.1
258	25.8	51.6	77.4	103.2	129.0	154.8	180.6	206.4	232.2
257	25.7	51.4	77.1	102.8	128.5	154.2	179.9	205.6	231.3
256	25.6	51.2	76.8	102.4	128.0	153.6	179.2	204.8	230.4
255	25.5	51.0	76.5	102.0	127.5	153.0	178.5	204.0	229.5

No. 170 Log. 230 — **Table VII—Continued** — No. 189 Log. 278

N.	0	1	2	3	4	5	6	7	8	9	Diff.
170	23 0449	0704	0960	1215	1470	1724	1979	2234	2488	2742	255
1	2996	3250	3504	3757	4011	4264	4517	4770	5023	5276	253
2	5528	5781	6033	6285	6537	6789	7041	7292	7544	7795	252
3	8046	8297	8548	8799	9049	9299	9550	9800	0050	0300	250
4	24 0549	0799	1048	1297	1546	1795	2044	2293	2541	2790	249
175	3038	3286	3534	3782	4030	4277	4525	4772	5019	5266	248
6	5513	5759	6006	6252	6499	6745	6991	7237	7482	7728	246
7	7973	8219	8464	8709	8954	9198	9443	9687	9932	0176	245
8	25 0420	0664	0908	1151	1395	1638	1881	2125	2368	2610	243
9	2853	3096	3338	3580	3822	4064	4306	4548	4790	5031	242
180	5273	5514	5755	5996	6237	6477	6718	6958	7198	7439	241
1	7679	7918	8158	8398	8637	8877	9116	9355	9594	9833	239
2	26 0071	0310	0548	0787	1025	1263	1501	1739	1976	2214	238
3	2451	2688	2925	3162	3399	3636	3873	4109	4346	4582	237
4	4818	5054	5290	5525	5761	5996	6232	6467	6702	6937	235
185	7172	7406	7641	7875	8110	8344	8578	8812	9046	9279	234
6	9513	9746	9980	0213	0446	0679	0912	1144	1377	1609	233
7	27 1842	2074	2306	2538	2770	3001	3233	3464	3696	3927	232
8	4158	4389	4620	4850	5081	5311	5542	5772	6002	6232	230
9	6462	6692	6921	7151	7380	7609	7838	8067	8296	8525	229

PROPORTIONAL PARTS

Diff.	1	2	3	4	5	6	7	8	9
255	25.5	51.0	76.5	102.0	127.5	153.0	178.5	204.0	229.5
254	25.4	50.8	76.2	101.6	127.0	152.4	177.8	203.2	228.6
253	25.3	50.6	75.9	101.2	126.5	151.8	177.1	202.4	227.7
252	25.2	50.4	75.6	100.8	126.0	151.2	176.4	201.6	226.8
251	25.1	50.2	75.3	100.4	125.5	150.6	175.7	200.8	225.9
250	25.0	50.0	75.0	100.0	125.0	150.0	175.0	200.0	225.0
249	24.9	49.8	74.7	99.6	124.5	149.4	174.3	199.2	224.1
248	24.8	49.6	74.4	99.2	124.0	148.8	173.6	198.4	223.2
247	24.7	49.4	74.1	98.8	123.5	148.2	172.9	197.6	222.3
246	24.6	49.2	73.8	98.4	123.0	147.6	172.2	196.8	221.4
245	24.5	49.0	73.5	98.0	122.5	147.0	171.5	196.0	220.5
244	24.4	48.8	73.2	97.6	122.0	146.4	170.8	195.2	219.6
243	24.3	48.6	72.9	97.2	121.5	145.8	170.1	194.4	218.7
242	24.2	48.4	72.6	96.8	121.0	145.2	169.4	193.6	217.8
241	24.1	48.2	72.3	96.4	120.5	144.6	168.7	192.8	216.9
240	24.0	48.0	72.0	96.0	120.0	144.0	168.0	192.0	216.0
239	23.9	47.8	71.7	95.6	119.5	143.4	167.3	191.2	215.1
238	23.8	47.6	71.4	95.2	119.0	142.8	166.6	190.4	214.2
237	23.7	47.4	71.1	94.8	118.5	142.2	165.9	189.6	213.3
236	23.6	47.2	70.8	94.4	118.0	141.6	165.2	188.8	212.4
235	23.5	47.0	70.5	94.0	117.5	141.0	164.5	188.0	211.5
234	23.4	46.8	70.2	93.6	117.0	140.4	163.8	187.2	210.6
233	23.3	46.6	69.9	93.2	116.5	139.8	163.1	186.4	209.7
232	23.2	46.4	69.6	92.8	116.0	139.2	162.4	185.6	208.8
231	23.1	46.2	69.3	92.4	115.5	138.6	161.7	184.8	207.9
230	23.0	46.0	69.0	92.0	115.0	138.0	161.0	184.0	207.0
229	22.9	45.8	68.7	91.6	114.5	137.4	160.3	183.2	206.1
228	22.8	45.6	68.4	91.2	114.0	136.8	159.6	182.4	205.2
227	22.7	45.4	68.1	90.8	113.5	136.2	158.9	181.6	204.3
226	22.6	45.2	67.8	90.4	113.0	135.6	158.2	180.8	203.4

No. 190 **No. 214**
Log. 278 **Table VII—Continued** **Log. 332**

N.	0	1	2	3	4	5	6	7	8	9	Diff.
190	27 8754	8982	9211	9439	9667	9895	0123	0351	0578	0806	228
1	28 1033	1261	1488	1715	1942	2169	2396	2622	2849	3075	227
2	3301	3527	3753	3979	4205	4431	4656	4882	5107	5332	226
3	5557	5782	6007	6232	6456	6681	6905	7130	7354	7578	225
4	7802	8026	8249	8473	8696	8920	9143	9366	9589	9812	223
195	29 0035	0257	0480	0702	0925	1147	1369	1591	1813	2034	222
6	2256	2478	2699	2920	3141	3363	3584	3804	4025	4246	221
7	4166	4687	4907	5127	5347	5567	5787	6007	6226	6446	220
8	6665	6884	7104	7323	7542	7761	7979	8198	8416	8635	219
9	8853	9071	9289	9507	9725	9943	0161	0378	0595	0813	218
200	30 1030	1247	1464	1681	1898	2114	2331	2547	2764	2980	217
1	3196	3412	3628	3844	4059	4275	4491	4706	4921	5136	216
2	5351	5566	5781	5996	6211	6425	6639	6854	7068	7282	215
3	7496	7710	7924	8137	8351	8564	8778	8991	9204	9417	213
4	9630	9843	0056	0268	0481	0693	0906	1118	1330	1542	212
205	31 1754	1966	2177	2389	2600	2812	3023	3234	3445	3656	211
6	3867	4078	4289	4499	4710	4920	5130	5340	5551	5760	210
7	5970	6180	6390	6599	6809	7018	7227	7436	7646	7854	209
8	8063	8272	8481	8689	8898	9106	9314	9522	9730	9938	208
9	32 0146	0354	0562	0769	0977	1184	1391	1598	1805	2012	207
210	2219	2426	2633	2839	3046	3252	3458	3665	3871	4077	206
1	4282	4488	4694	4899	5105	5310	5516	5721	5926	6131	205
2	6336	6541	6745	6950	7155	7359	7563	7767	7972	8176	204
3	8380	8583	8787	8991	9194	9398	9601	9805	0008	0211	203
4	33 0414	0617	0819	1022	1225	1427	1630	1832	2034	2236	202

PROPORTIONAL PARTS

Diff.	1	2	3	4	5	6	7	8	9
225	22.5	45.0	67.5	90.0	112.5	135.0	157.5	180.0	202.5
224	22.4	44.8	67.2	89.6	112.0	134.4	156.8	179.2	201.6
223	22.3	44.6	66.9	89.2	111.5	133.8	156.1	178.4	200.7
222	22.2	44.4	66.6	88.8	111.0	133.2	155.4	177.6	199.8
221	22.1	44.2	66.3	88.4	110.5	132.6	154.7	176.8	198.9
220	22.0	44.0	66.0	88.0	110.0	132.0	154.0	176.0	198.0
219	21.9	43.8	65.7	87.6	109.5	131.4	153.3	175.2	197.1
218	21.8	43.6	65.4	87.2	109.0	130.8	152.6	174.4	196.2
217	21.7	43.4	65.1	86.8	108.5	130.2	151.9	173.6	195.3
216	21.6	43.2	64.8	86.4	108.0	129.6	151.2	172.8	194.4
215	21.5	43.0	64.5	86.0	107.5	129.0	150.5	172.0	193.5
214	21.4	42.8	64.2	85.6	107.0	128.4	149.8	171.2	192.6
213	21.3	42.6	63.9	85.2	106.5	127.8	149.1	170.4	191.7
212	21.2	42.4	63.6	84.8	106.0	127.2	148.4	169.6	190.8
211	21.1	42.2	63.3	84.4	105.5	126.6	147.7	168.8	189.9
210	21.0	42.0	63.0	84.0	105.0	126.0	147.0	168.0	189.0
209	20.9	41.8	62.7	83.6	104.5	125.4	146.3	167.2	188.1
208	20.8	41.6	62.4	83.2	104.0	124.8	145.6	166.4	187.2
207	20.7	41.4	62.1	82.8	103.5	124.2	144.9	165.6	186.3
206	20.6	41.2	61.8	82.4	103.0	123.6	144.2	164.8	185.4
205	20.5	41.0	61.5	82.0	102.5	123.0	143.5	164.0	184.5
204	20.4	40.8	61.2	81.6	102.0	122.4	142.8	163.2	183.6
203	20.3	40.6	60.9	81.2	101.5	121.8	142.1	162.4	182.7
202	20.2	40.4	60.6	80.8	101.0	121.2	141.4	161.6	181.8

No. 215 Log. 332 — **Table VII—Continued** — No. 239 Log. 380

N.	0	1	2	3	4	5	6	7	8	9	Diff.
215	33 2438	2640	2842	3044	3246	3447	3649	3850	4051	4253	202
6	4454	4655	4856	5057	5257	5458	5658	5859	6059	6260	201
7	6460	6660	6860	7060	7260	7459	7659	7858	8058	8257	200
8	8456	8656	8855	9054	9253	9451	9650	9849	0047	0246	199
9	34 0444	0642	0841	1039	1237	1435	1632	1830	2028	2225	198
220	2423	2620	2817	3014	3212	3409	3606	3802	3999	4196	197
1	4392	4589	4785	4981	5178	5374	5570	5766	5962	6157	196
2	6353	6549	6744	6939	7135	7330	7525	7720	7915	8110	195
3	8305	8500	8694	8889	9083	9278	9472	9666	9860	0054	194
4	35 0248	0442	0636	0829	1023	1216	1410	1603	1796	1989	193
225	2183	2375	2568	2761	2954	3147	3339	3532	3724	3916	193
6	4108	4301	4493	4685	4876	5068	5260	5452	5643	5834	192
7	6026	6217	6408	6599	6790	6981	7172	7363	7554	7744	191
8	7935	8125	8316	8506	8696	8886	9076	9266	9456	9646	190
9	9835	0025	0215	0404	0593	0783	0972	1161	1350	1539	189
230	36 1728	1917	2105	2294	2482	2671	2859	3048	3236	3424	188
1	3612	3800	3988	4176	4363	4551	4739	4926	5113	5301	188
2	5488	5675	5862	6049	6236	6423	6610	6796	6983	7169	187
3	7356	7542	7729	7915	8101	8287	8473	8659	8845	9030	186
4	9216	9401	9587	9772	9958	0143	0328	0513	0698	0883	185
235	37 1068	1253	1437	1622	1806	1991	2175	2360	2544	2728	184
6	2912	3096	3280	3464	3647	3831	4015	4198	4382	4565	184
7	4748	4932	5115	5298	5481	5664	5846	6029	6212	6394	183
8	6577	6759	6942	7124	7306	7488	7670	7852	8034	8216	182
9	8398	8580	8761	8943	9124	9306	9487	9668	9849	0030	181
	38										

PROPORTIONAL PARTS

Diff.	1	2	3	4	5	6	7	8	9
202	20.2	40.4	60.6	80.8	101.0	121.2	141.4	161.6	181.8
201	20.1	40.2	60.3	80.4	100.5	120.6	140.7	160.8	180.9
200	20.0	40.0	60.0	80.0	100.0	120.0	140.0	160.0	180.0
199	19.9	39.8	59.7	79.6	99.5	119.4	139.3	159.2	179.1
198	19.8	39.6	59.4	79.2	99.0	118.8	138.6	158.4	178.2
197	19.7	39.4	59.1	78.8	98.5	118.2	137.9	157.6	177.3
196	19.6	39.2	58.8	78.4	98.0	117.6	137.2	156.8	176.4
195	19.5	39.0	58.5	78.0	97.5	117.0	136.5	156.0	175.5
194	19.4	38.8	58.2	77.6	97.0	116.4	135.8	155.2	174.6
193	19.3	38.6	57.9	77.2	96.5	115.8	135.1	154.4	173.7
192	19.2	38.4	57.6	76.8	96.0	115.2	134.4	153.6	172.8
191	19.1	38.2	57.3	76.4	95.5	114.6	133.7	152.8	171.9
190	19.0	38.0	57.0	76.0	95.0	114.0	133.0	152.0	171.0
189	18.9	37.8	56.7	75.6	94.5	113.4	132.3	151.2	170.1
188	18.8	37.6	56.4	75.2	94.0	112.8	131.6	150.4	169.2
187	18.7	37.4	56.1	74.8	93.5	112.2	130.9	149.6	168.3
186	18.6	37.2	55.8	74.4	93.0	111.6	130.2	148.8	167.4
185	18.5	37.0	55.5	74.0	92.5	111.0	129.5	148.0	166.5
184	18.4	36.8	55.2	73.6	92.0	110.4	128.8	147.2	165.6
183	18.3	36.6	54.9	73.2	91.5	109.8	128.1	146.4	164.7
182	18.2	36.4	54.6	72.8	91.0	109.2	127.4	145.6	163.8
181	18.1	36.2	54.3	72.4	90.5	108.6	126.7	144.8	162.9
180	18.0	36.0	54.0	72.0	90.0	108.0	126.0	144.0	162.0
179	17.9	35.8	53.7	71.6	89.5	107.4	125.3	143.2	161.1

No. 240 **No. 269**
Log. 380 **Table VII—Continued** **Log. 431**

N.	0	1	2	3	4	5	6	7	8	9	Diff.
240	38 0211	0392	0573	0754	0934	1115	1296	1476	1656	1837	181
1	2017	2197	2377	2557	2737	2917	3097	3277	3456	3636	180
2	3815	3995	4174	4353	4533	4712	4891	5070	5249	5428	179
3	5606	5785	5964	6142	6321	6499	6677	6856	7034	7212	178
4	7390	7568	7746	7924	8101	8279	8456	8634	8811	8989	178
245	9166	9343	9520	9698	9875	0051	0228	0405	0582	0759	177
6	39 0935	1112	1288	1464	1641	1817	1993	2169	2345	2521	176
7	2697	2873	3048	3224	3400	3575	3751	3926	4101	4277	176
8	4452	4627	4802	4977	5152	5326	5501	5676	5850	6025	175
9	6199	6374	6548	6722	6896	7071	7245	7419	7592	7766	174
250	7940	8114	8287	8461	8634	8808	8981	9154	9328	9501	173
1	9674	9847	0020	0192	0365	0538	0711	0883	1056	1228	173
2	40 1401	1573	1745	1917	2089	2261	2433	2605	2777	2949	172
3	3121	3292	3464	3635	3807	3978	4149	4320	4492	4663	171
4	4834	5005	5176	5346	5517	5688	5858	6029	6199	6370	171
255	6540	6710	6881	7051	7221	7391	7561	7731	7901	8070	170
6	8240	8410	8579	8749	8918	9087	9257	9426	9595	9764	169
7	9933	0102	0271	0440	0609	0777	0946	1114	1283	1451	169
8	41 1620	1788	1956	2124	2293	2461	2629	2796	2964	3132	168
9	3300	3467	3635	3803	3970	4137	4305	4472	4639	4806	167
260	4973	5140	5307	5474	5641	5808	5974	0141	6308	6474	167
1	6641	6807	6973	7139	7306	7472	7638	7804	7970	8135	166
2	8301	8467	8633	8798	8964	9129	9295	9460	0625	0701	165
3	9956	0121	0286	0451	0616	0781	0945	1110	1275	1439	165
4	42 1604	1768	1933	2007	2261	2426	2590	2754	2918	3082	164
265	3246	3410	3574	3737	3901	4065	4228	4392	4555	4718	164
6	4882	5045	5208	5371	5534	5697	5860	6023	6186	6349	163
7	6511	6674	6836	6999	7161	7324	7486	7648	7811	7973	162
8	8135	8297	8459	8621	8783	8944	9106	9268	0420	0591	162
9	9752 43	9914	0075	0236	0398	0559	0720	0881	1042	1203	161

PROPORTIONAL PARTS

Diff.	1	2	3	4	5	6	7	8	9
178	17.8	35.6	53.4	71.2	89.0	106.8	124.6	142.4	160.2
177	17.7	35.4	53.1	70.8	88.5	106.2	123.9	141.6	159.3
176	17.6	35.2	52.8	70.4	88.0	105.6	123.2	140.8	158.4
175	17.5	35.0	52.5	70.0	87.5	105.0	122.5	140.0	157.5
174	17.4	34.8	52.2	69.6	87.0	104.4	121.8	139.2	156.6
173	17.3	34.6	51.9	69.2	86.5	103.8	121.1	138.4	155.7
172	17.2	34.4	51.6	68.8	86.0	103.2	120.4	137.6	154.8
171	17.1	34.2	51.3	68.4	85.5	102.6	119.7	136.8	153.9
170	17.0	34.0	51.0	68.0	85.0	102.0	119.0	136.0	153.0
169	16.9	33.8	50.7	67.6	84.5	101.4	118.3	135.2	152.1
168	16.8	33.6	50.4	67.2	84.0	100.8	117.6	134.4	151.2
167	16.7	33.4	50.1	66.8	83.5	100.2	116.9	133.6	150.3
166	16.6	33.2	49.8	66.4	83.0	99.6	116.2	132.8	149.4
165	16.5	33.0	49.5	66.0	82.5	99.0	115.5	132.0	148.5
164	16.4	32.8	49.2	65.6	82.0	98.4	114.8	131.2	147.6
163	16.3	32.6	48.9	65.2	81.5	97.8	114.1	130.4	146.7
162	16.2	32.4	48.5	64.8	81.0	97.2	113.4	129.6	145.8
161	16.1	32.2	48.3	64.4	80.5	96.6	112.7	128.8	144.9

No. 270
Log. 431

Table VII—Continued

No. 299
Log. 476

N.	0	1	2	3	4	5	6	7	8	9	Diff.
270	43 1364	1525	1685	1846	2007	2167	2328	2488	2649	2809	161
1	2969	3130	3290	3450	3610	3770	3930	4090	4249	4409	160
2	4569	4729	4888	5048	5207	5367	5526	5685	5844	6004	159
3	6163	6322	6481	6640	6799	6957	7116	7275	7433	7592	159
4	7751	7909	8067	8226	8384	8542	8701	8859	9017	9175	158
275	9333	9491	9648	9806	9964	0122	0279	0437	0594	0752	158
6	44 0909	1066	1224	1381	1538	1695	1852	2009	2166	2323	157
7	2480	2637	2793	2950	3106	3263	3419	3576	3732	3889	157
8	4045	4201	4357	4513	4669	4825	4981	5137	5293	5449	156
9	5604	5760	5915	6071	6226	6382	6537	6692	6848	7003	155
280	7158	7313	7468	7623	7778	7933	8088	8242	8397	8552	155
1	8706	8861	9015	9170	9324	9478	9633	9787	9941	0095	154
2	45 0249	0403	0557	0711	0865	1018	1172	1326	1479	1633	154
3	1786	1940	2093	2247	2400	2553	2706	2859	3012	3165	153
4	3318	3471	3624	3777	3930	4082	4235	4387	4540	4692	153
285	4845	4997	5150	5302	5454	5606	5758	5910	6062	6214	152
6	6366	6518	6670	6821	6973	7125	7276	7428	7579	7731	152
7	7882	8033	8184	8336	8487	8638	8789	8940	9091	9242	151
8	9392	9543	9694	9845	9995	0146	0296	0447	0597	0748	151
9	46 0898	1048	1198	1348	1499	1649	1799	1948	2098	2248	150
290	2398	2548	2697	2847	2997	3146	3296	3445	3594	3744	150
1	3893	4042	4191	4340	4490	4639	4788	4936	5085	5234	149
2	5383	5532	5680	5829	5977	6126	6274	6423	6571	6719	149
3	6868	7016	7164	7312	7460	7608	7756	7904	8052	8200	148
4	8347	8495	8643	8790	8938	9085	9233	9380	9527	9675	148
295	9822	9969	0116	0263	0410	0557	0704	0851	0998	1145	147
6	47 1292	1438	1585	1732	1878	2025	2171	2318	2464	2610	146
7	2756	2903	3049	3195	3341	3487	3633	3779	3925	4071	146
8	4216	4362	4508	4653	4799	4944	5090	5235	5381	5526	146
9	5671	5816	5962	6107	6252	6397	6542	6687	6832	6976	145

PROPORTIONAL PARTS

Diff.	1	2	3	4	5	6	7	8	9
161	16.1	32.2	48.3	64.4	80.5	96.6	112.7	128.8	144.9
160	16.0	32.0	48.0	64.0	80.0	96.0	112.0	128.0	144.0
159	15.9	31.8	47.7	63.6	79.5	95.4	111.3	127.2	143.1
158	15.8	31.6	47.4	63.2	79.0	94.8	110.6	126.4	142.2
157	15.7	31.4	47.1	62.8	78.5	94.2	109.9	125.6	141.3
156	15.6	31.2	46.8	62.4	78.0	93.6	109.2	124.8	140.4
155	15.5	31.0	46.5	62.0	77.5	93.0	108.5	124.0	139.5
154	15.4	30.8	46.2	61.6	77.0	92.4	107.8	123.2	138.6
153	15.3	30.6	45.9	61.2	76.5	91.8	107.1	122.4	137.7
152	15.2	30.4	45.6	60.8	76.0	91.2	106.4	121.6	136.8
151	15.1	30.2	45.3	60.4	75.5	90.6	105.7	120.8	135.9
150	15.0	30.0	45.0	60.0	75.0	90.0	105.0	120.0	135.0
149	14.9	29.8	44.7	59.6	74.5	89.4	104.3	119.2	134.1
148	14.8	29.6	44.4	59.2	74.0	88.8	103.6	118.4	133.2
147	14.7	29.4	44.1	58.8	73.5	88.2	102.9	117.6	132.3
146	14.6	29.2	43.8	58.4	73.0	87.6	102.2	116.8	131.4
145	14.5	29.0	43.5	58.0	72.5	87.0	101.5	116.0	130.5
144	14.4	28.8	43.2	57.6	72.0	86.4	100.8	115.2	129.6
143	14.3	28.6	42.9	57.2	71.5	85.8	100.1	114.4	128.7
142	14.2	28.4	42.6	56.8	71.0	85.2	99.4	113.6	127.8
141	14.1	28.2	42.3	56.4	70.5	84.6	98.7	112.8	126.9
140	14.0	28.0	42.0	56.0	70.0	84.0	98.0	112.0	126.0

No. 300 **Log. 477**

Table VII—Continued

No. 339 **Log. 531**

N.	0	1	2	3	4	5	6	7	8	9	Diff.
300	47 7121	7266	7411	7555	7700	7844	7989	8133	8278	8422	145
1	8566	8711	8855	8999	9143	9287	9431	9575	9719	9863	144
2	48 0007	0151	0294	0438	0582	0725	0869	1012	1156	1299	144
3	1443	1586	1729	1872	2016	2159	2302	2445	2588	2731	143
4	2874	3016	3159	3302	3445	3587	3730	3872	4015	4157	143
305	4300	4442	4585	4727	4869	5011	5153	5295	5437	5579	142
6	5721	5863	6005	6147	6289	6430	6572	6714	6855	6997	142
7	7138	7280	7421	7563	7704	7845	7986	8127	8269	8410	141
8	8551	8692	8833	8974	9114	9255	9396	9537	9677	9818	141
9	9958	0099	0239	0380	0520	0661	0801	0941	1081	1222	140
310	49 1362	1502	1642	1782	1922	2062	2201	2341	2481	2621	140
1	2760	2900	3040	3179	3319	3458	3597	3737	3876	4015	139
2	4155	4294	4433	4572	4711	4850	4989	5128	5267	5406	139
3	5544	5683	5822	5960	6099	6238	6376	6515	6653	6791	139
4	6930	7068	7206	7344	7483	7621	7759	7897	8035	8173	138
315	8311	8448	8586	8724	8862	8999	9137	9275	9412	9550	138
6	9687	9824	9962	0099	0236	0374	0511	0648	0785	0922	137
7	50 1059	1196	1333	1470	1607	1744	1880	2017	2154	2291	137
8	2427	2564	2700	2837	2973	3109	3246	3382	3518	3655	136
9	3791	3927	4063	4199	4335	4471	4607	4743	4878	5014	136
320	5150	5286	5421	5557	5693	5828	5964	6099	6234	6370	136
1	6505	6640	6776	6911	7046	7181	7316	7451	7586	7721	135
2	7856	7991	8126	8260	8395	8530	8664	8799	8934	9068	135
3	9203	9337	9471	9606	9740	9874	0009	0143	0277	0411	134
4	51 0545	0679	0813	0947	1081	1215	1349	1482	1616	1750	134
325	1883	2017	2151	2284	2418	2551	2684	2816	2951	3084	133
6	3218	3351	3484	3617	3750	3883	4016	4149	4282	4415	133
7	4548	4681	4813	4946	5079	5211	5344	5476	5609	5741	133
8	5874	6006	6139	6271	6403	6535	6668	6800	6932	7064	132
9	7196	7328	7460	7592	7724	7855	7987	8119	8251	8382	132
330	8514	8646	8777	8909	9040	9171	9303	9434	9566	9697	131
1	9828	9959	0090	0221	0353	0484	0615	0745	0876	1007	131
2	52 1138	1269	1400	1530	1661	1792	1922	2053	2183	2314	131
3	2444	2575	2705	2835	2966	3096	3226	3356	3486	3616	130
4	3740	3870	4006	4136	4266	4396	4526	4656	4785	4915	130
335	5045	5174	5304	5434	5563	5693	5822	5951	6081	6210	129
6	6339	6469	6598	6727	6856	6985	7114	7243	7372	7501	129
7	7630	7759	7888	8016	8145	8274	8402	8531	8660	8788	129
8	8917	9045	9174	9302	9430	9559	9687	9815	9943	0072	128
9	53 0200	0328	0456	0584	0712	0840	0968	1096	1223	1351	128

PROPORTIONAL PARTS

Diff.	1	2	3	4	5	6	7	8	9
139	13.9	27.8	41.7	55.6	69.5	83.4	97.3	111.2	125.1
138	13.8	27.6	41.4	55.2	69.0	82.8	96.6	110.4	124.2
137	13.7	27.4	41.1	54.8	68.5	82.2	95.9	109.6	123.3
136	13.6	27.2	40.8	54.4	68.0	81.6	95.2	108.8	122.4
135	13.5	27.0	40.5	54.0	67.5	81.0	94.5	108.0	121.5
134	13.4	26.8	40.2	53.6	67.0	80.4	93.8	107.2	120.6
133	13.3	26.6	39.9	53.2	66.5	79.8	93.1	106.4	119.7
132	13.2	26.4	39.6	52.8	66.0	79.2	92.4	105.6	118.8
131	13.1	26.2	39.3	52.4	65.5	78.6	91.7	104.8	117.9
130	13.0	26.0	39.0	52.0	65.0	78.0	91.0	104.0	117.0
129	12.9	25.8	38.7	51.6	64.5	77.4	90.3	103.2	116.1
128	12.8	25.6	38.4	51.2	64.0	76.8	89.6	102.4	115.2
127	12.7	25.4	38.1	50.8	63.5	76.2	88.9	101.6	114.3

No. 340
Log. 531

Table VII—Continued

No. 379
Log. 579

N.	0	1	2	3	4	5	6	7	8	9	Diff.
340	53 1479	1607	1734	1862	1990	2117	2245	2372	2500	2627	128
1	2754	2882	3009	3136	3264	3391	3518	3645	3772	3899	127
2	4026	4153	4280	4407	4534	4661	4787	4914	5041	5167	127
3	5294	5421	5547	5674	5800	5927	6053	6180	6306	6432	126
4	6558	6685	6811	6937	7063	7189	7315	7441	7567	7693	126
345	7819	7945	8071	8197	8322	8448	8574	8699	8825	8951	126
6	9076	9202	9327	9452	9578	9703	9829	9954	0079	0204	125
7	54 0329	0455	0580	0705	0830	0955	1080	1205	1330	1454	125
8	1579	1704	1829	1953	2078	2203	2327	2452	2576	2701	125
9	2825	2950	3074	3199	3323	3447	3571	3696	3820	3944	124
350	4068	4192	4316	4440	4564	4688	4812	4936	5060	5183	124
1	5307	5431	5555	5678	5802	5925	6049	6172	6296	6419	124
2	6543	6666	6789	6913	7036	7159	7282	7405	7529	7652	123
3	7775	7898	8021	8144	8267	8389	8512	8635	8758	8881	123
4	9003	9126	9249	9371	9494	9616	9739	9861	9984	0106	123
355	55 0228	0351	0473	0595	0717	0840	0962	1084	1206	1328	122
6	1450	1572	1694	1816	1938	2060	2181	2303	2425	2547	122
7	2668	2790	2911	3033	3155	3276	3398	3519	3640	3762	121
8	3883	4004	4126	4247	4368	4489	4610	4731	4852	4973	121
9	5094	5215	5336	5457	5578	5699	5820	5940	6061	6182	121
360	6303	6423	6544	6664	6785	6905	7026	7146	7267	7387	120
1	7507	7627	7748	7868	7988	8108	8228	8349	8469	8589	120
2	8709	8829	8948	9068	9188	9308	9428	9548	9667	9787	120
3	9907	0026	0146	0265	0385	0504	0624	0743	0863	0982	119
4	56 1101	1221	1340	1459	1578	1698	1817	1936	2055	2174	119
365	2293	2412	2531	2650	2769	2887	3006	3125	3244	3362	119
6	3481	3600	3718	3837	3955	4074	4192	4311	4429	4548	119
7	4666	4784	4903	5021	5139	5257	5376	5494	5612	5730	118
8	5848	5966	6084	6202	6320	6437	6555	6673	6791	6909	118
9	7026	7144	7262	7379	7497	7614	7732	7849	7967	8084	118
370	8202	8319	8436	8554	8671	8788	8905	9023	9140	9257	117
1	9374	9491	9608	9725	9842	9959	0076	0193	0309	0426	117
2	57 0543	0660	0776	0893	1010	1126	1243	1359	1476	1592	117
3	1709	1825	1942	2058	2174	2291	2407	2523	2639	2755	116
4	2872	2988	3104	3220	3336	3452	3568	3684	3800	3915	116
375	4031	4147	4263	4379	4494	4610	4726	4841	4957	5072	116
6	5188	5303	5419	5534	5650	5765	5880	5996	6111	6226	115
7	6341	6457	6572	6687	6802	6917	7032	7147	7262	7377	115
8	7492	7607	7722	7836	7951	8066	8181	8295	8410	8525	115
9	8639	8754	8868	8983	9097	9212	9326	9441	9555	9669	114

PROPORTIONAL PARTS

Diff.	1	2	3	4	5	6	7	8	9
128	12.8	25.6	38.4	51.2	64.0	76.8	89.6	102.4	115.2
127	12.7	25.4	38.1	50.8	63.5	76.2	88.9	101.6	114.3
126	12.6	25.2	37.8	50.4	63.0	75.6	88.2	100.8	113.4
125	12.5	25.0	37.5	50.0	62.5	75.0	87.5	100.0	112.5
124	12.4	24.8	37.2	49.6	62.0	74.4	86.8	99.2	111.6
123	12.3	24.6	36.9	49.2	61.5	73.8	86.1	98.4	110.7
122	12.2	24.4	36.6	48.8	61.0	73.2	85.4	97.6	109.8
121	12.1	24.2	36.3	48.4	60.5	72.6	84.7	96.8	108.9
120	12.0	24.0	36.0	48.0	60.0	72.0	84.0	96.0	108.0
119	11.9	23.8	35.7	47.6	59.5	71.4	83.3	95.2	107.1

No. 380 **No. 414**
Log. 579 **Table VII—Continued** **Log. 617**

N.	0	1	2	3	4	5	6	7	8	9	Diff.
380	57 9784	9898	0012	0126	0241	0355	0469	0583	0697	0811	114
1	58 0925	1039	1153	1267	1381	1495	1608	1722	1836	1950	
2	2063	2177	2291	2404	2518	2631	2745	2858	2972	3085	
3	3199	3312	3426	3539	3652	3765	3879	3992	4105	4218	
4	4331	4444	4557	4670	4783	4896	5009	5122	5235	5348	113
385	5461	5574	5686	5799	5912	6024	6137	6250	6362	6475	
6	6587	6700	6812	6925	7037	7149	7262	7374	7486	7599	
7	7711	7823	7935	8047	8160	8272	8384	8496	8608	8720	112
8	8832	8944	9056	9167	9279	9391	9503	9615	9726	9838	
9	9950	0061	0173	0284	0396	0507	0619	0730	0842	0953	
390	59 1065	1176	1287	1399	1510	1621	1732	1843	1955	2066	
1	2177	2288	2399	2510	2621	2732	2843	2954	3064	3175	111
2	3286	3397	3508	3618	3729	3840	3950	4061	4171	4282	
3	4393	4503	4614	4724	4834	4945	5055	5165	5276	5386	
4	5496	5606	5717	5827	5937	6047	6157	6267	6377	6487	
395	6597	6707	6817	6927	7037	7146	7256	7366	7476	7586	110
6	7695	7805	7914	8024	8134	8243	8353	8462	8572	8681	
7	8791	8900	9009	9119	9228	9337	9446	9556	9665	9774	
8	9883	9992	0101	0210	0319	0428	0537	0646	0755	0864	109
9	60 0973	1082	1191	1299	1408	1517	1625	1734	1843	1951	
400	2000	2169	2277	2386	2494	2603	2711	2819	2928	3036	
1	3144	3253	3361	3469	3577	3686	3794	3902	4010	4118	108
2	4226	4334	4442	4550	4658	4766	4874	4982	5089	5197	
3	5305	5413	5521	5628	5736	5844	5951	6059	6166	6274	
4	6381	6489	6596	6704	6811	6919	7026	7133	7241	7348	
405	7455	7562	7669	7777	7884	7991	8098	8205	8312	8419	107
6	8526	8633	8740	8847	8954	9061	9167	9274	9381	9488	
7	9594	9701	9808	9914	0021	0128	0234	0341	0447	0554	
8	01 0660	0767	0873	0979	1086	1192	1298	1405	1511	1617	
9	1723	1829	1936	2042	2148	2254	2360	2466	2572	2678	106
410	2784	2890	2996	3102	3207	3313	3419	3525	3630	3736	
1	0842	3947	4053	4159	4264	4370	4475	4581	4686	4792	
2	4897	5003	5108	5213	5319	5424	5529	5634	5740	5845	
3	5950	6055	6160	6265	6370	6476	6581	0086	6790	6895	105
4	7000	7105	7210	7315	7420	7525	7629	7734	7839	7943	

PROPORTIONAL PARTS

Diff.	1	2	3	4	5	6	7	8	9
118	11.8	23.6	35.4	47.2	59.0	70.8	82.6	94.4	106.2
117	11.7	23.4	35.1	46.8	58.5	70.2	81.9	93.6	105.3
116	11.6	23.2	34.8	46.4	58.0	69.6	81.2	92.8	104.4
115	11.5	23.0	34.5	46.0	57.5	69.0	80.5	92.0	103.5
114	11.4	22.8	34.2	45.6	57.0	68.4	79.8	91.2	102.6
113	11.3	22.6	33.9	45.2	56.5	67.8	79.1	90.4	101.7
112	11.2	22.4	33.6	44.8	56.0	67.2	78.4	89.6	100.8
111	11.1	22.2	33.3	44.4	55.5	66.6	77.7	88.8	99.9
110	11.0	22.0	33.0	44.0	55.0	66.0	77.0	88.0	99.0
109	10.9	21.8	32.7	43.6	54.5	65.4	76.3	87.2	98.1
108	10.8	21.6	32.4	43.2	54.0	64.8	75.6	86.4	97.2
107	10.7	21.4	32.1	42.8	53.5	64.2	74.9	85.6	96.3
106	10.6	21.2	31.8	42.4	53.0	63.6	74.2	84.8	95.4
105	10.5	21.0	31.5	42.0	52.5	63.0	73.5	84.0	94.5
104	10.4	20.8	31.2	41.6	52.0	62.4	72.8	83.2	93.6

No. 415 Log. 618 — Table VII—Continued — No. 459 Log. 662

N.	0	1	2	3	4	5	6	7	8	9	Diff.
415	61 8048	8153	8257	8362	8466	8571	8676	8780	8884	8989	105
6	9093	9198	9302	9406	9511	9615	9719	9824	9928	0032	
7	62 0136	0240	0344	0448	0552	0656	0760	0864	0968	1072	104
8	1176	1280	1384	1488	1592	1695	1799	1903	2007	2110	
9	2214	2318	2421	2525	2628	2732	2835	2939	3042	3146	
420	3249	3353	3456	3559	3663	3766	3869	3973	4076	4179	
1	4282	4385	4488	4591	4695	4798	4901	5004	5107	5210	103
2	5312	5415	5518	5621	5724	5827	5929	6032	6135	6238	
3	6340	6443	6546	6648	6751	6853	6956	7058	7161	7263	
4	7366	7468	7571	7673	7775	7878	7980	8082	8185	8287	
425	8389	8491	8593	8695	8797	8900	9002	9104	9206	9308	102
6	9410	9512	9613	9715	9817	9919	0021	0123	0224	0326	
7	63 0428	0530	0631	0733	0835	0936	1038	1139	1241	1342	
8	1444	1545	1647	1748	1849	1951	2052	2153	2255	2356	
9	2457	2559	2660	2761	2862	2963	3064	3165	3266	3367	
430	3468	3569	3670	3771	3872	3973	4074	4175	4276	4376	101
1	4477	4578	4679	4779	4880	4981	5081	5182	5283	5383	
2	5484	5584	5685	5785	5886	5986	6087	6187	6287	6388	
3	6488	6588	6688	6789	6889	6989	7089	7189	7290	7390	
4	7490	7590	7690	7790	7890	7990	8090	8190	8290	8389	100
435	8489	8589	8689	8789	8888	8988	9088	9188	9287	9387	
6	9486	9586	9686	9785	9885	9984	0084	0183	0283	0382	
7	64 0481	0581	0680	0779	0879	0978	1077	1177	1276	1375	
8	1474	1573	1672	1771	1871	1970	2069	2168	2267	2366	
9	2465	2563	2662	2761	2860	2959	3058	3156	3255	3354	99
440	3453	3551	3650	3749	3847	3946	4044	4143	4242	4340	
1	4439	4537	4636	4734	4832	4931	5029	5127	5226	5324	
2	5422	5521	5619	5717	5815	5913	6011	6110	6208	6306	
3	6404	6502	6600	6698	6796	6894	6992	7089	7187	7285	98
4	7383	7481	7579	7676	7774	7872	7969	8067	8165	8262	
445	8360	8458	8555	8653	8750	8848	8945	9043	9140	9237	
6	9335	9432	9530	9627	9724	9821	9919	0016	0113	0210	
7	65 0308	0405	0502	0599	0696	0793	0890	0987	1084	1181	
8	1278	1375	1472	1569	1666	1762	1859	1956	2053	2150	97
9	2246	2343	2440	2536	2633	2730	2826	2923	3019	3116	
450	3213	3309	3405	3502	3598	3695	3791	3888	3984	4080	
1	4177	4273	4369	4465	4562	4658	4754	4850	4946	5042	
2	5138	5235	5331	5427	5523	5619	5715	5810	5906	6002	96
3	6098	6194	6290	6386	6482	6577	6673	6769	6864	6960	
4	7056	7152	7247	7343	7438	7534	7629	7725	7820	7916	
455	8011	8107	8202	8298	8393	8488	8584	8679	8774	8870	
6	8965	9060	9155	9250	9346	9441	9536	9631	9726	9821	
7	9916	0011	0106	0201	0296	0391	0486	0581	0676	0771	95
8	66 0865	0960	1055	1150	1245	1339	1434	1529	1623	1718	
9	1813	1907	2002	2096	2191	2286	2380	2475	2569	2663	

PROPORTIONAL PARTS

Diff.	1	2	3	4	5	6	7	8	9
105	10.5	21.0	31.5	42.0	52.5	63.0	73.5	84.0	94.5
104	10.4	20.8	31.2	41.6	52.0	62.4	72.8	83.2	93.6
103	10.3	20.6	30.9	41.2	51.5	61.8	72.1	82.4	92.7
102	10.2	20.4	30.6	40.8	51.0	61.2	71.4	81.6	91.8
101	10.1	20.2	30.3	40.4	50.5	60.6	70.7	80.8	90.9
100	10.0	20.0	30.0	40.0	50.0	60.0	70.0	80.0	90.0
99	9.9	19.8	29.7	39.6	49.5	59.4	69.3	79.2	89.1

No. 460 Log. 662 — **Table VII—Continued** — **No. 499 Log. 698**

N.	0	1	2	3	4	5	6	7	8	9	Diff.
460	66 2758	2852	2947	3041	3135	3230	3324	3418	3512	3607	
1	3701	3795	3889	3983	4078	4172	4266	4360	4454	4548	
2	4642	4736	4830	4924	5018	5112	5206	5299	5393	5487	94
3	5581	5675	5769	5862	5956	6050	6143	6237	6331	6424	
4	6518	6612	6705	6799	6892	6986	7079	7173	7266	7360	
465	7453	7546	7640	7733	7826	7920	8013	8106	8199	8293	
6	8386	8479	8572	8665	8759	8852	8945	9038	9131	9224	
7	9317	9410	9503	9596	9689	9782	9875	9967	0060	0153	93
8	67 0246	0339	0431	0524	0617	0710	0802	0895	0988	1080	
9	1173	1265	1358	1451	1543	1636	1728	1821	1913	2005	
470	2098	2190	2283	2375	2467	2560	2652	2744	2836	2929	
1	3021	3113	3205	3297	3390	3482	3574	3666	3758	3850	
2	3942	4034	4126	4218	4310	4402	4494	4586	4677	4769	92
3	4861	4953	5045	5137	5228	5320	5412	5503	5595	5687	
4	5778	5870	5962	6053	6145	6236	6328	6419	6511	6602	
475	6694	6785	6876	6968	7059	7151	7242	7333	7424	7516	
6	7607	7698	7789	7881	7972	8063	8154	8245	8336	8427	
7	8518	8609	8700	8791	8882	8973	9064	9155	9246	9337	91
8	9428	9519	9610	9700	9791	9882	9973	0063	0154	0245	
9	68 0336	0426	0517	0607	0698	0789	0879	0970	1060	1151	
480	1241	1332	1422	1513	1603	1693	1784	1874	1964	2055	
1	2145	2235	2326	2416	2506	2596	2686	2777	2867	2957	
2	3047	3137	3227	3317	3407	3497	3587	3677	3767	3857	90
3	3947	4037	4127	4217	4307	4396	4486	4576	4666	4756	
4	4845	4935	5025	5114	5204	5294	5383	5473	5563	5652	
485	5742	5831	5921	6010	6100	6189	6279	6368	6458	6547	
6	6636	6726	6815	6904	6994	7083	7172	7261	7351	7440	
7	7529	7618	7707	7796	7886	7975	8064	8153	8242	8331	89
8	8420	8509	8598	8687	8776	8865	8953	9042	9131	9220	
9	9309	9398	9486	9575	9664	9753	9841	9930	0019	0107	
490	69 0196	0285	0373	0462	0550	0639	0728	0816	0905	0993	
1	1081	1170	1258	1347	1435	1524	1612	1700	1789	1877	
2	1965	2053	2142	2230	2318	2406	2494	2583	2671	2759	
3	2847	2935	3023	3111	3199	3287	3375	3463	3551	3639	88
4	3727	3815	3903	3991	4078	4166	4254	4342	4430	4517	
495	4605	4693	4781	4868	4956	5044	5131	5219	5307	5394	
6	5482	5569	5657	5744	5832	5919	6007	6094	6182	6269	
7	6356	6444	6531	6618	6706	6793	6880	6968	7055	7142	
8	7229	7317	7404	7491	7578	7665	7752	7839	7926	8014	
9	8100	8188	8275	8362	8449	8535	8622	8709	8796	8883	87

PROPORTIONAL PARTS

Diff.	1	2	3	4	5	6	7	8	9
98	9.8	19.6	29.4	39.2	49.0	58.8	68.6	78.4	88.2
97	9.7	19.4	29.1	38.8	48.5	58.2	67.9	77.6	87.3
96	9.6	19.2	28.8	38.4	48.0	57.6	67.2	76.8	86.4
95	9.5	19.0	28.5	38.0	47.5	57.0	66.5	76.0	85.5
94	9.4	18.8	28.2	37.6	47.0	56.4	65.8	75.2	84.6
93	9.3	18.6	27.9	37.2	46.5	55.8	65.1	74.4	83.7
92	9.2	18.4	27.6	36.8	46.0	55.2	64.4	73.6	82.8
91	9.1	18.2	27.3	36.4	45.5	54.6	63.7	72.8	81.9
90	9.0	18.0	27.0	36.0	45.0	54.0	63.0	72.0	81.0
89	8.9	17.8	26.7	35.6	44.5	53.4	62.3	71.2	80.1
88	8.8	17.6	26.4	35.2	44.0	52.8	61.6	70.4	79.2
87	8.7	17.4	26.1	34.8	43.5	52.2	60.9	69.6	78.3
86	8.6	17.2	25.8	34.4	43.0	51.6	60.2	68.8	77.4

Table VII—Continued

N.	0	1	2	3	4	5	6	7	8	9	Diff.
500	69 8970	9057	9144	9231	9317	9404	9491	9578	9664	9751	
1	9838	9924	0011	0098	0184	0271	0358	0444	0531	0617	
2	70 0704	0790	0877	0963	1050	1136	1222	1309	1395	1482	
3	1568	1654	1741	1827	1913	1999	2086	2172	2258	2344	
4	2431	2517	2603	2689	2775	2861	2947	3033	3119	3205	
505	3291	3377	3463	3549	3635	3721	3807	3893	3979	4065	86
6	4151	4236	4322	4408	4494	4579	4665	4751	4837	4922	
7	5008	5094	5179	5265	5350	5436	5522	5607	5693	5778	
8	5864	5949	6035	6120	6206	6291	6376	6462	6547	6632	
9	6718	6803	6888	6974	7059	7144	7229	7315	7400	7485	
510	7570	7655	7740	7826	7911	7996	8081	8166	8251	8336	85
1	8421	8506	8591	8676	8761	8846	8931	9015	9100	9185	
2	9270	9355	9440	9524	9609	9694	9779	9863	9948	0033	
3	71 0117	0202	0287	0371	0456	0540	0625	0710	0794	0879	
4	0963	1048	1132	1217	1301	1385	1470	1554	1639	1723	
515	1807	1892	1976	2060	2144	2229	2313	2397	2481	2566	
6	2650	2734	2818	2902	2986	3070	3154	3238	3323	3407	84
7	3491	3575	3659	3742	3826	3910	3994	4078	4162	4246	
8	4330	4414	4497	4581	4665	4749	4833	4916	5000	5084	
9	5167	5251	5335	5418	5502	5586	5669	5753	5836	5920	
520	6003	6087	6170	6254	6337	6421	6504	6588	6671	6754	
1	6838	6921	7004	7088	7171	7254	7338	7421	7504	7587	
2	7671	7754	7837	7920	8003	8086	8169	8253	8336	8419	
3	8502	8585	8668	8751	8834	8917	9000	9083	9165	9248	83
4	9331	9414	9497	9580	9663	9745	9828	9911	9994	0077	
525	72 0159	0242	0325	0407	0490	0573	0655	0738	0821	0903	
6	0986	1068	1151	1233	1316	1398	1481	1563	1646	1728	
7	1811	1893	1975	2058	2140	2222	2305	2387	2469	2552	
8	2634	2716	2798	2881	2963	3045	3127	3209	3291	3374	82
9	3456	3538	3620	3702	3784	3866	3948	4030	4112	4194	
530	4276	4358	4440	4522	4604	4685	4767	4849	4931	5013	
1	5095	5176	5258	5340	5422	5503	5585	5667	5748	5830	
2	5912	5993	6075	6156	6238	6320	6401	6483	6564	6646	
3	6727	6809	6890	6972	7053	7134	7216	7297	7379	7460	
4	7541	7623	7704	7785	7866	7948	8029	8110	8191	8273	
535	8354	8435	8516	8597	8678	8759	8841	8922	9003	9084	81
6	9165	9246	9327	9408	9489	9570	9651	9732	9813	9893	
7	9974	0055	0136	0217	0298	0378	0459	0540	0621	0702	
8	73 0782	0863	0944	1024	1105	1186	1266	1347	1428	1508	
9	1589	1669	1750	1830	1911	1991	2072	2152	2233	2313	
540	2394	2474	2555	2635	2715	2796	2876	2956	3037	3117	
1	3197	3278	3358	3438	3518	3598	3679	3759	3839	3919	
2	3999	4079	4160	4240	4320	4400	4480	4560	4640	4720	80
3	4800	4880	4960	5040	5120	5200	5279	5359	5439	5519	
4	5599	5679	5759	5838	5918	5998	6078	6157	6237	6317	

PROPORTIONAL PARTS

Diff.	1	2	3	4	5	6	7	8	9
87	8.7	17.4	26.1	34.8	43.5	52.2	60.9	69.6	78.3
86	8.6	17.2	25.8	34.4	43.0	51.6	60.2	68.8	77.4
85	8.5	17.0	25.5	34.0	42.5	51.0	59.5	68.0	76.5
84	8.4	16.8	25.2	33.6	42.0	50.4	58.8	67.2	75.6

No. 545 **Log. 736** — **Table VII—Continued** — **No. 584** **Log. 767**

N.	0	1	2	3	4	5	6	7	8	9	Diff.
545	73 6397	6476	6556	6635	6715	6795	6874	6954	7034	7113	
6	7193	7272	7352	7431	7511	7590	7670	7749	7829	7908	
7	7987	8067	8146	8225	8305	8384	8463	8543	8622	8701	
8	8781	8860	8939	9018	9097	9177	9256	9335	9414	9493	
9	9572	9651	9731	9810	9889	9968	0047	0126	0205	0284	79
550	74 0363	0442	0521	0600	0678	0757	0836	0915	0994	1073	
1	1152	1230	1309	1388	1467	1546	1624	1703	1782	1860	
2	1939	2018	2096	2175	2254	2332	2411	2489	2508	2647	
3	2725	2804	2882	2961	3039	3118	3196	3275	3353	3431	
4	3510	3588	3667	3745	3823	3902	3980	4058	4136	4215	
555	4293	4371	4449	4528	4606	4684	4762	4840	4919	4997	
6	5075	5153	5231	5309	5387	5465	5543	5621	5699	5777	78
7	5855	5933	6011	6089	6167	6245	6323	6401	6479	6556	
8	6634	6712	6790	6868	6945	7023	7101	7179	7256	7334	
9	7412	7489	7567	7645	7722	7800	7878	7955	8033	8110	
560	8188	8266	8343	8421	8498	8576	8653	8731	8808	8885	
1	8963	9040	9118	9195	9272	9350	9427	9504	9582	9659	
2	9736	9814	9891	9968	0045	0123	0200	0277	0354	0431	
3	75 0508	0586	0663	0740	0817	0894	0971	1048	1125	1202	
4	1279	1356	1433	1510	1587	1664	1741	1818	1895	1972	
565	2048	2125	2202	2279	2356	2433	2509	2586	2663	2740	77
6	2816	2893	2970	3047	3123	3200	3277	3353	3430	3506	
7	3583	3660	3736	3813	3889	3966	4042	4119	4195	4272	
8	4348	4425	4501	4578	4654	4730	4807	4883	4960	5036	
9	5112	5189	5265	5341	5417	5494	5570	5646	5722	5799	
570	5875	5951	6027	6103	6180	6256	6332	6408	6484	6560	
1	6636	6712	6788	6864	6940	7016	7092	7168	7244	7320	76
2	7396	7472	7548	7624	7700	7775	7851	7927	8003	8079	
3	8155	8230	8306	8382	8458	8533	8609	8685	8761	8836	
4	8912	8988	9063	9139	9214	9290	9366	9441	9517	9592	
575	9668	9743	9819	9894	9970	0045	0121	0196	0272	0347	
6	76 0422	0498	0573	0649	0724	0799	0875	0950	1025	1101	
7	1176	1251	1326	1402	1477	1552	1627	1702	1778	1853	
8	1928	2003	2078	2153	2228	2303	2378	2453	2529	2604	
9	2679	2754	2829	2004	2978	3053	3128	3203	3278	3353	75
580	3428	3503	3578	3653	3727	3802	3877	3952	4027	4101	
1	4176	[illegible]	[illegible]	[illegible]	[illegible]	[illegible]	[illegible]	[illegible]	[illegible]	[illegible]	
2	4923	4998	5072	5147	5221	5296	5370	5445	5520	5594	
3	5669	5743	5818	5892	5966	6041	6115	6190	6264	6338	
4	6413	6487	6562	6636	6710	6785	6859	6933	7007	7082	

PROPORTIONAL PARTS

Diff.	1	2	3	4	5	6	7	8	9
83	8.3	16.6	24.9	33.2	41.5	49.8	58.1	66.4	74.7
82	8.2	16.4	24.6	32.8	41.0	49.2	57.4	65.6	73.8
81	8.1	16.2	24.3	32.4	40.5	48.6	56.7	64.8	72.9
80	8.0	16.0	24.0	32.0	40.0	48.0	56.0	64.0	72.0
79	7.9	15.8	23.7	31.6	39.5	47.4	55.3	63.2	71.1
78	7.8	15.6	23.4	31.2	39.0	46.8	54.6	62.4	70.2
77	7.7	15.4	23.1	30.8	38.5	46.2	53.9	61.6	69.3
76	7.6	15.2	22.8	30.4	38.0	45.6	53.2	60.8	68.4
75	7.5	15.0	22.5	30.0	37.5	45.0	52.5	60.0	67.5
74	7.4	14.8	22.2	29.6	37.0	44.4	51.8	59.2	66.6

Table VII—Continued

N.	0	1	2	3	4	5	6	7	8	9	Diff.
585	76 7156	7230	7304	7379	7453	7527	7601	7675	7749	7823	
6	7898	7972	8046	8120	8194	8268	8342	8416	8490	8564	74
7	8638	8712	8786	8860	8934	9008	9082	9156	9230	9303	
8	9377	9451	9525	9599	9673	9746	9820	9894	9968	0042	
9	77 0115	0189	0263	0336	0410	0484	0557	0631	0705	0778	
590	0852	0926	0999	1073	1146	1220	1293	1367	1440	1514	
1	1587	1661	1734	1808	1881	1955	2028	2102	2175	2248	
2	2322	2395	2468	2542	2615	2688	2762	2835	2908	2981	
3	3055	3128	3201	3274	3348	3421	3494	3567	3640	3713	
4	3786	3860	3933	4006	4079	4152	4225	4298	4371	4444	73
595	4517	4590	4663	4736	4809	4882	4955	5028	5100	5173	
6	5246	5319	5392	5465	5538	5610	5683	5756	5829	5902	
7	5974	6047	6120	6193	6265	6338	6411	6483	6556	6629	
8	6701	6774	6846	6919	6992	7064	7137	7209	7282	7354	
9	7427	7499	7572	7644	7717	7789	7862	7934	8006	8079	
600	8151	8224	8296	8368	8441	8513	8585	8658	8730	8802	
1	8874	8947	9019	9091	9163	9236	9308	9380	9452	9524	
2	9596	9669	9741	9813	9885	9957	0029	0101	0173	0245	72
3	78 0317	0389	0461	0533	0605	0677	0749	0821	0893	0965	
4	1037	1109	1181	1253	1324	1396	1468	1540	1612	1684	
605	1755	1827	1899	1971	2042	2114	2186	2258	2329	2401	
6	2473	2544	2616	2688	2759	2831	2902	2974	3046	3117	
7	3189	3260	3332	3403	3475	3546	3618	3689	3761	3832	
8	3904	3975	4046	4118	4189	4261	4332	4403	4475	4546	
9	4617	4689	4760	4831	4902	4974	5045	5116	5187	5259	
610	5330	5401	5472	5543	5615	5686	5757	5828	5899	5970	
1	6041	6112	6183	6254	6325	6396	6467	6538	6609	6680	71
2	6751	6822	6893	6964	7035	7106	7177	7248	7319	7390	
3	7460	7531	7602	7673	7744	7815	7885	7956	8027	8098	
4	8168	8239	8310	8381	8451	8522	8593	8663	8734	8804	
615	8875	8946	9016	9087	9157	9228	9299	9369	9440	9510	
6	9581	9651	9722	9792	9863	9933	0004	0074	0144	0215	
7	79 0285	0356	0426	0496	0567	0637	0707	0778	0848	0918	
8	0988	1059	1129	1199	1269	1340	1410	1480	1550	1620	
9	1691	1761	1831	1901	1971	2041	2111	2181	2252	2322	
620	2392	2462	2532	2602	2672	2742	2812	2882	2952	3022	70
1	3092	3162	3231	3301	3371	3441	3511	3581	3651	3721	
2	3790	3860	3930	4000	4070	4139	4209	4279	4349	4418	
3	4488	4558	4627	4697	4767	4836	4906	4976	5045	5115	
4	5185	5254	5324	5393	5463	5532	5602	5672	5741	5811	
625	5880	5949	6019	6088	6158	6227	6297	6366	6436	6505	
6	6574	6644	6713	6782	6852	6921	6990	7060	7129	7198	
7	7268	7337	7406	7475	7545	7614	7683	7752	7821	7890	
8	7960	8029	8098	8167	8236	8305	8374	8443	8513	8582	
9	8651	8720	8789	8858	8927	8996	9065	9134	9203	9272	69

PROPORTIONAL PARTS

Diff.	1	2	3	4	5	6	7	8	9
75	7.5	15.0	22.5	30.0	37.5	45.0	52.5	60.0	67.5
74	7.4	14.8	22.2	29.6	37.0	44.4	51.8	59.2	66.6
73	7.3	14.6	21.9	29.2	36.5	43.8	51.1	58.4	65.7
72	7.2	14.4	21.6	28.8	36.0	43.2	50.4	57.6	64.8
71	7.1	14.2	21.3	28.4	35.5	42.6	49.7	56.8	63.9
70	7.0	14.0	21.0	28.0	35.0	42.0	49.0	56.0	63.0
69	6.9	13.8	20.7	27.6	34.5	41.4	48.3	55.2	62.1

No. 630 **Log. 799** **Table VII—Continued** **No. 674** **Log. 829**

N.	0	1	2	3	4	5	6	7	8	9	Diff.
630	79 9341	9409	9478	9547	9616	9685	9754	9823	9892	9961	
1	80 0029	0098	0167	0236	0305	0373	0442	0511	0580	0648	
2	0717	0786	0854	0923	0992	1061	1129	1198	1266	1335	
3	1404	1472	1541	1609	1678	1747	1815	1884	1952	2021	
4	2089	2158	2226	2295	2363	2432	2500	2568	2637	2705	
635	2774	2842	2910	2979	3047	3116	3184	3252	3321	3389	
6	3457	3525	3594	3662	3730	3798	3867	3935	4003	4071	
7	4139	4208	4276	4344	4412	4480	4548	4616	4685	4753	
8	4821	4889	4957	5025	5093	5161	5229	5297	5365	5433	**68**
9	5501	5569	5637	5705	5773	5841	5908	5976	6044	6112	
640	80 6180	6248	6316	6384	6451	6519	6587	6655	6723	6790	
1	6858	6926	6994	7061	7129	7197	7264	7332	7400	7467	
2	7535	7603	7670	7738	7806	7873	7941	8008	8076	8143	
3	8211	8279	8346	8414	8481	8549	8616	8684	8751	8818	
4	8886	8953	9021	9088	9156	9223	9290	9358	9425	9492	
645	9560	9627	9694	9762	9829	9896	9964	0031	0098	0165	
6	81 0233	0300	0367	0434	0501	0569	0636	0703	0770	0837	
7	0904	0971	1039	1106	1173	1240	1307	1374	1441	1508	67
8	1575	1642	1709	1776	1843	1910	1977	2044	2111	2178	
9	2245	2312	2379	2445	2512	2579	2646	2713	2780	2847	
650	2913	2980	3047	3114	3181	3247	3314	3381	3448	3514	
1	3581	3648	3714	3781	3848	3914	3981	4048	4114	4181	
2	4248	4314	4381	4447	4514	4581	4647	4714	4780	4847	
3	4913	4980	5046	5113	5179	5246	5312	5378	5445	5511	
4	5578	5644	5711	5777	5843	5910	5976	6042	6109	6175	
655	6241	6308	6374	6440	6506	6573	6639	6705	6771	6838	
6	0004	6970	7036	7102	7169	7235	7001	7367	7433	7499	
7	7565	7631	7698	7764	7830	7896	7962	8028	8094	8160	
8	8226	8292	8358	8424	8400	8556	8022	8688	8754	8820	
9	8885	8951	9017	9083	9149	9215	9281	9346	9412	9478	66
660	9544	9610	9676	9741	9807	9873	9939	0004	0070	0136	
1	82 0201	0267	0333	0399	0464	0530	0595	0661	0727	0792	
2	0858	0924	0989	1055	1120	1186	1251	1317	1002	1448	
3	1514	1579	1645	1710	1775	1841	1906	1972	2037	2103	
4	2168	2233	2299	2364	2430	2495	2560	2626	2691	2756	
665	2822	2887	2952	3018	3083	3148	3213	3279	3344	3409	
6	3474	3539	3605	3670	3735	3800	3865	3930	3996	4061	
7	4126	4191	4256	4321	4386	4451	4516	4581	4646	4711	
8	4776	4841	4906	4071	5000	5101	5100	5231	5296	5361	65
9	5420	5491	5556	5621	5686	5751	5815	5880	5945	6010	
670	6075	6140	6204	6269	6334	6399	6464	6528	0593	6658	
1	6723	6787	6852	6917	6981	7046	7111	7175	7240	7305	
2	7369	7434	7499	7563	7628	7602	7757	7821	7886	7951	
3	8015	8080	8144	8209	8273	8338	8402	8467	8531	8595	
4	8660	8724	8789	8853	8918	8982	9046	9111	9175	9239	

PROPORTIONAL PARTS

Diff.	1	2	3	4	5	6	7	8	9
68	6.8	13.6	20.4	27.2	34.0	40.8	47.6	54.4	61.2
67	6.7	13.4	20.1	26.8	33.5	40.2	46.9	53.6	60.3
66	6.6	13.2	19.8	26.4	33.0	39.6	46.2	52.8	59.4
65	**6.5**	13.0	19.5	26.0	32.5	39.0	45.5	52.0	58.5
64	6.4	12.8	19.2	25.6	32.0	38.4	44.8	51.2	57.6

No. 675
Log. 829

Table VII—Continued

No. 719
Log. 857

N.	0	1	2	3	4	5	6	7	8	9	Diff.
675	82 9304	9368	9432	9497	9561	9625	9690	9754	9818	9882	
6	9947	0011	0075	0139	0204	0268	0332	0396	0460	0525	
7	83 0589	0653	0717	0781	0845	0909	0973	1037	1102	1166	
8	1230	1294	1358	1422	1486	1550	1614	1678	1742	1806	64
9	1870	1934	1998	2062	2126	2189	2253	2317	2381	2445	
680	2509	2573	2637	2700	2764	2828	2892	2956	3020	3083	
1	3147	3211	3275	3338	3402	3466	3530	3593	3657	3721	
2	3784	3848	3912	3975	4039	4103	4166	4230	4294	4357	
3	4421	4484	4548	4611	4675	4739	4802	4866	4929	4993	
4	5056	5120	5183	5247	5310	5373	5437	5500	5564	5627	
685	5691	5754	5817	5881	5944	6007	6071	6134	6197	6261	
6	6324	6387	6451	6514	6577	6641	6704	6767	6830	6894	
7	6957	7020	7083	7146	7210	7273	7336	7399	7462	7525	
8	7588	7652	7715	7778	7841	7904	7967	8030	8093	8156	
9	8219	8282	8345	8408	8471	8534	8597	8660	8723	8786	63
690	8849	8912	8975	9038	9101	9164	9227	9289	9352	9415	
1	9478	9541	9604	9667	9729	9792	9855	9918	9981	0043	
2	84 0106	0169	0232	0294	0357	0420	0482	0545	0608	0671	
3	0733	0796	0859	0921	0984	1046	1109	1172	1234	1297	
4	1359	1422	1485	1547	1610	1672	1735	1797	1860	1922	
695	1985	2047	2110	2172	2235	2297	2360	2422	2484	2547	
6	2609	2672	2734	2796	2859	2921	2983	3046	3108	3170	
7	3233	3295	3357	3420	3482	3544	3606	3669	3731	3793	
8	3855	3918	3980	4042	4104	4166	4229	4291	4353	4415	
9	4477	4539	4601	4664	4726	4788	4850	4912	4974	5036	
700	5098	5160	5222	5284	5346	5408	5470	5532	5594	5656	62
1	5718	5780	5842	5904	5966	6028	6090	6151	6213	6275	
2	6337	6399	6461	6523	6585	6646	6708	6770	6832	6894	
3	6955	7017	7079	7141	7202	7264	7326	7388	7449	7511	
4	7573	7634	7696	7758	7819	7881	7943	8004	8066	8128	
705	8189	8251	8312	8374	8435	8497	8559	8620	8682	8743	
6	8805	8866	8928	8989	9051	9112	9174	9235	9297	9358	
7	9419	9481	9542	9604	9665	9726	9788	9849	9911	9972	
8	85 0033	0095	0156	0217	0279	0340	0401	0462	0524	0585	
9	0646	0707	0769	0830	0891	0952	1014	1075	1136	1197	
710	1258	1320	1381	1442	1503	1564	1625	1686	1747	1809	
1	1870	1931	1992	2053	2114	2175	2236	2297	2358	2419	
2	2480	2541	2602	2663	2724	2785	2846	2907	2968	3029	61
3	3090	3150	3211	3272	3333	3394	3455	3516	3577	3637	
4	3698	3759	3820	3881	3941	4002	4063	4124	4185	4245	
715	4306	4367	4428	4488	4549	4610	4670	4731	4792	4852	
6	4913	4974	5034	5095	5156	5216	5277	5337	5398	5459	
7	5519	5580	5640	5701	5761	5822	5882	5943	6003	6064	
8	6124	6185	6245	6306	6366	6427	6487	6548	6608	6668	
9	6729	6789	6850	6910	6970	7031	7091	7152	7212	7272	

PROPORTIONAL PARTS

Diff.	1	2	3	4	5	6	7	8	9
65	6.5	13.0	19.5	26.0	32.5	39.0	45.5	52.0	58.5
64	6.4	12.8	19.2	25.6	32.0	38.4	44.8	51.2	57.6
63	6.3	12.6	18.9	25.2	31.5	37.8	44.1	50.4	56.7
62	6.2	12.4	18.6	24.8	31.0	37.2	43.4	49.6	55.8
61	6.1	12.2	18.3	24.4	30.5	36.6	42.7	48.8	54.9
60	6.0	12.0	18.0	24.0	30.0	36.0	42.0	48.0	54.0

No. 720 **No. 764**

Log. 857 **Table VII—Continued** **Log. 883**

N.	0	1	2	3	4	5	6	7	8	9	Diff.
720	85 7332	7393	7453	7513	7574	7634	7694	7755	7815	7875	
1	7935	7995	8056	8116	8176	8236	8297	8357	8417	8477	
2	8537	8597	8657	8718	8778	8838	8898	8958	9018	9078	
3	9138	9198	9258	9318	9379	9439	9499	9559	9619	9679	60
4	9739	9799	9859	9918	9978	0038	0098	0158	0218	0278	
725	86 0338	0398	0458	0518	0578	0637	0697	0757	0817	0877	
6	0937	0996	1056	1116	1176	1236	1295	1355	1415	1475	
7	1534	1594	1654	1714	1773	1833	1893	1952	2012	2072	
8	2131	2191	2251	2310	2370	2430	2489	2549	2608	2668	
9	2728	2787	2847	2906	2966	3025	3085	3144	3204	3263	
730	3323	3382	3442	3501	3561	3620	3680	3739	3799	3858	
1	3917	3977	4036	4096	4155	4214	4274	4333	4392	4452	
2	4511	4570	4630	4689	4748	4808	4867	4926	4985	5045	
3	5104	5163	5222	5282	5341	5400	5459	5519	5578	5637	
4	5696	5755	5814	5874	5933	5992	6051	6110	6169	6228	
735	6287	6346	6405	6465	6524	6583	6642	6701	6760	6819	
6	6878	6937	6996	7055	7114	7173	7232	7291	7350	7409	59
7	7467	7526	7585	7644	7703	7762	7821	7880	7939	7998	
8	8056	8115	8174	8233	8292	8350	8409	8468	8527	8586	
9	8644	8703	8762	8821	8879	8938	8997	9056	9114	9173	
740	9232	9290	9349	9408	9466	9525	9584	9642	9701	9760	
1	9818	9877	9935	9994	0053	0111	0170	0228	0287	0345	
2	87 0404	0462	0521	0579	0638	0696	0755	0813	0872	0930	
3	0989	1047	1106	1164	1223	1281	1339	1398	1456	1515	
4	1573	1631	1690	1748	1806	1865	1923	1981	2040	2098	
745	2156	2215	2273	2331	2389	2448	2506	2564	2622	2681	
6	2739	2797	2855	2913	2972	3030	3088	3146	3204	3262	
7	3321	3379	3437	3495	3553	3611	3669	3727	3785	3844	
8	3902	3960	4018	4076	4134	4192	4250	4308	4366	4424	58
9	4482	4540	4598	4656	4714	4772	4830	4888	4945	5003	
750	5061	5119	5177	5235	5293	5351	5409	5466	5524	5582	
1	5640	5698	5756	5813	5871	5929	5987	6045	6102	6160	
2	6218	6276	6333	6391	6449	6507	6564	6622	6680	6737	
3	6795	6853	6910	6968	7026	7083	7141	7199	7256	7314	
4	7371	7429	7487	7544	7602	7659	7717	7774	7832	7889	
755	7947	8004	8062	8119	8177	8234	8292	8349	8407	8464	
6	8522	8579	8637	8694	8752	8809	8866	8924	8981	9039	
7	9096	9153	9211	9268	9325	9383	9440	9497	9555	9612	
8	9669	9726	9784	9841	9898	9956	0013	0070	0127	0185	
9	88 0242	0299	0356	0413	0471	0528	0585	0642	0699	0756	
760	0814	0871	0928	0985	1042	1099	1156	1213	1271	1328	
1	1385	1442	1499	1556	1613	1670	1727	1784	1841	1898	
2	1955	2012	2069	2126	2183	2240	2297	2354	2411	2468	57
3	2525	2581	2638	2695	2752	2809	2866	2923	2980	3037	
4	3093	3150	3207	3264	3321	3377	3434	3491	3548	3605	

PROPORTIONAL PARTS

Diff.	1	2	3	4	5	6	7	8	9
59	5.9	11.8	17.7	23.6	29.5	35.4	41.3	47.2	53.1
58	5.8	11.6	17.4	23.2	29.0	34.8	40.6	46.4	52.2
57	5.7	11.4	17.1	22.8	28.5	34.2	39.9	45.6	51.3
56	5.6	11.2	16.8	22.4	28.0	33.6	39.2	44.8	50.4

No. 765
Log. 883

Table VII—Continued

No. 809
Log. 908

N.	0	1	2	3	4	5	6	7	8	9	Diff.
765	88 3661	3718	3775	3832	3888	3945	4002	4059	4115	4172	
6	4229	4285	4342	4399	4455	4512	4569	4625	4682	4739	
7	4795	4852	4909	4965	5022	5078	5135	5192	5248	5305	
8	5361	5418	5474	5531	5587	5644	5700	5757	5813	5870	
9	5926	5983	6039	6096	6152	6209	6265	6321	6378	6434	
770	6491	6547	6604	6660	6716	6773	6829	6885	6942	6998	
1	7054	7111	7167	7223	7280	7336	7392	7449	7505	7561	
2	7617	7674	7730	7786	7842	7898	7955	8011	8067	8123	
3	8179	8236	8292	8348	8404	8460	8516	8573	8629	8685	
4	8741	8797	8853	8909	8965	9021	9077	9134	9190	9246	
775	9302	9358	9414	9470	9526	9582	9638	9694	9750	9806	56
6	9862	9918	9974	0030	0086	0141	0197	0253	0309	0365	
7	89 0421	0477	0533	0589	0645	0700	0756	0812	0868	0924	
8	0980	1035	1091	1147	1203	1259	1314	1370	1426	1482	
9	1537	1593	1649	1705	1760	1816	1872	1928	1983	2039	
780	2095	2150	2206	2262	2317	2373	2429	2484	2540	2595	
1	2651	2707	2762	2818	2873	2929	2985	3040	3096	3151	
2	3207	3262	3318	3373	3429	3484	3540	3595	3651	3706	
3	3762	3817	3873	3928	3984	4039	4094	4150	4205	4261	
4	4316	4371	4427	4482	4538	4593	4648	4704	4759	4814	
785	4870	4925	4980	5036	5091	5146	5201	5257	5312	5367	
6	5423	5478	5533	5588	5644	5699	5754	5809	5864	5920	
7	5975	6030	6085	6140	6195	6251	6306	6361	6416	6471	
8	6526	6581	6636	6692	6747	6802	6857	6912	6967	7022	
9	7077	7132	7187	7242	7297	7352	7407	7462	7517	7572	
790	7627	7682	7737	7792	7847	7902	7957	8012	8067	8122	55
1	8176	8231	8286	8341	8396	8451	8506	8561	8615	8670	
2	8725	8780	8835	8890	8944	8999	9054	9109	9164	9218	
3	9273	9328	9383	9437	9492	9547	9602	9656	9711	9766	
4	9821	9875	9930	9985	0039	0094	0149	0203	0258	0312	
795	90 0367	0422	0476	0531	0586	0640	0695	0749	0804	0859	
6	0913	0968	1022	1077	1131	1186	1240	1295	1349	1404	
7	1458	1513	1567	1622	1676	1731	1785	1840	1894	1948	
8	2003	2057	2112	2166	2221	2275	2329	2384	2438	2492	
9	2547	2601	2655	2710	2764	2818	2873	2927	2981	3036	
800	3090	3144	3199	3253	3307	3361	3416	3470	3524	3578	
1	3633	3687	3741	3795	3849	3904	3958	4012	4066	4120	
2	4174	4229	4283	4337	4391	4445	4499	4553	4607	4661	
3	4716	4770	4824	4878	4932	4986	5040	5094	5148	5202	
4	5256	5310	5364	5418	5472	5526	5580	5634	5688	5742	54
805	5796	5850	5904	5958	6012	6066	6119	6173	6227	6281	
6	6335	6389	6443	6497	6551	6604	6658	6712	6766	6820	
7	6874	6927	6981	7035	7089	7143	7196	7250	7304	7358	
8	7411	7465	7519	7573	7626	7680	7734	7787	7841	7895	
9	7949	8002	8056	8110	8163	8217	8270	8324	8378	8431	

PROPORTIONAL PARTS

Diff.	1	2	3	4	5	6	7	8	9
57	5.7	11.4	17.1	22.8	28.5	34.2	39.9	45.6	51.3
56	5.6	11.2	16.8	22.4	28.0	33.6	39.2	44.8	50.4
55	5.5	11.0	16.5	22.0	27.5	33.0	38.5	44.0	49.5
54	5.4	10.8	16.2	21.6	27.0	32.4	37.8	43.2	48.6

No. 810 **Log. 908**

Table VII—Continued

No. 854 **Log. 931**

N.	0	1	2	3	4	5	6	7	8	9	Diff.
810	90 8485	8539	8592	8646	8699	8753	8807	8860	8914	8967	
1	9021	9074	9128	9181	9235	9289	9342	9396	9449	9503	
2	9556	9610	9663	9716	9770	9823	9877	9930	9984	0037	
3	91 0091	0144	0197	0251	0304	0358	0411	0464	0518	0571	
4	0624	0678	0731	0784	0838	0891	0944	0998	1051	1104	
815	1158	1211	1264	1317	1371	1424	1477	1530	1584	1637	
6	1690	1743	1797	1850	1903	1056	2009	2063	2116	2169	
7	2222	2275	2328	2381	2435	2488	2541	2594	2647	2700	
8	2753	2806	2859	2913	2966	3019	3072	3125	3178	3231	
9	3284	3337	3390	3443	3496	3549	3602	3655	3708	3761	53
820	3814	3867	3920	3973	4026	4079	4132	4184	4237	4290	
1	4343	4396	4449	4502	4555	4608	4660	4713	4766	4819	
2	4872	4925	4977	5030	5083	5136	5189	5241	5294	5347	
3	5400	5453	5505	5558	5611	5664	5716	5769	5822	5875	
4	5927	5980	6033	6085	6138	6191	6243	6296	6349	6401	
825	6454	6507	6559	6612	6664	6717	6770	6822	6875	6927	
6	6980	7033	7085	7138	7190	7243	7295	7348	7400	7453	
7	7506	7558	7611	7663	7716	7768	7820	7873	7925	7978	
8	8030	8083	8135	8188	8240	8293	8345	8397	8450	8502	
9	8555	8607	8659	8712	8764	8816	8869	8921	8973	9026	
830	9078	9130	9183	9235	9287	9340	9392	9444	9496	9549	
1	9601	9653	9706	9758	9810	9862	9914	9967	0019	0071	
2	92 0123	0176	0228	0280	0332	0384	0436	0489	0541	0593	
3	0645	0697	0749	0801	0853	0906	0958	1010	1062	1114	
4	1166	1218	1270	1322	1374	1426	1478	1530	1582	1634	52
835	1686	1738	1790	1842	1894	1946	1998	2050	2102	2154	
6	2206	2258	2310	2362	2414	2466	2518	2570	2622	2674	
7	2725	2777	2829	2881	2933	2985	3037	3089	3140	3192	
8	3244	3296	3348	3399	3451	3503	3555	3607	3658	3710	
9	3762	3814	3865	3917	3969	4021	4072	4124	4176	4228	
840	4279	4331	4383	4434	4486	4538	4589	4641	4693	4744	
1	4706	4040	4899	4951	5003	5054	5106	5157	5209	5261	
2	5312	5364	5415	5467	5518	5570	5621	5673	5725	5776	
3	5828	5879	5931	5982	6074	6085	0107	0100	0240	0291	
4	6342	6394	6445	6497	6548	6600	6651	6702	6754	6805	
845	6857	6908	6959	7011	7062	7114	7165	7216	7268	7319	
6	7370	7422	7473	7524	7576	7627	7678	7730	7781	7832	
7	7883	7935	7986	8037	8088	8140	8191	8242	8293	8345	
8	8396	8447	8498	8549	8601	8652	8703	8754	8805	8857	
9	8908	8959	9010	9061	9112	9163	0215	9266	9317	9368	
850	9419	9470	9521	9572	9623	9674	9725	9776	9827	9879	51
1	9930	9981	0032	0083	0134	0185	0236	0287	0338	0389	
2	93 0440	0491	0542	0592	0643	0694	0745	0796	0847	0898	
3	0949	1000	1051	1102	1153	1204	1254	1305	1356	1407	
4	1458	1509	1560	1610	1661	1712	1763	1814	1865	1915	

PROPORTIONAL PARTS

Diff.	1	2	3	4	5	6	7	8	9
53	5.3	10.6	15.9	21.2	26.5	31.8	37.1	42.4	47.7
52	5.2	10.4	15.6	20.8	26.0	31.2	36.4	41.6	46.8
51	5.1	10.2	15.3	20.4	25.5	30.6	35.7	40.8	45.9
50	5.0	10.0	15.0	20.0	25.0	30.0	35.0	40.0	45.0

No. 855 Log. 931 | Table VII—Continued | No. 899 Log. 954

N.	0	1	2	3	4	5	6	7	8	9	Diff.
855	93 1966	2017	2068	2118	2169	2220	2271	2322	2372	2423	
6	2474	2524	2575	2626	2677	2727	2778	2829	2879	2930	
7	2981	3031	3082	3133	3183	3234	3285	3335	3386	3437	
8	3487	3538	3589	3639	3690	3740	8791	3841	3892	3943	
9	3993	4044	4094	4145	4195	4246	4296	4347	4397	4448	
860	4498	4549	4599	4650	4700	4751	4801	4852	4902	4953	
1	5003	5054	5104	5154	5205	5255	5306	5356	5406	5457	
2	5507	5558	5608	5658	5709	5759	5809	5860	5910	5960	
3	6011	6061	6111	6162	6212	6262	6313	6363	6413	6463	
4	6514	6564	6614	6665	6715	6765	6315	6865	6916	6966	
865	7016	7066	7116	7167	7217	7267	7317	7367	7418	7468	
6	7518	7568	7618	7668	7718	7769	7819	7869	7919	7969	
7	8019	8069	8119	8169	8219	8269	8320	8370	8420	8470	50
8	8520	8570	8620	8670	8720	8770	8820	8870	8920	8970	
9	9020	9070	9120	9170	9220	9270	9320	9369	9419	9469	
870	9519	9569	9619	9669	9719	9769	9819	9869	9918	9968	
1	94 0018	0068	0118	0168	0218	0267	0317	0367	0417	0467	
2	0516	0566	0616	0666	0716	0765	0815	0865	0915	0964	
3	1014	1064	1114	1163	1213	1263	1313	1362	1412	1462	
4	1511	1561	1611	1660	1710	1760	1809	1859	1909	1958	
875	2008	2058	2107	2157	2207	2256	2306	2355	2405	2455	
6	2504	2554	2603	2653	2702	2752	2801	2851	2901	2950	
7	3000	3049	3099	3148	3198	3247	3297	3346	3396	3445	
8	3495	3544	3593	3643	3692	3742	3791	3841	3890	3939	
9	3989	4038	4088	4137	4186	4236	4285	4335	4384	4433	
880	4483	4532	4581	4631	4680	4729	4779	4828	4877	4927	
1	4976	5025	5074	5124	5173	5222	5272	5321	5370	5419	
2	5469	5518	5567	5616	5665	5715	5764	5813	5862	5912	
3	5961	6010	6059	6108	6157	6207	6256	6305	6354	6403	
4	6452	6501	6551	6600	6649	6698	6747	6796	6845	6894	
885	6943	6992	7041	7090	7139	7189	7238	7287	7336	7385	
6	7434	7483	7532	7581	7630	7679	7728	7777	7826	7875	49
7	7924	7973	8022	8070	8119	8168	8217	8266	8315	8364	
8	8413	8462	8511	8560	8608	8657	8706	8755	8804	8853	
9	8902	8951	8999	9048	9097	9146	9195	9244	9292	9341	
890	9390	9439	9488	9536	9585	9634	9683	9731	9780	9829	
1	9878	9926	9975	0024	0073	0121	0170	0219	0267	0316	
2	95 0365	0414	0462	0511	0560	0608	0657	0706	0754	0803	
3	0851	0900	0949	0997	1046	1095	1143	1192	1240	1289	
4	1338	1386	1435	1483	1532	1580	1629	1677	1726	1775	
895	1823	1872	1920	1969	2017	2066	2114	2163	2211	2260	
6	2308	2356	2405	2453	2502	2550	2599	2647	2696	2744	
7	2792	2841	2889	2938	2986	3034	3083	3131	3180	3228	
8	3276	3325	3373	3421	3470	3518	3566	3615	3663	3711	
9	3760	3808	3856	3905	3953	4001	4049	4098	4146	4194	

PROPORTIONAL PARTS

Diff.	1	2	3	4	5	6	7	8	9
51	5.1	10.2	15.3	20.4	25.5	30.6	35.7	40.8	45.9
50	5.0	10.0	15.0	20.0	25.0	30.0	35.0	40.0	45.0
49	4.9	9.8	14.7	19.6	24.5	29.4	34.3	39.2	44.1
48	4.8	9.6	14.4	19.2	24.0	28.8	33.6	38.4	43.2

No. 900 **No. 944**
Log. 954 **Table VII—Continued** **Log. 975**

N.	0	1	2	3	4	5	6	7	8	9	Diff.
900	95 4243	4291	4339	4387	4435	4484	4532	4580	4628	4677	
1	4725	4773	4821	4869	4918	4966	5014	5062	5110	5158	
2	5207	5255	5303	5351	5399	5447	5495	5543	5592	5640	
3	5688	5736	5784	5832	5880	5928	5976	6024	6072	6120	
4	6168	6216	6265	6313	6361	6409	6457	6505	6553	6601	48
905	6649	6697	6745	6793	6840	6888	6936	6984	7032	7080	
6	7128	7176	7224	7272	7320	7368	7416	7464	7512	7559	
7	7607	7655	7703	7751	7799	7847	7894	7942	7990	8038	
8	8086	8134	8181	8229	8277	8325	8373	8421	8468	8516	
9	8564	8612	8659	8707	8755	8803	8850	8898	8946	8994	
910	9041	9089	9137	9185	9232	9280	9328	9375	9423	9471	
1	9518	9566	9614	9661	9709	9757	9804	9852	9900	9947	
2	9995	0042	0090	0138	0185	0233	0280	0328	0376	0423	
3	96 0471	0518	0566	0613	0661	0709	0756	0804	0851	0899	
4	0946	0994	1041	1089	1136	1184	1231	1279	1326	1374	
915	1421	1469	1516	1563	1611	1658	1706	1753	1801	1848	
6	1895	1943	1990	2038	2085	2132	2180	2227	2275	2322	
7	2369	2417	2464	2511	2559	2606	2653	2701	2748	2795	
8	2843	2890	2937	2985	3032	3079	3126	3174	3221	3268	
9	3316	3363	3410	3457	3504	3552	3599	3646	3693	3741	
920	3788	3835	3882	3929	3977	4024	4071	4118	4165	4212	
1	4260	4307	4354	4401	4448	4495	4542	4590	4637	4684	
2	4731	4778	4825	4872	4919	4966	5013	5061	5108	5155	
3	5202	5249	5296	5343	5390	5437	5484	5531	5578	5625	
4	5672	5719	5766	5813	5860	5907	5954	6001	6048	6095	47
925	6142	6189	6236	6283	6329	6376	6423	6470	6517	6564	
6	6611	6658	6705	6752	6799	6845	6892	6939	6986	7033	
7	7080	7127	7173	7220	7267	7314	7361	7408	7454	7501	
8	7548	7595	7642	7688	7735	7782	7829	7875	7922	7969	
9	8016	8062	8109	8156	8203	8249	8296	8343	8390	8436	
930	8483	8530	8576	8623	8670	8716	8763	8810	8856	8903	
1	8950	8996	9043	9090	9136	9183	9229	9276	9323	9369	
2	9416	9463	9509	9556	9602	9649	9695	9742	9789	9835	
3	9882	9928	9975	0021	0068	0114	0161	0207	0254	0300	
4	97 0347	0393	0440	0486	0533	0579	0626	0672	0719	0765	
935	0812	0858	0904	0951	0997	1044	1090	1137	1183	1229	
6	1276	1322	1369	1415	1461	1508	1554	1601	1647	1693	
7	1740	1786	1832	1879	1925	1971	2018	2064	2110	2157	
8	2203	2249	2295	2342	2388	2434	2481	2527	2573	2619	
9	2666	2712	2758	2804	2851	2897	2943	2989	3035	3082	
940	3128	3174	3220	3266	3313	3359	3405	3451	3497	3543	
1	3590	3636	3682	3728	3774	3820	3866	3913	3959	4005	
2	4051	4097	4143	4189	4235	4281	4327	4374	4420	4466	
3	4512	4558	4604	4650	4696	4742	4788	4834	4880	4926	
4	4972	5018	5064	5110	5156	5202	5248	5294	5340	5386	46

PROPORTIONAL PARTS

Diff.	1	2	3	4	5	6	7	8	9
47	4.7	9.4	14.1	18.8	23.5	28.2	32.9	37.6	42.3
46	4.6	9.2	13.8	18.4	23.0	27.6	32.2	36.8	41.4

No. 945 Log. 975 — Table VII—Continued — No. 989 Log. 995

N.	0	1	2	3	4	5	6	7	8	9	Diff.
945	97 5432	5478	5524	5570	5616	5662	5707	5753	5799	5845	
6	5891	5937	5983	6029	6075	6121	6167	6212	6258	6304	
7	6350	6396	6442	6488	6533	6579	6625	6671	6717	6763	
8	6808	6854	6900	6946	6992	7037	7083	7129	7175	7220	
9	7266	7312	7358	7403	7449	7495	7541	7586	7632	7678	
950	7724	7769	7815	7861	7906	7952	7998	8043	8089	8135	
1	8181	8226	8272	8317	8363	8409	8454	8500	8546	8591	
2	8637	8683	8728	8774	8819	8865	8911	8956	9002	9047	
3	9093	9138	9184	9230	9275	9321	9366	9412	9457	9503	
4	9548	9594	9639	9685	9730	9776	9821	9867	9912	9958	
955	98 0003	0049	0094	0140	0185	0231	0276	0322	0367	0412	
6	0458	0503	0549	0594	0640	0685	0730	0776	0821	0867	
7	0912	0957	1003	1048	1093	1139	1184	1229	1275	1320	
8	1366	1411	1456	1501	1547	1592	1637	1683	1728	1773	
9	1819	1864	1909	1954	2000	2045	2090	2135	2181	2226	
960	2271	2316	2362	2407	2452	2497	2543	2588	2633	2678	
1	2723	2769	2814	2859	2904	2949	2994	3040	3085	3130	
2	3175	3220	3265	3310	3356	3401	3446	3491	3536	3581	
3	3626	3671	3716	3762	3807	3852	3897	3942	3987	4032	
4	4077	4122	4167	4212	4257	4302	4347	4392	4437	4482	
965	4527	4572	4617	4662	4707	4752	4797	4842	4887	4932	45
6	4977	5022	5067	5112	5157	5202	5247	5292	5337	5382	
7	5426	5471	5516	5561	5606	5651	5696	5741	5786	5830	
8	5875	5920	5965	6010	6055	6100	6144	6189	6234	6279	
9	6324	6369	6413	6458	6503	6548	6593	6637	6682	6727	
970	6772	6817	6861	6906	6951	6996	7040	7085	7130	7175	
1	7219	7264	7309	7353	7398	7443	7488	7532	7577	7622	
2	7666	7711	7756	7800	7845	7890	7934	7979	8024	8068	
3	8113	8157	8202	8247	8291	8336	8381	8425	8470	8514	
4	8559	8604	8648	8693	8737	8782	8826	8871	8916	8960	
975	9005	9049	9094	9138	9183	9227	9272	9316	9361	9405	
6	9450	9494	9539	9583	9628	9672	9717	9761	9806	9850	
7	9895	9939	9983	0028	0072	0117	0161	0206	0250	0294	
8	99 0339	0383	0428	0472	0516	0561	0605	0650	0694	0738	
9	0783	0827	0871	0916	0960	1004	1049	1093	1137	1182	
980	1226	1270	1315	1359	1403	1448	1492	1536	1580	1625	
1	1669	1713	1758	1802	1846	1890	1935	1979	2023	2067	
2	2111	2156	2200	2244	2288	2333	2377	2421	2465	2509	
3	2554	2598	2642	2686	2730	2774	2819	2863	2907	2951	
4	2995	3039	3083	3127	3172	3216	3260	3304	3348	3392	
985	3436	3480	3524	3568	3613	3657	3701	3745	3789	3833	
6	3877	3921	3965	4009	4053	4097	4141	4185	4229	4273	
7	4317	4361	4405	4449	4493	4537	4581	4625	4669	4713	44
8	4757	4801	4845	4889	4933	4977	5021	5065	5108	5152	
9	5196	5240	5284	5328	5372	5416	5460	5504	5547	5591	

PROPORTIONAL PARTS

Diff.	1	2	3	4	5	6	7	8	9
46	4.6	9.2	13.8	18.4	23.0	27.6	32.2	36.8	41.4
45	4.5	9.0	13.5	18.0	22.5	27.0	31.5	36.0	40.5
44	4.4	8.8	13.2	17.6	22.0	26.4	30.8	35.2	39.6
43	4.3	8.6	12.9	17.2	21.5	25.8	30.1	34.4	38.7

No. 990 **No. 999**
Log. 995 **Table VII—Concluded** **Log. 999**

N.	0	1	2	3	4	5	6	7	8	9	Diff.
990	99 5635	5679	5723	5767	5811	5854	5898	5942	5986	6030	
1	6074	6117	6161	6205	6249	6293	6337	6380	6424	6468	44
2	6512	6555	6599	6643	6687	6731	6774	6818	6862	6906	
3	6949	6993	7037	7080	7124	7168	7212	7255	7299	7343	
4	7386	7430	7474	7517	7561	7605	7648	7692	7736	7779	
995	7823	7867	7910	7954	7998	8041	8085	8129	8172	8216	
6	8259	8303	8347	8390	8434	8477	8521	8564	8608	8652	
7	8695	8739	8782	8826	8869	8913	8956	9000	9043	9087	
8	9131	9174	9218	9261	9305	9348	9392	9435	9479	9522	
9	9565	9609	9652	9696	9739	9783	9826	9870	9913	9957	43

Table VIII(a) Values of S, T, and C in Table VIII, for angles between 0° and 2° and between 88° and 90°

If we were to plot the values of the logarithmic functions given in Table VIII as ordinates and corresponding minutes as abscissas, it would be found that the points for each function were on a curve with variable radius. It would be noted further that the curves for sines, tangents, and cotangents were of comparatively small radii when the angles were small; that the curves for cosines, cotangents, and tangents of angles near 90°, respectively, had the same shape as the curves for sines, tangents, and cotangents of the complements of the angles; and that other than the portions of the curves just mentioned were nearly straight lines for short distances.

When seconds are involved, it will be sufficiently accurate to interpolate in the ordinary manner between adjacent values in the tables —in other words, to assume that the curve joining two adjacent points is a straight line—for all functions between 2° and 88°, and also for sines of angles between 88° and 90° and for cosines of angles between 0° and 2°. The values in the columns headed S, T, and C provide a means (1) of accurately determining for any given angle between 0° and 2° the logarithmic sine, tangent, or cotangent, and for any given angle between 88° and 90°, the logarithmic cosine, cotangent, or tangent; or (2) for a given value of the logarithmic sine, tangent, or cotangent of accurately determining the angle when it lies between 0° and 2°, and for any given value of the cosine, cotangent, or tangent, the angle when it lies between 88° and 90°.

Table VIII(a)—Concluded

Given: angle. Required: logarithmic function.

$$\left.\begin{aligned}\log\sin\alpha &= \log\alpha\ (\text{in seconds}) + S\\ \log\tan\alpha &= \log\alpha\ (\text{in seconds}) + T\\ \log\cot\alpha &= C - \log\alpha\ (\text{in seconds})\end{aligned}\right\}\ \text{In which } \alpha \text{ is less than } 2^\circ.$$

$$\left.\begin{aligned}\log\cos\beta &= \log(90^\circ-\beta)\ (\text{in seconds}) + S\\ \log\tan\beta &= C - \log(90^\circ-\beta)\ (\text{in seconds})\\ \log\cot\beta &= \log(90^\circ-\beta)\ (\text{in seconds}) + T\end{aligned}\right\}\ \text{In which } \beta \text{ lies between } 88^\circ \text{ and } 90^\circ.$$

Given: logarithmic function. Required: angle.

$$\left.\begin{aligned}\log\alpha\ (\text{in seconds}) &= \log\sin\alpha - S\\ &= \log\tan\alpha - T\\ &= C - \log\cot\alpha\end{aligned}\right\}\ \text{In which } \alpha \text{ is less than } 2^\circ.$$

$$\left.\begin{aligned}\log(90^\circ-\beta)\ (\text{in seconds}) &= \log\cos\beta - S\\ &= C - \log\tan\beta\\ &= \log\cot\beta - T\end{aligned}\right\}\ \text{In which } \beta \text{ lies between } 88^\circ \text{ and } 90^\circ.$$

Examples

Given: angle.

Required: logarithmic function.

$$\text{When}\left\{\begin{matrix}\alpha\\ 90^\circ-\beta\end{matrix}\right\} = 19'\,22'' = 1162''$$

$$\log 1162'' = 3.065206$$

$$S\ (\text{for } 19') = 4.685573$$

$$\left.\begin{matrix}\log\sin 19'\,22''\\ \log\cos 89^\circ\,40'\,38''\end{matrix}\right\} = 7.750779$$

$$\text{When}\left\{\begin{matrix}\alpha\\ 90^\circ-\beta\end{matrix}\right\} = 23'\,21'' = 1401''$$

$$\log 1401'' = 3.146438$$

$$C\ (\text{for } 23') = 15.314419$$

$$\left.\begin{matrix}\log\cot 23'\,21''\\ \log\tan 89^\circ 36' 39''\end{matrix}\right\} = 12.167981$$

Given: logarithmic function.

Required: angle.

$$\text{When}\left\{\begin{matrix}\log\sin\alpha\\ \log\cos\beta\end{matrix}\right\} = 7.750779$$

$$S\ (\text{for } 19') = 4.685573$$

$$\log\left\{\begin{matrix}\alpha\\ 90^\circ-\beta\end{matrix}\right\}(\text{in seconds}) = 3.065206$$

$$\text{or}\quad \begin{aligned}\alpha &= 1162'' = 19'\,22''\\ \beta &= 89^\circ\,40'\,38''\end{aligned}$$

$$\text{When}\left\{\begin{matrix}\log\cot\alpha\\ \log\tan\beta\end{matrix}\right\} = 12.167981$$

$$C\ (\text{for } 23') = 15.314419$$

$$\log\left\{\begin{matrix}\alpha\\ 90^\circ-\beta\end{matrix}\right\}(\text{in seconds}) = 3.146438$$

$$\text{or}\quad \begin{aligned}\alpha &= 1401'' = 23'\,21''\\ \beta &= 89^\circ\,36'\,39''\end{aligned}$$

Table VIII Logarithmic sines, cosines, tangents, and cotangents

0° 179°

″	′	Sine.	S.*	T.*	Tang.	Cotang.	C.*	D. 1″.	Cosine.	′
			4.685				15.314			
0	**0**	Inf. neg.	575	575	Inf. neg.	Inf. pos.	425		**10.00 0000**	**60**
60	1	6.46 3726	575	575	6.46 3726	13.53 6274	425		0000	59
120	2	.76 4756	575	575	.76 4756	.23 5244	425		0000	58
180	3	6.94 0847	575	575	6.94 0847	13.05 9153	425		0000	57
240	4	7.06 5786	575	575	7.06 5786	12.93 4214	425		0000	56
300	**5**	**7.16 2696**	575	575	**7.16 2696**	**12.83 7304**	425	.02	**10.00 0000**	**55**
360	6	.24 1877	575	575	.24 1878	.75 8122	425	.00	9.99 9999	54
420	7	.30 8824	575	575	.30 8825	.69 1175	425	.00	9999	53
480	8	.36 6816	574	576	.36 6817	.63 3183	424	.00	9999	52
540	9	.41 7968	574	576	.41 7970	.58 2030	424	.02	9999	51
600	**10**	**7.46 3726**	574	576	**7.46 3727**	**12.53 6273**	424	.00	**9.99 9998**	**50**
660	11	.50 5118	574	576	.50 5120	.49 4880	424	.02	9998	49
720	12	.54 2906	574	577	.54 2909	.45 7091	423	.00	9997	48
780	13	.57 7668	574	577	.57 7672	.42 2328	423	.02	9997	47
840	14	.60 9853	574	577	.60 9857	.39 0143	423	.00	9996	46
900	**15**	**7.63 9816**	573	578	**7.63 9820**	**12.36 0180**	422	.02	**9.99 9996**	**45**
960	16	.66 7845	573	578	.66 7849	.33 2151	422	.00	9995	44
1020	17	.69 4173	573	578	.69 4179	.30 5821	422	.02	9995	43
1080	18	.71 8997	573	579	.71 9003	.28 0997	421	.02	9994	42
1140	19	.74 2478	573	579	.74 2484	.25 7516	421	.00	9993	41
1200	**20**	**7.76 4754**	572	580	**7.76 4761**	**12.23 5239**	420	.02	**9.99 9993**	**40**
1260	21	.78 5943	572	580	.78 5951	.21 4049	420	.02	9992	39
1320	22	.80 6146	572	581	.80 6155	.19 3845	419	.02	9991	38
1380	23	.82 5451	572	581	.82 5460	.17 4540	419	.02	9990	37
1440	24	.84 3934	571	582	.84 3944	.15 6056	418	.00	9989	36
1500	**25**	**7.86 1662**	571	583	**7.86 1674**	**12.13 8326**	417	.02	**9.99 9989**	**35**
1560	26	.87 8695	571	583	.87 8708	.12 1292	417	.02	9988	34
1620	27	.89 5085	570	584	.89 5099	.10 4901	416	.02	9987	33
1680	28	.91 0879	570	584	.91 0894	.08 9106	416	.02	9986	32
1740	29	.92 6119	570	585	.92 6134	.07 3866	415	.03	9985	31
1800	**30**	**7.94 0842**	569	586	**7.94 0858**	**12.05 9142**	414	.02	**9.99 9983**	**30**
1860	31	.95 5082	569	587	.95 5100	.04 4900	413	.02	9982	29
1920	32	.96 8870	569	587	.96 8889	.03 1111	413	.02	9981	28
1980	33	.98 2233	568	588	.98 2253	.01 7747	412	.02	9980	27
2040	34	7.99 5198	568	589	7.99 5219	12.00 4781	411	.03	9979	26
2100	35	**8.00 7787**	567	590	**8.00 7809**	**11.99 2191**	410	.02	**9.99 9977**	**25**
2160	36	.02 0021	567	591	.02 0044	.97 9956	409	.02	9976	24
2220	37	.03 1919	566	592	.03 1945	.96 8055	408	.03	9975	23
2280	38	.04 3501	566	593	.04 3527	.95 6473	407	.02	9973	22
2340	39	.05 4781	566	593	.05 4809	.94 5191	407	.02	9972	21
2400	**40**	**8.06 5776**	565	594	**8.06 5806**	**11.93 4194**	406	.03	**9.99 9971**	**20**
2460	41	.07 6500	565	595	.07 6531	.92 3460	405	.02	9969	19
2520	42	.08 6965	564	596	.08 6997	.91 3003	404	.03	9968	18
2580	43	.09 7183	564	598	.09 7217	.90 2783	402	.03	9966	17
2640	44	.10 7167	563	599	.10 7203	.89 2797	401	.02	9964	16
2700	**45**	**8.11 6926**	562	600	**8.11 6963**	**11.88 3037**	400	.03	**9.99 9963**	**15**
2760	46	.12 6471	562	601	.12 6510	.87 3490	399	.03	9961	14
2820	47	.13 5810	561	602	.13 5851	.86 4149	398	.02	9959	13
2880	48	.14 4953	561	603	.14 4996	.85 5004	397	.03	9958	12
2940	49	.15 3907	560	604	.15 3952	.84 6048	396	.03	9956	11
3000	**50**	**8.16 2681**	560	605	**8.16 2727**	**11.83 7273**	395	.03	**9.99 9954**	**10**
3060	51	.17 1280	559	607	.17 1328	.82 8672	393	.03	9952	9
3120	52	.17 9713	558	608	.17 9763	.82 0237	392	.03	9950	8
3180	53	.18 7985	558	609	.18 8036	.81 1964	391	.03	9948	7
3240	54	.19 6102	557	611	.19 6156	.80 3844	389	.03	9946	6
3300	**55**	**8.20 4070**	556	612	**8.20 4126**	**11.79 5874**	388	.03	**9.99 9944**	**5**
3360	56	.21 1895	556	613	.21 1953	.78 8047	387	.03	9942	4
3420	57	.21 9581	555	615	.21 9641	.78 0359	385	.03	9940	3
3480	58	.22 7134	554	616	.22 7195	.77 2805	384	.03	9938	2
3540	59	.23 4557	554	618	.23 4621	.76 5379	382	.03	9936	1
3600	**60**	**8.24 1855**	553	619	**8.24 1921**	**11.75 8079**	381		**9.99 9934**	**0**
			4.685				15.314			
″	′	Cosine.	S.*	T.*	Cotang.	Tang.	C.*	D. 1″.	Sine.	′

90° 89°

* For use of *S*, *T*, and *C* see Table VIII(*a*), preceding this table.

1° **Table VIII—Continued** **178°**

″	′	Sine.	S.*	T.*	Tang.	Cotang.	C.*	D. 1″.	Cosine.	′
			4.685				15.314			
3600	**0**	**8.24 1855**	553	619	**8.24 1921**	**11.75 8079**	381	.03	**9.99 9934**	**60**
3660	1	.24 9033	552	620	.24 9102	.75 0898	380	.05	9932	59
3720	2	.25 6094	551	622	.25 6165	.74 3835	378	.03	9929	58
3780	3	.26 3042	551	623	.26 3115	.73 6885	377	.03	9927	57
3840	4	.26 9881	550	625	.26 9956	.73 0044	375	.05	9925	56
3900	**5**	**8.27 6614**	549	627	**8.27 6691**	**11.72 3309**	373	.03	**9.99 9922**	**55**
3960	6	.28 3243	548	628	.28 3323	.71 6677	372	.03	9920	54
4020	7	.28 9773	547	630	.28 9856	.71 0144	370	.05	9918	53
4080	8	.29 6207	546	632	.29 6292	.70 3708	368	.03	9915	52
4140	9	.30 2546	546	633	.30 2634	.69 7366	367	.05	9913	51
4200	**10**	**8.30 8794**	545	635	**8.30 8884**	**11.69 1116**	365	.05	**9.99 9910**	**50**
4260	11	.31 4954	544	637	.31 5046	.68 4954	363	.03	9907	49
4320	12	.32 1027	543	638	.32 1122	.67 8878	362	.05	9905	48
4380	13	.32 7016	542	640	.32 7114	.67 2886	360	.05	9902	47
4440	14	.33 2924	541	642	.33 3025	.66 6975	358	.03	9899	46
4500	**15**	**8.33 8753**	540	644	**8.33 8856**	**11.66 1144**	356	.05	**9.99 9897**	**45**
4560	16	.34 4504	539	646	.34 4610	.65 5390	354	.05	9894	44
4620	17	.35 0181	539	648	.35 0289	.64 9711	352	.05	9891	43
4680	18	.35 5783	538	649	.35 5895	.64 4105	351	.05	9888	42
4740	19	.36 1315	537	651	.36 1430	.63 8570	349	.05	9885	41
4800	**20**	**8.36 6777**	536	653	**8.36 6895**	**11.63 3105**	347	.05	**9.99 9882**	**40**
4860	21	.37 2171	535	655	.37 2292	.62 7708	345	.05	9879	39
4920	22	.37 7499	534	657	.37 7622	.62 2378	343	.05	9876	38
4980	23	.38 2762	533	659	.38 2889	.61 7111	341	.05	9873	37
5040	24	.38 7062	532	661	.38 8092	.61 1908	339	.05	9870	36
5100	**25**	**8.39 3101**	531	663	**8.39 3234**	**11.60 6766**	337	.05	**9.99 9867**	**35**
5160	26	.39 8179	530	666	.39 8315	.60 1685	334	.05	9864	34
5220	27	.40 3199	529	668	.40 3338	.59 6662	332	.05	9861	33
5280	28	.40 8161	527	670	.40 8304	.59 1696	330	.07	9858	32
5340	29	.41 3068	526	672	.41 3213	.58 6787	328	.05	9854	31
5400	**30**	**8.41 7919**	525	674	**8.41 8068**	**11.58 1932**	326	.05	**9.99 9851**	**30**
5460	31	.42 2717	524	676	.42 2869	.57 7131	324	.07	9848	29
5520	32	.42 7462	523	679	.42 7618	.57 2382	321	.05	9844	28
5580	33	.43 2156	522	681	.43 2315	.56 7685	319	.05	9841	27
5610	31	.40 0800	521	083	.43 6962	.56 3038	317	.07	9838	26
5700	**35**	**8.44 1394**	520	685	**8.44 1560**	**11.55 8440**	315	.05	**9.99 9834**	**25**
5760	36	.11 5041	518	088	.44 6110	.55 3890	312	.07	9831	24
5820	37	.45 0440	517	690	.45 0613	.54 9387	310	.05	9827	23
5880	38	4893	516	693	5070	4930	307	.07	9824	22
5940	39	.45 9301	515	695	.45 0481	.51 0510	005	.07	9820	21
6000	**40**	**8.46 3665**	514	697	**8.46 3849**	**11.53 6151**	303	.05	**9.99 9816**	**20**
6060	41	.46 7985	512	700	.46 8172	.53 1828	300	.07	9813	19
6120	42	.47 2263	511	702	.47 2454	.52 7546	298	.07	9809	18
6180	43	.47 6498	510	705	.47 6693	.52 3307	295	.07	9805	17
6240	44	.48 0693	509	707	.48 0892	.51 9108	293	.07	9801	16
6300	**45**	**8.48 4848**	507	710	**8.48 5050**	**11.51 4950**	290	.05	**9.99 9797**	**15**
6360	46	.48 8963	506	713	.48 9170	.51 0830	287	.07	9794	14
6420	47	.49 3040	505	715	.49 3250	.50 6750	285	.07	9790	13
6480	18	.10 7078	503	718	.49 7293	.50 2707	282	.07	9786	12
6540	49	.50 1080	502	720	.50 1298	.40 8702	280	.07	9782	11
6600	**50**	**8.50 5045**	501	723	**8.50 5267**	**11.49 4733**	277	.07	**9.99 9778**	**10**
6660	51	.50 8974	499	726	.50 9200	.49 0800	274	.08	9774	9
6720	52	.51 2867	498	729	.51 3098	.48 6902	271	.07	9769	8
6780	53	.51 6726	497	731	.51 6961	.48 3039	269	.07	9765	7
6840	54	.52 0551	495	734	.52 0790	.47 9210	266	.07	9761	6
6900	**55**	**8.52 4343**	494	737	**8.52 4586**	**11.47 5414**	263	.07	**9.99 9757**	**5**
6960	56	.52 8102	492	740	.52 8349	.47 1651	260	.08	9753	4
7020	57	.53 1828	491	743	.53 2080	.46 7920	257	.07	9748	3
7080	58	5523	490	745	5779	4221	255	.07	9744	2
7140	59	.53 9186	488	748	.53 9447	.46 0553	252	.08	9740	1
7200	**60**	**8.54 2819**	487	751	**8.54 3084**	**11.45 6916**	249		**9.99 9735**	**0**
			4.685				15.314			
″	′	Cosine.	S.*	T.*	Cotang.	Tang.	C.*	D. 1″.	Sine.	′

91° **88°**

* For use of *S*, *T*, and *C* see Table VIII(*a*), preceding this table.

2° **Table VIII—Continued** 177°

′	Sine.	D. 1″.	Cosine.	D.1″.	Tang.	D.1″.	Cotang.	′
0	8.54 2819	60.05	9.99 9735	.07	8.54 3084	60.12	11.45 6916	60
1	6422	59.55	9731	.08	.54 6691	59.62	.45 3309	59
2	.54 9995	59.07	9726	.07	.55 0268	59.15	.44 9732	58
3	.55 3539	58.58	9722	.08	3817	58.65	6183	57
4	.55 7054	58.10	9717	.07	.55 7336	58.20	.44 2664	56
5	8.56 0540	57.65	9.99 9713	.08	8.56 0828	57.72	11.43 9172	55
6	3999	57.20	9708	.07	4291	57.27	5709	54
7	.56 7431	56.75	9704	.08	.56 7727	56.83	.43 2273	53
8	.57 0836	56.30	9699	.08	.57 1137	56.38	.42 8863	52
9	4214	55.87	9694	.08	4520	55.95	5480	51
10	8.57 7566	55.43	9.99 9689	.07	8.57 7877	55.52	11.42 2123	50
11	.58 0892	55.02	9685	.08	.58 1208	55.10	.41 8792	49
12	4193	54.60	9680	.08	4514	54.68	5486	48
13	.58 7469	54.20	9675	.08	.58 7795	54.27	.41 2205	47
14	.59 0721	53.78	9670	.08	.59 1051	53.87	.40 8949	46
15	8.59 3948	53.40	9.99 9665	.08	8.59 4283	53.48	11.40 5717	45
16	.59 7152	53.00	9660	.08	.59 7492	53.08	.40 2508	44
17	.60 0332	52.62	9655	.08	.60 0677	52.70	.39 9323	43
18	3489	52.23	9650	.08	3839	52.32	6161	42
19	6623	51.85	9645	.08	.60 6978	51.93	.39 3022	41
20	8.60 9734	51.48	9.99 9640	.08	8.61 0094	51.58	11.38 9906	40
21	.61 2823	51.13	9635	.10	3189	51.22	6811	39
22	5891	50.77	9629	.08	6262	50.85	3738	38
23	.61 8937	50.42	9624	.08	.61 9313	50.50	.38 0687	37
24	.62 1962	50.05	9619	.08	.62 2343	50.15	.37 7657	36
25	8.62 4965	49.72	9.99 9614	.10	8.62 5352	49.80	11.37 4648	35
26	.62 7948	49.38	9608	.08	.62 8340	49.47	.37 1660	34
27	.63 0911	49.05	9603	.10	.63 1308	49.13	.36 8692	33
28	3854	48.70	9597	.08	4256	48.80	5744	32
29	6776	48.40	9592	.10	.63 7184	48.48	.36 2816	31
30	8.63 9680	48.05	9.99 9586	.08	8.64 0093	48.15	11.35 9907	30
31	.64 2563	47.75	9581	.10	2982	47.85	7018	29
32	5428	47.43	9575	.08	5853	47.52	4147	28
33	.64 8274	47.13	9570	.10	.64 8704	47.22	.35 1296	27
34	.65 1102	46.82	9564	.10	.65 1537	46.92	.34 8463	26
35	8.65 3911	46.52	9.99 9558	.08	8.65 4352	46.62	11.34 5648	25
36	6702	46.22	9553	.10	7149	46.32	2851	24
37	.65 9475	45.92	9547	.10	.65 9928	46.02	.34 0072	23
38	.66 2230	45.63	9541	.10	.66 2689	45.73	.33 7311	22
39	4968	45.35	9535	.10	5433	45.45	4567	21
40	8.66 7689	45.07	9.99 9529	.08	8.66 8160	45.17	11.33 1840	20
41	.67 0393	44.78	9524	.10	.67 0870	44.88	.32 9130	19
42	3080	44.52	9518	.10	3563	44.60	6437	18
43	5751	44.23	9512	.10	6239	44.35	3761	17
44	.67 8405	43.97	9506	.10	.67 8900	44.07	.32 1100	16
45	8.68 1043	43.70	9.99 9500	.12	8.68 1544	43.80	11.31 8456	15
46	3665	43.45	9493	.10	4172	43.53	5828	14
47	6272	43.18	9487	.10	6784	43.28	3216	13
48	.68 8863	42.92	9481	.10	.68 9381	43.03	.31 0619	12
49	.69 1438	42.67	9475	.10	.69 1963	42.77	.30 8037	11
50	8.69 3998	42.42	9.99 9469	.10	8.69 4529	42.53	11.30 5471	10
51	6543	42.17	9463	.12	7081	42.27	2919	9
52	.69 9073	41.93	9456	.10	.69 9617	42.03	.30 0383	8
53	.70 1589	41.68	9450	.12	.70 2139	41.78	.29 7861	7
54	4090	41.45	9443	.10	4646	41.57	5354	6
55	8.70 6577	41.20	9.99 9437	.10	8.70 7140	41.30	11.29 2860	5
56	.70 9049	40.97	9431	.12	.70 9618	41.08	.29 0382	4
57	.71 1507	40.75	9424	.10	.71 2083	40.85	.28 7917	3
58	3952	40.52	9418	.12	4534	40.63	5466	2
59	6383	40.28	9411	.12	6972	40.40	3028	1
60	8.71 8800		9.99 9404		8.71 9396		11.28 0604	0
′	Cosine.	D.1″.	Sine.	D.1″.	Cotang.	D.1″.	Tang.	′

92° 87°

3° **Table VIII—Continued** 176°

′	Sine.	D.1″.	Cosine.	D.1″.	Tang.	D.1″.	Cotang.	′
0	**8.71 8800**	40.07	**9.99 9404**	.10	**8.71 9396**	40.17	**11.28 0604**	**60**
1	.72 1204	39.85	9398	.12	.72 1806	39.97	.27 8194	59
2	3595	39.62	9391	.12	4204	39.73	5796	58
3	5972	39.42	9384	.10	6588	39.52	3412	57
4	.72 8337	39.18	9378	.12	.72 8959	39.30	.27 1041	56
5	**8.73 0688**	38.98	**9.99 9371**	.12	**8.73 1317**	39.10	**11.26 8683**	**55**
6	3027	38.78	9364	.12	3663	38.88	6337	54
7	5354	38.55	9357	.12	5990	38.68	4004	53
8	7667	38.37	9350	.12	.73 8317	38.48	.26 1683	52
9	.73 9969	38.17	9343	.12	.74 0626	38.27	.25 9374	51
10	**8.74 2259**	37.95	**9.99 9336**	.12	**8.74 2922**	38.08	**11.25 7078**	**50**
11	4536	37.77	9329	.12	5207	37.87	4793	49
12	6802	37.55	9322	.12	7479	37.68	2521	48
13	.74 9055	37.37	9315	.12	.74 9740	37.48	.25 0260	47
14	.75 1297	37.18	9308	.12	.75 1989	37.30	.24 8011	46
15	**8.75 3528**	36.98	**9.99 9301**	.12	**8.75 4227**	37.10	**11.24 5773**	**45**
16	5747	36.80	9294	.12	6453	36.92	3547	44
17	.75 7955	36.60	9287	.13	.75 8668	36.73	.24 1332	43
18	.76 0151	36.43	9279	.12	.76 0872	36.55	.23 9128	42
19	2337	36.23	9272	.12	3065	36.35	6935	41
20	**8.76 4511**	36.07	**9.99 9265**	.13	**8.76 5246**	36.18	**11.23 4754**	**40**
21	6675	35.88	9257	.12	7417	36.02	2583	39
22	.76 8828	35.70	9250	.13	.76 9578	35.82	.23 0422	38
23	.77 0970	35.52	9242	.12	.77 1727	35.65	.22 8273	37
24	3101	35.37	9235	.13	3866	35.48	6134	36
25	**8.77 5223**	35.17	**9.99 9227**	.12	**8.77 5995**	35.32	**11.22 4005**	**35**
26	7333	35.02	9220	.13	.77 8114	35.13	.22 1886	34
27	.77 9434	34.83	9212	.12	.78 0222	34.97	.21 9778	33
28	.78 1524	34.68	9205	.13	2320	34.80	7680	32
29	3605	34.50	9197	.13	4408	34.63	5592	31
30	**8.78 5675**	34.35	**9.99 9189**	.13	**8.78 6486**	34.47	**11.21 3514**	**30**
31	7736	34.18	9181	.12	.78 8554	34.32	.21 1446	29
32	.78 9787	34.02	9174	.13	.79 0613	34.15	.20 9387	28
33	.79 1828	33.85	9166	.13	2662	33.98	7338	27
34	3859	33.70	9158	.13	4701	33.83	5200	26
35	**8.79 5881**	33.55	**9.99 9150**	.13	**8.79 6731**	33.68	**11.20 3269**	**25**
36	7894	33.38	9142	.13	.79 8752	33.52	.20 1248	24
37	.79 9897	33.25	9134	.13	.80 0763	33.37	.19 9237	23
38	.80 1892	33.07	9126	.13	2765	33.22	7235	22
39	3876	32.93	9118	.13	4758	33.07	5242	21
40	**8.80 5852**	32.78	**9.99 9110**	.13	**8.80 6742**	32.92	**11.19 3258**	**20**
41	7819	32.63	9102	.13	.80 8717	32.77	.19 1283	19
42	.80 9777	32.48	9094	.13	.81 0683	32.63	.18 9317	18
43	.81 1726	32.35	9086	.15	2641	32.47	7359	17
44	3667	32.20	9077	.13	4589	32.33	5411	16
45	**8.81 5599**	32.05	**9.99 9069**	.13	**8.81 6529**	32.20	**11.18 3471**	**15**
46	7522	31.90	9061	.13	.81 8461	32.05	.18 1539	14
47	.81 9436	31.78	9053	.15	.82 0384	31.90	.17 0616	13
48	.82 1343	31.62	9044	.13	2298	31.78	7702	12
49	3240	31.50	9036	.15	4205	31.63	5795	11
50	**8.82 5130**	31.35	**9.99 9027**	.13	**8.82 6103**	31.48	**11.17 3897**	**10**
51	7011	31.22	9019	.15	7992	31.37	2008	9
52	.82 8884	31.08	9010	.13	.82 9874	31.23	.17 0126	8
53	.83 0749	30.97	9002	.15	.83 1748	31.08	.16 8252	7
54	2607	30.82	8993	.15	3613	30.97	6387	6
55	**8.83 4456**	30.68	**9.99 8984**	.13	**8.83 5471**	30.83	**11.16 4529**	**5**
56	6297	30.55	8976	.15	7321	30.70	2679	4
57	8130	30.43	8967	.15	.83 9163	30.58	.16 0837	3
58	.83 9956	30.30	8958	.13	.84 0998	30.45	.15 9002	2
59	.84 1774	30.18	8950	.15	2825	30.32	7175	1
60	**8.84 3585**		**9.99 8941**		**8.84 4644**		**11.15 5356**	**0**
′	Cosine.	D.1″.	Sine.	D.1″.	Cotang.	D.1″.	Tang.	′

93° 86°

4° **Table VIII—Continued** 175°

′	Sine.	D.1″.	Cosine.	D.1″.	Tang.	D.1″.	Cotang.	′
0	**8.84 3585**	30.03	**9.99 8941**	.15	**8.84 4644**	30.18	**11.15 5356**	**60**
1	5387	29.93	8932	.15	6455	30.08	3545	59
2	7183	29.80	8923	.15	.84 8260	29.95	.15 1740	58
3	.84 8971	29.67	8914	.15	.85 0057	29.82	.14 9943	57
4	.85 0751	29.57	8905	.15	1846	29.70	8154	56
5	**8.85 2525**	29.43	**9.99 8896**	.15	**8.85 3628**	29.58	**11.14 6372**	**55**
6	4291	29.30	8887	.15	5403	29.47	4597	54
7	6049	29.20	8878	.15	7171	29.35	2829	53
8	7801	29.08	8869	.15	.85 8932	29.23	.14 1068	52
9	.85 9546	28.95	8860	.15	.86 0686	29.12	.13 9314	51
10	**8.86 1283**	28.85	**9.99 8851**	.17	**8.86 2433**	29.00	**11.13 7567**	**50**
11	3014	28.73	8841	.15	4173	28.88	5827	49
12	4738	28.62	8832	.15	5906	28.77	4094	48
13	6455	28.50	8823	.17	7632	28.65	2368	47
14	8165	28.38	8813	.15	.86 9351	28.55	.13 0649	46
15	**8.86 9868**	28.28	**9.99 8804**	.15	**8.87 1064**	28.43	**11.12 8936**	**45**
16	.87 1565	28.17	8795	.17	2770	28.32	7230	44
17	3255	28.05	8785	.15	4469	28.22	5531	43
18	4938	27.95	8776	.17	6162	28.12	3838	42
19	6615	27.83	8766	.15	7849	28.00	2151	41
20	**8.87 8285**	27.73	**9.99 8757**	.17	**8.87 9529**	27.88	**11.12 0471**	**40**
21	.87 9949	27.63	8747	.15	.88 1202	27.78	.11 8798	39
22	.88 1607	27.52	8738	.17	2869	27.68	7131	38
23	3258	27.42	8728	.17	4530	27.58	5470	37
24	4903	27.32	8718	.17	6185	27.47	3815	36
25	**8.88 6542**	27.20	**9.99 8708**	.15	**8.88 7833**	27.38	**11.11 2167**	**35**
26	8174	27.12	8699	.17	.88 9476	27.27	.11 0524	34
27	.88 9801	27.00	8689	.17	.89 1112	27.17	.10 8888	33
28	.89 1421	26.90	8679	.17	2742	27.07	7258	32
29	3035	26.80	8669	.17	4366	26.97	5634	31
30	**8.89 4643**	26.72	**9.99 8659**	.17	**8.89 5984**	26.87	**11.10 4016**	**30**
31	6246	26.60	8649	.17	7596	26.78	2404	29
32	7842	26.50	8639	.17	.89 9203	26.67	.10 0797	28
33	.89 9432	26.42	8629	.17	.90 0803	26.58	.09 9197	27
34	.90 1017	26.32	8619	.17	2398	26.48	7602	26
35	**8.90 2596**	26.22	**9.99 8609**	.17	**8.90 3987**	26.38	**11.09 6013**	**25**
36	4169	26.12	8599	.17	5570	26.28	4430	24
37	5736	26.02	8589	.18	7147	26.20	2853	23
38	7297	25.93	8578	.17	.90 8719	26.10	.09 1281	22
39	.90 8853	25.85	8568	.17	.91 0285	26.02	.08 9715	21
40	**8.91 0404**	25.75	**9.99 8558**	.17	**8.91 1846**	25.92	**11.08 8154**	**20**
41	1949	25.65	8548	.18	3401	25.83	6599	19
42	3488	25.57	8537	.17	4951	25.73	5049	18
43	5022	25.47	8527	.18	6495	25.63	3505	17
44	6550	25.38	8516	.17	8034	25.57	1966	16
45	**8.91 8073**	25.30	**9.99 8506**	.18	**8.91 9568**	25.47	**11.08 0432**	**15**
46	.91 9591	25.20	8495	.17	.92 1096	25.38	.07 8904	14
47	.92 1103	25.12	8485	.18	2619	25.28	7381	13
48	2610	25.03	8474	.17	4136	25.22	5864	12
49	4112	24.95	8464	.18	5649	25.12	4351	11
50	**8.92 5609**	24.85	**9.99 8453**	.18	**8.92 7156**	25.03	**11.07 2844**	**10**
51	7100	24.78	8442	.18	.92 8658	24.95	.07 1342	9
52	.92 8587	24.68	8431	.17	.93 0155	24.87	.06 9845	8
53	.93 0068	24.60	8421	.18	1647	24.78	8353	7
54	1544	24.52	8410	.18	3134	24.70	6866	6
55	**8.93 3015**	24.43	**9.99 8399**	.18	**8.93 4616**	24.62	**11.06 5384**	**5**
56	4481	24.35	8388	.18	6093	24.53	3907	4
57	5942	24.27	8377	.18	7565	24.45	2435	3
58	7398	24.20	8366	.18	.93 9032	24.37	.06 0968	2
59	.93 8850	24.10	8355	.18	.94 0494	24.30	.05 9506	1
60	**8.94 0296**		**9.99 8344**		**8.94 1952**		**11.05 8048**	**0**
′	Cosine.	D.1″.	Sine.	D.1″.	Cotang.	D.1″.	Tang.	′

94° 85°

5° **Table VIII—Continued** 174°

′	Sine.	D.1″.	Cosine.	D.1″.	Tang.	D.1″.	Cotang.	′
0	**8.94 0296**	24.03	**9.99 8344**	.18	**8.94 1952**	24.20	**11.05 8048**	**60**
1	1738	23.93	8333	.18	3404	24.13	6596	59
2	3174	23.87	8322	.18	4852	24.05	5148	58
3	4606	23.80	8311	.18	6295	23.98	3705	57
4	6034	23.70	8300	.18	7734	23.90	2266	56
5	**8.94 7456**	23.63	**9.99 8289**	.20	**8.94 9168**	23.82	**11.05 0832**	**55**
6	.94 8874	23.55	8277	.18	.95 0597	23.73	.04 9403	54
7	.95 0287	23.48	8266	.18	2021	23.67	7979	53
8	1696	23.40	8255	.20	3441	23.58	6550	52
9	3100	23.32	8243	.18	4856	23.52	5144	51
10	**8.95 4499**	23.25	**9.99 8232**	.20	**8.95 6267**	23.45	**11.04 3733**	**50**
11	5894	23.17	8220	.18	7674	23.35	2326	49
12	7284	23.10	8209	.20	.95 9075	23.30	.04 0925	48
13	.95 8670	23.03	8197	.18	.96 0473	23.22	.03 9527	47
14	.96 0052	22.95	8186	.20	1866	23.15	8134	46
15	**8.96 1429**	22.87	**9.99 8174**	.18	**8.96 3255**	23.07	**11.03 6745**	**45**
16	2801	22.82	8163	.20	4639	23.00	5361	44
17	4170	22.73	8151	.20	6019	22.92	3981	43
18	5534	22.65	8139	.18	7394	22.87	2606	42
19	6893	22.60	8128	.20	.96 8766	22.78	.00 1234	41
20	**8.96 8249**	22.52	**9.99 8116**	.20	**8.97 0133**	22.72	**11.02 9867**	**40**
21	.96 9600	22.45	8104	.20	1496	22.65	8504	39
22	.97 0947	22.37	8092	.20	2855	22.57	7145	38
23	2289	22.32	8080	.20	4209	22.52	5791	37
24	3628	22.23	8068	.20	5560	22.43	4440	36
25	**8.97 4962**	22.18	**9.99 8056**	.20	**8.97 6906**	22.37	**11.02 3094**	**35**
26	6293	22.10	8044	.20	8248	22.30	1752	34
27	7619	22.03	8032	.20	.97 9586	22.25	.02 0414	33
28	.97 8941	21.97	8020	.20	.98 0921	22.17	.01 9079	32
29	.98 0259	21.90	8008	.20	2251	22.10	7749	31
30	**8.98 1573**	21.83	**9.99 7996**	.20	**8.98 3577**	22.03	**11.01 6423**	**30**
31	2883	21.77	7984	.20	4899	21.97	5101	29
32	4189	21.72	7972	.22	6217	21.92	3783	28
33	5491	21.63	7959	.20	7532	21.83	2468	27
34	6789	21.57	7947	.20	.98 8842	21.78	.01 1158	26
35	**8.98 8083**	21.52	**9.99 7935**	.22	**8.99 0149**	21.70	**11.00 9851**	**25**
36	.98 9374	21.43	7922	.20	1451	21.65	8549	24
37	.99 0660	21.38	7910	.22	2750	21.58	7250	23
38	1943	21.32	7897	.20	4045	21.53	5955	22
39	3222	21.25	7885	.22	5337	21.45	4663	21
40	**8.99 4497**	21.18	**9.99 7872**	.20	**8.99 6624**	21.40	**11.00 3376**	**20**
41	5768	21.13	7860	.22	7908	21.33	2092	19
42	7036	21.05	7847	.20	8.99 9188	21.28	11.00 0812	18
43	8299	21.02	7835	.22	9.00 0465	21.22	10.99 9535	17
44	8.99 9560	20.93	7822	.22	1738	21.15	8262	16
45	**9.00 0816**	20.88	**9.99 7809**	.20	**9.00 3007**	21.08	**10.99 6993**	**15**
46	2069	20.82	7797	.22	4272	21.03	5728	14
47	3318	20.75	7784	.22	5534	20.97	4466	13
48	4563	20.70	7771	.22	6792	20.92	3208	12
49	5805	20.65	7758	.22	8047	20.85	1953	11
50	**9.00 7044**	20.57	**9.99 7745**	.22	**9.00 9298**	20.80	**10.99 0702**	**10**
51	8278	20.53	7732	.22	.01 0546	20.73	.98 9454	9
52	.00 9510	20.45	7719	.22	1790	20.68	8210	8
53	.01 0737	20.42	7706	.22	3031	20.62	6969	7
54	1962	20.33	7693	.22	4268	20.57	5732	6
55	**9.01 3182**	20.30	**9.99 7680**	.22	**9.01 5502**	20.50	**10.98 4498**	**5**
56	4400	20.22	7667	.22	6732	20.45	3268	4
57	5613	20.18	7654	.22	7959	20.40	2041	3
58	6824	20.12	7641	.22	.01 9183	20.33	.98 0817	2
59	8031	20.07	7628	.23	.02 0403	20.28	.97 9597	1
60	**9.01 9235**		**9.99 7614**		**9.02 1620**		**10.97 8380**	**0**
′	Cosine.	D.1″.	Sine.	D.1″.	Cotang.	D.1″.	Tang.	′

95° 84°

6° **Table VIII—Continued** 173°

′	Sine.	D.1″.	Cosine.	D.1″.	Tang.	D.1″.	Cotang.	′
0	**9.01 9235**	20.00	**9.99 7614**	.22	**9.02 1620**	20.23	**10.97 8380**	**60**
1	.02 0435	19.95	7601	.22	2834	20.17	7166	59
2	1632	19.88	7588	.23	4044	20.12	5956	58
3	2825	19.85	7574	.22	5251	20.07	4749	57
4	4016	19.78	7561	.23	6455	20.00	3545	56
5	**9.02 5203**	19.72	**9.99 7547**	.22	**9.02 7655**	19.95	**10.97 2345**	**55**
6	6386	19.68	7534	.23	.02 8852	19.90	.97 1148	54
7	7567	19.62	7520	.22	.03 0046	19.85	.96 9954	53
8	8744	19.57	7507	.23	1237	19.80	8763	52
9	.02 9918	19.52	7493	.22	2425	19.73	7575	51
10	**9.03 1089**	19.47	**9.99 7480**	.23	**9.03 3609**	19.70	**10.96 6391**	**50**
11	2257	19.40	7466	.23	4791	19.63	5209	49
12	3421	19.35	7452	.22	5969	19.58	4031	48
13	4582	19.32	7439	.23	7144	19.53	2856	47
14	5741	19.25	7425	.23	8316	19.48	1684	46
15	**9.03 6896**	19.20	**9.99 7411**	.23	**9.03 9485**	19.43	**10.96 0515**	**45**
16	8048	19.15	7397	.23	.04 0651	19.37	.95 9349	44
17	.03 9197	19.08	7383	.23	1813	19.33	8187	43
18	.04 0342	19.05	7369	.23	2973	19.28	7027	42
19	1485	19.00	7355	.23	4130	19.23	5870	41
20	**9.04 2625**	18.95	**9.99 7341**	.23	**9.04 5284**	19.17	**10.95 4716**	**40**
21	3762	18.88	7327	.23	6434	19.13	3566	39
22	4895	18.85	7313	.23	7582	19.08	2418	38
23	6026	18.80	7299	.23	8727	19.03	1273	37
24	7154	18.75	7285	.23	.04 9869	18.98	.95 0131	36
25	**9.04 8279**	18.68	**9.99 7271**	.23	**9.05 1008**	18.93	**10.94 8992**	**35**
26	.04 9400	18.65	7257	.25	2144	18.88	7856	34
27	.05 0519	18.60	7242	.23	3277	18.83	6723	33
28	1635	18.57	7228	.23	4407	18.80	5593	32
29	2749	18.50	7214	.25	5535	18.73	4465	31
30	**9.05 3859**	18.45	**9.99 7199**	.23	**9.05 6659**	18.70	**10.94 3341**	**30**
31	4966	18.42	7185	.25	7781	18.65	2219	29
32	6071	18.35	7170	.23	.05 8900	18.60	.94 1100	28
33	7172	18.32	7156	.25	.06 0016	18.57	.93 9984	27
34	8271	18.27	7141	.23	1130	18.50	8870	26
35	**9.05 9367**	18.22	**9.99 7127**	.25	**9.06 2240**	18.47	**10.93 7760**	**25**
36	.06 0460	18.18	7112	.23	3348	18.42	6652	24
37	1551	18.13	7098	.25	4453	18.38	5547	23
38	2639	18.08	7083	.25	5556	18.32	4444	22
39	3724	18.03	7068	.25	6655	18.28	3345	21
40	**9.06 4806**	17.98	**9.99 7053**	.23	**9.06 7752**	18.25	**10.93 2248**	**20**
41	5885	17.95	7039	.25	8846	18.20	1154	19
42	6962	17.90	7024	.25	.06 9938	18.15	.93 0062	18
43	8036	17.85	7009	.25	.07 1027	18.10	.92 8973	17
44	.06 9107	17.82	6994	.25	2113	18.07	7887	16
45	**9.07 0176**	17.77	**9.99 6979**	.25	**9.07 3197**	18.02	**10.92 6803**	**15**
46	1242	17.73	6964	.25	4278	17.97	5722	14
47	2306	17.67	6949	.25	5356	17.93	4644	13
48	3366	17.63	6934	.25	6432	17.88	3568	12
49	4424	17.60	6919	.25	7505	17.85	2495	11
50	**9.07 5480**	17.55	**9.99 6904**	.25	**9.07 8576**	17.80	**10.92 1424**	**10**
51	6533	17.50	6889	.25	.07 9644	17.77	.92 0356	9
52	7583	17.47	6874	.27	.08 0710	17.72	.91 9290	8
53	8631	17.42	6858	.25	1773	17.67	8227	7
54	.07 9676	17.38	6843	.27	2833	17.63	7167	6
55	**9.08 0719**	17.33	**9.99 6828**	.27	**9.08 3891**	17.60	**10.91 6109**	**5**
56	1759	17.30	6812	.25	4947	17.55	5053	4
57	2797	17.25	6797	.25	6000	17.50	4000	3
58	3832	17.20	6782	.27	7050	17.47	2950	2
59	4864	17.17	6766	.25	8098	17.43	1902	1
60	**9.08 5894**		**9.99 6751**		**9.08 9144**		**10.91 0856**	**0**
′	Cosine.	D.1″.	Sine.	D.1″.	Cotang.	D.1″.	Tang.	′

96° 83°

7° **Table VIII—Continued** **172°**

′	Sine.	D.1″.	Cosine.	D.1″.	Tang.	D.1″.	Cotang.	′
0	9.08 5894	17.13	9.99 6751	.27	9.08 9144	17.38	10.91 0856	60
1	6922	17.08	6735	.25	.09 0187	17.35	.90 9813	59
2	7947	17.05	6720	.27	1228	17.30	8772	58
3	8970	17.00	6704	.27	2266	17.27	7734	57
4	.08 9990	16.97	6688	.25	3302	17.23	6698	56
5	9.09 1008	16.93	9.99 6673	.27	9.09 4336	17.18	10.90 5664	55
6	2024	16.88	6657	.27	5367	17.13	4633	54
7	3037	16.83	6641	.27	6395	17.12	3605	53
8	4047	16.82	6625	.25	7422	17.07	2578	52
9	5056	16.77	6610	.27	8446	17.03	1554	51
10	9.09 6062	16.72	9.99 6594	.27	9.09 9468	16.98	10.90 0532	50
11	7065	16.68	6578	.27	.10 0487	16.95	89 9513	49
12	8066	16.65	6562	.27	1504	16.92	8496	48
13	.09 9065	16.62	6546	.27	2519	16.88	7481	47
14	.10 0062	16.57	6530	.27	3532	16.83	6468	46
15	9.10 1056	16.53	9.99 6514	.27	9.10 4542	16.80	10.89 5458	45
16	2048	16.48	6498	.27	5550	16.77	4450	44
17	3037	16.47	6482	.28	6556	16.72	3444	43
18	4025	16.42	6465	.27	7559	16.68	2441	42
19	5010	16.37	6449	.27	8560	16.65	1440	41
20	9.10 5992	16.35	9.99 6433	.27	9.10 9559	16.62	10.89 0441	40
21	6973	16.30	6417	.28	.11 0556	16.58	.88 9444	39
22	7951	16.27	6400	.27	1551	16.53	8449	38
23	8927	16.23	6384	.27	2543	16.50	7457	37
24	.10 9901	16.20	6368	.28	3533	16.47	6467	36
25	9.11 0873	16.15	9.99 6351	.27	9.11 4521	16.43	10.88 5479	35
26	1842	16.12	6335	.28	5507	16.40	4493	34
27	2809	16.08	6318	.27	6491	16.35	3509	33
28	3774	16.05	6302	.28	7472	16.33	2528	32
29	4737	16.02	6285	.27	8452	16.28	1548	31
30	9.11 5698	15.97	9.99 6269	.28	9.11 9429	16.25	10.88 0571	30
31	0056	15.95	6252	.28	.12 0404	16.22	.87 9596	29
32	7613	15.90	6235	.27	1377	16.18	8623	28
33	8567	15.87	6219	.28	2348	16.15	7652	27
34	.11 9519	15.83	6202	.28	3317	16.12	6683	26
35	9.12 0469	15.80	9.99 6185	.28	9.12 4284	16.08	10.87 5716	25
36	1417	15.75	6168	.28	5249	16.05	4751	24
37	2362	15.73	6151	.28	6211	16.02	3789	23
38	3306	15.70	6134	.28	7172	15.97	2828	22
39	4248	15.65	6117	.28	8130	15.95	1870	21
40	9.12 5187	15.63	9.99 6100	.28	9.12 9087	15.90	10.87 0913	20
41	6125	15.58	6083	.28	.13 0041	15.88	.86 9959	19
42	7060	15.55	6066	.28	0994	15.83	9006	18
43	7993	15.53	6049	.28	1944	15.82	8056	17
44	8925	15.48	6032	.28	2893	15.77	7107	16
45	9.12 9854	15.45	9.99 6015	.28	9.13 3839	15.75	10.86 6161	15
46	.13 0781	15.42	5998	.30	4784	15.70	5216	14
47	1706	15.40	5980	.28	5726	15.68	4274	13
48	2630	15.35	5963	.28	6667	15.63	3333	12
49	3551	15.32	5946	.30	7605	15.62	2395	11
50	9.13 4470	15.28	9.99 5928	.28	9.13 8542	15.57	10.86 1458	10
51	5387	15.27	5911	.28	.13 9476	15.55	.86 0524	9
52	6303	15.22	5894	.30	.14 0409	15.52	.85 9591	8
53	7216	15.20	5876	.28	1340	15.48	8660	7
54	8128	15.15	5859	.30	2269	15.45	7731	6
55	9.13 9037	15.12	9.99 5841	.30	9.14 3196	15.42	10.85 6804	5
56	.13 9944	15.10	5823	.28	4121	15.38	5879	4
57	.14 0850	15.07	5806	.30	5044	15.37	4956	3
58	1754	15.02	5788	.28	5966	15.32	4034	2
59	2655	15.00	5771	.30	6885	15.30	3115	1
60	9.14 3555		9.99 5753		9.14 7803		10.85 2197	0
′	Cosine.	D.1″.	Sine.	D.1″.	Cotang.	D.1″.	Tang.	′

97° **82°**

8° **Table VIII—Continued** 171°

′	Sine.	D.1″.	Cosine.	D.1″.	Tang.	D.1″.	Cotang.	′
0	9.14 3555	14.97	9.99 5753	.30	9.14 7803	15.25	10.85 2197	60
1	4453	14.93	5735	.30	8718	15.23	1282	59
2	5349	14.90	5717	.30	.14 9632	15.20	.85 0368	58
3	6243	14.88	5699	.30	.15 0544	15.17	.84 9456	57
4	7136	14.83	5681	.28	1454	15.15	8546	56
5	9.14 8026	14.82	9.99 5664	.30	9.15 2363	15.10	10.84 7637	55
6	8915	14.78	5646	.30	3269	15.08	6731	54
7	.14 9802	14.73	5628	.30	4174	15.05	5826	53
8	.15 0686	14.72	5610	.32	5077	15.02	4923	52
9	1569	14.70	5591	.30	5978	14.98	4022	51
10	9.15 2451	14.65	9.99 5573	.30	9.15 6877	14.97	10.84 3123	50
11	3330	14.63	5555	.30	7775	14.93	2225	49
12	4208	14.58	5537	.30	8671	14.90	1329	48
13	5083	14.57	5519	.30	.15 9565	14.87	.84 0435	47
14	5957	14.55	5501	.32	.16 0457	14.83	.83 9543	46
15	9.15 6830	14.50	9.99 5482	.30	9.16 1347	14.82	10.83 8653	45
16	7700	14.48	5464	.30	2236	14.78	7764	44
17	8569	14.43	5446	.32	3123	14.75	6877	43
18	.15 9435	14.43	5427	.30	4008	14.73	5992	42
19	.16 0301	14.38	5409	.32	4892	14.70	5108	41
20	9.16 1164	14.35	9.99 5390	.30	9.16 5774	14.67	10.83 4226	40
21	2025	14.33	5372	.32	6654	14.63	3346	39
22	2885	14.30	5353	.32	7532	14.62	2468	38
23	3743	14.28	5334	.30	8409	14.58	1591	37
24	4600	14.23	5316	.32	.16 9284	14.55	.83 0716	36
25	9.16 5454	14.22	9.99 5297	.32	9.17 0157	14.53	10.82 9843	35
26	6307	14.20	5278	.30	1029	14.50	8971	34
27	7159	14.15	5260	.32	1899	14.47	8101	33
28	8008	14.13	5241	.32	2767	14.45	7233	32
29	8856	14.10	5222	.32	3634	14.42	6366	31
30	9.16 9702	14.08	9.99 5203	.32	9.17 4499	14.38	10.82 5501	30
31	.17 0547	14.03	5184	.32	5362	14.37	4638	29
32	1389	14.02	5165	.32	6224	14.33	3776	28
33	2230	14.00	5146	.32	7084	14.30	2916	27
34	3070	13.97	5127	.32	7942	14.28	2058	26
35	9.17 3908	13.93	9.99 5108	.32	9.17 8799	14.27	10.82 1201	25
36	4744	13.90	5089	.32	.17 9655	14.22	.82 0345	24
37	5578	13.88	5070	.32	.18 0508	14.20	.81 9492	23
38	6411	13.85	5051	.32	1360	14.18	8640	22
39	7242	13.83	5032	.32	2211	14.13	7789	21
40	9.17 8072	13.80	9.99 5013	.33	9.18 3059	14.13	10.81 6941	20
41	8900	13.77	4993	.32	3907	14.08	6093	19
42	.17 9726	13.75	4974	.32	4752	14.08	5248	18
43	.18 0551	13.72	4955	.33	5597	14.03	4403	17
44	1374	13.70	4935	.32	6439	14.02	3561	16
45	9.18 2196	13.67	9.99 4916	.33	9.18 7280	14.00	10.81 2720	15
46	3016	13.63	4896	.32	8120	13.97	1880	14
47	3834	13.62	4877	.33	8958	13.93	1042	13
48	4651	13.58	4857	.32	.18 9794	13.92	.81 0206	12
49	5466	13.57	4838	.33	.19 0629	13.88	.80 9371	11
50	9.18 6280	13.53	9.99 4818	.33	9.19 1462	13.87	10.80 8538	10
51	7092	13.52	4798	.32	2294	13.83	7706	9
52	7903	13.48	4779	.33	3124	13.82	6876	8
53	8712	13.45	4759	.33	3953	13.78	6047	7
54	.18 9519	13.43	4739	.32	4780	13.77	5220	6
55	9.19 0325	13.42	9.99 4720	.33	9.19 5606	13.73	10.80 4394	5
56	1130	13.38	4700	.33	6430	13.72	3570	4
57	1933	13.35	4680	.33	7253	13.68	2747	3
58	2734	13.33	4660	.33	8074	13.67	1926	2
59	3534	13.30	4640	.33	8894	13.65	1106	1
60	9.19 4332		9.99 4620		9.19 9713		10.80 0287	0
′	Cosine.	D.1″.	Sine.	D.1″.	Cotang.	D.1″.	Tang.	′

98° 81°

9° **Table VIII—Continued** 170°

′	Sine.	D.1″.	Cosine.	D.1″.	Tang.	D.1″.	Cotang.	′
0	**9.19 4332**	13.28	**9.99 4620**	.33	**9.19 9713**	13.60	**10.80 0287**	**60**
1	5129	13.27	4600	.33	.20 0529	13.60	.79 9471	59
2	5925	13.23	4580	.33	1345	13.57	8655	58
3	6719	13.20	4560	.33	2159	13.53	7841	57
4	7511	13.18	4540	.35	2971	13.52	7029	56
5	**9.19 8302**	13.15	**9.99 4519**	.33	**9.20 3782**	13.50	**10.79 6218**	**55**
6	9091	13.13	4499	.33	4592	13.47	5408	54
7	.19 9879	13.12	4479	.33	5400	13.45	4600	53
8	.20 0666	13.08	4459	.35	6207	13.43	3793	52
9	1451	13.05	4438	.33	7013	13.40	2987	51
10	**9.20 2234**	13.05	**9.99 4418**	.33	**9.20 7817**	13.37	**10.79 2183**	**50**
11	3017	13.00	4398	.35	8619	13.35	1381	49
12	3797	13.00	4377	.33	.20 9420	13.33	.79 0580	48
13	4577	12.95	4357	.35	.21 0220	13.30	.78 9780	47
14	5354	12.95	4336	.33	1018	13.28	8982	46
15	**9.20 6131**	12.92	**9.99 4316**	.35	**9.21 1815**	13.27	**10.78 8185**	**45**
16	6906	12.88	4295	.35	2611	13.23	7389	44
17	7679	12.88	4274	.33	3405	13.22	6595	43
18	8452	12.83	4254	.35	4198	13.18	5802	42
19	9222	12.83	4233	.35	4989	13.18	5011	41
20	**9.20 9992**	12.80	**9.99 4212**	.35	**9.21 5780**	13.13	**10.78 4220**	**40**
21	.21 0760	12.77	4191	.33	6568	13.13	3432	39
22	1526	12.75	4171	.35	7356	13.10	2644	38
23	2291	12.73	4150	.35	8142	13.07	1858	37
24	3055	12.72	4129	.35	8926	13.07	1074	36
25	**9.21 3818**	12.68	**9.99 4108**	.35	**9.21 9710**	13.03	**10.78 0290**	**35**
26	4579	12.65	4087	.35	.22 0492	13.00	.77 9508	34
27	5338	12.65	4066	.35	1272	13.00	8728	33
28	6097	12.62	4045	.35	2052	12.97	7948	32
29	6854	12.58	4024	.35	2830	12.95	7170	31
30	**9.21 7609**	12.57	**9.99 4003**	.35	**9.22 3607**	12.92	**10.77 6393**	**30**
31	8363	12.55	3982	.37	4382	12.90	5618	29
32	9116	12.53	3960	.35	5156	12.88	4844	28
33	.21 9868	12.50	3939	.35	5929	12.85	4071	27
34	.22 0618	12.48	3918	.35	6700	12.85	3300	26
35	**9.22 1367**	12.47	**9.99 3897**	.37	**9.22 7471**	12.80	**10.77 2529**	**25**
36	2115	12.43	3875	.35	8239	12.80	1761	24
37	2861	12.42	3854	.37	9007	12.77	0993	23
38	3606	12.38	3832	.35	.22 9773	12.77	.77 0227	22
39	4349	12.38	3811	.37	.23 0539	12.72	.76 9461	21
40	**9.22 5092**	12.35	**9.99 3789**	.35	**9.23 1302**	12.72	**10.76 8698**	**20**
41	5833	12.33	3768	.37	2065	12.68	7935	19
42	6573	12.30	3746	.35	2826	12.67	7174	18
43	7311	12.28	3725	.37	3586	12.65	6414	17
44	8048	12.27	3703	.37	4345	12.63	5655	16
45	**9.22 8784**	12.23	**9.99 3681**	.35	**9.23 5103**	12.60	**10.76 4897**	**15**
46	.22 9518	12.23	3660	.37	5859	12.58	4141	14
47	.23 0252	12.20	3638	.37	6614	12.57	3386	13
48	0984	12.18	3616	.37	7368	12.53	2632	12
49	1715	12.15	3594	.37	8120	12.53	1880	11
50	**9.23 2444**	12.13	**9.99 3572**	.37	**9.23 8872**	12.50	**10.76 1128**	**10**
51	3172	12.12	3550	.37	.23 9622	12.48	.76 0378	9
52	3899	12.10	3528	.37	.24 0371	12.45	.75 9629	8
53	4625	12.07	3506	.37	1118	12.45	8882	7
54	5349	12.07	3484	.37	1865	12.42	8135	6
55	**9.23 6073**	12.03	**9.99 3462**	.37	**9.24 2610**	12.40	**10.75 7390**	**5**
56	6795	12.00	3440	.37	3354	12.38	6646	4
57	7515	12.00	3418	.37	4097	12.37	5903	3
58	8235	11.97	3396	.37	4839	12.33	5161	2
59	8953	11.95	3374	.38	5579	12.33	4421	1
60	**9.23 9670**		**9.99 3351**		**9.24 6319**		**10.75 3681**	**0**
′	Cosine.	D.1″.	Sine.	D.1″.	Cotang.	D.1″.	Tang.	′

99° 80°

10° **Table VIII—Continued** 169°

′	Sine.	D.1″.	Cosine.	D.1″.	Tang.	D.1″.	Cotang.	′
0	**9.23 9670**	11.93	**9.99 3351**	.37	**9.24 6319**	12.30	**10.75 3681**	**60**
1	.24 0386	11.92	3329	.37	7057	12.28	2943	59
2	1101	11.88	3307	.38	7794	12.27	2206	58
3	1814	11.87	3284	.37	8530	12.23	1470	57
4	2526	11.85	3262	.37	9264	12.23	0736	56
5	**9.24 3237**	11.83	**9.99 3240**	.38	**9.24 9998**	12.20	**10.75 0002**	**55**
6	3947	11.82	3217	.37	.25 0730	12.18	.74 9270	54
7	4656	11.78	3195	.38	1461	12.17	8539	53
8	5363	11.77	3172	.38	2191	12.15	7809	52
9	6069	11.77	3149	.37	2920	12.13	7080	51
10	**9.24 6775**	11.72	**9.99 3127**	.38	**9.25 3648**	12.10	**10.74 6352**	**50**
11	7478	11.72	3104	.38	4374	12.10	5626	49
12	8181	11.70	3081	.37	5100	12.07	4900	48
13	8883	11.67	3059	.38	5824	12.05	4176	47
14	.24 9583	11.65	3036	.38	6547	12.03	3453	46
15	**9.25 0282**	11.63	**9.99 3013**	.38	**9.25 7269**	12.02	**10.74 2731**	**45**
16	0980	11.62	2990	.38	7990	12.00	2010	44
17	1677	11.60	2967	.38	8710	11.98	1290	43
18	2373	11.57	2944	.38	.25 9429	11.95	.74 0571	42
19	3067	11.57	2921	.38	.26 0146	11.95	.73 9854	41
20	**9.25 3761**	11.53	**9.99 2898**	.38	**9.26 0863**	11.92	**10.73 9137**	**40**
21	4453	11.52	2875	.38	1578	11.90	8422	39
22	5144	11.50	2852	.38	2292	11.88	7708	38
23	5834	11.48	2829	.38	3005	11.87	6995	37
24	6523	11.47	2806	.38	3717	11.85	6283	36
25	**9.25 7211**	11.45	**9.99 2783**	.40	**9.26 4428**	11.83	**10.73 5572**	**35**
26	7898	11.42	2759	.38	5138	11.82	4862	34
27	8583	11.42	2736	.38	5847	11.80	4153	33
28	9268	11.38	2713	.38	6555	11.77	3445	32
29	.25 9951	11.37	2690	.40	7261	11.77	2739	31
30	**9.26 0633**	11.35	**9.99 2666**	.38	**9.26 7967**	11.73	**10.73 2033**	**30**
31	1314	11.33	2643	.40	8671	11.73	1329	29
32	1994	11.32	2619	.38	.26 9375	11.70	.73 0625	28
33	2673	11.30	2596	.40	.27 0077	11.70	.72 9923	27
34	3351	11.27	2572	.38	0779	11.67	9221	26
35	**9.26 4027**	11.27	**9.99 2549**	.40	**9.27 1479**	11.65	**10.72 8521**	**25**
36	4703	11.23	2525	.40	2178	11.63	7822	24
37	5377	11.23	2501	.38	2876	11.62	7124	23
38	6051	11.20	2478	.40	3573	11.60	6427	22
39	6723	11.20	2454	.40	4269	11.58	5731	21
40	**9.26 7395**	11.17	**9.99 2430**	.40	**9.27 4964**	11.57	**10.72 5036**	**20**
41	8065	11.15	2406	.40	5658	11.55	4342	19
42	8734	11.13	2382	.38	6351	11.53	3649	18
43	.26 9402	11.12	2359	.40	7043	11.52	2957	17
44	.27 0069	11.10	2335	.40	7734	11.50	2266	16
45	**9.27 0735**	11.08	**9.99 2311**	.40	**9.27 8424**	11.48	**10.72 1576**	**15**
46	1400	11.07	2287	.40	9113	11.47	0887	14
47	2064	11.03	2263	.40	.27 9801	11.45	.72 0199	13
48	2726	11.03	2239	.42	.28 0488	11.43	.71 9512	12
49	3388	11.02	2214	.40	1174	11.40	8826	11
50	**9.27 4049**	10.98	**9.99 2190**	.40	**9.28 1858**	11.40	**10.71 8142**	**10**
51	4708	10.98	2166	.40	2542	11.38	7458	9
52	5367	10.97	2142	.40	3225	11.37	6775	8
53	6025	10.93	2118	.42	3907	11.35	6093	7
54	6681	10.93	2093	.40	4588	11.33	5412	6
55	**9.27 7337**	10.90	**9.99 2069**	.42	**9.28 5268**	11.32	**10.71 4732**	**5**
56	7991	10.90	2044	.40	5947	11.28	4053	4
57	8645	10.87	2020	.40	6624	11.28	3376	3
58	9297	10.85	1996	.42	7301	11.27	2699	2
59	.27 9948	10.85	1971	.40	7977	11.25	2023	1
60	**9.28 0599**		**9.99 1947**		**9.28 8652**		**10.71 1348**	**0**
′	Cosine.	D.1″.	Sine.	D.1″.	Cotang.	D.1″.	Tang.	′

100° 79°

11° **Table VIII—Continued** **168°**

′	Sine.	D.1″.	Cosine.	D.1″.	Tang.	D.1″.	Cotang.	′
0	**9.28 0599**	10.82	**9.99 1947**	.42	**9.28 8652**	11.23	**10.71 1348**	**60**
1	1248	10.82	1922	.42	9326	11.22	0674	59
2	1897	10.78	1897	.40	.28 9999	11.20	.71 0001	58
3	2544	10.77	1873	.42	.29 0671	11.18	.70 9329	57
4	3190	10.77	1848	.42	1342	11.18	8658	56
5	**9.28 3836**	10.73	**9.99 1823**	.40	**9.29 2013**	11.15	**10.70 7987**	**55**
6	4480	10.73	1799	.42	2682	11.13	7318	54
7	5124	10.70	1774	.42	3350	11.12	6650	53
8	5766	10.70	1749	.42	4017	11.12	5983	52
9	6408	10.67	1724	.42	4684	11.08	5316	51
10	**9.28 7048**	10.67	**9.99 1699**	.42	**9.29 5349**	11.07	**10.70 4651**	**50**
11	7688	10.63	1674	.42	6013	11.07	3987	49
12	8326	10.63	1649	.42	6677	11.03	3323	48
13	8964	10.60	1624	.42	7339	11.03	2661	47
14	.28 9600	10.60	1599	.42	8001	11.02	1999	46
15	**9.29 0236**	10.57	**9.99 1574**	.42	**9.29 8662**	11.00	**10.70 1338**	**45**
16	0870	10.57	1549	.42	9322	10.97	0678	44
17	1504	10.55	1524	.43	.29 9980	10.97	.70 0020	43
18	2137	10.52	1498	.42	.30 0638	10.95	.69 9362	42
19	2768	10.52	1473	.42	1295	10.93	8705	41
20	**9.29 3399**	10.50	**9.99 1448**	.43	**9.30 1951**	10.93	**10.69 8049**	**40**
21	4029	10.48	1422	.42	2607	10.90	7393	39
22	4658	10.47	1397	.42	3261	10.88	6739	38
23	5286	10.45	1372	.43	3914	10.88	6086	37
24	5913	10.43	1346	.42	4567	10.85	5433	36
25	**9.29 6539**	10.42	**9.99 1321**	.43	**9.30 5218**	10.85	**10.69 4782**	**35**
26	7164	10.40	1295	.42	5869	10.83	4131	34
27	7788	10.40	1270	.43	6519	10.82	3481	33
28	8412	10.37	1244	.43	7168	10.80	2832	32
29	9034	10.35	1218	.42	7816	10.78	2184	31
30	**9.29 9655**	10.35	**9.99 1193**	.43	**9.30 8463**	10.77	**10.69 1537**	**30**
31	.30 0276	10.32	1167	.43	9109	10.75	0891	29
32	0895	10.32	1141	.43	.30 9754	10.75	.69 0246	28
33	1514	10.30	1115	.42	.31 0399	10.72	.68 9601	27
34	2132	10.27	1090	.43	1042	10.72	8958	26
35	**9.30 2748**	10.27	**9.99 1064**	.43	**9.31 1685**	10.70	**10.68 8315**	**25**
36	3364	10.25	1038	.43	2327	10.68	7673	24
37	3979	10.23	1012	.43	2968	10.67	7032	23
38	4593	10.23	0986	.43	3608	10.65	6392	22
39	5207	10.20	0960	.43	4247	10.63	5753	21
40	**9.30 5819**	10.18	**9.99 0934**	.43	**9.31 4885**	10.63	**10.68 5115**	**20**
41	6430	10.18	0908	.43	5523	10.60	4477	19
42	7041	10.15	0882	.45	6159	10.60	3841	18
43	7650	10.15	0855	.43	6795	10.58	3205	17
44	8259	10.13	0829	.43	7430	10.57	2570	16
45	**9.30 8867**	10.12	**9.99 0803**	.43	**9.31 8064**	10.55	**10.68 1936**	**15**
46	.30 9474	10.10	0777	.45	8697	10.55	1303	14
47	.31 0080	10.08	0750	.43	9330	10.52	0670	13
48	0685	10.07	0724	.45	.31 9961	10.52	.68 0039	12
49	1289	10.07	0697	.43	.32 0592	10.50	.67 9408	11
50	**9.31 1893**	10.03	**9.99 0671**	.43	**9.32 1222**	10.48	**10.67 8778**	**10**
51	2495	10.03	0645	.45	1851	10.47	8149	9
52	3097	10.02	0618	.45	2479	10.45	7521	8
53	3698	9.98	0591	.43	3106	10.45	6894	7
54	4297	10.00	0565	.45	3733	10.42	6267	6
55	**9.31 4897**	9.97	**9.99 0538**	.45	**9.32 4358**	10.42	**10.67 5642**	**5**
56	5495	9.95	0511	.43	4983	10.40	5017	4
57	6092	9.95	0485	.45	5607	10.40	4393	3
58	6689	9.92	0458	.45	6231	10.37	3769	2
59	7284	9.92	0431	.45	6853	10.37	3147	1
60	**9.31 7879**		**9.99 0404**		**9.32 7475**		**10.67 2525**	**0**
′	Cosine.	D.1″.	Sine.	D.1″.	Cotang.	D.1″.	Tang.	′

101° **78°**

12° **Table VIII—Continued** **167°**

′	Sine.	D.1″.	Cosine.	D.1″.	Tang.	D.1″.	Cotang.	′
0	**9.31 7879**	9.90	**9.99 0404**	.43	**9.32 7475**	10.33	**10.67 2525**	**60**
1	8473	9.88	0378	.45	8095	10.33	1905	59
2	9066	9.87	0351	.45	8715	10.32	1285	58
3	.31 9658	9.85	0324	.45	9334	10.32	0666	57
4	.32 0249	9.85	0297	.45	.32 9953	10.28	.67 0047	56
5	**9.32 0840**	9.83	**9.99 0270**	.45	**9.33 0570**	10.28	**10.66 9430**	**55**
6	1430	9.82	0243	.47	1187	10.27	8813	54
7	2019	9.80	0215	.45	1803	10.25	8197	53
8	2607	9.78	0188	.45	2418	10.25	7582	52
9	3194	9.77	0161	.45	3033	10.22	6967	51
10	**9.32 3780**	9.77	**9.99 0134**	.45	**9.33 3646**	10.22	**10.66 6354**	**50**
11	4366	9.73	0107	.47	4259	10.20	5741	49
12	4950	9.73	0079	.45	4871	10.18	5129	48
13	5534	9.72	0052	.45	5482	10.18	4518	47
14	6117	9.72	.99 0025	.47	6093	10.15	3907	46
15	**9.32 6700**	9.68	**9.98 9997**	.45	**9.33 6702**	10.15	**10.66 3298**	**45**
16	7281	9.68	9970	.47	7311	10.13	2689	44
17	7862	9.67	9942	.45	7919	10.13	2081	43
18	8442	9.65	9915	.47	8527	10.10	1473	42
19	9021	9.63	9887	.45	9133	10.10	0867	41
20	**9.32 9599**	9.62	**9.98 9860**	.47	**9.33 9739**	10.08	**10.66 0261**	**40**
21	.33 0176	9.62	9832	.47	.34 0344	10.07	.65 9656	39
22	0753	9.60	9804	.45	0948	10.07	9052	38
23	1329	9.57	9777	.47	1552	10.05	8448	37
24	1903	9.58	9749	.47	2155	10.03	7845	36
25	**9.33 2473**	9.55	**9.98 9721**	.47	**9.34 2757**	10.02	**10.65 7243**	**35**
26	3051	9.55	9693	.47	3358	10.00	6642	34
27	3624	9.52	9665	.47	3958	10.00	6042	33
28	4195	9.53	9637	.45	4558	9.98	5442	32
29	4767	9.50	9610	.47	5157	9.97	4843	31
30	**9.33 5337**	9.48	**9.98 9582**	.48	**9.34 5755**	9.97	**10.65 4245**	**30**
31	5906	9.48	9553	.47	6353	9.93	3647	29
32	6475	9.47	9525	.47	6949	9.93	3051	28
33	7043	9.45	9497	.47	7545	9.93	2455	27
34	7610	9.43	9469	.47	8141	9.90	1859	26
35	**9.33 8176**	9.43	**9.98 9441**	.47	**9.34 8735**	9.90	**10.65 1265**	**25**
36	8742	9.42	9413	.47	9329	9.88	0671	24
37	9307	9.40	9385	.48	.34 9922	9.87	.65 0078	23
38	.33 9871	9.38	9356	.47	.35 0514	9.87	.64 9486	22
39	.34 0434	9.37	9328	.47	1106	9.85	8894	21
40	**9.34 0996**	9.37	**9.98 9300**	.48	**9.35 1697**	9.83	**10.64 8303**	**20**
41	1558	9.35	9271	.47	2287	9.82	7713	19
42	2119	9.33	9243	.48	2876	9.82	7124	18
43	2679	9.33	9214	.47	3465	9.80	6535	17
44	3239	9.30	9186	.48	4053	9.78	5947	16
45	**9.34 3797**	9.30	**9.98 9157**	.48	**9.35 4640**	9.78	**10.64 5360**	**15**
46	4355	9.28	9128	.47	5227	9.77	4773	14
47	4912	9.28	9100	.48	5813	9.75	4187	13
48	5469	9.25	9071	.48	6398	9.73	3602	12
49	6024	9.25	9042	.47	6982	9.73	3018	11
50	**9.34 6579**	9.25	**9.98 9014**	.48	**9.35 7566**	9.72	**10.64 2434**	**10**
51	7134	9.22	8985	.48	8149	9.70	1851	9
52	7687	9 22	8956	.48	8731	9.70	1269	8
53	8240	9.20	8927	.48	9313	9.67	0687	7
54	8792	9.18	8898	.48	.35 9893	9.68	.64 0107	6
55	**9.34 9343**	9.17	**9.98 8869**	.48	**9.36 0474**	9.65	**10.63 9526**	**5**
56	.34 9893	9.17	8840	.48	1053	9.65	8947	4
57	.35 0443	9.15	8811	.48	1632	9.63	8368	3
58	0992	9.13	8782	.48	2210	9.62	7790	2
59	1540	9.13	8753	.48	2787	9.62	7213	1
60	**9.35 2088**		**9.98 8724**		**9.36 3364**		**10.63 6636**	**0**
′	Cosine.	D.1″.	Sine.	D.1″.	Cotang.	D.1″.	Tang.	′

102° **77°**

13° **Table VIII—Continued** **166°**

′	Sine.	D.1″.	Cosine.	D.1″.	Tang.	D.1″.	Cotang.	′
0	**9.35 2088**	9.12	**9.98 8724**	.48	**9.36 3364**	9.60	**10.63 6636**	**60**
1	2635	9.10	8695	.48	3940	9.58	6060	59
2	3181	9.08	8666	.50	4515	9.58	5485	58
3	3726	9.08	8636	.48	5090	9.57	4910	57
4	4271	9.07	8607	.48	5664	9.55	4336	56
5	**9.35 4815**	9.05	**9.98 8578**	.50	**9.36 6237**	9.55	**10.63 3763**	**55**
6	5358	9.05	8548	.48	6810	9.53	3190	54
7	5901	9.03	8519	.50	7382	9.52	2618	53
8	6443	9.02	8489	.48	7953	9.52	2047	52
9	6984	9.00	8460	.50	8524	9.50	1476	51
10	**9.35 7524**	9.00	**9.98 8430**	.48	**9.36 9094**	9.48	**10.63 0906**	**50**
11	8064	8.98	8401	.50	.36 9663	9.48	.63 0337	49
12	8603	8.97	8371	.48	.37 0232	9.45	.62 9768	48
13	9141	8.95	8342	.50	0799	9.47	9201	47
14	.35 9678	8.95	8312	.50	1367	9.43	8633	46
15	**9.36 0215**	8.95	**9.98 8282**	.50	**9.37 1933**	9.43	**10.62 8067**	**45**
16	0752	8.92	8252	.48	2499	9.42	7501	44
17	1287	8.92	8223	.50	3064	9.42	6936	43
18	1822	8.90	8193	.50	3629	9.40	6371	42
19	2356	8.88	8163	.50	4193	9.38	5807	41
20	**9.36 2889**	8.88	**9.98 8133**	.50	**9.37 4756**	9.38	**10.62 5244**	**40**
21	3422	8.87	8103	.50	5319	9.37	4681	39
22	3954	8.85	8073	.50	5881	9.35	4119	38
23	4485	8.85	8043	.50	6442	9.35	3558	37
24	5016	8.83	8013	.50	7003	9.33	2997	36
25	**9.36 5546**	8.82	**9.98 7983**	.50	**9.37 7563**	9.32	**10.62 2437**	**35**
26	6075	8.82	7953	.52	8122	9.32	1878	34
27	6604	8.78	7922	.50	8681	9.30	1319	33
28	7131	8.80	7892	.50	9239	9.30	0761	32
29	7659	8.77	7862	.50	.37 9797	9.28	.62 0203	31
30	**9.36 8185**	8.77	**9.98 7832**	.52	**9.38 0354**	9.27	**10.61 9646**	**30**
31	8711	8.75	7801	.50	0910	9.27	9090	29
32	9236	8.75	7771	.52	1466	9.23	8534	28
33	.36 9761	8.72	7740	.50	2020	9.25	7980	27
34	.37 0285	8.72	7710	.52	2575	9.23	7425	26
35	**9.37 0808**	8.70	**9.98 7679**	.50	**9.38 3129**	9.22	**10.61 6871**	**25**
36	1330	8.70	7649	.52	3682	9.20	6318	24
37	1852	8.68	7618	.50	4234	9.20	5766	23
38	2373	8.68	7588	.52	4786	9.18	5214	22
39	[illegible]	8.67	7557	.52	5337	9.18	4663	21
40	**9.37 3414**	8.65	**9.98 7526**	.50	**9.38 5888**	9.17	**10.61 4112**	**20**
41	3933	8.65	7496	.52	6438	9.15	3562	19
42	4452	8.63	7465	.52	6987	9.15	3013	18
43	4970	8.62	7434	.52	7536	9.13	2464	17
44	5487	8.60	7403	.52	8084	9.12	1916	16
45	**9.37 6003**	8.60	**9.98 7372**	.52	**9.38 8631**	9.12	**10.61 1369**	**15**
46	6519	8.60	7341	.52	9178	9.10	0822	14
47	7035	8.57	7310	.52	.38 9724	9.10	.61 0276	13
48	7549	8.57	7279	.52	.39 0270	9.08	.60 9730	12
49	8063	8.57	7248	.52	0815	9.08	9185	11
50	**9.37 8577**	8.53	**9.98 7217**	.52	**9.39 1360**	9.05	**10.60 8640**	**10**
51	9089	8.53	7186	.52	1903	9.07	8097	9
52	.37 9601	8.53	7155	.52	2447	9.03	7553	8
53	.38 0113	8.52	7124	.53	2989	9.03	7011	7
54	0624	8.50	7092	.52	3531	9.03	6469	6
55	**9.38 1134**	8.48	**9.98 7061**	.52	**9.39 4073**	9.02	**10.60 5927**	**5**
56	1643	8.48	7030	.53	4614	9.00	5386	4
57	2152	8.48	6998	.52	5154	9.00	4846	3
58	2661	8.45	6967	.52	5694	8.98	4306	2
59	3168	8.45	6936	.53	6233	8.97	3767	1
60	**9.38 3675**		**9.98 6904**		**9.39 6771**		**10.60 3229**	**0**
′	Cosine.	D.1″.	Sine.	D.1″.	Cotang.	D.1″.	Tang.	′

103° **76°**

14° **Table VIII—Continued** 165°

′	Sine.	D.1″.	Cosine.	D.1″.	Tang.	D.1″.	Cotang.	′
0	**9.38 3675**	8.45	**9.98 6904**	.52	**9.39 6771**	8.97	**10.60 3229**	**60**
1	4182	8.42	6873	.53	7309	8.95	2691	59
2	4687	8.42	6841	.53	7846	8.95	2154	58
3	5192	8.42	6809	.52	8383	8.93	1617	57
4	5697	8.40	6778	.53	8919	8.93	1081	56
5	**9.38 6201**	8.38	**9.98 6746**	.53	**9.39 9455**	8.92	**10.60 0545**	**55**
6	6704	8.38	6714	.52	.39 9990	8.90	.60 0010	54
7	7207	8.37	6683	.53	.40 0524	8.90	.59 9476	53
8	7709	8.35	6651	.53	1058	8.88	8942	52
9	8210	8.35	6619	.53	1591	8.88	8409	51
10	**9.38 8711**	8.33	**9.98 6587**	.53	**9.40 2124**	8.87	**10.59 7876**	**50**
11	9211	8.33	6555	.53	2656	8.85	7344	49
12	.38 9711	8.32	6523	.53	3187	8.85	6813	48
13	.39 0210	8.30	6401	.53	3718	8.85	6282	47
14	0708	8.30	6459	.53	4249	8.82	5751	46
15	**9.39 1206**	8.28	**9.98 6427**	.53	**9.40 4778**	8.83	**10.59 5222**	**45**
16	1703	8.27	6395	.53	5308	8.80	4692	44
17	2199	8.27	6363	.53	5836	8.80	4164	43
18	2695	8.27	6331	.53	6364	8.80	3636	42
19	3191	8.23	6299	.55	6892	8.78	3108	41
20	**9.39 3685**	8.23	**9.98 6266**	.53	**9.40 7419**	8.77	**10.59 2581**	**40**
21	4179	8.23	6234	.53	7945	8.77	2055	39
22	4673	8.22	6202	.55	8471	8.75	1529	38
23	5166	8.20	6169	.53	8996	8.75	1004	37
24	5658	8.20	6137	.55	.40 9521	8.73	.59 0479	36
25	**9.39 6150**	8.18	**9.98 6104**	.53	**9.41 0045**	8.73	**10.58 9955**	**35**
26	6641	8.18	6072	.55	0569	8.72	9431	34
27	7132	8.15	6039	.53	1092	8.72	8908	33
28	7621	8.17	6007	.55	1615	8.70	8385	32
29	8111	8.15	5974	.53	2137	8.68	7863	31
30	**9.39 8600**	8.13	**9.98 5942**	.55	**9.41 2658**	8.68	**10.58 7342**	**30**
31	9088	8.12	5909	.55	3179	8.67	6821	29
32	.39 9575	8.12	5876	.55	3699	8.67	6301	28
33	.40 0062	8.12	5843	.53	4219	8.65	5781	27
34	0549	8.10	5811	.55	4738	8.65	5262	26
35	**9.40 1035**	8.08	**9.98 5778**	.55	**9.41 5257**	8.63	**10.58 4743**	**25**
36	1520	8.08	5745	.55	5775	8.63	4225	24
37	2005	8.07	5712	.55	6293	8.62	3707	23
38	2489	8.05	5679	.55	6810	8.60	3190	22
39	2972	8.05	5646	.55	7326	8.60	2674	21
40	**9.40 3455**	8.05	**9.98 5613**	.55	**9.41 7842**	8.60	**10.58 2158**	**20**
41	3938	8.03	5580	.55	8358	8.58	1642	19
42	4420	8.02	5547	.55	8873	8.57	1127	18
43	4901	8.02	5514	.57	9387	8.57	0613	17
44	5382	8.00	5480	.55	.41 9901	8.57	.58 0099	16
45	**9.40 5862**	7.98	**9.98 5447**	.55	**9.42 0415**	8.55	**10.57 9585**	**15**
46	6341	7.98	5414	.55	0927	8.55	9073	14
47	6820	7.98	5381	.57	1440	8.53	8560	13
48	7299	7.97	5347	.55	1952	8.52	8048	12
49	7777	7.95	5314	.57	2463	8.52	7537	11
50	**9.40 8254**	7.95	**9.98 5280**	.55	**9.42 2974**	8.50	**10.57 7026**	**10**
51	8731	7.93	5247	.57	3484	8.48	6516	9
52	9207	7.93	5213	.55	3993	8.50	6007	8
53	.40 9682	7.92	5180	.57	4503	8.47	5497	7
54	.41 0157	7.92	5146	.55	5011	8.47	4989	6
55	**9.41 0632**	7.90	**9.98 5113**	.57	**9.42 5519**	8.47	**10.57 4481**	**5**
56	1106	7.88	5079	.57	6027	8.45	3973	4
57	1579	7.88	5045	.57	6534	8.45	3466	3
58	2052	7.87	5011	.55	7041	8.43	2959	2
59	2524	7.87	4978	.57	7547	8.42	2453	1
60	**9.41 2996**		**9.98 4944**		**9.42 8052**		**10.57 1948**	**0**
′	Cosine.	D.1″.	Sine.	D.1″.	Cotang.	D.1″.	Tang.	′

104° 75°

15° **Table VIII—Continued** **164°**

'	Sine.	D.1″.	Cosine.	D.1″.	Tang.	D.1″.	Cotang.	'
0	**9.41 2996**	7.85	**9.98 4944**	.57	**9.42 8052**	8.43	**10.57 1948**	**60**
1	3467	7.85	4910	.57	8558	8.40	1442	59
2	3938	7.83	4876	.57	9062	8.40	0938	58
3	4408	7.83	4842	.57	.42 9566	8.40	.57 0434	57
4	4878	7.82	4808	.57	.43 0070	8.38	.56 9930	56
5	**9.41 5347**	7.80	**9.98 4774**	.57	**9.43 0573**	8.37	**10.56 9427**	**55**
6	5815	7.80	4740	.57	1075	8.37	8925	54
7	6283	7.80	4706	.57	1577	8.37	8423	53
8	6751	7.77	4672	.57	2079	8.35	7921	52
9	7217	7.78	4638	.58	2580	8.33	7420	51
10	**9.41 7684**	7.77	**9.98 4603**	.57	**9.43 3080**	8.33	**10.56 6920**	**50**
11	8150	7.75	4569	.57	3580	8.33	6420	49
12	8615	7.73	4535	.58	4080	8.32	5920	48
13	9079	7.75	4500	.57	4579	8.32	5421	47
14	.41 9544	7.72	4466	.57	5078	8.30	4922	46
15	**9.42 0007**	7.72	**9.98 4432**	.58	**9.43 5576**	8.28	**10.56 4424**	**45**
16	0470	7.72	4397	.57	6073	8.28	3927	44
17	0933	7.70	4363	.58	6570	8.28	3430	43
18	1395	7.70	4328	.57	7067	8.27	2933	42
19	1857	7.68	4294	.58	7563	8.27	2437	41
20	**9.42 2318**	7.67	**9.98 4259**	.58	**9.43 8059**	8.25	**10.56 1941**	**40**
21	2778	7.67	4224	.57	8554	8.23	1446	39
22	3238	7.65	4190	.58	9048	8.25	0952	38
23	3697	7.65	4155	.58	.43 9543	8.22	.56 0457	37
24	4156	7.65	4120	.58	.44 0036	8.22	.55 9964	36
25	**9.42 4615**	7.63	**9.98 4085**	.58	**9.44 0529**	8.22	**10.55 9471**	**35**
26	5073	7.62	4050	.58	1022	8.20	8978	34
27	5530	7.62	4015	.57	1514	8.20	8486	33
28	5987	7.60	3981	.58	2006	8.18	7994	32
29	6443	7.60	3946	.58	2497	8.18	7503	31
30	**9.42 6899**	7.58	**9.98 3911**	.60	**9.44 2988**	8.18	**10.55 7012**	**30**
31	7354	7.58	3875	.58	3479	8.15	6521	29
32	7809	7.57	3840	.58	3968	8.17	6032	28
33	8263	7.57	3805	.58	4458	8.15	5542	27
34	8717	7.55	3770	.58	4947	8.13	5053	26
35	**9.42 9170**	7.55	**9.98 3735**	.58	**9.44 5435**	8.13	**10.55 4565**	**25**
36	.42 9623	7.53	3700	.60	5923	8.13	4077	24
37	.43 0075	7.53	3664	.58	6411	8.12	3589	23
38	0527	7.52	3629	.58	6898	8.10	3102	22
39	0978	7.52	3594	.60	7384	8.10	2616	21
40	**9.43 1429**	7.50	**9.98 3558**	.58	**9.44 7870**	8.10	**10.55 2130**	**20**
41	1879	7.50	3523	.60	8356	8.08	1644	19
42	2329	7.48	3487	.58	8841	8.08	1159	18
43	2778	7.47	3452	.60	9326	8.07	0674	17
44	3226	7.48	3416	.58	.44 9810	8.07	.55 0190	16
45	**9.43 3675**	7.45	**9.98 3381**	.60	**9.45 0294**	8.05	**10.54 9706**	**15**
46	4122	7.45	3345	.60	0777	8.05	9223	14
47	4569	7.45	3309	.60	1260	8.05	8740	13
48	5016	7.43	3273	.58	1743	8.03	8257	12
49	5462	7.43	3238	.60	2225	8.02	7775	11
50	**9.43 5908**	7.42	**9.98 3202**	.60	**9.45 2706**	8.02	**10.54 7294**	**10**
51	6353	7.42	3166	.60	3187	8.02	6813	9
52	6798	7.40	3130	.60	3668	8.00	6332	8
53	7242	7.40	3094	.60	4148	8.00	5852	7
54	7686	7.38	3058	.60	4628	7.98	5372	6
55	**9.43 8129**	7.38	**9.98 3022**	.60	**9.45 5107**	7.98	**10.54 4893**	**5**
56	8572	7.37	2986	.60	5586	7.97	4414	4
57	9014	7.37	2950	.60	6064	7.97	3936	3
58	9456	7.35	2914	.60	6542	7.95	3458	2
59	.43 9897	7.35	2878	.60	7019	7.95	2981	1
60	**9.44 0338**		**9.98 2842**		**9.45 7496**		**10.54 2404**	**0**
'	Cosine.	D.1″.	Sine.	D.1″.	Cotang.	D.1″.	Tang.	'

105° **74°**

16° **Table VIII—Continued** **163°**

′	Sine.	D.1″.	Cosine.	D.1″.	Tang.	D.1″.	Cotang.	′
0	**9.44 0338**	7.33	**9.98 2842**	.62	**9.45 7496**	7.95	**10.54 2504**	**60**
1	0778	7.33	2805	.60	7973	7.93	2027	59
2	1218	7.33	2769	.60	8449	7.93	1551	58
3	1658	7.30	2733	.62	8925	7.92	1075	57
4	2096	7.32	2696	.60	9400	7.92	0600	56
5	**9.44 2535**	7.30	**9.98 2660**	.60	**9.45 9875**	7.90	**10.54 0125**	**55**
6	2973	7.28	2624	.62	.46 0349	7.90	.53 9651	54
7	3410	7.28	2587	.60	0823	7.90	9177	53
8	3847	7.28	2551	.62	1297	7.88	8703	52
9	4284	7.27	2514	.62	1770	7.87	8230	51
10	**9.44 4720**	7.25	**9.98 2477**	.60	**9.46 2242**	7.88	**10.53 7758**	**50**
11	5155	7.25	2441	.62	2715	7.85	7285	49
12	5590	7.25	2404	.62	3186	7.87	6814	48
13	6025	7.23	2367	.60	3658	7.83	6342	47
14	6459	7.23	2331	.62	4128	7.85	5872	46
15	**9.44 6893**	7.22	**9.98 2294**	.62	**9.46 4599**	7.83	**10.53 5401**	**45**
16	7326	7.22	2257	.62	5069	7.83	4931	44
17	7759	7.20	2220	.62	5539	7.82	4461	43
18	8191	7.20	2183	.62	6008	7.82	3992	42
19	8623	7.18	2146	.62	6477	7.80	3523	41
20	**9.44 9054**	7.18	**9.98 2109**	.62	**9.46 6945**	7.80	**10.53 3055**	**40**
21	9485	7.17	2072	.62	7413	7.78	2587	39
22	.44 9915	7.17	2035	.62	7880	7.78	2120	38
23	.45 0345	7.17	1998	.62	8347	7.78	1653	37
24	0775	7.15	1961	.62	8814	7.77	1186	36
25	**9.45 1204**	7.13	**9.98 1924**	.63	**9.46 9280**	7.77	**10.53 0720**	**35**
26	1632	7.13	1886	.62	.46 9746	7.75	.53 0254	34
27	2060	7.13	1849	.62	.47 0211	7.75	.52 9789	33
28	2488	7.12	1812	.63	0676	7.75	9324	32
29	2915	7.12	1774	.62	1141	7.73	8859	31
30	**9.45 3342**	7.10	**9.98 1737**	.62	**9.47 1605**	7.73	**10.52 8395**	**30**
31	3768	7.10	1700	.63	2069	7.72	7931	29
32	4194	7.08	1662	.62	2532	7.72	7468	28
33	4619	7.08	1625	.63	2995	7.70	7005	27
34	5044	7.08	1587	.63	3457	7.70	6543	26
35	**9.45 5469**	7.07	**9.98 1549**	.62	**9.47 3919**	7.70	**10.52 6081**	**25**
36	5893	7.05	1512	.63	4381	7.68	5619	24
37	6316	7.05	1474	.63	4842	7.68	5158	23
38	6739	7.05	1436	.62	5303	7.67	4697	22
39	7162	7.03	1399	.63	5763	7.67	4237	21
40	**9.45 7584**	7.03	**9.98 1361**	.63	**9.47 6223**	7.67	**10.52 3777**	**20**
41	8006	7.02	1323	.63	6683	7.65	3317	19
42	8427	7.02	1285	.63	7142	7.65	2858	18
43	8848	7.00	1247	.63	7601	7.63	2399	17
44	9268	7.00	1209	.63	8059	7.63	1941	16
45	**9.45 9688**	7.00	**9.98 1171**	.63	**9.47 8517**	7.63	**10.52 1483**	**15**
46	.46 0108	6.98	1133	.63	8975	7.62	1025	14
47	0527	6.98	1095	.63	9432	7.62	0568	13
48	0946	6.97	1057	.63	.47 9889	7.60	.52 0111	12
49	1364	6.97	1019	.63	.48 0345	7.60	.51 9655	11
50	**9.46 1782**	6.95	**9.98 0981**	.65	**9.48 0801**	7.60	**10.51 9199**	**10**
51	2199	6.95	0942	.63	1257	7.58	8743	9
52	2616	6.93	0904	.63	1712	7.58	8288	8
53	3032	6.93	0866	.65	2167	7.57	7833	7
54	3448	6.93	0827	.63	2621	7.57	7379	6
55	**9.46 3864**	6.92	**9.98 0789**	.65	**9.48 3075**	7.57	**10.51 6925**	**5**
56	4279	6.92	0750	.63	3529	7.55	6471	4
57	4694	6.90	0712	.65	3982	7.55	6018	3
58	5108	6.90	0673	.63	4435	7.53	5565	2
59	5522	6.88	0635	.65	4887	7.53	5113	1
60	**9.46 5935**		**9.98 0596**		**9.48 5339**		**10.51 4661**	**0**
′	Cosine.	D.1″.	Sine.	D.1″.	Cotang.	D.1″.	Tang.	′

106° **73°**

17° **Table VIII—Continued** **162°**

′	Sine.	D.1″.	Cosine.	D.1″.	Tang.	D.1″.	Cotang.	′
0	**9.46 5935**	6.88	**9.98 0596**	.63	**9.48 5339**	7.53	**10.51 4661**	**60**
1	6348	6.88	0558	.65	5791	7.52	4209	59
2	6761	6.87	0519	.65	6242	7.52	3758	58
3	7173	6.87	0480	.63	6693	7.50	3307	57
4	7585	6.85	0442	.65	7143	7.50	2857	56
5	**9.46 7996**	6.85	**9.98 0403**	.65	**9.48 7593**	7.50	**10.51 2407**	**55**
6	8407	6.83	0364	.65	8043	7.48	1957	54
7	8817	6.83	0325	.65	8492	7.48	1508	53
8	9227	6.83	0286	.65	8941	7.48	1059	52
9	.46 9637	6.82	0247	.65	9300	7.47	0610	51
10	**9.47 0046**	6.82	**9.98 0208**	.65	**9.48 9838**	7.47	**10.51 0162**	**50**
11	0455	6.80	0169	.65	.49 0286	7.45	.50 9714	49
12	0863	6.80	0130	.65	0733	7.45	9267	48
13	1271	6.80	0091	.65	1180	7.45	8820	47
14	1679	6.78	0052	.67	1627	7.43	8373	46
15	**9.47 2086**	6.77	**9.98 0012**	.65	**9.49 2073**	7.43	**10.50 7927**	**45**
16	2492	6.77	.97 9973	.65	2519	7.43	7481	44
17	2898	6.77	9934	.65	2965	7.42	7035	43
18	3304	6.77	9895	.67	3410	7.40	6590	42
19	3710	6.75	9855	.65	3854	7.42	6146	41
20	**9.47 4115**	6.73	**9.97 9816**	.67	**9.49 4299**	7.40	**10.50 5701**	**40**
21	4519	6.73	9776	.65	4743	7.38	5257	39
22	4923	6.73	9737	.67	5186	7.40	4814	38
23	5327	6.72	9697	.65	5630	7.38	4370	37
24	5730	6.72	9658	.67	6073	7.37	3927	36
25	**9.47 6133**	6.72	**9.97 9618**	.65	**9.49 6515**	7.37	**10.50 3485**	**35**
26	6536	6.70	9579	.67	6957	7.37	3043	34
27	6938	6.70	9539	.67	7399	7.37	2601	33
28	7340	6.68	9499	.67	7841	7.35	2159	32
29	7741	6.68	9459	.65	8282	7.33	1718	31
30	**9.47 8142**	6.67	**9.97 9420**	.67	**9.49 8722**	7.35	**10.50 1278**	**30**
31	8542	6.67	9380	.67	9163	7.33	0837	29
32	8942	6.67	9340	.67	.49 9603	7.33	.50 0397	28
33	9342	6.65	9300	.67	.50 0042	7.32	.49 9958	27
34	.47 9741	6.65	9260	.67	0481	7.32	9519	26
35	**9.48 0140**	6.65	**9.97 9220**	.67	**9.50 0920**	7.32	**10.49 9080**	**25**
36	0539	6.63	9180	.67	1359	7.30	8641	24
37	0937	6.62	9140	.67	1797	7.30	8203	23
38	1334	6.62	9100	.68	2235	7.28	7765	22
39	1731	6.62	9059	.67	2672	7.28	7328	21
40	**9.48 2128**	6.62	**9.97 9019**	.67	**9.50 3109**	7.28	**10.49 6891**	**20**
41	2525	6.60	8979	.67	3546	7.27	6454	19
42	2921	6.58	8939	.68	3982	7.27	6018	18
43	3316	6.60	8898	.67	4418	7.27	5582	17
44	3712	6.58	8858	.68	4854	7.25	5146	16
45	**9.48 4107**	6.57	**9.97 8817**	.67	**9.50 5289**	7.25	**10.49 4711**	**15**
46	4501	6.57	8777	.67	5724	7.25	4276	14
47	4895	6.57	8737	.68	6159	7.23	3841	13
48	5289	6.55	8696	.68	6593	7.23	3407	12
49	5682	6.55	8655	.67	7027	7.23	2973	11
50	**9.48 6075**	6.53	**9.97 8615**	.68	**9.50 7460**	7.22	**10.49 2540**	**10**
51	6467	6.55	8574	.68	7893	7.22	2107	9
52	6860	6.52	8533	.67	8326	7.22	1674	8
53	7251	6.53	8493	.68	8759	7.20	1241	7
54	7643	6.52	8452	.68	9191	7.18	0809	6
55	**9.48 8034**	6.50	**9.97 8411**	.68	**9.50 9622**	7.20	**10.49 0378**	**5**
56	8424	6.50	8370	.68	.51 0054	7.18	.48 9946	4
57	8814	6.50	8329	.68	0485	7.18	9515	3
58	9204	6.48	8288	.68	0916	7.17	9084	2
59	9593	6.48	8247	.68	1346	7.17	8654	1
60	**9.48 9982**		**9.97 8206**		**9.51 1776**		**10.48 8224**	**0**
′	Cosine.	D.1″.	Sine.	D.1″.	Cotang.	D.1″.	Tang.	′

107° **72°**

18° **Table VIII—Continued** 161°

'	Sine.	D.1″.	Cosine.	D.1″.	Tang.	D.1″.	Cotang.	'
0	**9.48 9982**	6.48	**9.97 8206**	.68	**9.51 1776**	7.17	**10.48 8224**	**60**
1	.49 0371	6.47	8165	.68	2206	7.15	7794	59
2	0759	6.47	8124	.68	2635	7.15	7365	58
3	1147	6.47	8083	.68	3064	7.15	6936	57
4	1535	6.45	8042	.68	3493	7.13	6507	56
5	**9.49 1922**	6.43	**9.97 8001**	.70	**9.51 3921**	7.13	**10.48 6079**	**55**
6	2308	6.45	7959	.68	4349	7.13	5651	54
7	2695	6.43	7918	.68	4777	7.12	5223	53
8	3081	6.42	7877	.70	5204	7.12	4796	52
9	3466	6.42	7835	.68	5631	7.10	4369	51
10	**9.49 3851**	6.42	**9.97 7794**	.70	**9.51 6057**	7.12	**10.48 3934**	**50**
11	4236	6.42	7752	.68	6484	7.10	3516	49
12	4621	6.40	7711	.70	6910	7.08	3090	48
13	5005	6.38	7669	.68	7335	7.10	2665	47
14	5388	6.40	7628	.70	7761	7.08	2239	46
15	**9.49 5772**	6.37	**9.97 7586**	.70	**9.51 8186**	7.07	**10.48 1814**	**45**
16	6154	6.38	7544	.68	8610	7.07	1390	44
17	6537	6.37	7503	.70	9034	7.07	0966	43
18	6919	6.37	7461	.70	9458	7.07	0542	42
19	7301	6.35	7419	.70	.51 9882	7.05	.48 0118	41
20	**9.49 7682**	6.35	**9.97 7377**	.70	**9.52 0305**	7.05	**10.47 9695**	**40**
21	8064	6.33	7335	.70	0728	7.05	9272	39
22	8444	6.35	7293	.70	1151	7.03	8849	38
23	8825	6.32	7251	.70	1573	7.03	8427	37
24	9204	6.33	7209	.70	1995	7.03	8005	36
25	**9.49 9584**	6.32	**9.97 7167**	.70	**9.52 2417**	7.02	**10.47 7583**	**35**
26	.49 9963	6.32	7125	.70	2838	7.02	7162	34
27	.50 0342	6.32	7083	.70	3259	7.02	6741	33
28	0721	6.30	7041	.70	3680	7.00	6320	32
29	1099	6.28	6999	.70	4100	7.00	5900	31
30	**9.50 1476**	6.30	**9.97 6957**	.72	**9.52 4520**	7.00	**10.47 5480**	**30**
31	1854	6.28	6914	.70	4940	6.98	5060	29
32	2231	6.27	6872	.70	5359	6.98	4641	28
33	2607	6.28	6830	.72	5778	6.98	4222	27
34	2984	6.27	6787	.70	6197	6.97	3803	26
35	**9.50 3360**	6.25	**9.97 6745**	.72	**9.52 6615**	6.97	**10.47 3385**	**25**
36	3735	6.25	6702	.70	7033	6.97	2967	24
37	4110	6.25	6660	.72	7451	6.95	2549	23
38	4485	6.25	6617	.72	7868	6.95	2132	22
39	4860	6.23	6574	.70	8285	6.95	1715	21
40	**9.50 5234**	6.23	**9.97 6532**	.72	**9.52 8702**	6.95	**10.47 1298**	**20**
41	5608	6.22	6489	.72	9119	6.93	0881	19
42	5981	6.22	6446	.70	9535	6.93	0465	18
43	6354	6.22	6404	.72	.52 9951	6.92	.47 0049	17
44	6727	6.20	6361	.72	.53 0366	6.92	.46 9634	16
45	**9.50 7099**	6.20	**9.97 6318**	.72	**9.53 0781**	6.92	**10.46 9219**	**15**
46	7471	6.20	6275	.72	1196	6.92	8804	14
47	7843	6.18	6232	.72	1611	6.90	8389	13
48	8214	6.18	6189	.72	2025	6.90	7975	12
49	8585	6.18	6146	.72	2439	6.90	7561	11
50	**9.50 8956**	6.17	**9.97 6103**	.72	**9.53 2853**	6.88	**10.46 7147**	**10**
51	9326	6.17	6060	.72	3266	6.88	6734	9
52	.50 9696	6.15	6017	.72	3679	6.88	6321	8
53	.51 0065	6.15	5974	.73	4092	6.87	5908	7
54	0434	6.15	5930	.72	4504	6.87	5496	6
55	**9.51 0803**	6.15	**9.97 5887**	.72	**9.53 4916**	6.87	**10.46 5084**	**5**
56	1172	6.13	5844	.73	5328	6.85	4672	4
57	1540	6.12	5800	.72	5739	6.85	4261	3
58	1907	6.13	5757	.72	6150	6.85	3850	2
59	2275	6.12	5714	.73	6561	6.85	3439	1
60	**9.51 2642**		**9.97 5670**		**9.53 6972**		**10.46 3028**	**0**
'	Cosine.	D.1″.	Sine.	D.1″.	Cotang.	D.1″.	Tang.	'

108° 71°

19° Table VIII—Continued 160°

′	Sine.	D.1″.	Cosine.	D.1″.	Tang.	D.1″.	Cotang.	′
0	9.51 2642	6.12	9.97 5670	.72	9.53 6972	6.83	10.46 3028	60
1	3009	6.10	5627	.73	7382	6.83	2618	59
2	3375	6.10	5583	.73	7792	6.83	2208	58
3	3741	6.10	5539	.72	8202	6.82	1798	57
4	4107	6.08	5496	.73	8611	6.82	1389	56
5	9.51 4472	6.08	9.97 5452	.73	9.53 9020	6.82	10.46 0980	55
6	4837	6.08	5408	.72	9429	6.80	0571	54
7	5202	6.07	5365	.73	.53 9837	6.80	.46 0163	53
8	5566	6.07	5321	.73	.54 0245	6.80	.45 9755	52
9	5930	6.07	5277	.73	0653	6.80	9347	51
10	9.51 6294	6.05	9.97 5233	.73	9.54 1061	6.78	10.45 8939	50
11	6657	6.05	5189	.73	1468	6.78	8532	49
12	7020	6.03	5145	.73	1875	6.77	8125	48
13	7382	6.05	5101	.73	2281	6.78	7719	47
14	7745	6.03	5057	.73	2688	6.77	7312	46
15	9.51 8107	6.02	9.97 5013	.73	9.54 3094	6.75	10.45 6906	45
16	8468	6.02	4969	.73	3499	6.77	6501	44
17	8829	6.02	4925	.75	3905	6.75	6095	43
18	9190	6.02	4880	.73	4310	6.75	5690	42
19	9551	6.00	4836	.73	4715	6.73	5285	41
20	9.51 9911	6.00	9.97 4792	.73	9.54 5119	6.75	10.45 4881	40
21	.52 0271	6.00	4748	.75	5524	6.73	4476	39
22	0631	5.98	4703	.73	5928	6.72	4072	38
23	0990	5.98	4659	.75	6331	6.73	3669	37
24	1349	5.97	4614	.73	6735	6.72	3265	36
25	9.52 1707	5.98	9.97 4570	.75	9.54 7138	6.70	10.45 2862	35
26	2066	5.97	4525	.73	7540	6.72	2460	34
27	2424	5.95	4481	.75	7943	6.70	2057	33
28	2781	5.95	4436	.75	8345	6.70	1655	32
29	3138	5.95	4391	.73	8747	6.70	1253	31
30	9.52 3495	5.95	9.97 4347	.75	9.54 9149	6.68	10.45 0851	30
31	3852	5.93	4302	.75	9550	6.68	0450	29
32	4208	5.93	4257	.75	.54 9951	6.68	.45 0049	28
33	4564	5.93	4212	.75	.55 0352	6.67	.44 9648	27
34	4920	5.92	4167	.75	0752	6.68	9248	26
35	9.52 5275	5.92	9.97 4122	.75	9.55 1153	6.65	10.44 8847	25
36	5630	5.90	4077	.75	1552	6.67	8448	24
37	5984	5.92	4032	.75	1952	6.65	8048	23
38	6339	5.90	3987	.75	2351	6.65	7649	22
39	6693	5.88	3942	.75	2750	6.65	7250	21
40	9.52 7046	5.90	9.97 3897	.75	9.55 3149	6.65	10.44 6851	20
41	7400	5.88	3852	.75	3548	6.63	6452	19
42	7753	5.87	3807	.77	3946	6.63	6054	18
43	8105	5.88	3761	.75	4344	6.62	5656	17
44	8458	5.87	3716	.75	4741	6.63	5259	16
45	9.52 8810	5.85	9.97 3671	.77	9.55 5139	6.62	10.44 4861	15
46	9161	5.87	3625	.75	5536	6.62	4464	14
47	9513	5.85	3580	.75	5933	6.60	4067	13
48	.52 9864	5.85	3535	.77	6329	6.60	3671	12
49	.53 0215	5.83	3489	.75	6725	6.60	3275	11
50	9.53 0565	5.83	9.97 3444	.77	9.55 7121	6.60	10.44 2879	10
51	0915	5.83	3398	.77	7517	6.60	2483	9
52	1265	5.82	3352	.75	7913	6.58	2087	8
53	1614	5.82	3307	.77	8308	6.58	1692	7
54	1963	5.82	3261	.77	8703	6.57	1297	6
55	9.53 2312	5.82	9.97 3215	.77	9.55 9097	6.57	10.44 0903	5
56	2661	5.80	3169	.75	9491	6.57	0509	4
57	3009	5.80	3124	.77	.55 9885	6.57	.44 0115	3
58	3357	5.78	3078	.77	.56 0279	6.57	.43 9721	2
59	3704	5.80	3032	.77	0673	6.55	9327	1
60	9.53 4052		9.97 2986		9.56 1066		10.43 8934	0
′	Cosine.	D.1″.	Sine.	D.1″.	Cotang.	D.1″.	Tang.	′

109° 70°

20° **Table VIII—Continued** **159°**

′	Sine.	D.1″.	Cosine.	D.1″.	Tang.	D.1″.	Cotang.	′
0	**9.53 4052**	5.78	**9.97 2986**	.77	**9.56 1066**	6.55	**10.43 8934**	**60**
1	4399	5.77	2940	.77	1459	6.53	8541	59
2	4745	5.78	2894	.77	1851	6.55	8149	58
3	5092	5.77	2848	.77	2244	6.53	7756	57
4	5438	5.75	2802	.78	2636	6.53	7364	56
5	**9.53 5783**	5.77	**9.97 2755**	.77	**9.56 3028**	6.52	**10.43 6972**	**55**
6	6129	5.75	2709	.77	3419	6.53	6581	54
7	6474	5.73	2663	.77	3811	6.52	6189	53
8	6818	5.75	2617	.78	4202	6.52	5798	52
9	7163	5.73	2570	.77	4593	6.50	5407	51
10	**9.53 7507**	5.73	**9.97 2524**	.77	**9.56 4983**	6.50	**10.43 5017**	**50**
11	7851	5.72	2478	.78	5373	6.50	4627	49
12	8194	5.73	2431	.77	5763	6.50	4237	48
13	8538	5.70	2385	.78	6153	6.48	3847	47
14	8880	5.72	2338	.78	6542	6.50	3458	46
15	**9.53 9223**	5.70	**9.97 2291**	.77	**9.56 6932**	6.47	**10.43 3068**	**45**
16	9565	5.70	2245	.78	7320	6.48	2680	44
17	.53 9907	5.68	2198	.78	7709	6.48	2291	43
18	.54 0249	5.68	2151	.77	8098	6.47	1902	42
19	0590	5.68	2105	.78	8486	6.45	1514	41
20	**9.54 0931**	5.68	**9.97 2058**	.78	**9.56 8873**	6.47	**10.43 1127**	**40**
21	1272	5.68	2011	.78	9261	6.45	0739	39
22	1613	5.67	1964	.78	.56 9648	6.45	.43 0352	38
23	1953	5.67	1917	.78	.57 0035	6.45	.42 9965	37
24	2293	5.65	1870	.78	0422	6.45	9578	36
25	**9.54 2632**	5.65	**9.97 1823**	.78	**9.57 0809**	6.43	**10.42 9191**	**35**
26	2971	5.65	1776	.78	1195	6.43	8805	34
27	3310	5.65	1729	.78	1581	6.43	8419	33
28	3649	5.63	1682	.78	1967	6.42	8033	32
29	3987	5.63	1635	.78	2352	6.43	7648	31
30	**9.54 4325**	5.63	**9.97 1588**	.80	**9.57 2738**	6.42	**10.42 7262**	**30**
31	4663	5.62	1540	.78	3123	6.40	6877	29
32	5000	5.63	1493	.78	3507	6.42	6493	28
33	5338	5.60	1446	.80	3892	6.40	6108	27
34	5674	5.62	1398	.78	4276	6.40	5724	26
35	**9.54 6011**	5.60	**9.97 1351**	.80	**9.57 4660**	6.40	**10.42 5340**	**25**
36	6347	5.60	1303	.78	5044	6.38	4956	24
37	6683	5.60	1256	.80	5427	6.38	4573	23
38	7019	5.58	1208	.78	5810	6.38	4190	22
39	7354	5.58	1161	.80	6193	6.38	3807	21
40	**9.54 7689**	5.58	**9.97 1113**	.78	**9.57 6576**	6.38	**10.42 3424**	**20**
41	8024	5.58	1066	.80	6959	6.37	3041	19
42	8359	5.57	1018	.80	7341	6.37	2659	18
43	8693	5.57	0970	.80	7723	6.35	2277	17
44	9027	5.55	0922	.80	8104	6.37	1896	16
45	**9.54 9360**	5.55	**9.97 0874**	.78	**9.57 8486**	6.35	**10.42 1514**	**15**
46	.54 9693	5.55	0827	.80	8867	6.35	1133	14
47	.55 0026	5.55	0779	.80	9248	6.35	0752	13
48	0359	5.55	0731	.80	.57 9629	6.33	.42 0371	12
49	0692	5.53	0683	.80	.58 0009	6.33	.41 9991	11
50	**9.55 1024**	5.53	**9.97 0635**	.82	**9.58 0389**	6.33	**10.41 9611**	**10**
51	1356	5.52	0586	.80	0769	6.33	9231	9
52	1687	5.52	0538	.80	1149	6.32	8851	8
53	2018	5.52	0490	.80	1528	6.32	8472	7
54	2349	5.52	0442	.80	1907	6.32	8093	6
55	**9.55 2680**	5.50	**9.97 0394**	.82	**9.58 2286**	6.32	**10.41 7714**	**5**
56	3010	5.52	0345	.80	2665	6.32	7335	4
57	3341	5.48	0297	.80	3044	6.30	6956	3
58	3670	5.50	0249	.82	3422	6.30	6578	2
59	4000	5.48	0200	.80	3800	6.28	6200	1
60	**9.55 4329**		**9.97 0152**		**9.58 4177**		**10.41 5823**	**0**
′	Cosine.	D.1″.	Sine.	D.1″.	Cotang.	D.1″.	Tang.	′

110° **69°**

21° **Table VIII—Continued** **158°**

′	Sine.	D.1″.	Cosine.	D.1″.	Tang.	D.1″.	Cotang.	′
0	**9.55 4329**	5.48	**9.97 0152**	.82	**9.58 4177**	6.30	**10.41 5823**	**60**
1	4658	5.48	0103	.80	4555	6.28	5445	59
2	4987	5.47	0055	.82	4932	6.28	5068	58
3	5315	5.47	.97 0006	.82	5309	6.28	4691	57
4	5643	5.47	.96 9957	.80	5686	6.27	4314	56
5	**9.55 5971**	5.47	**9.96 9909**	.82	**9.58 6062**	6.28	**10.41 3938**	**55**
6	6299	5.45	9860	.82	6439	6.27	3561	54
7	6626	5.45	9811	.82	6815	6.25	3185	53
8	6953	5.45	9762	.80	7190	6.27	2810	52
9	7280	5.43	9714	.82	7566	6.25	2434	51
10	**9.55 7606**	5.43	**9.96 9665**	.82	**9.58 7941**	6.25	**10.41 2059**	**50**
11	7932	5.43	9616	.82	8316	6.25	1684	49
12	8258	5.42	9567	.82	8691	6.25	1309	48
13	8583	5.43	9518	.82	9066	6.23	0934	47
14	8909	5.42	9469	.82	9440	6.23	0560	46
15	**9.55 9234**	5.40	**9.96 9420**	.83	**9.58 9814**	6.23	**10.41 0186**	**45**
16	9558	5.42	9370	.82	.59 0188	6.23	.40 9812	44
17	.55 9883	5.40	9321	.82	0562	6.22	9438	43
18	.56 0207	5.40	9272	.82	0935	6.22	9065	42
19	0531	5.40	9223	.83	1308	6.22	8692	41
20	**9.56 0855**	5.38	**9.96 9173**	.82	**9.59 1681**	6.22	**10.40 8319**	**40**
21	1178	5.38	9124	.82	2054	6.20	7946	39
22	1501	5.38	9075	.83	2426	6.22	7574	38
23	1824	5.37	9025	.82	2799	6.20	7201	37
24	2146	5.37	8976	.83	3171	6.18	6829	36
25	**9.56 2468**	5.37	**9.96 8926**	.82	**9.59 3542**	6.20	**10.40 6458**	**35**
26	2790	5.37	8877	.83	3914	6.18	6086	34
27	3112	5.35	8827	.83	4285	6.18	5715	33
28	3433	5.37	8777	.82	4656	6.18	5344	32
29	3755	5.33	8728	.83	5027	6.18	4973	31
30	**9.56 4075**	5.35	**9.96 8678**	.83	**9.59 5398**	6.17	**10.40 4602**	**30**
31	4396	5.33	8628	.83	5768	6.17	4232	29
32	4716	5.33	8578	.83	6138	6.17	3862	28
33	5036	5.33	8528	.82	6508	6.17	3492	27
34	5356	5.33	8479	.83	6878	6.15	3122	26
35	**9.56 5676**	5.32	**9.96 8429**	.83	**9.59 7247**	6.15	**10.40 2753**	**25**
36	5995	5.32	8379	.83	7616	6.15	2384	24
37	6314	5.30	8329	.85	7985	6.15	2015	23
38	6632	5.32	8278	.83	8354	6.13	1646	22
39	6951	5.30	8228	.83	8722	6.15	1278	21
40	**9.56 7269**	5.30	**9.96 8178**	.83	**9.59 9091**	6.13	**10.40 0909**	**20**
41	7587	5.28	8128	.83	9459	6.13	0541	19
42	7904	5.30	8078	.85	.59 9827	6.12	.40 0173	18
43	8222	5.28	8027	.83	.60 0194	6.13	.39 9806	17
44	8539	5.28	7977	.83	0562	6.12	9438	16
45	**9.56 8856**	5.27	**9.96 7927**	.85	**9.60 0929**	6.12	**10.39 9071**	**15**
46	9172	5.27	7876	.83	1296	6.12	8704	14
47	9488	5.27	7826	.85	1663	6.10	8337	13
48	.56 9804	5.27	7775	.83	2029	6.10	7971	12
49	.57 0120	5.25	7725	.85	2395	6.10	7605	11
50	**9.57 0435**	5.27	**9.96 7674**	.83	**9.60 2761**	6.10	**10.39 7239**	**10**
51	0751	5.25	7624	.85	3127	6.10	6873	9
52	1066	5.23	7573	.85	3493	6.08	6507	8
53	1380	5.25	7522	.85	3858	6.08	6142	7
54	1695	5.23	7471	.83	4223	6.08	5777	6
55	**9.57 2009**	5.23	**9.96 7421**	.85	**9.60 4588**	6.08	**10.39 5412**	**5**
56	2323	5.22	7370	.85	4953	6.07	5047	4
57	2636	5.23	7319	.85	5317	6.08	4683	3
58	2950	5.22	7268	.85	5682	6.07	4318	2
59	3263	5.20	7217	.85	6046	6.07	3954	1
60	**9.57 3575**		**9.96 7166**		**9.60 6410**		**10.39 3590**	**0**
′	Cosine.	D.1″.	Sine.	D.1″.	Cotang.	D.1″.	Tang.	′

111° **68°**

22° Table VIII—Continued 157°

′	Sine.	D.1″.	Cosine.	D.1″.	Tang.	D.1″.	Cotang.	′
0	9.57 3575	5.22	9.96 7166	.85	9.60 6410	6.05	10.39 3590	60
1	3888	5.20	7115	.85	6773	6.07	3227	59
2	4200	5.20	7064	.85	7137	6.05	2863	58
3	4512	5.20	7013	.87	7500	6.05	2500	57
4	4824	5.20	6961	.85	7863	6.03	2137	56
5	9.57 5136	5.18	9.96 6910	.85	9.60 8225	6.05	10.39 1775	55
6	5447	5.18	6859	.85	8588	6.03	1412	54
7	5758	5.18	6808	.87	8950	6.03	1050	53
8	6069	5.17	6756	.85	9312	6.03	0688	52
9	6379	5.17	6705	.87	.60 9674	6.03	.39 0326	51
10	9.57 6689	5.17	9.96 6653	.85	9.61 0036	6.02	10.38 9964	50
11	6999	5.17	6602	.87	0397	6.03	9603	49
12	7309	5.15	6550	.85	0759	6.02	9241	48
13	7618	5.15	6499	.87	1120	6.00	8880	47
14	7927	5.15	6447	.87	1480	6.02	8520	46
15	9.57 8236	5.15	9.96 6395	.85	9.61 1841	6.00	10.38 8159	45
16	8545	5.13	6344	.87	2201	6.00	7799	44
17	8853	5.15	6292	.87	2561	6.00	7439	43
18	9162	5.13	6240	.87	2921	6.00	7079	42
19	9470	5.12	6188	.87	3281	6.00	6719	41
20	9.57 9777	5.13	9.96 6136	.85	9.61 3641	5.98	10.38 6359	40
21	.58 0085	5.12	6085	.87	4000	5.98	6000	39
22	0392	5.12	6033	.87	4359	5.98	5641	38
23	0699	5.10	5981	.87	4718	5.98	5282	37
24	1005	5.12	5929	.88	5077	5.97	4923	36
25	9.58 1312	5.10	9.96 5876	.87	9.61 5435	5.97	10.38 4565	35
26	1618	5.10	5824	.87	5793	5.97	4207	34
27	1924	5.08	5772	.87	6151	5.97	3849	33
28	2229	5.10	5720	.87	6509	5.97	3491	32
29	2535	5.08	5668	.88	6867	5.95	3133	31
30	9.58 2840	5.08	9.96 5615	.87	9.61 7224	5.97	10.38 2776	30
31	3145	5.07	5563	.87	7582	5.95	2418	29
32	3449	5.08	5511	.88	7939	5.93	2061	28
33	3754	5.07	5458	.87	8295	5.95	1705	27
34	4058	5.05	5406	.88	8652	5.93	1348	26
35	9.58 4361	5.07	9.96 5353	.87	9.61 9008	5.93	10.38 0992	25
36	4665	5.05	5301	.88	9364	5.93	0636	24
37	4968	5.07	5248	.88	.61 9720	5.93	.38 0280	23
38	5272	5.03	5195	.87	.62 0076	5.93	.37 9924	22
39	5574	5.05	5143	.88	0432	5.92	9568	21
40	9.58 5877	5.03	9.96 5090	.88	9.62 0787	5.92	10.37 9213	20
41	6179	5.05	5037	.88	1142	5.92	8858	19
42	6482	5.02	4984	.88	1497	5.92	8503	18
43	6783	5.03	4931	.87	1852	5.92	8148	17
44	7085	5.02	4879	.88	2207	5.90	7793	16
45	9.58 7386	5.03	9.96 4826	.88	9.62 2561	5.90	10.37 7439	15
46	7688	5.02	4773	.88	2915	5.90	7085	14
47	7989	5.00	4720	.90	3269	5.90	6731	13
48	8289	5.02	4666	.88	3623	5.88	6377	12
49	8590	5.00	4613	.88	3976	5.90	6024	11
50	9.58 8890	5.00	9.96 4560	.88	9.62 4330	5.88	10.37 5670	10
51	9190	4.98	4507	.88	4683	5.88	5317	9
52	9489	5.00	4454	.90	5036	5.87	4964	8
53	.58 9789	4.98	4400	.88	5388	5.88	4612	7
54	.59 0088	4.98	4347	.88	5741	5.87	4259	6
55	9.59 0387	4.98	9.96 4294	.90	9.62 6093	5.87	10.37 3907	5
56	0686	4.97	4240	.88	6445	5.87	3555	4
57	0984	4.97	4187	.90	6797	5.87	3203	3
58	1282	4.97	4133	.88	7149	5.87	2851	2
59	1580	4.97	4080	.90	7501	5.85	2499	1
60	9.59 1878		9.96 4026		9.62 7852		10.37 2148	0
′	Cosine.	D.1″.	Sine.	D.1″.	Cotang.	D.1″.	Tang.	′

112° 67°

23° **Table VIII—Continued** **156°**

′	Sine.	D.1″.	Cosine.	D.1″.	Tang.	D.1″.	Cotang.	′
0	**9.59 1878**	4.97	**9.96 4026**	.90	**9.62 7825**	5.85	**10.37 2148**	**60**
1	2176	4.95	3972	.88	8203	5.85	1797	59
2	2473	4.95	3919	.90	8554	5.85	1446	58
3	2770	4.95	3865	.90	8905	5.83	1095	57
4	3067	4.93	3811	.90	9255	5.85	0745	56
5	**9.59 3363**	4.93	**9.96 3757**	.88	**9.62 9606**	5.83	**10.37 0394**	**55**
6	3659	4.93	3704	.90	.62 9956	5.83	.37 0044	54
7	3955	4.93	3650	.90	.63 0306	5.83	.36 9694	53
8	4251	4.93	3596	.90	0656	5.82	9344	52
9	4547	4.92	3542	.90	1005	5.83	8995	51
10	**9.59 4842**	4.92	**9.96 3488**	.90	**9.63 1355**	5.82	**10.36 8645**	**50**
11	5137	4.92	3434	.92	1704	5.82	8296	49
12	5432	4.92	3379	.90	2053	5.82	7947	48
13	5727	4.90	3325	.90	2402	5.80	7598	47
14	6021	4.90	3271	.90	2750	5.82	7250	46
15	**9.59 6315**	4.90	**9.96 3217**	.90	**9.63 3099**	5.80	**10.36 6901**	**45**
16	6609	4.90	3163	.92	3447	5.80	6553	44
17	6903	4.88	3108	.90	3795	5.80	6205	43
18	7196	4.90	3054	.92	4143	5.78	5857	42
19	7490	4.88	2999	.90	4490	5.80	5510	41
20	**9.59 7783**	4.87	**9.96 2945**	.92	**9.63 4838**	5.78	**10.36 5162**	**40**
21	8075	4.88	2890	.90	5185	5.78	4815	39
22	8368	4.87	2836	.92	5532	5.78	4468	38
23	8660	4.87	2781	.90	5879	5.78	4121	37
24	8952	4.87	2727	.92	6226	5.77	3774	36
25	**9.59 9244**	4.87	**9.96 2672**	.92	**9.63 6572**	5.78	**10.36 3428**	**35**
26	9536	4.85	2617	.92	6919	5.77	3081	34
27	.59 9827	4.85	2562	.90	7265	5.77	2735	33
28	.60 0118	4.85	2508	.92	7611	5.75	2389	32
29	0409	4.85	2453	.92	7956	5.77	2044	31
30	**9.60 0700**	4.83	**9.96 2398**	.92	**9.63 8302**	5.75	**10.36 1698**	**30**
31	0990	4.83	2343	.92	8647	5.75	1353	29
32	1280	4.83	2288	.92	8992	5.75	1008	28
33	1570	4.83	2233	.92	9337	5.75	0663	27
34	1860	4.83	2178	.92	.63 9682	5.75	.36 0318	26
35	**9.60 2150**	4.82	**9.96 2123**	.93	**9.64 0027**	5.73	**10.35 9973**	**25**
36	2439	4.82	2067	.92	0371	5.75	9629	24
37	2728	4.82	2012	.92	0716	5.73	9284	23
38	3017	4.80	1957	.92	1060	5.73	8940	22
39	3305	4.82	1902	.93	1404	5.72	8596	21
40	**9.60 3594**	4.80	**9.96 1846**	.92	**9.64 1747**	5.73	**10.35 8253**	**20**
41	3882	4.80	1791	.93	2091	5.72	7909	19
42	4170	4.78	1735	.92	2434	5.72	7566	18
43	4457	4.80	1680	.93	2777	5.72	7223	17
44	4745	4.78	1624	.92	3120	5.72	6880	16
45	**9.60 5032**	4.78	**9.96 1569**	.93	**9.64 3463**	5.72	**10.35 6537**	**15**
46	5319	4.78	1513	.92	3806	5.70	6194	14
47	5606	4.77	1458	.93	4148	5.70	5852	13
48	5892	4.78	1402	.93	4490	5.70	5510	12
49	6179	4.77	1346	.93	4832	5.70	5168	11
50	**9.60 6465**	4.77	**9.96 1290**	.92	**9.64 5174**	5.70	**10.35 4826**	**10**
51	6751	4.75	1235	.93	5516	5.68	4484	9
52	7036	4.77	1179	.93	5857	5.70	4143	8
53	7322	4.75	1123	.93	6199	5.68	3801	7
54	7607	4.75	1067	.93	6540	5.68	3460	6
55	**9.60 7892**	4.75	**9.96 1011**	.93	**9.64 6881**	5.68	**10.35 3119**	**5**
56	8177	4.73	0955	.93	7222	5.67	2778	4
57	8461	4.73	0899	.93	7562	5.68	2438	3
58	8745	4.73	0843	.95	7903	5.67	2097	2
59	9029	4.73	0786	.93	8243	5.67	1757	1
60	**9.60 9313**		**9.96 0730**		**9.64 8583**		**10.35 1417**	**0**
′	Cosine.	D.1″.	Sine.	D.1″.	Cotang.	D.1″.	Tang.	′

113° **66°**

24° **Table VIII—Continued** 155°

′	Sine.	D.1″.	Cosine.	D.1″.	Tang.	D.1″.	Cotang.	′
0	**9.60 9313**	4.73	**9.96 0730**	.93	**9.64 8583**	5.67	**10.35 1417**	**60**
1	9597	4.72	0674	.93	8923	5.67	1077	59
2	.60 9880	4.73	0618	.95	9263	5.65	0737	58
3	.61 0164	4.72	0561	.93	9602	5.67	0398	57
4	0447	4.70	0505	.95	.64 9942	5.65	.35 0058	56
5	**9.61 0729**	4.72	**9.96 0448**	.93	**9.65 0281**	5.65	**10.34 9719**	**55**
6	1012	4.70	0392	.95	0620	5.65	9380	54
7	1294	4.70	0335	.93	0959	5.63	9041	53
8	1576	4.70	0279	.95	1297	5.65	8703	52
9	1858	4.70	0222	.95	1636	5.63	8364	51
10	**9.61 2140**	4.68	**9.96 0165**	.93	**9.65 1974**	5.63	**10.34 8026**	**50**
11	2421	4.68	0109	.95	2312	5.63	7688	49
12	2702	4.68	.96 0052	.95	2650	5.63	7350	48
13	2983	4.68	.95 9995	.95	2988	5.63	7012	47
14	3264	4.68	9038	.93	3326	5.62	6674	40
15	**9.61 3545**	4.67	**9.95 9882**	.95	**9.65 3663**	5.62	**10.34 6337**	**45**
16	3825	4.67	9825	.95	4000	5.62	6000	44
17	4105	4.67	9768	.95	4337	5.62	5663	43
18	4385	4.67	9711	.95	4674	5.62	5326	42
19	4665	4.65	9654	.97	5011	5.62	4989	41
20	**9.61 4944**	4.65	**9.95 9596**	.95	**9.65 5348**	5.60	**10.34 4652**	**40**
21	5223	4.65	9539	.95	5684	5.60	4316	39
22	5502	4.65	9482	.95	6020	5.60	3980	38
23	5781	4.65	9425	.95	6356	5.60	3644	37
24	6060	4.63	9368	.97	6692	5.60	3308	36
25	**9.61 6338**	4.63	**9.95 9310**	.95	**9.65 7028**	5.60	**10.34 2972**	**35**
26	6616	4.63	9253	.97	7364	5.58	2636	34
27	6894	4.63	9195	.95	7699	5.58	2301	33
28	7172	4.63	9138	.97	8034	5.58	1966	32
29	7450	4.62	9080	.95	8369	5.58	1631	31
30	**9.61 7727**	4.62	**9.95 9023**	.97	**9.65 8704**	5.58	**10.34 1296**	**30**
31	8004	4.62	8965	.95	9039	5.57	0961	29
32	8281	4.62	8908	.97	9373	5.58	0627	28
33	8558	4.60	8850	.97	.65 9708	5.57	.34 0292	27
34	8834	4.60	8792	.97	.66 0042	5.57	.33 9958	26
35	**9.61 9110**	4.60	**9.95 8734**	.95	**9.66 0376**	5.57	**10.33 9624**	**25**
36	9386	4.60	8677	.97	0710	5.55	9290	24
37	9662	4.60	8619	.97	1043	5.57	8957	23
38	.61 9938	4.58	8561	.97	1377	5.55	8623	22
39	.62 0213	4.58	8503	.97	1710	5.55	8290	21
40	**9.62 0488**	4.58	**9.95 8445**	.97	**9.66 2043**	5.55	**10.33 7957**	**20**
41	0763	4.58	8387	.97	2376	5.55	7624	19
42	1038	4.58	8329	.97	2709	5.55	7291	18
43	1313	4.57	8271	.97	3042	5.55	6958	17
44	1587	4.57	8213	.98	3375	5.53	6625	16
45	**9.62 1861**	4.57	**9.95 8154**	.97	**9.66 3707**	5.53	**10.33 6293**	**15**
46	2135	4.57	8096	.97	4039	5.53	5961	14
47	2409	4.55	8038	.98	4371	5.53	5629	13
48	2682	4.57	7979	.97	4703	5.53	5297	12
49	2956	4.55	7921	.97	5035	5.52	4965	11
50	**9.62 3229**	4.55	**9.95 7863**	.98	**9.66 5366**	5.53	**10.33 4634**	**10**
51	3502	4.53	7804	.97	5698	5.52	4302	9
52	3774	4.55	7746	.98	6029	5.52	3971	8
53	4047	4.53	7687	.98	6360	5.52	3640	7
54	4319	4.53	7628	.97	6691	5.50	3309	6
55	**9.62 4591**	4.53	**9.95 7570**	.98	**9.66 7021**	5.52	**10.33 2979**	**5**
56	4863	4.53	7511	.98	7352	5.50	2648	4
57	5135	4.52	7452	.98	7682	5.52	2318	3
58	5406	4.52	7393	.97	8013	5.50	1987	2
59	5677	4.52	7335	.98	8343	5.50	1657	1
60	**9.62 5948**		**9.95 7276**		**9.66 8673**		**10.33 1327**	**0**
′	Cosine.	D.1″.	Sine.	D.1″.	Cotang.	D.1″.	Tang.	′

114° 65°

25° **Table VIII—Continued** **154°**

′	Sine.	D.1″.	Cosine.	D.1″.	Tang.	D.1″.	Cotang.	′
0	**9.62 5948**	4.52	**9.95 7276**	.98	**9.66 8673**	5.48	**10.33 1327**	**60**
1	6219	4.52	7217	.98	9002	5.50	0998	59
2	6490	4.50	7158	.98	9332	5.48	0668	58
3	6760	4.50	7099	.98	9661	5.50	0339	57
4	7030	4.50	7040	.98	.66 9991	5.48	.33 0009	56
5	**9.62 7300**	4.50	**9.95 6981**	1.00	**9.67 0320**	5.48	**10.32 9680**	**55**
6	7570	4.50	6921	.98	0649	5.47	9351	54
7	7840	4.48	6862	.98	0977	5.48	9023	53
8	8109	4.48	6803	.98	1306	5.48	8694	52
9	8378	4.48	6744	1.00	1635	5.47	8365	51
10	**9.62 8647**	4.48	**9.95 6684**	.98	**9.67 1963**	5.47	**10.32 8037**	**50**
11	8916	4.48	6625	.98	2291	5.47	7709	49
12	9185	4.47	6566	1.00	2619	5.47	7381	48
13	9453	4.47	6506	.98	2947	5.45	7053	47
14	9721	4.47	6447	1.00	3274	5.47	6726	46
15	**9.62 9989**	4.47	**9.95 6387**	1.00	**9.67 3602**	5.45	**10.32 6398**	**45**
16	.63 0257	4.45	6327	.98	3929	5.47	6071	44
17	0524	4.47	6268	1.00	4257	5.45	5743	43
18	0792	4.45	6208	1.00	4584	5.45	5416	42
19	1059	4.45	6148	.98	4911	5.43	5089	41
20	**9.63 1326**	4.45	**9.95 6089**	1.00	**9.67 5237**	5.45	**10.32 4763**	**40**
21	1593	4.43	6029	1.00	5564	5.43	4436	39
22	1859	4.43	5969	1.00	5890	5.45	4110	38
23	2125	4.45	5000	1.00	6217	5.43	3783	37
24	2392	4.43	5849	1.00	6543	5.43	3457	36
25	**9.63 2658**	4.42	**9.95 5789**	1.00	**9.67 6869**	5.42	**10.32 3131**	**35**
26	2923	4.43	5729	1.00	7194	5.43	2806	34
27	3189	4.43	5669	1.00	7520	5.43	2480	33
28	3454	4.42	5609	.98	7846	5.42	2154	32
29	3719	4.42	5548	1.00	8171	5.42	1829	31
30	**9.63 3984**	4.42	**9.95 5488**	1.00	**9.67 8496**	5.42	**10.32 1504**	**30**
31	4249	4.42	5428	1.00	8821	5.42	1179	29
32	4514	4.40	5368	1.02	9146	5.42	0854	28
33	4778	4.40	5307	1.00	9471	5.40	0529	27
34	5042	4.40	5247	1.00	.67 9795	5.40	.32 0205	26
35	**9.63 5306**	4.40	**9.95 5186**	1.00	**9.68 0120**	5.40	**10.31 9880**	**25**
36	5570	4.40	5126	1.00	0444	5.40	9556	24
37	5834	4.38	5065	1.00	0768	5.40	9232	23
38	6097	4.38	5005	1.02	1092	5.40	8908	22
39	6360	4.38	4944	1.00	1416	5.40	8584	21
40	**9.63 6623**	4.38	**9.95 4883**	1.00	**9.68 1740**	5.38	**10.31 8260**	**20**
41	6886	4.37	4823	1.02	2063	5.40	7937	19
42	7148	4.38	4762	1.02	2387	5.38	7613	18
43	7411	4.37	4701	1.02	2710	5.38	7290	17
44	7673	4.37	4640	1.02	3033	5.38	6967	16
45	**9.63 7935**	4.37	**9.95 4579**	1.02	**9.68 3356**	5.38	**10.31 6644**	**15**
46	8197	4.35	4518	1.02	3679	5.37	6321	14
47	8458	4.37	4457	1.02	4001	5.38	5999	13
48	8720	4.35	4396	1.02	4324	5.37	5676	12
49	8981	4.35	4335	1.02	4646	5.37	5354	11
50	**9.63 9242**	4.35	**9.95 4274**	1.02	**9.68 4968**	5.37	**10.31 5032**	**10**
51	9503	4.35	4213	1.02	5290	5.37	4710	9
52	.63 9764	4.33	4152	1.03	5612	5.37	4388	8
53	.64 0024	4.33	4090	1.02	5934	5.35	4066	7
54	0284	4.33	4029	1.02	6255	5.37	3745	6
55	**9.64 0544**	4.33	**9.95 3963**	1.03	**9.68 6577**	5.35	**10.31 3423**	**5**
56	0804	4.33	3906	1.02	6898	5.35	3102	4
57	1064	4.33	3845	1.03	7219	5.35	2781	3
58	1324	4.32	3783	1.02	7540	5.35	2460	2
59	1583	4.32	3722	1.03	7861	5.35	2139	1
60	**9.64 1842**		**9.95 3660**		**9.68 8182**		**10.31 1818**	**0**
′	Cosine.	D.1″.	Sine.	D.1″.	Cotang.	D.1″.	Tang.	′

115° **64°**

26° **Table VIII—Continued** **153°**

′	Sine.	D.1″.	Cosine.	D.1″.	Tang.	D.1″.	Cotang.	′
0	**9.64 1842**	4.32	**9.95 3660**	1.02	**9.68 8182**	5.33	**10.31 1818**	**60**
1	2101	4.32	3599	1.03	8502	5.32	1498	59
2	2360	4.30	3537	1.03	8823	5.33	1177	58
3	2618	4.32	3475	1.03	9143	5.33	0857	57
4	2877	4.30	3413	1.02	9463	5.33	0537	56
5	**9.64 3135**	4.30	**9.95 3352**	1.03	**9.68 9783**	5.33	**10.31 0217**	**55**
6	3393	4.28	3290	1.03	.69 0103	5.33	.30 9897	54
7	3650	4.30	3228	1.03	0423	5.32	9577	53
8	3908	4.28	3166	1.03	0742	5.33	9258	52
9	4165	4.30	3104	1.03	1062	5.32	8938	51
10	**9.64 4423**	4.28	**9.95 3042**	1.03	**9.69 1381**	5.32	**10.30 8619**	**50**
11	4680	4.27	2980	1.03	1700	5.32	8300	49
12	4936	4.28	2918	1.05	2019	5.32	7981	48
13	5193	4.28	2855	1.03	2338	5.30	7662	47
14	5450	4.27	2793	1.03	2656	5.32	7344	46
15	**9.64 5706**	4.27	**9.95 2731**	1.03	**9.69 2975**	5.30	**10.30 7025**	**45**
16	5962	4.27	2669	1.05	3293	5.32	6707	44
17	6218	4.27	2606	1.03	3612	5.30	6388	43
18	6474	4.25	2544	1.05	3930	5.30	6070	42
19	6729	4.25	2481	1.03	4228	5.30	5752	41
20	**9.64 6984**	4.27	**9.95 2419**	1.05	**9.69 4566**	5.28	**10.30 5434**	**40**
21	7240	4.23	2356	1.03	4883	5.30	5117	39
22	7494	4.25	2294	1.05	5201	5.28	4799	38
23	7749	4.25	2231	1.05	5518	5.30	4482	37
24	8004	4.23	2168	1.03	5836	5.28	4164	36
25	**9.64 8258**	4.23	**9.95 2106**	1.05	**9.69 6153**	5.28	**10.30 3847**	**35**
26	8512	4.23	2043	1.05	6470	5.28	3530	34
27	8766	4.23	1980	1.05	6787	5.27	3213	33
28	9020	4.23	1917	1.05	7103	5.28	2897	32
29	9274	4.22	1854	1.05	7420	5.27	2580	31
30	**9.64 9527**	4.23	**9.95 1791**	1.05	**9.69 7736**	5.28	**10.30 2264**	**30**
31	.64 9781	4.22	1728	1.05	8053	5.27	1947	29
32	.65 0034	4.22	1665	1.05	8369	5.27	1631	28
33	0287	4.20	1602	1.05	8685	5.27	1315	27
34	0539	4.22	1539	1.05	9001	5.25	0999	26
35	**9.65 0792**	4.20	**9.95 1476**	1.07	**9.69 9316**	5.27	**10.30 0684**	**25**
36	1044	4.22	1412	1.05	9632	5.25	0368	24
37	1297	4.20	1349	1.05	.69 9947	5.27	.30 0053	23
38	1549	4.18	1286	1.07	.70 0263	5.25	.29 9737	22
39	1800	4.20	1222	1.05	0578	5.25	9422	21
40	**9.65 2052**	4.20	**9.95 1159**	1.05	**9.70 0893**	5.25	**10.29 9107**	**20**
41	2304	4.18	1096	1.07	1208	5.25	8792	19
42	2555	4.18	1032	1.07	1523	5.23	8477	18
43	2806	4.18	0968	1.05	1837	5.25	8163	17
44	3057	4.18	0905	1.07	2152	5.23	7848	16
45	**9.65 3308**	4.17	**9.95 0841**	1.05	**9.70 2466**	5.25	**10.29 7534**	**15**
46	3558	4.17	0778	1.07	2781	5.23	7219	14
47	3808	4.18	0714	1.07	3095	5.23	6905	13
48	4059	4.17	0650	1.07	3409	5.22	6591	12
49	4309	4.15	0586	1.07	3722	5.23	6278	11
50	**9.65 4558**	4.17	**9.95 0522**	1.07	**9.70 4036**	5.23	**10.29 5964**	**10**
51	4808	4.17	0458	1.07	4350	5.22	5650	9
52	5058	4.15	0394	1.07	4663	5.22	5337	8
53	5307	4.15	0330	1.07	4976	5.23	5024	7
54	5556	4.15	0266	1.07	5290	5.22	4710	6
55	**9.65 5805**	4.15	**9.95 0202**	1.07	**9.70 5603**	5.22	**10.29 4397**	**5**
56	6054	4.13	0138	1.07	5916	5.20	4084	4
57	6302	4.15	0074	1.07	6228	5.22	3772	3
58	6551	4.13	.95 0010	1.08	6541	5.22	3459	2
59	6799	4.13	.94 9945	1.07	6854	5.20	3146	1
60	**9.65 7047**		**9.94 9881**		**9.70 7166**		**10.29 2834**	**0**
′	Cosine.	D.1″.	Sine.	D.1″.	Cotang.	D.1″.	Tang.	′

116° **63°**

27° **Table VIII—Continued** 152°

′	Sine.	D.1″.	Cosine.	D.1″.	Tang.	D.1″.	Cotang.	′
0	9.65 7047	4.13	9.94 9881	1.08	9.70 7166	5.20	10.29 2834	60
1	7295	4.12	9816	1.07	7478	5.20	2522	59
2	7542	4.13	9752	1.07	7790	5.20	2210	58
3	7790	4.12	9688	1.08	8102	5.20	1898	57
4	8037	4.12	9623	1.08	8414	5.20	1586	56
5	9.65 8284	4.12	9.94 9558	1.07	9.70 8726	5.18	10.29 1274	55
6	8531	4.12	9494	1.08	9037	5.20	0963	54
7	8778	4.12	9429	1.08	9349	5.18	0651	53
8	9025	4.10	9364	1.07	9660	5.18	0340	52
9	9271	4.10	9300	1.08	9971	5.18	.29 0029	51
10	9.65 9517	4.10	9.94 9235	1.08	9.71 0282	5.18	10.28 9718	50
11	.65 9763	4.10	9170	1.08	0593	5.18	9407	49
12	.66 0009	4.10	9105	1.08	0904	5.18	9096	48
13	0255	4.10	9040	1.08	1215	5.17	8785	47
14	0501	4.08	8975	1.08	1525	5.18	8475	46
15	9.66 0746	4.08	9.94 8910	1.08	9.71 1836	5.17	10.28 8164	45
16	0991	4.08	8845	1.08	2146	5.17	7854	44
17	1236	4.08	8780	1.08	2456	5.17	7544	43
18	1481	4.08	8715	1.08	2766	5.17	7234	42
19	1726	4.07	8650	1.10	3076	5.17	6924	41
20	9.66 1970	4.07	9.94 8584	1.08	9.71 3386	5.17	10.28 6614	40
21	2214	4.08	8519	1.08	3696	5.15	6304	39
22	2459	4.07	8454	1.10	4005	5.15	5995	38
23	2703	4.05	8388	1.08	4314	5.17	5686	37
24	2946	4.07	8323	1.10	4624	5.15	5376	36
25	9.66 3190	4.05	9.94 8257	1.08	9.71 4933	5.15	10.28 5067	35
26	3433	4.07	8192	1.10	5242	5.15	4758	34
27	3677	4.05	8126	1.10	5551	5.15	4449	33
28	3920	4.05	8060	1.08	5860	5.13	4140	32
29	4163	4.05	7995	1.10	6168	5.15	3832	31
30	9.66 4406	4.03	9.94 7929	1.10	9.71 6477	5.13	10.28 3523	30
31	4648	4.05	7863	1.10	6785	5.13	3215	29
32	4891	4.03	7797	1.10	7093	5.13	2907	28
33	5133	4.03	7731	1.10	7401	5.13	2599	27
34	5375	4.03	7665	1.08	7709	5.13	2291	26
35	9.66 5617	4.03	9.94 7600	1.12	9.71 8017	5.13	10.28 1983	25
36	5850	4.02	7533	1.10	8325	5.13	1675	24
37	6100	4.03	7467	1.10	8633	5.12	1367	23
38	6342	4.02	7401	1.10	8940	5.13	1060	22
39	6583	4.02	7335	1.10	9248	5.12	0752	21
40	9.66 6824	4.02	9.94 7269	1.10	9.71 9555	5.12	10.28 0445	20
41	7065	4.00	7203	1.12	.71 9862	5.12	.28 0138	19
42	7305	4.02	7136	1.10	.72 0169	5.12	.27 9831	18
43	7546	4.00	7070	1.10	0476	5.12	9524	17
44	7786	4.02	7004	1.12	0783	5.10	9217	16
45	9.66 8027	4.00	9.94 6937	1.10	9.72 1089	5.12	10.27 8911	15
46	8267	3.98	6871	1.12	1396	5.10	8604	14
47	8506	4.00	6804	1.10	1702	5.12	8298	13
48	8746	4.00	6738	1.12	2009	5.10	7991	12
49	8986	3.98	6671	1.12	2315	5.10	7685	11
50	9.66 9225	3.98	9.94 6604	1.10	9.72 2621	5.10	10.27 7379	10
51	9464	3.98	6538	1.12	2927	5.08	7073	9
52	9703	3.98	6471	1.12	3232	5.10	6768	8
53	.66 9942	3.98	6404	1.12	3538	5.10	6462	7
54	.67 0181	3.97	6337	1.12	3844	5.08	6156	6
55	9.67 0419	3.98	9.94 6270	1.12	9.72 4149	5.08	10.27 5851	5
56	0658	3.97	6203	1.12	4454	5.10	5546	4
57	0896	3.97	6136	1.12	4760	5.08	5240	3
58	1134	3.97	6069	1.12	5065	5.08	4935	2
59	1372	3.95	6002	1.12	5370	5.07	4630	1
60	9.67 1609		9.94 5935		9.72 5674		10.27 4326	0
′	Cosine.	D.1″.	Sine.	D.1″.	Cotang.	D.1″.	Tang.	′

117° 62°

28° **Table VIII—Continued** **151°**

′	Sine.	D.1″.	Cosine.	D.1″.	Tang.	D.1″.	Cotang.	′
0	**9.67 1609**	3.97	**9.94 5935**	1.12	**9.72 5674**	5.08	**10.27 4326**	**60**
1	1847	3.95	5868	1.13	5979	5.08	4021	59
2	2084	3.95	5800	1.12	6284	5.07	3716	58
3	2321	3.95	5733	1.12	6588	5.07	3412	57
4	2558	3.95	5666	1.13	6892	5.05	3108	56
5	**9.67 2795**	3.95	**9.94 5598**	1.12	**9.72 7197**	5.07	**10.27 2803**	**55**
6	3032	3.93	5531	1.12	7501	5.07	2499	54
7	3268	3.95	5464	1.13	7805	5.07	2195	53
8	3505	3.93	5396	1.13	8109	5.05	1891	52
9	3741	3.93	5328	1.12	8412	5.07	1588	51
10	**9.67 3977**	3.93	**9.94 5261**	1.13	**9.72 8716**	5.07	**10.27 1284**	**50**
11	4213	3.92	5193	1.13	9020	5.05	0980	49
12	4448	3.93	5125	1.12	9323	5.05	0677	48
13	4684	3.92	5058	1.13	9626	5.05	0374	47
14	4919	3.93	4990	1.13	.72 9929	5.07	.27 0071	46
15	**9.67 5155**	3.92	**9.94 4922**	1.13	**9.73 0233**	5.03	**10.26 9767**	**45**
16	5390	3.90	4854	1.13	0535	5.05	9465	44
17	5624	3.92	4786	1.13	0838	5.05	9162	43
18	5859	3.92	4718	1.13	1141	5.05	8859	42
19	6094	3.90	4650	1.13	1444	5.03	8556	41
20	**9.67 6328**	3.90	**9.94 4582**	1.13	**9.73 1746**	5.03	**10.26 8254**	**40**
21	6562	3.90	4514	1.13	2048	5.05	7952	39
22	6796	3.90	4446	1.15	2351	5.03	7649	38
23	7030	3.90	4377	1.13	2653	5.03	7347	37
24	7264	3.90	4309	1.13	2955	5.03	7045	36
25	**9.67 7498**	3.88	**9.94 4241**	1.15	**9.73 3257**	5.02	**10.26 6743**	**35**
26	7731	3.88	4172	1.13	3558	5.03	6442	34
27	7964	3.88	4104	1.13	3860	5.03	6140	33
28	8197	3.88	4036	1.15	4162	5.02	5838	32
29	8430	3.88	3967	1.13	4463	5.02	5537	31
30	**9.67 8663**	3.87	**9.94 3899**	1.15	**9.73 4764**	5.03	**10.26 5236**	**30**
31	8895	3.88	3830	1.15	5066	5.02	4934	29
32	9128	3.87	3761	1.13	5367	5.02	4633	28
33	9360	3.87	3693	1.15	5668	5.02	4332	27
34	9592	3.87	3624	1.15	5969	5.00	4031	26
35	**9.67 9824**	3.87	**9.94 3555**	1.15	**9.73 6269**	5.02	**10.26 3731**	**25**
36	.68 0056	3.87	3486	1.15	6570	5.00	3430	24
37	0288	3.85	3417	1.15	6870	5.02	3130	23
38	0519	3.85	3348	1.15	7171	5.00	2829	22
39	0750	3.87	3279	1.15	7471	5.00	2529	21
40	**9.68 0982**	3.85	**9.94 3210**	1.15	**9.73 7771**	5.00	**10.26 2229**	**20**
41	1213	3.83	3141	1.15	8071	5.00	1929	19
42	1443	3.85	3072	1.15	8371	5.00	1629	18
43	1674	3.85	3003	1.15	8671	5.00	1329	17
44	1905	3.83	2934	1.17	8971	5.00	1029	16
45	**9.68 2135**	3.83	**9.94 2864**	1.15	**9.73 9271**	4.98	**10.26 0729**	**15**
46	2365	3.83	2795	1.15	9570	5.00	0430	14
47	2595	3.83	2726	1.17	.73 9870	4.98	.26 0130	13
48	2825	3.83	2656	1.15	.74 0169	4.98	.25 9831	12
49	3055	3.82	2587	1.17	0468	4.98	9532	11
50	**9.68 3284**	3.83	**9.94 2517**	1.15	**9.74 0767**	4.98	**10.25 9233**	**10**
51	3514	3.82	2448	1.17	1066	4.98	8934	9
52	3743	3.82	2378	1.17	1365	4.98	8635	8
53	3972	3.82	2308	1.15	1664	4.97	8336	7
54	4201	3.82	2239	1.17	1962	4.98	8038	6
55	**9.68 4430**	3.80	**9.94 2169**	1.17	**9.74 2261**	4.97	**10.25 7739**	**5**
56	4658	3.82	2099	1.17	2559	4.98	7441	4
57	4887	3.80	2029	1.17	2858	4.97	7142	3
58	5115	3.80	1959	1.17	3156	4.97	6844	2
59	5343	3.80	1889	1.17	3454	4.97	6546	1
60	**9.68 5571**		**9.94 1819**		**9.74 3752**		**10.25 6248**	**0**
′	Cosine.	D.1″.	Sine.	D.1″.	Cotang.	D.1″.	Tang.	′

118° **61°**

29° **Table VIII—Continued** **150°**

′	Sine.	D.1″.	Cosine.	D.1″.	Tang.	D.1″.	Cotang.	′
0	9.68 **5571**	3.80	9.94 **1819**	1.17	9.74 **3752**	4.97	10.25 **6248**	**60**
1	5799	3.80	1749	1.17	4050	4.97	5950	59
2	6027	3.78	1679	1.17	4348	4.95	5652	58
3	6254	3.80	1609	1.17	4645	4.97	5355	57
4	6482	3.78	1539	1.17	4943	4.95	5057	56
5	9.68 **6709**	3.78	9.94 **1469**	1.18	9.74 **5240**	4.97	10.25 **4760**	**55**
6	6936	3.78	1398	1.17	5538	4.95	4462	54
7	7163	3.77	1328	1.17	5835	4.95	4165	53
8	7389	3.78	1258	1.18	6132	4.95	3868	52
9	7616	3.78	1187	1.17	6429	4.95	3571	51
10	9.68 **7843**	3.77	9.94 **1117**	1.18	9.74 **6726**	4.95	10.25 **3274**	**50**
11	8069	3.77	1046	1.18	7023	4.93	2977	49
12	8295	3.77	0975	1.17	7319	4.95	2681	48
13	8521	3.77	0905	1.18	7616	4.95	2384	47
14	8747	3.75	0834	1.18	7913	4.93	2087	46
15	9.68 **8972**	3.77	9.94 **0763**	1.17	9.74 **8209**	4.93	10.25 **1791**	**45**
16	9198	3.75	0693	1.18	8505	4.93	1495	44
17	9423	3.75	0622	1.18	8801	4.93	1199	43
18	9648	3.75	0551	1.18	9097	4.93	0903	42
19	.68 9873	3.75	0480	1.18	9393	4.93	0607	41
20	9.69 **0098**	3.75	9.94 **0409**	1.18	9.74 **9689**	4.93	10.25 **0311**	**40**
21	0323	3.75	0338	1.18	.74 9985	4.93	.25 0015	39
22	0548	3.73	0267	1.18	.75 0281	4.92	.24 9719	38
23	0772	3.73	0196	1.18	0576	4.93	9424	37
24	0996	3.73	0125	1.18	0872	4.92	9128	36
25	9.69 **1220**	3.73	9.94 **0054**	1.20	9.75 **1167**	4.92	10.24 **8833**	**35**
26	1444	3.73	.93 9982	1.18	1462	4.92	8538	34
27	1668	3.73	9911	1.18	1757	4.92	8243	33
28	1892	3.72	9840	1.20	2052	4.92	7948	32
29	2115	3.73	9768	1.18	2347	4.92	7653	31
30	9.69 **2339**	3.72	9.93 **9697**	1.20	9.75 **2642**	4.92	10.24 **7358**	**30**
31	2562	3.72	9625	1.18	2937	4.90	7063	29
32	2785	3.72	9554	1.20	3231	4.92	6769	28
33	3008	3.72	9482	1.20	3526	4.90	6474	27
34	3231	3.70	9410	1.18	3820	4.92	6180	26
35	9.69 **3453**	3.72	9.93 **9339**	1.20	9.75 **4115**	4.90	10.24 **5885**	**25**
36	3676	3.70	9267	1.20	4409	4.90	5591	24
37	3898	3.70	9195	1.20	4703	4.90	5207	23
38	4120	3.70	9123	1.18	4997	4.90	5003	22
39	4342	3.70	9052	1.20	5291	4.90	4709	21
40	9.69 **4564**	3.70	[illegible]	1.20	[illegible]	4.88	[illegible]	**20**
41	4786	3.68	8908	1.20	5878	4.90	4122	19
42	5007	3.70	8836	1.22	6172	4.88	3828	18
43	5229	3.68	8763	1.20	6465	4.90	3535	17
44	5450	3.68	8691	1.20	6759	4.88	3241	16
45	9.69 **5671**	3.68	9.93 **8619**	1.20	9.75 **7052**	4.88	10.24 **2948**	**15**
46	5892	3.68	8547	1.20	7345	4.88	2655	14
47	6113	3.68	8475	1.22	7638	4.88	2362	13
48	6334	3.67	8402	1.20	7931	4.88	2069	12
49	6554	3.68	8330	1.20	8224	4.88	1776	11
50	9.69 **6775**	3.67	9.93 **8258**	1.22	9.75 **8517**	4.88	10.24 **1483**	**10**
51	6995	3.67	8185	1.20	8810	4.87	1190	9
52	7215	3.67	8113	1.22	9102	4.88	0898	8
53	7435	3.65	8040	1.22	9395	4.87	0605	7
54	7654	3.67	7967	1.20	9687	4.87	0313	6
55	9.69 **7874**	3.67	9.93 **7895**	1.22	9.75 **9979**	4.88	10.24 **0021**	**5**
56	8094	3.65	7822	1.22	.76 0272	4.87	.23 9728	4
57	8313	3.65	7749	1.22	0564	4.87	9436	3
58	8532	3.65	7676	1.20	0856	4.87	9144	2
59	8751	3.65	7604	1.22	1148	4.85	8852	1
60	9.69 **8970**		9.93 **7531**		9.76 **1439**		10.23 **8561**	**0**
′	Cosine.	D.1″.	Sine.	D.1″.	Cotang.	D.1″.	Tang.	′

119° **60°**

30° **Table VIII—Continued** 149°

′	Sine.	D.1″.	Cosine.	D.1″.	Tang.	D.1″.	Cotang.	′
0	9.69 8970	3.65	9.93 7531	1.22	9.76 1439	4.87	10.23 8561	60
1	9189	3.63	7458	1.22	1731	4.87	8269	59
2	9407	3.65	7385	1.22	2023	4.85	7977	58
3	9626	3.63	7312	1.23	2314	4.87	7686	57
4	.69 9844	3.63	7238	1.22	2606	4.85	7394	56
5	9.70 0062	3.63	9.93 7165	1.22	9.76 2897	4.85	10.23 7103	55
6	0280	3.63	7092	1.22	3188	4.85	6812	54
7	0498	3.63	7019	1.22	3479	4.85	6521	53
8	0716	3.62	6946	1.23	3770	4.85	6230	52
9	0933	3.63	6872	1.22	4061	4.85	5939	51
10	9.70 1151	3.62	9.93 6799	1.23	9.76 4352	4.85	10.23 5648	50
11	1368	3.62	6725	1.22	4643	4.83	5357	49
12	1585	3.62	6652	1.23	4933	4.85	5067	48
13	1802	3.62	6578	1.22	5224	4.83	4776	47
14	2019	3.62	6505	1.23	5514	4.85	4486	46
15	9.70 2236	3.60	9.93 6431	1.23	9.76 5805	4.83	10.23 4195	45
16	2452	3.62	6357	1.22	6095	4.83	3905	44
17	2669	3.60	6284	1.23	6385	4.83	3615	43
18	2885	3.60	6210	1.23	6675	4.83	3325	42
19	3101	3.60	6136	1.23	6965	4.83	3035	41
20	9.70 3317	3.60	9.93 6062	1.23	9.76 7255	4.83	10.23 2745	40
21	3533	3.60	5988	1.23	7545	4.82	2455	39
22	3749	3.58	5914	1.23	7834	4.83	2166	38
23	3964	3.58	5840	1.23	8124	4.83	1876	37
24	4179	3.60	5766	1.23	8414	4.82	1586	36
25	9.70 4395	3.58	9.93 5692	1.23	9.76 8703	4.82	10.23 1297	35
26	4610	3.58	5618	1.25	8992	4.82	1008	34
27	4825	3.58	5543	1.23	9281	4.83	0719	33
28	5040	3.57	5469	1.23	9571	4.82	0429	32
29	5254	3.58	5395	1.25	.76 9860	4.80	.23 0140	31
30	9.70 5469	3.57	9.93 5320	1.23	9.77 0148	4.82	10.22 9852	30
31	5683	3.58	5246	1.25	0437	4.82	9563	29
32	5898	3.57	5171	1.23	0726	4.82	9274	28
33	6112	3.57	5097	1.25	1015	4.80	8985	27
34	6326	3.55	5022	1.23	1303	4.82	8697	26
35	9.70 6539	3.57	9.93 4948	1.25	9.77 1592	4.80	10.22 8408	25
36	6753	3.57	4873	1.25	1880	4.80	8120	24
37	6967	3.55	4798	1.25	2168	4.82	7832	23
38	7180	3.55	4723	1.23	2457	4.80	7543	22
39	7393	3.55	4649	1.25	2745	4.80	7255	21
40	9.70 7606	3.55	9.93 4574	1.25	9.77 3033	4.80	10.22 6967	20
41	7819	3.55	4499	1.25	3321	4.78	6679	19
42	8032	3.55	4424	1.25	3608	4.80	6392	18
43	8245	3.55	4349	1.25	3896	4.80	6104	17
44	8458	3.53	4274	1.25	4184	4.78	5816	16
45	9.70 8670	3.53	9.93 4199	1.27	9.77 4471	4.80	10.22 5529	15
46	8882	3.53	4123	1.25	4759	4.78	5241	14
47	9094	3.53	4048	1.25	5046	4.78	4954	13
48	9306	3.53	3973	1.25	5333	4.80	4667	12
49	9518	3.53	3898	1.27	5621	4.78	4379	11
50	9.70 9730	3.52	9.93 3822	1.25	9.77 5908	4.78	10.22 4092	10
51	.70 9941	3.53	3747	1.27	6195	4.78	3805	9
52	.71 0153	3.52	3671	1.25	6482	4.77	3518	8
53	0364	3.52	3596	1.27	6768	4.78	3232	7
54	0575	3.52	3520	1.25	7055	4.78	2945	6
55	9.71 0786	3.52	9.93 3445	1.27	9.77 7342	4.77	10.22 2658	5
56	0997	3.52	3369	1.27	7628	4.78	2372	4
57	1208	3.52	3293	1.27	7915	4.77	2085	3
58	1419	3.50	3217	1.27	8201	4.78	1799	2
59	1629	3.50	3141	1.25	8488	4.77	1512	1
60	9.71 1839		9.93 3066		9.77 8774		10.22 1226	0
′	Cosine.	D.1″.	Sine.	D.1″.	Cotang.	D.1″.	Tang.	′

120° 59°

31° **Table VIII—Continued** **148°**

′	Sine.	D.1″.	Cosine.	D.1″.	Tang.	D.1″.	Cotang.	′
0	**9.71 1839**	3.52	**9.93 3066**	1.27	**9.77 8774**	4.77	**10.22 1226**	**60**
1	2050	3.50	2990	1.27	9060	4.77	0940	59
2	2260	3.48	2914	1.27	9346	4.77	0654	58
3	2469	3.50	2838	1.27	9632	4.77	0368	57
4	2679	3.50	2762	1.28	.77 9918	4.75	.22 0082	56
5	**9.71 2889**	3.48	**9.93 2685**	1.27	**9.78 0203**	4.77	**10.21 9797**	**55**
6	3098	3.50	2609	1.27	0489	4.77	9511	54
7	3308	3.48	2533	1.27	0775	4.75	9225	53
8	3517	3.48	2457	1.28	1060	4.77	8940	52
9	3726	3.48	2380	1.27	1346	4.75	8654	51
10	**9.71 3935**	3.48	**9.93 2304**	1.27	**9.78 1631**	4.75	**10.21 8369**	**50**
11	4144	3.47	2228	1.28	1916	4.75	8084	49
12	4352	3.48	2151	1.27	2201	4.75	7799	48
13	4561	3.47	2075	1.28	2486	4.75	7514	47
14	4769	3.48	1998	1.28	2771	4.75	7229	46
15	**9.71 4978**	3.47	**9.93 1921**	1.27	**9.78 3056**	4.75	**10.21 6944**	**45**
16	5186	3.47	1845	1.28	3341	4.75	6659	44
17	5394	3.47	1768	1.28	3626	4.73	6374	43
18	5602	3.45	1691	1.28	3910	4.75	6090	42
19	5809	3.47	1614	1.28	4195	4.73	5805	41
20	**9.71 6017**	3.45	**9.93 1537**	1.28	**9.78 4479**	4.75	**10.21 5521**	**40**
21	6224	3.47	1460	1.28	4764	4.73	5236	39
22	6432	3.45	1383	1.28	5048	4.73	4952	38
23	6639	3.45	1306	1.28	5332	4.73	4668	37
24	6846	3.45	1229	1.28	5616	4.73	4384	36
25	**9.71 7053**	3.43	**9.93 1152**	1.28	**9.78 5900**	4.73	**10.21 4100**	**35**
26	7259	3.45	1075	1.28	6184	4.73	3816	34
27	7466	3.45	0998	1.28	6468	4.73	3532	33
28	7673	3.43	0921	1.30	6752	4.73	3248	32
29	7879	3.43	0843	1.28	7036	4.72	2964	31
30	**9.71 8085**	3.43	**9.93 0766**	1.30	**9.78 7319**	4.73	**10.21 2681**	**30**
31	8201	3.43	0688	1.28	7603	4.72	2397	29
32	8497	3.43	0611	1.30	7886	4.73	2114	28
33	8703	3.43	0533	1.28	8170	4.72	1830	27
34	8909	3.42	0456	1.30	8453	4.72	1547	26
35	**9.71 9114**	3.43	**9.93 0378**	1.30	**9.78 8736**	4.72	**10.21 1264**	**25**
36	9320	3.42	0300	1.28	9019	4.72	0981	24
37	9525	3.42	0223	1.30	9302	4.72	0698	23
38	9730	3.42	0145	1.30	9585	4.72	0415	22
39	.71 9935	3.42	.93 0067	1.30	.78 9868	4.72	.21 0132	21
40	**9.72 0140**	3.42	**9.92 9989**	1.30	**9.79 0151**	4.72	**10.20 9849**	**20**
41	0345	3.40	9911	1.30	0434	4.70	9566	19
42	0549	3.42	9833	1.30	0716	4.72	9284	18
43	0754	3.40	9755	1.30	0999	4.70	9001	17
44	0958	3.40	9677	1.30	1281	4.70	8719	16
45	**9.72 1162**	3.40	**9.92 9599**	1.30	**9.79 1563**	4.72	**10.20 8437**	**15**
46	1366	3.40	9521	1.32	1846	4.70	8154	14
47	1570	3.40	9442	1.30	2128	4.70	7872	13
48	1774	3.40	9364	1.30	2410	4.70	7590	12
49	1978	3.38	9286	1.32	2692	4.70	7308	11
50	**9.72 2181**	3.40	**9.92 9207**	1.30	**9.79 2974**	4.70	**10.20 7026**	**10**
51	2385	3.38	9129	1.32	3256	4.70	6744	9
52	2588	3.38	9050	1.30	3538	4.68	6462	8
53	2791	3.38	8972	1.32	3819	4.70	6181	7
54	2994	3.38	8893	1.30	4101	4.70	5899	6
55	**9.72 3197**	3.38	**9.92 8815**	1.32	**9.79 4383**	4.68	**10.20 5617**	**5**
56	3400	3.38	8736	1.32	4664	4.70	5336	4
57	3603	3.37	8657	1.32	4946	4.68	5054	3
58	3805	3.37	8578	1.32	5227	4.68	4773	2
59	4007	3.38	8499	1.32	5508	4.68	4492	1
60	**9.72 4210**		**9.92 8420**		**9.79 5789**		**10.20 4211**	**0**
′	Cosine.	D.1″.	Sine.	D.1″.	Cotang.	D.1″.	Tang.	′

121° **58°**

32° **Table VIII—Continued** 147°

′	Sine.	D.1″.	Cosine.	D.1″.	Tang.	D.1″.	Cotang.	′
0	**9.72 4210**	3.37	**9.92 8420**	1.30	**9.79 5789**	4.68	**10.20 4211**	**60**
1	4412	3.37	8342	1.32	6070	4.68	3930	59
2	4614	3.37	8263	1.33	6351	4.68	3649	58
3	4816	3.35	8183	1.32	6632	4.68	3368	57
4	5017	3.37	8104	1.32	6913	4.68	3087	56
5	**9.72 5219**	3.35	**9.92 8025**	1.32	**9.79 7194**	4.67	**10.20 2806**	**55**
6	5420	3.37	7946	1.32	7474	4.68	2526	54
7	5622	3.35	7867	1.33	7755	4.68	2245	53
8	5823	3.35	7787	1.32	8036	4.67	1964	52
9	6024	3.35	7708	1.32	8316	4.67	1684	51
10	**9.72 6225**	3.35	**9.92 7629**	1.33	**9.79 8596**	4.68	**10.20 1404**	**50**
11	6426	3.33	7549	1.32	8877	4.67	1123	49
12	6626	3.35	7470	1.33	9157	4.67	0843	48
13	6827	3.33	7390	1.33	9437	4.67	0563	47
14	7027	3.35	7310	1.32	9717	4.67	0283	46
15	**9.72 7228**	3.33	**9.92 7231**	1.33	**9.79 9997**	4.67	**10.20 0003**	**45**
16	7428	3.33	7151	1.33	.80 0277	4.67	.19 9723	44
17	7628	3.33	7071	1.33	0557	4.65	9443	43
18	7828	3.32	6991	1.33	0836	4.67	9164	42
19	8027	3.33	6911	1.33	1116	4.67	8884	41
20	**9.72 8227**	3.33	**9.92 6831**	1.33	**9.80 1396**	4.65	**10.19 8604**	**40**
21	8427	3.32	6751	1.33	1675	4.67	8325	39
22	8626	3.32	6671	1.33	1955	4.65	8045	38
23	8825	3.32	6591	1.33	2234	4.65	7766	37
24	9024	3.32	6511	1.33	2513	4.65	7487	36
25	**9.72 9223**	3.32	**9.92 6431**	1.33	**9.80 2792**	4.67	**10.19 7208**	**35**
26	9422	3.32	6351	1.35	3072	4.65	6928	34
27	9621	3.32	6270	1.33	3351	4.65	6649	33
28	.72 9820	3.30	6190	1.33	3630	4.65	6370	32
29	.73 0018	3.32	6110	1.35	3909	4.63	6091	31
30	**9.73 0217**	3.30	**9.92 6029**	1.33	**9.80 4187**	4.65	**10.19 5813**	**30**
31	0415	3.30	5949	1.35	4466	4.65	5534	29
32	0613	3.30	5868	1.33	4745	4.63	5255	28
33	0811	3.30	5788	1.35	5023	4.65	4977	27
34	1009	3.28	5707	1.35	5302	4.63	4698	26
35	**9.73 1206**	3.30	**9.92 5626**	1.35	**9.80 5580**	4.65	**10.19 4420**	**25**
36	1404	3.30	5545	1.33	5859	4.63	4141	24
37	1602	3.28	5465	1.35	6137	4.63	3863	23
38	1799	3.28	5384	1.35	6415	4.63	3585	22
39	1996	3.28	5303	1.35	6693	4.63	3307	21
40	**9.73 2193**	3.28	**9.92 5222**	1.35	**9.80 6971**	4.63	**10.19 3029**	**20**
41	2390	3.28	5141	1.35	7249	4.63	2751	19
42	2587	3.28	5060	1.35	7527	4.63	2473	18
43	2784	3.27	4979	1.37	7805	4.63	2195	17
44	2980	3.28	4897	1.35	8083	4.63	1917	16
45	**9.73 3177**	3.27	**9.92 4816**	1.35	**9.80 8361**	4.62	**10.19 1639**	**15**
46	3373	3.27	4735	1.35	8638	4.63	1362	14
47	3569	3.27	4654	1.37	8916	4.62	1084	13
48	3765	3.27	4572	1.35	9193	4.63	0807	12
49	3961	3.27	4491	1.37	9471	4.62	0529	11
50	**9.73 4157**	3.27	**9.92 4409**	1.35	**9.80 9748**	4.62	**10.19 0252**	**10**
51	4353	3.27	4328	1.37	.81 0025	4.62	.18 9975	9
52	4549	3.25	4246	1.37	0302	4.63	9698	8
53	4744	3.25	4164	1.35	0580	4.62	9420	7
54	4939	3.27	4083	1.37	0857	4.62	9143	6
55	**9.73 5135**	3.25	**9.92 4001**	1.37	**9.81 1134**	4.60	**10.18 8866**	**5**
56	5330	3.25	3919	1.37	1410	4.62	8590	4
57	5525	3.23	3837	1.37	1687	4.62	8313	3
58	5719	3.25	3755	1.37	1964	4.62	8036	2
59	5914	3.25	3673	1.37	2241	4.60	7759	1
60	**9.73 6109**		**9.92 3591**		**9.81 2517**		**10.18 7483**	**0**
′	Cosine.	D.1″.	Sine.	D.1″.	Cotang.	D.1″.	Tang.	′

122° 57°

33° **Table VIII—Continued** 146°

′	Sine.	D.1″.	Cosine.	D.1″.	Tang.	D.1″.	Cotang.	′
0	**9.73 6109**	3.23	**9.92 3591**	1.37	**9.81 2517**	4.62	**10.18 7483**	**60**
1	6303	3.25	3509	1.37	2794	4.60	7206	59
2	6498	3.23	3427	1.37	3070	4.62	6930	58
3	6692	3.23	3345	1.37	3347	4.60	6653	57
4	6886	3.23	3263	1.37	3623	4.60	6377	56
5	**9.73 7080**	3.23	**9.92 3181**	1.38	**9.81 3899**	4.62	**10.18 6101**	**55**
6	7274	3.22	3098	1.37	4176	4.60	5824	54
7	7467	3.23	3016	1.38	4452	4.60	5548	53
8	7661	3.23	2933	1.37	4728	4.60	5272	52
9	7855	3.22	2851	1.38	5004	4.60	4996	51
10	**9.73 8048**	3.22	**9.92 2768**	1.37	**9.81 5280**	4.58	**10.18 4720**	**50**
11	8241	3.22	2686	1.38	5555	4.60	4445	49
12	8434	3.22	2603	1.38	5831	4.60	4169	48
13	8627	3.22	2520	1.37	6107	4.58	3893	47
14	8820	3.22	2438	1.38	6382	4.60	3618	46
15	**9.73 9013**	3.22	**9.92 2355**	1.38	**9.81 6658**	4.58	**10.18 3342**	**45**
16	9206	3.20	2272	1.38	6933	4.60	3067	44
17	9398	3.20	2189	1.38	7209	4.58	2791	43
18	9590	3.22	2106	1.38	7484	4.58	2516	42
19	9783	3.20	2023	1.38	7759	4.60	2241	41
20	**9.73 9975**	3.20	**9.92 1940**	1.38	**9.81 8035**	4.58	**10.18 1965**	**40**
21	.74 0167	3.20	1857	1.38	8310	4.58	1690	39
22	0359	3.18	1774	1.38	8585	4.58	1415	38
23	0550	3.20	1691	1.40	8860	4.58	1140	37
24	0742	3.20	1607	1.38	9135	4.58	0865	36
25	**9.74 0934**	3.18	**9.92 1524**	1.38	**9.81 9410**	4.57	**10.18 0590**	**35**
26	1125	3.18	1441	1.40	9684	4.58	0316	34
27	1316	3.20	1357	1.38	.81 9959	4.58	.18 0041	33
28	1508	3.18	1274	1.40	.82 0234	4.57	.17 9766	32
29	1699	3.17	1190	1.38	0508	4.58	9492	31
30	**9.74 1889**	3.18	**9.92 1107**	1.40	**9.82 0783**	4.57	**10.17 9217**	**30**
31	2080	3.18	1023	1.40	1057	4.58	8943	29
32	2271	3.18	0939	1.38	1332	4.57	8668	28
33	2462	3.17	0856	1.40	1606	4.57	8394	27
34	2652	3.17	0772	1.40	1880	4.57	8120	26
35	**9.74 2842**	3.18	**9.92 0688**	1.40	**9.82 2154**	4.58	**10.17 7846**	**25**
36	3033	3.17	0604	1.40	2429	4.57	7571	24
37	3223	3.17	0520	1.40	2703	4.57	7297	23
38	3413	3.15	0436	1.40	2977	4.57	7023	22
39	3602	3.17	0352	1.40	3251	4.55	6749	21
40	**9.74 3792**	3.17	**9.92 0268**	1.40	[illegible]	4.57	[illegible]	**20**
41	3982	3.15	0184	1.42	3798	4.57	6202	19
42	4171	3.17	0099	1.40	4072	4.55	5928	18
43	4361	3.15	.92 0015	1.40	4345	4.57	5655	17
44	4550	3.15	.91 9931	1.42	4619	4.57	5381	16
45	**9.74 4739**	3.15	**9.91 9846**	1.40	**9.82 4893**	4.55	**10.17 5107**	**15**
46	4928	3.15	9762	1.42	5166	4.55	4834	14
47	5117	3.15	9677	1.40	5439	4.57	4561	13
48	5306	3.13	9593	1.42	5713	4.55	4287	12
49	5494	3.15	9508	1.40	5986	4.55	4014	11
50	**9.74 5683**	3.13	**9.91 9424**	1.42	**9.82 6254**	4.55	**10.17 3741**	**10**
51	5871	3.15	9339	1.42	6532	4.55	3468	9
52	6060	3.13	9254	1.42	6805	4.55	3195	8
53	6248	3.13	9169	1.40	7078	4.55	2922	7
54	6436	3.13	9085	1.42	7351	4.55	2649	6
55	**9.74 6624**	3.13	**9.91 9000**	1.42	**9.82 7624**	4.55	**10.17 2376**	**5**
56	6812	3.12	8915	1.42	7897	4.55	2103	4
57	6999	3.13	8830	1.42	8170	4.53	1830	3
58	7187	3.12	8745	1.43	8442	4.55	1558	2
59	7374	3.13	8659	1.42	8715	4.53	1285	1
60	**9.74 7562**		**9.91 8574**		**9.82 8987**		**10.17 1013**	**0**
′	Cosine.	D.1″.	Sine.	D.1″.	Cotang.	D.1″.	Tang.	′

123° 56°

34° **Table VIII—Continued** 145°

′	Sine.	D.1″	Cosine.	D.1″	Tang.	D.1″	Cotang.	′
0	**9.74 7562**	3.12	**9.91 8574**	1.42	**9.82 8987**	4.55	**10.17 1013**	**60**
1	7749	3.12	8489	1.42	9260	4.53	0740	59
2	7936	3.12	8404	1.43	9532	4.55	0468	58
3	8123	3.12	8318	1.42	.82 9805	4.53	.17 0195	57
4	8310	3.12	8233	1.43	.83 0077	4.53	.16 9923	56
5	**9.74 8497**	3.10	**9.91 8147**	1.42	**9.83 0349**	4.53	**10.16 9651**	**55**
6	8683	3.12	8062	1.43	0621	4.53	9379	54
7	8870	3.10	7976	1.42	0893	4.53	9107	53
8	9056	3.12	7891	1.43	1165	4.53	8835	52
9	9243	3.10	7805	1.43	1437	4.53	8563	51
10	**9.74 9429**	3.10	**9.91 7719**	1.42	**9.83 1709**	4.53	**10.16 8291**	**50**
11	9615	3.10	7634	1.43	1981	4.53	8019	49
12	9801	3.10	7548	1.43	2253	4.53	7747	48
13	.74 9987	3.08	7462	1.43	2525	4.52	7475	47
14	.75 0172	3.10	7376	1.43	2796	4.53	7204	46
15	**9.75 0358**	3.08	**9.91 7290**	1.43	**9.83 3068**	4.52	**10.16 6932**	**45**
16	0543	3.10	7204	1.43	3339	4.53	6661	44
17	0729	3.08	7118	1.43	3611	4.52	6389	43
18	0914	3.08	7032	1.43	3882	4.53	6118	42
19	1099	3.08	6946	1.45	4154	4.52	5846	41
20	**9.75 1284**	3.08	**9.91 6859**	1.43	**9.83 4425**	4.52	**10.16 5575**	**40**
21	1469	3.08	6773	1.43	4696	4.52	5304	39
22	1654	3.08	6687	1.45	4967	4.52	5033	38
23	1839	3.07	6600	1.43	5238	4.52	4762	37
24	2023	3.08	6514	1.45	5509	4.52	4491	36
25	**9.75 2208**	3.07	**9.91 6427**	1.43	**9.83 5780**	4.52	**10.16 4220**	**35**
26	2392	3.07	6341	1.45	6051	4.52	3949	34
27	2576	3.07	6254	1.45	6322	4.52	3678	33
28	2760	3.07	6167	1.43	6593	4.52	3407	32
29	2944	3.07	6081	1.45	6864	4.50	3136	31
30	**9.75 3128**	3.07	**9.91 5994**	1.45	**9.83 7134**	4.52	**10.16 2866**	**30**
31	3312	3.05	5907	1.45	7405	4.50	2595	29
32	3495	3.07	5820	1.45	7675	4.52	2325	28
33	3679	3.07	5733	1.45	7946	4.50	2054	27
34	3862	3.07	5646	1.45	8216	4.52	1784	26
35	**9.75 4046**	3.05	**9.91 5559**	1.45	**9.83 8487**	4.50	**10.16 1513**	**25**
36	4229	3.05	5472	1.45	8757	4.50	1243	24
37	4412	3.05	5385	1.47	9027	4.50	0973	23
38	4595	3.05	5297	1.45	9297	4.52	0703	22
39	4778	3.03	5210	1.45	9568	4.50	0432	21
40	**9.75 4960**	3.05	**9.91 5123**	1.47	**9.83 9838**	4.50	**10.16 0162**	**20**
41	5143	3.05	5035	1.45	.84 0108	4.50	.15 9892	19
42	5326	3.03	4948	1.47	0378	4.50	9622	18
43	5508	3.03	4860	1.45	0648	4.48	9352	17
44	5690	3.03	4773	1.47	0917	4.50	9083	16
45	**9.75 5872**	3.03	**9.91 4685**	1.45	**9.84 1187**	4.50	**10.15 8813**	**15**
46	6054	3.03	4598	1.47	1457	4.50	8543	14
47	6236	3.03	4510	1.47	1727	4.48	8273	13
48	6418	3.03	4422	1.47	1996	4.50	8004	12
49	6600	3.03	4334	1.47	2266	4.48	7734	11
50	**9.75 6782**	3.02	**9.91 4246**	1.47	**9.84 2535**	4.50	**10.15 7465**	**10**
51	6963	3.02	4158	1.47	2805	4.48	7195	9
52	7144	3.03	4070	1.47	3074	4.48	6926	8
53	7326	3.02	3982	1.47	3343	4.48	6657	7
54	7507	3.02	3894	1.47	3612	4.50	6388	6
55	**9.75 7688**	3.02	**9.91 3806**	1.47	**9.84 3882**	4.48	**10.15 6118**	**5**
56	7869	3.02	3718	1.47	4151	4.48	5849	4
57	8050	3.00	3630	1.48	4420	4.48	5580	3
58	8230	3.02	3541	1.47	4689	4.48	5311	2
59	8411	3.00	3453	1.47	4958	4.48	5042	1
60	**9.75 8591**		**9.91 3365**		**9.84 5227**		**10.15 4773**	**0**
′	Cosine.	D.1″.	Sine.	D.1″.	Cotang.	D.1″.	Tang.	′

124° 55°

35° **Table VIII—Continued** **144°**

′	Sine.	D.1″.	Cosine.	D.1″.	Tang.	D.1″.	Cotang.	′
0	**9.75 8591**	3.02	**9.91 3365**	1.48	**9.84 5227**	4.48	**10 15 4773**	**60**
1	8772	3.00	3276	1.48	5496	4.47	4504	59
2	8952	3.00	3187	1.47	5764	4.48	4236	58
3	9132	3.00	3099	1.48	6033	4.48	3967	57
4	9312	3.00	3010	1.47	6302	4.47	3698	56
5	**9.75 9492**	3.00	**9.91 2922**	1.48	**9.84 6570**	4.48	**10.15 3430**	**55**
6	9672	3.00	2833	1.48	6839	4.48	3161	54
7	.75 9852	2.98	2744	1.48	7108	4.47	2892	53
8	.76 0031	3.00	2655	1.48	7376	4.47	2624	52
9	0211	2.98	2566	1.48	7644	4.48	2356	51
10	**9.76 0390**	2.98	**9.91 2477**	1.48	**9.84 7913**	4.47	**10.15 2087**	**50**
11	0569	2.98	2388	1.48	8181	4.47	1819	49
12	0748	2.98	2299	1.48	8449	4.47	1551	48
13	0927	2.98	2210	1.48	8717	4.48	1283	47
14	1106	2.98	2121	1.50	8986	4.47	1014	46
15	**9.76 1285**	2.98	**9.91 2031**	1.48	**9.84 9254**	4.47	**10.15 0746**	**45**
16	1464	2.97	1942	1.48	9522	4.47	0478	44
17	1642	2.98	1853	1.50	.84 9790	4.45	.15 0210	43
18	1821	2.97	1763	1.48	.85 0057	4.47	.14 9943	42
19	1999	2.97	1674	1.50	0325	4.47	9675	41
20	**9.76 2177**	2.98	**9.91 1584**	1.48	**9.85 0593**	4.47	**10.14 9407**	**40**
21	2356	2.97	1495	1.50	0861	4.47	9139	39
22	2534	2.97	1405	1.50	1129	4.45	8871	38
23	2712	2.95	1315	1.48	1396	4.47	8604	37
24	2889	2.97	1226	1.50	1664	4.45	8336	36
25	**9.76 3067**	2.97	**9.91 1136**	1.50	**9.85 1931**	4.47	**10.14 8069**	**35**
26	3245	2.95	1046	1.50	2199	4.45	7801	34
27	3422	2.97	0956	1.50	2466	4.45	7534	33
28	[illegible]	2.95	[illegible]	1.50	[illegible]	4.47	[illegible]	[illegible]
29	3777	2.95	0776	1.50	3001	4.45	6999	31
30	**9.76 3954**	2.95	**9.91 0686**	1.50	**9.85 3268**	4.45	**10.14 6732**	**30**
31	4131	2.95	0596	1.50	3535	4.45	6465	29
32	4308	2.95	0506	1.52	3802	4.45	6198	28
33	4485	2.95	0415	1.50	4069	4.45	5931	27
34	4662	2.93	0325	1.50	4336	4.45	5664	26
35	**9.76 4838**	2.95	**9.91 0235**	1.52	**9.85 4603**	4.45	**10.14 5397**	**25**
36	5015	2.93	0144	1.50	4870	4.45	5130	24
37	5191	2.93	.91 0054	1.52	5137	4.45	4863	23
38	5367	2.95	.90 9963	1.50	5404	4.45	4596	22
39	5544	2.93	9873	1.52	5671	4.45	4329	21
40	**9.76 5720**	2.93	**9.90 9782**	1.52	**9.85 5938**	4.43	**10.14 4062**	**20**
41	5896	2.93	9691	1.50	6204	4.45	3796	19
42	6072	2.92	9601	1.52	6471	4.43	3529	18
43	6247	2.93	9510	1.52	6737	4.45	3263	17
44	6423	2.92	9419	1.52	7004	4.43	2996	16
45	**9.76 6598**	2.93	**9.90 9328**	1.53	**9.85 7270**	4.45	**10.14 2730**	**15**
46	6774	2.92	9237	1.52	7537	4.43	2463	14
47	6949	2.92	9146	1.52	7803	4.43	2197	13
48	7124	2.93	9055	1.52	8069	4.45	1931	12
49	7300	2.92	8964	1.52	8336	4.43	1664	11
50	**9.76 7475**	2.90	**9.90 8873**	1.53	**9.85 8602**	4.43	**10.14 1398**	**10**
51	7649	2.92	8781	1.52	8868	4.43	1132	9
52	7824	2.92	8690	1.52	9134	4.43	0866	8
53	7999	2.90	8599	1.53	9400	4.43	0600	7
54	8173	2.92	8507	1.52	9666	4.43	0334	6
55	**9.76 8348**	2.90	**9.90 8416**	1.53	**9.85 9932**	4.43	**10.14 0068**	**5**
56	8522	2.92	8324	1.52	.86 0198	4.43	.13 9802	4
57	8697	2.90	8233	1.53	0464	4.43	9536	3
58	8871	2.90	8141	1.53	0730	4.42	9270	2
59	9045	2.90	8049	1.52	0995	4.43	9005	1
60	**9.76 9219**		**9.90 7958**		**9.86 1261**		**10.13 8739**	**0**
′	Cosine.	D.1″.	Sine.	D.1″.	Cotang.	D.1″.	Tang.	′

125° **54°**

36° **Table VIII—Continued** 143°

′	Sine.	D.1″.	Cosine.	D.1″.	Tang.	D.1″.	Cotang.	′
0	9.76 9219	2.90	9.90 7958	1.53	9.86 1261	4.43	10.13 8739	60
1	9393	2.88	7866	1.53	1527	4.42	8473	59
2	9566	2.90	7774	1.53	1792	4.43	8208	58
3	9740	2.88	7682	1.53	2058	4.42	7942	57
4	.76 9913	2.90	7590	1.53	2323	4.43	7677	56
5	9.77 0087	2.88	9.90 7498	1.53	9.86 2589	4.42	10.13 7411	55
6	0260	2.88	7406	1.53	2854	4.42	7146	54
7	0433	2.88	7314	1.53	3119	4.43	6881	53
8	0606	2.88	7222	1.55	3385	4.42	6615	52
9	0779	2.88	7129	1.53	3650	4.42	6350	51
10	9.77 0952	2.88	9.90 7037	1.53	9.86 3915	4.42	10.13 6085	50
11	1125	2.88	6945	1.55	4180	4.42	5820	49
12	1298	2.87	6852	1.53	4445	4.42	5555	48
13	1470	2.88	6760	1.55	4710	4.42	5290	47
14	1643	2.87	6667	1.53	4975	4.42	5025	46
15	9.77 1815	2.87	9.90 6575	1.55	9.86 5240	4.42	10.13 4760	45
16	1987	2.87	6482	1.55	5505	4.42	4495	44
17	2159	2.87	6389	1.55	5770	4.42	4230	43
18	2331	2.87	6296	1.53	6035	4.42	3965	42
19	2503	2.87	6204	1.55	6300	4.40	3700	41
20	9.77 2675	2.87	9.90 6111	1.55	9.86 6564	4.42	10.13 3436	40
21	2847	2.85	6018	1.55	6829	4.42	3171	39
22	3018	2.87	5925	1.55	7094	4.40	2906	38
23	3190	2.85	5832	1.55	7358	4.42	2642	37
24	3361	2.87	5739	1.57	7623	4.40	2377	36
25	9.77 3533	2.85	9.90 5645	1.55	9.86 7887	4.42	10.13 2113	35
26	3704	2.85	5552	1.55	8152	4.40	1848	34
27	3875	2.85	5459	1.55	8416	4.40	1584	33
28	4046	2.85	5366	1.57	8680	4.42	1320	32
29	4217	2.85	5272	1.55	8945	4.40	1055	31
30	9.77 4388	2.83	9.90 5179	1.57	9.86 9209	4.40	10.13 0791	30
31	4558	2.85	5085	1.55	9473	4.40	0527	29
32	4729	2.83	4992	1.57	.86 9737	4.40	.13 0263	28
33	4899	2.85	4898	1.57	.87 0001	4.40	.12 9999	27
34	5070	2.83	4804	1.55	0265	4.40	9735	26
35	9.77 5240	2.83	9.90 4711	1.57	9.87 0529	4.40	10.12 9471	25
36	5410	2.83	4617	1.57	0793	4.40	9207	24
37	5580	2.83	4523	1.57	1057	4.40	8943	23
38	5750	2.83	4429	1.57	1321	4.40	8679	22
39	5920	2.83	4335	1.57	1585	4.40	8415	21
40	9.77 6090	2.82	9.90 4241	1.57	9.87 1849	4.38	10.12 8151	20
41	6259	2.83	4147	1.57	2112	4.40	7888	19
42	6429	2.82	4053	1.57	2376	4.40	7624	18
43	6598	2.83	3959	1.58	2640	4.38	7360	17
44	6768	2.82	3864	1.57	2903	4.40	7097	16
45	9.77 6937	2.82	9.90 3770	1.57	9.87 3167	4.38	10.12 6833	15
46	7106	2.82	3676	1.58	3430	4.40	6570	14
47	7275	2.82	3581	1.57	3694	4.38	6306	13
48	7444	2.82	3487	1.58	3957	4.38	6043	12
49	7613	2.80	3392	1.57	4220	4.40	5780	11
50	9.77 7781	2.82	9.90 3298	1.58	9.87 4484	4.38	10.12 5516	10
51	7950	2.82	3203	1.58	4747	4.38	5253	9
52	8119	2.80	3108	1.57	5010	4.38	4990	8
53	8287	2.80	3014	1.58	5273	4.40	4727	7
54	8455	2.82	2919	1.58	5537	4.38	4463	6
55	9.77 8624	2.80	9.90 2824	1.58	9.87 5800	4.38	10.12 4200	5
56	8792	2.80	2729	1.58	6063	4.38	3937	4
57	8960	2.80	2634	1.58	6326	4.38	3674	3
58	9128	2.78	2539	1.58	6589	4.38	3411	2
59	9295	2.80	2444	1.58	6852	4.37	3148	1
60	9.77 9463		9.90 2349		9.87 7114		10.12 2886	0
′	Cosine.	D.1″.	Sine.	D.1″.	Cotang.	D.1″.	Tang.	′

126° 53°

37° **Table VIII—Continued** **142°**

′	Sine.	D.1″.	Cosine.	D.1″.	Tang.	D.1″.	Cotang.	′
0	**9.77 9463**	2.80	**9.90 2349**	1.60	**9.87 7114**	4.38	**10.12 2886**	**60**
1	9631	2.78	2253	1.58	7377	4.38	2623	59
2	9798	2.80	2158	1.58	7640	4.38	2360	58
3	.77 9966	2.78	2063	1.60	7903	4.37	2097	57
4	.78 0133	2.78	1967	1.58	8165	4.38	1835	56
5	**9.78 0300**	2.78	**9.90 1872**	1.60	**9.87 8428**	4.38	**10.12 1572**	**55**
6	0467	2.78	1776	1.58	8691	4.37	1309	54
7	0634	2.78	1681	1.60	8953	4.38	1047	53
8	0801	2.78	1585	1.58	9216	4.37	0784	52
9	0968	2.77	1490	1.60	9478	4.38	0522	51
10	**9.78 1134**	2.78	**9.90 1394**	1.60	**9.87 9741**	4.37	**10.12 0259**	**50**
11	1301	2.78	1298	1.60	.88 0003	4.37	.11 9997	49
12	1468	2.77	1202	1.60	0265	4.38	9735	48
13	1634	2.77	1106	1.60	0528	4.37	9472	47
14	1800	2.77	1010	1.60	0790	4.37	9210	46
15	**9.78 1966**	2.77	**9.90 0914**	1.60	**9.88 1052**	4.37	**10.11 8948**	**45**
16	2132	2.77	0818	1.60	1314	4.38	8686	44
17	2298	2.77	0722	1.60	1577	4.37	8423	43
18	2464	2.77	0626	1.62	1839	4.37	8161	42
19	2630	2.77	0529	1.60	2101	4.37	7899	41
20	**9.78 2796**	2.75	**9.90 0433**	1.60	**9.88 2363**	4.37	**10.11 7637**	**40**
21	2961	2.77	0337	1.62	2625	4.37	7375	39
22	3127	2.75	0240	1.60	2887	4.35	7113	38
23	3292	2.77	0144	1.62	3148	4.37	6852	37
24	3458	2.75	.90 0047	1.60	3410	4.37	6590	36
25	**9.78 3623**	2.75	**9.89 9951**	1.62	**9.88 3672**	4.37	**10.11 6328**	**35**
26	3788	2.75	9854	1.62	3934	4.37	6066	34
27	3953	2.75	9757	1.62	4196	4.35	5804	33
28	4118	2.73	9660	1.60	4457	4.37	5543	32
29	4282	2.75	9564	1.62	4719	4.35	5281	31
30	**9.78 4447**	2.75	**9.89 9467**	1.62	**9.88 4980**	4.37	**10.11 5020**	**30**
31	4612	2.73	9370	1.62	5242	4.37	4758	29
32	4776	2.75	9273	1.62	5504	4.35	4496	28
33	4941	2.73	9176	1.63	5765	4.35	4235	27
34	5105	[illegible]	9078	[illegible]	6026	4.37	3974	26
35	**9.78 5269**	2.73	**9.89 8981**	1.62	**9.88 6288**	4.35	**10.11 3712**	**25**
36	5433	[illegible]	8884	[illegible]	6549	4.37	3451	24
37	5597	2.73	8787	1.63	6811	4.35	3189	23
38	5761	2.73	8689	1.62	7072	4.35	2928	22
[illegible]	[illegible]	[illegible]	[illegible]	[illegible]	[illegible]	[illegible]	2667	21
40	**9.78 6089**	2.72	**9.89 8494**	1.62	**9.88 7594**	4.35	**10.11 2406**	**20**
41	6252	2.73	8397	1.63	7855	4.35	2145	19
42	6416	2.72	8299	1.62	8116	4.37	1884	18
43	6579	2.72	8202	1.63	8378	4.35	1622	17
44	6742	2.73	8104	1.63	8639	4.35	1361	16
45	**9.78 6906**	2.72	**9.89 8006**	1.63	**9.88 8900**	4.35	**10.11 1100**	**15**
46	7069	2.72	7908	1.63	9161	4.33	0839	14
47	7232	2.72	7810	1.63	9421	4.35	0579	13
48	7395	2.70	7712	1.63	9682	4.35	0318	12
49	7557	2.72	7614	1.63	.88 9943	4.35	.11 0057	11
50	**9.78 7720**	2.72	**9.89 7516**	1.63	**9.89 0204**	4.35	**10.10 9796**	**10**
51	7883	2.70	7418	1.63	0465	4.33	9535	9
52	8045	2.72	7320	1.63	0725	4.35	9275	8
53	8208	2.70	7222	1.65	0986	4.35	9014	7
54	8370	2.70	7123	1.63	1247	4.33	8753	6
55	**9.78 8532**	2.70	**9.89 7025**	1.65	**9.89 1507**	4.35	**10.10 8493**	**5**
56	8694	2.70	6926	1.63	1768	4.33	8232	4
57	8856	2.70	6828	1.65	2028	4.35	7972	3
58	9018	2.70	6729	1.63	2289	4.33	7711	2
59	9180	2.70	6631	1.65	2549	4.35	7451	1
60	**9.78 9342**		**9.89 6532**		**9.89 2810**		**10.10 7190**	**0**
′	Cosine.	D.1″.	Sine.	D.1″.	Cotang.	D.1″.	Tang.	′

127° **52°**

38° **Table VIII—Continued** 141°

′	Sine.	D.1″.	Cosine.	D.1″.	Tang.	D.1″.	Cotang.	′
0	**9.78 9342**	2.70	**9.89 6532**	1.65	**9.89 2810**	4.33	**10.10 7190**	**60**
1	9504	2.68	6433	1.63	3070	4.35	6930	59
2	9665	2.70	6335	1.65	3331	4.33	6669	58
3	9827	2.68	6236	1.65	3591	4.33	6409	57
4	.78 9988	2.68	6137	1.65	3851	4.33	6149	56
5	**9.79 0149**	2.68	**9.89 6038**	1.65	**9.89 4111**	4.35	**10.10 5889**	**55**
6	0310	2.68	5939	1.65	4372	4.33	5628	54
7	0471	2.68	5840	1.65	4632	4.33	5368	53
8	0632	2.68	5741	1.67	4892	4.33	5108	52
9	0793	2.68	5641	1.65	5152	4.33	4848	51
10	**9.79 0954**	2.68	**9.89 5542**	1.65	**9.89 5412**	4.33	**10.10 4588**	**50**
11	1115	2.67	5443	1.67	5672	4.33	4328	49
12	1275	2.68	5343	1.65	5932	4.33	4068	48
13	1436	2.67	5244	1.65	6192	4.33	3808	47
14	1596	2.68	5145	1.67	6452	4.33	3548	46
15	**9.79 1757**	2.67	**9.89 5045**	1.67	**9.89 6712**	4.32	**10.10 3288**	**45**
16	1917	2.67	4945	1.65	6971	4.33	3029	44
17	2077	2.67	4846	1.67	7231	4.33	2769	43
18	2237	2.67	4746	1.67	7491	4.33	2509	42
19	2397	2.67	4646	1.67	7751	4.32	2249	41
20	**9.79 2557**	2.65	**9.89 4546**	1.67	**9.89 8010**	4.33	**10.10 1990**	**40**
21	2716	2.67	4446	1.67	8270	4.33	1730	39
22	2876	2.65	4346	1.67	8530	4.32	1470	38
23	3035	2.67	4246	1.67	8789	4.33	1211	37
24	3195	2.65	4146	1.67	9049	4.32	0951	36
25	**9.79 3354**	2.67	**9.89 4046**	1.67	**9.89 9308**	4.33	**10.10 0692**	**35**
26	3514	2.65	3946	1.67	9568	4.32	0432	34
27	3673	2.65	3846	1.68	.89 9827	4.33	.10 0173	33
28	3832	2.65	3745	1.67	.90 0087	4.32	.09 9913	32
29	3991	2.65	3645	1.68	0346	4.32	9654	31
30	**9.79 4150**	2.63	**9.89 3544**	1.67	**9.90 0605**	4.32	**10.09 9395**	**30**
31	4308	2.65	3444	1.68	0864	4.33	9136	29
32	4467	2.65	3343	1.67	1124	4.32	8876	28
33	4626	2.63	3243	1.68	1383	4.32	8617	27
34	4784	2.63	3142	1.68	1642	4.32	8358	26
35	**9.79 4942**	2.65	**9.89 3041**	1.68	**9.90 1901**	4.32	**10.09 8099**	**25**
36	5101	2.63	2940	1.68	2160	4.33	7840	24
37	5259	2.63	2839	1.67	2420	4.32	7580	23
38	5417	2.63	2739	1.68	2679	4.32	7321	22
39	5575	2.63	2638	1.70	2938	4.32	7062	21
40	**9.79 5733**	2.63	**9.89 2536**	1.68	**9.90 3197**	4.32	**10.09 6803**	**20**
41	5891	2.63	2435	1.68	3456	4.30	6544	19
42	6049	2.62	2334	1.68	3714	4.32	6286	18
43	6206	2.63	2233	1.68	3973	4.32	6027	17
44	6364	2.62	2132	1.70	4232	4.32	5768	16
45	**9.79 6521**	2.63	**9.89 2030**	1.68	**9.90 4491**	4.32	**10.09 5509**	**15**
46	6679	2.62	1929	1.70	4750	4.30	5250	14
47	6836	2.62	1827	1.68	5008	4.32	4992	13
48	6993	2.62	1726	1.70	5267	4.32	4733	12
49	7150	2.62	1624	1.68	5526	4.32	4474	11
50	**9.79 7307**	2.62	**9.89 1523**	1.70	**9.90 5785**	4.30	**10.09 4215**	**10**
51	7464	2.62	1421	1.70	6043	4.32	3957	9
52	7621	2.60	1319	1.70	6302	4.30	3698	8
53	7777	2.62	1217	1.70	6560	4.32	3440	7
54	7934	2.62	1115	1.70	6819	4.30	3181	6
55	**9.79 8091**	2.60	**9.89 1013**	1.70	**9.90 7077**	4.32	**10.09 2923**	**5**
56	8247	2.60	0911	1.70	7336	4.30	2664	4
57	8403	2.62	0809	1.70	7594	4.32	2406	3
58	8560	2.60	0707	1.70	7853	4.30	2147	2
59	8716	2.60	0605	1.70	8111	4.30	1889	1
60	**9.79 8872**		**9.89 0503**		**9.90 8369**		**10.09 1631**	**0**
′	Cosine.	D.1″.	Sine.	D.1″.	Cotang.	D.1″.	Tang.	′

128° 51°

39° **Table VIII—Continued** **140°**

′	Sine.	D.1″.	Cosine.	D.1″.	Tang.	D.1″.	Cotang.	′
0	9.79 8872	2.60	9.89 0503	1.72	9.90 8369	4.32	10.09 1631	60
1	9028	2.60	0400	1.70	8628	4.30	1372	59
2	9184	2.58	0298	1.72	8886	4.30	1114	58
3	9339	2.60	0195	1.70	9144	4.30	0856	57
4	9495	2.60	.89 0093	1.72	9402	4.30	0598	56
5	9.79 9651	2.58	9.88 9990	1.70	9.90 9660	4.30	10.09 0340	55
6	9806	2.60	9888	1.72	.90 9918	4.32	.09 0082	54
7	.79 9962	2.58	9785	1.72	.91 0177	4.30	.08 9823	53
8	.80 0117	2.58	9682	1.72	0435	4.30	9565	52
9	0272	2.58	9579	1.70	0693	4.30	9307	51
10	9.80 0427	2.58	9.88 9477	1.72	9.91 0951	4.30	10.08 9049	50
11	0582	2.58	9374	1.72	1209	4.30	8791	49
12	0737	2.58	9271	1.72	1467	4.30	8533	48
13	0892	2.58	9168	1.73	1725	4.28	8275	47
14	1047	2.57	9064	1.72	1982	4.30	8018	46
15	9.80 1201	2.58	9.88 8961	1.72	9.91 2240	4.30	10.08 7760	45
16	1356	2.58	8858	1.72	2498	4.30	7502	44
17	1511	2.57	8755	1.73	2756	4.30	7244	43
18	1665	2.57	8651	1.72	3014	4.28	6986	42
19	1819	2.57	8548	1.73	3271	4.30	6729	41
20	9.80 1973	2.58	9.88 8444	1.72	9.91 3529	4.30	10.08 6471	40
21	2128	2.57	8341	1.73	3787	4.28	6213	39
22	2282	2.57	8237	1.72	4044	4.30	5956	38
23	2436	2.55	8134	1.73	4302	4.30	5698	37
24	2589	2.57	8030	1.73	4560	4.28	5440	36
25	9.80 2743	2.57	9.88 7926	1.73	9.91 4817	4.30	10.08 5183	35
26	2897	2.55	7822	1.73	5075	4.28	4925	34
27	3050	2.57	7718	1.73	5332	4.30	4668	33
28	3204	2.55	7614	1.73	5590	4.28	4410	32
29	3357	2.57	7510	1.73	5847	4.28	4153	31
30	9.80 3511	2.55	9.88 7406	1.73	9.91 6104	4.30	10.08 3896	30
31	3664	2.55	7302	1.73	6362	4.28	3638	29
32	3817	2.55	7198	1.75	6619	4.30	3381	28
33	3970	2.55	7093	1.73	6877	4.28	3123	27
34	4123	2.55	6989	1.73	7134	4.28	2866	26
35	9.80 4276	2.55	9.88 6885	1.75	9.91 7391	4.28	10.08 2609	25
36	4428	2.55	6780	1.73	7648	4.30	2352	24
37	4581	2.55	6676	1.75	7906	4.28	2094	23
38	4734	2.53	6571	1.75	8163	4.28	1837	22
39	4886	2.55	6466	1.73	8420	4.28	1580	21
40	9.80 5039	2.53	9.88 6362	1.75	9.91 8677	4.28	10.08 1323	20
41	5191	2.53	6257	1.75	8934	4.28	1066	19
42	5343	2.53	6152	1.75	9191	4.28	0809	18
43	5495	2.53	6047	1.75	9448	4.28	0552	17
44	5647	2.53	5942	1.75	9705	4.28	0295	16
45	9.80 5799	2.53	9.88 5837	1.75	9.91 9962	4.28	10.08 0038	15
46	5951	2.53	5732	1.75	.92 0219	4.28	.07 9781	14
47	6103	2.52	5627	1.75	0476	4.28	9524	13
48	6254	2.53	5522	1.77	0733	4.28	9267	12
49	6406	2.52	5416	1.75	0990	4.28	9010	11
50	9.80 6557	2.52	9.88 5311	1.77	9.92 1247	4.27	10.07 8753	10
51	6709	2.52	5205	1.75	1503	4.28	8497	9
52	6860	2.52	5100	1.77	1760	4.28	8240	8
53	7011	2.53	4994	1.75	2017	4.28	7983	7
54	7163	2.52	4889	1.77	2274	4.27	7726	6
55	9.80 7314	2.52	9.88 4783	1.77	9.92 2530	4.28	10.07 7470	5
56	7465	2.50	4677	1.75	2787	4.28	7213	4
57	7615	2.52	4572	1.77	3044	4.27	6956	3
58	7766	2.52	4466	1.77	3300	4.28	6700	2
59	7917	2.50	4360	1.77	3557	4.28	6443	1
60	9.80 8067		9.88 4254		9.92 3814		10.07 6186	0
′	Cosine.	D.1″.	Sine.	D.1″.	Cotang.	D.1″.	Tang.	′

129° **50°**

40° **Table VIII—Continued** 139°

′	Sine.	D.1″.	Cosine.	D.1″.	Tang.	D.1″.	Cotang.	′
0	**9.80 8067**	2.52	**9.88 4254**	1.77	**9.92 3814**	4.27	**10.07 6186**	**60**
1	8218	2.50	4148	1.77	4070	4.28	5930	59
2	8368	2.52	4042	1.77	4327	4.27	5673	58
3	8519	2.50	3936	1.78	4583	4.28	5417	57
4	8669	2.50	3829	1.77	4840	4.27	5160	56
5	**9.80 8819**	2.50	**9.88 3723**	1.77	**9.92 5096**	4.27	**10.07 4904**	**55**
6	8969	2.50	3617	1.78	5352	4.28	4648	54
7	9119	2.50	3510	1.77	5609	4.27	4391	53
8	9269	2.50	3404	1.78	5865	4.28	4135	52
9	9419	2.50	3297	1.77	6122	4.27	3878	51
10	**9.80 9569**	2.48	**9.88 3191**	1.78	**9.92 6378**	4.27	**10.07 3622**	**50**
11	9718	2.50	3084	1.78	6634	4.27	3366	49
12	.80 9868	2.48	2977	1.77	6890	4.28	3110	48
13	.81 0017	2.50	2871	1.78	7147	4.27	2853	47
14	0167	2.48	2764	1.78	7403	4.27	2597	46
15	**9.81 0316**	2.48	**9.88 2657**	1.78	**9.92 7659**	4.27	**10.07 2341**	**45**
16	0465	2.48	2550	1.78	7915	4.27	2085	44
17	0614	2.48	2443	1.78	8171	4.27	1829	43
18	0763	2.48	2336	1.78	8427	4.28	1573	42
19	0912	2.48	2229	1.80	8684	4.27	1316	41
20	**9.81 1061**	2.48	**9.88 2121**	1.78	**9.92 8940**	4.27	**10.07 1060**	**40**
21	1210	2.47	2014	1.78	9196	4.27	0804	39
22	1358	2.48	1907	1.80	9452	4.27	0548	38
23	1507	2.47	1799	1.78	9708	4.27	0292	37
24	1655	2.48	1692	1.80	.92 9964	4.27	.07 0036	36
25	**9.81 1804**	2.47	**9.88 1584**	1.78	**9.93 0220**	4.25	**10.06 9780**	**35**
26	1952	2.47	1477	1.80	0475	4.27	9525	34
27	2100	2.47	1369	1.80	0731	4.27	9269	33
28	2248	2.47	1261	1.80	0987	4.27	9013	32
29	2396	2.47	1153	1.78	1243	4.27	8757	31
30	**9.81 2544**	2.47	**9.88 1046**	1.80	**9.93 1499**	4.27	**10.06 8501**	**30**
31	2692	2.47	0938	1.80	1755	4.25	8245	29
32	2840	2.47	0830	1.80	2010	4.27	7990	28
33	2988	2.45	0722	1.82	2266	4.27	7734	27
34	3135	2.47	0613	1.80	2522	4.27	7478	26
35	**9.81 3283**	2.45	**9.88 0505**	1.80	**9.93 2778**	4.25	**10.06 7222**	**25**
36	3430	2.47	0397	1.80	3033	4.27	6967	24
37	3578	2.45	0289	1.82	3289	4.27	6711	23
38	3725	2.45	0180	1.80	3545	4.25	6455	22
39	3872	2.45	.88 0072	1.82	3800	4.27	6200	21
40	**9.81 4019**	2.45	**9.87 9963**	1.80	**9.93 4056**	4.25	**10.06 5944**	**20**
41	4166	2.45	9855	1.82	4311	4.27	5689	19
42	4313	2.45	9746	1.82	4567	4.25	5433	18
43	4460	2.45	9637	1.80	4822	4.27	5178	17
44	4607	2.43	9529	1.82	5078	4.25	4922	16
45	**9.81 4753**	2.45	**9.87 9420**	1.82	**9.93 5333**	4.27	**10.06 4667**	**15**
46	4900	2.43	9311	1.82	5589	4.25	4411	14
47	5046	2.45	9202	1.82	5844	4.27	4156	13
48	5193	2.43	9093	1.82	6100	4.25	3900	12
49	5339	2.43	8984	1.82	6355	4.27	3645	11
50	**9.81 5485**	2.45	**9.87 8875**	1.82	**9.93 6611**	4.25	**10.06 3389**	**10**
51	5632	2.43	8766	1.83	6866	4.25	3134	9
52	5778	2.43	8656	1.82	7121	4.27	2879	8
53	5924	2.42	8547	1.82	7377	4.25	2623	7
54	6069	2.43	8438	1.83	7632	4.25	2368	6
55	**9.81 6215**	2.43	**9.87 8328**	1.82	**9.93 7887**	4.25	**10.06 2113**	**5**
56	6361	2.43	8219	1.83	8142	4.27	1858	4
57	6507	2.42	8109	1.83	8398	4.25	1602	3
58	6652	2.43	7999	1.82	8653	4.25	1347	2
59	6798	2.42	7890	1.83	8908	4.25	1092	1
60	**9.81 6943**		**9.87 7780**		**9.93 9163**		**10.06 0837**	**0**
′	Cosine.	D.1″.	Sine.	D.1″.	Cotang.	D.1″.	Tang.	′

130° 49°

41° **Table VIII—Continued** 138°

′	Sine.	D.1″.	Cosine.	D.1″.	Tang.	D.1″.	Cotang.	′
0	**9.81 6943**	2.42	**9.87 7780**	1.83	**9.93 9163**	4.25	**10.06 0837**	**60**
1	7088	2.42	7670	1.83	9418	4.25	0582	59
2	7233	2.43	7560	1.83	9673	4.25	0327	58
3	7379	2.42	7450	1.83	.93 9928	4.25	.06 0072	57
4	7524	2.40	7340	1.83	.94 0183	4.27	.05 9817	56
5	**9.81 7668**	2.42	**9.87 7230**	1.83	**9.94 0439**	4.25	**10.05 9561**	**55**
6	7813	2.42	7120	1.83	0694	4.25	9306	54
7	7958	2.42	7010	1.85	0949	4.25	9051	53
8	8103	2.40	6899	1.83	1204	4.25	8796	52
9	8247	2.42	6789	1.85	1459	4.23	8541	51
10	**9.81 8392**	2.40	**9.87 6678**	1.83	**9.94 1713**	4.25	**10.05 8287**	**50**
11	8536	2.42	6568	1.85	1968	4.25	8032	49
12	8681	2.40	6457	1.83	2223	4.25	7777	48
13	8825	2.40	6347	1.85	2478	4.25	7522	47
14	8969	2.40	6236	1.85	2733	4.25	7267	46
15	**9.81 9113**	2.40	**9.87 6125**	1.85	**9.94 2988**	4.25	**10.05 7012**	**45**
16	9257	2.40	6014	1.83	3243	4.25	6757	44
17	9401	2.40	5904	1.85	3498	4.23	6502	43
18	9545	2.40	5793	1.85	3752	4.25	6248	42
19	9689	2.38	5682	1.85	4007	4.25	5993	41
20	**9.81 9832**	2.40	**9.87 5571**	1.87	**9.94 4262**	4.25	**10.05 5738**	**40**
21	.81 9976	2.40	5459	1.85	4517	4.23	5483	39
22	.82 0120	2.38	5348	1.85	4771	4.25	5229	38
23	0263	2.38	5237	1.85	5026	4.25	4974	37
24	0406	2.40	5126	1.87	5281	4.23	4719	36
25	**9.82 0550**	2.38	**9.87 5014**	1.85	**9.94 5535**	4.25	**10.05 4465**	**35**
26	0693	2.38	4903	1.87	5790	4.25	4210	34
27	0836	2.38	4791	1.85	6045	4.23	3955	33
28	0979	2.38	4680	1.87	6299	4.25	3701	32
29	1122	2.38	4568	1.87	6554	4.23	3446	31
30	**9.82 1265**	2.37	**9.87 4456**	1.87	**9.94 6808**	4.25	**10.05 3192**	**30**
31	1407	2.38	4344	1.87	7063	4.25	2937	29
32	1550	2.38	4232	1.85	7318	4.23	2682	28
33	1693	2.37	4121	1.87	7572	4.25	2428	27
34	1835	2.37	4009	1.88	7827	4.23	2173	26
35	**9.82 1977**	2.38	**9.87 3896**	1.87	**9.94 8081**	4.23	**10.05 1919**	**25**
36	2120	2.37	3784	1.87	8335	4.25	1665	24
37	2262	2.37	3672	1.87	8590	4.23	1410	23
38	2404	2.37	3560	1.87	8844	4.25	1156	22
39	2546	2.37	3448	1.88	9099	4.23	0901	21
40	**9.82 2688**	2.37	**9.87 3335**	1.87	**9.94 9353**	4.25	**10.05 0647**	**20**
41	2830	2.37	3223	1.88	9608	4.23	0392	19
42	2972	2.37	3110	1.87	.94 9862	4.23	.05 0138	18
43	3114	2.35	2998	1.88	.95 0116	4.25	.04 9884	17
44	3255	2.37	2885	1.88	0371	4.23	9629	16
45	**9.82 3397**	2.37	**9.87 2772**	1.88	**9.95 0625**	4.23	**10.04 9375**	**15**
46	3539	2.35	2659	1.87	0879	4.23	9121	14
47	3680	2.35	2547	1.88	1133	4.25	8867	13
48	3821	2.37	2434	1.88	1388	4.23	8612	12
49	3963	2.35	2321	1.88	1642	4.23	8358	11
50	**9.82 4104**	2.35	**9.87 2208**	1.88	**9.95 1896**	4.23	**10.04 8104**	**10**
51	4245	2.35	2095	1.90	2150	4.25	7850	9
52	4386	2.35	1981	1.88	2405	4.23	7595	8
53	4527	2.35	1868	1.88	2659	4.23	7341	7
54	4668	2.33	1755	1.90	2913	4.23	7087	6
55	**9.82 4808**	2.35	**9.87 1641**	1.88	**9.95 3167**	4.23	**10.04 6833**	**5**
56	4949	2.35	1528	1.90	3421	4.23	6579	4
57	5090	2.33	1414	1.88	3675	4.23	6325	3
58	5230	2.35	1301	1.90	3929	4.23	6071	2
59	5371	2.33	1187	1.90	4183	4.23	5817	1
60	**9.82 5511**		**9.87 1073**		**9.95 4437**		**10.04 5563**	**0**
′	Cosine.	D.1″.	Sine.	D.1″.	Cotang.	D.1″.	Tang.	′

131° 48°

42° **Table VIII—Continued** 137°

′	Sine.	D.1″.	Cosine.	D.1″.	Tang.	D.1″.	Cotang.	′
0	**9.82 5511**	2.33	**9.87 1073**	1.88	**9.95 4437**	4.23	**10.04 5563**	**60**
1	5651	2.33	0960	1.90	4691	4.25	5309	59
2	5791	2.33	0846	1.90	4946	4.23	5054	58
3	5931	2.33	0732	1.90	5200	4.23	4800	57
4	6071	2.33	0618	1.90	5454	4.23	4546	56
5	**9.82 6211**	2.33	**9.87 0504**	1.90	**9.95 5708**	4.22	**10.04 4292**	**55**
6	6351	2.33	0390	1.90	5961	4.23	4039	54
7	6491	2.33	0276	1.92	6215	4.23	3785	53
8	6631	2.32	0161	1.90	6469	4.23	3531	52
9	6770	2.33	.87 0047	1.90	6723	4.23	3277	51
10	**9.82 6910**	2.32	**9.86 9933**	1.92	**9.95 6977**	4.23	**10.04 3023**	**50**
11	7049	2.33	9818	1.90	7231	4.23	2769	49
12	7189	2.32	9704	1.92	7485	4.23	2515	48
13	7328	2.32	9589	1.92	7739	4.23	2261	47
14	7467	2.32	9474	1.90	7993	4.23	2007	46
15	**9.82 7606**	2.32	**9.86 9360**	1.92	**9.95 8247**	4.22	**10.04 1753**	**45**
16	7745	2.32	9245	1.92	8500	4.23	1500	44
17	7884	2.32	9130	1.92	8754	4.23	1246	43
18	8023	2.32	9015	1.92	9008	4.23	0992	42
19	8162	2.32	8900	1.92	9262	4.23	0738	41
20	**9.82 8301**	2.30	**9.86 8785**	1.92	**9.95 9516**	4.22	**10.04 0484**	**40**
21	8439	2.32	8670	1.92	.95 9769	4.23	.04 0231	39
22	8578	2.30	8555	1.92	.96 0023	4.23	.03 9977	38
23	8716	2.32	8440	1.93	0277	4.22	9723	37
24	8855	2.30	8324	1.92	0530	4.23	9470	36
25	**9.82 8993**	2.30	**9.86 8209**	1.93	**9.96 0784**	4.23	**10.03 9216**	**35**
26	9131	2.30	8093	1.92	1038	4.23	8962	34
27	9269	2.30	7978	1.93	1292	4.22	8708	33
28	9407	2.30	7862	1.92	1545	4.23	8455	32
29	9545	2.30	7747	1.93	1799	4.22	8201	31
30	**9.82 9683**	2.30	**9.86 7631**	1.93	**9.96 2052**	4.23	**10.03 7948**	**30**
31	9821	2.30	7515	1.93	2306	4.23	7694	29
32	.82 9959	2.30	7399	1.93	2560	4.22	7440	28
33	.83 0097	2.28	7283	1.93	2813	4.23	7187	27
34	0234	2.30	7167	1.93	3067	4.22	6933	26
35	**9.83 0372**	2.28	**9.86 7051**	1.93	**9.96 3320**	4.23	**10.03 6680**	**25**
36	0509	2.28	6935	1.93	3574	4.23	6426	24
37	0646	2.30	6819	1.93	3828	4.22	6172	23
38	0784	2.28	6703	1.95	4081	4.23	5919	22
39	0921	2.28	6586	1.93	4335	4.22	5665	21
40	**9.83 1058**	2.28	**9.86 6470**	1.95	**9.96 4588**	4.23	**10.03 5412**	**20**
41	1195	2.28	6353	1.93	4842	4.22	5158	19
42	1332	2.28	6237	1.95	5095	4.23	4905	18
43	1469	2.28	6120	1.93	5349	4.22	4651	17
44	1606	2.27	6004	1.95	5602	4.22	4398	16
45	**9.83 1742**	2.28	**9.86 5887**	1.95	**9.96 5855**	4.23	**10.03 4145**	**15**
46	1879	2.27	5770	1.95	6109	4.22	3891	14
47	2015	2.28	5653	1.95	6362	4.23	3638	13
48	2152	2.27	5536	1.95	6616	4.22	3384	12
49	2288	2.28	5419	1.95	6869	4.23	3131	11
50	**9.83 2425**	2.27	**9.86 5302**	1.95	**9.96 7123**	4.22	**10.03 2877**	**10**
51	2561	2.27	5185	1.95	7376	4.22	2624	9
52	2697	2.27	5068	1.97	7629	4.23	2371	8
53	2833	2.27	4950	1.95	7883	4.22	2117	7
54	2969	2.27	4833	1.95	8136	4.22	1864	6
55	**9.83 3105**	2.27	**9.86 4716**	1.97	**9.96 8389**	4.23	**10.03 1611**	**5**
56	3241	2.27	4598	1.95	8643	4.22	1357	4
57	3377	2.25	4481	1.97	8896	4.22	1104	3
58	3512	2.27	4363	1.97	9149	4.23	0851	2
59	3648	2.25	4245	1.97	9403	4.22	0597	1
60	**9.83 3783**		9.86 4127		9.96 9656		10.03 0344	**0**
′	Cosine.	D.1″.	Sine.	D.1″.	Cotang.	D.1″.	Tang.	′

132° 47°

43° **Table VIII—Continued** 136°

′	Sine.	D.1″.	Cosine.	D.1″.	Tang.	D.1″.	Cotang.	′
0	9.83 3783	2.27	9.86 4127	1.95	9.96 9656	4.22	10.03 0344	60
1	3919	2.25	4010	1.97	.96 9909	4.22	.03 0091	59
2	4054	2.25	3892	1.97	.97 0162	4.23	.02 9838	58
3	4189	2.27	3774	1.97	0416	4.22	9584	57
4	4325	2.25	3656	1.97	0669	4.22	9331	56
5	9.83 4460	2.25	9.86 3538	1.98	9.97 0922	4.22	10.02 9078	55
6	4595	2.25	3419	1.97	1175	4.23	8825	54
7	4730	2.25	3301	1.97	1429	4.22	8571	53
8	4865	2.23	3183	1.98	1682	4.22	8318	52
9	4999	2.25	3064	1.97	1935	4.22	8065	51
10	9.83 5134	2.25	9.86 2946	1.98	9.97 2188	4.22	10.02 7812	50
11	5269	2.23	2827	1.97	2441	4.23	7559	49
12	5403	2.25	2709	1.98	2695	4.22	7305	48
13	5538	2.23	2590	1.98	2948	4.22	7052	47
14	5672	2.25	2471	1.97	3201	4.22	6799	46
15	9.83 5807	2.23	9.86 2353	1.98	9.97 3454	4.22	10.02 6546	45
16	5941	2.23	2234	1.98	3707	4.22	6293	44
17	6075	2.23	2115	1.98	3960	4.22	6040	43
18	6209	2.23	1996	1.98	4213	4.22	5787	42
19	6343	2.23	1877	1.98	4466	4.23	5534	41
20	9.83 6477	2.23	9.86 1758	2.00	9.97 4720	4.22	10.02 5280	40
21	6611	2.23	1638	1.98	4973	4.22	5027	39
22	6745	2.22	1519	1.98	5226	4.22	4774	38
23	6878	2.23	1400	2.00	5479	4.22	4521	37
24	7012	2.23	1280	1.98	5732	4.22	4268	36
25	9.83 7146	2.22	9.86 1161	2.00	9.97 5985	4.22	10.02 4015	35
26	7279	2.22	1041	1.98	6238	4.22	3762	34
27	7412	2.23	0922	2.00	6491	4.22	3509	33
28	7546	2.22	0802	2.00	6744	4.22	3250	32
29	7679	2.22	0682	2.00	6997	4.22	3003	31
30	9.83 7812	2.22	9.86 0562	2.00	9.97 7250	4.22	10.02 2750	30
31	7945	2.22	0442	2.00	7503	4.22	2497	29
32	8078	2.22	0322	2.00	7756	4.22	2244	28
33	8211	2.22	0202	2.00	8009	4.22	1991	27
34	8344	2.22	.86 0082	2.00	8262	4.22	1738	26
35	9.83 8477	2.22	9.85 9962	2.00	9.97 8515	4.22	10.02 1485	25
36	8610	2.20	9842	2.02	8768	4.22	1232	24
37	8742	2.22	9721	2.00	9021	4.22	0979	20
38	8875	2.20	9601	2.02	9274	4.22	0726	22
39	9007	2.22	9480	2.00	9527	4.22	0473	21
40	9.83 9140	2.20	9.85 9360	2.02	9.97 9780	4.22	10.02 0220	20
41	9272	2.20	9239	2.00	.98 0033	4.22	.01 9967	19
42	9404	2.20	9119	2.02	0286	4.20	9714	18
43	9536	2.20	8998	2.02	0538	4.22	9462	17
44	9668	2.20	8877	2.02	0791	4.22	9209	16
45	9.83 9800	2.20	9.85 8756	2.02	9.98 1044	4.22	10.01 8956	15
46	.83 9932	2.20	8635	2.02	1297	4.22	8703	14
47	.84 0064	2.20	8514	2.02	1550	4.22	8450	13
48	0196	2.20	8393	2.02	1803	4.22	8197	12
49	0328	2.18	8272	2.02	2056	4.22	7944	11
50	9.84 0459	2.20	9.85 8151	2.03	9.98 2309	4.22	10.01 7691	10
51	0591	2.18	8029	2.02	2562	4.20	7438	9
52	0722	2.20	7908	2.03	2814	4.22	7186	8
53	0854	2.18	7786	2.02	3067	4.22	6933	7
54	0985	2.18	7665	2.03	3320	4.22	6680	6
55	9.84 1116	2.18	9.85 7543	2.02	9.98 3573	4.22	10.01 6427	5
56	1247	2.18	7422	2.03	3826	4.22	6174	4
57	1378	2.18	7300	2.03	4079	4.22	5921	3
58	1509	2.18	7178	2.03	4332	4.20	5668	2
59	1640	2.18	7056	2.03	4584	4.22	5416	1
60	9.84 1771		9.85 6934		9.98 4837		10.01 5163	0
′	Cosine.	D.1″.	Sine.	D.1″.	Cotang.	D.1″.	Tang.	′

133° 46°

44° **Table VIII—Concluded** **135°**

′	Sine.	D.1″.	Cosine.	D.1″.	Tang.	D.1″.	Cotang.	′
0	9.84 1771	2.18	9.85 6934	2.03	9.98 4837	4.22	10.01 5163	60
1	1902	2.18	6812	2.03	5090	4.22	4910	59
2	2033	2.17	6690	2.03	5343	4.22	4657	58
3	2163	2.18	6568	2.03	5596	4.20	4404	57
4	2294	2.17	6446	2.05	5848	4.22	4152	56
5	9.84 2424	2.18	9.85 6323	2.03	9.98 6101	4.22	10.01 3899	55
6	2555	2.17	6201	2.05	6354	4.22	3646	54
7	2685	2.17	6078	2.03	6607	4.22	3393	53
8	2815	2.18	5956	2.05	6860	4.20	3140	52
9	2946	2.17	5833	2.03	7112	4.22	2888	51
10	9.84 3076	2.17	9.85 5711	2.05	9.98 7365	4.22	10.01 2635	50
11	3206	2.17	5588	2.05	7618	4.22	2382	49
12	3336	2.17	5465	2.05	7871	4.20	2129	48
13	3466	2.15	5342	2.05	8123	4.22	1877	47
14	3595	2.17	5219	2.05	8376	4.22	1624	46
15	9.84 3725	2.17	9.85 5096	2.05	9.98 8629	4.22	10.01 1371	45
16	3855	2.15	4973	2.05	8882	4.20	1118	44
17	3984	2.17	4850	2.05	9134	4.22	0866	43
18	4114	2.15	4727	2.07	9387	4.22	0613	42
19	4243	2.15	4603	2.05	9640	4.22	0360	41
20	9.84 4372	2.17	9.85 4480	2.07	9.98 9893	4.20	10.01 0107	40
21	4502	2.15	4356	2.05	.99 0145	4.22	.00 9855	39
22	4631	2.15	4233	2.07	0398	4.22	9602	38
23	4760	2.15	4109	2.05	0651	4.20	9349	37
24	4889	2.15	3986	2.07	0903	4.22	9097	36
25	9.84 5018	2.15	9.85 3862	2.07	9.99 1156	4.22	10.00 8844	35
26	5147	2.15	3738	2.07	1409	4.22	8591	34
27	5276	2.15	3614	2.07	1662	4.20	8338	33
28	5405	2.13	3490	2.07	1914	4.22	8086	32
29	5533	2.15	3366	2.07	2167	4.22	7833	31
30	9.84 5662	2.13	9.85 3242	2.07	9.99 2420	4.20	10.00 7580	30
31	5790	2.15	3118	2.07	2672	4.22	7328	29
32	5919	2.13	2994	2.08	2925	4.22	7075	28
33	6047	2.13	2869	2.07	3178	4.22	6822	27
34	6175	2.15	2745	2.08	3431	4.20	6569	26
35	9.84 6304	2.13	9.85 2620	2.07	9.99 3683	4.22	10.00 6317	25
36	6432	2.13	2496	2.08	3936	4.22	6064	24
37	6560	2.13	2371	2.07	4189	4.20	5811	23
38	6688	2.13	2247	2.08	4441	4.22	5559	22
39	6816	2.13	2122	2.08	4694	4.22	5306	21
40	9.84 6944	2.12	9.85 1997	2.08	9.99 4947	4.20	10.00 5053	20
41	7071	2.13	1872	2.08	5199	4.22	4801	19
42	7199	2.13	1747	2.08	5452	4.22	4548	18
43	7327	2.12	1622	2.08	5705	4.20	4295	17
44	7454	2.13	1497	2.08	5957	4.22	4043	16
45	9.84 7582	2.12	9.85 1372	2.10	9.99 6210	4.22	10.00 3790	15
46	7709	2.12	1246	2.08	6463	4.20	3537	14
47	7836	2.13	1121	2.08	6715	4.22	3285	13
48	7964	2.12	0996	2.10	6968	4.22	3032	12
49	8091	2.12	0870	2.08	7221	4.20	2779	11
50	9.84 8218	2.12	9.85 0745	2.10	9.99 7473	4.22	10.00 2527	10
51	8345	2.12	0619	2.10	7726	4.22	2274	9
52	8472	2.12	0493	2.08	7979	4.20	2021	8
53	8599	2.12	0368	2.10	8231	4.22	1769	7
54	8726	2.10	0242	2.10	8484	4.22	1516	6
55	9.84 8852	2.12	9.85 0116	2.10	9.99 8737	4.20	10.00 1263	5
56	8979	2.12	84 9990	2.10	8989	4.22	1011	4
57	9106	2.10	9864	2.10	9242	4.22	0758	3
58	9232	2.12	9738	2.12	9495	4.20	0505	2
59	9359	2.10	9611	2.10	9.99 9747	4.22	0253	1
60	9.84 9485		9.84 9485		10.00 0000		10.00 0000	0
′	Cosine.	D.1″.	Sine.	D.1″.	Cotang.	D.1″.	Tang.	′

134° **45°**

Table IX Natural sines and cosines

′	0° Sine	0° Cosine	′	′	0° Sine	0° Cosine	′	′	0° Sine	0° Cosine	′
0	.00000	1	60	21	.00611	.99998	39	41	.01193	.99993	19
1	.00029	1	59	22	.00640	.99998	38	42	.01222	.99993	18
2	.00058	1	58	23	.00669	.99998	37	43	.01251	.99992	17
3	.00087	1	57	24	.00698	.99998	36	44	.01280	.99992	16
4	.00116	1	56	25	.00727	.99997	35	45	.01309	.99991	15
5	.00145	1	55	26	.00756	.99997	34	46	.01338	.99991	14
6	.00175	1	54	27	.00785	.99997	33	47	.01367	.99991	13
7	.00204	1	53	28	.00814	.99997	32	48	.01396	.99990	12
8	.00233	1	52	29	.00844	.99996	31	49	.01425	.99990	11
9	[illegible]	1	51	30	.00873	.99996	30	50	.01454	[illegible]	10
10	.00291	1	50	31	.00902	.99996	29	51	.01483	.99989	9
11	.00320	.99999	49	32	.00931	.99996	28	52	.01513	.99989	8
12	.00349	.99999	48	33	.00960	.99995	27	53	.01542	.99988	7
13	.00378	.99999	47	34	.00989	.99995	26	54	.01571	.99988	6
14	.00407	.99999	46	35	.01018	.99995	25	55	.01600	.99987	5
15	.00436	.99999	45	36	.01047	.99995	24	56	.01629	.99987	4
16	.00465	.99999	44	37	.01076	.99994	23	57	.01658	.99986	3
17	.00495	.99999	43	38	.01105	.99994	22	58	.01687	.99986	2
18	.00524	.99999	42	39	.01134	.99994	21	59	.01716	.99985	1
19	.00553	.99998	41	40	.01164	.99993	20	60	.01745	.99985	0
20	.00582	.99998	40								
′	Cosine	Sine	′	′	Cosine	Sine	′	′	Cosine	Sine	′
	89°				89°				89°		

Table IX—Continued

′	1° Sine	1° Cosine	2° Sine	2° Cosine	3° Sine	3° Cosine	4° Sine	4° Cosine	′
0	.01745	.99985	.03490	.99939	.05234	.99863	.06976	.99756	60
1	.01774	.99984	.03519	.99938	.05263	.99861	.07005	.99754	59
2	.01803	.99984	.03548	.99937	.05292	.99860	.07034	.99752	58
3	.01832	.99983	.03577	.99936	.05321	.99858	.07063	.99750	57
4	.01862	.99983	.03606	.99935	.05350	.99857	.07092	.99748	56
5	.01891	.99982	.03635	.99934	.05379	.99855	.07121	.99746	55
6	.01920	.99982	.03664	.99933	.05408	.99854	.07150	.99744	54
7	.01949	.99981	.03693	.99932	.05437	.99852	.07179	.99742	53
8	.01978	.99980	.03723	.99931	.05466	.99851	.07208	.99740	52
9	.02007	.99980	.03752	.99930	.05495	.99849	.07237	.99738	51
10	.02036	.99979	.03781	.99929	.05524	.99847	.07266	.99736	50
11	.02065	.99979	.03810	.99927	.05553	.99846	.07295	.99734	49
12	.02094	.99978	.03839	.99926	.05582	.99844	.07324	.99731	48
13	.02123	.99977	.03868	.99925	.05611	.99842	.07353	.99729	47
14	.02152	.99977	.03897	.99924	.05640	.99841	.07382	.99727	46
15	.02181	.99976	.03926	.99923	.05669	.99839	.07411	.99725	45
16	.02211	.99976	.03955	.99922	.05698	.99838	.07440	.99723	44
17	.02240	.99975	.03984	.99921	.05727	.99836	.07469	.99721	43
18	.02269	.99974	.04013	.99919	.05756	.99834	.07498	.99719	42
19	.02298	.99974	.04042	.99918	.05785	.99833	.07527	.99716	41
20	.02327	.99973	.04071	.99917	.05814	.99831	.07556	.99714	40
21	.02356	.99972	.04100	.99916	.05844	.99829	.07585	.99712	39
22	.02385	.99972	.04129	.99915	.05873	.99827	.07614	.99710	38
23	.02414	.99971	.04159	.99913	.05902	.99826	.07643	.99708	37
24	.02443	.99970	.04188	.99912	.05931	.99824	.07672	.99705	36
25	.02472	.99969	.04217	.99911	.05960	.99822	.07701	.99703	35
26	.02501	.99969	.04246	.99910	.05989	.99821	.07730	.99701	34
27	.02530	.99968	.04275	.99909	.06018	.99819	.07759	.99699	33
28	.02560	.99967	.04304	.99907	.06047	.99817	.07788	.99696	32
29	.02589	.99966	.04333	.99906	.06076	.99815	.07817	.99694	31
30	.02618	.99966	.04362	.99905	.06105	.99813	.07846	.99692	30
31	.02647	.99965	.04391	.99904	.06134	.99812	.07875	.99689	29
32	.02676	.99964	.04420	.99902	.06163	.99810	.07904	.99687	28
33	.02705	.99963	.04449	.99901	.06192	.99808	.07933	.99685	27
34	.02734	.99963	.04478	.99900	.06221	.99806	.07962	.99683	26
35	.02763	.99962	.04507	.99898	.06250	.99804	.07991	.99680	25
36	.02792	.99961	.04536	.99897	.06279	.99803	.08020	.99678	24
37	.02821	.99960	.04565	.99896	.06308	.99801	.08049	.99676	23
38	.02850	.99959	.04594	.99894	.06337	.99799	.08078	.99673	22
39	.02879	.99959	.04623	.99893	.06366	.99797	.08107	.99671	21
40	.02908	.99958	.04653	.99892	.06395	.99795	.08136	.99668	20
41	.02938	.99957	.04682	.99890	.06424	.99793	.08165	.99666	19
42	.02967	.99956	.04711	.99889	.06453	.99792	.08194	.99664	18
43	.02996	.99955	.04740	.99888	.06482	.99790	.08223	.99661	17
44	.03025	.99954	.04769	.99886	.06511	.99788	.08252	.99659	16
45	.03054	.99953	.04798	.99885	.06540	.99786	.08281	.99657	15
46	.03083	.99952	.04827	.99883	.06569	.99784	.08310	.99654	14
47	.03112	.99952	.04856	.99882	.06598	.99782	.08339	.99652	13
48	.03141	.99951	.04885	.99881	.06627	.99780	.08368	.99649	12
49	.03170	.99950	.04914	.99879	.06656	.99778	.08397	.99647	11
50	.03199	.99949	.04943	.99878	.06685	.99776	.08426	.99644	10
51	.03228	.99948	.04972	.99876	.06714	.99774	.08455	.99642	9
52	.03257	.99947	.05001	.99875	.06743	.99772	.08484	.99639	8
53	.03286	.99946	.05030	.99873	.06773	.99770	.08513	.99637	7
54	.03316	.99945	.05059	.99872	.06802	.99768	.08542	.99635	6
55	.03345	.99944	.05088	.99870	.06831	.99766	.08571	.99632	5
56	.03374	.99943	.05117	.99869	.06860	.99764	.08600	.99630	4
57	.03403	.99942	.05146	.99867	.06889	.99762	.08629	.99627	3
58	.03432	.99941	.05175	.99866	.06918	.99760	.08658	.99625	2
59	.03461	.99940	.05205	.99864	.06947	.99758	.08687	.99622	1
60	.03490	.99939	.05234	.99863	.06976	.99756	.08716	.99619	0
′	88° Cosine	88° Sine	87° Cosine	87° Sine	86° Cosine	86° Sine	85° Cosine	85° Sine	′

Table IX—Continued

′	5°		6°		7°		8°		′
	Sine	Cosine	Sine	Cosine	Sine	Cosine	Sine	Cosine	
0	.08716	.99619	.10453	.99452	.12187	.99255	.13917	.99027	60
1	.08745	.99617	.10482	.99449	.12216	.99251	.13946	.99023	59
2	.08774	.99614	.10511	.99446	.12245	.99248	.13975	.99019	58
3	.08803	.99612	.10540	.99443	.12274	.99244	.14004	.99015	57
4	.08831	.99609	.10569	.99440	.12302	.99240	.14033	.99011	56
5	.08860	.99607	.10597	.99437	.12331	.99237	.14061	.99006	55
6	.08889	.99604	.10626	.99434	.12360	.99233	.14090	.99002	54
7	.08918	.99602	.10655	.99431	.12389	.99230	.14119	.98998	53
8	.08947	.99599	.10684	.99428	.12418	.99226	.14148	.98994	52
9	.08976	.99596	.10713	.99424	.12447	.99222	.14177	.98990	51
10	.09005	.99594	.10742	.99421	.12476	.99219	.14205	.98986	50
11	.09034	.99591	.10771	.99418	.12504	.99215	.14234	.98982	49
12	.09063	.99588	.10800	.99415	.12533	.99211	.14263	.98978	48
13	.09092	.99586	.10829	.99412	.12562	.99208	.14292	.98973	47
14	.09121	.99583	.10858	.99409	.12591	.99204	.14320	.98969	46
15	.09150	.99580	.10887	.99406	.12620	.99200	.14349	.98965	45
16	.09179	.99578	.10916	.99402	.12649	.99197	.14378	.98961	44
17	.09208	.99575	.10945	.99399	.12678	.99193	.14407	.98957	43
18	.09237	.99572	.10973	.99396	.12706	.99189	.14436	.98953	42
19	.09266	.99570	.11002	.99393	.12735	.99186	.14464	.98948	41
20	.09295	.99567	.11031	.99390	.12764	.99182	.14493	.98944	40
21	.09324	.99564	.11060	.99386	.12793	.99178	.14522	.98940	39
22	.09353	.99562	.11089	.99383	.12822	.99175	.14551	.98936	38
23	.09382	.99559	.11118	.99380	.12851	.99171	.14580	.98931	37
24	.09411	.99556	.11147	.99377	.12880	.99167	.14608	.98927	36
25	.09440	.99553	.11176	.99374	.12908	.99163	.14637	.98923	35
26	.09469	.99551	.11205	.99370	.12937	.99160	.14666	.98919	34
27	.09498	.99548	.11234	.99367	.12966	.99156	.14695	.98914	33
28	.09527	.99545	.11263	.99364	.12995	.99152	.14723	.98910	32
29	.09556	.99542	.11291	.99360	.13024	.99148	.14752	.98906	31
30	.09585	.99540	.11320	.99357	.13053	.99144	.14781	.98902	30
31	.09614	.99537	.11349	.99354	.13081	.99141	.14810	.98897	29
32	.09642	.99534	.11378	.99351	.13110	.99137	.14838	.98893	28
33	.09671	.99531	.11407	.99347	.13139	.99133	.14867	.98889	27
34	.09700	.99528	.11436	.99344	.13168	.99129	.14896	.98884	26
35	.09729	.99526	.11465	.99341	.13197	.99125	.14925	.98880	25
36	.09758	.99523	.11494	.99337	.13226	.99122	.14954	.98876	24
37	.09787	.99520	.11523	.99334	.13254	.99118	.14982	.98871	23
38	.09816	.99517	.11552	.99331	.13283	.99114	.15011	.98867	22
39	.09845	.99514	.11580	.99327	.13312	.99110	.15040	.98863	21
40	.09874	.99511	.11609	.99324	.13341	.99106	.15069	.98858	20
41	.09903	.99508	.11638	.99320	.13370	.99102	.15097	.98854	19
42	.09932	.99506	.11667	.99317	.13399	.99098	.15126	.98849	18
43	.09961	.99503	.11696	.99314	.13427	.99094	.15155	.98845	17
44	.09990	.99500	.11725	.99310	.13456	.99091	.15184	.98841	16
45	.10019	.99497	.11754	.99307	.13485	.99087	.15212	.98836	15
46	.10048	.99494	.11783	.99303	.13514	.99083	.15241	.98832	14
47	.10077	.99491	.11812	.99300	.13543	.99079	.15270	.98827	13
48	.10106	.99488	.11840	.99297	.13572	.99075	.15299	.98823	12
49	.10135	.99485	.11869	.99293	.13600	.99071	.15327	.98818	11
50	.10164	.99482	.11898	.99290	.13629	.99067	.15356	.98814	10
51	.10192	.99479	.11927	.99286	.13658	.99063	.15385	.98809	9
52	.10221	.99476	.11956	.99283	.13687	.99059	.15414	.98805	8
53	.10250	.99473	.11985	.99279	.13716	.99055	.15442	.98800	7
54	.10279	.99470	.12014	.99276	.13744	.99051	.15471	.98796	6
55	.10308	.99467	.12043	.99272	.13773	.99047	.15500	.98791	5
56	.10337	.99464	.12071	.99269	.13802	.99043	.15529	.98787	4
57	.10366	.99461	.12100	.99265	.13831	.99039	.15557	.98782	3
58	.10395	.99458	.12129	.99262	.13860	.99035	.15586	.98778	2
59	.10424	.99455	.12158	.99258	.13889	.99031	.15615	.98773	1
60	.10453	.99452	.12187	.99255	.13917	.99027	.15643	.98769	0
′	Cosine	Sine	Cosine	Sine	Cosine	Sine	Cosine	Sine	′
	84°		83°		82°		81°		

Table IX—Continued

′	9° Sine	9° Cosine	10° Sine	10° Cosine	11° Sine	11° Cosine	12° Sine	12° Cosine	′
0	.15643	.98769	.17365	.98481	.19081	.98163	.20791	.97815	60
1	.15672	.98764	.17393	.98476	.19109	.98157	.20820	.97809	59
2	.15701	.98760	.17422	.98471	.19138	.98152	.20848	.97803	58
3	.15730	.98755	.17451	.98466	.19167	.98146	.20877	.97797	57
4	.15758	.98751	.17479	.98461	.19195	.98140	.20905	.97791	56
5	.15787	.98746	.17508	.98455	.19224	.98135	.20933	.97784	55
6	.15816	.98741	.17537	.98450	.19252	.98129	.20962	.97778	54
7	.15845	.98737	.17565	.98445	.19281	.98124	.20990	.97772	53
8	.15873	.98732	.17594	.98440	.19309	.98118	.21019	.97766	52
9	.15902	.98728	.17623	.98435	.19338	.98112	.21047	.97760	51
10	.15931	.98723	.17651	.98430	.19366	.98107	.21076	.97754	50
11	.15959	.98718	.17680	.98425	.19395	.98101	.21104	.97748	49
12	.15988	.98714	.17708	.98420	.19423	.98096	.21132	.97742	48
13	.16017	.98709	.17737	.98414	.19452	.98090	.21161	.97735	47
14	.16046	.98704	.17766	.98409	.19481	.98084	.21189	.97729	46
15	.16074	.98700	.17794	.98404	.19509	.98079	.21218	.97723	45
16	.16103	.98695	.17823	.98399	.19538	.98073	.21246	.97717	44
17	.16132	.98690	.17852	.98394	.19566	.98067	.21275	.97711	43
18	.16160	.98686	.17880	.98389	.19595	.98061	.21303	.97705	42
19	.16189	.98681	.17909	.98383	.19623	.98056	.21331	.97698	41
20	.16218	.98676	.17937	.98378	.19652	.98050	.21360	.97692	40
21	.16246	.98671	.17966	.98373	.19680	.98044	.21388	.97686	39
22	.16275	.98667	.17995	.98368	.19709	.98039	.21417	.97680	38
23	.16304	.98662	.18023	.98362	.19737	.98033	.21445	.97673	37
24	.16333	.98657	.18052	.98357	.19766	.98027	.21474	.97667	36
25	.16361	.98652	.18081	.98352	.19794	.98021	.21502	.97661	35
26	.16390	.98648	.18109	.98347	.19823	.98016	.21530	.97655	34
27	.16419	.98643	.18138	.98341	.19851	.98010	.21559	.97648	33
28	.16447	.98638	.18166	.98336	.19880	.98004	.21587	.97642	32
29	.16476	.98633	.18195	.98331	.19908	.97998	.21616	.97636	31
30	.16505	.98629	.18224	.98325	.19937	.97992	.21644	.97630	30
31	.16533	.98624	.18252	.98320	.19965	.97987	.21672	.97623	29
32	.16562	.98619	.18281	.98315	.19994	.97981	.21701	.97617	28
33	.16591	.98614	.18309	.98310	.20022	.97975	.21729	.97611	27
34	.16620	.98609	.18338	.98304	.20051	.97969	.21758	.97604	26
35	.16648	.98604	.18367	.98299	.20079	.97963	.21786	.97598	25
36	.16677	.98600	.18395	.98294	.20108	.97958	.21814	.97592	24
37	.16706	.98595	.18424	.98288	.20136	.97952	.21843	.97585	23
38	.16734	.98590	.18452	.98283	.20165	.97946	.21871	.97579	22
39	.16763	.98585	.18481	.98277	.20193	.97940	.21899	.97573	21
40	.16792	.98580	.18509	.98272	.20222	.97934	.21928	.97566	20
41	.16820	.98575	.18538	.98267	.20250	.97928	.21956	.97560	19
42	.16849	.98570	.18567	.98261	.20279	.97922	.21985	.97553	18
43	.16878	.98565	.18595	.98256	.20307	.97916	.22013	.97547	17
44	.16906	.98561	.18624	.98250	.20336	.97910	.22041	.97541	16
45	.16935	.98556	.18652	.98245	.20364	.97905	.22070	.97534	15
46	.16964	.98551	.18681	.98240	.20393	.97899	.22098	.97528	14
47	.16992	.98546	.18710	.98234	.20421	.97893	.22126	.97521	13
48	.17021	.98541	.18738	.98229	.20450	.97887	.22155	.97515	12
49	.17050	.98536	.18767	.98223	.20478	.97881	.22183	.97508	11
50	.17078	.98531	.18795	.98218	.20507	.97875	.22212	.97502	10
51	.17107	.98526	.18824	.98212	.20535	.97869	.22240	.97496	9
52	.17136	.98521	.18852	.98207	.20563	.97863	.22268	.97489	8
53	.17164	.98516	.18881	.98201	.20592	.97857	.22297	.97483	7
54	.17193	.98511	.18910	.98196	.20620	.97851	.22325	.97476	6
55	.17222	.98506	.18938	.98190	.20649	.97845	.22353	.97470	5
56	.17250	.98501	.18967	.98185	.20677	.97839	.22382	.97463	4
57	.17279	.98496	.18995	.98179	.20706	.97833	.22410	.97457	3
58	.17308	.98491	.19024	.98174	.20734	.97827	.22438	.97450	2
59	.17336	.98486	.19052	.98168	.20763	.97821	.22467	.97444	1
60	.17365	.98481	.19081	.98163	.20791	.97815	.22495	.97437	0
′	Cosine	Sine	Cosine	Sine	Cosine	Sine	Cosine	Sine	′
	80°		79°		78°		77°		

Table IX—Continued

′	13° Sine	13° Cosine	14° Sine	14° Cosine	15° Sine	15° Cosine	16° Sine	16° Cosine	′
0	.22495	.97437	.24192	.97030	.25882	.96593	.27564	.96126	60
1	.22523	.97430	.24220	.97023	.25910	.96585	.27592	.96118	59
2	.22552	.97424	.24249	.97015	.25938	.96578	.27620	.96110	58
3	.22580	.97417	.24277	.97008	.25966	.96570	.27648	.96102	57
4	.22608	.97411	.24305	.97001	.25994	.96562	.27676	.96094	56
5	.22637	.97404	.24333	.96994	.26022	.96555	.27704	.96086	55
6	.22665	.97398	.24362	.96987	.26050	.96547	.27731	.96078	54
7	.22693	.97391	.24390	.96980	.26079	.96540	.27759	.96070	53
8	.22722	.97384	.24418	.96973	.26107	.96532	.27787	.96062	52
9	.22750	.97378	.24446	.96966	.26135	.96524	.27815	.96054	51
10	.22778	.97371	.24474	.96959	.26163	.96517	.27843	.96046	50
11	.22807	.97365	.24503	.96952	.26191	.96509	.27871	.96037	49
12	.22835	.97358	.24531	.96945	.26219	.96502	.27899	.96029	48
13	.22863	.97351	.24559	.96937	.26247	.96494	.27927	.96021	47
14	.22892	.97345	.24587	.96930	.26275	.96486	.27955	.96013	46
15	.22920	.97338	.24615	.96923	.26303	.96479	.27983	.96005	45
16	.22948	.97331	.24644	.96916	.26331	.96471	.28011	.95997	44
17	.22977	.97325	.24672	.96909	.26359	.96463	.28039	.95989	43
18	.23005	.97318	.24700	.96902	.26387	.96456	.28067	.95981	42
19	.23033	.97311	.24728	.96894	.26415	.96448	.28095	.95972	41
20	.23062	.97304	.24756	.96887	.26443	.96440	.28123	.95964	40
21	.23090	.97298	.24784	.96880	.26471	.96433	.28150	.95956	39
22	.23118	.97291	.24813	.96873	.26500	.96425	.28178	.95948	38
23	.23146	.97284	.24841	.96866	.26528	.96417	.28206	.95940	37
24	.23175	.97278	.24869	.96858	.26556	.96410	.28234	.95931	36
25	.23203	.97271	.24897	.96851	.26584	.96402	.28262	.95923	35
26	.23231	.97264	.24925	.96844	.26612	.96394	.28290	.95915	34
27	.23260	.97257	.24954	.96837	.26640	.96386	.28318	.95907	33
28	.23288	.97251	.24982	.96829	.26668	.96379	.28346	.95898	32
29	.23316	.97244	.25010	.96822	.26696	.96371	.28374	.95890	31
30	.23345	.97237	.25038	.96815	.26724	.96363	.28402	.95882	30
31	.23373	.97230	.25066	.96807	.26752	.96355	.28429	.95874	29
32	.23401	.97223	.25094	.96800	.26780	.96347	.28457	.95865	28
33	.23429	.97217	.25122	.96793	.26808	.96340	.28485	.95857	27
34	.23458	.97210	.25151	.96786	.26836	.96332	.28513	.95849	26
35	.23486	.97203	.25179	.96778	.26864	.96324	.28541	.95841	25
36	.23514	.97196	.25207	.96771	.26892	.96316	.28569	.95832	24
37	.23542	.97189	.25235	.96764	.26920	.96308	.28597	.95824	23
38	.23571	.97182	.25263	.96756	.26948	.96301	.28625	.95816	22
39	.23599	.97176	.25291	.96749	.26976	.96293	.28652	.95807	21
40	.23627	.97169	.25320	.96742	.27004	.96285	.28680	.95799	20
41	.23656	.97162	.25348	.96734	.27032	.96277	.28708	.95791	19
42	.23684	.97155	.25376	.96727	.27060	.96269	.28736	.95782	18
43	.23712	.97148	.25404	.96719	.27088	.96261	.28764	.95774	17
44	.23740	.97141	.25432	.96712	.27116	.96253	.28792	.95766	16
45	.23769	.97134	.25460	.96705	.27144	.96246	.28820	.95757	15
46	.23797	.97127	.25488	.96697	.27172	.96238	.28847	.95749	14
47	.23825	.97120	.25516	.96690	.27200	.96230	.28875	.95740	13
48	.23853	.97113	.25545	.96682	.27228	.96222	.28903	.95732	12
49	.23882	.97106	.25573	.96675	.27256	.96214	.28931	.95724	11
50	.23910	.97100	.25601	.96667	.27284	.96206	.28959	.95715	10
51	.23938	.97093	.25629	.96660	.27312	.96198	.28987	.95707	9
52	.23966	.97086	.25657	.96653	.27340	.96190	.29015	.95698	8
53	.23995	.97079	.25685	.96645	.27368	.96182	.29042	.95690	7
54	.24023	.97072	.25713	.96638	.27396	.96174	.29070	.95681	6
55	.24051	.97065	.25741	.96630	.27424	.96166	.29098	.95673	5
56	.24079	.97058	.25769	.96623	.27452	.96158	.29126	.95664	4
57	.24108	.97051	.25798	.96615	.27480	.96150	.29154	.95656	3
58	.24136	.97044	.25826	.96608	.27508	.96142	.29182	.95647	2
59	.24164	.97037	.25854	.96600	.27536	.96134	.29209	.95639	1
60	.24192	.97030	.25882	.96593	.27564	.96126	.29237	.95630	0
′	76° Cosine	76° Sine	75° Cosine	75° Sine	74° Cosine	74° Sine	73° Cosine	73° Sine	′

Table IX—Continued

′	17° Sine	17° Cosine	18° Sine	18° Cosine	19° Sine	19° Cosine	20° Sine	20° Cosine	′
0	.29237	.95630	.30902	.95106	.32557	.94552	.34202	.93969	60
1	.29265	.95622	.30929	.95097	.32584	.94542	.34229	.93959	59
2	.29293	.95613	.30957	.95088	.32612	.94533	.34257	.93949	58
3	.29321	.95605	.30985	.95079	.32639	.94523	.34284	.93939	57
4	.29348	.95596	.31012	.95070	.32667	.94514	.34311	.93929	56
5	.29376	.95588	.31040	.95061	.32694	.94504	.34339	.93919	55
6	.29404	.95579	.31068	.95052	.32722	.94495	.34366	.93909	54
7	.29432	.95571	.31095	.95043	.32749	.94485	.34393	.93899	53
8	.29460	.95562	.31123	.95033	.32777	.94476	.34421	.93889	52
9	.29487	.95554	.31151	.95024	.32804	.94466	.34448	.93879	51
10	.29515	.95545	.31178	.95015	.32832	.94457	.34475	.93869	50
11	.29543	.95536	.31206	.95006	.32859	.94447	.34503	.93859	49
12	.29571	.95528	.31233	.94997	.32887	.94438	.34530	.93849	48
13	.29599	.95519	.31261	.94988	.32914	.94428	.34557	.93839	47
14	.29626	.95511	.31289	.94979	.32942	.94418	.34584	.93829	46
15	.29654	.95502	.31316	.94970	.32969	.94409	.34612	.93819	45
16	.29682	.95493	.31344	.94961	.32997	.94399	.34639	.93809	44
17	.29710	.95485	.31372	.94952	.33024	.94390	.34666	.93799	43
18	.29737	.95476	.31399	.94943	.33051	.94380	.34694	.93789	42
19	.29765	.95467	.31427	.94933	.33079	.94370	.34721	.93779	41
20	.29793	.95459	.31454	.94924	.33106	.94361	.34748	.93769	40
21	.29821	.95450	.31482	.94915	.33134	.94351	.34775	.93759	39
22	.29849	.95441	.31510	.94906	.33161	.94342	.34803	.93748	38
23	.29876	.95433	.31537	.94897	.33189	.94332	.34830	.93738	37
24	.29904	.95424	.31565	.94888	.33216	.94322	.34857	.93728	36
25	.29932	.95415	.31593	.94878	.33244	.94313	.34884	.93718	35
26	.29960	.95407	.31620	.94869	.33271	.94303	.34912	.93708	34
27	.29987	.95398	.31648	.94860	.33298	.94293	.34939	.93698	33
28	.30015	.95389	.31675	.94851	.33326	.94284	.34966	.93688	32
29	.30043	.95380	.31703	.94842	.33353	.94274	.34993	.93677	31
30	.30071	.95372	.31730	.94832	.33381	.94264	.35021	.93667	30
31	.30098	.95363	.31758	.94823	.33408	.94254	.35048	.93657	29
32	.30126	.95354	.31786	.94814	.33436	.94245	.35075	.93647	28
33	.30154	.95345	.31813	.94805	.33463	.94235	.35102	.93637	27
34	.30182	.95337	.31841	.94795	.33490	.94225	.35130	.93626	26
35	.30209	.95328	.31868	.94786	.33518	.94215	.35157	.93616	25
36	.30237	.95319	.31896	.94777	.33545	.94206	.35184	.93606	24
37	.30265	.95310	.31923	.94768	.33573	.94196	.35211	.93596	23
38	.30292	.95301	.31951	.94758	.33600	.94186	.35239	.93585	22
39	.30320	.95293	.31979	.94749	.33627	.94176	.35266	.93575	21
40	.30348	.95284	.32006	.94740	.33655	.94167	.35293	.93565	20
41	.30376	.95275	.32034	.94730	.33682	.94157	.35320	.93555	19
42	.30403	.95266	.32061	.94721	.33710	.94147	.35347	.93544	18
43	.30431	.95257	.32089	.94712	.33737	.94137	.35375	.93534	17
44	.30459	.95248	.32116	.94702	.33764	.94127	.35402	.93524	16
45	.30486	.95240	.32144	.94693	.33792	.94118	.35429	.93514	15
46	.30514	.95231	.32171	.94684	.33819	.94108	.35456	.93503	14
47	.30542	.95222	.32199	.94674	.33846	.94098	.35484	.93493	13
48	.30570	.95213	.32227	.94665	.33874	.94088	.35511	.93483	12
49	.30597	.95204	.32254	.94656	.33901	.94078	.35538	.93472	11
50	.30625	.95195	.32282	.94646	.33929	.94068	.35565	.93462	10
51	.30653	.95186	.32309	.94637	.33956	.94058	.35592	.93452	9
52	.30680	.95177	.32337	.94627	.33983	.94049	.35619	.93441	8
53	.30708	.95168	.32364	.94618	.34011	.94039	.35647	.93431	7
54	.30736	.95159	.32392	.94609	.34038	.94029	.35674	.93420	6
55	.30763	.95150	.32419	.94599	.34065	.94019	.35701	.93410	5
56	.30791	.95142	.32447	.94590	.34093	.94009	.35728	.93400	4
57	.30819	.95133	.32474	.94580	.34120	.93999	.35755	.93389	3
58	.30846	.95124	.32502	.94571	.34147	.93989	.35782	.93379	2
59	.30874	.95115	.32529	.94561	.34175	.93979	.35810	.93368	1
60	.30902	.95106	.32557	.94552	.34202	.93969	.35837	.93358	0
′	72° Cosine	72° Sine	71° Cosine	71° Sine	70° Cosine	70° Sine	69° Cosine	69° Sine	′

Table IX—Continued

′	21° Sine	21° Cosine	22° Sine	22° Cosine	23° Sine	23° Cosine	24° Sine	24° Cosine	′
0	.35837	.93358	.37461	.92718	.39073	.92050	.40674	.91355	60
1	.35864	.93348	.37488	.92707	.39100	.92039	.40700	.91343	59
2	.35891	.93337	.37515	.92697	.39127	.92028	.40727	.91331	58
3	.35918	.93327	.37542	.92686	.39153	.92016	.40753	.91319	57
4	.35945	.93316	.37569	.92675	.39180	.92005	.40780	.91307	56
5	.35973	.93306	.37595	.92664	.39207	.91994	.40806	.91295	55
6	.36000	.93295	.37622	.92653	.39234	.91982	.40833	.91283	54
7	.36027	.93285	.37649	.92642	.39260	.91971	.40860	.91272	53
8	.36054	.93274	.37676	.92631	.39287	.91959	.40886	.91260	52
9	.36081	.93264	.37703	.92620	.39314	.91948	.40913	.91248	51
10	.36108	.93253	.37730	.92609	.39341	.91936	.40939	.91236	50
11	.36135	.93243	.37757	.92598	.39367	.91925	.40966	.91224	49
12	.36162	.93232	.37784	.92587	.39394	.91914	.40992	.91212	48
13	.36190	.93222	.37811	.92576	.39421	.91902	.41019	.91200	47
14	.36217	.93211	.37838	.92565	.39448	.91891	.41045	.91188	46
15	.36244	.93201	.37865	.92554	.39474	.91879	.41072	.91176	45
16	.36271	.93190	.37892	.92543	.39501	.91868	.41098	.91164	44
17	.36298	.93180	.37919	.92532	.39528	.91856	.41125	.91152	43
18	.36325	.93169	.37946	.92521	.39555	.91845	.41151	.91140	42
19	.36352	.93159	.37973	.92510	.39581	.91833	.41178	.91128	41
20	.36379	.93148	.37999	.92499	.39608	.91822	.41204	.91116	40
21	.36406	.93137	.38026	.92488	.39635	.91810	.41231	.91104	39
22	.36434	.93127	.38053	.92477	.39661	.91799	.41257	.91092	38
23	.36461	.93116	.38080	.92466	.39688	.91787	.41284	.91080	37
24	.36488	.93106	.38107	.92455	.39715	.91775	.41310	.91068	36
25	.36515	.93095	.38134	.92444	.39741	.91764	.41337	.91056	35
26	.36542	.93084	.38161	.92432	.39768	.91752	.41363	.91044	34
27	.36569	.93074	.38188	.92421	.39795	.91741	.41390	.91032	33
28	.36596	.93063	.38215	.92410	.39822	.91729	.41416	.91020	32
29	.36623	.93052	.38241	.92399	.39848	.91718	.41443	.91008	31
30	.36650	.93042	.38268	.92388	.39875	.91706	.41469	.90996	30
31	.36677	.93031	.38295	.92377	.39902	.91694	.41496	.90984	29
32	.36704	.93020	.38322	.92366	.39928	.91683	.41522	.90972	28
33	.36731	.93010	.38349	.92355	.39955	.91671	.41549	.90960	27
34	.36758	.92999	.38376	.92343	.39982	.91660	.41575	.90948	26
35	.36785	.92988	.38403	.92332	.40008	.91648	.41602	.90936	25
36	.36812	.92978	.38430	.92321	.40035	.91636	.41628	.90924	24
37	.36839	.92967	.38456	.92310	.40062	.91625	.41655	.90911	23
38	.36867	.92956	.38483	.92299	.40088	.91613	.41681	.90899	22
39	.36894	.92945	.38510	.92287	.40115	.91601	.41707	.90887	21
40	.36921	.92935	.38537	.92276	.40141	.91590	.41734	.90875	20
41	.36948	.92924	.38564	.92265	.40168	.91578	.41760	.90863	19
42	.36975	.92913	.38591	.92254	.40195	.91566	.41787	.90851	18
43	.37002	.92902	.38617	.92243	.40221	.91555	.41813	.90839	17
44	.37029	.92892	.38644	.92231	.40248	.91543	.41840	.90826	16
45	.37056	.92881	.38671	.92220	.40275	.91531	.41866	.90814	15
46	.37083	.92870	.38698	.92209	.40301	.91519	.41892	.90802	14
47	.37110	.92859	.38725	.92198	.40328	.91508	.41919	.90790	13
48	.37137	.92849	.38752	.92186	.40355	.91496	.41945	.90778	12
49	.37164	.92838	.38778	.92175	.40381	.91484	.41972	.90766	11
50	.37191	.92827	.38805	.92164	.40408	.91472	.41998	.90753	10
51	.37218	.92816	.38832	.92152	.40434	.91461	.42024	.90741	9
52	.37245	.92805	.38859	.92141	.40461	.91449	.42051	.90729	8
53	.37272	.92794	.38886	.92130	.40488	.91437	.42077	.90717	7
54	.37299	.92784	.38912	.92119	.40514	.91425	.42104	.90704	6
55	.37326	.92773	.38939	.92107	.40541	.91414	.42130	.90692	5
56	.37353	.92762	.38966	.92096	.40567	.91402	.42156	.90680	4
57	.37380	.92751	.38993	.92085	.40594	.91390	.42183	.90668	3
58	.37407	.92740	.39020	.92073	.40621	.91378	.42209	.90655	2
59	.37434	.92729	.39046	.92062	.40647	.91366	.42235	.90643	1
60	.37461	.92718	.39073	.92050	.40674	.91355	.42262	.90631	0
′	Cosine	Sine	Cosine	Sine	Cosine	Sine	Cosine	Sine	′
	68°		67°		66°		65°		

Table IX—Continued

′	25° Sine	25° Cosine	26° Sine	26° Cosine	27° Sine	27° Cosine	28° Sine	28° Cosine	′
0	.42262	.90631	.43837	.89879	.45399	.89101	.46947	.88295	60
1	.42288	.90618	.43863	.89867	.45425	.89087	.46973	.88281	59
2	.42315	.90606	.43889	.89854	.45451	.89074	.46999	.88267	58
3	.42341	.90594	.43916	.89841	.45477	.89061	.47024	.88254	57
4	.42367	.90582	.43942	.89828	.45503	.89048	.47050	.88240	56
5	.42394	.90569	.43968	.89816	.45529	.89035	.47076	.88226	55
6	.42420	.90557	.43994	.89803	.45554	.89021	.47101	.88213	54
7	.42446	.90545	.44020	.89790	.45580	.89008	.47127	.88199	53
8	.42473	.90532	.44046	.89777	.45606	.88995	.47153	.88185	52
9	.42499	.90520	.44072	.89764	.45632	.88981	.47178	.88172	51
10	.42525	.90507	.44098	.89752	.45658	.88968	.47204	.88158	50
11	.42552	.90495	.44124	.89739	.45684	.88955	.47229	.88144	49
12	.42578	.90483	.44151	.89726	.45710	.88942	.47255	.88130	48
13	.42604	.90470	.44177	.89713	.45736	.88928	.47281	.88117	47
14	.42631	.90458	.44203	.89700	.45762	.88915	.47306	.88103	46
15	.42657	.90446	.44229	.89687	.45787	.88902	.47332	.88089	45
16	.42683	.90433	.44255	.89674	.45813	.88888	.47358	.88075	44
17	.42709	.90421	.44281	.89662	.45839	.88875	.47383	.88062	43
18	.42736	.90408	.44307	.89649	.45865	.88862	.47409	.88048	42
19	.42762	.90396	.44333	.89636	.45891	.88848	.47434	.88034	41
20	.42788	.90383	.44359	.89623	.45917	.88835	.47460	.88020	40
21	.42815	.90371	.44385	.89610	.45942	.88822	.47486	.88006	39
22	.42841	.90358	.44411	.89597	.45968	.88808	.47511	.87993	38
23	.42867	.90346	.44437	.89584	.45994	.88795	.47537	.87979	37
24	.42894	.90334	.44464	.89571	.46020	.88782	.47562	.87965	36
25	.42920	.90321	.44490	.89558	.46046	.88768	.47588	.87951	35
26	.42946	.90309	.44516	.89545	.46072	.88755	.47614	.87937	34
27	.42972	.90296	.44542	.89532	.46097	.88741	.47639	.87923	33
28	.42999	.90284	.44568	.89519	.46123	.88728	.47665	.87909	32
29	.43025	.90271	.44594	.89506	.46149	.88715	.47690	.87896	31
30	.43051	.90259	.44620	.89493	.46175	.88701	.47716	.87882	30
31	.43077	.90246	.44646	.89480	.46201	.88688	.47741	.87868	29
32	.43104	.90233	.44672	.89467	.46226	.88674	.47767	.87854	28
33	.43130	.90221	.44698	.89454	.46252	.88661	.47793	.87840	27
34	.43156	.90208	.44724	.89441	.46278	.88647	.47818	.87826	26
35	.43182	.90196	.44750	.89428	.46304	.88634	.47844	.87812	25
36	.43209	.90183	.44776	.89415	.46330	.88620	.47869	.87798	24
37	.43235	.90171	.44802	.89402	.46355	.88607	.47895	.87784	23
38	.43261	.90158	.44828	.89389	.46381	.88593	.47920	.87770	22
39	.43287	.90146	.44854	.89376	.46407	.88580	.47946	.87756	21
40	.43313	.90133	.44880	.89363	.46433	.88566	.47971	.87743	20
41	.43340	.90120	.44906	.89350	.46458	.88553	.47997	.87729	19
42	.43366	.90108	.44932	.89337	.46484	.88539	.48022	.87715	18
43	.43392	.90095	.44958	.89324	.46510	.88526	.48048	.87701	17
44	.43418	.90082	.44984	.89311	.46536	.88512	.48073	.87687	16
45	.43445	.90070	.45010	.89298	.46561	.88499	.48099	.87673	15
46	.43471	.90057	.45036	.89285	.46587	.88485	.48124	.87659	14
47	.43497	.90045	.45062	.89272	.46613	.88472	.48150	.87645	13
48	.43523	.90032	.45088	.89259	.46639	.88458	.48175	.87631	12
49	.43549	.90019	.45114	.89245	.46664	.88445	.48201	.87617	11
50	.43575	.90007	.45140	.89232	.46690	.88431	.48226	.87603	10
51	.43602	.89994	.45166	.89219	.46716	.88417	.48252	.87589	9
52	.43628	.89981	.45192	.89206	.46742	.88404	.48277	.87575	8
53	.43654	.89968	.45218	.89193	.46767	.88390	.48303	.87561	7
54	.43680	.89956	.45243	.89180	.46793	.88377	.48328	.87546	6
55	.43706	.89943	.45269	.89167	.46819	.88363	.48354	.87532	5
56	.43733	.89930	.45295	.89153	.46844	.88349	.48379	.87518	4
57	.43759	.89918	.45321	.89140	.46870	.88336	.48405	.87504	3
58	.43785	.89905	.45347	.89127	.46896	.88322	.48430	.87490	2
59	.43811	.89892	.45373	.89114	.46921	.88308	.48456	.87476	1
60	.43837	.89879	.45399	.89101	.46947	.88295	.48481	.87462	0
′	Cosine 64°	Sine 64°	Cosine 63°	Sine 63°	Cosine 62°	Sine 62°	Cosine 61°	Sine 61°	′

Table IX—Continued

	29°		30°		31°		32°		
′	Sine	Cosine	Sine	Cosine	Sine	Cosine	Sine	Cosine	′
0	.48481	.87462	.50000	.86603	.51504	.85717	.52992	.84805	60
1	.48506	.87448	.50025	.86588	.51529	.85702	.53017	.84789	59
2	.48532	.87434	.50050	.86573	.51554	.85687	.53041	.84774	58
3	.48557	.87420	.50076	.86559	.51579	.85672	.53066	.84759	57
4	.48583	.87406	.50101	.86544	.51604	.85657	.53091	.84743	56
5	.48608	.87391	.50126	.86530	.51628	.85642	.53115	.84728	55
6	.48634	.87377	.50151	.86515	.51653	.85627	.53140	.84712	54
7	.48659	.87363	.50176	.86501	.51678	.85612	.53164	.84697	53
8	.48684	.87349	.50201	.86486	.51703	.85597	.53189	.84681	52
9	.48710	.87335	.50227	.86471	.51728	.85582	.53214	.84666	51
10	.48735	.87321	.50252	.86457	.51753	.85567	.53238	.84650	50
11	.48761	.87306	.50277	.86442	.51778	.85551	.53263	.84635	49
12	.48786	.87292	.50302	.86427	.51803	.85536	.53288	.84619	48
13	.48811	.87278	.50327	.86413	.51828	.85521	.53312	.84604	47
14	.48837	.87264	.50352	.86398	.51852	.85506	.53337	.84588	46
15	.48862	.87250	.50377	.86384	.51877	.85491	.53361	.84573	45
16	.48888	.87235	.50403	.86369	.51902	.85476	.53386	.84557	44
17	.48913	.87221	.50428	.86354	.51927	.85461	.53411	.84542	43
18	.48938	.87207	.50453	.86340	.51952	.85446	.53435	.84526	42
19	.48964	.87193	.50478	.86325	.51977	.85431	.53460	.84511	41
20	.48989	.87178	.50503	.86310	.52002	.85416	.53484	.84495	40
21	.49014	.87164	.50528	.86295	.52026	.85401	.53509	.84480	39
22	.49040	.87150	.50553	.86281	.52051	.85385	.53534	.84464	38
23	.49065	.87136	.50578	.86266	.52076	.85370	.53558	.84448	37
24	.49090	.87121	.50603	.86251	.52101	.85355	.53583	.84433	36
25	.49116	.87107	.50628	.86237	.52126	.85340	.53607	.84417	35
26	.49141	.87093	.50654	.86222	.52151	.85325	.53632	.84402	34
27	.49166	.87079	.50679	.86207	.52175	.85310	.53656	.84386	33
28	.49192	.87064	.50704	.86192	.52200	.85294	.53681	.84370	32
29	.49217	.87050	.50729	.86178	.52225	.85279	.53705	.84355	31
30	.49242	.87036	.50754	.86163	.52250	.85264	.53730	.84339	30
31	.49268	.87021	.50779	.86148	.52275	.85249	.53754	.84324	29
32	.49293	.87007	.50804	.86133	.52299	.85234	.53779	.84308	28
33	.49318	.86993	.50829	.86119	.52324	.85218	.53804	.84292	27
34	.49344	.86978	.50854	.86104	.52349	.85203	.53828	.84277	26
35	.49369	.86964	.50879	.86089	.52374	.85188	.53853	.84261	25
36	.49394	.86949	.50904	.86074	.52399	.85173	.53877	.84245	24
37	.49419	.86935	.50929	.86059	.52423	.85157	.53902	.84230	23
38	.49445	.86921	.50954	.86045	.52448	.85142	.53926	.84214	22
39	.49470	.86906	.50979	.86030	.52473	.85127	.53951	.84198	21
40	.49495	.86892	.51004	.86015	.52498	.85112	.53975	.84182	20
41	.49521	.86878	.51029	.86000	.52522	.85096	.54000	.84167	19
42	.49546	.86863	.51054	.85985	.52547	.85081	.54024	.84151	18
43	.49571	.86849	.51079	.85970	.52572	.85066	.54049	.84135	17
44	.49596	.86834	.51104	.85956	.52597	.85051	.54073	.84120	16
45	.49622	.86820	.51129	.85941	.52621	.85035	.54097	.84104	15
46	.49647	.86805	.51154	.85926	.52646	.85020	.54122	.84088	14
47	.49672	.86791	.51179	.85911	.52671	.85005	.54146	.84072	13
48	.49697	.86777	.51204	.85896	.52696	.84989	.54171	.84057	12
49	.49723	.86762	.51229	.85881	.52720	.84974	.54195	.84041	11
50	.49748	.86748	.51254	.85866	.52745	.84959	.54220	.84025	10
51	.49773	.86733	.51279	.85851	.52770	.84943	.54244	.84009	9
52	.49798	.86719	.51304	.85836	.52794	.84928	.54269	.83994	8
53	.49824	.86704	.51329	.85821	.52819	.84913	.54293	.83978	7
54	.49849	.86690	.51354	.85806	.52844	.84897	.54317	.83962	6
55	.49874	.86675	.51379	.85792	.52869	.84882	.54342	.83946	5
56	.49899	.86661	.51404	.85777	.52893	.84866	.54366	.83930	4
57	.49924	.86646	.51429	.85762	.52918	.84851	.54391	.83915	3
58	.49950	.86632	.51454	.85747	.52943	.84836	.54415	.83899	2
59	.49975	.86617	.51479	.85732	.52967	.84820	.54440	.83883	1
60	.50000	.86603	.51504	.85717	.52992	.84805	.54464	.83867	0
′	Cosine	Sine	Cosine	Sine	Cosine	Sine	Cosine	Sine	′
	60°		59°		58°		57°		

Table IX—Continued

′	33°		34°		35°		36°		′
	Sine	Cosine	Sine	Cosine	Sine	Cosine	Sine	Cosine	
0	.54464	.83867	.55919	.82904	.57358	.81915	.58779	.80902	60
1	.54488	.83851	.55943	.82887	.57381	.81899	.58802	.80885	59
2	.54513	.83835	.55968	.82871	.57405	.81882	.58826	.80867	58
3	.54537	.83819	.55992	.82855	.57429	.81865	.58849	.80850	57
4	.54561	.83804	.56016	.82839	.57453	.81848	.58873	.80833	56
5	.54586	.83788	.56040	.82822	.57477	.81832	.58896	.80816	55
6	.54610	.83772	.56064	.82806	.57501	.81815	.58920	.80799	54
7	.54635	.83756	.56088	.82790	.57524	.81798	.58943	.80782	53
8	.54659	.83740	.56112	.82773	.57548	.81782	.58967	.80765	52
9	.54683	.83724	.56136	.82757	.57572	.81765	.58990	.80748	51
10	.54708	.83708	.56160	.82741	.57596	.81748	.59014	.80730	50
11	.54732	.83692	.56184	.82724	.57619	.81731	.59037	.80713	49
12	.54756	.83676	.56208	.82708	.57643	.81714	.59061	.80696	48
13	.54781	.83660	.56232	.82692	.57667	.81698	.59084	.80679	47
14	.54805	.83645	.56256	.82675	.57691	.81681	.59108	.80662	46
15	.54829	.83629	.56280	.82659	.57715	.81664	.59131	.80644	45
16	.54854	.83613	.56305	.82643	.57738	.81647	.59154	.80627	44
17	.54878	.83597	.56329	.82626	.57762	.81631	.59178	.80610	43
18	.54902	.83581	.56353	.82610	.57786	.81614	.59201	.80593	42
19	.54927	.83565	.56377	.82593	.57810	.81597	.59225	.80576	41
20	.54951	.83549	.56401	.82577	.57833	.81580	.59248	.80558	40
21	.54975	.83533	.56425	.82561	.57857	.81563	.59272	.80541	39
22	.54999	.83517	.56449	.82544	.57881	.81546	.59295	.80524	38
23	.55024	.83501	.56473	.82528	.57904	.81530	.59318	.80507	37
24	.55048	.83485	.56497	.82511	.57928	.81513	.59342	.80489	36
25	.55072	.83469	.56521	.82495	.57952	.81496	.59365	.80472	35
26	.55097	.83453	.56545	.82478	.57976	.81479	.59389	.80455	34
27	.55121	.83437	.56569	.82462	.57999	.81462	.59412	.80438	33
28	.55145	.83421	.56593	.82446	.58023	.81445	.59436	.80420	32
29	.55169	.83405	.56617	.82429	.58047	.81428	.59459	.80403	31
30	.55194	.83389	.56641	.82413	.58070	.81412	.59482	.80386	30
31	.55218	.83373	.56665	.82396	.58094	.81395	.59506	.80368	29
32	.55242	.83356	.56689	.82380	.58118	.81378	.59529	.80351	28
33	.55266	.83340	.56713	.82363	.58141	.81361	.59552	.80334	27
34	.55291	.83324	.56736	.82347	.58165	.81344	.59576	.80316	26
35	.55315	.83308	.56760	.82330	.58189	.81327	.59599	.80299	25
36	.55339	.83292	.56784	.82314	.58212	.81310	.59622	.80282	24
37	.55363	.83276	.56808	.82297	.58236	.81293	.59646	.80264	23
38	.55388	.83260	.56832	.82281	.58260	.81276	.59669	.80247	22
39	.55412	.83244	.56856	.82264	.58283	.81259	.59693	.80230	21
40	.55436	.83228	.56880	.82248	.58307	.81242	.59716	.80212	20
41	.55460	.83212	.56904	.82231	.58330	.81225	.59739	.80195	19
42	.55484	.83195	.56928	.82214	.58354	.81208	.59763	.80178	18
43	.55509	.83179	.5695[illegible]	.82198	.58378	.81191	.59786	.80160	17
44	.55533	.83163	.56976	.82181	.58401	.81174	.59809	.80143	16
45	.55557	.83147	.57000	.82165	.58425	.81157	.59832	.80125	15
46	.55581	.83131	.57024	.82148	.58449	.81140	.59856	.80108	14
47	.55605	.83115	.57047	.82132	.58472	.81123	.59879	.80091	13
48	.55630	.83098	.57071	.82115	.58496	.81106	.59902	.80073	12
49	.55654	.83082	.57095	.82098	.58519	.81089	.59926	.80056	11
50	.55678	.83066	.57119	.82082	.58543	.81072	.59949	.80038	10
51	.55702	.83050	.57143	.82065	.58567	.81055	.59972	.80021	9
52	.55726	.83034	.57167	.82048	.58590	.81038	.59995	.80003	8
53	.55750	.83017	.57191	.82032	.58614	.81021	.60019	.79986	7
54	.55775	.83001	.57215	.82015	.58637	.81004	.60042	.79968	6
55	.55799	.82985	.57238	.81999	.58661	.80987	.60065	.79951	5
56	.55823	.82969	.57262	.81982	.58684	.80970	.60089	.79934	4
57	.55847	.82953	.57286	.81965	.58708	.80953	.60112	.79916	3
58	.55871	.82936	.57310	.81949	.58731	.80936	.60135	.79899	2
59	.55895	.82920	.57334	.81932	.58755	.80919	.60158	.79881	1
60	.55919	.82904	.57358	.81915	.58779	.80902	.60182	.79864	0
′	Cosine	Sine	Cosine	Sine	Cosine	Sine	Cosine	Sine	′
	56°		55°		54°		53°		

Table IX—Continued

′	37° Sine	37° Cosine	38° Sine	38° Cosine	39° Sine	39° Cosine	40° Sine	40° Cosine	′
0	.60182	.79864	.61566	.78801	.62932	.77715	.64279	.76604	60
1	.60205	.79846	.61589	.78783	.62955	.77696	.64301	.76586	59
2	.60228	.79829	.61612	.78765	.62977	.77678	.64323	.76567	58
3	.60251	.79811	.61635	.78747	.63000	.77660	.64346	.76548	57
4	.60274	.79793	.61658	.78729	.63022	.77641	.64368	.76530	56
5	.60298	.79776	.61681	.78711	.63045	.77623	.64390	.76511	55
6	.60321	.79758	.61704	.78694	.63068	.77605	.64412	.76492	54
7	.60344	.79741	.61726	.78676	.63090	.77586	.64435	.76473	53
8	.60367	.79723	.61749	.78658	.63113	.77568	.64457	.76455	52
9	.60390	.79706	.61772	.78640	.63135	.77550	.64479	.76436	51
10	.60414	.79688	.61795	.78622	.63158	.77531	.64501	.76417	50
11	.60437	.79671	.61818	.78604	.63180	.77513	.64524	.76398	49
12	.60460	.79653	.61841	.78586	.63203	.77494	.64546	.76380	48
13	.60483	.79635	.61864	.78568	.63225	.77476	.64568	.76361	47
14	.60506	.79618	.61887	.78550	.63248	.77458	.64590	.76342	46
15	.60529	.79600	.61909	.78532	.63271	.77439	.64612	.76323	45
16	.60553	.79583	.61932	.78514	.63293	.77421	.64635	.76304	44
17	.60576	.79565	.61955	.78496	.63316	.77402	.64657	.76286	43
18	.60599	.79547	.61978	.78478	.63338	.77384	.64679	.76267	42
19	[illegible]	[illegible]	.62001	.78460	.63361	.77366	.64701	.76248	41
20	.60645	.79512	.62024	.78442	.63383	.77347	.64723	[illegible]	40
21	.60668	.79494	.62046	.78424	.63406	.77329	.64746	.76210	39
22	.60691	.79477	.62069	.78405	.63428	.77310	.64768	.76192	38
23	.60714	.79459	.62092	.78387	.63451	.77292	.64790	.76173	37
24	.60738	.79441	.62115	.78369	.63473	.77273	.64812	.76154	36
25	.60761	.79424	.62138	.78351	.63496	.77255	.64834	.76135	35
26	.60784	.79406	.62160	.78333	.63518	.77236	.64856	.76116	34
27	.60807	.79388	.62183	.78315	.63540	.77218	.64878	.76097	33
28	[illegible]	.79371	[illegible]	[illegible]	[illegible]	[illegible]	[illegible]	[illegible]	32
29	.60853	.79353	.62229	.78279	.63585	.77181	.64923	.76059	31
30	.60876	.79335	.62251	.78261	.63608	.77162	.64945	.76041	30
31	.60899	.79318	.62274	.78243	.63630	.77144	.64967	.76022	29
32	.60922	.79300	.62297	.78225	.63653	.77125	.64989	.76003	28
33	.60945	.79282	.62320	.78206	.63675	.77107	.65011	.75984	27
34	.60968	.79264	.62342	.78188	.63698	.77088	.65033	.75965	26
35	.60991	.79247	.62365	.78170	.63720	.77070	.65055	.75946	25
36	.61015	.79229	.62388	.78152	.63742	.77051	.65077	.75927	24
37	[illegible]	[illegible]	[illegible]	.78134	[illegible]	.77033	.65100	.75908	23
38	.61061	.79193	.62433	.78116	.63787	.77014	.65122	.75889	22
39	.61084	.79176	.62456	.78098	.63810	.76996	.65144	.75870	21
40	[illegible]	[illegible]	[illegible]	[illegible]	[illegible]	[illegible]	.65166	.75851	20
41	.61130	.79140	.62502	.78061	.63854	.76959	.65188	.75832	19
42	.61153	.79122	.62524	.78043	.63877	.76940	.65210	.75813	18
43	.61176	.79105	.62547	.78025	.63899	.76921	.65232	.75794	17
44	.61199	.79087	.62570	.78007	.63922	.76903	.65254	.75775	16
45	.61222	.79069	.62592	.77988	.63944	.76884	.65276	.75756	15
46	.61245	.79051	.62615	.77970	.63966	.76866	.65298	.75738	14
47	.61268	.79033	.62638	.77952	.63989	.76847	.65320	.75719	13
48	.61291	.79016	.62660	.77934	.64011	.76828	.65342	.75700	12
49	.61314	.78998	.62683	.77916	.64033	.76810	.65364	.75680	11
50	.61337	.78980	.62706	.77897	.64056	.76791	.65386	.75661	10
51	.61360	.78962	.62728	.77879	.64078	.76772	.65408	.75642	9
52	.61383	.78944	.62751	.77861	.64100	.76754	.65430	.75623	8
53	.61406	.78926	.62774	.77843	.64123	.76735	.65452	.75604	7
54	.61429	.78908	.62796	.77824	.64145	.76717	.65474	.75585	6
55	.61451	.78891	.62819	.77806	.64167	.76698	.65496	.75566	5
56	.61474	.78873	.62842	.77788	.64190	.76679	.65518	.75547	4
57	.61497	.78855	.62864	.77769	.64212	.76661	.65540	.75528	3
58	.61520	.78837	.62887	.77751	.64234	.76642	.65562	.75509	2
59	.61543	.78819	.62909	.77733	.64256	.76623	.65584	.75490	1
60	.61566	.78801	.62932	.77715	.64279	.76604	.65606	.75471	0
′	52° Cosine	52° Sine	51° Cosine	51° Sine	50° Cosine	50° Sine	49° Cosine	49° Sine	′

Table IX—Concluded

′	41° Sine	41° Cosine	42° Sine	42° Cosine	43° Sine	43° Cosine	44° Sine	44° Cosine	′
0	.65606	.75471	.66913	.74314	.68200	.73135	.69466	.71934	60
1	.65628	.75452	.66935	.74295	.68221	.73116	.69487	.71914	59
2	.65650	.75433	.66956	.74276	.68242	.73096	.69508	.71894	58
3	.65672	.75414	.66978	.74256	.68264	.73076	.69529	.71873	57
4	.65694	.75395	.66999	.74237	.68285	.73056	.69549	.71853	56
5	.65716	.75375	.67021	.74217	.68306	.73036	.69570	.71833	55
6	.65738	.75356	.67043	.74198	.68327	.73016	.69591	.71813	54
7	.65759	.75337	.67064	.74178	.68349	.72996	.69612	.71792	53
8	.65781	.75318	.67086	.74159	.68370	.72976	.69633	.71772	52
9	.65803	.75299	.67107	.74139	.68391	.72957	.69654	.71752	51
10	.65825	.75280	.67129	.74120	.68412	.72937	.69675	.71732	50
11	.65847	.75261	.67151	.74100	.68434	.72917	.69696	.71711	49
12	.65869	.75241	.67172	.74080	.68455	.72897	.69717	.71691	48
13	.65891	.75222	.67194	.74061	.68476	.72877	.69737	.71671	47
14	.65913	.75203	.67215	.74041	.68497	.72857	.69758	.71650	46
15	.65935	.75184	.67237	.74022	.68518	.72837	.69779	.71630	45
16	.65956	.75165	.67258	.74002	.68539	.72817	.69800	.71610	44
17	.65978	.75146	.67280	.73983	.68561	.72797	.69821	.71590	43
18	.66000	.75126	.67301	.73963	.68582	.72777	.69842	.71569	42
19	.66022	.75107	.67323	.73944	.68603	.72757	.69862	.71549	41
20	.66044	.75088	.67344	.73924	.68624	.72737	.69883	.71529	40
21	.66066	.75069	.67366	.73904	.68645	.72717	.69904	.71508	39
22	.66088	.75050	.67387	.73885	.68666	.72697	.69925	.71488	38
23	.66109	.75030	.67409	.73865	.68688	.72677	.69946	.71468	37
24	.66131	.75011	.67430	.73846	.68709	.72657	.69966	.71447	36
25	.66153	.74992	.67452	.73826	.68730	.72637	.69987	.71427	35
26	.66175	.74973	.67473	.73806	.68751	.72617	.70008	.71407	34
27	.66197	.74953	.67495	.73787	.68772	.72597	.70029	.71386	33
28	.66218	.74934	.67516	.73767	.68793	.72577	.70049	.71366	32
29	.66240	.74915	.67538	.73747	.68814	.72557	.70070	.71345	31
30	.66262	.74896	.67559	.73728	.68835	.72537	.70091	.71325	30
31	.66284	.74876	.67580	.73708	.68857	.72517	.70112	.71305	29
32	.66306	.74857	.67602	.73688	.68878	.72497	.70132	.71284	28
33	.66327	.74838	.67623	.73669	.68899	.72477	.70153	.71264	27
34	.66349	.74818	.67645	.73649	.68920	.72457	.70174	.71243	26
35	.66371	.74799	.67666	.73629	.68941	.72437	.70195	.71223	25
36	.66393	.74780	.67688	.73610	.68962	.72417	.70215	.71203	24
37	.66414	.74760	.67709	.73590	.68983	.72397	.70236	.71182	23
38	.66436	.74741	.67730	.73570	.69004	.72377	.70257	.71162	22
39	.66458	.74722	.67752	.73551	.69025	.72357	.70277	.71141	21
40	.66480	.74703	.67773	.73531	.69046	.72337	.70298	.71121	20
41	.66501	.74683	.67795	.73511	.69067	.72317	.70319	.71100	19
42	.66523	.74664	.67816	.73491	.69088	.72297	.70339	.71080	18
43	.66545	.74644	.67837	.73472	.69109	.72277	.70360	.71059	17
44	.66566	.74625	.67859	.73452	.69130	.72257	.70381	.71039	16
45	.66588	.74606	.67880	.73432	.69151	.72236	.70401	.71019	15
46	.66610	.74586	.67901	.73413	.69172	.72216	.70422	.70998	14
47	.66632	.74567	.67923	.73393	.69193	.72196	.70443	.70978	13
48	.66653	.74548	.67944	.73373	.69214	.72176	.70463	.70957	12
49	.66675	.74528	.67965	.73353	.69235	.72156	.70484	.70937	11
50	.66697	.74509	.67987	.73333	.69256	.72136	.70505	.70916	10
51	.66718	.74489	.68008	.73314	.69277	.72116	.70525	.70896	9
52	.66740	.74470	.68029	.73294	.69298	.72095	.70546	.70875	8
53	.66762	.74451	.68051	.73274	.69319	.72075	.70567	.70855	7
54	.66783	.74431	.68072	.73254	.69340	.72055	.70587	.70834	6
55	.66805	.74412	.68093	.73234	.69361	.72035	.70608	.70813	5
56	.66827	.74392	.68115	.73215	.69382	.72015	.70628	.70793	4
57	.66848	.74373	.68136	.73195	.69403	.71995	.70649	.70772	3
58	.66870	.74353	.68157	.73175	.69424	.71974	.70670	.70752	2
59	.66891	.74334	.68179	.73155	.69445	.71954	.70690	.70731	1
60	.66913	.74314	.68200	.73135	.69466	.71934	.70711	.70711	0
′	48° Cosine	48° Sine	47° Cosine	47° Sine	46° Cosine	46° Sine	45° Cosine	45° Sine	′

Table X Natural tangents and cotangents

′	0°		1°		2°		3°		′
	TAN.	CO-TAN.	TAN.	CO-TAN.	TAN.	CO-TAN.	TAN.	CO-TAN.	
0	.00000	Infinite.	.01746	57.2900	.03492	28.6363	.05241	19.0811	60
1	.00029	3437.750	.01775	56.3506	.03521	28.3994	.05270	18.9755	59
2	.00058	1718.870	.01804	55.4415	.03550	28.1664	.05299	18.8711	58
3	.00087	1145.920	.01833	54.5613	.03579	27.9372	.05328	18.7678	57
4	.00116	859.436	.01862	53.7086	.03609	27.7117	.05357	18.6656	56
5	.00145	687.549	.01891	52.8821	.03638	27.4899	.05387	18.5645	55
6	.00175	572.957	.01920	52.0807	.03667	27.2715	.05416	18.4645	54
7	.00204	491.106	.01949	51.3032	.03696	27.0566	.05445	18.3655	53
8	.00233	429.718	.01978	50.5485	.03725	26.8450	.05474	18.2677	52
9	.00262	381.971	.02007	49.8157	.03754	26.6367	.05503	18.1708	51
10	.00291	343.774	.02036	49.1039	.03783	26.4316	.05533	18.0750	50
11	.00320	312.521	.02066	48.4121	.03812	26.2296	.05562	17.9802	49
12	.00349	286.478	.02095	47.7395	.03842	26.0307	.05591	17.8863	48
13	.00378	264.441	.02124	47.0853	.03871	25.8348	.05620	17.7934	47
14	.00407	245.552	.02153	46.4489	.03900	25.6418	.05649	17.7015	46
15	.00436	229.182	.02182	45.8294	.03929	25.4517	.05678	17.6106	45
16	.00465	214.858	.02211	45.2261	.03958	25.2644	.05708	17.5205	44
17	.00495	202.219	.02240	44.6386	.03987	25.0798	.05737	17.4314	43
18	.00524	190.984	.02269	44.0661	.04016	24.8978	.05766	17.3432	42
19	.00553	180.932	.02298	43.5081	.04046	24.7185	.05795	17.2558	41
20	.00582	171.885	.02328	42.9641	.04075	24.5418	.05824	17.1693	40
21	.00611	163.700	.02357	42.4335	.04104	24.3675	.05854	17.0837	39
22	.00640	156.259	.02386	41.9158	.04133	24.1957	.05883	16.9990	38
23	.00669	149.465	.02415	41.4106	.04162	24.0263	.05912	16.9150	37
24	.00698	143.237	.02444	40.9174	.04191	23.8593	.05941	16.8319	36
25	.00727	137.507	.02473	40.4358	.04220	23.6945	.05970	16.7496	35
26	.00756	132.219	.02502	39.9655	.04250	23.5321	.05999	16.6681	34
27	.00785	127.321	.02531	39.5059	.04279	23.3718	.06029	16.5874	33
28	.00814	122.774	.02560	39.0568	.04308	23.2137	.06058	16.5075	32
29	.00844	118.540	.02589	38.6177	.04337	23.0577	.06087	16.4283	31
30	.00873	114.589	.02619	38.1885	.04366	22.9038	.06116	16.3499	30
31	.00902	110.892	.02648	37.7686	.04395	22.7519	.06145	16.2722	29
32	.00931	107.426	.02677	37.3579	.04424	22.6020	.06175	16.1952	28
33	.00960	104.171	.02706	36.9560	.04454	22.4541	.06204	16.1190	27
34	.00989	101.107	.02735	36.5627	.04483	22.3081	.06233	16.0435	26
35	.01018	98.2179	.02764	36.1776	.04512	22.1640	.06262	15.9687	25
36	.01047	95.4895	.02793	35.8006	.04541	22.0217	.06291	15.8945	24
37	.01076	92.9085	.02822	35.4313	.04570	21.8813	.06321	15.8211	23
38	.01105	90.4633	.02851	35.0695	.04599	21.7426	.06350	15.7483	22
39	.01135	88.1436	.02881	34.7151	.04628	21.6056	.06379	15.6762	21
40	.01164	85.9398	.02910	34.3678	.04658	21.4704	[illegible]	15.6048	20
41	.01193	83.8435	.02939	34.0273	.04687	21.3369	.06437	15.5340	19
42	.01222	81.8470	.02968	33.6935	.04716	21.2049	.06467	15.4638	18
43	.01251	79.9434	.02997	33.3662	.04745	21.0747	.06496	15.3943	17
44	.01280	78.1263	.03026	33.0452	.04774	20.9460	.06525	15.3254	16
45	.01309	76.3900	.03055	32.7303	.04803	20.8188	.06554	15.2571	15
46	.01338	74.7292	.03084	32.4213	.04832	20.6932	.06584	15.1893	14
47	.01367	73.1390	.03114	32.1181	.04862	20.5691	.06613	15.1222	13
48	.01396	71.6151	.03143	31.8205	.04891	20.4465	.06642	15.0557	12
49	.01425	70.1533	.03172	31.5284	.04920	20.3253	.06671	14.9898	11
50	.01455	68.7501	.03201	31.2416	.04949	20.2056	.06700	14.9244	10
51	.01484	67.4019	.03230	30.9599	.04978	20.0872	.06730	14.8596	9
52	.01513	66.1055	.03259	30.6833	.05007	19.9702	.06759	14.7954	8
53	.01542	64.8580	.03288	30.4116	.05037	19.8546	.06788	14.7317	7
54	.01571	63.6567	.03317	30.1446	.05066	19.7403	.06817	14.6685	6
55	.01600	62.4992	.03346	29.8823	.05095	19.6273	.06847	14.6059	5
56	.01629	61.3829	.03376	29.6245	.05124	19.5156	.06876	14.5438	4
57	.01658	60.3058	.03405	29.3711	.05153	19.4051	.06905	14.4823	3
58	.01687	59.2659	.03434	29.1220	.05182	19.2959	.06934	14.4212	2
59	.01716	58.2612	.03463	28.8771	.05212	19.1879	.06963	14.3607	1
60	.01746	57.2900	.03492	28.6363	.05241	19.0811	.06993	14.3007	0
′	CO-TAN.	TAN.	CO-TAN.	TAN.	CO-TAN.	TAN.	CO-TAN.	TAN.	′
	89°		88°		87°		86°		

Table X—Continued

′	4° Tan.	4° Co-tan.	5° Tan.	5° Co-tan.	6° Tan.	6° Co-tan.	7° Tan.	7° Co-tan.	′
0	.06993	14.3007	.08749	11.4301	.10510	9.51436	.12278	8.14435	60
1	.07022	14.2411	.08778	11.3919	.10540	9.48781	.12308	8.12481	59
2	.07051	14.1821	.08807	11.3540	.10569	9.46141	.12338	8.10536	58
3	.07080	14.1235	.08837	11.3163	.10599	9.43515	.12367	8.08600	57
4	.07110	14.0655	.08866	11.2789	.10628	9.40904	.12397	8.06674	56
5	.07139	14.0079	.08895	11.2417	.10657	9.38307	.12426	8.04756	55
6	.07168	13.9507	.08925	11.2048	.10687	9.35724	.12456	8.02848	54
7	.07197	13.8940	.08954	11.1681	.10716	9.33154	.12485	8.00948	53
8	.07227	13.8378	.08983	11.1316	.10746	9.30599	.12515	7.99058	52
9	.07256	13.7821	.09013	11.0954	.10775	9.28058	.12544	7.97176	51
10	.07285	13.7267	.09042	11.0594	.10805	9.25530	.12574	7.95302	50
11	.07314	13.6719	.09071	11.0237	.10834	9.23016	.12603	7.93438	49
12	.07344	13.6174	.09101	10.9882	.10863	9.20516	.12633	7.91582	48
13	.07373	13.5634	.09130	10.9529	.10893	9.18028	.12662	7.89734	47
14	.07402	13.5098	.09159	10.9178	.10922	9.15554	.12692	7.87895	46
15	.07431	13.4566	.09189	10.8829	.10952	9.13093	.12722	7.86064	45
16	.07461	13.4039	.09218	10.8483	.10981	9.10646	.12751	7.84242	44
17	.07490	13.3515	.09247	10.8139	.11011	9.08211	.12781	7.82428	43
18	.07519	13.2996	.09277	10.7797	.11040	9.05789	.12810	7.80622	42
19	.07548	13.2480	.09306	10.7457	.11070	9.03379	.12840	7.78825	41
20	.07578	13.1969	.09335	10.7119	.11099	9.00983	.12869	7.77035	40
21	.07607	13.1461	.09365	10.6783	.11128	8.98598	.12899	7.75254	39
22	.07636	13.0958	.09394	10.6450	.11158	8.96227	.12929	7.73480	38
23	.07665	13.0458	.09423	10.6118	.11187	8.93867	.12958	7.71715	37
24	.07695	12.9962	.09453	10.5789	.11217	8.91520	.12988	7.69957	36
25	.07724	12.9469	.09482	10.5462	.11246	8.89185	.13017	7.68208	35
26	.07753	12.8981	.09511	10.5136	.11276	8.86862	.13047	7.66466	34
27	.07782	12.8496	.09541	10.4813	.11305	8.84551	.13076	7.64732	33
28	.07812	12.8014	.09570	10.4491	.11335	8.82252	.13106	7.63005	32
29	.07841	12.7536	.09600	10.4172	.11364	8.79964	.13136	7.61287	31
30	.07870	12.7062	.09629	10.3854	.11394	8.77689	.13165	7.59575	30
31	.07899	12.6591	.09658	10.3538	.11423	8.75425	.13195	7.57872	29
32	.07929	12.6124	.09688	10.3224	.11452	8.73172	.13224	7.56176	28
33	.07958	12.5660	.09717	10.2913	.11482	8.70931	.13254	7.54487	27
34	.07987	12.5199	.09746	10.2602	.11511	8.68701	.13284	7.52806	26
35	.08017	12.4742	.09776	10.2294	.11541	8.66482	.13313	7.51132	25
36	.08046	12.4288	.09805	10.1988	.11570	8.64275	.13343	7.49465	24
37	.08075	12.3838	.09834	10.1683	.11600	8.62078	.13372	7.47806	23
38	.08104	12.3390	.09864	10.1381	.11629	8.59893	.13402	7.46154	22
39	.08134	12.2946	.09893	10.1080	.11659	8.57718	.13432	7.44509	21
40	.08163	12.2505	.09923	10.0780	.11688	8.55555	.13461	7.42871	20
41	.08192	12.2067	.09952	10.0483	.11718	8.53402	.13491	7.41240	19
42	.08221	12.1632	.09981	10.0187	.11747	8.51259	.13521	7.39616	18
43	.08251	12.1201	.10011	9.98931	.11777	8.49128	.13550	7.37999	17
44	.08280	12.0772	.10040	9.96007	.11806	8.47007	.13580	7.36389	16
45	.08309	12.0346	.10069	9.93101	.11836	8.44896	.13609	7.34786	15
46	.08339	11.9923	.10099	9.90211	.11865	8.42795	.13639	7.33190	14
47	.08368	11.9504	.10128	9.87338	.11895	8.40705	.13669	7.31600	13
48	.08397	11.9087	.10158	9.84482	.11924	8.38625	.13698	7.30018	12
49	.08427	11.8673	.10187	9.81641	.11954	8.36555	.13728	7.28442	11
50	.08456	11.8262	.10216	9.78817	.11983	8.34496	.13758	7.26873	10
51	.08485	11.7853	.10246	9.76009	.12013	8.32446	.13787	7.25310	9
52	.08514	11.7448	.10275	9.73217	.12042	8.30406	.13817	7.23754	8
53	.08544	11.7045	.10305	9.70441	.12072	8.28376	.13846	7.22204	7
54	.08573	11.6645	.10334	9.67680	.12101	8.26355	.13876	7.20661	6
55	.08602	11.6248	.10363	9.64935	.12131	8.24345	.13906	7.19125	5
56	.08632	11.5853	.10393	9.62205	.12160	8.22344	.13935	7.17594	4
57	.08661	11.5461	.10422	9.59490	.12190	8.20352	.13965	7.16071	3
58	.08690	11.5072	.10452	9.56791	.12219	8.18370	.13995	7.14553	2
59	.08720	11.4685	.10481	9.54106	.12249	8.16398	.14024	7.13042	1
60	.08749	11.4301	.10510	9.51436	.12278	8.14435	.14054	7.11537	0
′	Co-tan.	Tan.	Co-tan.	Tan.	Co-tan.	Tan.	Co-tan.	Tan.	′
	85°		84°		83°		82°		

Table X—Continued

′	8° Tan.	8° Co-tan.	9° Tan.	9° Co-tan.	10° Tan.	10° Co-tan.	11° Tan.	11° Co-tan.	′
0	.14054	7.11537	.15838	6.31375	.17633	5.67128	.19438	5.14455	60
1	.14084	7.10038	.15868	6.30189	.17663	5.66165	.19468	5.13658	59
2	.14113	7.08546	.15898	6.29007	.17693	5.65205	.19498	5.12862	58
3	.14143	7.07059	.15928	6.27829	.17723	5.64248	.19529	5.12069	57
4	.14173	7.05579	.15958	6.26655	.17753	5.63295	.19559	5.11279	56
5	.14202	7.04105	.15988	6.25486	.17783	5.62344	.19589	5.10490	55
6	.14232	7.02637	.16017	6.24321	.17813	5.61397	.19619	5.09704	54
7	.14262	7.01174	.16047	6.23160	.17843	5.60452	.19649	5.08921	53
8	.14291	6.99718	.16077	6.22003	.17873	5.59511	.19680	5.08139	52
9	.14321	6.98268	.16107	6.20851	.17903	5.58573	.19710	5.07360	51
10	.14351	6.96823	.16137	6.19703	.17933	5.57638	.19740	5.06584	50
11	.14381	6.95385	.16167	6.18559	.17963	5.56706	.19770	5.05809	49
12	.14410	6.93952	.16196	6.17419	.17993	5.55777	.19801	5.05037	48
13	.14440	6.92525	.16226	6.16283	.18023	5.54851	.19831	5.04267	47
14	.14470	6.91104	.16256	6.15151	.18053	5.53927	.19861	5.03499	46
15	.14499	6.89688	.16286	6.14023	.18083	5.53007	.19891	5.02734	45
16	.14529	6.88278	.16316	6.12899	.18113	5.52090	.19921	5.01971	44
17	.14559	6.86874	.16346	6.11779	.18143	5.51176	.19952	5.01210	43
18	.14588	6.85475	.16376	6.10664	.18173	5.50264	.19982	5.00451	42
19	.14618	6.84082	.16405	6.09552	.18203	5.49356	.20012	4.99695	41
20	.14648	6.82694	.16435	6.08444	.18233	5.48451	.20042	4.98940	40
21	.14678	6.81312	.16465	6.07340	.18263	5.47548	.20073	4.98188	39
22	.14707	6.79936	.16495	6.06240	.18293	5.46648	.20103	4.97438	38
23	.14737	6.78564	.16525	6.05143	.18323	5.45751	.20133	4.96690	37
24	.14767	6.77199	.16555	6.04051	.18353	5.44857	.20164	4.95945	36
25	.14796	6.75838	.16585	6.02962	.18383	5.43966	.20194	4.95201	35
26	.14826	6.74483	.16615	6.01878	.18414	5.43077	.20224	4.94460	34
27	.14856	6.73133	.16645	6.00797	.18444	5.42192	.20254	4.93721	33
28	.14886	6.71789	.16674	5.99720	.18474	5.41309	.20285	4.92984	32
29	.14915	6.70450	.16704	5.98646	.18504	5.40429	.20315	4.92249	31
30	.14945	6.69116	.16734	5.97576	.18534	5.39552	.20345	4.91516	30
31	.14975	6.67787	.16764	5.96510	.18564	5.38677	.20376	4.90785	29
32	.15005	6.66463	.16794	5.95448	.18594	5.37805	.20406	4.90056	28
33	.15034	6.65144	.16824	5.94390	.18624	5.36936	.20436	4.89330	27
34	.15064	6.63831	.16854	5.93335	.18654	5.36070	.20466	4.83605	26
35	.15094	6.62523	.16884	5.92283	.18684	5.35206	.20497	4.87882	25
36	.15124	6.61219	.16914	5.91235	.18714	5.34345	.20527	4.87162	24
37	.15153	6.59921	.16944	5.90191	.18745	5.33487	.20557	4.86444	23
38	.15183	6.58627	.16974	5.89151	.18775	5.32631	.20588	4.85727	22
39	.15213	6.57339	.17004	5.88114	.18805	5.31778	.20618	4.85013	21
40	.15243	6.56055	.17033	5.87080	.18835	5.30928	.20648	4.84300	20
41	.15272	6.54777	.17063	5.86051	.18865	5.30080	.20679	4.83590	19
42	.15302	6.53503	.17093	5.85024	.18895	5.29235	.20709	4.82882	18
43	.15332	6.52234	.17123	5.84001	.18925	5.28393	.20739	4.82175	17
44	.15362	6.50970	.17153	5.82982	.18955	5.27553	.20770	4.81471	16
45	.15391	6.49710	.17183	5.81966	.18986	5.26715	.20800	4.80769	15
46	.15421	6.48456	.17213	5.80953	.19016	5.25880	.20830	4.80068	14
47	.15451	6.47206	.17243	5.79944	.19046	5.25048	.20861	4.79370	13
48	.15481	6.45961	.17273	5.78938	.19076	5.24218	.20891	4.78673	12
49	.15511	6.44720	.17303	5.77936	.19106	5.23391	.20921	4.77978	11
50	.15540	6.43484	.17333	5.76937	.19136	5.22566	.20952	4.77286	10
51	.15570	6.42253	.17363	5.75941	.19166	5.21744	.20982	4.76595	9
52	.15600	6.41026	.17393	5.74949	.19197	5.20925	.21013	4.75906	8
53	.15630	6.39804	.17423	5.73960	.19227	5.20107	.21043	4.75219	7
54	.15660	6.38587	.17453	5.72974	.19257	5.19293	.21073	4.74534	6
55	.15689	6.37374	.17483	5.71992	.19287	5.18480	.21104	4.73851	5
56	.15719	6.36165	.17513	5.71013	.19317	5.17671	.21134	4.73170	4
57	.15749	6.34961	.17543	5.70037	.19347	5.16863	.21164	4.72490	3
58	.15779	6.33761	.17573	5.69064	.19378	5.16058	.21195	4.71813	2
59	.15809	6.32566	.17603	5.68094	.19408	5.15256	.21225	4.71137	1
60	.15838	6.31375	.17633	5.67128	.19438	5.14455	.21256	4.70463	0
′	81° Co-tan.	81° Tan.	80° Co-tan.	80° Tan.	79° Co-tan.	79° Tan.	78° Co-tan.	78° Tan.	′

Table X—Continued

′	12°		13°		14°		15°		′
	Tan.	Co-tan.	Tan.	Co-tan.	Tan.	Co-tan.	Tan.	Co-tan.	
0	.21256	4.70463	.23087	4.33148	.24933	4.01078	.26795	3.73205	60
1	.21286	4.69791	.23117	4.32573	.24964	4.00582	.26826	3.72771	59
2	.21316	4.69121	.23148	4.32001	.24995	4.00086	.26857	3.72338	58
3	.21347	4.68452	.23179	4.31430	.25026	3.99592	.26888	3.71907	57
4	.21377	4.67786	.23209	4.30860	.25056	3.99099	.26920	3.71476	56
5	.21408	4.67121	.23240	4.30291	.25087	3.98607	.26951	3.71046	55
6	.21438	4.66458	.23271	4.29724	.25118	3.98117	.26982	3.70616	54
7	.21469	4.65797	.23301	4.29159	.25149	3.97627	.27013	3.70188	53
8	.21499	4.65138	.23332	4.28595	.25180	3.97139	.27044	3.69761	52
9	.21529	4.64480	.23363	4.28032	.25211	3.96651	.27076	3.69335	51
10	.21560	4.63825	.23393	4.27471	.25242	3.96165	.27107	3.68909	50
11	.21590	4.63171	.23424	4.26911	.25273	3.95680	.27138	3.68485	49
12	.21621	4.62518	.23455	4.26352	.25304	3.95196	.27169	3.68061	48
13	.21651	4.61868	.23485	4.25795	.25335	3.94713	.27201	3.67638	47
14	.21682	4.61219	.23516	4.25239	.25366	3.94232	.27232	3.67217	46
15	.21712	4.60572	.23547	4.24685	.25397	3.93751	.27263	3.66796	45
16	.21743	4.59927	.23578	4.24132	.25428	3.93271	.27294	3.66376	44
17	.21773	4.59283	.23608	4.23580	.25459	3.92793	.27326	3.65957	43
18	.21804	4.58641	.23639	4.23030	.25490	3.92316	.27357	3.65538	42
19	.21834	4.58001	.23670	4.22481	.25521	3.91839	.27388	3.65121	41
20	.21864	4.57363	.23700	4.21933	.25552	3.91364	.27419	3.64705	40
21	.21895	4.56726	.23731	4.21387	.25583	3.90890	.27451	3.64289	39
22	.21925	4.56091	.23762	4.20842	.25614	3.90417	.27482	3.63874	38
23	.21956	4.55458	.23793	4.20298	.25645	3.89945	.27513	3.63461	37
24	.21986	4.54826	.23823	4.19756	.25676	3.89474	.27545	3.63048	36
25	.22017	4.54196	.23854	4.19215	.25707	3.89004	.27576	3.62636	35
26	.22047	4.53568	.23885	4.18675	.25738	3.88536	.27607	3.62224	34
27	.22078	4.52941	.23916	4.18137	.25769	3.88068	.27638	3.61814	33
28	.22108	4.52316	.23946	4.17600	.25800	3.87601	.27670	3.61405	32
29	.22139	4.51693	.23977	4.17064	.25831	3.87136	.27701	3.60996	31
30	.22169	4.51071	.24008	4.16530	.25862	3.86671	.27732	3.60588	30
31	.22200	4.50451	.24039	4.15997	.25893	3.86208	.27764	3.60181	29
32	.22231	4.49832	.24069	4.15465	.25924	3.85745	.27795	3.59775	28
33	.22261	4.49215	.24100	4.14934	.25955	3.85284	.27826	3.59370	27
34	.22292	4.48600	.24131	4.14405	.25986	3.84824	.27858	3.58966	26
35	.22322	4.47986	.24162	4.13877	.26017	3.84364	.27889	3.58562	25
36	.22353	4.47374	.24193	4.13350	.26048	3.83906	.27920	3.58160	24
37	.22383	4.46764	.24223	4.12825	.26079	3.83449	.27952	3.57758	23
38	.22414	4.46155	.24254	4.12301	.26110	3.82992	.27983	3.57357	22
39	.22444	4.45548	.24285	4.11778	.26141	3.82537	.28015	3.56957	21
40	.22475	4.44942	.24316	4.11256	.26172	3.82083	.28046	3.56557	20
41	.22505	4.44338	.24347	4.10736	.26203	3.81630	.28077	3.56159	19
42	.22536	4.43735	.24377	4.10216	.26235	3.81177	.28109	3.55761	18
43	.22567	4.43134	.24408	4.09699	.26266	3.80726	.28140	3.55364	17
44	.22597	4.42534	.24439	4.09182	.26297	3.80276	.28172	3.54968	16
45	.22628	4.41936	.24470	4.08666	.26328	3.79827	.28203	3.54573	15
46	.22658	4.41340	.24501	4.08152	.26359	3.79378	.28234	3.54179	14
47	.22689	4.40745	.24532	4.07639	.26390	3.78931	.28266	3.53785	13
48	.22719	4.40152	.24562	4.07127	.26421	3.78485	.28297	3.53393	12
49	.22750	4.39560	.24593	4.06616	.26452	3.78040	.28329	3.53001	11
50	.22781	4.38969	.24624	4.06107	.26483	3.77595	.28360	3.52609	10
51	.22811	4.38381	.24655	4.05599	.26515	3.77152	.28391	3.52219	9
52	.22842	4.37793	.24686	4.05092	.26546	3.76709	.28423	3.51829	8
53	.22872	4.37207	.24717	4.04586	.26577	3.76268	.28454	3.51441	7
54	.22903	4.36623	.24747	4.04081	.26608	3.75828	.28486	3.51053	6
55	.22934	4.36040	.24778	4.03578	.26639	3.75388	.28517	3.50666	5
56	.22964	4.35459	.24809	4.03075	.26670	3.74950	.28549	3.50279	4
57	.22995	4.34879	.24840	4.02574	.26701	3.74512	.28580	3.49894	3
58	.23026	4.34300	.24871	4.02074	.26733	3.74075	.28612	3.49509	2
59	.23056	4.33723	.24902	4.01576	.26764	3.73640	.28643	3.49125	1
60	.23087	4.33148	.24933	4.01078	.26795	3.73205	.28675	3.48741	0
′	Co-tan.	Tan.	Co-tan.	Tan.	Co-tan.	Tan.	Co-tan.	Tan.	′
	77°		76°		75°		74°		

Table X—Continued

′	16° Tan.	16° Co-tan.	17° Tan.	17° Co-tan.	18° Tan.	18° Co-tan.	19° Tan.	19° Co-tan.	′
0	.28675	3.48741	.30573	3.27085	.32492	3.07768	.34433	2.90421	60
1	.28706	3.48359	.30605	3.26745	.32524	3.07464	.34465	2.90147	59
2	.28738	3.47977	.30637	3.26406	.32556	3.07160	.34498	2.89873	58
3	.28769	3.47596	.30669	3.26067	.32588	3.06857	.34530	2.89600	57
4	.28800	3.47216	.30700	3.25729	.32621	3.06554	.34563	2.89327	56
5	.28832	3.46837	.30732	3.25392	.32653	3.06252	.34596	2.89055	55
6	.28864	3.46458	.30764	3.25055	.32685	3.05950	.34628	2.88783	54
7	.28895	3.46080	.30796	3.24719	.32717	3.05649	.34661	2.88511	53
8	.28927	3.45703	.30828	3.24383	.32749	3.05349	.34693	2.88240	52
9	.28958	3.45327	.30860	3.24049	.32782	3.05049	.34726	2.87970	51
10	.28990	3.44951	.30891	3.23714	.32814	3.04749	.34758	2.87700	50
11	.29021	3.44576	.30923	3.23381	.32846	3.04450	.34791	2.87430	49
12	.29053	3.44202	.30955	3.23048	.32878	3.04152	.34824	2.87161	48
13	.29084	3.43829	.30987	3.22715	.32911	3.03854	.34856	2.86892	47
14	.29116	3.43456	.31019	3.22384	.32943	3.03556	.34889	2.86624	46
15	.29147	3.43084	.31051	3.22053	.32975	3.03260	.34922	2.86356	45
16	.29179	3.42713	.31083	3.21722	.33007	3.02963	.34954	2.86089	44
17	.29210	3.42343	.31115	3.21392	.33040	3.02667	.34987	2.85822	43
18	.29242	3.41973	.31147	3.21063	.33072	3.02372	.35019	2.85555	42
19	.29274	3.41604	.31178	3.20734	.33104	3.02077	.35052	2.85289	41
20	.29305	3.41236	.31210	3.20406	.33136	3.01783	.35085	2.85023	40
21	.29337	3.40869	.31242	3.20079	.33160	3.01489	.35117	2.84758	39
22	.29368	3.40502	.31274	3.19752	.33201	3.01196	.35150	2.84494	38
23	.29400	3.40136	.31306	3.19426	.33233	3.00903	.35183	2.84229	37
24	.29432	3.39771	.31338	3.19100	.33266	3.00611	.35216	2.83965	36
25	.29463	3.39406	.31370	3.18775	.33298	3.00319	.35248	2.83702	35
26	.29495	3.39042	.31402	3.18451	.33330	3.00028	.35281	2.83439	34
27	.29526	3.38679	.31434	3.18127	.33363	2.99738	.35314	2.83176	33
28	.29558	3.38317	.31466	3.17804	.33395	2.99447	.35346	2.82914	32
29	.29590	3.37955	.31498	3.17481	.33427	2.99158	.35379	2.82653	31
30	.29621	3.37594	.31530	3.17159	.33460	2.98868	.35412	2.82391	30
31	.29653	3.37234	.31562	3.16838	.33492	2.98580	.35445	2.82130	29
32	.29685	3.36875	.31594	3.16517	.33524	2.98292	.35477	2.81870	28
33	.29716	3.36516	.31626	3.16197	.33557	2.98004	.35510	2.81610	27
34	.29748	3.36158	.31658	3.15877	.33589	2.97717	.35543	2.81350	26
35	.29780	3.35800	.31690	3.15558	.33621	2.97430	.35576	2.81091	25
36	.29811	3.35443	.31722	3.15240	.33654	2.97144	.35608	2.80833	24
37	.29843	3.35087	.31754	3.14922	.33686	2.96858	.35641	2.80574	23
38	.29875	3.34732	.31786	3.14605	.33718	2.96573	.35674	2.80316	22
39	.29906	3.34377	.31818	3.14288	.33751	2.96288	.35707	2.80059	21
40	.29938	3.34023	.31850	3.13972	.33783	2.96004	.35740	2.79802	20
41	.29970	3.33670	.31882	3.13656	.33816	2.95721	.35772	2.79545	19
42	.30001	3.33317	.31914	3.13341	.33848	2.95437	.35805	2.79289	18
43	.30033	3.32965	.31946	3.13027	.33881	2.95155	.35838	2.79033	17
44	.30065	3.32614	.31978	3.12713	.33913	2.94872	.35871	2.78778	16
45	.30097	3.32264	.32010	3.12400	.33945	2.94590	.35904	2.78523	15
46	.30128	3.31914	.32042	3.12087	.33978	2.94309	.35937	2.78269	14
47	.30160	3.31565	.32074	3.11775	.34010	2.94028	.35969	2.78014	13
48	.30192	3.31216	.32106	3.11464	.34043	2.93748	.36002	2.77761	12
49	.30224	3.30868	.32139	3.11153	.34075	2.93468	.36035	2.77507	11
50	.30255	3.30521	.32171	3.10842	.34108	2.93189	.36068	2.77254	10
51	.30287	3.30174	.32203	3.10532	.34140	2.92910	.36101	2.77002	9
52	.30319	3.29829	.32235	3.10223	.34173	2.92632	.36134	2.76750	8
53	.30351	3.29483	.32267	3.09914	.34205	2.92354	.36167	2.76498	7
54	.30382	3.29139	.32299	3.09606	.34238	2.92076	.36199	2.76247	6
55	.30414	3.28795	.32331	3.09298	.34270	2.91799	.36232	2.75996	5
56	.30446	3.28452	.32363	3.08991	.34303	2.91523	.36265	2.75746	4
57	.30478	3.28109	.32396	3.08685	.34335	2.91246	.36298	2.75496	3
58	.30509	3.27767	.32428	3.08379	.34368	2.90971	.36331	2.75246	2
59	.30541	3.27426	.32460	3.08073	.34400	2.90696	.36364	2.74997	1
60	.30573	3.27085	.32492	3.07768	.34433	2.90421	.36397	2.74748	0
′	Co-tan.	Tan.	Co-tan.	Tan.	Co-tan.	Tan.	Co-tan.	Tan.	′
	73°		72°		71°		70°		

Table X—Continued

′	20° Tan.	20° Co-tan.	21° Tan.	21° Co-tan.	22° Tan.	22° Co-tan.	23° Tan.	23° Co-tan.	′
0	.36397	2.74748	.38386	2.60509	.40403	2.47509	.42447	2.35585	60
1	.36430	2.74499	.38420	2.60283	.40436	2.47302	.42482	2.35395	59
2	.36463	2.74251	.38453	2.60057	.40470	2.47095	.42516	2.35205	58
3	.36496	2.74004	.38487	2.59831	.40504	2.46888	.42551	2.35015	57
4	.36529	2.73756	.38520	2.59606	.40538	2.46682	.42585	2.34825	56
5	.36562	2.73509	.38553	2.59381	.40572	2.46476	.42619	2.34636	55
6	.36595	2.73263	.38587	2.59156	.40606	2.46270	.42654	2.34447	54
7	.36628	2.73017	.38620	2.58932	.40640	2.46065	.42688	2.34258	53
8	.36661	2.72771	.38654	2.58708	.40674	2.45860	.42722	2.34069	52
9	.36694	2.72526	.38687	2.58484	.40707	2.45655	.42757	2.33881	51
10	.36727	2.72281	.38721	2.58261	.40741	2.45451	.42791	2.33693	50
11	.36760	2.72036	.38754	2.58038	.40775	2.45246	.42826	2.33505	49
12	.36793	2.71792	.38787	2.57815	.40809	2.45043	.42860	2.33317	48
13	.36826	2.71548	.38821	2.57593	.40843	2.44839	.42894	2.33130	47
14	.36859	2.71305	.38854	2.57371	.40877	2.44636	.42929	2.32943	46
15	.36892	2.71062	.38888	2.57150	.40911	2.44433	.42963	2.32756	45
16	.36925	2.70819	.38921	2.56928	.40945	2.44230	.42998	2.32570	44
17	.36958	2.70577	.38955	2.56707	.40979	2.44027	.43032	2.32383	43
18	.36991	2.70335	.38988	2.56487	.41013	2.43825	.43067	2.32197	42
19	.37024	2.70094	.39022	2.56266	.41047	2.43623	.43101	2.32012	41
20	.37057	2.69853	.39055	2.56046	.41081	2.43422	.43136	2.31826	40
21	.37090	2.69612	.39089	2.55827	.41115	2.43220	.43170	2.31641	39
22	.37124	2.69371	.39122	2.55608	.41149	2.43019	.43205	2.31456	38
23	.37157	2.69131	.39156	2.55389	.41183	2.42819	.43239	2.31271	37
24	.37190	2.68892	.39190	2.55170	.41217	2.42618	.43274	2.31086	36
25	.37223	2.68653	.39223	2.54952	.41251	2.42418	.43308	2.30902	35
26	.37256	2.68414	.39257	2.54734	.41285	2.42218	.43343	2.30718	34
27	.37289	2.68175	.39290	2.54516	.41319	2.42019	.43378	2.30534	33
28	.37322	2.67937	.39324	2.54299	.41353	2.41819	.43412	2.30351	32
29	.37355	2.67700	.39357	2.54082	.41387	2.41620	.43447	2.30167	31
30	.37388	2.67462	.39391	2.53865	.41421	2.41421	.43481	2.29984	30
31	.37422	2.67225	.39425	2.53648	.41455	2.41223	.43516	2.29801	29
32	.37455	2.66989	.39458	2.53432	.41490	2.41025	.43550	2.29619	28
33	.37488	2.66752	.39492	2.53217	.41524	2.40827	.43585	2.29437	27
34	.37521	2.66516	.39526	2.53001	.41558	2.40629	.43620	2.29254	26
35	.37554	2.66281	.39559	2.52786	.41592	2.40432	.43654	2.29073	25
36	.37588	2.66046	.39593	2.52571	.41626	2.40235	.43689	2.28891	24
37	.37621	2.65811	.39626	2.52357	.41660	2.40038	.43724	2.28710	23
38	.37654	2.65576	.39660	2.52142	.41694	2.39841	.43758	2.28528	22
39	.37687	2.65342	.39694	2.51929	.41728	2.39645	.43793	2.28348	21
40	.37720	2.65109	.39727	2.51715	.41763	2.39449	.43828	2.28167	20
41	.37754	2.64875	.39761	2.51502	.41797	2.39253	.43862	2.27987	19
42	.37787	2.64642	.39795	2.51289	.41831	2.39058	.43897	2.27806	18
43	.37820	2.64410	.39829	2.51076	.41865	2.38862	.43932	2.27626	17
44	.37853	2.64177	.39862	2.50864	.41899	2.38668	.43966	2.27447	16
45	.37887	2.63945	.39896	2.50652	.41933	2.38473	.44001	2.27267	15
46	.37920	2.63714	.39930	2.50440	.41968	2.38279	.44036	2.27088	14
47	.37953	2.63483	.39963	2.50229	.42002	2.38084	.44071	2.26909	13
48	.37986	2.63252	.39997	2.50018	.42036	2.37891	.44105	2.26730	12
49	.38020	2.63021	.40031	2.49807	.42070	2.37697	.44140	2.26552	11
50	.38053	2.62791	.40065	2.49597	.42105	2.37504	.44175	2.26374	10
51	.38086	2.62561	.40098	2.49386	.42139	2.37311	.44210	2.26196	9
52	.38120	2.62332	.40132	2.49177	.42173	2.37118	.44244	2.26018	8
53	.38153	2.62103	.40166	2.48967	.42207	2.36925	.44279	2.25840	7
54	.38186	2.61874	.40200	2.48758	.42242	2.36733	.44314	2.25663	6
55	.38220	2.61646	.40234	2.48549	.42276	2.36541	.44349	2.25486	5
56	.38253	2.61418	.40267	2.48340	.42310	2.36349	.44384	2.25309	4
57	.38286	2.61190	.40301	2.48132	.42345	2.36158	.44418	2.25132	3
58	.38320	2.60963	.40335	2.47924	.42379	2.35967	.44453	2.24956	2
59	.38353	2.60736	.40369	2.47716	.42413	2.35776	.44488	2.24780	1
60	.38386	2.60509	.40403	2.47509	.42447	2.35585	.44523	2.24604	0
′	69° Co-tan.	69° Tan.	68° Co-tan.	68° Tan.	67° Co-tan.	67° Tan.	66° Co-tan.	66° Tan.	′

Table X—Continued

′	24° Tan.	24° Co-tan.	25° Tan.	25° Co-tan.	26° Tan.	26° Co-tan.	27° Tan.	27° Co-tan.	′
0	.44523	2.24604	.46631	2.14451	.48773	2.05030	.50953	1.96261	60
1	.44558	2.24428	.46666	2.14288	.48809	2.04879	.50989	1.96120	59
2	.44593	2.24252	.46702	2.14125	.48845	2.04728	.51026	1.95979	58
3	.44627	2.24077	.46737	2.13963	.48881	2.04577	.51063	1.95838	57
4	.44662	2.23902	.46772	2.13801	.48917	2.04426	.51099	1.95698	56
5	.44697	2.23727	.46808	2.13639	.48953	2.04276	.51136	1.95557	55
6	.44732	2.23553	.46843	2.13477	.48989	2.04125	.51173	1.95417	54
7	.44767	2.23378	.46879	2.13316	.49026	2.03975	.51209	1.95277	53
8	.44802	2.23204	.46914	2.13154	.49062	2.03825	.51246	1.95137	52
9	.44837	2.23030	.46950	2.12993	.49098	2.03675	.51283	1.94997	51
10	.44872	2.22857	.46985	2.12832	.49134	2.03526	.51319	1.94858	50
11	.44907	2.22683	.47021	2.12671	.49170	2.03376	.51356	1.94718	49
12	.44942	2.22510	.47056	2.12511	.49206	2.03227	.51393	1.94579	48
13	.44977	2.22337	.47092	2.12350	.49242	2.03078	.51430	1.94440	47
14	.45012	2.22164	.47128	2.12190	.49278	2.02929	.51467	1.94301	46
15	.45047	2.21992	.47163	2.12030	.49315	2.02780	.51503	1.94162	45
16	.45082	2.21819	.47199	2.11871	.49351	2.02631	.51540	1.94023	44
17	.45117	2.21647	.47234	2.11711	.49387	2.02483	.51577	1.93885	43
18	.45152	2.21475	.47270	2.11552	.49423	2.02335	.51614	1.93746	42
19	.45187	2.21304	.47305	2.11392	.49459	2.02187	.51651	1.93608	41
20	.45222	2.21132	.47341	2.11233	.49495	2.02039	.51688	1.93470	40
21	.45257	2.20961	.47377	2.11075	.49532	2.01891	.51724	1.93332	39
22	.45292	2.20790	.47412	2.10916	.49568	2.01743	.51761	1.93195	38
23	.45327	2.20619	.47448	2.10758	.49604	2.01596	.51798	1.93057	37
24	.45362	2.20449	.47483	2.10600	.49640	2.01449	.51835	1.92920	36
25	.45397	2.20278	.47519	2.10442	.49677	2.01302	.51872	1.92782	35
26	.45432	2.20108	.47555	2.10284	.49713	2.01155	.51909	1.92645	34
27	.45467	2.19938	.47590	2.10126	.49749	2.01008	.51946	1.92508	33
28	.45502	2.19769	.47626	2.09969	.49786	2.00862	.51983	1.92371	32
29	.45537	2.19599	.47662	2.09811	.49822	2.00715	.52020	1.92235	31
30	.45573	2.19430	.47698	2.09654	.49858	2.00569	.52057	1.92098	30
31	.45608	2.19261	.47733	2.09498	.49894	2.00423	.52094	1.91962	29
32	.45643	2.19092	.47769	2.09341	.49931	2.00277	.52131	1.91826	28
33	.45678	2.18923	.47805	2.09184	.49967	2.00131	.52168	1.91690	27
34	.45713	2.18755	.47840	2.09028	.50004	1.99986	.52205	1.91554	26
35	.45748	2.18587	.47876	2.08872	.50040	1.99841	.52242	1.91418	25
36	.45784	2.18419	.47912	2.08716	.50076	1.99695	.52279	1.91282	24
37	.45819	2.18251	.47948	2.08560	.50113	1.99550	.52316	1.91147	23
38	.45854	2.18084	.47984	2.08405	.50149	1.99406	.52353	1.91012	22
39	.45889	2.17916	.48019	2.08250	.50185	1.99261	.52390	1.90876	21
40	.45924	2.17749	.48055	2.08094	.50222	1.99116	.52427	1.90741	20
41	.45960	2.17582	.48091	2.07939	.50258	1.98972	.52464	1.90607	19
42	.45995	2.17416	.48127	2.07785	.50295	1.98828	.52501	1.90472	18
43	.46030	2.17249	.48163	2.07630	.50331	1.98684	.52538	1.90337	17
44	.46065	2.17083	.48198	2.07476	.50368	1.98540	.52575	1.90203	16
45	.46101	2.16917	.48234	2.07321	.50404	1.98396	.52613	1.90069	15
46	.46136	2.16751	.48270	2.07167	.50441	1.98253	.52650	1.89935	14
47	.46171	2.16585	.48306	2.07014	.50477	1.98110	.52687	1.89801	13
48	.46206	2.16420	.48342	2.06860	.50514	1.97966	.52724	1.89667	12
49	.46242	2.16255	.48378	2.06706	.50550	1.97823	.52761	1.89533	11
50	.46277	2.16090	.48414	2.06553	.50587	1.97680	.52798	1.89400	10
51	.46312	2.15925	.48450	2.06400	.50623	1.97538	.52836	1.89266	9
52	.46348	2.15760	.48486	2.06247	.50660	1.97395	.52873	1.89133	8
53	.46383	2.15596	.48521	2.06094	.50696	1.97253	.52910	1.89000	7
54	.46418	2.15432	.48557	2.05942	.50733	1.97111	.52947	1.88867	6
55	.46454	2.15268	.48593	2.05790	.50769	1.96969	.52984	1.88734	5
56	.46489	2.15104	.48629	2.05637	.50806	1.96827	.53022	1.88602	4
57	.46525	2.14940	.48665	2.05485	.50843	1.96685	.53059	1.88469	3
58	.46560	2.14777	.48701	2.05333	.50879	1.96544	.53096	1.88337	2
59	.46595	2.14614	.48737	2.05182	.50916	1.96402	.53134	1.88205	1
60	.46631	2.14451	.48773	2.05030	.50953	1.96261	.53171	1.88073	0
′	Co-tan.	Tan. 65°	Co-tan.	Tan. 64°	Co-tan.	Tan. 63°	Co-tan.	Tan. 62°	′

Table X—Continued

′	28° Tan.	28° Co-tan.	29° Tan.	29° Co-tan.	30° Tan.	30° Co-tan.	31° Tan.	31° Co-tan.	′
0	.53171	1.88073	.55431	1.80405	.57735	1.73205	.60086	1.66428	60
1	.53208	1.87941	.55469	1.80281	.57774	1.73089	.60126	1.66318	59
2	.53246	1.87809	.55507	1.80158	.57813	1.72973	.60165	1..6209	58
3	.53283	1.87677	.55545	1.80034	.57851	1.72857	.60205	1.66099	57
4	.53320	1.87546	.55583	1.79911	.57890	1.72741	.60245	1.65990	56
5	.53358	1.87415	.55621	1.79788	.57929	1.72625	.60284	1.65881	55
6	.53395	1.87283	.55659	1.79665	.57968	1.72509	.60324	1.65772	54
7	.53432	1.87152	.55697	1.79542	.58007	1.72393	.60364	1.65663	53
8	.53470	1.87021	.55736	1.79419	.58046	1.72278	.60403	1.65534	52
9	.53507	1.86891	.55774	1.79296	.58085	1.72163	.60443	1.65445	51
10	.53545	1.86760	.55812	1.79174	.58124	1.72047	.60483	1.65337	50
11	.53582	1.86630	.55850	1.79051	.58162	1.71932	.60522	1.65228	49
12	.53620	1.86499	.55888	1.78929	.58201	1.71817	.60562	1.65120	48
13	.53657	1.86369	.55926	1.78807	.58240	1.71702	.60602	1.65011	47
14	.53694	1.86239	.55964	1.78685	.58279	1.71588	.60642	1.64903	46
15	.53732	1.86109	.56003	1.78563	.58318	1.71473	.60681	1.64795	45
16	.53769	1.85979	.56041	1.78441	.58357	1.71358	.60721	1.64687	44
17	.53807	1.85850	.56079	1.78319	.58396	1.71244	.60761	1.64579	43
18	.53844	1.85720	.56117	1.78198	.58435	1.71129	.60801	1.64471	42
19	.53882	1.85591	.56156	1.78077	.58474	1.71015	.60841	1.64363	41
20	.53920	1.85462	.56194	1.77955	.58513	1.70901	.60881	1.64256	40
21	.53957	1.85333	.56232	1.77834	.58552	1.70787	.60921	1.64148	39
22	.53995	1.85204	.56270	1.77713	.58591	1.70673	.60960	1.64041	38
23	.54032	1.85075	.56309	1.77592	.58631	1.70560	.61000	1.63934	37
24	.54070	1.84946	.56347	1.77471	.58670	1.70446	.61040	1.63826	36
25	.54107	1.84818	.56385	1.77351	.58709	1.70332	.61080	1.63719	35
26	.54145	1.84689	.56424	1.77230	.58748	1.70219	.61120	1.63612	34
27	.54183	1.84561	.56462	1.77110	.58787	1.70106	.61160	1.63505	33
28	.54220	1.84433	.56500	1.76990	.58826	1.69992	.61200	1.63398	32
29	.54258	1.84305	.56539	1.76869	.58865	1.69879	.61240	1.63292	31
30	.54296	1.84177	.56577	1.76749	.58904	1.69766	.61280	1.63185	30
31	.54333	1.84049	.56616	1.76630	.58944	1.69653	.61320	1.63079	29
32	.54371	1.83922	.56654	1.76510	.58983	1.69541	.61360	1.62972	28
33	.54409	1.83794	.56693	1.76390	.59022	1.69428	.61400	1.62866	27
34	.54446	1.83667	.56731	1.76271	.59061	1.69316	.61440	1.62760	26
35	.54484	1.83540	.56769	1.76151	.59101	1.69203	.61480	1.62654	25
36	.54522	1.83413	.56808	1.76032	.59140	1.69091	.61520	1.62548	24
37	.54560	1.83286	.56846	1.75913	.59179	1.68979	.61561	1.62442	23
38	.54597	1.83159	.56885	1.75794	.59218	1.68866	.61601	1.62336	22
39	.54635	1.83033	.56923	1.75675	.59258	1.68754	.61641	1.62230	21
40	.54673	1.82906	.56962	1.75556	.59297	1.68643	.61681	1.62125	20
41	.54711	1.82780	.57000	1.75437	.59336	1.68531	.61721	1.62019	19
42	.54748	1.82654	.57039	1.75319	.59376	1.68419	.61761	1.61914	18
43	.54786	1.82528	.57078	1.75200	.59415	1.68308	.61801	1.61808	17
44	.54824	1.82402	.57116	1.75082	.59454	1.68196	.61842	1.61703	16
45	.54862	1.82276	.57155	1.74964	.59494	1.68085	.61882	1.61598	15
46	.54900	1.82150	.57193	1.74846	.59533	1.67974	.61922	1.61493	14
47	.54938	1.82025	.57232	1.74728	.59573	1.67863	.61962	1.61388	13
48	.54975	1.81899	.57271	1.74610	.59612	1.67752	.62003	1.61283	12
49	.55013	1.81774	.57309	1.74492	.59651	1.67641	.62043	1.61179	11
50	.55051	1.81649	.57348	1.74375	.59691	1.67530	.62083	1.61074	10
51	.55089	1.81524	.57386	1.74257	.59730	1.67419	.62124	1.60970	9
52	.55127	1.81399	.57425	1.74140	.59770	1.67309	.62164	1.60865	8
53	.55165	1.81274	.57464	1.74022	.59809	1.67198	.62204	1.60761	7
54	.55203	1.81150	.57503	1.73905	.59849	1.67088	.62245	1.60657	6
55	.55241	1.81025	.57541	1.73788	.59888	1.66978	.62285	1.60553	5
56	.55279	1.80901	.57580	1.73671	.59928	1.66867	.62325	1.60449	4
57	.55317	1.80777	.57619	1.73555	.59967	1.66757	.62366	1.60345	3
58	.55355	1.80653	.57657	1.73438	.60007	1.66647	.62406	1.60241	2
59	.55393	1.80529	.57696	1.73321	.60046	1.66538	.62446	1.60137	1
60	.55431	1.80405	.57735	1.73205	.60086	1.66428	.62487	1.60033	0
′	Co-tan. 61°	Tan. 61°	Co-tan. 60°	Tan. 60°	Co-tan. 59°	Tan. 59°	Co-tan. 58°	Tan. 58°	′

Table X—Continued

	32°		33°		34°		35°		
′	Tan.	Co-tan.	Tan.	Co-tan.	Tan.	Co-tan.	Tan.	Co-tan.	′
0	.62487	1.60033	.64941	1.53986	.67451	1.48256	.70021	1.42815	60
1	.62527	1.59930	.64982	1.53888	.67493	1.48163	.70064	1.42726	59
2	.62568	1.59826	.65023	1.53791	.67536	1.48070	.70107	1.42638	58
3	.62608	1.59723	.65065	1.53693	.67578	1.47977	.70151	1.42550	57
4	.62649	1.59620	.65106	1.53595	.67620	1.47885	.70194	1.42462	56
5	.62689	1.59517	.65148	1.53497	.67663	1.47792	.70238	1.42374	55
6	.62730	1.59414	.65189	1.53400	.67705	1.47699	.70281	1.42286	54
7	.62770	1.59311	.65231	1.53302	.67748	1.47607	.70325	1.42198	53
8	.62811	1.59208	.65272	1.53205	.67790	1.47514	.70368	1.42110	52
9	.62852	1.59105	.65314	1.53107	.67832	1.47422	.70412	1.42022	51
10	.62892	1.59002	.65355	1.53010	.67875	1.47330	.70455	1.41934	50
11	.62933	1.58900	.65397	1.52913	.67917	1.47238	.70499	1.41847	49
12	.62973	1.58797	.65438	1.52816	.67960	1.47146	.70542	1.41759	48
13	.63014	1.58695	.65480	1.52719	.68002	1.47053	.70586	1.41672	47
14	.63055	1.58593	.65521	1.52622	.68045	1.46962	.70629	1.41584	46
15	.63095	1.58490	.65563	1.52525	.68088	1.46870	.70673	1.41497	45
16	.63136	1.58388	.65604	1.52429	.68130	1.46778	.70717	1.41409	44
17	.63177	1.58286	.65646	1.52332	.68173	1.46686	.70760	1.41322	43
18	.63217	1.58184	.65688	1.52235	.68215	1.46595	.70804	1.41235	42
19	.63258	1.58083	.65729	1.52139	.68258	1.46503	.70848	1.41148	41
20	.63299	1.57981	.65771	1.52043	.68301	1.46411	.70891	1.41061	40
21	.63340	1.57879	.65813	1.51946	.68343	1.46320	.70935	1.40974	39
22	.63380	1.57778	.65854	1.51850	.68386	1.46229	.70979	1.40887	38
23	.63421	1.57676	.65896	1.51754	.68429	1.46137	.71023	1.40800	37
24	.63462	1.57575	.65938	1.51658	.68471	1.46046	.71066	1.40714	36
25	.63503	1.57474	.65980	1.51562	.68514	1.45955	.71110	1.40627	35
26	.63544	1.57372	.66021	1.51466	.68557	1.45864	.71154	1.40540	34
27	.63584	1.57271	.66063	1.51370	.68600	1.45773	.71198	1.40454	33
28	.63625	1.57170	.66105	1.51275	.68642	1.45682	.71242	1.40367	32
29	.63666	1.57069	.66147	1.51179	.68685	1.45592	.71285	1.40281	31
30	.63707	1.56969	.66189	1.51084	.68728	1.45501	.71329	1.40195	30
31	.63748	1.56868	.66230	1.50988	.68771	1.45410	.71373	1.40109	29
32	.63789	1.56767	.66272	1.50893	.68814	1.45320	.71417	1.40022	28
33	.63830	1.56667	.66314	1.50797	.68857	1.45229	.71461	1.39936	27
34	.63871	1.56566	.66356	1.50702	.68900	1.45139	.71505	1.39850	26
35	.63912	1.56466	.66398	1.50607	.68942	1.45049	.71549	1.39764	25
36	.63953	1.56366	.66440	1.50512	.68985	1.44958	.71593	1.39679	24
37	.63994	1.56265	.66482	1.50417	.69028	1.44868	.71637	1.39593	23
38	.64035	1.56165	.66524	1.50322	.69071	1.44778	.71681	1.39507	22
39	.64076	1.56065	.66566	1.50228	.69114	1.44688	.71725	1.39421	21
40	[illegible]	[illegible]	[illegible]	[illegible]	[illegible]	[illegible]	[illegible]	[illegible]	20
41	.64158	1.55866	.66650	1.50038	.69200	1.44508	.71813	1.39250	19
42	.64199	1.55766	.66692	1.49944	.69243	1.44418	.71857	1.39165	18
43	.64240	1.55666	.66734	1.49849	.69286	1.44329	.71901	1.39079	17
44	.64281	1.55567	.66776	1.49755	.69329	1.44239	.71946	1.38994	16
45	.64322	1.55467	.66818	1.49661	.69372	1.44149	.71990	1.38909	15
46	.64363	1.55368	.66860	1.49566	.69416	1.44060	.72034	1.38824	14
47	.64404	1.55269	.66902	1.49472	.69459	1.43970	.72078	1.38738	13
48	.64446	1.55170	.66944	1.49378	.69502	1.43881	.72122	1.38653	12
49	.64487	1.55071	.66986	1.49284	.69545	1.43792	.72166	1.38568	11
50	.64528	1.54972	.67028	1.49190	.69588	1.43703	.72211	1.38484	10
51	.64569	1.54873	.67071	1.49097	.69631	1.43614	.72255	1.38399	9
52	.64610	1.54774	.67113	1.49003	.69675	1.43525	.72299	1.38314	8
53	.64652	1.54675	.67155	1.48909	.69718	1.43436	.72344	1.38229	7
54	.64693	1.54576	.67197	1.48816	.69761	1.43347	.72388	1.38145	6
55	.64734	1.54478	.67239	1.48722	.69804	1.43258	.72432	1.38060	5
56	.64775	1.54379	.67282	1.48629	.69847	1.43169	.72477	1.37976	4
57	.64817	1.54281	.67324	1.48536	.69891	1.43080	.72521	1.37891	3
58	.64858	1.54183	.67366	1.48442	.69934	1.42992	.72565	1.37807	2
59	.64899	1.54085	.67409	1.48349	.69977	1.42903	.72610	1.37722	1
60	.64941	1.53986	.67451	1.48256	.70021	1.42815	.72654	1.37638	0
′	Co-tan.	Tan.	Co-tan.	Tan.	Co-tan.	Tan.	Co-tan.	Tan.	′
	57°		56°		55°		54°		

Table X—Continued

′	36°		37°		38°		39°		′
	Tan.	Co-tan.	Tan.	Co-tan.	Tan.	Co-tan.	Tan.	Co-tan.	
0	.72654	1.37638	.75355	1.32704	.78129	1.27994	.80978	1.23490	60
1	.72699	1.37554	.75401	1.32624	.78175	1.27917	.81027	1.23416	59
2	.72743	1.37470	.75447	1.32544	.78222	1.27841	.81075	1.23343	58
3	.72788	1.37386	.75492	1.32464	.78269	1.27764	.81123	1.23270	57
4	.72832	1.37302	.75538	1.32384	.78316	1.27688	.81171	1.23196	56
5	.72877	1.37218	.75584	1.32304	.78363	1.27611	.81220	1.23123	55
6	.72921	1.37134	.75629	1.32224	.78410	1.27535	.81268	1.23050	54
7	.72966	1.37050	.75675	1.32144	.78457	1.27458	.81316	1.22977	53
8	.73010	1.36967	.75721	1.32064	.78504	1.27382	.81364	1.22904	52
9	.73055	1.36883	.75767	1.31984	.78551	1.27306	.81413	1.22831	51
10	.73100	1.36800	.75812	1.31904	.78598	1.27230	.81461	1.22758	50
11	.73144	1.36716	.75858	1.31825	.78645	1.27153	.81510	1.22685	49
12	.73189	1.36633	.75904	1.31745	.78692	1.27077	.81558	1.22612	48
13	.73234	1.36549	.75950	1.31666	.78739	1.27001	.81606	1.22539	47
14	.73278	1.36466	.75996	1.31586	.78786	1.26925	.81655	1.22467	46
15	.73323	1.36383	.76042	1.31507	.78834	1.26849	.81703	1.22394	45
16	.73368	1.36300	.76088	1.31427	.78881	1.26774	.81752	1.22321	44
17	.73413	1.36217	.76134	1.31348	.78928	1.26698	.81800	1.22249	43
18	.73457	1.36133	.76180	1.31269	.78975	1.26622	.81849	1.22176	42
19	.73502	1.36051	.76226	1.31190	.79022	1.26546	.81898	1.22104	41
20	.73547	1.35968	.76272	1.31110	.79070	1.26471	.81946	1.22031	40
21	.73592	1.35885	.76318	1.31031	.79117	1.26395	.81995	1.21959	39
22	.73637	1.35802	.76364	1.30952	.79164	1.26319	.82044	1.21886	38
23	.73681	1.35719	.76410	1.30873	.79212	1.26244	.82092	1.21814	37
24	.73726	1.35637	.76456	1.30795	.79259	1.26169	.82141	1.21742	36
25	.73771	1.35554	.76502	1.30716	.79306	1.26093	.82190	1.21670	35
26	.73816	1.35472	.76548	1.30637	.79354	1.26018	.82238	1.21598	34
27	.73861	1.35389	.76594	1.30558	.79401	1.25943	.82287	1.21526	33
28	.73906	1.35307	.76640	1.30480	.79449	1.25867	.82336	1.21454	32
29	.73951	1.35224	.76686	1.30401	.79496	1.25792	.82385	1.21382	31
30	.73996	1.35142	.76733	1.30323	.79544	1.25717	.82434	1.21310	30
31	.74041	1.35060	.76779	1.30244	.79591	1.25642	.82483	1.21238	29
32	.74086	1.34978	.76825	1.30166	.79639	1.25567	.82531	1.21166	28
33	.74131	1.34896	.76871	1.30087	.79686	1.25492	.82580	1.21094	27
34	.74176	1.34814	.76918	1.30009	.79734	1.25417	.82629	1.21023	26
35	.74221	1.34732	.76964	1.29931	.79781	1.25343	.82678	1.20951	25
36	.74267	1.34650	.77010	1.29853	.79829	1.25268	.82727	1.20879	24
37	.74312	1.34568	.77057	1.29775	.79877	1.25193	.82776	1.20808	23
38	.74357	1.34487	.77103	1.29696	.79924	1.25118	.82825	1.20736	22
39	.74402	1.34405	.77149	1.29618	.79972	1.25044	.82874	1.20665	21
40	.74447	1.34323	.77196	1.29541	.80020	1.24969	.82923	1.20593	20
41	.74492	1.34242	.77242	1.29463	.80067	1.24895	.82972	1.20522	19
42	.74538	1.34160	.77289	1.29385	.80115	1.24820	.83022	1.20451	18
43	.74583	1.34079	.77335	1.29307	.80163	1.24746	.83071	1.20379	17
44	.74628	1.33998	.77382	1.29229	.80211	1.24672	.83120	1.20308	16
45	.74674	1.33916	.77428	1.29152	.80258	1.24597	.83169	1.20237	15
46	.74719	1.33835	.77475	1.29074	.80306	1.24523	.83218	1.20166	14
47	.74764	1.33754	.77521	1.28997	.80354	1.24449	.83268	1.20095	13
48	.74810	1.33673	.77568	1.28919	.80402	1.24375	.83317	1.20024	12
49	.74855	1.33592	.77615	1.28842	.80450	1.24301	.83366	1.19953	11
50	.74900	1.33511	.77661	1.28764	.80498	1.24227	.83415	1.19882	10
51	.74946	1.33430	.77708	1.28687	.80546	1.24153	.83465	1.19811	9
52	.74991	1.33349	.77754	1.28610	.80594	1.24079	.83514	1.19740	8
53	.75037	1.33268	.77801	1.28533	.80642	1.24005	.83564	1.19669	7
54	.75082	1.33187	.77848	1.28456	.80690	1.23931	.83613	1.19599	6
55	.75128	1.33107	.77895	1.28379	.80738	1.23858	.83662	1.19528	5
56	.75173	1.33026	.77941	1.28302	.80786	1.23784	.83712	1.19457	4
57	.75219	1.32946	.77988	1.28225	.80834	1.23710	.83761	1.19387	3
58	.75264	1.32865	.78035	1.28148	.80882	1.23637	.83811	1.19316	2
59	.75310	1.32785	.78082	1.28071	.80930	1.23563	.83860	1.19246	1
60	.75355	1.32704	.78129	1.27994	.80978	1.23490	.83910	1.19175	0
′	Co-tan.	Tan.	Co-tan.	Tan.	Co-tan.	Tan.	Co-tan.	Tan	′
	53°		52°		51°		50°		

Table X—Continued

′	40° Tan.	40° Co-tan.	41° Tan.	41° Co-tan.	42° Tan.	42° Co-tan.	43° Tan.	43° Co-tan.	′
0	.83910	1.19175	.86929	1.15037	.90040	1.11061	.93252	1.07237	60
1	.83960	1.19105	.86980	1.14969	.90093	1.10996	.93306	1.07174	59
2	.84009	1.19035	.87031	1.14902	.90146	1.10931	.93360	1.07112	58
3	.84059	1.18964	.87082	1.14834	.90199	1.10867	.93415	1.07049	57
4	.84108	1.18894	.87133	1.14767	.90251	1.10802	.93469	1.06987	56
5	.84158	1.18824	.87184	1.14699	.90304	1.10737	.93524	1.06925	55
6	.84208	1.18754	.87236	1.14632	.90357	1.10672	.93578	1.06862	54
7	.84258	1.18684	.87287	1.14565	.90410	1.10607	.93633	1.06800	53
8	.84307	1.18614	.87338	1.14498	.90463	1.10543	.93688	1.06738	52
9	.84357	1.18544	.87389	1.14430	.90516	1.10478	.93742	1.06676	51
10	.84407	1.18474	.87441	1.14363	.90569	1.10414	.93797	1.06613	50
11	.84457	1.18404	.87492	1.14296	.90621	1.10349	.93852	1.06551	49
12	.84507	1.18334	.87543	1.14229	.90674	1.10285	.93906	1.06489	48
13	.84556	1.18264	.87595	1.14162	.90727	1.10220	.93961	1.06427	47
14	.84606	1.18194	.87646	1.14095	.90781	1.10156	.94016	1.06365	46
15	.84656	1.18125	.87698	1.14028	.90834	1.10091	.94071	1.06303	45
16	.84706	1.18055	.87749	1.13961	.90887	1.10027	.94125	1.06241	44
17	.84756	1.17986	.87801	1.13894	.90940	1.09963	.94180	1.06179	43
18	.84806	1.17916	.87852	1.13828	.90993	1.09899	.94235	1.06117	42
19	.84856	1.17846	.87904	1.13761	.91046	1.09834	.94290	1.06056	41
20	.84906	1.17777	.87955	1.13694	.91099	1.09770	.94345	1.05994	40
21	.84956	1.17708	.88007	1.13627	.91153	1.09706	.94400	1.05932	39
22	.85006	1.17638	.88059	1.13561	.91206	1.09642	.94455	1.05870	38
23	.85057	1.17569	.88110	1.13494	.91259	1.09578	.94510	1.05809	37
24	.85107	1.17500	.88162	1.13428	.91313	1.09514	.94565	1.05747	36
25	.85157	1.17430	.88214	1.13361	.91366	1.09450	.94620	1.05685	35
26	.85207	1.17361	.88265	1.13295	.91419	1.09386	.94676	1.05624	34
27	[illegible]	[illegible]	.88317	1.13228	.91473	1.09322	.94731	1.05562	33
28	.85307	1.17223	.88369	1.13162	[illegible]	[illegible]	[illegible]	[illegible]	32
29	.85358	1.17154	.88421	1.13096	.91580	1.09195	.94841	1.05439	31
30	.85408	1.17085	.88473	1.13029	.91633	1.09131	.94896	1.05378	30
31	.85458	1.17016	.88524	1.12963	.91687	1.09067	.94952	1.05317	29
32	.85509	1.16947	.88576	1.12897	.91740	1.09003	.95007	1.05255	28
33	.85559	1.16878	.88628	1.12831	.91794	1.08940	.95062	1.05194	27
34	.85609	1.16809	.88680	1.12765	.91847	1.08876	.95118	1.05133	26
35	.85660	1.16741	.88732	1.12699	.91901	1.08813	.95173	1.05072	25
36	.85710	1.16672	.88784	1.12633	.91955	1.08749	.95229	1.05010	24
37	.85761	1.16603	[illegible]	1.12567	.92008	1.08686	[illegible]	[illegible]	23
38	.85811	1.16535	.88888	1.12501	.92062	1.08622	.95340	1.04888	22
39	.85862	1.16466	.88940	1.12435	.92116	1.08559	.95395	1.04827	21
40	.85912	1.16398	[illegible]	[illegible]	[illegible]	[illegible]	[illegible]	[illegible]	20
41	.85963	1.16329	.89045	1.12303	.92224	1.08432	.95506	1.04705	19
42	.86014	1.16261	.89097	1.12238	.92277	1.08369	.95562	1.04644	18
43	.86064	1.16192	.89149	1.12172	.92331	1.08306	.95618	1.04583	17
44	.86115	1.16124	.89201	1.12106	.92385	1.08243	.95673	1.04522	16
45	.86166	1.16056	.89253	1.12041	.92439	1.08179	.95729	1.04461	15
46	.86216	1.15987	.89306	1.11975	.92493	1.08116	.95785	1.04401	14
47	.86267	1.15919	.89358	1.11909	.92547	1.08053	.95841	1.04340	13
48	.86318	1.15851	.89410	1.11844	.92601	1.07990	.95897	1.04279	12
49	.86368	1.15783	.89463	1.11778	.92655	1.07927	.95952	1.04218	11
50	.86419	1.15715	.89515	1.11713	.92709	1.07864	.96008	1.04158	10
51	.86470	1.15647	.89567	1.11648	.92763	1.07801	.96064	1.04097	9
52	.86521	1.15579	.89620	1.11582	.92817	1.07738	.96120	1.04036	8
53	.86572	1.15511	.89672	1.11517	.92872	1.07676	.96176	1.03976	7
54	.86623	1.15443	.89725	1.11452	.92926	1.07613	.96232	1.03915	6
55	.86674	1.15375	.89777	1.11387	.92980	1.07550	.96288	1.03855	5
56	.86725	1.15308	.89830	1.11321	.93034	1.07487	.96344	1.03794	4
57	.86776	1.15240	.89883	1.11256	.93088	1.07425	.96400	1.03734	3
58	.86827	1.15172	.89935	1.11191	.93143	1.07362	.96457	1.03674	2
59	.86878	1.15104	.89988	1.11126	.93197	1.07299	.96513	1.03613	1
60	.86929	1.15037	.90040	1.11061	.93252	1.07237	.96569	1.03553	0
′	49° Co-tan.	49° Tan.	48° Co-tan.	48° Tan.	47° Co-tan.	47° Tan.	46° Co-tan.	46° Tan.	′

Table X—Concluded

′	44° Tan.	44° Co-tan.	′	′	44° Tan.	44° Co-tan.	′	′	44° Tan.	44° Co-tan.	′
0	.96569	1.03553	60	21	.97756	1.02295	39	41	.98901	1.01112	19
1	.96625	1.03493	59	22	.97813	1.02236	38	42	.98958	1.01053	18
2	.96681	1.03433	58	23	.97870	1.02176	37	43	.99016	1.00994	17
3	.96738	1.03372	57	24	.97927	1.02117	36	44	.99073	1.00935	16
4	.96794	1.03312	56	25	.97984	1.02057	35	45	.99131	1.00876	15
5	.96850	1.03252	55	26	.98041	1.01998	34	46	.99189	1.00818	14
6	.96907	1.03192	54	27	.98098	1.01939	33	47	.99247	1.00759	13
7	.96963	1.03132	53	28	.98155	1.01879	32	48	.99304	1.00701	12
8	.97020	1.03072	52	29	.98213	1.01820	31	49	.99362	1.00642	11
9	.97076	1.03012	51	30	.98270	1.01761	30	50	.99420	1.00583	10
10	.97133	1.02952	50	31	.98327	1.01702	29	51	.99478	1.00525	9
11	.97189	1.02892	49	32	.98384	1.01642	28	52	.99536	1.00467	8
12	.97246	1.02832	48	33	.98441	1.01583	27	53	.99594	1.00408	7
13	.97302	1.02772	47	34	.98499	1.01524	26	54	.99652	1.00350	6
14	.97359	1.02713	46	35	.98556	1.01465	25	55	.99710	1.00291	5
15	.97416	1.02653	45	36	.98613	1.01406	24	56	.99768	1.00233	4
16	.97472	1.02593	44	37	.98671	1.01347	23	57	.99826	1.00175	3
17	.97529	1.02533	43	38	.98728	1.01288	22	58	.99884	1.00116	2
18	.97586	1.02474	42	39	.98786	1.01229	21	59	.99942	1.00058	1
19	.97643	1.02414	41	40	.98843	1.01170	20	60	1	1	0
20	.97700	1.02355	40								
′	45° Co-tan.	45° Tan.	′	′	45° Co-tan.	45° Tan.	′	′	45° Co-tan.	45° Tan.	′

TABLE XI

FACTORS FOR DETERMINING STRENGTH OF FIGURE

Table XI Factors for determining strength of figure

Values of $(\delta_A^2 + \delta_A\delta_B + \delta_B^2)$ for various combinations of distance angles *A* and *B* of a triangle

	10°	12°	14°	16°	18°	20°	22°	24°	26°	28°	30°	35°	40°	45°	50°	55°	60°	65°	70°	75°	80°	85°	90°
10°	428	359																					
12	359	295	253																				
14	315	253	214	187																			
16	284	225	187	162	143																		
18	262	204	168	143	126	113																	
20°	245	189	153	130	113	100	91																
22	232	177	142	119	103	91	81	74															
24	221	167	134	111	95	83	74	67	61														
26	213	160	126	104	89	77	68	61	56	51													
28	206	153	120	99	83	72	63	57	51	47	43												
30°	199	148	115	94	79	68	59	53	48	43	40	33											
35	188	137	106	85	71	60	52	46	41	37	33	27	23										
40	179	129	99	79	65	54	47	41	36	32	29	23	19	16									
45	172	124	93	74	60	50	43	37	32	28	25	20	16	13	11								
50°	167	119	89	70	57	47	39	34	29	26	23	18	14	11	9	8							
55	162	115	86	67	54	44	37	32	27	24	21	16	12	10	8	7	5						
60	159	112	83	64	51	42	35	30	25	22	19	14	11	9	7	5	4	4					
65	155	109	80	62	49	40	33	28	24	21	18	13	10	7	6	5	4	3	2				
70°	152	106	78	60	48	38	32	27	23	19	17	12	9	7	5	4	3	2	2	1			
75	150	104	76	58	46	37	30	25	21	18	16	11	8	6	4	3	2	2	1	1	1		
80	147	102	74	57	45	36	29	24	20	17	15	10	7	5	4	3	2	1	1	1	0	0	
85	145	100	73	55	43	34	28	23	19	16	14	10	7	5	3	2	2	1	1	0	0	0	0

90°	143	98	71	54	42	33	27	22	19	16	13	9	6	4	3	2	1	1	1	0	0	0	0
95°	140	96	70	53	41	32	26	22	18	15	13	9	6	4	3	2	1	1	0	0	0	0	
100	138	95	68	51	40	31	25	21	17	14	12	8	6	4	3	2	1	1	0	0	0		
105	136	93	67	50	39	30	25	20	17	14	12	8	5	4	2	2	1	1	0	0			
110	134	91	65	49	38	30	24	19	16	13	11	7	5	3	2	2	1	1	1				
115°	132	89	64	48	37	29	23	19	15	13	11	7	5	3	2	2	1	1					
120	129	88	62	46	36	28	22	18	15	12	10	7	5	3	2	2	1						
125	127	86	61	45	35	27	22	18	14	12	10	7	5	4	3	2							
130	125	84	59	44	34	26	21	17	14	12	10	7	5	4	3								
135°	122	82	58	43	33	26	21	17	14	12	10	7	5	4									
140	119	80	56	42	32	25	20	17	14	12	10	8	6										
145	116	77	55	41	32	25	21	17	15	13	11	9											
150	112	75	54	40	32	26	21	18	16	15	13												
152°	111	75	53	40	32	26	22	19	17	16													
154	110	74	53	41	33	27	23	21	19														
156	108	74	54	42	34	28	25	22															
158	107	74	54	43	35	30	27																
160	107	74	56	45	38	33																	
162°	107	76	59	48	42																		
164	109	79	63	54																			
166	113	86	71																				
168	122	98																					
170	143																						

Table XII Trigonometric formulas

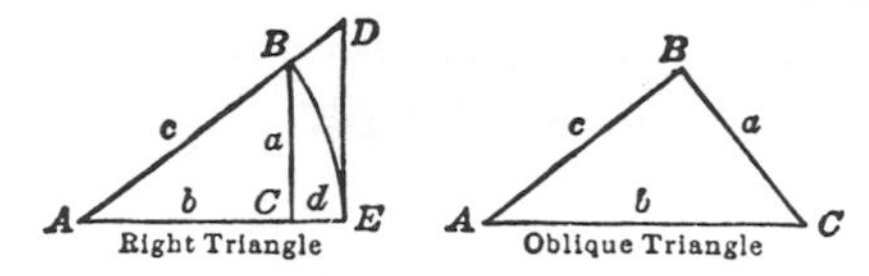

Right Triangles

$$\sin A = \frac{a}{c} = \cos B \qquad \sec A = \frac{c}{b} = \operatorname{cosec} B$$

$$\cos A = \frac{b}{c} = \sin B \qquad \operatorname{cosec} A = \frac{c}{a} = \sec B$$

$$\tan A = \frac{a}{b} = \cot B \qquad \operatorname{vers} A = \frac{c-b}{c} = \frac{d}{c}$$

$$\cot A = \frac{b}{a} = \tan B \qquad \operatorname{exsec} A = \frac{e}{c}$$

$$a = c \sin A = c \cos B = b \tan A = b \cot B = \sqrt{c^2 - b^2}$$

$$b = c \cos A = c \sin B = a \cot A = a \tan B = \sqrt{c^2 - a^2}$$

$$c = \frac{a}{\sin A} = \frac{a}{\cos B} = \frac{b}{\sin B} = \frac{b}{\cos A} = \frac{d}{\operatorname{vers} A} = \frac{e}{\operatorname{exsec} A} = \sqrt{a^2 + b^2}$$

$$d = c \operatorname{vers} A \qquad e = c \operatorname{exsec} A$$

Oblique Triangles

Given	Sought	Formulas
A, B, a	b, c	$b = \frac{a}{\sin A} \cdot \sin B$ $\quad c = \frac{a}{\sin A} \cdot \sin (A + B)$
A, a, b	B, c	$\sin B = \frac{\sin A}{a} \cdot b$ $\quad c = \frac{a}{\sin A} \cdot \sin C$
C, a, b	$\frac{1}{2}(A + B)$	$\frac{1}{2}(A + B) = 90^\circ - \frac{1}{2}C$
	$\frac{1}{2}(A - B)$	$\tan \frac{1}{2}(A - B) = \frac{a - b}{a + b} \cdot \tan \frac{1}{2}(A + B)$
a, b, c	A	If $s = \frac{1}{2}(a + b + c)$, $\sin \frac{1}{2}A = \sqrt{\frac{(s-b)(s-c)}{bc}}$
		$\cos \frac{1}{2}A = \sqrt{\frac{s(s-a)}{bc}}$, $\tan \frac{1}{2}A = \sqrt{\frac{(s-b)(s-c)}{s(s-a)}}$
		$\sin A = 2 \frac{\sqrt{s(s-a)(s-b)(s-c)}}{bc}$
		$\operatorname{vers} A = \frac{2(s-b)(s-c)}{bc}$
	area	$area = \sqrt{s(s-a)(s-b)(s-c)}$
C, a, b	*area*	$area = \frac{1}{2}ab \sin C$
A, B, C, a	*area*	$area = \frac{a^2 \sin B \cdot \sin C}{2 \sin A}$

INDEX